Jutta B. Sperber
Anthropological Aspects in the Christian-Muslim Dialogues of the Vatican

Judaism, Christianity, and Islam – Tension, Transmission, Transformation

Edited by Patrice Brodeur, Alexandra Cuffel, Assaad Elias Kattan, and Georges Tamer

Volume 14

Jutta B. Sperber

Anthropological Aspects in the Christian-Muslim Dialogues of the Vatican

—

DE GRUYTER

ISBN 978-3-11-058997-9
e-ISBN (PDF) 978-3-11-059091-3
e-ISBN (EPUB) 978-3-11-058973-3
ISSN 2196-405X

Library of Congress Control Number: 2019932234

Bibliographic information published by the Deutsche Nationalbibliothek
The Deutsche Nationalbibliothek lists this publication in the Deutsche Nationalbibliografie;
detailed bibliographic data are available on the Internet at http://dnb.dnb.de.

© 2019 Walter de Gruyter GmbH, Berlin/Boston
Published in German as *Die anthropologischen Aspekte in den christlich-muslimischen Dialogen des Vatikan* © 2018 Vandenhoeck & Ruprecht, Göttingen.
Printing and binding: CPI books GmbH, Leck

www.degruyter.com

Acknowledgements

It's all done and the English edition is within reach! I still remember the casual way in which Dr. Andrea Pacini, who was then my direct contact with the Fondazione Giovanni Agnelli in Turin, said, "We shall help you with this project. Just draw up a list of what it will cost and then we'll see where we go. And of course you must include a calculation of the price for an English translation." I believe that neither he nor those who were standing invisibly but could be clearly sensed behind him – and certainly not I myself – had any idea how long and stony and also often complicated the path before us would be. Now we can all only breathe a deep sigh of relief and say, Thank God that we have now finally got there. But, since support is always needed for good and important ideas, like that of Father Maurice Borrmans M.Afr., that, after the Christian-Muslim dialogues of the World Council of Churches, I should also produce a book on the dialogues of the Vatican, an idea with which his fellow brother, H.E. Archbishop Michael Louis Fitzgerald, who was then the President of the Papal Council for Interreligious Dialogue, also concurred without much hesitation, in retrospect that was outwardly the decisive step. Maurice Borrmans' support opened doors for me in Rome and also to the Fondazione Piero Rossano in Vezza d'Alba and to the archives of the World Council of Churches in Geneva. The support from Andrea Pacini and the Fondazione Giovanni Agnelli also led my regional church, the Evangelical Lutheran Church in Bavaria, to back the project with a habilitation scholarship and the requisite leave of absence and again, at the end, with a contribution to the translation costs. All of this amounts to a proud sum of money and time from that source alone, coupled with patience as well, the justification for which can perhaps only be grasped in its profundity by someone who has not only written a second encyclopaedic work – twice the size – for submission to a German university, but also taken on the burden of seeing to a second English translation so that the subject matter can enjoy the international attention which it deserves. In this connection, my special thanks go to Dr. Dorothea Greiner, initially the head of personnel in the Bavarian Church and later my regional bishop in Bayreuth, who has given me great and personal support to this very day. I can only voice the hope that not only this book but also its further repercussions will prove her and all my other benefactors right.

These benefactors include, not least, the German Catholic Bishops' Conference. Although it was not able to grant a scholarship to a Protestant pastor doing research on a Vatican subject, it has covered the lion's share of the translation costs. The Harms Foundation also contributed to my living expenses during the intensive efforts at the beginning of my research, thanks to the intermediary of Prof. Dr. Klaus Hock of the University of Rostock who supervised this undertaking as a habilitation thesis with both persistence and patience. Prof. Dr. Perry Schmidt-Leukel of the University of Münster generously included me in the Excellence Cluster "Religion and Politics" where I was able to complete my work and, in the process, received a number of different suggestions from that wide circle of colleagues. And, finally, I also believe that the German Society for Mission Studies will bear a share of the expenditure which the de Gruyter publishing house will incur with this publication in English. But what could money alone achieve without the work and cooperation of my two translators, Margaret A. Pater who had already translated my doctoral thesis into English and Neville Williamson who assisted her with the final section of this mammoth enterprise? Infinite thanks to both of them for their work.

An enterprise like this naturally depends not only on professional support but also on personal encouragement, and here there may be overlap as there certainly was in my case. Family and friends have to share the burden of such an undertaking if it is to have a chance of success and, on the other hand, colleagues in the field become personal friends. The best way of thanking them all is to celebrate with them as I was already able to do to some extent on the occasion of the habilitation ceremony at the University of Rostock and still hope to do with others later. So here I merely wish to mention those who, to my great regret, are no longer able to celebrate with me. They include, first and foremost, Father Maurice Borrmans himself. Without him this book would never have come about. I was profoundly grieved when I learnt that he would never be able to hold it in his hands. I should also have wished to give a copy to Monsignor Dr. Johannes Paffhausen of the Überwasser Church in Münster in whose residence I was allowed to live for a year. As someone who had worked for many years under Pope John Paul II in the Vatican Secretariat of State, he was an excellent interlocutor with whom I could discuss questions and impressions about my research work in a most uncomplicated way.

In addition to these theological companions along the way, three other persons are now no longer at my side who were close family members or whom I count as such in the case of Italy. The first who had to leave me was my mother and the next my motherly friend Barbara von Ihne in Arcore. How much both of them, each in her own way without ever having met one another personally, were concerned for me and my welfare, in view of what I had taken upon myself with this project, is something I am only now realising bit by bit as I look back, as well as the extent of the support which Ernst-Christoph von Ihne has given and offered to me over the past ten or more years. For the former I should like to express my heartfelt thanks to him and for the latter I can unfortunately no longer say "Yes, thank you" since January 17th this year. So relief, gratitude and joy are mixed with sadness and regret, but may gratitude and joy gain the upper hand with a view to the future and, not least, the motto of the family which he bequeathed to me: *nulla dies sine linea.*

Tremestieri Etneo/Catania,   Dr.theol.habil. Jutta Sperber
St. Valentine's Day 2019

Contents

Introduction — 1
1   Preliminary Remarks — 1
2   The research situation — 2
3   Aids for the Reader — 5
            Notes — 6

I   The theological foundations for the dialogue with Muslims: Statements by the magisterium — 7
1   The Second Vatican Council — 7
1.1   *Nostra Aetate* – a first look at other religions — 7
1.2   *Dignitatis Humanae* – Religious liberty in the name of human dignity — 8
1.3   Other declarations of the Council – humankind comes into view — 9
2   The official statements of the magisterium — 9
2.1   The pontificate of Paul VI — 9
2.2   The pontificate of John Paul II — 12
3   Further statements of the magisterium — 18
3.1   The magisterium of Paul VI — 18
3.2   The magisterium of John Paul I – few words on a permanent issue — 22
3.3   The magisterium of John Paul II – more than 25 years of involvement in dialogue with Muslims — 23
4   Documents from the Curia — 54
4.1   The Secretariat for Non-Christians – many legal issues between people — 54
4.2   Dominus Iesus: no more than humanly equal on any account — 55
4.3   And in conclusion human rights again — 56
            Notes — 57

II   The Secretariat's Publications — 93
1   The beginnings of the Secretariat's work — 94
2   Publications of the Commission pour les Relations Religieuses avec les Musulmans — 98
2.1   Harmony and conflict — 98

2.2	A reference work on Islam – Information for decision-makers —— 113	
2.3	Joint prayer – with limitations —— 116	
2.4	Politics (and Law): Insights into Islamic Anthropology —— 117	
2.5	Experiences with dialogue on the eve of the third millennium —— 140	
2.6	Religious liberty as a topical theme —— 142	
3	Guides to dialogue —— 185	
3.1	Orientations pour un dialogue entre chrétiens et musulmans —— 185	
3.2	Guidelines for dialogue between Christians and Muslims —— 197	
4	Religions – Thèmes fondamentaux pour une connaissance dialogique —— 214	
4.1	Introduction through the "homo religiosus" —— 214	
4.2	Humankind's striving for salvation – the basis of interreligious dialogue —— 220	
4.3	Humankind – open to a relationship with God —— 223	
4.4	Good and evil —— 225	
5	Religions in the World – a publication for international organisations —— 231	
6	Chiesa e islam – early John Paul II and his focus on Islam —— 233	
7	Meeting in Friendship – Messages to Muslims for the End of Ramadan —— 234	
7.1	The beginning under Father Joseph Cuoq (1967–1972) – burdened by the past —— 236	
7.2	Cardinal Sergio Pignedoli (1973–1979) – against the backdrop of real encounters —— 237	
7.3	Archbishop Jean Jadot (1980–1983) – materialism as a common opponent? —— 238	
7.4	Cardinal Francis Arinze and the great influence of Pope John Paul II (2000) —— 238	
7.5	Cardinal Francis Arinze (1984–2001): all about life worthy of humans —— 239	
7.6	Archbishop Michael Fitzgerald (2002–2005) – many unusual events —— 241	
8	Questionnaire on Mankind: too difficult to be made into a book —— 242	
8.1	Humanity's calling: Servant of God – but how? —— 242	
8.2	Peace – difficult in many respects —— 243	
8.3	Social inequality – for Islam an evil —— 245	

8.4	Family issues: the controversial problem of birth control —— **245**	
8.5	Industrialisation and the dispute about secularisation —— **246**	
8.6	Human freedom – differing priorities —— **247**	
8.7	Individuals and transcendental collectivism in the Islamic community —— **248**	
8.8	All people are brothers – or not entirely? —— **249**	
8.9	Life after death: a certainty —— **250**	
9	Not anthropology is problematic, but its concrete effects —— **250**	
	Notes —— **251**	
III	**The Christian-Muslim Dialogues —— 272**	
1	The beginnings and basic structures —— **272**	
1.1	Rome 1972 – an unintentional dialogue —— **272**	
1.2	Grottaferrata 1975: a key role for the dialogue with Islam —— **273**	
1.3	Challenges of the dialogue and how it was presented —— **276**	
1.4	After some time: wishes addressed to the Muslims and small steps backwards —— **277**	
2	Contacts with Saudi Arabia – working session on human rights —— **278**	
3	Contacts with the World Islamic Call Society, Tripoli, Libya —— **279**	
3.1	The very first dialogue: Tripoli 1976 —— **279**	
3.2	A new beginning from 1989 onwards: also many anthropological elements —— **292**	
3.3	Malta 1990: the importance of tolerance —— **293**	
3.4	Tripoli 1993 – religion and mass media: a joint supervisory committee —— **300**	
3.5	Continuation in Vienna in 1994: still more common commitment —— **301**	
3.6	Rome 1997: the subject of mission raises many questions —— **301**	
3.7	Tripoli 2002: dialogues between evaluation and concepts of the future —— **303**	
4	Muslims in Europe – familiar demands in a difficult situation —— **304**	
5	Dialogue on the question of spirituality: the thirst for God —— **306**	
5.1	Holiness from a Muslim point of view —— **307**	
5.2	Holiness from a Christian point of view —— **312**	
5.3	The mutual criticism —— **314**	
6	Relationships with Turkey – strong academic cooperation —— **317**	

7	Cooperation with the Al Albait Foundation —— **319**	
7.1	Beginning with religious education —— **319**	
7.2	Amman 1990 on children's rights: from the unborn to school children —— **344**	
7.2.1	The rights of the unborn —— **344**	
7.3	Rome 1992: women in society —— **365**	
7.4	Amman 1994: nationalism —— **386**	
7.5	Rome 1996 – using the resources of the earth rightly —— **412**	
7.6	The last dialogue conference for the time being – Amman 1997 —— **436**	
8	Contacts with the Secretariat for Inter-religious Dialogue, Tehran —— **463**	
8.1	The beginnings still under the Shah —— **463**	
8.2	A dialogue on the theological evaluation of modernity —— **464**	
8.3	A dialogue on religious plurality – continuation needed —— **467**	
8.4	A dialogue on youth and many old questions —— **468**	
9	Islamic-Catholic Liaison Committee —— **469**	
9.1	The creation of the Islamic-Catholic Liaison Committee —— **469**	
9.2	The first meeting in Cairo – issues can only be touched on —— **470**	
9.3	The second meeting in Rabat: many delicate points —— **470**	
9.4	The third meeting: very anthropological issues —— **471**	
9.5	The fourth meeting: the culture of dialogue and religious values —— **472**	
9.6	The further development: less issues, more discussion —— **472**	
9.7	Surprising unanimity on the Holy Land: "two state" solution —— **474**	
9.8	Human dignity and human rights in armed conflicts —— **474**	
10	Joint Dialogue Committee with the Al-Azhar University —— **475**	
10.1	The first contacts with Al-Azhar —— **475**	
10.2	The founding agreement —— **477**	
10.3	The development of the cooperation: better and better despite difficulties —— **477**	
10.4	The witness of the Al-Azhar at the prayers for peace in Assisi in 2002 —— **479**	
10.5	Meeting in 2002 – a start with activities against religious extremism —— **479**	
11	Regional dialogues —— **481**	
11.1	The regional dialogues: an activity of the whole Church from the very beginning —— **481**	

11.2	North Africa in 1988 – also possible stimuli for others —— **481**	
11.3	West Africa in 1991 – the decisive point: democracy —— **483**	
11.4	South East Asia in 1994 —— **484**	
12	Future prospects: difficult but still hopeful —— **488**	
	Notes —— **489**	
IV	**Evaluation —— 529**	
1	The statements of the magisterium —— **529**	
1.1	The Second Vatican Council and its time —— **529**	
1.2	The official statements by the magisterium —— **530**	
1.3	The magisterium of Paul VI —— **531**	
1.4	The magisterium of John Paul II —— **531**	
1.5	Documents from the Curia —— **537**	
2	The Secretariat's own publications —— **538**	
2.1	The basic starting point —— **538**	
2.2	Publications of the Commission for Religious Relations with Muslims —— **539**	
2.3	Additional publications of the Pontifical Council for Interreligious Dialogue —— **557**	
3	The Christian-Muslim dialogues —— **558**	
3.1	The beginnings —— **558**	
3.2	Dialogues with Libya —— **560**	
3.3	The European situation in the dialogue —— **567**	
3.4	Dialogue on holiness —— **567**	
3.5	Dialogues with the Royal Academy for Islamic Civilization Research, Jordan —— **569**	
3.6	Dialogue with Iran —— **596**	
3.7	Islamic-Catholic Liaison Committee —— **597**	
3.8	Dialogues with the Al-Azhar University —— **598**	
3.9	Regional dialogues —— **600**	
3.10	After 40 years of dialogue and September 11th —— **604**	
4	Final remarks —— **604**	

Bibliography —— 606

Index —— 641

Introduction

1 Preliminary Remarks

This study has a significant pre-history. Almost as soon as my dissertation *Christians and Muslims. The Dialogue Activities of the World Council of Churches and their Theological Foundation*[1] appeared, which reviewed the subjects of the Christian-Muslim dialogue of the World Council of Churches from its beginnings to the early 1990s, and also described the theological debates about the bases for interreligious dialogue in general, I received a noteworthy letter from Rome from Father Maurice Borrmans, responsible at that time for the periodical *Islamochristiana*. He wrote "that German girl" should now come to Rome to undertake a similar treatment of the Christian-Muslim dialogues of the Vatican as well. It was a very spontaneous letter; later on, he could not say exactly what had led him to write like that. As far as I was concerned, although I had sworn to myself three months earlier that I should never again get involved in battling with so much material of this kind, I was immediately enchanted by the idea. The implementation took a long time, however, because it was more difficult to finance a Protestant than to entrust the material to her. So I was only able to begin the research work quite a number of years later, supported initially by my own Bavarian Protestant Church which granted me a scholarship to work on a habilitation thesis for this purpose.

The scope of my study was determined to a large extent by an agreement with a Finnish colleague, Risto Jukko, who was also interested in these dialogues. He undertook the theological part while I concentrated on the anthropological aspects which were considerably more extensive, as could be expected. In Christian-Muslim dialogue, anthropology must also take account of the Islamic approach which is considerably more juridical and also comprises a vast wealth of detail. The major emphasis of the dialogues lay on such anthropological issues in the broadest sense. Despite repeated efforts on the Catholic side, the Muslim participants refused to get involved in theological questions. For this reason, above all, Risto Jukkko's study *Trinity in Unity in Christian-Muslim Relations*[2] was available long before mine. The title describes the programme; Jukko was concerned with the theological approach of the Catholic Church in relation to the dialogue, with matters such as Christology, pneumatology and the doctrine of the kingdom of God in this context. In so far as questions of anthropology arose, the question was how they fitted into this framework of Christian theology which was certainly important, but it was not a question of the details, nor of the practical consequences apart from the dialogue.

2 The research situation

As the spontaneous inquiry from Maurice Borrmans demonstrates, there was (and is) so far no study of this kind although studies related to these subjects have been made, including some which he himself produced such as *Dialoguer avec les Musulmans*[3] to name just one recent one. But the studies which already exist have other main emphases, namely theological ones. The first characteristic of my work is its consistent anthropological approach, rather than starting from theology which has been the primary focus elsewhere and also in Risto Jukko's study, although it is clearly an anthropology rooted in and based on theology on *both* sides. As a result, this study ranges from theology to practice, from theological statements via human beings to the question what humans should do, may do or may not. So it is of interest not only for theologians but for everyone who is concerned about the practical and, indeed, very human coexistence of Christians and Muslims. The differences between Christians and Muslims relating to their image of God are probably well known but are all the problems really solved if one excludes theology? Or, to put it another way, why are there so many discussions about Islam and Muslims in society and politics which go beyond theology? These questions lead in to the central issue of this study, to theological anthropology when comparing Christians and Muslims.

The practical starting point of my investigation was the documents of the Roman Catholic Church which I had been invited to examine. My examination and the study to which it led comprised a series of steps; one could also say that it proceeded in concentric circles from outside to inside, starting with the magisterial approach of the Catholic Church to these questions and in this case again with the Second Vatican Council. This approach is set out in the first of the three main chapters and in a hierarchical and chronological order as is appropriate to documents of this kind. The second main characteristic of this study is that it consistently pursues how the issue was dealt with beyond the magisterium as such by looking at the Secretariat for Non-Christians, as it then was, now the Pontifical Council for Interreligious Dialogue; so, to begin with, the publications of the then Secretariat for Non-Christians which relate to Islam were also examined to see how the treatment of these questions developed. That was the second step which was undertaken historically and which then logically became the second step of my analysis to which the second main chapter has been devoted. The third main chapter then deals logically with how the Christian-Muslim dialogues themselves handled the same issues. Here, on the one hand, the main lines which were followed become recognisable, mainly on the Catholic side and then also on the Islamic side. But certain individual publications and the course of certain dialogue meetings are also presented because the request addressed to me was precisely to make things available which were otherwise not accessible to a broader public and which had also not been taken up in the work of Risto Jukko because of its main focus. (I have deliberately maintained this basic structure right through to the evaluation precisely because, in contrast to the German study as a whole, it should also be accessible in translation in Italy.) The third main chapter follows the line of the various different dialogue partners which the Vatican had in the course of time. This was fairly natural because there was a whole series of different Muslim participants with whom dialogues were conducted over longer periods with the clear desire, which was never stated explicitly in the documents consulted (cf. below), to guarantee as broad a representation as possible of dialogue partners on the Muslim side. This kind of sub-division according to dialogue partners makes it possible to focus directly on individual strands of the dialogue. At the same time, when combining the two previous chapters and against that background, it becomes evident where the Muslim emphases lay in relation to the question of anthropology, to what extent these were identical with the Catholic approach, where they differed typically and beyond the limits of individual dialogue partners, and with what consequences. Finally, my investigation pursues the dual aim of making Vatican material on the Christian-Muslim dialogue accessible and of analysing it from the point of view of anthropology and its practical expressions and consequences. That also meant that other Catholic dialogue activities, such as those of regional bodies, religious orders or universities, could not be included because that would inevitably have exceeded the limits of what one study can cover. This was a regrettably necessary limitation in order to keep the subject manageable. The interreligious dialogue of the Catholic Church was naturally more comprehensive, not only with regard to the Catholic participants worldwide but also with reference to the dialogue with religions apart from Islam. That applies especially to the fundamental theological decisions taken at the beginning, but it was precisely not the aim of this study to pursue those lines further. Anyone interested in that can refer, for example, to the book published by Ernst Fürlinger *Der Dialog muss weitergehen*[4] which also deals with the dialogue with Judaism, Buddhism, Hinduism and traditional religions during the long pontificate of John Paul II, although it is clear that the main focus was on Islam.

When looking at the Christian-Muslim dialogues which I have analysed, one can sense that an effort was made by the Vatican to ensure that the Muslim partners were representative to some degree, which meant speaking with a whole series of different representatives, on the one hand, and, on the other, there was always the awareness that its own representative participation would never be matched by any other partner in any interreligious dialogue. Who is really prepared to join a dialogue, which subjects are discussed and with what intensity, and how this will be documented or not, all inevitably depends very much on the partners in the dialogue. In practice, especially at the beginning, there were also various random features, to put it mildly (cf. situation of sources). Apart from that, there was an attempt on the Vatican side from the very beginning to include all the expertise available on the subject and to link it up through the Secretariat's own bulletin, going beyond the dialogue with just one religion, although the special standing of Islam as a post-Christian religion was repeatedly emphasised because this made dialogue with Muslims particularly difficult. (The dialogue with Judaism is treated *de facto* as related to Christian ecumenism and has therefore not been included in the work of the Pontifical Council for Interreligious dialogue from the start, however one may judge this approach.) Since it was necessary to select a date for the conclusion of the analysis, I chose the end of the pontificate of John Paul II. The practical reason was that I was able to consult most of those relevant documents myself during the research I did in Rome. But it has also become evident that, with the coming of Benedict XVI, there was a certain change of direction, at least for some time. Therefore this period also constitutes a unity of content.

The situation of the sources varies, as is evident also in the study. I had access to the library of the Pontifical Council for Interreligious Dialogue which is normally totally closed to the public. I tried also to gain access to the archives, arguing that, especially after September 11th, it was appropriate to disclose all the positive and negative aspects of the relationships with Islam in order to create a solid basis for further trustful dialogue. But this argument did not help me to overcome the strict rules of the Vatican's archive policy. Even the publication of material from an unofficial guideline for action from the early days of the Secretariat, which I found in its library and was able to consult, was finally prohibited so that my study had to be revised again during the final stage. Presenting the work of the Secretariat, and especially the reconstruction of dialogues on which the extent of the publications varies considerably, would have been made much easier if I had been granted access and reference to archival material. This can only be regretted but not changed at the moment. I have tried to fill this gap by reference to the archives of the World Council of Churches in Geneva and of the Fondazione Pietro Rossano in Vezza d'Alba in Piedmont which has tried to preserve the inheritance of this man who was so important for the early years of the Secretariat. I have included this material as background information. On one decisive point, the very first Christian-Muslim dialogue of the Vatican on invitation from Libya, I received detailed insights from the point of view of a participant in a conversation which I had in Rome, but I had to commit myself not to quote anything directly. My knowledge has merely influenced the form of the presentation which may point the reader to this or that conclusion without anything being stated explicitly.

3 Aids for the Reader

In connection with this study, I was able to draw extensively on the experience I gained during my doctoral study of the Christian-Muslim dialogues of the World Council of Churches and their theological basis. However, there are also some characteristic differences. Since this analysis concentrated from the beginning on anthropological issues, a further sub-division of the questions in line with the individual subjects discussed did not seem to me helpful. In addition, on the Catholic side, far more lines were found that were deliberately chosen and followed up than was the case with the World Council of Churches. I considered it important to point this out time and again. And, precisely in the field of the dialogues, the situation of the sources is so different, as was stated above, that it was simply clearer to set the individual strands side by side. A summary of content drawing them together would present a false picture of the weight of the different dialogues with different partners. But it is again possible, as it was with the dissertation, to read the individual chapters with their comments in isolation although they are related to one another in content. This is made immediately clear by the script of the remarks belonging to the different chapters. It is indeed also understandable that a reader may be interested only in the dialogues or only in the publications of the Secretariat which are difficult to access, and the reader should be able to do this as comfortably as possible. Therefore the presentation remains relatively close to the documents, ranging from literal or indirect quotations to summaries in order to provide an overview. That is what would be expected in this case. But often it appeared necessary to add relevant comments in order to make particular issues clear and comprehensible when they belonged to a broader context. These can be recognised either by a brief comment in brackets or, if they are longer, by the use of italics. These comments have then also usually been included in the evaluation.

In addition, in order to make the study less burdensome, I have again adopted the system of multiple endnotes. The sources of literal quotations are always stated at the beginning, other sources in the order in which they were used and further references then follow with specific mention. For the transcription of Arabic terms, I have followed the system in the documents available to me apart from a few terms which I would consider to have become part of the English language by now. Since they come from what can be considered a uniform source, I felt this was justified. For the sake of brevity, the term "Catholic" is used in the text although it would really be more accurate to write "Roman Catholic". However, when reference is made to the Eastern Churches, this does not mean the Uniates but the Orthodox Churches. But this is very rarely the case. Abbreviations in the bibliography are based on the list provided by Siegfried Schwertner in the *Theologische Realenzyklopädie*. In the bibliography, the same documents in different editions are combined; the same applies to authors even when their names are spelled differently. To this end, I had sometimes to make considerable additions at this point – it is not easy to fit a prince into a normal scheme. Where the names and titles of meetings occur, I have followed the alphabetical order strictly without giving special importance to any article except that they follow one another consistently and words that happen also to start with the Arabic *Al* are only listed afterwards.

Notes

1 Sperber, Jutta, Christians and Muslims, The Dialogue Activities of the World Council of Churches and their Theological Foundation, Theologische Bibliothek Töpelmann 107 (also ev. theol. Diss. Neuendettelsau); Berlin/New York 2000.
2 Jukko, Risto, Trinity in Unity in Christian-Muslim Relations, The Work of the Pontifical Council for Interreligious Dialogue, History of Christian-Muslim Relations 7, Leiden/Boston 2007.
3 Borrmans, Maurice, Dialoguer avec les Musulmans, Une cause perdue ou une cause à gagner? Paris 2011.
4 Fürlinger, Ernst (ed.), Der Dialog muss weitergehen, Ausgewählte vatikanische Dokumente zum interreligiösen Dialog (1964–2008), Freiburg/Basel/Wien 2009.

I The theological foundations for the dialogue with Muslims: Statements by the magisterium

Intended addressees: inside and outside

The work of a Vatican institution such as the Pontifical Council for Interreligious Dialogue is naturally influenced to a high degree by the official statements of the church's magisterium on the issues with which it is concerned. The best example is that this Council itself commissioned a compendium in the Italian language for the thirtieth anniversary of its existence which compiled all the statements related to its work in chronological order, and this volume then appeared in the following years in English and French with the relevant subsequent additions. The immediate aim was twofold: on the one hand, to give its own believers easier access to what their church thinks about non-Christian religions and interreligious dialogue and thus to prepare them for the latter, and, on the other hand, in some way to offer this thinking in a condensed form to the followers of other religions as well – and both, furthermore, on the basis of the relevant statements by this same magisterium[1].

1 The Second Vatican Council

1.1 *Nostra Aetate* – a first look at other religions

As the Secretariat for Non-Christians, today's Pontifical Council for Interreligious Dialogue was a creation of the Second Vatican Council. So the new decisions of that Council on the relationship with non-Christian religions and their followers were naturally of very special significance. At many points, however, they are still of a very general nature; there are few passages which deal concretely with particular religions and their followers. One exception is *Nostra Aetate* (28 October 1965) which, in relation to Islam in particular, identifies agreements in the content of doctrine starting with faith in God, the Creator, which is also emphasised elsewhere. Mention is also made of the hostilities between Christians and Muslims as well as the Muslims' high appreciation for a moral life. The paragraph concludes with a call jointly to protect and promote social justice, moral values, peace and freedom in the interest of all people.[2]

One of the main intentions of *Nostra Aetate* was indeed to examine what people have in common and what urges them to live out their purpose together – clearly an anthropological issue. The answer given by *Nostra Aetate* is that all nations have a common origin and a common goal, namely God, and therefore form one single community. Against this background, no discrimination against any person was possible. People expected answers from the various religions to the obscure and timeless riddles of their existence, for example: what is the human being? What is the meaning and goal of human life? What is good and what is sin? What is the cause and the meaning of suffering? What is death? What is finally the ultimate mystery surrounding our existence?[3]

1.2 *Dignitatis Humanae* – Religious liberty in the name of human dignity

As the name itself reveals, *Dignitatis Humanae* (7 December 1965) relates especially to human beings and the dignity given to them. The dignity of the individual, which was based on God's word but also on reason, and was expressed in reason and free will and, consequently, also in personal responsibility, included the right to religious liberty as well. Human nature compelled people to seek the truth, precisely in the religious realm, and because of their divine disposition human beings also had access to the truth; special emphasis is laid here on the human conscience and divine law which governs the world and hence also human beings. But this also means that people have a duty to seek the truth. The social nature of human beings (and of religion), in turn, implied that this religious liberty could not only be the inner freedom of conscience but had to be freedom for the individual to practise religion in community. This then also comprised worship, teaching, religious authorities, activities of a charitable or cultural nature, the necessary buildings and, not least, propagation also outside of one's own community – naturally, always taking due account of the justified requirements of public order. The right to active freedom of religion had to become a civil right, however; to refuse it constituted a violation of the human person and of the divine order for humanity. A secular authority should recognise and support the religious life of its citizens, but a desire to control or hinder it exceeded its competence. It is striking that the footnotes do not only mention someone like Thomas Aquinas to show how deeply this image of human beings and their capabilities is rooted in the Catholic tradition, but also, for example, the encyclical of Pius XI *With Burning Anxiety* which refers to the concrete threat to this same religious liberty during the period of National Socialism.[4]

What is not mentioned is the problem of a secular authority which simultaneously understands itself as religious and has therefore developed a somewhat different view of religious liberty as is the case with Islam.

1.3 Other declarations of the Council – humankind comes into view

In the other statements of the Council, the anthropological references are less clear, at least in an interreligious framework. *Lumen Gentium* (21 November 1964) emphasised the Church also as a sign of and means to the unity of the human race but underlined, at the same time, that traces of goodness are also present in people's hearts and minds and in their rites and cultures, and the Church can incorporate, purify and perfect them.[5] *Ad Gentes* (7 December 1965) went further into this with reference to proclamation: all Christians and especially priests and missionaries had to (get to) know the culture of the country (also on the spot) in which they were living and working in order then to be able to recognise and relate to these traces. In this missionary context, immigrants were definitely seen positively. Fundamentally, Christ, and with him the Church which witnessed to him, was not a stranger anywhere, because Christ was the model of the renewed humankind according to which Christians should live also in the religious culture of each particular country. Thus mission was explicitly closely connected with human nature. From a purely human point of view, as is also mentioned, the gospel had been the ferment for freedom and progress and was still the ferment for fraternity, unity and peace.[6] *Gaudium et Spes* (7 December 1965) again emphasised, on the one hand, the humanity of Jesus and, on the other, also the possible function of the Church as a link between different human communities and nations. The Church explicitly recognised everything which promoted the unity of humankind; among the human merits mentioned, the will to create better living conditions for all can be given special mention here. The document ends with a sentence which certainly applies particularly to the Muslims with whom, as stated elsewhere in a similar passage, we do not share the same faith but many values.

> "Et puisque nous sommes destinés à une seule et même vocation divine, nous pouvons aussi et nous devons coopérer, sans violence et sans arrière-pensée, à la construction du monde dans une paix véritable."[7]

2 The official statements of the magisterium

2.1 The pontificate of Paul VI

2.1.1 Paul VI and the founding of the Secretariat for Non-Christians
The most important documents initially for the Secretariat for Non-Christians were certainly the two which determined its establishment and goal, *Progre-*

diente Concilio (19 May 1964) and *Regimini Ecclesiae Universae*. The first document emphasised explicitly that this step was based both on the progress of the Second Vatican Council and also on the personal considerations of Pope Paul VI. Cardinal Paul Marella was appointed president of the newly established Secretariat because of his wisdom/caution and his knowledge of other religions – in that order. The second document of 15 August 1967 is more detailed as far as the organisational provisions are concerned. It stipulated that the president was to be assisted by a secretary and an under-secretary. The members of the Secretariat, with the exception of the Prefect of the Congregation for the Evangelisation of the Peoples, were to be chosen by the Pope, either from the ranks of the College of Cardinals or from proposals made by the bishops' conferences in mission countries or other countries with a high proportion of non-Christians. The Secretariat was to be advised by a group of experts approved by the Pope. Attached to the secretariat was an office for relationships with Muslims (founded on 1 March 1965), the only one of this kind which clearly indicated the importance attributed to Muslims. The goals, however, were formulated in a much more open way. The Secretariat was not only to promote studies on the issue – explicitly without limiting the competence of the Congregation for the Evangelisation of the Peoples – but also to maintain contacts with non-Christians so that mutual appreciation could develop. Special value was attached to this human aspect. The Secretariat was responsible for finding ways of doing this and was also the partner to be approached and the competent body for initiatives and training in the field of dialogue.[8]

2.1.2 Ecclesiam Suam – the first step towards interreligious dialogue

In this field of dialogue there are also instructions, which probably provided strong inspiration even for the Council documents, to be found mainly in the encyclical *Ecclesiam Suam* (6 August 1964), which was really concerned about the attitudes which the Catholic Church should adopt in the world of today. These included the task of proclamation, of mission. For its inner dynamic, love of one's neighbour, the term "dialogue" was used which had recently been introduced. The further comments made clear how this was to be understood. The whole of divine revelation can be understood as dialogue. God offered the Church a dialogical relationship through Jesus Christ in the Holy Spirit and the Church in turn had to establish such a relationship with humankind. But, seen more anthropologically, dialogue with its variability can also be described as the best form of relationship with the people of the modern world. The goal of conversion was not lost sight of here, but before one could convert someone, indeed in order to be able to convert them, one had first to take a step towards them and talk with them. So dialogue was also an instrument of mission. But conversion was not the direct focus and hence the freedom and dignity of the other person could be respected; the focus was on the advantage of one's partner (one should not understand oneself as civilisation but certainly as its agent) and on a broader sense of common feelings and convictions. Hence dialogue was also an art of spiritual communion and not merely conversation. As such, dialogue should be clear, i.e. dialogue presupposed, indeed required, that people understood one another. It had to be gentle and not proud, offensive or polemical. It should be marked by trust because that was the prerequisite for friendship. And it should be (pedagogically) skilful and take account of the psychological and moral condition and the sensitivities of the listeners. In a true dialogue, there could be no prior condemnation; it was a combination of truth and mercy, intelligence and love. The climate for dialogue was friendship, indeed, service of the other. A dialogue of this kind is in itself a declaration in favour of honest peace in freedom. The partner in this dialogue was also identified more clearly, although it was naturally emphasised that the message of the Church was addressed to all people and especially to the poor. Muslims, as worshippers of the one God, were counted within the inner circle and were thus given special emphasis. They deserved admiration for what was true and good in their worship of God. On the very general list of spiritual and moral values, which the Church can promote and defend jointly with other religions, religious liberty (again), the fraternity of all human beings, a healthy culture, social welfare and civil order are mentioned. But another sentence was significant as a reason for dialogue from a theological and anthropological point of view.

> "S'il existe dans l'homme une *âme naturellement chrétienne*, nous voulons lui rendre hommage de notre estime et de notre conversation."[9]

2.1.3 Other encyclicals: help for human beings – both inward and outward

Evangelii Nuntiandi (8 December 1975) again made clear that the new developments of the Second Vatican Council should in no sense be understood as a reason for abandoning the proclamation of the gospel to all people. Naturally, this should not use means like forcing people's conscience or talking them into something but, as such, preaching the gospel was not an attack on religious liberty but rather a form of respect for it, because that was what provided the possibility to chose. Indeed, there was even a human right to the Good News. Finally, and this was the starting point, while religions were certainly the living expression of the soul of many people and bore within them the echo of many centuries of an honest and genuine search for God – here mention is made of prayer and elsewhere of penitence, this search had not reached its goal. For this reason, people – and their culture(s) – needed the gospel.[10]

But here, as already underlined by *Populorum Progressio* (26 March 1967), it was not only a matter of an economic but also, above all, of a human development which also had moral and spiritual aspects and these constituted the real goal. The central focus of dialogue should be the human being and not goods or techniques. In the long run, that was also a more lasting aid to development, which again made clear how closely the basic decisions and aims could be related to quite concrete results:

> "Entre les civilisations comme entre les personnes, un dialogue sincère est en effet créateur de fraternité. (...) Passée l'assistance, les relations ainsi établies dureront. Qui ne voit de quel poids elles seront pour la paix du monde?"[11]

2.2 The pontificate of John Paul II

2.2.1 The early encyclicals – little directly related to Islam

Pope John Paul II continued to follow the line which had been laid down by the Second Vatican Council and his predecessor. Very little was said about Islam, especially from the anthropological point of view. Here and there, human dignity was emphasised and also respect for people of other faiths, for example in *Familiaris Consortio* (22 November 1981) in connection with interreligious marriages. In this context, the emphasis was again that love of one's neighbour implied seeing each person as a brother, especially when the latter was poor and weak, suffering and treated unjustly – a statement which can be of great significance in practical situations. Despite the awareness of human sin(s), according to *Sollicitudo Rei Socialis* (30 December 1987), the main focus was on human beings and their abilities based on their being in God's image. There were no grounds for pessimism, passivity or hopelessness; that could be as much a sin as egoism or the desire for power. The list of different forms of dialogue in *Redemptoris Missio* (7 December 1990) is also of practical interest. They range from an interchange between religious experts or official representatives to cooperation for the sake of holistic development and the protection of religious values, from a mutual exchange of religious experiences to a dialogue of life in which there is mutual witness to one another's human and religious values and mutual assistance in building up a more just and fraternal society. *Dominum et Vivificantem* (18 May 1986), on the other hand, argued along more religious, theological lines and referred to the fact that human beings had only one and the same ultimate vocation and this was one theological reason for assuming that the Holy Spirit offered all people a share in the Pascal mystery in a way known only to God. But the argumentation became most personal when the Pope (repeatedly) discussed the prayer of the religions for peace in Assisi which he had brought into being. Peace and its prerequisite, the development of the human being as an individual and in community, were also a religious matter and, according to his personal conviction stated many times, every genuine prayer was the work of the Holy Spirit who was present in a mysterious way in the heart of every human being.[12]

2.2.2 Redemptionis Anno –focus on Jerusalem and the Holy Land

The encyclical *Redemptionis Anno*, published for good reasons on Good Friday, 20 April 1984, occupies a special place. A pilgrimage of the Pope to the Holy Land, like that of Paul VI 20 years earlier, had unfortunately not been possible. Nevertheless, the encyclical stated and demonstrated that the Pope was very familiar with the problems of the Holy Land and its inhabitants. There is hardly any other place connected with dialogue between Christians and Muslims where human suffering is so immediate and seems so tragically inescapable. The most disturbing thing, when reading this document about two decades later, is how frighteningly up to date it is. Two peoples and three religions still continue, using well-founded religious and historical arguments, to claim this country and especially Jerusalem as their holy city. It was maintained that Jerusalem was a city of peace for the whole of humankind and naturally particularly for those who worship one God. But, in reality, Jerusalem was the cause of constant rivalries, of violence and special demands – a problem which affected not only those who live there but all who have brothers in the faith there. They should stand up for the preservation of the unique character of this city. The Jerusalem question was decisive for establishing peace with justice in the Near East. A concrete, just solution had to be found which could safeguard and balance the various interests and hopes in a lasting way. This needed to take the form of a special status guaranteed by international law and agreement between the nations so that no party could even challenge it again. According to the Pope's conviction, the religious nature of the city and, above all, the practice of a common, monotheistic faith could help to achieve this. On the other hand, even at that time, he made overall reference to the efforts undertaken along these lines especially by the popes in the twentieth century and spoke of peace there as his own personal dream. As far as Israelis and Palestinians were concerned, he could only speak of an antagonism which nothing seemed able to resolve. The only recourse was prayer:

"Et puisque le fruit de la rédemption est la réconciliation de l'homme avec Dieu et de chaque homme avec ses frères, il nous faut donc prier pour que, à Jérusalem aussi, dans la Terre Sainte de Jésus, ceux qui croient en Dieu, après tant de divisions et de désaccords douloureux, trouvent la réconciliation et la paix."[13]

2.2.3 The New Catechism – a summary of the previous one

The new *Catechism* published on 11 October 1992 also has a special standing, although of a different kind. As various commentators have underlined, it does not contain much relating to Muslims, Islam or followers of other religions in general, and especially nothing new, but this was also hardly to be expected considering the intention of a catechism. As in *Lumen Gentium*, Islam was mentioned on the theological and not the anthropological level. It was generally emphasised that the relation of the Church to non-Christian religions was based (anthropologically) on the origin (e.g. creation myths) and purpose of all persons being the same, so that they also had a yearning for God in their hearts that was expressed above all in prayer but also in sacrifice, cult or meditation. These expressions were considered so universal that human beings could certainly be described as religious beings. It was emphasised explicitly that the Church defended the ability of human reason to recognise God – and this was indeed the foundation for dialogue with followers of other religions. The standing and mandate of the (Catholic) Church remained unaffected by this; all that was mentioned in a quotation from *Lumen Gentium* was the possibility of salvation for non-Christians who, through no fault of their own, did not know Christ and the Church but obeyed their conscience – although no details were given. Freedom of conscience and thus also of religion were included as part of human dignity. The details include the fact that God's covenant with Noah (and hence with all of humanity) is mentioned in the catechism and apparently considered still valid, and there is reference to interreligious marriages and the special problem these raise for the education of the children. There was also a very interesting appeal to rich countries as far as they were able to welcome strangers who needed security and material welfare for survival. On the other hand, immigrants should gratefully respect the material and spiritual heritage of their host countries. This is one of the many points at which followers of other religions or even Muslims were not mentioned explicitly but can be understood, as pointed out several times by Bishop Michael Fitzgerald, the secretary of the Pontifical Council for Interreligious Dialogue at the time, in his commentary on the Catechism. And one can only agree with his concluding opinion on the question of other religions in this Catechism.

"It may well be that a later catechism will have more to say about the relations of the Church with other religions."[14]

2.2.4 Other documents: considering common and specific situations and tasks

Tertio Millenio Adveniente (10 November 1944) then mentioned two essential areas for the work of the church: confrontation with secularism and dialogue with the major religions – in itself a telling formulation. And, within this interreligious dialogue, the dialogue with the Jews and that with the Muslims was to have a special place. Here the Pope expressed the wish that God would grant that such encounters could also be held in particularly symbolic places, and he immediately made this more concrete by referring to Bethlehem, Jerusalem and Mount Sinai.[15]

Vita Consecrata (25 March 1986) occupies a special place dealing with the role of the religious orders in interreligious dialogue among other things, although it did not mention Islam by name. But, when it emphasised e. g. the special importance of the presence of the contemplative life in places where other religions were more widespread than Christianity, Islamic countries naturally come to mind. However, where it discussed interreligious dialogue between different forms of monasticism, Islam was relatively uninteresting in comparison. The presence of religious orders, it stated generally, encouraged interreligious dialogue – in the sense that, in the context of the mandate to evangelise, interreligious dialogue had now also become a task which the orders could not avoid, starting with the education of their members. The free spirit characteristic of them could especially promote a dialogue of life. But it was also a matter of sharing physical and spiritual suffering and of a common commitment to justice, peace and the integrity of creation. The active orders could develop a dialogue of works and the contribution of nuns was especially important when it was a matter of promoting the dignity of women and of the equality and right understanding of the mutuality of the sexes.[16]

But at this level there were also statements on the situation in particular regions, for example in Africa. In *Ecclesia in Africa* (14 September 1995), Christians were again exhorted not to forget that the Muslims imitated the faith of Abraham and tried to live according to the demands of the Decalogue. It was God's will that we should witness to them while respecting the values and religious traditions on each side. But God did not want to be one in whose name other people were killed. Special attention should be paid to religious liberty, explicitly including the outward, public manifestations of faith (which constitute a particular problem for Islam). The concluding words on dialogue with the Muslims in Africa could really apply everywhere.

> "Chrétiens et musulmans sont appelés à promouvoir un dialogue exempt de tous les dangers qu'entraînent un irénisme de mauvais aloi ou un fondamentalisme militant, et à s'élever contre des politiques déloyales, ainsi que contre tout manque de réciprocité en matière de liberté religieuse."[17]

Another statement of this kind was the *Exhortation Post-Synodale pour le Liban* of May 1997. It thanked the Muslim and Druse delegates for their presence at the synod and for their active participation in the dialogue and underlined the long tradition of relations between Christians and Muslims in that country. Precisely at that time of reconstruction, dialogue, which must always be concerned about moral values, social justice, peace and freedom, but also about the protection of life and the family, in addition to concrete cooperation, played an important part in rebuilding mutual trust. Religious dialogue was indispensible for this although dialogue was not merely a dialogue of intellectuals but a dialogue in normal, everyday life and should promote coexistence with its various tasks and challenges. It was important (precisely in Lebanon) to work for the common good and not to concentrate on particular interests, whether of specific individuals or a specific group. The Catholic Church explicitly indicated its openness for dialogue and cooperation with other Arab countries. Because they belonged to the Arab cultural context, Lebanese Christians, together with the Christians in other Arab countries, were considered to have a special opportunity for engaging in an authentic, profound dialogue. Lebanon, in particular, could have a pioneering role in dialogue and cooperation, but in the context of this document there were again statements which went far beyond that country and even the Christian-Muslim dialogue to which reference was being made.

> "Pour des hommes de bonne volonté, il est impensable que des membres d'une même communauté humaine, vivant sur la même terre, en viennent à se méfier les uns des autres, à s'opposer et à s'exclure au nom de leurs religions respectives."[18]

The later *Apostolic Exhortation: Ecclesia in Asia* (6 November 1999) naturally did not deal so exclusively with Islam but often mentioned the religions and cultures of Asia and their traditions and values, indeed, spirituality in a very general sense. Where Islam was mentioned explicitly, it was often more a matter of problematic situations: the difficulties of coexistence in certain, predominantly Muslim states and, of course, the situation in the Holy Land and especially in Jerusalem which gave rise to a new papal appeal to keep the peace and maintain the unity of the city. The valuable experiences of the Eastern Catholic churches particularly in dialogue with Islam were also mentioned and the great opportunity which the forthcoming turn of the millennium offered for interreligious dialogue generally and for encounters with the leaders of the major world religions in particular. Even where it was not mentioned, Islam can be clearly recognised in the background when it was pointed out that in various countries religious liberty was systematically restricted or not granted at all. This was where the key term of theocracy came in, but it was also a matter of evangelisation being prohibited so that only a silent witness of life (especially of the religious orders) was possible, and of the problem of discrimination because of conversion to Christianity. There were even two explicit appeals for the recognition of the right to freedom of conscience and religious liberty and other basic human rights. Mention was also made of positive experiences gained, e.g. with Catholic schools which promoted interreligious understanding, with interreligious dialogue at the expert level, joint activities for holistic human development and the defence of human and religious values (the field of the mass media was referred to here particularly as an important future task) and the interreligious gathering for prayer on 7 October 1986 in Assisi – prayer and contemplation in interreligious dialogue were especially important for the Pope; dialogue was not simply a strategy for more peaceful coexistence between nations. The Church saw itself in a much more profound way as the sacrament of unity for the whole of humanity, but this also included commitment to peace, justice and reconciliation on an international and interreligious level and based on two explicit basic principles.

> "She <i.e. the Church> continues to insist on the negotiated and non-military resolution of conflicts, and she looks to the day when nations will abandon war as a way of vindicating claims or means of resolving differences. She is convinced that war creates more problems than it ever solves, that dialogue is the only just and noble path to agreement and reconciliation, and that the patient and wise art of peacemaking is especially blessed by God."[19]

Finally, *Novo Millenio Ineunte* made several references to the new millennium having a mixture of different religions and cultures also in originally Christian countries, and stated as the challenges of this millennium the ecological crisis, the problems of peace, contempt for the fundamental human rights of many people (followed by an excursus on respect for life especially at its beginning and end and in view of the potential of genetic engineering, explicitly emphasising that it was not a matter here of forcing an approach based on faith on unbelievers but of interpreting and defending values which were inherent in human nature itself) and, as a major challenge, interreligious dialogue.[20]

In conclusion, mention must be made of the *Exhortation apostolique post-synodale "Ecclesia in Europa"* (28 June 2003). Under the heading of dialogue with other religions, it underlined the extreme importance of the dialogue with Judaism and then also with Islam, and in the latter case the expression *"juste rapport"*[21] was used which, as can be deduced from the following sections, related mainly to mutuality in questions of religious liberty for which the Church was an advocate – in line with its previous commitment to this question, but also because this mutuality was not yet a reality and in Europe, in particular, the Church's contributions to it had so far been unilateral. Certain sentences were quite critical, and critical also in the original sense of the term meaning distinguishing; they introduced a very new tone and should probably be understood at least partially in connection with the debate in the European Union either on a reference to God in the constitution or on the issue of the potential membership of Turkey, both of which were also questions of Europe's identity. One must be aware, it stated, "de la divergence notable entre la culture européenne, qui a de profondes racines chrétiennes, et la pensée musulmane."[22]

3 Further statements of the magisterium

3.1 The magisterium of Paul VI

The declarations by Paul VI (and all the more those of John Paul II), which can be seen as in any way relevant to anthropology in the Christian-Muslim dialogue, are so numerous that they cannot merely be ordered chronologically if one wishes to grasp them in a meaningful way. It is helpful to identify to whom they are addressed. Which statements are directed (more) to the inside, primarily of course to the Secretariat for Non-Christians itself, but also to a broader circle of church officials and faithful? In this case, different emphases and clarifications are to be expected from statements which are addressed more or less directly to the Muslims themselves. The presentation of the magisterial statements of Paul VI follows this line from inside to outside in considerable detail. Both aspects certainly reflect the development and importance of this initial phase.

3.1.1 For those responsible for the dialogue: much praise and encouragement

On 25 September 1968, almost five years after the establishment of the *Secretariat for Non-Christians*, Paul VI received its members and advisers. In his address, he began by praising their competence, experience – and caution. He again repeated that the work of the Secretariat was a matter of knowledge and understanding of and of dialogue and cooperation with other religions. Respect and appreciation for their moral and spiritual values would enable brotherhood to increase in the family of humankind and the ideal of the unity of all people in the light of God would become effective. It was a matter of the *homo religiosus* – and here explicit reference was made both to Thomas Aquinas and to *Nostra Aetate*. After further, significant praise both for the cooperation with other dicasteries of the Roman curia and for the necessary independence of the Secretariat, Paul VI emphasised that the Secretariat had to establish itself within the Church as a visible, institutional sign of the dialogue with non-Christians, a delicate area in which there were various dangers. It is fairly clear that the address as a whole was not dealing with new insights but with confirmation of and encouragement for the chosen path of dialogue. The Pope's closing wish, as confirmed by many of those concerned, was not merely an empty phrase but intended as seriously as it was necessary: "Que l'assurance de Notre estime et de Notre intérêt vous encourage!"[23]

The next address of this kind, given four years later on the occasion of a congress of the Secretariat, also stated that the work of the Secretariat was noble and demanding and had to overcome many barriers. But at the same time there was high praise: the Secretariat had contributed a lot to a new climate between the Catholic Church and the major religions (thus including Islam). The focus was also on human beings in these religions when Paul VI emphasised, e.g. that knowledge of other religions (a recognised aim of the Secretariat) had as its aim love, dialogue and serious, fraternal cooperation. The religions naturally contributed to peace, brotherliness and justice, inspired morality and awakened hope, and social contacts became difficult without religion. Every person shared in the unfathomable mystery of God because as God's creature he/she had the likeness of God and represented the humanity of Christ.[24] *These last two reasons, which go far beyond human beings being God's creatures in a clearly Christian understanding, would naturally not have been communicable to Muslims and thus demonstrate in another way that this is also a purely internal document.*

The address on the tenth anniversary mentioned the visible progress (organisation of encounters, link with about 25 dialogue commissions in all the bishops' conferences), but again emphasised that the meeting point with humanity was precisely the search for the likeness of God which had left its traces in all people and also provided the urge for the unity of all. With numerous quotations and strong words, it underlined what a milestone the Council had been in this respect. e.g. "En fait, les Pères Conciliaires ont vu dans ces Confessions une expression très significative – même si elle est incomplète – du génie religieuse de l'humanité".[25]

Still more immediately connected with the papal promotion of Christian-Muslim dialogue was the transfer of the Institute for Arabic Studies (Africa missionaries) to Rome as the Pontificio Istituto di Studi Arabi e d'Islamistica. The often mentioned idea behind it was to facilitate access in general by means of access through language and, inter alia, thus bring about better coexistence and cooperation and come closer to the more just, more peaceful and more brotherly world referred to so often. This was underlined particularly in an address to the oriental patriarchs who lived where religions, cultures and peoples collided concretely, the difficulties could not be denied and peace efforts were especially necessary. The papal speech offered some interesting suggestions. Cultural differences were seen as divine providence which should serve to supplement and enrich and not lead to hostility and an attitude of conqueror and conquered, and naturally Western culture owed much to Arab culture. With regard to the refugee problem, there was a suggestion of concern for development and especially education because this commitment would finally also lead to freedom.[26]

The address to the College of Cardinals on 22 December 1967 was of a rather special nature, six months after the Six-Day War. It was particularly concerned with Jerusalem and the holy sites of Judaism, Christianity and Islam. The emphasis, in line with earlier and later papal pronouncements, was on freedom of worship, respect, the preservation of and access to the holy sites which had to be guaranteed and (the latter) protected by special privileges in the form of a special status that had to be guaranteed by an international organisation. But there was also reference to the free exercise of the legitimate religious and civil rights of all persons, places and activities of all communities on Palestinian territory.[27] *The indirect, internal churchly character of this statement is interesting because one would rather have expected this in another framework before a different audience.*

3.1.2 To his own faithful – cautious publicity for the new approach

It is significant that the clearest statements on anthropology in an interreligious (if not Christian-Muslim) framework were made in contexts which had nothing directly to do with the institutional dialogue or the Middle East. Usually, during a general audience or on some other occasion, they were simply addressed to the faithful and emphasised, almost apologetically sometimes, that others also had moral and religious values, that they were better than they appeared, and so it was important to awaken this residue of goodness first and foremost through dialogue – or (later) that this had already happened. One very neutral but impressive statement, on the other hand, was that "l'homme qui pense, qui agit, qui commande, qui souffre, qui s'exprime artistiquement, saisit quelque chose de Dieu".[28] The counterpart to the apologetic, coaxing tones was strong words which made clear that no wrong conclusion should be drawn from this. The unity of humankind corresponded to the unity of truth, namely of religion which linked human beings with God; a positive, objective religion corresponded to the subjective religiosity of human beings. This was not just a fact but a necessity. Paul VI became still more direct when speaking to a general assembly of the synod of bishops which was to discuss the problem of dialogue versus mission when he referred to the "Pasteur suprême successeur de Pierre, principe perpétuel et visible et fondement de l'unité".[29] *That is extremely clear and confirmatory addressed to a Catholic audience; even representatives of other Christian confessions would be unable to share such a theology/anthropology, let alone Muslims.*

3.1.3 Addressed to the Muslims: taking account of many situations

Similar statements, although less forceful, were made only to the diplomatic corps which is made up both of Christians and of representatives of other religions. There was reference to the Catholic Church giving individuals and also the nation or civilisation what they still lack for perfection or, in a less anthropological and more theological vein, to the idea of a progression of the light which guides us from natural reason via religious traditions (that which is true and good in them) to Christianity.[30] But the speech to the delegations of the nations at the Council mentioned nothing of this kind. There it was a matter of the moral law in the human heart and, looking ahead, of the positive effects which the declaration on non-Christian religions would have on Catholics and thus indirectly for peace. Moreover, the reference to the declaration on religious liberty pointed out that this was usually acknowledged in theory but less in practice. Religious liberty, and often more precisely the freedom for minorities to evangelise, raised an issue which Paul VI had already discussed before in a prominent place and which had become something of a perpetual issue, usually with reminders and exhortations, whether the addressee was the Ugandan parliament, or the Tunisian or Iranian ambassador (whose connection with the ancient Persian king Cyrus offered an excellent point of contact in relation to minorities). Among the commendatory references in this connection, we find the consciously theological but still neutral constitution of Indonesia and the religious policy of Pakistan since the establishment of that republic.[31] Overall, many of the papal statements in relation to particular Muslims were also quite concretely determined by the situation and therefore less abstract, praising peaceful Christian-Muslim coexistence where this existed, e.g. in Senegal, or expressing concern about fraternal coexistence like in connection with the civil war in Lebanon. Naturally, a peace initiative for Palestine received papal support and there was reference once again to the status of the holy sites and Jerusalem and to religions working for unity on these questions and not divisively. Only once was this effort of the Church for justice and peace explicitly justified in relation to a Muslim, the ambassador of Pakistan, with an explicit justification that both faiths worship God together who will judge humanity at the last judgement. Otherwise, the basis was simply given as a common faith or, before a more general audience, there was talk of the goal of civilisation being to organise solidarity between human beings in such a way than no one was deprived of bread or dignity.[32] But some statements and occasions do not fit into this framework in comparison; for example the Pope's pilgrimage to Uganda where he celebrated Catholic, Anglican and explicitly also Muslim martyrs. An address in the Philippines also went very far when the Pope addressed the peoples of Asia about the spiritual vision in their own tradition by means of which they were able to resist materialism and even also help Western civilisation to deal with the dangers which progress can bring even when viewed positively. But, after a lapse of nearly 30 years, there was another anthropological or simply human question which still provided food for thought.

> "Si la conscience de la fraternité universelle arrive à pénétrer vraiment le coeur des hommes, auront-ils encore besoin de s'armer au point de devenir assassins aveugles et fanatiques de leurs propres frères, innocents en soi (...)?"[33]

3.2 The magisterium of John Paul I – few words on a permanent issue

Pope *John Paul I* gave the beginnings of an answer during his brief pontificate but without going into details when he stated, in a speech to an assembly of the World Conference on Religion and Peace, "Pour que la paix, en effet, se réalise, sa nécessité doit être profondément ressentie par la conscience, car elle naît d'une conception fondamentalement spirituelle de l'humanité. Cet aspect religieux pousse non seulement au pardon et à la réconciliation mais aussi à l'engagement pour favoriser l'amitié et la collaboration entre les individus et les peuples."[34]

What was not considered is the extent to which other religions, and in our case Islam in particular, share this spiritual conception of humanity that was just mentioned in passing, and thus also a comparable conviction about the necessity of peace. The fact that, three decades later, this question is all the more relevant, especially with regard to Islam and its sphere of influence, might indicate that there are greater divergences on this point that the general reference to God as creator and judge and an equally general mention of common values might lead one to believe.

3.3 The magisterium of John Paul II – more than 25 years of involvement in dialogue with Muslims

For more than 25 years John Paul II continually took a stand in his public speeches in favour of interreligious dialogue in general and also for Christian-Muslim dialogue in particular. The effect of this extraordinary continuity, which followed on almost directly from the 15 years of similar activity by Paul VI, should not be underestimated either internally or externally. The fact alone was of great significance, especially in the Islamic world, although this says nothing about the extent to which the details contained in written statements were really perceived. While maintaining the usual sequence, because of the large number of papal statements produced by John Paul II we shall look here more closely at the issues raised in connection with Christian-Muslim dialogue and the anthropological aspects in particular.

3.3.1 Casablanca 1985 – more than just a speech to young Muslims

The *Speech to Young Muslims in Casablanca on 19 August 1985* must receive special treatment in this connection because, in the Catholic context, it is in fact almost on the same level as *Nostra Aetate*.[35] It really did display great thoroughness and a lot of detail and tried, in addition, to use expressions understandable to Muslims and appropriate to the way young people think (as will be demonstrated more precisely). This predestined it for the influential role in the history of theology which it enjoys. The outward occasion for this speech was twofold: a return invitation from the King of Morocco which coincided with the Year of Youth. The combination of the two, as the Pope himself stated explicitly at the beginning of his speech, led to the first encounter of John Paul II with young Muslims.

The Pope began with what they had in common: life in this world with its hopes and fears, via Abraham as an example of faith in God and reliance on his goodness and – at this point one can sense an adaptation to Muslim formulations – of submission to the will of God. Only subsequently was there mention of faith in one God who created the world. Then he outlined what the content of his speech was to be: he wished firstly to speak about God and then about the human values which had their basis in God and were important particularly for young people faced with decisions about their lives. Then the Pope introduced himself, or rather his office and function, but he concluded by saying that he came as a believer and wanted to give a witness of faith about things that were useful according to his experience, and here he spoke about human beings generally as his brothers.

The next section, according to the sub-heading, was to be about obedience to the will of God (the wording and sequence were again perhaps a conscious gesture towards the Muslim understanding) but initially it retold almost the whole story of creation. This constituted the reason why we human beings belonged to God and his holy law directed our path (a formulation which is a clear reminder of sharia and right guidance as fundamental Islamic entities), and we were finally also responsible to him as our judge who was, however, merciful (again a divine attribute which is almost standard in Islamic usage) when someone returned to God. Respect for every other human being was also grounded in the will of God and in creation. Then he moved on from God having created humankind and waiting for it because he wanted to gather everyone together (an audible reference to the way Islam speaks about the last judgement) to the question whether people and the nations who formed a community (that too is a word with an Islamic background) should not prefer friendship and unity between them. Only then did he quote briefly a few relevant statements from *Nostra Aetate* into which he also introduced the idea of the church's special interest in Muslims. After a few sentences about the necessity and basis of Christian-Muslim dialogue in faith, and greater detail about the necessity of prayer, he came to the point that faith was evident in respect for other religious convictions and gave an anthropological reason: every human being expected to be respected for what he/she was and what he/she conscientiously believed. This brought freedom of conscience and religion into play again of course – external pressure in these matters was not worthy of the voluntary respect of reason and the heart which constituted human dignity. That, in turn, was the true meaning of religious liberty which respected God and human beings simultaneously. These statements were a clear attempt to build a bridge for the Muslims to understand how we view religious liberty, although the idea that Muslims gave one-sided emphasis to God's dignity and tended to lose sight of human dignity was still perceptible for those who knew the situation.

The next section also underlined this reciprocity but in a different way when the Pope quoted *Nostra Aetate's* reference to our being unable to appeal to God if we did not behave in a brotherly way to human beings (the basis for this being that human beings are in the likeness of God). This argument was taken up but less forcefully: the reason for respect, love and assistance to every human being resided in their being God's creatures and "dans un certain sens, son image et son représentant".[36] This was a clear juxtaposition of the Christian and Muslim views where both were presented in advance as being open to interpretation – again a bridge to understanding but which was not and could not be spelled out more precisely. Instead, the argument returned initially to human rights and again emphasised that they were an expression of the will of God and a necessity of human nature as created by God.

But then a further section dealt with the idea that human beings in general, and young people in particular, had to do something. (One has to build the world and not just dream about it.) Here, work was seen as service to one's neighbour which incidentally also established bonds of solidarity and could become a purifying human experience. The focus is again constantly on the dignity and freedom of every person. The next step broadened the approach to include the world-wide community in which, even though God intended it otherwise (humanity and the world were compared with a body in which everything had its own function – a New Testament image of the church which is generalised here), cultures and races existed which were not respected. Here the Pope appealed directly to the young people to change this in the interest of peace in the world and praised the tolerance of Morocco especially in relation to the Jews. One can clearly sense and hear that the Pope saw young people, also generally, as allies in the struggle for peace and also for a just world. The theological, anthropological basis given for the latter was that God gave the earth to humanity as a whole for all to draw sustenance from it in solidarity. But this material aspect was not sufficient; human beings, in reference to the New Testament, did not live by bread alone. However, before there was any talk of the word of God, a difficult issue in a Christian-Muslim context, the quotation (for which no source is mentioned) was interrupted and replaced by the necessity of an intellectual and spiritual life. Human beings had to develop their minds and consciences. This inner meaning and direction were often lacking today, also in the commitment to development, that is, to our fellow human beings. Here again there was a commendatory mention of Arab traditions which led on to the mutual recognition of religious values. Explicit reference was made here to the importance of prayer, fasting, alms, penitence and forgiveness, faith in a merciful judge and the hope of resurrection. The theological differences were also stated and not minimised but, according to the conviction expressed in this speech, they can be accepted with humility and respect in mutual tolerance. Christians and Muslims had generally not understood one another properly and opposed one another to the point of exhaustion in polemics and wars. He (personally) believed that God invited us to change our old habits, to respect one another and encourage one another to perform good works along God's path. Then words of thanks followed, especially to God who had placed in our hearts the feelings of mercy and understanding, of forgiveness and reconciliation in the service of cooperation, and a further word about his personal conviction that, along the road marked out, a world can come about in which men and women with a living faith endeavoured to construct a human society corresponding to God's will.

The speech closed with a prayer which again combined Christian and Muslim similarities. The basic attitude of the speaker to the young Muslims in the audience and his own claim were summarised very well in the last paragraph before the concluding section.

> "Je souhaite, chers jeunes, que vous puissiez contribuer à construire ainsi un monde où Dieu ait la première place pour aider et sauver l'homme. Sur ce chemin, vous êtes assurés de l'estime et de la collaboration de vos frères et soeurs catholiques que je représente parmi vous ce soir."[37]

3.3.2 Addressing the office bearers: peace, religious liberty and – spirituality

John Paul II also addressed the *Secretariat for Non-Christians* more or less directly on a number of occasions, the first being shortly after his election on 27 April 1979 at the time of the first general assembly of the Secretariat shortly before the fifteenth anniversary of the solemn announcement of the founding of this Secretariat in St. Peter's basilica. The address consciously follows the line of Paul VI who, as has been shown, actively promoted interreligious dialogue. At the same time, this address like a number of similar ones reflects the progress made in the meantime, in this case the network of local organisations which really made worldwide work possible, forms of dialogue well adapted to individual groups and cooperation e.g. with the World Council of Churches. For the future work, he described the Secretariat as an expression of the will of the Church to establish contact with the many followers of non-Christian religions who were looking for something to give their lives meaning and orientation. What was also important for the dialogue was respect and appreciation for others and a knowledge of them that cannot just be obtained from books. The only really perfect language was that of love. The Pope's wish was that dialogue would also develop in places where Christians were in the majority.[38]

The next similar speech was given at the next general assembly five years later and mentioned e.g. that parts of the Church had established serious and constructive relations with the followers of other religions in their cultural context. The Secretariat for Non-Christians had provided the stimulus for this and now it needed to go deeper and into greater detail, preferably by an exchange of ideas and reflection and with a special commission in each bishops' conference. There was also a positive reference to the presence of the director of the parallel sub-unit at the World Council of Churches, but nothing particular about the Muslims. In some of the statements, one can sense their presence in the background, for example when the Pope spoke of the damage the communities had undergone in the course of the centuries, when he referred to behaviour adapted to specific situations (here there is even explicit mention of Charles de

Foucauld) or when, in an anthropological way, the goal of dialogue, of all religions and all believers is said to be "que chaque homme puisse atteindre son but transcendant et réaliser sa croissance authentique".[39] From time to time the Pope also received the representatives of the World Council of Churches who met annually with their colleagues of the Secretariat for Non-Christians to discuss questions of dialogue, exchange experiences and coordinate their future activities. On the occasion of one such audience in April 1986, he emphasised that the good will of individuals was not enough; the large number of normal faithful had to understand followers of other religions and accept them as brothers and sisters with whom they could share life peacefully, and they should therefore be encouraged to show mutual respect and appreciation and to cooperate in society. In addition, he already spoke of the multi-religious prayer gathering planned in Assisi and underlined prayer as the best means for uniting humankind because it also changed the relationships between those who prayed together: they discovered that they were pilgrims looking for the same goal, brothers and sisters who shared responsibility for humankind and children of the same God and Father.[40] He himself saw his role as being to underline the firm, unchangeable will of the Church to engage in dialogue; dialogue was not to be forgotten even where the Church's proclamation was successful. But, at least when addressing such a general assembly of the Secretariat for Non-Christians, he quite openly named questions which, in his opinion, still needed clarification, e.g. how God is at work in the life of nations with different religions.

But with *Pastor bonus* he also intervened in a juridical way and the Secretariat for Non-Christians became the Pontifical Council for Interreligious Dialogue. Apart from the name, nothing much changed. The aim specified was the promotion of human dignity as well as spiritual and moral values. When required or necessary, the newly created Pontifical Council was to cooperate with the Congregation for the Faith or the Congregation for the Oriental Churches and the Evangelisation of the Peoples on an equal footing. The Muslims still had a special standing. "Auprès du Conseil est constituée une Commission pour la promotion des relations avec les musulmans du point de vue religieux, sous la direction du président du Conseil."[41] Other statements were not so concretely directed to Muslims but the priorities (studies and encounters) listed or directly demanded by the Pope and the issues (media, transmission of human and spiritual values to new generations, human rights and duties, safeguarding creation, call for justice, aid for the poor, hungry, sick and homeless in their struggle for a life in human dignity) were certainly important in this context and recurred as issues in the dialogues. But the problem of peace clearly had priority. In the past, different nations had not known one another at all; now they had to learn to live together in harmony and peace in pluralist societies. But this was explicitly not a unilateral matter. "Non seulement les chrétiens mais les peuples de toutes religions doivent faire face au défi d'apprendre à comprendre les autres croyances et pratiques religieuses, à résoudre les conflits dans la paix, à développer l'estime et le respect entre ceux dont les voies et les valeurs sont différentes."[42] An address at about the time of the Gulf war showed clear signs of our subject when it stated that the dialogue between Jews, Christians and Muslims reminded us of the same creator God who has made his will known to us and calls us to eternal salvation. The basic foundation for this dialogue was mutual respect which led to religious liberty. The latter was a permanent issue, just as prayer for peace was for John Paul II. At the general assembly of the Pontifical Council for Interreligious Dialogue in 1992, the Pope referred to an inquiry conducted by the Pontifical Council itself to identify the reactions to the church's dialogue documents both inside the church and among followers of other religions. His reaction and instruction were unambiguous. "En réaffirmant la valeur de ces enseignements du Magistère, je vous encourage à faire connaître le message qu'ils contiennent."[43]

At the 30th anniversary, the questions quoted above from *Nostra Aetate* were repeated in slightly different language. But the following statement was very clear. "[L]a recherche de perfection, de purification et de conformité à la volonté divine n'est pas limitée aux chrétiens. Elle interpelle chaque être humain. Il n'est donc guère étonnant de trouver dans les traditions religieuses de l'humanité une conscience claire de l'appel aux plus hautes valeurs."[44] In addition, it is repeatedly clear that John Paul II set great store by spiritual interchange in order to give the dialogue depth and quality and protect it from the danger of mere activism.[45]

But precisely in the Christian-Muslim dialogue there were other questions and areas in the foreground, especially after 11 September 2001John Paul II stated more clearly than ever before that peace was something which human beings could not achieve despite all their own necessary efforts; it could only be received by humble hearts as a divine gift. Two statements in his address to the assembly of the Pontifical Council for Interreligious Dialogue are particularly characteristic.

> "It has been suggested that we are witnessing a veritable clash of religions. But, as I have already said on numerous occasions, this would be to falsify religion itself. Believers know that, far from doing evil, they are obliged to do good, to work to alleviate human suffering, to build together a just and harmonious world."[46]
>
> "Misunderstandings arise, prejudice can stand in the way of common accord, and the hand offered in friendship may even be refused. A true spirituality of dialogue has to take such situations into account and provide the motivation for persevering, even in the face of opposition or when the results appear to be meagre."[47]

The *Pontificio Istituto per I Studi Arabi e d'Islamistica* was also praised for its contribution to Christian-Muslim relations and the further importance of its work was underlined, albeit in a very general way. Another statement which was really directed to the missionary orders is of greater anthropological relevance. "Celui qui propose la Bonne Nouvelle invite les religions non chrétiennes à découvrir le Christ, mais il est aussi appelé, par les signes de la présence de Dieu dans ces religions, à recevoir des éclairages nouveaux sur des façons différentes de vivre en homme, et donc avec Dieu."[48]

John Paul II also showed a special interest in the work of the *Franciscans* in the field of dialogue with Muslims, a tradition which can be traced back to the founder of the order himself. Together with members of other orders and with specialists, they had gathered in August 1995 for a congress on forms of fundamentalism in Islam and Christianity. The Pope emphasised that their regional work had given them direct experience of the effects of the Islamic fundamentalism that had been in evidence especially in the past few years. In order to be able to provide a truly lasting response to this phenomenon, however, it was necessary also to deal with the unsolved problems it had created and was still perpetrating. On closer examination, it was in fact not a purely religious phenomenon; in many cases religion was being exploited for political aims or to detract from social or economic difficulties. The Pope made a clear distinction between condemning the phenomenon as such and its effects on the persons who represented it. "Si l'on doit condamner l'intolérance et la violence suscitées par l'intégrisme *(sic!)*, il importe au plus haut point de poser un regard de foi et d'amour sur les personnes qui prennent de telles attitudes et qui en souffrent fréquemment."[49]

3.3.3 The Near East – the speeches are as concrete as the difficulties
As a result of his pastoral visits and pilgrimages, and still more through the obligatory visits of the bishops to the Vatican, John Paul II was also in direct contact with Christians who had more or less daily experience of Christian-Muslim dialogue. Here one thinks geographically of the Near East and North Africa, to begin with, and then also of countries further east and south in Asia and Africa. However, Arabia received little direct mention in the papal discourses dealing with Muslims; only once in all the years are the bishops of the Arab region mentioned generally as partners, but without going into detail about their countries; this could be seen as a precautionary measure because the limitations of dialogue, namely its rejection on the Muslim side, and still more the problem of the necessary but inadequate religious liberty which then led to discrimination and living as a restricted minority, were immediately given a prominent place in

the Pope's address. On the anthropological level, the basis quoted for dialogue was that God is the common father of the whole human family so that all that was necessary was to promote the development of the human person according to God's plan. In two further addresses, a distinction was made between the Latin and Oriental rites. In the speech to the Latin rite bishops, there was also a hint of the difficult social situation of Christians in those countries; it mentioned victims of violence and dealt with peace in the Near East. John Paul II underlined very strongly that the core of peace negotiations had to be the respect for and dignity of each individual who must have the right to live In his/her own country in peace and security, and that it was therefore all the more necessary to appeal for dialogue, encounter and for the love which each person must feel for his/her brother/sister and for every human being. Apart from appreciating what had been achieved especially on the level of a dialogue of life (school/education, health/social services), the effort for a genuine, trustful dialogue with Islam and also with Judaism was one of the major demands from which the Church could not escape and also a contribution to the true religious liberty so often mentioned (no discrimination or marginalisation on the basis of faith; the special status of a religion should not be to the disadvantage of or harm other religions). Here again he came full circle, linking fundamental statements on the status and rights of persons with the specific, problematic situation. In contrast, the address to the Chaldean bishops, namely the bishops of the Nestorian churches in union with Rome in Iraq, Iran, Lebanon, Egypt, Syria, Turkey and the United States, adopted a purely pragmatic approach. It dealt with the Iraq crisis, the embargo, good relations with the religious authorities, and papal initiatives such as a day of fasting and a further invitation to Assisi, all with a view to solidarity and peace in the world. Here, in a sense, the reasoning related (almost exclusively) to implementation; it was not essential always to quote the bases although it was possible.

A summary of more or less everything necessary in this connection and for the region, and of what was also said sooner or later, is found in the two speeches of the Pope at the beginning and end of March 1991, namely immediately after the second Gulf War, at the gathering of the patriarchs and bishops of the Near and Middle East called by the Pope, and in the closing statement to this gathering which was strongly influenced by them and consisted of taking stock and outlining prospects and measures for the region. The special importance attached to this event by the Pontifical Council for Interreligious Dialogue is evidenced by the separate small publication of all the documents in French, English and Arabic. Indeed, all the crisis points of the region were touched on with detailed knowledge and in a balanced way, as well as the problem that they were so often over-simplified or presented in a prejudiced way, and that the Gulf war was even described as a religious war. There was emphatic rejection of any supposedly "holy" war and an appeal for peace with justice because poverty (and also pressure on the conscience) were the main causes of war; a reference back to the first prayer for peace in Assisi was not lacking, nor was a reference to the overwhelmingly minority situation of the Catholic Christians locally, which naturally restricted their ability to be effective in a dialogue which was recognised as absolutely and immediately necessary, and in relation to other, concrete measures, especially when it was linked with limitations on religious liberty, or the latter did not exist at all – a direct reference to Saudi Arabia. Two separate statements on anthropology are particularly striking because of their fundamental nature and their positive character.

> "With humble means, in conformity with her spiritual nature, the Church tries to *evoke or arouse the sense of truth, justice and fraternity* which the Creator placed in the heart of each individual, of each person always considered in his or her transcendent and social dimension. These basic considerations motivated my many recent interventions when the peace in the Gulf and, in a certain sense, the peace of the world were threatened. It seemed necessary to me, indeed, to recall the great principles of a morality and law which challenge in a like manner the conscience of every person and which are to be applied everywhere and are applicable to each of the members of the international community."[50]
>
> "It will be necessary to overcome the rancour and cultural divisions and especially those which have been created between diverse religious worlds. It is a hope which finds its deepest foundation in the common faith of these peoples in God the Creator and in trust in mankind, God's creature, called by him to preserve the world and improve it."[51]

A much clearer tendency to almost exclusive, concrete language was evident when the Pope was speaking himself in a region or reacting directly to one of its specific problems. His arguments were still relatively basic in a sermon in Damascus where, although he touched on the Holy Land and on rights and peace for all the peoples there, he also spoke more fundamentally about constructing a fraternal, just society with solidarity in which every person's human dignity and fundamental rights would be fully recognised. At a meeting with patriarchs and bishops, there was a more general reference to the positive effects of dialogue and cooperation between Christians and Muslims, even including common witness for the full recognition of human dignity, however that might be understood in practice. In the speeches in Bethlehem and Jerusalem, on the contrary, the choice of words was more adapted to the special solemnity of that pilgrimage. But John Paul II again became more concrete in a message to the Latin patriarch of Jerusalem where it was a matter of the coexistence of Israelis and Palestinians and of this only being possible peacefully and lastingly if fundamental rights existed for all, namely the right for both peoples to live in dignity and security. That was followed in this case by an appeal to the spiritual leaders of both religions to draw from their faith the energy needed for internal and external peace, because that was what the people really wanted.[52] *It was relatively clear and certainly understandable that it was less a question of thorough theological arguments than of placing certain "simple", fundamental issues on the table, with all the personal and institutional weight of the speaker, in order perhaps to be able to get something moving in the stalemate of the political situation.*

3.3.4 North Africa – here the same words have a quite different ring

A lot of scope is devoted in the speeches of John Paul II to the situation among Christians and Muslims in *North Africa*, although most of these speeches, whether given in the Vatican or locally, were addressed to the bishops and have a predominantly pastoral tone, not least because of the really difficult situation especially in Algeria where attacks by Muslims, as was mentioned directly and more often indirectly, had killed 19 monks and nuns, including the Bishop of Oran and the seven Trappist monks of Notre-Dame de l'Atlas. Whereas, initially, he stated quite generally that dialogue was ultimately a question of friendship which needed time, gradual, mutual contact, differentiation and discretion in order to take account of a hesitant mentality, he also referred to the unavoidable tension between respect for the person and convictions of one's partners in dialogue and one's own faith convictions, and he even stated that, for many Muslims, the nuns were the Church and the Muslims were happy about what had happened. But, in view of the later tragic developments, these statements look very different and it becomes clear how profound the differences and tensions can really be and that some things are not always what they initially seem to be. The limits to dialogue have simply been reached when a partner in the dialogue is no longer prepared to perceive the other religion as it really is and allows him/herself to be influenced by prejudices broadcast by the media and which distort reality. It is certainly of special significance when there is a call in such a situation for openness and good neighbourliness, for sharing joys and sorrows, for cooperation and deeds of mercy for the holistic development and complete liberation of human beings. It also becomes clearer why it was so important to John Paul II to win young people for dialogue; that is where the long-term, lasting and also necessary developments lie. The reference to the vulnerability of the small flock in relation to the Muslims, the new chapter of dialogue and cooperation between followers of different religions in the history of the Church – all of this had a more concrete meaning here than elsewhere. And the statement that God was the God of life and willed the life of people and not their death meant quite practically in these circumstances that no one should kill in the name of God or put his brother to death. Even a quotation from another context (address to the UNO in 1995), that the Church's respect for culture was based on respect for the search for answers to the problem of human life, had a special meaning in this context, and naturally also when the good will of the civil authorities was mentioned or other sentences stated that unofficial and official dialogue helped to overcome false images from the media. Even relatively frequent statements like the importance of mutual respect and appreciation for religious liberty, or the claim that values were rooted in human nature and there were therefore common values in all cultures, ceased to be mere phrases in this context and could be seen for what they were worth with their whole weight. It is indeed true that, for peace and interreligious dialogue, to overcome mutual mistrust and to be able at all to learn to serve the common good of humankind, patience and resolve are needed simultaneously – and in no small measure.

Several times, the papal addresses to the North African bishops mentioned – and this was unusual – the marriages between Christian women and Muslim men where the women lived in a Muslim family, an environment totally determined by Islam. Often this produced an inner change when the question of the children's education arose and there was a wish to anticipate this way of life which was completely imbued with Islamic religious practices. Naturally, sacraments were a problem but the organs of the curia existed to provide competent help and the women could be sure of the concern of the Church and the prayers of the Pope. The bishops themselves saw mixed marriages as practical contexts for dialogue and the Pope wished this situation to develop so that there was increasing respect for the freedom of conscience of every individual, and parents generally should help their children to seek the truth and live accordingly, to s eek the good and further it.[53]

The speeches addressed to the bishops or faithful of individual countries did not differ significantly from the general tenor with its emphasis on religious liberty and human dignity in the background and the emphasis on what Christian social service contributed to the development of human beings including the furthering of women. What is more interesting is that, during the last few years against the background of the most recent events, the exemplary importance of the dialogue with Islam was underlined. It was "particulièrement requis après les événements tragiques liés au terrorisme qui ont marqués le début du troisième millénaire et que l'opinion peut être tentée d'imputer à des causes d'origine religieuse."[54]

3.3.5 Turkey – another programmatic speech

As far as *Turkey* was concerned, the addresses to the bishops, in particular, again fitted predominantly into the framework of emphasising religious liberty and the commitment to it, and also dealt with the possibilities of an everyday dialogue alongside the institutional one. The only specific reference was to the relations between the University of Ankara and the Gregorian University. The address to the Catholic congregation in Ankara was different and can be considered programmatic in comparison. It began – certainly with the whole Islamic world in mind – by emphasising the Church's appreciation for the Muslims' religious values which were a valuable spiritual heritage for human beings and society, provided orientation especially for the lives of young people and could fill the gap caused by materialism. These values were also a firm foundation for social and legal organisation – a statement which, in a country where Islamic law has not become the state law, should not be overestimated but was nevertheless unusual. Here, too, John Paul II emphasised that a new age had begun for Christians and Muslims in the sense that it was necessary to recognise and develop the spiritual bonds which united the two for the sake of the protection and development of all people, meaning concretely social justice, moral values, peace and freedom. The serious practice of the Abrahamic faith – the implication here is of a faith which permeates the whole of life – offered a secure foundation for human dignity, brotherhood and freedom, and there is explicit reference here to belief in creation according to which human beings were the crown of creation. As God's creatures, human beings had inviolable rights but were also bound by the law of good and evil which had its origin in the divine order. All these common elements had to be combined in the principle of cooperation for the progress of humankind, for competition in doing what is good, for the extension of peace and brotherliness, and all this naturally in a context of free proclamation of the faith. This concludes with an encouraging, "Ayez aussi le sentiment d'être en communion avec l'Église universelle que le Pape représente devant vous, dans son humble personne."[55]

3.3.6 Religious liberty time and again

In a general way, too, religious liberty was in the foreground with its fundamental importance for human beings and therefore for the state, again with special emphasis in the states which were more or less firmly in Communist hands before 1989. There, John Paul II particularly underlined the link between religious liberty and democracy. He called on the bishops of Bosnia-Herzegovina, for example, explicitly to love their neighbours, and later also to a dialogue with the Islamic community, while in the Ukraine and in Kazakhstan he went into greater detail, e.g. about the deportation of the (Muslim) Tartars from Crimea to the Asian Soviet republics and their wish to return; he again spoke about young people and, naturally, especially envisaged a common front against materialism, communism, atheism, consumerism and hedonism. It was also striking that, in the angelus message in Kazakhstan, he commended everyone to Mary; since her son was the redeemer, she was the mother of all. But there were also other, very telling anthropological references, for example during a mass there. "Dearly beloved, *humanity's homeland is the Kingdom of heaven!*"[56] Whereas in the Ukraine the wording was: "If God is removed from the world, nothing human remains. (...) At the root of every authentic humanism there is always the humble and trusting acknowledgement of the primacy of God."[57]

The above mentioned question of religious liberty was also carried over into Pakistan and Bangladesh or rather into the addresses to the bishops of those countries. It is evident that there was still some work of persuasion and dialogue needed here. The questions about the relations with the Muslims which required clarification in a serious and enlightened dialogue were unfortunately not mentioned specifically; instead, the focus was on areas of potential cooperation: social justice, moral values, peace, development, freedom and human dignity, the role of the family in society and promotion of the common good. There was also discussion of the basis for such cooperation, from the rather superficial place of God in public opinion in these countries to the universal and unchangeable moral norms which went back to the created order, were written on the human heart and thus became the common treasure of humanity and the real point where followers of different religious traditions could meet.[58]

3.3.7 South-East Asia: a historical review

With regard to the situation in South-East Asia, he again stated that faith in God the Creator offered a solid foundation for mutual understanding and peaceful dialogue. Indeed, when speaking to the faithful in Indonesia, John Paul II went on to say that the question of the rich young man, namely the question of eternal life and of the future of humanity after death, was raised by people of all generations, nations, cultures and languages. Men and women of all times and in all places, hence also the Muslims, recognised that their future after death depended on their earthly life. In general, hope of eternal life was widespread everywhere there (the address was given in Sumatra) and it was important to preserve this religious character of life, this openness to transcendent values and this society, which was marked by respect for God and his commands, for future generations. In view of the problems in some regions of the Philippines and in Malaysia, however, the focus was on religious liberty again, and John Paul II began by underlining that religious differences did not necessarily prevent living together; both Christians and Muslims could be partners in building up society according to the values taught by God. In so doing, tolerance, peace and concern for the poorer and weaker were the task and the goal. The similarities with the speech in Indonesia are quite obvious. As violence escalated in Mindanao, the tone changed; here it was a question of campaigning against terrorism and violence and the decisive role of religious leaders in this context. Religions (and confessions) had to cooperate to eliminate the social and cultural causes of terrorism. Again, (John Paul II quoted himself), it was a matter of the greatness and dignity of humanity and of the unity of the human family which needed to become more obvious to serve this purpose. The Pope was still more explicit in view of the Islamisation of social and civil life in Malaysia where religious liberty seemed in the meantime to exist only on paper. His language was clear: without an honest debate there could be neither peace nor progress, and it was also not possible to separate religion from morality and banish morality to the private realm. Then John Paul II almost took stock of the international developments relating to human rights. "Par bonheur, il y a maintenant à travers le monde une conscience croissante de l'importance des droits fondamentaux pour édifier des sociétés justes et stables, capables de traduire les aspirations des peuples pour vivre dans la dignité et la liberté. De plus, les citoyens qui redoutent les réactions adverses quand ils expriment leurs convictions ne peuvent prendre part pleinement à la construction de la société dans laquelle ils vivent."[59]

3.3.8 Black Africa: Mali as a positive example

The situation in *Africa* is extremely diverse (with the exception of North Africa where the culture is quite different). The African synod of bishops had made a general offer to the Muslims to cooperate on peace and justice in mutual respect for the religious liberty of the individual and of the community because that was God's will – a God, moreover, who also did not wish to be an idol in whose name one could kill other people. What form such cooperation could or should take in practice had naturally not yet been specified.[60] Mali played a special part in the addresses of John Paul II which is surprising since the number of Catholics in that country, the northern part of which lies in the Sahara, is extremely small. The number of Muslims, on the other hand, is all the greater. One can sense in the Pope's addresses that relations were not always unproblematic but seem to have improved in the course of the years (between 1981, 1990 and 1996). He again spoke very openly about freedom of conscience and worship. Otherwise, he referred to the Muslims as the bearers of authentic religious values which the Christians should recognise and respect, and to the many different dimensions of Islam which had deep roots in many African nations. Dialogue in general was not always easy, was not desired by everyone and it was sometimes difficult to find a common language and representative partners for dialogue. Here, the Christians had to be realistic and simultaneously generous, and John Paul II saw the initiative for this in relation to the Muslims clearly on the side of the Catholic Church. The solidarity desired was much more than vague tolerance or simple acceptance of others; it should lead to a community of brothers who valued and loved one another.[61] From the point of view of anthropology, he gave a speech of special interest to 1,500 Christian and Muslim young people in the palace of culture of Bamako on the occasion of his pastoral visit in January 1990; it is somewhat reminiscent of the Casablanca speech which was also addressed to young people. The papal visitor began by observing that all people belonged to one human family. The reference to creation was followed by the idea of the unique dignity and superior value of human beings. Since humankind was created by God, it followed that one had to respect, love and help every human being. Whether one described human beings as images of God or as his representatives (i.e. adopted a Christian or Muslim view), in any case the human being was a sign which pointed to God. And his/her rights (i.e. human rights) were an expression of the will of God and of the existence of human nature as created by God. On the other hand, human beings as God's creatures were also marked by a dependency (the radical nature of which must probably be described in German theological, philosophical terms as "schlechthinnig" – pure and simple) which might perhaps destroy their pride but, provided that they recognised and accepted it voluntarily, gave their lives meaning and an unlimited breadth. It was more or less assumed by John Paul II in his speech that Christians and Muslims had this in common, because he then continued that each certainly had different motives and means for achieving this ideal; on the one side, they needed to become perfect representatives of God on earth and to witness to the good attributes of God as expressed in the beautiful names of God, while the others had to become true children of God according to Christ's example. This kind of human dignity was seen in any case as contrasting with ideologies and passing happiness. It was linked with a high morality and did not merely seek material wellbeing. Coupled with a call for solidarity also in the realm of work, he moved on to speak about the necessity of a spiritual life because man simply did not live by bread alone. It was this spiritual life which gave meaning to development and orientation towards the wellbeing of humankind – of the whole human being and of all of humankind.[62]

It seems that John Paul II gave his most fundamentally anthropological speeches with the clearest references to Muslims when speaking to young people, perhaps with the idea that young people need the most guidance about who they are and who other people are in general, about where they stand in life, what they should do with life and the right way of going about it. The Pope's answer to this was simple, clear, comprehensible for both sides and well-founded without wasting a lot of words. On the other hand, it was a major demonstration of trust by the host country, Mali, that the Pope was allowed to give such a speech before Muslim young people as well, although young people are generally considered especially susceptible to influences and by Muslims in particular,. This again indicated that Mali and Christian-Muslim relations there had a special status, namely a specially harmonious one.

Along the Gulf of Guinea, the proportion of Muslims compared with Christians is very high in some places, so it is no real surprise that religious liberty should again have been an issue, naturally always in the interest of the whole nation. However, these addresses did not offer any special new reasons or particularly telling formulations, apart from a definition of the common values frequently mentioned, namely peace, solidarity, brotherliness, justice and freedom.[63] The neighbouring states along the same latitude also received papal attention. They too are more or less strongly marked by Islam. The speeches contained little that was new in content; they referred to efforts to get to know one another, for the sake of respect and religious liberty and for the common good. One relatively basic statement was that fear of others often came "de la méconnaissance profonde des valeurs religieuses qui l'animent".[64] Among the states along the Gulf of Guinea, *Nigeria* received special attention, unfortunately as a country in which there had repeatedly been disputes with bloodshed, among other things because some Muslims were becoming increasingly militant to the point of wanting to force their understanding of Islamic law on followers of other religions as well. It is obvious that here again the question of religious liberty was at issue. John Paul II repeated an urgent appeal, both before the bishops and before the Muslim leaders of the country. "The Creator of the one great human family to which we all belong desires that we bear witness to the divine image in every human being by respecting each person with his or her values and religious traditions, and by working together for human progress and development at all levels."[65] But in these speeches there were no special, new reasons or particularly telling formulations, apart from a definition of the common values so often mentioned before, namely peace, solidarity, brotherliness, justice and freedom. The conversations with the bishops of other African countries also followed similar lines. Where Muslims were in the majority or difficulties existed, the focus was mainly on the importance of religious liberty as the core and foundation for all human rights and hence as the basis for the development of the individual and of society. On the other hand, it was clear that the Catholic

Church was moving into dialogue with the Muslims (but without abandoning its missionary dimension) and explicitly recognising common values even though common conceptions of dialogue were sometimes lacking.[66]

3.3.9 Europe: encouragement for dialogue

The Muslims in a society could also become an issue, as shown by the example of the papal addresses to the bishops of Northern France. It was the large number of Muslims among the immigrants which caused anxiety here. John Paul II referred to the possibility of defending essential values jointly with Muslim believers and exhorted the bishops to continue to encourage dialogue with Muslims. Again he made special mention of young people who needed to learn to coexist in friendship, and also to human rights, but this time it was the rights of foreigners to protection which should not suddenly be forgotten because the economic situation had deteriorated. This address was therefore very concrete whereas another to the Curia did little more than quote *Nostra Aetate* which had to be read in the light of *Lumen Gentium*, and both were seen as the cognitive framework for the prayers for peace in Assisi. A very similar collection of quotations made up the first speech at a general audience in which Muslims were discussed, in addition to an appeal for intercession for and cooperation in the realm of interreligious dialogue.[67] At a general audience almost 15 years later the message sounded quite different. There was greater detail about the Muslims' attitude to God (submissiveness and openness for God's will) which reflected the existential condition of every person before God. It went on to say that Christians were happy about common values; in view of a world which had forgotten God, people were called in a spirit of love to defend human dignity, moral values and freedom. John Paul II even spoke about a common pilgrimage to eternity expressed in prayer, fasting and charitable works but also in common efforts for peace and justice, for human progress and environmental protection. He also referred to refraining from any form of violence in humble submission to the will of God, speaking of God as creator and guide of the human family. In summary, "The Church has a high regard for them <i.e. the Muslims>, convinced that their faith in the transcendent God contributes to building a new human family based on the highest aspirations of the human heart."[68] Sometimes at general audiences John Paul II also spoke about experiences he had had during his pastoral visits or pilgrimages, among other things also with Muslims. So on one occasion he talked at greater length about the situation in Lebanon having actually been marked by centuries of harmonious coexistence between very different Christian and Muslim groups, an open society without any extremism such as was often found in political and social life in other places where there was unjustified reference to religion. In the Pope's eyes, Lebanon had a peace mission in the region provided it received support from outside (and not the opposite like during the civil war). On another occasion he stated that dialogue with Islam was becoming increasingly important and necessary at the beginning of the third millennium, and then said that the warm reception by the Grand Mufti of Syria and his accompaniment when visiting the large Umayyad Mosque in Damascus (a sanctuary of John the Baptist revered by both Christians and Muslims) had encouraged him – not impressive theology but an important human gesture. The message for the World Day of the Sick (11 February 1997) was particularly unusual. John Paul II used the occasion to refer to his pilgrimage of reconciliation to Hebron and to his wish that Jerusalem and Rome might become focal points for a pilgrimage for peace throughout the world. To this end he asked the sick for their prayers, seeing them as strengthened by the sacrifice of their suffering.[69]

3.3.10 Before the United Nations – important reflections

Because of its fundamental, anthropological nature and also because of the significance attached to it by the occasion and the audience, special emphasis must be given to the speech by John Paul II on 5 October 1995 in New York to mark the 50th anniversary of the United Nations, even though Muslims are not mentioned there specifically but only the basic problem of the diversity of humankind. As he said when introducing this point, he had experienced this himself during his pastoral journeys in the course of 17 years, and he immediately added an analysis of the situation.

> "Malheureusement, il faut encore que le monde apprenne à vivre dans la diversité, ainsi que l'ont douloureusement rappelé les événements récents des Balkans et d'Afrique centrale. La réalité de la <différence> et la particularité de l'<autre> peuvent parfois être ressenties comme un poids, ou même comme une menace. Amplifiée par des ressentiments d'origine historique et exacerbée par les manipulations de personnages sans scrupules, la peur de la <différence> peut conduire à nier l'humanité même de l'<autre>"[70]

– which leads to an unending spiral of violence. Then the argumentation went a stage deeper: despite all the differences, there were also fundamental similarities between individuals and nations because, in reality, cultures were only different ways of handling the question of the meaning of personal existence or, in other words, different ways of expressing the transcendental dimension of human life. This, in turn, led to respect for cultures and thus somehow logically to respect for religious liberty. "Dans cette perspective, il nous est possible de reconnaître l'importance de préserver le droit fondamental à la liberté de religion et à la liberté de conscience, colonnes essentielles sur lesquelles repose la structure des droits humains et fondement de toute société réellement libre. Il n'est permis à personne d'annihiler ces droits en faisant usage de coercition pour imposer une réponse au mystère de l'homme."[71] He did not specify which religion or system he may have had in mind but expanded the idea along another line. To ignore or even abolish the differences would deprive one of the depth of the mystery of human life. He still maintained that there was an unchangeable truth about human beings, but it was so complex that every culture could teach one something about one dimension or another of this truth by means of respectful dialogue.[72] *Here indeed, lines of thought which can be found time and again ever since Vatican II were summarised clearly and spelled out further. The last point about the complexity of the truth, in particular, can be seen as a step in an established direction (one can also recall what was said about a common pilgrimage) but it is a new step which was taken almost incidentally.* The speech itself then switched to the concrete level of nationalism which was rejected as contempt for others even if it had a religious basis (which then constituted fundamentalism) and to patriotism which was legitimate as love of one's homeland, and it then came back, after a fundamental, anthropological discussion, to the everyday business of the United Nations as an institution which should be honoured.

It would certainly be interesting if, in relation to Islamic culture (because the reference here is explicitly to cultures and one is almost compelled to see the religions in the background although it is significant that they are not even mentioned generally), it were to be said in some other place or more concretely in future which dimension of this very complex truth about human beings the Christian culture could teach.

In this connection, the message for the World Day of Peace in 1999, delivered on 7 January, must be mentioned; it refers back again to the 50th anniversary of the adoption of the Universal Declaration on Human Rights. Human dignity was seen here as a transcendent value which was related to creation and linked human beings with their creator and also with all their fellow creatures. The Human Rights Declaration was affirmed explicitly because it only proclaimed the rights of human beings and was not the body which granted people these rights, because the rights resided in the human person, in human dignity. The message went on to mention the traditional subdivision of human rights into civil and political rights, on the one hand, and economic, social and cultural rights, on the other, but emphasised that all human rights were endangered if one was violated. There was no specific mention of Islam or of convergences or divergences over human rights, and this was understandable in view of the supposedly worldwide audience, but these can be understood at various points. Among the convergences one can count the rejection of various ideologies, the effects of which had also threatened human rights: in addition to Marxism, racism and nationalism mention was also made of materialistic consumerism in which the individual and his/her wishes were seen as the final goal and which was indifferent to the negative consequences of this attitude for others. Muslims would certainly also welcome the statement that the universality and indivisibility of human rights did not exclude legitimate cultural and political differences; here unfortunately there are only melodious sentences but no precise details about how far such differences can go. Divergence can arise already over the question of the extent to which the individual should be emphasised and what the exact relation is between people's rights and duties; this, too, is a problem which was formulated and expressed but finally left in the air. The address became more specific when dealing with individual human rights that were under special threat. These included the right to life and the right to freedom of religion. On the former, it must be said that the threat to life was definitely not only seen in its initial and final phase but in every form of violence against life, and expressly also the violence embodied in poverty and hunger, armed conflicts, illegal traffic in drugs and armaments and in thoughtless destruction of the natural environment. That faith and violence had nothing to do with one another sounded like an entreaty when dealing with threats to freedom of religion. The hints of divergences with Islam were very obvious here when the demand was made for the right to change one's religion, for the public practice of religion and for the abolition of discrimination against religious minorities. It was immediately clear even without mentioning it that there were major differences about that. More agreement could certainly be reached on another point which was mentioned more with a view to the future: the ambivalent role of the mass media between glorifying violence and promoting mutual understanding and peace.[73]

3.3.11 On the international stage – general values and religious liberty in particular

Elsewhere, naturally, John Paul II had an audience with a more religious approach and less international than during his appearance before the United Nations in particular. Therefore his statements took greater account of specific situations, especially specific situations of dialogue and coexistence. As far as Christians and Muslims were concerned, these also covered a wide field, beginning with an audience in which a third religion predominated. This often led to an explicit or implicit passing mention of Muslims and their religion.

When, for example, John Paul II was addressing the faithful in India and speaking about the typical Indian spiritual view of human beings and its significance for the whole of humanity, it was easy to perceive Hinduism in the background. There was only one explicit mention of everyone (Hindus, Muslims, Sikhs, Buddhists, Jains, Parsees and Christians) when he talked about the similarity of all in brotherly love. He also declared as a kind of common truth that the human being's striving for material and social wellbeing and full dignity was related to the deepest needs of his/her spiritual nature. The argument jumped somewhat to working together for human rights, especially for religious liberty understood as freedom of conscience and religious propagation and, in a further step, to the struggle against hunger, poverty, ignorance, persecution, discrimination and every kind of enslavement of the human spirit. This could not be achieved without religion and a spiritual understanding of humankind.

There was also only marginal reference to Muslims in the message to the peoples of Asia. It is interesting that this relatively early speech (21 February 1981), which referred back explicitly to Paul VI and spoke with similar clarity about the truth, did not start anthropologically from creation but from Christology. Every individual and every nation, it stated, had become children of God through the cross and resurrection of Christ, shared in the divine nature and were heirs of eternal life. Every person received the full measure of his/her dignity and his/her final purpose in Christ. Christians extended their hands to all men and women of good will who believed in this inestimable dignity of every human person. The special bond between all religions, and that too was unusual in this message, was described as "une reconnaissance du besoin de prière, en tant qu'expression de la spiritualité de l'homme orientée vers un Absolu."[74]

An address to representatives of various religions in Indonesia also remained very general although Indonesia is predominantly Muslim. As a solid basis for mutual understanding and peaceful interchange, it did not immediately name faith in God, the creator, but firstly the view of the universe as an organic whole. That on its own led to sensitivity in human relationships, to a great appreciation for love and cooperation within the family and to a very strong concern for justice and the recognition of the rights of every individual. The latter point naturally was again very important with clear references to the Catholic teaching that the right to religious liberty is rooted in the dignity of every person created by God and that religious liberty in general is especially necessary for minorities. The question of truth was no obstacle to mutual openness because believers in particular could not expect other believers to act against their own consciences.

But these very general formulations were certainly not just adapted for listeners whose religious views meant they were further removed from Christians than the Muslims because, in a speech before Hindu and Muslim representatives in Kenya, John Paul II stated explicitly that what he had said nearly five years earlier only to the Hindus of the country applied equally to relations with the Muslims. The point at issue was human needs on various levels (spiritual, material, social) and the meeting point was in the worship of God and in spiritual values. "Dieu veut que tous ceux qui l'adorent, même s'ils ne sont pas unis par le même culte, soient néanmoins unis dans la fraternité et dans le service commun pour le bien de tous."[75]

But elsewhere the Muslims were definitely emphasised, e. g. in Tanzania and the Sudan. The former address to religious representatives was of interest because it showed, on the one hand, that Muslims certainly had a special standing compared with Hindus and Buddhists, since John Paul II considered the dialogue between Christians and Muslims to be increasingly important in today's world; but, on the other hand, this did not seem to make much difference on the anthropological level. On a general level, he claimed that interreligious dialogue led to better mutual understanding and thus to furthering common ideals in the realm of religious liberty, human fraternity and social progress. The background presupposition was a particular view of and expectation attached to religion in the world, namely that religion would promote harmony and peace. The section which was addressed especially to Muslims listed, on the one hand, the values taught by God – tolerance, justice, peace and concern for the weaker and poorer – and emphasised the general experience that religious differences alone did not necessarily destroy common life; on the other hand, more directly than in *Dignitatis Humanae*, he strongly underlined religious liberty as the basis for mutual respect and cooperation and made the point very clearly.

> "C'est la reconnaissance que chaque personne a un droit inaliénable et un droit solennel de suivre sa conscience en cherchant la vérité et en lui obéissant. Le Seigneur du ciel et de la terre ne peut aimer une observance religieuse imposée de l'extérieur. Que deviendraient alors les merveilleux dons de la raison et de la liberté qui donnent aux individus le privilège de prendre leurs responsabilités personnelles et qui constituent la valeur et la gloire des fils et des filles aimés du Créateur (...)?"[76]

Here, the image of reasonable, free human beings, created and hence responsible, was the argument for religious liberty, one of the most sensitive issues in the Islamic realm, and behind it one can certainly understand that reason played a very important part in Islam. There are grounds for doubt, however, as to whether sufficient account was taken of the Islamic view that a truly reasonable person can really only become a Muslim. The very fact that this subject continues to be so important between Christians and Muslims indicates that it raises fundamental issues which are naturally not expressed so clearly if only one side presents its view and attempts to convince the other. The difficulties of whatever kind become all the clearer the more direct and especially hostile the confrontation is.

As has already become clear, there have been plenty of such situations and from time to time John Paul II also made statements directly on specific situations.

Where things were going quite well and a lot of cooperation already existed on various different levels, e. g. in Tunisia, it could almost be taken for granted that statements would be made on faith in the creator as a basis for recognising the dignity of the creatures and one could move on to other questions, for example what one is responsible for in society and what not. According to the Pope, not for the technical problems of the modern economy or of international cooperation, but definitely for the life of society in the sense of a social conscience which reminded people of ethical principles that should be considered when taking concrete decisions (e. g. respect for life, dignity of the individual, decency). In addition, the existing differences could also be seen positively; they might not be directly God's will but God did at least allow them. "L'ouverture à l'autre est, en quelque sorte, une réponse à Dieu qui permet nos différences et qui veut que nous nous connaissions plus profondément."[77]

But, in the face to face encounter, such situations seemed to be fairly rare. During an address in Bangladesh, John Paul II even mentioned explicitly and generally that, as a result of past conflicts and a lack of understanding, Christians and Muslims were sometimes afraid of and killed one another. In Bosnia-Herzegovina, the situation had already escalated to warfare and now the major task was to find the way back to a society in which diversity was positive and could perhaps even be exemplary. To this end, human values were fundamentally important. "Il ne s'agit pas seulement de la reconstruction matérielle; il est tout d'abord nécessaire de pourvoir à la réédification spirituelle des âmes dans lesquelles la fureur dévastatrice de la guerre a souvent ébranlé, voire même compromis, les valeurs sur lesquelles se fonde toute coexistence civile. C'est précisément à partir de là, des fondements spirituels de la coexistence humaine, qu'il faut recommencer."[78] The situation in the Holy Land seemed to be a real tangle; the tone of the papal address to the authorities there can be described as pleading or even despairing. In view of the oppressive situation for the people concerned, there was no room for anthropological analyses, only for appeals for peace, although one can hear how even the appellant doubted their effect.[79]

3.3.12 Dialogue participants as main addressees – immaterial values common to all

The situation is simpler and more pleasant when the representatives of the religions are used to coming together for *dialogues* within whatever framework. John Paul II himself had invited people to such encounters, not so much for dialogue but to Assisi to pray for peace. Numerically, as far as representation and content were concerned, Muslims only played a real part starting with the second encounter which was limited to the religions well established in Europe, namely Judaism, Christianity and Islam, because the aim was prayer and fasting for peace in Europe or, more precisely, in the Balkans. Looking back to Assisi and forward to an encounter on the dialogue between religions and cultures, John Paul II said about the religious leaders who had gathered in Assisi, "They wished to show that genuine religious belief is an inexhaustible wellspring of mutual respect and harmony among peoples; indeed it is the chief antidote to violence and conflict."[80] That was also emphasised at the Assisi gathering in January 2002 at which Islam was indirectly still more the focal point, of course. And the annual papal addresses for the World Day of Peace (January 1st) must also be quoted here. In practice, the addresses never mentioned Muslims directly and were naturally also not addressed exclusively to Muslims or to situations where Christians and Muslims meet, so they are, at most, only indirectly significant for this study when quoted in the contexts examined here, which is frequently the case. One exception to the excessive "transparency", despite the lack of reference to a particular religion, was the address for the New Year 2002, only a few months after the attacks of September 11th. The wording is clearly programmatic.

> "Instead, even when the truth has been reached – and this can happen only in a limited and imperfect way – it can never be imposed. Respect for a person's conscience, where the image of God himself is reflected (cf. *Gen* 1:26–27), means that we can only propose the truth to others, who are then responsible for accepting it. To try to impose on others by violent means what we consider to be the truth is an offence against human dignity, and ultimately an offence against God whose image that person bears. For this reason, what is usually referred to as fundamentalism is an attitude radically opposed to belief in God. *Terrorism exploits not just people, it exploits God:* it ends by making him an idol to be used for one's own purposes."[81]

More detailed reflection on anthropology can be found in the address given at the audience for the participants in a Christian-Muslim conference on religious education in modern society (organised by the Pontifical Council for Interreligious Dialogue, among others). John Paul II explicitly welcomed the theme which was also, as he said, one of his favourite issues, since indeed young people were very important to him generally. The young, as he stated at the end of the speech, had to learn to distinguish between progress and progress, to rediscover the social character of human life, namely the inalienable rights and responsibilities of humanity. He was concerned about a better transmission of religious values, first and foremost respect for others and openness to all the children of God irrespective of race, religion, economic status, gender, language or ethnic group. The background was the critical debate on progress in many different areas and especially on modern affluence, with the attendant danger of forgetting or ignoring the transcendent, spiritual aspect of the individual before God, because it resulted in excessive individualism, addiction to pleasure, a feeling of loneliness and in violence and self-destruction, and thus led implicitly to a refusal to recognise God as creator and law giver (!) whose will humans as humans had to respect and obey. The will of God alone was the key to true human happiness. However, progress was not rejected wholesale in view of considerations based on the Christian theology of creation. "D'un autre côté, ce ne serait pas réaliste de retourner en arrière en rejetant le progrès et ce serait faire montre de manque de confiance dans les puissances intellectuelles dont Dieu a doté l'humanité. Cela équivaudrait à abdiquer la vocation que Dieu a donnée à l'homme – la vocation de collaborer avec lui dans le travail de la création."[82]

Another interesting address was the one John Paul II gave to the participants in a dialogue dealing with the situation of women. This subject was also suited to fundamental remarks, in this case that God had created human beings as men and women and therefore all people deserved to enjoy the same dignity. The differences between women and men were not to be exploited to oppress and discriminate against the ones and to declare the superiority of the others, which often happened in practice and against which both Christians and Muslims should act together in order to promote respect for the right role of and freedom for women. Naturally, biblical characters were also mentioned, ranging from the strong women in the Old Testament to the women in the Gospels, and first and foremost Mary, of course, as well as the respect and appreciation Jesus himself showed for women. That there might be some need for clarification here between Christians and Muslims can at least be read between the lines.[83]

In contrast, comparable addresses on the occasion of a Christian-Muslim dialogue on holiness in the two religions, or a colloquium on reconciliation between those who believe in one God, offered nothing on anthropology; nor did a message to a colloquium on religion and the earth's resources in which there was only very brief reference to one's duties to the coming generations and the moral duty of solidarity. An address given in Lisbon to an ecumenical meeting in which apparently Jews and Muslims also participated is more interesting in this respect. There was talk (now almost naturally) of the spiritual value of human beings, of their inviolable dignity and consequently their inalienable rights (in contrast to merely practical usefulness) and of human solidarity also beyond the boundaries of religions. The religious dimension of human beings was related to prayer, obedience and justice. It was expressed, for instance, in such diverse forms as personal honour, moral discipline in public and private life (here there was even reference to the need to defy the relaxation of the principles of morality and justice and the plague of ethical permissiveness), respect for life, for the family and its values, for wise and courageous social and political involvement and for generous work. There was mention of the emptiness of a purely anthropocentric humanism; clearly the image they jointly opposed was again the more material approach of modern people. "Sans dimension religieuse l'homme se trouve appauvri, car frustré d'un de ses droits fondamentaux. Et nous désirons tous éviter cet appauvrissement de l'homme."[84]

3.3.13 Speaking directly to the Muslims: peace is a major issue

Then there is still – apart from the Casablanca speech already discussed – the wide range of speeches which, as far as one can tell, John Paul II addressed directly to *Muslims* and in which he also referred directly to Islam and Muslims themselves. There is a particularly close link between the work of the Pontifical Council for Interreligious Dialogue and three addresses, one to the professors of Ankara University, one to a delegation of the World Islamic Call Society and one to a delegation from Al-Azhar. All three meetings with the Pope took place following events in which the Pontifical Council was at least involved, like the colloquium on theological education (Passing on religious values to the youth of today) in the context of which the cooperation agreement (i.e. mutual visits, exchange of professors) between the University of Ankara and the Gregorian University was renewed. The basic condition for this cooperation stated by John Paul II was freedom of conscience and of religion; at the same time he referred to the means for promoting them as an important part of cooperation in dialogue and research. Values such as the dignity of the individual and the equality of all human beings lay behind the questions and problems generally considered urgent today. Human beings needed to develop their minds and consciences in order to understand their role in a constantly changing world. The remarks to the delegation of the World Islamic Call Society (which had actually participated in a dialogue on the subject of mission) were very similar. In this connection as well, respect for the inalienable dignity and freedom of human beings created and loved by God was important, and both sides were to work for freedom of belief and of religious practice. There was certainly the idea of a common front against modern atheism in the statement that modern society was also unable to lead people to the goal for which they were created without a reference to God. And the address to the delegation from Al-Azhar, on the day after the establishment of a Joint Committee for Dialogue between the Pontifical Council for Interreligious Dialogue and the Permanent Committee of Al-Azhar for Dialogue between the Monotheistic Religions, also included a reference to respect, knowledge and acceptance which needed to be mutual if dialogue was to be convincing. The final goal which was apparently not easily reached was peace.[85]

This was often the subject when the Pope addressed Muslims in a way that went beyond the usual encounters during his journeys or at the accreditation of diplomats. Thus peace was the real concern in the message to all Muslims for the sake of Lebanon on 7 September 1989, namely after more than 14 years of civil war. It is striking that there was a very strong emphasis on God as the judge of all humanity over against people's indifference to this tragedy. People had no option to act or not to act; on the contrary, "[i]l s'agit d'un devoir de solidarité humaine que votre conscience d'homme et votre appartenance à la grande famille des croyants imposent à chacun de vous".[86] It was again a special situation of war and conflict which led John Paul II to write his message for the end of the Ramadan fast himself and not to leave it to the Pontifical Council for Interreligious Dialogue. It is very interesting that, in this connection, the otherwise usual sequence of knowledge, dialogue and cooperation was reversed. Aid for the war victims (and for lasting peace in the Near East) was to be the concrete basis for a serious, profound and lasting dialogue between Catholics and Muslims, and this would lead in turn to greater mutual understanding, trust and religious liberty. It gave the impression that all these things somehow belonged together without it being compulsory to lay down a particular sequence. The latter might result from the concrete situation and perhaps certain processes can or must take place simultaneously. But naturally the message did not dwell on this. It referred to the common values in both religions which could serve as an alternative to the attractions of power, money and material pleasures. The message became explicitly anthropological when dealing with human beings' inner peace which was related to the knowledge of where one came from, why one existed on earth and to whom one would return one day. The other points were too directly and exclusively linked with the particular occasion to be of interest from an anthropological point of view.[87]

3.3.14 Pastoral journeys: main emphasis on Africa

We now come to the vast arena of the encounters between John Paul II and Muslims during his *pastoral journeys* that he often said would have been incomplete without these encounters which he explicitly requested. When one considers to whom he was speaking, it becomes clear why the Casablanca speech was so exceptional even with regard to the situation. Africa was very much in focus while Europe and America North and South received much less attention, together with Asia. What was completely lacking, with the exception of Casablanca, before his journey in the anniversary year were the key countries of Arabian

Islam. This again makes clear why that encounter was of such importance to John Paul II and why that speech acquired such canonical significance.[88]

The other main emphases on the *African continent* corresponded more or less to those of the encounters with the Catholic bishops and concentrated mainly on *states along the Gulf of Guinea*. From the very beginning, human dignity and human rights were a key issue in the addresses John Paul II gave to Muslims, clearly also following on from Paul VI as demonstrated by a relatively early speech in Ghana. In Senegal, which is predominantly Muslim, John Paul II emphasised the dialogue of life in general: mutual acceptance, mutual respect for freedom of conscience and culture, sharing and cooperation. Dialogue was seen here as a possibility of making modern society more human, more just and more respectful with regard to the freedom, rights and dignity of the individual. Indirectly, taking account of the context, the address also dealt with economic problems like poverty and just cooperation, the question of equal opportunities in the fields of education and health and stated that everyone should really enjoy a healthy, dignified family life. Almost as a passing reference it mentioned that modern people were also seeking a faith which could give meaning and direction to their lives. Speaking to the Muslim representatives of Guinea about the common characteristics which are derived from faith, reference was also made to the sense of human dignity which formed the basis for fundamental human rights. It is interesting that prayer was also mentioned in connection with interreligious dialogue. It prepared people for encounter with their neighbours and helped to form relationships of respect, understanding, appreciation and love without any discrimination. An address to Muslim representatives in Benin underlined this again. It is "le sens de la dignité de la personne humaine ouverte à la transcendance"[89], together with the appreciation of morality which both religions had in common and these also constituted some of the sources of human rights. The references to prayer were practically identical to those already mentioned, but it was also stated that respect and brotherliness should provide the motivation for abolishing poverty. What was new was the detailed discussion of the role of the family in forming the conscience. Parents prepared their children for the exercise of freedom of conscience and worship and this, in turn, was an essential prerequisite for the common life of the nation. In other places, the importance of religious liberty in the upbringing of children had been interpreted more as a right to the religious education of children and this was considered especially important in a world which tended to forget or destroy the spiritual aspect of human beings, as stated in a speech to the Muslim leaders in Ghana. There was also reference there to sacred respect for the dignity of human beings as servants of God and to the fact that both religions confessed complete submission to God (which sounds to Christians more like a Muslim expression) which was also the real meaning of being brothers and sisters in faith. The explanation of what Christians and Muslims could achieve together in society was relatively precise. "Nous pouvons promouvoir plus d'honnêteté et de discipline dans la vie publique et privée, plus de courage et de sagesse en politique, l'élimination des antagonismes politiques, l'élimination de toute discrimination à cause de la race, de la couleur, de l'origine ethnique, de la religion ou du sexe".[90]

Naturally, many of these statements were not to be seen as mutually exclusive; so, in an address to Muslim representatives in Cameroon, John Paul II himself based the expression "brothers" simply on creation and then went into more detail about the Christian way of speaking about human beings as God's representatives compared with the Muslim language about God having created human beings as his delegates. Here, too, it was a question of deriving human dignity from this and, in addition, this dignity became the criterion for the aspects of modern technology and science that should be integrated into society. But in this speech John Paul II also made clear that not everyone saw things in this way. One of the major challenges of the present time was therefore to live together peacefully and constructively. Not all religious (and other) groups recognised the rights of others. They resisted proposals for cooperation and human brotherliness. But the Catholic way was dialogue which also implied tolerance of differences. And, vice versa, in an address to the Muslim leaders of Kenya, the desire for genuine service to humanity compelled Christians and Muslims to unite their efforts.

An address to the president of *Sudan* during a pastoral visit was especially emphatic with regard to human rights, religious liberty and legal equality – for good reasons. But there was also emphasis on the argument that human beings were rational beings capable of solving any conflict if there was a will to do so. *This was a skilfully veiled slap in the face based on creation theology for many in the audience and was almost unparalleled.*

On the other hand, speaking to Muslims in the *Philippines*, John Paul II also used reasons drawn from creation theology for addressing them as brothers, although this was usually reserved for Christians in other interreligious contexts, and went on to say that the efforts of human beings were directed to God and to the truth which came from God whether or not they were aware of it.[91]

All the other speeches by John Paul II were addressed to Muslims living in *Western countries*. The common reference to God as creator was the indisputable basis but even human rights were no longer mentioned because there was no question about their recognition in these countries although there were undoubtedly difficulties of integration. The most important statements were certainly those related to the Balkans, whether in connection with the second encounter for prayer for peace specifically in that region, "Puisque tous les êtres humains ont été créés par Dieu et sont tous membres de l'unique famille humaine, nous avons le devoir d'apporter notre aide à tous"[92], or at the meeting with Muslim representatives in Sarajevo itself. "Dieu a placé tous les êtres humains sur la terre afin qu'ils parcourent un pèlerinage de paix chacun dans la situation et la culture qui sont les siennes."[93].

3.3.15 Speaking to diplomats: religious liberty

A very different context was the addresses to *ambassadors* on the occasion of their accreditation or at the traditional New Year reception. In this context, when there was explicit reference to Muslim concerns, it related more to countries with a high or very high proportion of Muslims in the population. On more or less every occasion, the now familiar series of items was quoted, although with varying completeness: belief in God the creator – the dignity of human beings as creatures who come from God and return to God – human rights, especially passive and active religious liberty – the effects at the level of coexistence and common action (justice and solidarity but also peace). The God given dignity of human beings was the pivotal focus, as John Paul II himself clearly stated.

> "Indeed, the Holy See's activity in the international forum stems from this specific vision of the human person, and from the conviction that when it is undermined or abandoned the very foundation of human society is threatened. This perspective of development calls for the advancement of freedom through the political recognition of the duty to guarantee human rights. Not least of these rights are: the freedom of authentic religious practice; the entitlement to build and maintain places of worship, including those for religious minorities; active participation of all citizens in democratic civic life; and access to education."[94]

John Paul II spoke unusually frankly about human rights and religious liberty in his speech to the New Year reception in 1990, namely shortly after the spectacular upheavals in the countries of Eastern Europe. He talked about the positive consequences of the Pancasila principle in Indonesia for the relation between Islam and other religions and for society in general. But this was not the case in all countries with a Muslim majority. On the contrary, Christians there felt that they were second-class citizens, constantly under suspicion, without churches and prevented from organizing religious education and charitable activities. His reaction was a masterly example of diplomacy, namely of recognition which made clear that this could surely not be Islamic thinking, underlined by the reference to the need (frequently requested) for mutuality and the practical realisation of religious liberty. "Je suis persuadé que les grandes traditions de l'Islam, telles que l'accueil de l'étranger, la fidélité en amitié, la patience en face de l'adversité, l'importance accordée à la foi en Dieu sont autant de principes qui devraient permettre de dépasser des attitudes sectaires inadmissibles. Je souhaite vivement que, si les fidèles musulmans trouvent justement aujourd'hui dans les pays de tradition chrétienne les facilités essentielles pour satisfaire les exigences de leur religion, les chrétiens puissent de même bénéficier d'un traitement comparable dans tous les pays de tradition islamique. La liberté religieuse ne saurait être limitée à une simple tolérance. Elle est une réalité civile et sociale, assortie de droits précis permettant aux croyants et à leurs communautés de *témoigner sans crainte de leur foi en Dieu* et d'en vivre toutes les exigences".[95] In places where religious liberty was no major problem, such remarks were kept correspondingly brief; they were often linked with a reference to the protective function of the Catholic Church for its members or as a kind of protective influence for the rights of individuals and groups, or they called for things which were guaranteed by national legislation but were in practice apparently not implemented. This demand was frequently coupled with a reference to the situation of Muslim minorities in countries with a Christian majority and the same treatment was demanded for Christian communities in countries with a Muslim majority. The assessment of individual countries can vary in the course of time. The details of the arguments also varied although it is striking that, in the very last speeches, the importance and usefulness of the separation of religion and state for religious liberty, and also for religious and state institutions and for society in general, was generally emphasised. Particularly when addressing the new ambassador of Bangladesh, who had himself raised the issue of religious liberty, John Paul II went into somewhat more detail, emphasising that faith generally made people more attentive to, responsible for and generous in their commitment to the public good because they saw others as their brothers, and linking religious liberty with the welfare of society. "[L]a garantie assurée à chaque individu de la liberté pour professer sa religion renforce l'intégrité morale d'un peuple et donc favorise une société plus juste au service du bien commun".[96] And addressing the ambassador of the Republic of Iran, of all people, John Paul II set up a common front against modern people, modern society and their spiritual crisis. The world needed a common witness to human dignity. The basis was the belief that, despite their weaknesses, human beings were able to seek and recognise the good and to distinguish it from evil which they could reject. This led on to statements like the following about freedom of conscience and religion and about peace. "[E]n suivant la loi de la conscience et les préceptes de leur propre religion, les croyants, bien qu'ayant des points de vue différents sur maint sujet, seront capables de travailler ensemble pour affronter les problèmes urgents que rencontre la famille humaine".[97] Later on, however, there were statements which somewhat relativised this view of the conscience when, referring to the constant violations of human rights everywhere, John Paul II said, "I appeal today to the consciences of our contemporaries. Indeed the human conscience must be educated".[98] But otherwise (for example speaking to the Pakistani ambassador) he established the link between intelligence and free will as gifts from the creator and part of human nature with the right to religious liberty. Apart from this, there was little in these addresses that went beyond the quasi normal subject of human rights or, more fundamentally, human dignity. This included praising the commitment of the Jordanian monarchy in the field of dialogue between the monotheistic religions and, even more, a positive statement, which sounds more Muslim than Christian, about the differences between people. "Les différences (...) manifestent au contraire la richesse et la grandeur de la nature humaine, créée par Dieu de telle sorte que tous ceux qui en participent forment une unique famille humaine".[99] When addressing all the accredited diplomats, on one occasion John Paul II also provided an indication of how important fundamental conceptions of natural law can be in the context of anthropological thinking and argumentation as well.

> "Pour éviter de tomber dans le chaos *deux exigences* me semblent s'imposer. D'abord *retrouver* au sein des États et entre les États *la valeur primordiale de la loi naturelle*, qui a inspiré jadis le droit des gens et les premiers penseurs du droit international. Même si certains remettent aujourd'hui en question sa validité, je suis convaincu que ses principes généraux et universels sont toujours capables de faire mieux percevoir l'unité du genre humain et de favoriser le perfectionnement de la conscience des gouvernants comme des gouvernés. Ensuite *l'action persévérante d'hommes d'États probes et désintéressé.*"[100]

4 Documents from the Curia

4.1 The Secretariat for Non-Christians – many legal issues between people

Among the documents from the Curia, the first of interest is the one on the attitude of the Catholic Church to the followers of other religions which was published by what was then the Secretariat for Non-Christians (3 November 1988). However, its main emphasis is the relationship between dialogue and mission, so the concrete references to Muslims are very brief and hardly go beyond what the Vatican Council said. The chapter on racism, on the other hand, is very clear or at least transparent. It speaks about discriminatory legislation affecting religious minorities, meaning the limitations on civil and religious rights, explicitly also when residence is granted but not citizenship or when one loses one's citizenship on becoming a Christian, and about second-class citizens, for example in relation to employment in public service. Under the sub-heading of current forms of racism, it states literally at the end, "Dans ce contexte il faut aussi mentionner les situations dans lesquelles on impose à d'autres communautés, à l'intérieur d'un même pays, la propre loi religieuse avec ses conséquences pour la vie quotidienne, comme par exemple la *sharia* dans quelques États à prédominance musulmane".[101] Only after starting in this way is an anthropological link established. It is very fundamental in character and begins by observing that the Christian doctrine of humankind was developed by referring back to and in the light of biblical revelation, but also in a permanent encounter with the hopes and experiences of the nations. Then the well-known statements follow about the equality of all persons because of their origin and destiny, but also the observation that differences certainly exist with regard to physical, intellectual and even moral abilities. The other major religions also saw things in this way and here there is a reference back to Paul VI who even claimed that the fatherhood of God implied the brotherhood of humankind, although it is also stated elsewhere that it is not always possible to harmonise all concrete, cultural and religious values. But, on the very basic level, a far-reaching agreement with the natural sciences is conjured up, with the reservation, however, that one is dealing with human beings at different levels and the natural sciences have nothing to say about the purpose of humanity or on moral norms. Overall, the agreement is considered to be so great that it would be possible to draw up common declarations, conventions and international agreements, for example on the protection of human rights. A twofold process is suggested, both at the level of the heart, namely of inner convictions, and at the level of the law. The aim is to reach agreement between legislation and the moral law by democratic means, and within one country it is essential that the same laws apply to everyone and that religious minorities are able to preserve their religious specificity. At another point, the importance of interreligious dialogue for achieving this aim is mentioned and it is stated that the local churches should support the rights of minorities also where they themselves are not affected. Respect for human beings is seen explicitly as respect for their basic rights, dignity and fundamental equality.

4.2 Dominus Iesus: no more than humanly equal on any account

Mention must be made here of *Dominus Iesus* which, as a document from the Congregation for the Doctrine of the Faith, concentrates mainly on the realm of the theology of religions and only touches briefly on anthropology. As far as the addressees and the line of argument are concerned, it is more – depending on one's approach – a document of reflection on one's own standpoint or of delimitation from others, and the language and style are clearly un-dialogical and hence at breach with the documents of the Second Vatican Council, even though the latter are quoted correctly and persistently. It is also critical, for example, that in interreligious dialogue preference is so often given to the subject of creation in comparison to the Christological question of redemption; it is simply easier to find a consensus on the former, whereas a specifically Christian witness is expressed in the latter and that is clearly the concern of this document.

The overall impression is precisely no longer that of a generally clear standpoint from which one can approach others in a masterly and open way and try to find common views, as in the new approach of the Second Vatican Council; on the contrary, the same statements and standpoints appear to have become uncertain, challenged by the development that they had caused and had undergone. They seem to have changed from pioneering positions into defensive bastions and this is the really negative connotation for interreligious (and also for internal ecumenical) dialogue.

Although comparatively weaker, the tendency for things that were positive and could be taken for granted to acquire a negative, defensive connotation is also perceptible in the most concrete sentence on anthropology in interreligious dialogue. "Die Parität, die Voraussetzung für den Dialog ist, bezieht sich auf die gleiche personale Würde der Partner, nicht auf die Lehrinhalte und noch weniger auf Jesus Christus, den menschgewordenen Sohn Gottes, im Vergleich zu den Gründern anderer Religionen."[102] (Parity, which is a prerequisite for inter-religious dialogue, refers to the equal personal dignity of the parties in dialogue, not to doctrinal content, and still less to the position of Jesus Christ – who is God himself made man – in comparison with the founders of other religions.)

4.3 And in conclusion human rights again

On the other hand, a very important document from the curia specifically in connection with Muslims and Christian-Muslim dialogue is *Erga Migrantes Caritas Christi* which contains a separate chapter on Muslim migrants. Among the common values which create bonds, faith in one creator is naturally listed as well as common religious practices and commitment to common goals. But then comes the big "but" of human rights in comparison with the Islamic legal conceptions which often collide, as is spelled out further on.

"Beside these points of agreement there are, however, also divergences, some of which have to do with legitimate acquisitions of modern life and thought. Thinking in particular of human rights, we hope that there will be, on the part of our Muslim brothers and sisters, a growing awareness that fundamental liberties, the inviolable rights of the person, the equal dignity of man and woman, the democratic principle of government and the healthy lay character of the State are principles that cannot be surrendered. It will likewise be necessary to reach harmony between the vision of faith and the just autonomy of creation".[103]

Immediately afterwards, the bitter experiences of Christian-Muslim mixed marriages are mentioned in which frequently the usually weakest members of Muslim families, the wives, are deprived even of the rights they do have – not to mention those to which they were accustomed – to the extent of unknowingly signing or reciting the Islamic confession of faith. The next problem is the baptism of the children because the two religions are diametrically opposed to one another on this as well. The same applies to the baptism or conversion of adult Muslims. As has been demonstrated, this problem had already been mentioned repeatedly, but never so clearly or with such far reaching effect as in this document, and despite decades of intensive efforts in dialogue. Overall, it boils down to saying that both Christians and Muslims emphasise that human beings are created by God and in due time will be raised to life again and judged by this God, and that God therefore plays the decisive part in human life and its orientation.

But as to what this means specifically for the human being as a man or woman and in the other details of their lives, and how precisely the dignity and rights of human beings or the demands addressed to them should be seen, there are obviously differences about this between Christians and Muslims which should not be underestimated and, in addition to their being reflected in church documents, they need further examination related to particular partners in dialogues which may also set other emphases in detail in addition to human rights although this is undeniably an urgent issue at the end of this chapter on fundamentals.

Notes

1 cf. Conseil Pontifical pour le Dialogue Interreligieux (éd.); Le dialogue interreligieux dans l'enseignement officiel de l'église catholique (1963–1997), Documents rassemblés par Francesco Gioia, s.l. 1998, p. 3–5. Here mention is also made of the earlier editions in Italian (Documents 1963–1993) and English (Documents 1963–1995) as well as an important element of content that the dialogue with the Jews has not been included because a separate commission is responsible for this and not the Pontifical Council for Interreligious Dialogue; incidentally, the commission for Jewish dialogue was established at the same time as the commission for Islam within the

Secretariat for Non-Christians on 23 October 1974 (cf. Fitzgerald, M[ichael] L[ouis], The Secretariat for Non-Christians is Ten Years Old, Islamochristiana 1 (1975), p.87). (On this and for other details of the book, cf. also the presentation by the author himself : Gioia, Francesco, Presentation of the Book Il dialogo inter-religioso nel magistero pontificio: Documenti 1963–1993, Pro Dialogo 87 (1994), p.214f). In what follows, decisions of the magisterium are quoted from this compendium where possible, namely from the most recent (French) edition at the beginning of this study, but references are made to the internal numeration (n.) to enable easy and quick identification in the other language editions as well. (In 2006, an updated Italian edition appeared which includes the documents up to 2005, but it was not possible to take this into account and it is probably only of interest in Catholic contexts where the use of Italian is more widespread. In the meantime, a German volume with a different basis also exists which even extends into the time of Benedict XVI: CIBEDO e.V. (ed.), Die offiziellen Dokumente der Katholischen Kirche zum Dialog mit dem Islam, Regensburg 2009. This volume can only be recommended but not included here.) For the documents of the Council and the encyclicals, Latin has been omitted since, in the period following the Council, the importance of that language has also declined dramatically in the Catholic Church, and in the interreligious dialogue in practice only French and English were of any significance. cf. also in this connection Secrétariat pour les Non-Chrétiens (éd.), Notes sur l'organisation et les règles d'action du Secrétariat, s.a. s.l. (À l'usage exclusif du Secrétariat), where p.25 states on the languages to be used in the necessary bulletin planned (now Pro Dialogo), "Il faudra évidemment affronter la question des langues, mais nous croyons qu'il suffira du français et de l'anglais pour répondre à notre but." Anyone who is looking only for a brief summary of the most important developments in interreligious dialogue and the Catholic Church can refer to: Kroeger, James H., Milestones in Interreligious Dialogue, Studies in Interreligious Dialogue 7 (1997), p. 232–239. For a brief review of questions and developments in Christian-Muslim dialogue, reference can be made to Michel, Thomas, Islamo-Christian Dialogue: Reflections on the Recent Teachings of the Church, BSNC 59 (1985), p. 172–193, with noteworthy results which are still (or again) valid today: "Neither community can await an idealized partner for dialogue to begin(…) [W]e cannot postpone our taking up the challenges (…) until the times change, for history shows us that the times will not change until we change them." (p.192) Another good summary is: Zago, Marcello, Le Magistère sur le Dialogue Islamo-Chrétien, BSNC 62 (1986), p. 171–181. The initial pages offer a good explanation of the nature and role of the church's magisterium in general and of the position adopted by the Secretariat towards it. "D'une manière générale le Secrétariat sert de caisse de résonance pour le magistère papal, universel, régional, local. C'est aussi un centre de partage des réflexions et des expériences dialogiques d'hommes et de groupes du monde entier. Il veut en particulier animer et encourager le dialogue, qui doit être assumé comme une tâche surtout par les églises particulières pour engager tout le monde." (p. 176–177)

2 cf. Conseil Pontifical pour le Dialogue Interreligieux, p. 16–17, n.7–8; cf. also *Lumen Gentium*, ibid., p.21, n.23. With regard to the evaluation of the creation of the Secretariat for Non-Christians, cf. ibid., p.125, n.196, where Paul VI himself says, "Il est bien certain que ce Secrétariat, étant donné sa destination précise, se situe hors du deuxième Concile du Vatican. Il est cependant né de l'atmosphère d'union et de bonne entente qui a nettement caractérisé le Concile." John Paul II, on its 25th anniversary, calls it "un des fruits les plus concrets et durables du Concile Vatican II" (ibid., p.425, n.584). For the background history of *Nostra Aetate* see Henry, A[ntonin]-M[arie] (ed.), Les relations de l'église avec les religions non chrétiennes, Unam Sanctam 61, Paris 1966. There one can find a very illuminating article by Robert Caspar (La religion musulmane, p. 201–236), together with a chronology of the events (p. 285–286) and a synoptic review of the various editions of the document (Les formes successives de la declaration, p. 297–305). Caspar emphasises p. 204–205 the role of Paul VI in the evolution of the document in this form and finally concludes (p.219), "l'intention du Concile était de rechercher les convergences doctrinales qui, en rendant justice à la grandeur et aux aspects de vérité de la foi musulmane, justifieraient un dialogue et une collaboration pratiques"; in comparison, other points were of lesser importance. This is in harmony with the uniqueness of this development in the history of theology which he describes clearly on p. 207–208, a development with which not all the fathers of the Council easily agreed (cf. p. 201–202, 207, 230). Reference is also made to notes 3 and 20. Bühlmann, Walbert, Alle haben denselben Gott, Begegnung mit den Menschen und Religionen Asiens, Frankfurt/Main 1978, p. 26–39, also provides a very readable survey of the developments concerning the decisive articles of the Council and of the Secretariat for Non-Christians. For interreligious dialogue and the Secretariat for Non-Christians in general, the importance of *Nostra Aetate* cannot be over-estimated, cf. Borrmans, Maurice, L'esprit de la déclaration *Nostra Aetate*: Paul VI et Jean Paul II en dialogue avec les Musulmans, in: Isizoh, Chidi Denis (ed.), Milestones in Interreligious Dialogue, A Reading of Selected Catholic Church Documents on Relations with People of Other Religions, Essays in Honour of Francis Cardinal Arinze, Rome/Lagos 2002, p.47, where right at the beginning there is reference to the charter of interreligious dialogue and the foundation document of the Secretariat. But on the very next page it is stated that the concern of the article is to show what use Paul VI and John Paul II made of the brief statements on Islam and Muslims and how they went on to interpret them. Especially John Paul II and his magisterium had proved to be very important, indeed even prophetic, at this point, according to p.56 f. One particular point which is considered to need further examination is called "common ethics" on p.60 from which *Nostra Aetate* had moved on very quickly to the necessary cooperation, so (p.61) John Paul II's involvement in human rights and especially in religious liberty is appreciated, as will also be made clear at the end of these comments. On the other hand, naturally *Nostra Aetate* had a preparatory history in theology and the church which points in the direction of North Africa, as is very well presented by Henri Tessier in his article L'expérience Missionnaire de l'Eglise au Maghreb et la Mission du Conseil pontifical pour le Dialogue inter-religieux, Pro Dialogo 72 (1989), p. 323–333. Nor should one underestimate the role of Cardinal Augustin Bea in the evolution of the wording of *Nostra Aetate* and in the establishment of the Secretariat for Non-Christians; on the contrary, he played a key role according to Schmidt, Stejpan, Cardinal Bea and Interreligious Dialogue, Pro Dialogo 74 (1990), p. 143–149. It is also interesting that it is stated ibid., p.145 f, that the three key ideas in the planning were that God is the father of all humankind, that all people are his children and that they are therefore all brothers. The automatic consequence of this is that any form of discrimination, violence or persecution for nationalist or racist reasons must be condemned. One of the best descriptions of the evolution of the Secretariat for Non-Christians is undoubtedly that in Zago, Marcello, The Life and Activity of the Pontifical Council for Inter-religious Dialogue, Pro Dialogo 74 (1990), p.151: "One might say that the Secretariat had Paul VI as founder, the Secretariat for Unity as Godfather, and (*sic*) the declaration 'Nostra Aetate' as its charter, giving it both inspiration and identity." Another person must not go without mention here, whom Fitzgerald, Michael L[ouis] / Borelli, John (eds.), Interfaith Dialogue, A Catholic View, London/Maryknoll 2006 consider a prophet of dialogue as far as Islam is concerned, namely Louis Massignon (1883–1962). Of him it is said (from p.229 onwards): "Maybe more than any other /p.230 individual, he prepared the way for the new attitude of the Catholic church towards Islam expressed in the conciliar declaration *Nostra Aetate*."

3 Quoted from Conseil Pontifical pour le Dialogue Interreligieux, p.15, n.3, cf. also n.1–2 and p.19, n.17–19. *Gaudium et Spes* also emphasises this common goal, cf. ibid., p.38, n.66 and especially p. 39–40, n.71. For further background on *Nostra Aetate* cf. Kaulig, Ludger, Ebenen des christlich-islamischen Dialogs, Beobachtungen und Analysen zu den Wegen einer Begegnung, Christentum und Islam im Dialog / Christian-Muslim Relations 3 (also Catholic theological dissertation Bochum), Münster 2004, p. 7–12. Kaulig points out that the real emphasis in the draft and the discussions was the relation with Judaism. Islam and the other religions were only introduced at relatively short notice and hence also briefly. It is also worth reading the article by Maurice Borrmans, Les évaluations en conflit autour de *Nostra Aetate*, Communio 25/5 (2000), p. 96–123, who also deals with the political conflicts related to this conciliar declaration (cf. also note 2). Precisely the statement about the unity of all human beings, because they owe their existence on this earth to God and also have God as the goal of their existence, was well received among Muslims, e.g. Dr. H. Munawir Sjadzali, Minister for Religious Affairs in the Government of Indonesia, Received the Pope in the Name of All Religious Believers in Indonesia, with the Following Address, 10 October 1989, Pro Dialogo 73 (1990), p.10. Afterwards, these statements were also seen as a reason for interreligious dialogue, as in Rossano, Pietro, Le cheminement du dialogue interreligieux de "*Nostra Aetate*" à nos jours, Pro Dialogo 74 (1990), p. 131–133, 137, whereas, as the following chapter clearly shows, the word dialogue as such was seen very sceptically at the beginning. Later it was understood as taking one's real partner seriously, not merely an object of the church's mission, and as an appropriate answer to the concept of the unity of all of humanity derived from God. NA 1a itself did not only point to what people have in common but to "ce qu'ils *sont en commun*" (ibid., p.132). Rossano not only sees the philosophical, theological roots to which the Council refers (dialogical or new thinking promoted by men like Buber, Heidegger, Jaspers or Guardini, to mention only a few); he also recognises that this was a major breakthrough in church history which (probably not by chance) coincided with the end of Eurocentrism in church and society. The extent to which this new approach had taken hold is shown clearly by the sermon of Cardinal Francis Arinze, the president of the Pontifical Council for Interreligious Dialogue, on the occasion of its silver jubilee where he said (In Faith and Gratitude, Pro Dialogo 74 (1990), p.117): "It is this (i.e. the Catholic) Church which promotes and must promote both the encounter with all other believers and the proclamation of the Gospel to every man and woman."

4 According to Conseil Pontifical pour le Dialogue Interreligieux, p. 23–27, n. 29–40; the notes are 17 and 18, p.25, and 15, p.24. As for the question of the special protection of minorities and their – also religious – rights, this is touched on briefly in *Gaudium et Spes*, cf. ibid., p. 41–24, n.79. In this connection mention can also be made of the article by Onah, Godfrey I., Reflections on *Dignitatis Humanae*, 2–4, in: Isizoh, Chidi Denis (ed.), Milestones in Interreligious Dialogue, A Reading of Selected Church Documents on Relations with People of Other Religions, Essays in Honour of Francis Cardinal Arinze, Rome/Lagos 2002, p. 83–110, the subtitle of which is: The Anthropological Basis for Interreligious Dialogue. For this reason, the article's approach is broader that this presentation but all the more fundamental. To begin with, it underlines dignity and freedom as the two pillars of the anthropology of Vatican II and emphasises that these only became interesting for everyone in the course of the 20th century when they were also seen as characteristic of everyone. He sees *Dignitatis Humanae* firmly within the tradition of the philosophy of being in God's image in the sense of *capax Dei* and thus, finally, in the categories of the *analogia entis* of Thomas Aquinas who is also mentioned, although on this question, and on the question of the consequences of sin for man's being in God's image, the divergent attitude of Protestant theology is also considered. In line with this ontological *capax Dei* as the basis for human dignity, the anthropology of Vatican II is certainly also seen as restorative compared with the approaches of existentialist philosophy and also compared with Descartes and Kant. It is also interesting that Joseph Ratzinger (now Pope Benedict XVI), in an early commentary on the Council, emphasised that starting from existence in the image of God, which is in many respects vaguer in the Old Testament than in the New, was typical for the spirit of dialogue (p.92f). It is certainly wise, in the author's view, to start from human reason which is at least potentially common to all people and then to move on to faith. The attitude of confrontation over against totalitarian states and societies is seen most clearly, according to Onah, in the definition of freedom which is understood much more strongly as freedom from than as freedom for, and that in the double sense that people should neither be forced to act against their conscience nor should they be prevented from acting according to their conscience. On the mention of public witness, p.100 states: "This was the nearest that the Council came to making a reference to the freedom of choice in religious freedom" – which is indeed the most sensitive question of all in relation to Islam. He states that the observation was also important that freedom, especially religious freedom, belonged to the fundamental, natural rights of human beings because it is rooted in human nature, in their being in God's image; (human beings cannot be open to God if they are not free, but this does not imply arbitrariness because the conscience must not close the door to transcendence of which it is the voice). Hence freedom as well as being created in God's image can finally not be lost. This brings Onah to some very telling statements about anthropology precisely in relation to (interreligious) dialogue: "The dignity of the human person, his freedom and his social nature – these are some of the anthropological factors that commend and even sometimes command dialogue as means of a healthy interaction between human beings. (...) Since religion has to do with man's relation with God and every human being is naturally capable of this relation, does it then mean that man is naturally religious? Precisely! Every human being is naturally religious, even when he does not profess any formal religion." (p.105) And the penultimate sentence states significantly (p.110): "Human beings are likely to agree more easily about human nature than about the nature of God." Although it is not stated as such here, that must surely be connected with the fact that the logical and the phenomenological levels are more easily accessible for human beings that the notorious ontological level. But at this juncture it must also be pointed out that this important document of the Council was much debated and criticised before it was finally adopted, and that its present wording scarcely hints that the church was far from being faithful always to this approach, indeed that it had condemned and combated it fiercely 100 years earlier, and that Pius XII stood for the ideal of Catholicism as the state religion, whereas here for the first time another ideal is brought into play, namely the independence of church and state while both cooperate simultaneously for the common good (something which, as this chapter will show, was increasingly emphasised later precisely against the background of the experiences of a lack of religious liberty in the Islamic context). This is discussed quite thoroughly by Caspar, Robert, Pour une théologie chrétienne du monde sécularisé: quelques perspectives récentes, in : Groupe de recherches islamo-chrétien (éd.), Pluralisme et laïcité. Chrétiens et musulmans proposent, Paris 1996, p. 188–191, who states expressly (p.190) : "Exactement un siècle plus tard, Vatican II reprend à son compte la proposition condamnée. Réjouissons-nous, mais ayons un peu de mémoire!"This is perhaps most true especially in relation to Muslims and the Islamic world.

5 cf. Conseil Pontifical pour le Dialogue Interreligieux, p.19, n.20, and p.22, n.24.

6 cf. p. 27–37, n.41–62, especially p. 28–29, n.43. In this connection, *Dei Verbum* must also be mentioned which suggests special editions of the bible for non-Christians, cf. ibid. p.23, n.27.

7 ibid. p.42, n.82, cf. p. 37–42, n 63–81 (esp. p.37, n.63, p.38/39, n.67–69, p.40, n.74)and *Apostolicam Actuositatem*, ibid. p.23, n.28. Kaulig emphasises on p.36 re *Gaudium et Spes* that it links human dignity with the vocation of human beings to communion with God, a link which certainly becomes current thinking. In general, the anthropological statements of Vatican II are seen as a theological codification of the *homo religiosus*, a concept which becomes central to the work of the Secretariat for Non-Christians, as stated e.g. by Rossano, P[ietro], Contents of Dialogue with Non-Christians, BSNC 26 (1974), p. 143–144, and idem, Le cheminement du dialogue interreligieux de *"Nostra Aetate"* à nos jours, Pro Dialogo 74 (1990), p.134, where he also says that in this way the anthropological pessimism was overcome in Catholic mission work. Another interesting statement is that of Fitzgerald Borelli, p.136: "Since the Vatican Council there has been little development in the official teaching of the Church about Islam."

8 According to Conseil Pontifical pour le Dialogue Interreligieux, p. 45–46, n.84 and p.66, n.139–140; cf. also Note 2. The statements of the Pope at Pentecost in 1964 follow this line as well but also emphasise the inner connection with that festival which should probably be understood as referring to the involvement of the Secretariat in the missionary task of the Church (ibid. p.124, n.194). In a speech of 23 June 1964 to the college of cardinals, on the other hand, Paul VI underlined that he intended this Secretariat to provide a clear demonstration of the catholic dimension of the Church, as quoted by Borrmans, Maurice, Le pape Paul VI et les musulmans, Islamochristiana 4 (1978), p.4This was reflected in the Secretariat's own understanding of itself, according to Rossano, Pietro, The Secretariat for Non-Christian Religions from the Beginnings to the Present Day: History, Ideas, Problems, BSNC 41/42 (1979), p.89: "[W]e represent the external sign of the interest of the Church for the immense multitude of the followers of religious traditions in the world." But this article indicates very honestly that up to that point there was a major difference between the claims and the practical reality. As stated e.g. on p.100: "Coresponsible with the Secretariat are the Members with whom we would like to work more closely. We also have, on paper, a distinguished group of Consultors. As I have said, we would like to call on them more than we have done in the past." Fitzgerald also provides very comprehensive and detailed information, p.87: on the structure of a typical organ of the curia with (as a rule) a cardinal as president, a secretary and an under-secretary, cardinals and bishops as members and up to eleven advisers who are specialists for Islam, not to mention the correspondents. On the under-secretariat for Islam cf. Caspar, p.204. It should be pointed out in general that the establishment of the Secretariat preceded the Council documents *Lumen Gentium* (21 November 1964) and *Nostra Aetate* (28 October 1965) and even the magna charta of dialogue, Ecclesiam Suam (6 August 1964). A letter of Paul VI indicates that he already had the plan to establish the Secretariat at the opening of the second session of the Council, according to Arinze, Francis, Reflections on the Silver Jubilee of the Pontifical Council for Inter-religious Dialogue, Pro Dialogo 72 (1989), p. 313–318. However, the special role of the Islam Commission is described by the first secretary of the Secretariat, Pierre Humbertclaude, in his essay, "Clarification of the Nature and Role of the Secretariat for non-Christians", BSNC 11 (1969), p.87 as follows: "Even before the Secretariat was set up, there already existed a Commission on Islam which had a life of its own and continued to operate on these lines after the official creation of the Secretariat, to which it was united only a year later. As it was already constituted with its own head and its own records, it remained distinct from the Secretariat while at the same time being united with it."

9 Conseil Pontifical pour le Dialogue Interreligieux, p.59, n.123, cf. also p.46, 48–51, 53–54, 58, 62–63, n.85, 91, 95, 97–98, 104, 106–108, 121, 124, 132–134. Willi Henkel in his essay Directions in Vatican Documents on Interreligious Dialogue, published in Daneel, Inus / Van Engen, Charles / Vroom, Hendrik (eds.), Fullness of Life for All, Challenges for Mission in Early

21ˢᵗCentury,Amsterdam / New York 2003, p.236, rightly speaks about this encyclical as the "'Magna Charta' of dialogue". The same expression was used by John Paul II almost 20 years earlier in an address to the Secretariat for Non-Christians (Conseil Pontifical pour le Dialogue Interreligieux, p.294, n.418, cf. also an address to bishops from Iran, p.618, n.817*). Borrmans, Le Pape Paul VI et les Musulmans, p. 5–6, refers expressly to the striking similarity with the later Council documents (which Caspar, p.213, takes for granted, cf. Note 19) and their basic attitude and puts forward the hypothesis that Paul VI was here presenting the programme for his pontificate (it was his first encyclical to which John Paul II also referred, cf. Conseil Pontifical pour le Dialogue Interreligieux, p.503, n.686), which he then also tried to implement despite all the difficulties. And right at the beginning of his article, on p.1, he mentions that the later pope had a close friendship with Louis Massignon over many years. Caspar, for his part, mentions, p.212, n.25, that *Ecclesiam Suam* was the first papal document to refer positively to Islam.

10 cf. Conseil Pontifical pour le Dialogue Interreligieux, p. 67–68, 71–72, n.141–143, 148–149, for penitence cf. *Paenitemini*, ibid. p.64, n.136.

11 ibid. p.65, n.137.

12 On human dignity cf. e.g. *Redemptor Hominis*, ibid. p.75, n.153–154, in connection with mixed marriages, ibid. p. 78–79, n.161, on love of neighbour, ibid. p.76, n.157, on appreciating people, ibid. p. 85–86, n.171, on forms of dialogue, ibid. p.93, n.180, on the theology of religions, ibid. p.85, n.170, on Assisi cf. *Sollicitudo Rei Socialis*, ibid. p. 87–88, n.174 and *Redemptoris Missio*, ibid. p. 89–90, n.177; moreover BSNC 64 (1987) is almost totally devoted to this event. The general evaluation of John Paul II in Rossano, Le cheminement du dialogue interreligieux de "*Nostra Aetate*" à nos jours, p.136, is interesting: "Jean Paul II a par suite confirmé le cheminement du dialogue par le puissant charisme de sa personnalité. Dès sa première encyclique *Redemptor Hominis* et dès son premier pèlerinage à Yasna Gora, la rencontre avec les religions non-chrétiennes apparaît comme l'une des priorités de son ministère universel. Ayant attentivement suivi de Rome les efforts de dialogue interreligieux, des ses origines à nos jours, je pense pouvoir affirmer que celui-ci se serait arrêté, qu'il se serait égaré et perdu dans les péripéties de mille et une discussions s'il n'y avait pas eu la main sûre des Papes pour en ouvrir et en soutenir le cheminement. C'est de cette impulsion et de ces directives du Pape qu'ont bénéficié les épiscopats et les Eglises catholiques d'Amérique, d'Asie et d'Afrique, et mêmes les communautés chrétiennes qui adhèrent au Conseil Œcuménique des Eglises."

13 Conseil Pontifical pour le Dialogue Interreligieux, p.83, n.168 and p. 79–82, n.162–167; for the background and more precise information cf. Kreutz, Andrej, The Vatican and the Palestinians : A Historical Overview, Islamochristiana 18 (1992), p. 109–125. The author, a specialist with a number of publications on the relations between the Vatican and the Near East, sets out clearly the emphases and developments during various phases. Up to the end of the first World War, the primary focus was clearly on concern about the holy sites for Christianity. It was only parallel to the developments which followed that the Vatican became more concerned about the Arab Christians and their problems as well and thus also about the Muslim Palestinians who had previously either been completely ignored or viewed with fear and mistrust. The Second Vatican Council strengthened the understanding both of the Jewish and of the Palestinian side, and in the case of the latter both the beginnings of a dialogue with Islam and the aspects which can be seen as generally related to "third world" problems played a part. Thus the view of Jerusalem mentioned above and of the rights of all the parties also reflected a development in the Vatican, and the tendency, whatever happened, not to make its proposals too concrete but rather to limit itself to appeals of general validity which would not offend anyone, had a longer tradition in connection with this issue as the author demonstrates. The article concludes (p. 123–125) with extracts from a press release of the Holy See of 25 January 1991 on relations with the state of Israel. Diplomatic relations between the Holy See and the state of Israel were only established in 1994 and the first Israeli ambassador to the Holy See was not accredited until 1997, cf. ISRAEL, *The New Ambassador to the Holy See Is Received by the Pope (10 April 1997)*, Islamochristiana 23 (1997), p. 199–200. Shortly afterwards, in June 1997, John Paul II made an appeal for peace to Benjamin Netanyahu and also to Yassir Arafat, cf. The Holy Father Appeals for Middle-East Peace (16 June 1007), Islamochristiana 23 (1997), p. 252–253. More direct reference to Jerusalem is made by Farhat, Edmond (ed.), Gerusalemme nei documenti pontifici, Città del Vaticano 1987, discussed by Silvio Ferrari in Islamochristiana 14 (1988), p. 362–364. Bethlehem: 22 March 2000, **To the Chairman, Palestine Autonomous Territory, and Palestinian Friends**, Pro Dialogo 104/105 (2000), p.171, referring back to *Redemptionis Anno*, emphasises that the Holy See had always recognised "that the Palestinian people have a natural right to a homeland". But at the same time John Paul II tried to safeguard the rights of the religious minorities, namely of the Palestinian Christians in particular, and adopted a clear stand against terrorism (Vatican City: 10 November 2003, **To a Delegation of the Organisation for the Liberation of Palestine** (Extract), Pro Dialogo 115 (2004), p.14. Various addresses come back repeatedly to the Middle East problem and underline the importance of a just, peaceful solution, e. g. Damascus, Syria: 5 May 2001, **Discourse upon Arrival at the International Airport** (Excerpt), Pro Dialogo 107 (2001), p. 167–168, and Damascus Syria: 8 May 2001, **Discourse of Farewell at the International Airport**, Pro Dialogo 107 (2001), p. 175–176, and deal in more or less detail with the difficulties of the situation, e. g. **Aux nouveaux Ambassadeurs de Népal, Tunisie, Estonie, Zambie, Guinée, Sri Lanka, Mongolie, Afrique du Sud et Gambie** (Extrait), Pro Dialogo 107 (2001), p.189, Cité du Vatican: 17 May 2002, **Au nouvel Ambassadeur de la Jordanie près le Saint-Siège** (Extrait), Pro Dialogo 110 (2002), p. 183–184. The wording is particularly good in Cité du Vatican: 12 décembre 2003, **A l'Ambassadeur du Qatar près le Saint-Siège** (Extrait), Pro Dialogo 115 (2004), p.22: "En effet, il n'y aura de paix véritable dans cette région que moyennant le renoncement aux violences réciproques et le recours à un dialogue courageux qui puisse aboutir à la reconnaissance du droit de chacun à vivre librement sur sa terre, dans le respect de la justice et de la sécurité pour tous, particulièrement autour des Lieux saints." But these beautiful, telling arguments do not exist in isolation; they also convey the urgent, almost desperate side of the recognition of how far removed this all is from the concrete reality and how necessary it is, indeed a duty, to convince people that peace is still possible at all (cf. Vatican City, 20 January 2004, **To the Participants in the Christian-Muslim Dialogue which Was Organized by the Islamic-Catholic Liaison Committee in Rome**, Pro Dialogo 115 (2004), p.32, referring back to the message for World Peace Day 2004).

14 Fitzgerald, Michael L[ouis], Other Religions in the Catechism of the Catholic Church, Islamochristiana 19 (1993), p.41, also p. 29–30, 32–40 (on p.35 he discusses how the catechism keeps a clear distance from revelations which go beyond the revelation in Christ or wish to correct it; Islam is not mentioned directly here but is clearly intended) and Conseil Pontifical pour le Dialogue Interreligieux, p. 92–97, n.181–184. (Fitzgerald, The Secretariat for Non-Christians Is Ten Years Old, p.91, mentioned however that the advisers to the Secretariat had requested a positive presentation of relations with Muslims in the catechism.) Immediately following the article by Michael Fitzgerald on the catechism, there is another article on it which absolutely deserves mention: Arkoun, Mohammed, Réflexions d'un musulman sur le "nouveau catéchisme", Islamochristiana 19 (1993), p. 43–54. The article appeared originally in April 1993 in the Revue des deux mondes and, for the reprint, a postscript was added at the request of the author which modifies the tone to some extent. The author is an Algerian and Frenchmen and spent many years as professor at the Sorbonne and director of the Institut d'Etudes Arabes et Islamiques de Paris III and a member of the Islamic-Christian research group. His criticism of the "new" Catechism is very fundamental. He considers the document, which naturally also attracted much attention outside of the Catholic Church, to be absolutely counterproductive as far as the dialogue with Muslims and with other religions and modernism in general was concerned. It constituted a self-contained, closed world of faith and revelation and could only provoke a comparable attitude, especially on the Muslim side, of shutting oneself up in the world of one's own revelation with its traditional interpretations – precisely what the author had been working to change for many years and was still attempting: "Les discours pour la paix par les religions, sur la tolérance, les secours humanitaires, la justice et la liberté pour tous les peuples fleuriront d'autant plus que les clôtures dogmatiques seront officiellement verrouillées et gardées pour chaque religion." (p.46) On the Christian, western side he saw a system at work that ignored the links between religion and politics and thus deceived itself about its own role in society and the world. But he also saw in it the (also typically French) secular, scientifically minded world of modernism with its closedness, powerful position and hence opposition to other worlds developed less along these lines in parallel to the mentally closed world of the catechism, only that the secular catechism was called human rights and the humanitarian attitude concealed not dogmas but one's own, initially colonial and now economic position of dominance which today helps people with problems it has itself brought about. Also particularly in the realm of science(s), he saw this unquestioned dominant power at work – Judaism alone still had thinkers of this kind to offer; all other religions "sont livrées au regard froid, distant, pas toujours désintéressé d'observateurs plus ou moins libérés d'une grille de perception et d'interprétation élaborée par la raison scientifique hégémonique."(p. 51) As far as Islam is concerned, in his view the consequences of this overwhelming position of dominance in all fields are especially drastic: innumerable false evaluations of modern "Islam" in particular and the prevention of a new definition within Islam, indeed of any discussion of the relation between the religious and the political. He envisaged an open, non-normative discourse between all people on the revelatory documents of the different religions which could e. g. also shed new light on the religious paradigms of Christianity for secularised people in the West. In the postscript, he emphasised yet again that he was not really interested in value judgements but in something quite different, namely religious anthropology which he however considered needed considerable expansion in this realm: "[I]l me semble possible et nécessaire de défendre, d'expliquer, de fonder intellectuellement le principe d'une recherche et d'un enseignement portant sur l'anthropologie religieuse, indissociable de l'anthropologie sociale et culturelle. Je m'efforce de faire découvrir aux musulmans cette discipline qu'ils ignorent encore dans leur écrasante majorité; mais je maintiens aussi que la pratique actuelle de l'anthropologie reste éloignée des problèmes soulevés justement par le type de regard intellectuel qui commande le discours du Nouveau Catéchisme." (p.53)

15 cf. Conseil Pontifical pour le Dialogue Interreligieux, p. 102–103, n.190*
16 cf. ibid. p.106, 110–111, n.194*, 198*
17 ibid. p.105, n.192*
18 ibid. p.113, n.200*, and p.112, 114–115, n.199*, 201*. Also Akasheh, Khaled, *Une espérance nouvelle pour le Liban*, Lecture de l'exhortation apostolique post-synodale à la lumière des relations islamo-chrétiennes, in: Isizoh, Chidi Denis (ed.), Milestones in Interreligious Dialogue, A Reading of Selected Catholic Church Documents on Relations with People of Other Religions, Essays in Honour of Francis Cardinal Arinze, Rome/Lagos 2002, p.248, where he underlines that the participation of official Muslim representatives of Lebanon at the Lebanese bishops'

synod remained a historic event, also for Christian-Muslim dialogue everywhere. The three Muslims had even been received by the Pope for a meal and were thus the first Muslims ever to have been invited to his table.

19 New Delhi, India: 6 November 1999, **Apostolic Exhortation: Ecclesia in Asia** (Extract), Pro Dialogo 103 (2000), p.45, and p. 28–30, 32, 37, 40–46. The document is also marked by many more points of contact; e.g. on p.41 it refers to *Nostra Aetate* as the magna charta of interreligious dialogue for our time (cf. Note 9), p. 41–42 quotes the encyclical *Redemptoris Missio* on the aims of and ways or false ways to interreligious dialogue and finally p.34 quotes an address given on 5 February 1986 in Madras before non-Christian representatives which outlined anthropological grounds for the Christian attitude of respect for other religions, on the one hand, respect for human beings in their search for answers to the deepest questions in their lives and, on the other, respect for the action of the Holy Spirit in other people.

20 cf. Vatican City: 6 January 2001, **Apostolic Letter, Novo Millennio Ineunte** (Excerpts), Pro Dialogo 106 (2001), p. 43–46. A detailed discussion of the background and effects of the Council documents on the issue of Islam or rather on the Islamic faith with detailed bibliographical references can be found in Renz, Andreas, Der Mensch unter dem An-Spruch Gottes, Offenbarungsverständnis und Menschenbild des Islam im Urteil gegenwärtiger christlicher Theologie, Christentum und Islam: Anthropologische Grundlagen und Entwicklungen 1, Würzburg 2002, p. 27–33. Renz presents the theological approach of the Council fathers as "emphatically *theocentric*" (p.29) and also underlines that, because of the problematic areas of polygamy, divorce legislation and the relation between the state and religion, it was decided not to mention Islamic family and social ethics (p.31). As far as the post-conciliar documents of the magisterium are concerned, Renz emphasizes that there is in fact no direct reference to Islam (p.37). So it is hardly surprising that it was and is necessary to widen the view somewhat in order to include what can be relevant in concrete Christian-Muslim dialogues with regard to human beings, their lives and their problems. Such detail can naturally not be expected for all the expressions of the magisterium because John Paul II in particular set great store by interreligious dialogue and the wealth of his statements easily exceeds the framework of individual studies (as already stated by Kaulig, p.38, Note 111) and especially of a study which wishes to allow Christian-Muslim dialogue to speak for itself.

21 Cité du Vatican: 28 June 2003, **Exhortation apostolique post-synodale, "Ecclesia in Europa"** (Extrait), Pro Dialogo 118 (2005), p.10.

22 ibid.

23 Conseil Pontifical pour le Dialogue Interreligieux, p.170, n.256, but also p. 168–169, n.253–255. This indication of support is already found in an address by the Pope to the Council fathers: "Nous avons le ferme propos de conduire à bon terme les conséquences qui découlent de la célébration du Concile œcuménique et de poursuivre les activités auxquelles il a donné naissance, comme celles des trois secrétariats qui déploient déjà une très heureuse activité" (ibid., p.147, n.224). Similarly also in an address to the College of Cardinals on 23 June 1970 (cf. ibid. p.183, n.273). As far as the *homo religiosus* is concerned, there is a good definition in a book by Cardinal Paul Poupard: "un homme qui croit à l'origine sacrée de la vie et au sens de l'existence humaine comme participation à une Réalité qui va au-delà de cette existence"(Les Religions, p.27, quoted from Goncalves, Teresa, Card. PAUL POUPARD, Le *Religioni nel Mondo*, I Edizione Italiana, Edizioni Piemme, 1990, p.122, Pro Dialogo 74 (1990), p.205).

24 Conseil Pontifical pour le Dialogue Interreligieux, p. 199–201, n.295–297. On the problems of the reasons cf. also Kaulig, p. 31–32, 41–43.

25 Conseil Pontifical pour le Dialogue Interreligieux, p.24, n.313, but also the whole document, p. 213–215, n.313–315.

26 cf. ibid. p.158, n.237 and especially p. 136–137, n.208–210. This positive view of cultural differences is somewhat reminiscent of the claim by the Prophet Mohammed that the legal disagreements between Muslims even constitute a divine favour. The hope for better relations between the Catholic Church and Islam is also mentioned in the address on the occasion of the canonisation of Nicolas Tavelic and his companions (cf. p. 182–183, n.272, with the sub-heading "L'amour des martyrs pour les musulmans").

27 cf. ibid. p.161, n.244, and as a broader framework for this also Note 13.

28 ibid. p.171, n.257, and p.134, n.206, p.167, n.252, p.206, n.303.

29 ibid. p.218, n.318, and p. 194–195, n.290. Kaulig deals with this phenomenon in a more comprehensive and fundamental way, not on a primarily anthropological but a theological level (the question of the true religion and the religions) and comes to the conclusion that such a strong emphasis could be understood as a deliberately chosen opposite pole intended to anticipate or combat uncertainties about this among the people (cf. p.61). The results of this investigation also point clearly in that direction.

30 cf. Conseil Pontifical pour le Dialogue Interreligieux, p.164, n.247, p.181, n.271.

31 cf. ibid. p.148, n.225, p.120, n.187, p.139, n.213, p.176, n.264, p.196, n.292, p. 190–191, n.284, p.189, n.280–281, p.198, n.294. Borrmans, Le Pape Paul VI et les Musulmans, also mentions that Paul VI was always concerned to establish diplomatic relations with countries where Islam was the state religion or with a more or less significant proportion of Muslim population, and provides a precise list with names and dates (p. 9–10, Note 19). For John Paul II such relations were also important, hence the mention of the presence for the first time of a Palestinian representative at the traditional New Year reception, Conseil Pontifical pour le Dialogue Interreligieux, p.655, n.846*. Very generally, the approaches made by Pope Paul VI to non-Christians during concrete encounters probably contributed more to the initial, fundamental decisions of the Secretariat for Non-Christians than any theological considerations about how the council documents, that were quite open in practice, should be interpreted in detail, cf. Rossano, Pietro, The Secretariat for Non-Christian Religions from the Beginnings to the Present Day: History, Ideas, Problems, p.92: "We must acknowledge that the way forward was shown less by questions and reflections around a table than by the Pope himself in his journeys to India and Palestine. The Pope spoke of other religions with theological esteem; he spoke to their adherents, calling them brothers. He underlined common spiritual and ethical values, and he expressed a desire to know them better and work with them."

32 cf. Conseil Pontifical pour le Dialogue Interreligieux, p.157, n.236, p.223, n.324, p. 180–181, n.270 (to the King of Morocco), p.180, n.269 (to the Pakistani ambassador), p.158, n.238, p.175, n.263, p.139, n.213 (Christmas message 1964). In some sense, another relevant aspect is when Muslims are simply addressed as brothers in the faith who believe in the one God (cf. p.223, n.325), when what they have in common is described as a basis for dialogue (cf. p.160, n.241) or when the message on the occasion of the independence of Mauritius is used for an appeal for dialogue (cf. p.166, n.250). Borrmans, Le Pape Paul VI et les Musulmans, p.10, mentions as almost the last words of Paul VI on the subject of Muslims an address on 13 February 1978 during the reception of the Egyptian president, Anwar el-Sadat, the content of which really refers again to everything that had been said before about peace in the Near East.

33 Conseil Pontifical pour le Dialogue Interreligieux, p.221, n.321, cf. p.175, n.263, p.186, n.276.

34 ibid. p.230, n.330.

35 According to information from Dr. Tarek Mitri, responsible for the dialogue programme of the World Council of Churches in Geneva and, as such, in constant, close contact with his Catholic colleagues and their work. A quotation from Kaulig, p.311, confirms this approach in a very striking way: "So fordert er [Claude Geffré] für den Islam einen zwar nicht mit dem jüdischen identischen, aber doch analogen Status ein: einen Platz (...) in der speziellen und nicht nur allgemeinen Heilsgeschichte, der über die Positionen von NA [*Nostra Aetate*] 3 und selbst der Ansprache Papst Johannes Pauls II vom 19.08.85 (...) hinausgeht." The same view is found in Lanfry, Jacques, A la mémoire du P. Joseph Cuoq, premier responsable pour l'Islam (1964–1974), Pro Dialogo 72 (1989), p.344. Kroeger, in his review of the milestones in interreligious dialogue and of dialogue with Muslims, also only quotes this speech of the Pope (p.238), and likewise Pontifical Council for Interreligious Dialogue / Maurice Borrmans, Guidelines for Dialogue between Christians and Muslims, Interreligious Documents 1, New York / Mahwah 1990, p.118, which lists only one papal speech in a list of dialogue meetings, namely this one. The book Recognize the Spiritual Bonds which Unite Us, 16 Years of Christian-Muslim Dialogue, published by Pontifical Council for Interreligious Dialogue (Vatican City 1994), devotes a whole chapter to the short visit (less than seven hours) under the heading "KING and POPE in Casablanca" (p.62), two pages of which (p. 64–65) contain excerpts from the speech to young Muslims, and thus sets it on the same level as all its own dialogue activities. Cardinal Francis Arinze as president of the Pontifical Council for Interreligious Dialogue also mentions this speech not before but together with the prayers for peace at Assisi as evidence of John Paul II's commitment to interreligious dialogue; cf. Cardinal Arinze Visits Pakistan, India, Malaysia, Libya, Pro Dialogo 71 (1989), p.286. A further indication is found in what Michael Louis Fitzgerald writes in his book review on *Seminarium, Joannes Paulus II e Islamismus*, Januario-Martio 1986, Città del Vaticano, 240 pp., published in Islamochristiana 13 (1987) (p.275): "Conversation turned to Pope John Paul II's visit to Morocco, its importance for Christian-Muslim relations, and the lessons that can be drawn from it. Why not have an issue of Seminarium, the Congregation's journal, dedicated to Islam? The idea clicked, and immediately a plan was drawn up. The result is no mere coffee-table book, but a collection of serious studies." The first three articles in this collection are devoted entirely to this speech and its importance. Thomas Michel, in charge of the Islam desk in the Secretariat for Non-Christians, describes this speech in Christlich-islamischer Dialog: Gedanken zu neueren Verlautbarungen der Kirche (Fortsetzung), CIBEDO Dokumentation 2 (1988), p.60, as the summary of all the previous addresses to Muslims which he had listed and explained. This speech, which had lasted a full hour, was the most comprehensive presentation of his conception of dialogue. The cornerstone for his approach to Muslims was that both sides, by virtue of their faith in the one God, claimed spiritual descent from Abraham and had many things in common anyway so they also had much to say to one another; similarly, idem. Pope John Paul II's Teaching about Islam in His Addresses to Muslims, BSNC 62 (1985), p.191. Ellul, Joseph, A Christian Understanding of Islam, in: Pontifical Council for Interreligious Dialogue / World Islamic Call Society (eds.), Co-existence between Religions, Reality and Horizons, Vatican City, p. 12–15, quotes only this speech alongside Council documents and *Ecclesiam Suam*. Among the Muslims too it is striking to observe the importance attributed to this speech by the other side which in turn led to its being at least translated and published; cf. Bayraktar, Mehmet, Le Pape Jean Paul II et la politique islamique de la papauté, Islamochristiana 15 (1898), p.239, where, despite all his questions and reservations, he speaks in favour of dialogue and mutual knowledge precisely in the current difficult situation and with special emphasis on the corrective effect of this dialogue: "Si les interlocuteurs ne corrigent pas réciproquement leurs erreurs, quel peut être le sens réel d'un dialogue authentique?" The special importance of this speech certainly related and relates to its being the first and only speech not only of this but of any pope in history before such a large Muslim audience in an Arab country. Further importance was attributed to it because the host King of Morocco was not just any monarch of a Muslim country but one of the direct descendants of the Prophet Mohammed and viewed more or less as a Caliph in his country (Pontifical Council for Interreligious Dialogue (ed.), Recognize the Spiritual Bonds which Unite, p.66). On both sides this speech was probably intended from the beginning to have the effect of uniting religions. Thus Hassan II, when introducing the Pope, quotes the conversation which had led to this encounter with the young people (ibid. p.67): "I recall that the Pope said to me with a sweet smile when he received me at the Vatican and I invited him to visit Morocco: 'But what will I do if I come to Morocco? I could not pray with the people since Your country is a purely Muslim state.' I answered: 'Holiness, your responsibility is not only religious, but equally educative and moral.' And I added, 'I am certain that tens and tens of thousands of Moroccans – especially the Youth – would be really happy if you would give them a talk on morality and relations between people, communities, nations and religions'." The speech can be found in French translation: Allocution de S.M. le Roi Hassan II à l'occasion de la rencontre du Pape avec la jeunesse marocaine, Islamochristiana 11 (1985), p. 228–229. BSNC 60 (1985) also discusses the event in more detail. John Paul II saw this speech initially as a possibility for contributing to the necessary broad effect of the dialogue via the youth (cf. Conseil Pontifical pour le Dialogue Interreligieux, p.373, n.515). Later on, (after Assisi which however, as an interreligious event and prayer for peace with initially very modest Muslim participation in all respects, had a quite different significance), he counted it among the great opportunities he had had to meet representatives of Islam (ibid. p.424, n.582) and considered that the encounter had given a new stimulus to the relationships and dialogue between Christians and Muslims (ibid. p.736, n.927*). Of the four pages he devotes to Islam in a conversation with the journalist Vittorio Messori, half a page deals with this visit alone, this speech and its significance. "Der auf Einladung von König Hassan II. erfolgte Besuch in Marokko kann zweifellos als historisches Ereignis gewertet werden. Es handelte sich nämlich nicht nur um einen Höflichkeitsbesuch, sondern um einen wirklichen Pastoralbesuch. Unvergeßlich bleibt mir die Begegnung mit der Jugend im Stadion von Casablanca (1985). Besonders beeindruckend war die Aufgeschlossenheit der Jugendlichen für das Wort des Papstes, als er den Glauben an den Einen Gott veranschaulichte. Dies war gewiß ein noch nie dagewesenes Ereignis." (Johannes Paul II, Die Schwelle der Hoffnung überschreiten, ed. Vittorio Messori, Hamburg 1994, p.122). Logically, he himself repeatedly quotes his speech when addressing bishops, ibid. p.343, n.478, p.453, n.619, p.459, n.628, in a letter to the Latin patriarchate of Jerusalem (ibid. p.739, n.929*), in the General Audience, cf. Vatican City: 5 May 1999, General Audience: Religious Dialogue with Islam, Pro Dialogo 102 (1999), p.303, before young people in Bangladesh (Conseil Pontifical pour le Dialogue Interreligieux, p.404, n.558) and in Indonesia (ibid. p.482, n.658), but also more specifically before Muslims, e.g. before the professors of Ankara University (ibid. p.462, n.633), before the Muslim leaders of Senegal (ibid. p.552, n.747), before the Muslim representative of Guinea (ibid. p.557, n.754) and of Benin (ibid. p.587, n.787), before the Iranian ambassador (ibid. p.721, n.13*) and in a message to all Muslims concerning Lebanon (ibid. p.477, n.652). The mention of this meeting to the Moroccan ambassador was more a matter of politeness (ibid. p.734, n.926*). On the Muslim side as well, the later quotation by the Pope was also seen as decisive for the importance of this speech; cf. Ayoub, Mahmoud, Pope John Paul II on Islam, in: Sharwin, Byron L./Kasimow, Harold (eds.), John Paul II and Interreligious Dialogue, Maryknoll 1999, p.176: "Since the Pope referred often to this address in later encounter with Muslims, it must be regarded as an especially significant statement of his position toward Islam and Muslims." Another external indication is that the Secretariat for Non-Christians published this speech as an offprint of Islamochristiana (11 (1985), p. 191–208) in French and in English and Arabic translation. Cardinal Francis Arinze in his preface emphasises that the Pope dealt in this speech with two subjects which were the cornerstones of his approach to the Muslims, namely the common features of the Christian and Islamic faiths in that Christians and Muslims believe in the same God, and their common heritage as spiritual descendents of Abraham. Arinze concludes: "For the Secretariat for non-Christians, which is charged with promoting dialogue between the Catholic Church and other religions, the papal visit to Morocco is of great significance and is an encouragement." (Secretariatus pro non-Christianis (ed.), Le discours de S.S. Jean-Paul II lors de sa rencontre avec les jeunes musulmans à Casablanca (Maroc), le 19 août 1985, The Speech of the Holy Father John Paul II to Young Muslims during His Meeting with them at Casablanca (Morocco), August the 19th 1985, s.l.s.a., s.p. The editors of Islamochristiana had stated (p.191): "un longue discours qui demeure un document important pour le dialogue islamo-chrétien." The foreword to the second, completely revised edition of the Guidelines for Dialogue between Christians and Muslims also quotes *Nostra Aetate* 3 (the passage most valued by the Muslims) and at the same time includes three quotations from the Casablanca speech (Rossano, Pietro, Foreword, in: Pontifical Council for Interreligious Dialogue / Borrmans, Maurice, Guidelines for Dialogue between Christians and Muslims, Interreligious Documents 1, New York / Mahwah 1990, p. 2–3). In the messages at the end of Ramadan, this papal speech is also quoted more frequently than any other; as early as 1986 it was the Pope's prayer in the context of this meeting in which God was referred to explicitly as the one who has given human beings ("us") an inner law for their lives; also in 1996 and in great detail for the special date of 2000, cf. Pontifical Council for Interreligious Dialogue (ed.), Meeting in Friendship, Messages to Muslims for the End of Ramadan (1967–2002), Vatican City 2003, p.41, 60, 68. An outward sign of the extreme importance attributed to the event is the fact that on its tenth anniversary there was a commemorative event organised by the papal nuntio in Morocco to which the president of the Pontifical Council for Interreligious Dialogue, Cardinal Francis Arinze, also came together with a whole series of Muslim dignitaries, and to which both the Pope and the King of Morocco sent words of greeting, cf. Akasheh, Khaled, Report on the Activities of the PCID: Relations with Muslims, Pro Dialogo 101 (1999), p.216, and *MAROC Commémoration du Xème anniversaire de la visite au Maroc de Sa Sainteté le Pape Jean Paul II, effectué en août 1985 (15 octobre 1995)*, Islamochristiana 22 (1996), p. 243–251. For Arinze's own speech cf. Arinze, Francis, The Way Ahead for Muslims and Christians, Pro Dialogo 91 (1996), p. 26–32. Mention must also be made here of an essay by Maurice Borrmans which deals with this speech in great detail and is therefore interesting also beyond the fundamental and anthropological aspects important here: Le discours de Jean-Paul II aux jeunes de Casablanca, Seminarium 26/1, p. 44–72. Zago also greatly values the anthropological aspects in which he sees one of the major novelties of this speech compared with the important Council documents. Le Magistère sur le Dialogue Islamo-Chrétien, p. 177–179, with discrete but clear emphasis on human rights: "Les droits de l'homme sont largement mis en relief et sont indiqué comme expression et réalisation de la volonté de Dieu. Dans cette lumière est présentée, certes avec discrétion, la réciprocité du respect et de la liberté religieuse." (p.179) Despite all the importance attributed to this visit and this speech on the Catholic side, it must be stated explicitly that the Arab-Islamic world outside of Morocco preferred to take no note of the speech or, at the most, only of its political aspects (statements on Jerusalem, Palestinian question) or reacted very negatively, cf. Demeerseman, Gérard, Echos de presse dans certain pays arabes el musulmans à propos de la visite du Pape, Islamochristiana 11 (1985), p. 229–230. It is striking that the evaluation on the two sides is radically different. The small volume Pontifical Council for Interreligious Dialogue (ed.), Journeying together. The Catholic Church in Dialogue with the Religious Traditions of the World, Città del Vaticano 1999, does however reproduce on p. 84–89 an address by John Paul II to the General Audience on 5 May 1999 under the heading "Muslims and Christians". The statements made there mainly reflect the content of *Nostra Aetate* except for his explicit reference to a "common pilgrimage to eternity" (p.88).

36 Conseil Pontifical pour le Dialogue Interreligieux, p.335, n.469; for the previous reasoning cf. p. 332–335, n.465–468.

37 ibid. p.340, n.474, cf. p. 337–341, n.471–475; Borrmans, L'esprit de la déclaration *Nostra Aetate:* Paul VI et Jean Paul II en dialogue avec les Musulmans, underlines on p.67, note 63, as the lesson of John Paul II concerning the pedagogy of dialogue, that his language in this speech is biblical without his quoting the Bible or the Quran directly. I would add to this analysis that his language also attempted to echo the Quran without quoting the Quran literally and that this is a necessary addition for dialogue (and the pedagogy of dialogue). The information on p.59, note 66, is also very interesting, namely that John Paul II used the closing prayer again with minor changes in February 1992 at the encounter with Senegalese Muslims. The more detailed article by Borrmans, Le discours de Jean-Paul II aux jeunes de Casablanca, also refers to the similarities with the Quran and other Islamic sources, and states that this skilfully solved the general problem that direct quotations from either of the holy writings could be counterproductive in the dialogue context of the speech, cf. especially p.57 f, notes 22 and 23. On p.48 f there is a very interesting and appropriate comparison of the structure and content with a Pauline letter and the indication at the end (p.69), which is also taken from Henri Teissier (p.70, note 33), that, with regard to the emphasis on practice in this speech – and elsewhere – the Charta for Christian-Muslim dialogue which should be mentioned is not *Nostra Aetate* but *Gaudium et Spes* which deals with the Church in today's world (cf. also Note 7). The last idea of special interest from an anthropological point of view is on p.53 that every authentic monotheism has its own personalism; here Borrmans refers for the Muslim side to the modern philosopher Muhammad Aziz Lahbabi. Elsewhere (Les évaluations en conflit autour de *Nostra Aetate*, p.120), Borrmans describes this speech as a commentary on the corresponding passage of the Council's declaration, but he has previously (p.113) explained explicitly that, when the Council's declaration was being produced, the term "personal" was deliberately avoided because it is untranslatable into Arabic. This leads one to conclude that the anthropological question of the concept of "person" in Islam is far from being clarified and one must be very careful not to overestimate the importance of Lahbabi in this connection.

38 cf. Conseil Pontifical pour le dialogue Interreligieux, p. 234–237, n.332–335. In the context of general audiences, John Paul II also called for intercession for and cooperation in the concerns of the Secretariat; cf. p.321, n.449.

39 ibid. p.296, n.419, cf. p. 297–299, n.420–422.

40 cf. ibid. p.372, n.515, p. 373–374, n.517, as is well stated by Fitzgerald/Borelli, p.167, concerning the monotheistic religions: "From human responsibility it is an easy step to the idea of service to humanity." Speaking to participants at a "jubilee dialogue" in Assisi, John Paul II once put it as follows (Jean Paul II reçoit les participants de la rencontre d'Assise, Islamochristiana 15 (1989), p.246): "My intention in inviting leaders from various world religions to come to that small town (...) to pray for peace in the world, was that we may be present in our common humanity before God."

41 Conseil Pontifical pour le Dialogue Interreligieux, p.857, §162, n.1035, cf. n.1034, §§ 159, 161 and p.425, n.584, p.426, n.585, p.427, n.589. This was promulgated on 28 June 1988 and came into force on 1 March 1989 according to Pontifical Council for Interreligious Dialogue/ Maurice Borrmans, p.120/121, note 2; ibid. p.155 had mentioned the formation of a Commission for Islam on 22 October 1974. John Paul II also confirmed the role of the Pontifical Council for Interreligious Dialogue before the Muslims themselves and underlined its merits (Damascus, Syria: 6 May 2001, Discourse during the Visit to the Umayyad Mosque, Pro Dialogo 107 (2001), p. 171–172): "At the highest level, the Pontifical Council for Interreligious Dialogue represents the Catholic Church in this task <to advance interreligious dialogue between the Catholic Church and Islam>. For more than thirty years the Council has sent a message to Muslims on the occasion of Îd al-Fitr at the close of Ramadan, and I /172 am very happy that this gesture has been welcomed by many Muslims as a sign of growing friendship between us. In recent years the Council has established a liaison committee with international Islamic Organizations, and also with al-Azhar in Egypt, which I had the pleasure of visiting last year." The change of name from a negative reference (non-Christians) had already been the wish of Cardinal Pignedoli in 1974, but apparently no satisfactory solution was found for a long time, since the description of the corresponding desk at the World Council of Churches, Working Unit for Dialogue with People of Living Faiths and Ideologies, was probably rightly felt to be too cumbersome, according to Fitzgerald, The Secretariat for Non-Christians Is Ten Years Old, p.88, note 1. idem, Twenty-five Years of Dialogue, The Pontifical Council for Interreligious Dialogue, Islamochristiana 15 (1989), p.109, sees the change of name positively. "The new title moreover is positive, rather that apparently discriminatory, and in fact corresponds much better to the role of the department." At the same time, in line with his superior, Cardinal Francis Arinze, he considers it a sign of the durability of this institution which can now no longer be seen as a temporary experiment. He ends the article with the words, "There is no reason why the Council should not pursue for many years its task of conscientizing, encouraging and guiding. There is still much to do."

42 Conseil Pontifical pour le Dialogue Interreligieux, p.493, n.672, cf. p.494, n.673–674, p.495, n.675, where he also says, however, "Le dialogue n'est pas surtout une idée à étudier mais il est une manière de vivre en relations avec les autres."

43 ibid. p.578, n.780, cf. p.576, n.779, p. 579–580, n.783 and earlier on p.527, n.714. A number of phenomena from the Islamic realm could also be included after the following statement although Muslims are not mentioned explicitly (n.780, p. 577–578). "Nous ne pouvons qu'être profondément peinés et attristés par l'apparition ou par la résurgence de préjugés et d'attitudes agressives qui sont parfois encouragées au nom de Dieu mais /p. 578 qui ne trouvent aucune base dans la foi au Dieu Créateur tout-puissant et miséricordieux." The clear stand of John Paul II against the Gulf war was very well received: "on behalf of all Arabs and Muslims, our sincerest thanks and appreciation for his noble stands, and to convey also our hearty congratulations and felicitations to His Holiness on the anniversary of his accession to the papal throne as well as our best wishes for a happy Christmas and a prosperous New Year" (El-Assad, Nassir El-Din, Inaugural Address, in: Pontifical Council for Interreligious Dialogue / Royal Academy for Islamic Civilization Research (eds.), The Rights and Education of Children in Islam and Christianity, Acts of a Muslim-Christian Colloquium, Vatican City s.a., p. 10).

44 Conseil Pontifical pour le Dialogue Interreligieux, p.652, n.843*, cf. p. 650–651, n.842*

45 cf. ibid. p. 652–653, n.844*, and also p. 691–692, n.884*. Moreover, in comparison it is striking that no other representative of the World Council of Churches is mentioned but during the concluding blessing Mary's intercession is invoked (p.653, n.844*), cf. Pope's Address to the Participants in the Plenary Assembly 1998, Pro Dialogo 101 (1999), p. 192 und Vatican City: 9 November 2001, **To the Participants in the Plenary Assembly of the Pontifical Council for Interreligious Dialogue**, Pro Dialogo 109 (2002), p. 12, whereas at the commemorative event for

Bishop Rossano the cooperation with the World Council of Churches is commended. "This is a significant collaboration, which Bishop Rossano started and encouraged. I would also like to pay him a tribute for this." And the blessing takes a different form (Vatican City: 16 June 2001, **To the Participants in the "Study and Reflection Days" on the Occasion of the Tenth Anniversary of the Death of H.E. Mgr. Piero Rossano**, Pro Dialogo 108 (2001), p. 293).
46 Vatican City: 9th November 2001, **To the Participants in the Plenary Assembly of the Pontifical Council for Interreligious Dialogue**, p.10, cf. Cité du Vatican: 18 novembre 2004, **Aux Chefs religieux d'Azerbaidjan**, Pro Dialogo 118 (2005), p.36: "Personne n'a le droit de présenter ou d'utiliser les religions comme instrument d'intolérance, comme un moyen d'agression, de violence et de mort."
47 Vatican City: 9th November 2001, **To the Participants in the Plenary Assembly of the Pontifical Council for Interreligious Dialogue**, p.11.
48 Conseil Pontifical pour le Dialogue Interreligieux, p.464, n.636, cf. p. 560–561, n.759. However, John Paul II also calls on the missionaries explicitly to engage in interreligious dialogue to which the Church should devote more attention in the new year, see Vatican City: 19 May 2001, **To the Participants in the General Chapter of the Society of the African Missions** (Excerpt), Pro Dialogo 107 (2001), p.192. A document of more general content although directed to a similar group of addressees is Vatican City: 26 May 2001, **Letter to Bishop François Blondel of Viviers on the Occasion of the Centenary of the Ordination to the Priesthood of Charles de Foucauld in 1901** (Extract), Pro Dialogo 108 (2001), p. 289–290. Mention must also be made of the work done by the Institute in the context of the Journées Romaines which was generally much appreciated by John Paul II, cf. Conseil Pontifical pour le Dialogue Interreligieux, p. 474–4766, n.648–650, although in this address the emphasis was more on witness and he did not deal directly with anthropological issues.
49 ibid. p.643, n.836*, cf. p.642.
50 Opening Discourse, in: **Rencontre** de S.S. le Pape Jean-Paul II avec les patriarches des églises orientales et les évêques des pays impliqués dans la guerre du golfe, **Meeting** of H.H. Pope John Paul II with the Patriarchs of the Oriental Churches and Bishops of Countries Involved in the Gulf War, 4–6 March 1991, s.l.s.a., p.20, also p. 19–21; on the importance cf. p.5, the preface by Cardinal Francis Arinze. The address can also be found in Message de Jean Paul II aux Patriarches et aux Evêques sur le Moyen-Orient (4 mars 1991), Islamochristiana 17 (1991), p. 287–289. And Conseil Pontifical pour le Dialogue Interreligieux, p. 451–453, n 618–621, Cité du Vatican: 17 mars 2001, **Aux Evêques du Rite Latin de la Région Arabe (C.E.L.R.A.)** (Extraits), Pro Dialogo 107 (2001), p. 162–164, Cité du Vatican: 11 décembre 2001, **Aux évêques de l'Eglise chaldéenne (d'Iraq, l'Iran, du Liban, d'Egypte, du Syrie, de la Turquie e** *(sic!)* **des Etats Unis d'Amérique) en visite "ad limina"**(extrait), Pro Dialogo 109 (2002), p.30 and Cité du Vatican: 1 août 2004, **À Sa Béatitude Emmanuel III Delly, Patriarche de Babylone des Chaldéens**, Pro Dialogo 118 (2005), p.25. In this connection mention must also be made of a joint declaration with His Holiness Aram I Keshishian (Conseil Pontifical pour le Dialogue Interreligieux, p.702, n.896*) which naturally deals mainly with the situation of the Armenian Church but also generally states how much Christians and Muslims in the Near East have in common by way of history, socio-economic problems and political destiny. Apart from purely diplomatic occasions, Yemen occurs only once in an address during the visit of the state president to the Vatican which remains relatively general, however (Cité du Vatican: 26 novembre 2004, **Au Président, S.E. Ali Abdullah Saleh, de la République du Yémen**, Pro Dialogo 118 (2005), p 38–39).
51 Solemn Closing, in: **Rencontre** de S.S. le Pape Jean-Paul II avec les patriarches des églises orientales et les évêqes des pays impliqués dans la guerre du golfe, **Meeting** of H.H. Pope John

Paul II with the Patriarchs of the Oriental Churches and Bishops of Countries Involved in the Gulf War, 4–6 March 1991, s.l.s.a., p.23, also p. 24–25. The idea is taken up again in: The Patriarchs of the Catholic Churches of the Middle East and the Presidents of the Bishops' Conferences of the Countries More Directly Involved in the Gulf War, in: **Rencontre** de S.S. le Pape Jean-Paul II avec les patriarches des églises orientales et les évêqes des pays impliqués dans la guerre du golfe, **Meeting** of H.H. Pope John Paul II with the Patriarchs of the Oriental Churches and Bishops of Countries Involved in the Gulf War, 4–6 March 1991, s.l.s.a., p.28: "Together with all believers, we are persuaded that with faith in God and trust in man, God's creature, the face of the world can indeed be changed."

52 cf. Damas Syrie: 6 mai 2001, **Homélie au Stade Abbassyne**, Pro Dialogo 107 (2001), p.169, Damas Syrie: 6 mai 2001, **Rencontre avec les Patriarches et les évêques au Patriarcat grec-melkite**, Pro Dialogo 107 (2001), p.170, Bethlehem: 22 March 2000, **Homily** (Extract), Pro Dialogo 104/105 (2000), p.172, Jerusalem: 25 March 2000, **To the Participants and Representatives of the Christian Churches and Confessions Present in the Holy Land** (Extract), Pro Dialogo 104/105, p.179, Cité du Vatican: 6 novembre 2000, **Message à Sa Béatitude Michel Sabbah**, Pro Dialogo 106 (2001), p. 18–19

53 cf. In the order of their mention: Conseil Pontifical pour le Dialogue Interreligieux, p.736, n.927*, p.270, n.382, p.271, n.385, p. 538–539, n.727, p 666–668, n.861*-863*, Cité du Vatican: 22 février 2003, **Aux évêques de la Région du Nord de l'Afrique pendant leur visite "ad limina"**(Extrait), Pro Dialogo 114 (2003), p. 314–316, Conseil Pontifical pour le Dialogue Interreligieux, p.271, n.384, p.539, n.728.

54 Castel Gandolfo: 30 août 2003, **Aux évêques de l'Eglise Catholique copte d'Egypte à l'occasion de leur visite "ad limina"** (Extrait), Pro Dialogo 114 (2003), p.344, cf. Conseil Pontifical pour le Dialogue Interreligieux, p.729, n.920*, p.662, n.857*.

55 ibid. p.241, n.341, cf. p. 239–240, n.338–339: According to Dr. Tarek Mitri, responsible for the Dialogue department of the World Council of Churches, this speech, given on 29 November 1979 in Ankara, does not have the almost canonical rank of the Casablanca speech but does occupy a significant place and is often quoted in internal Catholic contexts. The addresses to the bishops are found ibid. p.454, n.622, p. 608–610, n.810* and Vatican City: 19 February 2001, **To the Bishops of Turkey, Ad Limina Apostolorum** (Excerpts), Pro Dialogo 107 (2001), p. 161–162.

56 Astana, Kazakhstan: 23 September 2001, **Homily at the Holy Mass**, Pro Dialogo 108 (2001), p.309, cf. p. 310–311. At this mass, representatives of other religions must have been present because they were greeted explicitly (p.308). cf. also Astana, Kazakhstan: 23 September 2001, **Angelus Message**, Pro Dialogo 108 (2001), p.311, Astana. Kazakhstan: 23 September 2001, **To the Ordinaries of Central Asia** (Extract), Pro Dialogo 108 (2001), p.313, Kiev, Ukraine: 24 June 2001,**To the Illustrious Representatives of the All-Ukrainian Council of Churches and Religious Organizations** (Extract), Pro Dialogo 108 (2001), p. 293–295, Conseil Pontifical pour le Dialogue Interreligieux, p.714, n.908* (on the general situation in Bosnia-Herzegovina cf. p.715, n.909*), Cité du Vatican: 15 Janvier 1999, **Discours du Pape aux évêques de la Conférence épiscopale de la Bosnie-Herzégovine à l'occasion de leur visite "ad limina"** (extrait) Pro Dialogo 102 (1999), p.289.

57 Kiev, Ukraine: 24 June 2001, **To the Illustrious Representatives of the All-Ukrainian Council of Churches and Religious Organizations** (Extract), p.295.

58 cf. Conseil Pontifical pour le Dialogue Interreligieux, p.468, n.641, Vatican City: 15 May 2001, **To the Bishops of Bangladesh,** Pro Dialogo 107 (2001), p.176, Vatican City: 19 May 2001, **To the Bishops of Pakistan** (Excerpts), Pro Dialogo 107 (2001), p. 190–191.Several interesting points in

this connection are also contained in: The Pope Receives the Bishops of Pakistan on the Occasion of Their "ad limina" Visit (21 October 1994), Islamochristiana 21 (1995), p.198.

59 Conseil Pontifical pour le Dialogue Interreligieux, p.500/501, n.683, cf. n.682, p.465, n.638, p.485, n. 660 – 661, p.510, n. 696, Castel Gandolfo: 25 September 2003, **To the Bishops from the Philippines (Ecclesiastical Provinces of Cagayan de Oro, Cotabato, Davao, Lipa, Ozamis and Zamboanga) on the Occasion of their "ad limina" Visit** (Extract), Pro Dialogo 114 (2003), p. 347–348, where the Pope also suggests a forum of bishops and Ulama at the local level.

60 cf. Conseil Pontifical pour le Dialogue Interreligieux, p.606/607, n.808*.

61 cf. ibid. p.272/273, n.386, p.488, n.665, p.489, n.666, p.490, n.668, p.661, n.856*. The information about the special position of Mali and the situation there comes from Prof. Maurice Borrmans, formerly of the Pontificio Istituto per i Studi Arabi e d'Islamistica, Rome. The response of John Paul II to the welcome by the state president of Mali confirms this. He speaks about the forthcoming meeting with followers of different religions "me réjouissant des rapports harmonieux qui existent ici entre les religions africaines traditionnelles, les communautés musulmanes et les communautés chrétiennes. Je vois dans cette attitude positive une garantie du respect de la dignité et du bien-être de chacun. La foi religieuse, en effet, doit rapprocher les hommes et les conduire à une solidarité plus profonde dans la recherche commune de tout ce qui est noble et bon." (La visite pastorale de Jean Paul II au Mali (28–29 janvier 1990), Islamochristiana 16 (1990), p.255).

62 According to: La visite pastorale de Jean Paul II au Mali (28–29 janvier 1990), p. 256–258. John Paul II speaks in an equally fundamental way, for example, when he talks to Catholic young people in Senegal about Christian-Muslim relations, cf. Les rencontres de Jean Paul II avec les divers groupes de catholiques sénégalais, Islamochristiana 18 (1992), p.290 : "Bref, que Chrétiens et Musulmans collaborent dans ce qui fait grandir la communauté humaine! Vous, les jeunes, bien que vous ne soyez pas unis dans vos croyances, apprenez à vous respecter et à vous tolérer. En effet, la Bible nous montre que l'être humain possède *une dignité unique:* il est une *créature de Dieu* et il a donc une relation privilégiée avec Celui qui lui a tout donné. L'homme est invité à devenir vraiment fils de Dieu dans un partage de vie et d'amour: il a une valeur souveraine. Pour les Musulmans, l'homme est appelé à être un parfait représentant de Dieu sur la terre, en y témoignant, pour le service de tous, de ce que signifient ces Très Beaux Noms: miséricorde et compréhension, pardon et réconciliation. Chers amis, grande est la dignité de l'homme! Il est une route qui mène au Seigneur, un 'signe' qui révèle Dieu."

63 cf. Cité du Vatican: 28 août 1999, **Aux évêques du Côte d'Ivoire à l'occasion de leur visite "Ad limina"** (Extrait), Pro Dialogo 103 (2000), p.17, Vatican City: 15 February 2003, **To the Bishops of the Gambia, Liberia and Sierra Leone during their "ad limina" Visit** (Extract), Pro Dialogo 114 (2003), p.308, Cité du Vatican: 15 février 2003, **Aux évêques de la Guinée Conakry à l'occasion de leur visite "ad limina"** (Extrait), Pro Dialogo 114 (2003), p.310.

64 Cité du Vatican: 17 juin 2003, **Aux évêques du Burkina Faso et du Niger à l'occasion de leur visite "ad limina"** (Extrait), Pro Dialogo 114 (2003), p.338, cf. Conseil Pontifical pour le Dialogue Interreligieux, p.254, n.358, p.274/275, n.388, p.694, n.887*. Another address fits into this picture, namely that to the bishops of Chad during their obligatory visit in the Vatican, cf. A Rome (Vatican), le Pape reçoit les Evêques du Tchad en visite "ad limina" (27 juin 1994), Islamochristiana 20 (1994), p.270 f.

65 Abuja – Nigeria: 23 March 1998, **Discourse of the Pope at Meeting with the Episcopal Conference of Nigeria in the Apostolic Nunciature** (Excerpt), Pro Dialogo 99 (1998), p.278, cf. Conseil Pontifical pour le Dialogue Interreligieux, p.428, n.590, Cairo, Egypt: 25 February

2000, **A la fin de la Sainte Messe**, Pro Dialogo 104/105 (2000), p.166, Vatican City: 20 April 2002, **To the Bishops of Nigeria on the Occasion of their Visit "ad limina"** (Extract), Pro Dialogo 110 (2002), p. 177–178, Vatican City: 27 May 2004, **To the New Ambassador of Nigeria to the Holy See** (Extract), p.18.

66 cf. Cité du Vatican: 28 août 1999, **Aux évêques du Côte d'Ivoire à l'occasion de leur visite "Ad limina"** (Extrait), Pro Dialogo 103 (2000), p.17, Vatican City: 15 February 2003, **To the Bishops of the Gambia, Liberia and Sierra Leone during their "ad limina" Visit** (Extract), Pro Dialogo 114 (2003), p.308, Cité du Vatican: 15 février 2003, **Aux évêques de la Guinée Conakry à l'occasion de leur visite "ad limina"** (Extrait), Pro Dialogo 114 (2003), p.310, Cité du Vatican: 9 Septembre 1999, **Aux évêques du Tchad à l'occasion de leur visite "Ad limina"** (Extrait), Pro Dialogo 103 (2000), p.21, Conseil Pontifical pour le Dialogue Interreligieux, p.595, n.796, Cité du Vatican: 27 Septembre 1999, **Aux évêques de la République Centrafricaine en visite "Ad limina"** (Extrait), Pro Dialogo 103 (2000), p.24, Conseil Pontifical pour le Dialogue Interreligieux, p.437, n.599, p.595, n.796, Vatican City: 3 September 1999, **To the Bishops of Zambia on the Occasion of their "Ad limina" Visit** (Extract), Pro Dialogo 103 (2000), p.19, Vatican City: 6 September 1999, **To the Bishops of Malawi on the Occasion of their "Ad limina" Visit** (Extract), Pro Dialogo 103 (2000), p.20.

67 cf. Conseil Pontifical pour le Dialogue Interreligieux, p.547, n.738–739, p.413/414, n.569, p. 320–321, n.448–449. The latter address unfortunately contains two errors. Firstly, it is claimed (p.320, n.449) that the Secretariat for Non-Christians was only established after the Second Vatican Council whereas the founding of the Secretariat on 19 May 1964 even predated the two determinative Council documents, *Lumen Gentium* and *Nostra Aetate* (of 21 November 1964 and 28 October 1965 respectively). (A similar error occurred in another speech, p.503, n.686, where the impression is given that the Council had created the structures for dialogue, i.e. established the Secretariat mentioned above.) But the most painful erroneous statement is on p.320, n.448, that "les disciples de Mahomet honorent aussi Jésus comme Dieu", whereas in *Nostra Aetate* (p.17, n.7) the correct quotation would have been: "Bien qu'ils ne reconnaissent pas Jésus comme Dieu." Since the speech discusses theological agreements, one could also wonder whether this was not wishful thinking, but another publication of the same speech in English demonstrates that it was simply a mistake, because there the wording is correct. "Though they do not acknowledge Jesus as God" (General Audience of 5 June 1985: **Relations with Non-Christian Religions**, BSNC 60 (1985), p.229). The address to the Curia in December 1986 referred to before, in a sense as an explanation for Assisi, had far-reaching consequences, as can be clearly seen e.g. in **Message aux chrétiens du Maghreb de la Conférence des Evêques de la Région Nord de l'Afrique (C.E.R.N.A.)**, Alger, le 8 juin 1990, Pro Dialogo 74 (1990), p.162: "Le Pape Jean Paul II a pris des initiatives historiques, en particulier en suscitant la prière pour la Paix à Assise (1986). Dans son discours à la curie romaine à Noel 1986 il a signalé les fondements théologiques de cette attitude. Tous les hommes ont la même origine. Ils sont tous créés à l'image de Dieu. Ils sont tous habités par l'Esprit de Dieu qui les conduit, de l'intérieur même de leurs traditions religieuses et de leurs idéologies, vers leur vocation commune. Les collaborations entre croyants de traditions religieuses différentes, ou entre hommes de références diverses, se fondent sur cette unité, plus profonde que leur divergence."

68 Vatican City, 5 May 1999, **General Audience: Religious Dialogue with Islam**, p.303, cf. p. 304–305.

69 cf. Conseil Pontifical pour le Dialogue Interreligieux, p.703, n 898*, p.724, n.916*, Cité du Vatican: 16 mai 2001, **L'Audience générale** (Extraits), Pro Dialogo 107 (2001), p.177.

70 Conseil Pontifical pour le Dialogue Interreligieux, p.647, n.840*; in connection with the problems of the Middle East, John Paul II had already sent a letter earlier on (1991) to the UN general secretary, cf. p.518, n.705. And there are other statements made before similar international institutions, e.g. the Organisation for Security and Cooperation in Europe, where of course human dignity and human rights play a major part, but things are also said which lead one to think directly of the situation in France which is not always easy, e.g. (Cité du Vatican: 10 octobre 2003, **Aux participants à l'Assemblée parlementaire de l'Organisation pour la Sécurité et la Coopération en Europe** (O.S.C.E.), Pro Dialogo 115 (2004), p.9): "Lorsque les Etats sont disciplinés et équilibrés dans l'expression de leur nature séculière (...)" or in a surprisingly direct way as far as the practical, legal aspect of religious liberty is concerned (p.10): "La promotion de la liberté religieuse peut également avoir lieu à travers des dispositions prises pour les différentes disciplines juridiques des diverses religions, à condition que l'identité et la liberté de chaque religion soient garanties." Along very similar lines cf. Vatican City: 31 October 2003, **To the Ministers for the Interior of the European Union**, Pro Dialogo 115 (2004), p.13: "The recognition of the specific religious patrimony of a society demands the recognition of the symbols that qualify it. If, in the name of an erroneous interpretation of the principle of equality, one gives up expressing such religious traditions and connected cultural values, the fragmentation of today's multiethnic and multicultural societies could easily transform itself into a factor of instability and, thus, of conflict. Social cohesion and peace cannot be built by eliminating the religious features of every people: otherwise, such a proposition would result in less democracy, because it would be contrary to the spirit of nations and the sentiments of the majority of their peoples." An address that is interesting and not so easy to classify is that on the occasion of the accreditation of the new US ambassador just two days after the attack on the Pentagon and the World Trade Center. In this context there is no reference to Islam, nor to religion or God, but only to an "inhuman act", to the "human family" and to the "highest ideals of solidarity, justice and peace" (Castel Gandolfo: 13 September 2001, **To the New Ambassador of the United States of America to the Holy See** (Extract), Pro Dialogo 108 (2001), p.302). At the next new year reception for the diplomatic corps, John Paul II expressed himself somewhat more clearly when referring to (Cité du Vatican: 10 janvier 2002, **Discours du Saint-Père aux Membres du Corps Diplomatique accrédité près le Saint-Siège** (Extrait), Pro Dialogo 109 (2002), p.40): "frères et soeurs, fidèles de l'Islam authentique, religion de paix et d'amour du prochain". (ibid. p.40/41 also has an interesting list of the worldwide tasks still to be carried out.) And one year later at the same event he again takes up the subject (Cité du Vatican: 13 janvier 2003, **Au Corps Diplomatique accrédité près le Saint Siège à l'occasion de l'échange des Voeux du Nouvel An** (Extrait), Pro Dialogo 112 (2003), p.20): "Le dialogue oecuménique entre chrétiens et les contacts respectueux avec les autres religions, en particulier avec l'Islam, sont le meilleur antidote aux dérives sectaires, au fanatisme ou au terrorisme religieux." Addressing the ambassador of Syria, terrorism is even described explicitly as a phenomenon (Cité du Vatican: 15 may *(sic!)* 2003, **Au nouvel Ambassadeur de Syrie près le Saint-Siège** (Extrait), Pro Dialogo 114 (2003), p.329, "qui met en danger de manière insupportable le bien commun de la paix, de la dignité des personnes et des peuples." Probably also referring to the attack on Bali, he states Vatican City: 10 January 2004, **To the Ambassador of the Republic of Indonesia to the Holy See** (Extract), Pro Dialogo 115 (2004), p.25: "I am convinced, moreover, that Islamic, Christian and Jewish religious leaders must be at the forefront in condemning terrorism and denying terrorists any form of religious or moral legitimacy."

71 Conseil Pontifical pour le Dialogue Interreligieux, p.647/648, n.840*, the rest of the summary is found on p.648, n.840*.

72 cf. Also Kaulig who writes on p.75 about the theological development in John Paul II: "'Jene, die die Wahrheit besitzen' und 'jene, die sie nicht gefunden haben' sind längst nicht mehr so eindeutig Christen und Andersgläubigen zuzuordnen, wie es Paul VI. nahelegt *(sic!)*, so daß *(sic!)* das päpstliche Lehramt nun intensiver das Potential ausschöpft, das z.B. In *(sic!)* DH <i.e. *Dignitatis Humanae*> 3 vorgelegt ist."

73 cf. Message of His Holiness Pope John Paul II for the Celebration of the World Day of Peace (7 January 1999), Islamochristiana 25 (1999), p. 240–242. The document particularly contains the following, very skillful formulations. "To promote the good of the individual is thus to serve the common good, which is that point where rights and duties converge and reinforce one another." "To affirm the universality and indivisibility of rights is not to exclude legitimate cultural and political differences in the exercise of individual rights, provided that in every case the levels set for the whole of humanity by the Universal Declaration are respected", (both p.240). "Only when a culture of human rights which respects different traditions becomes an integral part of humanity's moral patrimony shall we be able to look to the future with serene confidence." (p.242)

74 Conseil Pontifical pour le Dialogue Interreligieux, p.262, n.371, cf. p.261, n.369, p.262, n.370, p.263, n.372, for the corresponding address in India cf. p.352, n.489, p.356, n.495–496. The following general sentence is also striking, p.354, n.492: "La culture n'est pas seulement une expression de la vie temporelle de l'homme mais une aide pour atteindre sa vie éternelle." The Muslims are also mentioned in an address in Benin (p.591, n.791) which is really directed to representatives of voodoo which has its African roots in that region (cf. Bellinger, Gerhard J., Knaurs großer Religionsführer, 670 Religionen, Kirchen und Kulte, weltanschaulich-religiöse Bewegungen und Gesellschaften sowie religionsphilosophische Schulen, Augsburg 1999, p.414).

75 Conseil Pontifical pour le Dialogue Interreligieux, p.331, n.463, cf. p.329, n.461, p. 481–482, n 657–658, p.484, n.659.The speech in Sri Lanka is also comparable where John Paul II referred to Buddhists, Hindus, Muslims and Christians in one breath and continued (p.621, n 820*), "Les croyances religieuses inspirent des valeurs communes telles que l'acceptation de l'autre, le dialogue, la compréhension dans la recherche de la vérité." Another similar reference is Vatican City: 22 January 2001, **To the New Ambassador of the Islamic Republic of Iran to the Holy See** (Excerpt), Pro Dialogo 106 (2001), p.49. "In the dialogue of cultures, people of good will come to see that there are values which are common to all cultures because they are rooted in the very nature of the human person. These are values which express humanity's most authentic distinctive features: the value of solidarity and peace; the value of education; the value of forgiveness and reconciliation; the value of life itself."

76 Conseil Pontifical pour le Dialogue Interreligieux, p.505/506, n.689, cf. p.503, n.686, p.506, n 690, p. 591–592, n.792.

77 ibid. p.671, n 865*, cf. p.670 and p.669, n.864*.

78 ibid. p.707, n 903*, cf. p.708, p.404, n 558.

79 cf. Vatican City: 13 December 2001, **To the Authorities Concerned about Crisis in Holy Land** (Extract), Pro Dialogo 109 (2002), p. 31–32; p.32: "How we wished that this message would be heard and promptly put into practice! How pleased we would have been not to have to repeat it!"

80 Vatican City: 6 March 2002, **To The Most Reverend Pietro Sambi Apostolic Nuncio to Cyprus**, Pro Dialogo 110 (2002), p.173. This statement refers concretely to the encounter on 24 January 2002. The European encounter took place on 9 January 1993, cf. Conseil Pontifical pour le Dialogue Interreligieux, p.583, n.785, and the first meeting on 27 October 1986, cf. ibid. p. 388–340, n.534–538 ; for the noteworthy general evaluation on the part of John Paul II cf. p. 409–417.

According to the declarations of Vatican II, the Catholic Church was the sacrament of the unity of humankind which was naturally given through creation and redemption in Jesus Christ but normally not yet visible. Despite all the continuing differences considered human and not of divine origin, this unity was visible in Assisi and thus the Catholic Church had done justice to its mandate from the Council. On p.415, n.570, he stated explicitly, "En présentant l'Église catholique lui tient par la main ses frères chrétiens et ceux-ci tous ensemble qui donnent la main aux frères des autres religions, la Journée d'Assise a été comme une expression visible de ces affirmations du Concile Vatican II. Avec elle et par elle, nous avons réussi, grâce à Dieu, à mettre en pratique, sans aucune ombre de confusion ni de syncrétisme, cette conviction qui est la nôtre, inculquée par le Concile, sur l'unité de principe et de fin de la famille humaine et sur le sens et la valeur des religions non chrétiennes." This evaluation was also included in documents of the Curia, cf. ibid. p.799, n.929 and p.809, n.952. With the first invitation to Assisi, John Paul II had taken an initiative which was taken up by others within the Catholic Church and thus gained momentum, also among Muslims. It would be quite impossible and is not the task of this presentation to include all of this material but, by way of example, reference can be made to Landousies, J., Algérie: Rapport sur le dialogue islamo-chrétien, Pro Dialogo 73 (1990), p.99f. Extracts from the Christian prayer for peace at Assisi already pointed in this direction (Christian Prayer for Peace, BSNC 64 (1987), p. 123): "Thus, peace is not only the absence of war but the state of fraternal concord and realized unity of the human family. (...) The *Spirit of God*, which we believe is God himself, is the Spirit which brings 'unity to all, through the bond of peace' (cf. *Eph* 4:3) because peace is one of his gifts (cf. *Gal* 5:22). (...) Our religious commitment to peace implies a simultaneous commitment to justice, namely not only in respect for but also in the promotion of the true dignity of all men and women, created 'in the image of God' (cf. *Gen* 1:26)." On the Muslim contribution to the first prayer for peace cf. Demeerseman, Gérard, Journée de prière pour la paix à Assise, Islamochristiana 13 (1987), p. 200–204. John Paul II continued to express concern for peace in the Balkans, driven as it were by the spirit of brotherliness, and in opposition to those who violated humanity, cf. Holy Father's New Appeal for Peace: "We Beg again, End this Terrible Slaughter!" (19 February 1994), Islamochristiana 20 (1994), p. 216–217. In this connection, the speeches are interesting which he had prepared for his pastoral visit to Sarajevo which was then cancelled; there he explained how the Declaration of Human Rights had evolved from the will for peace after the experiences of the Second World War, and that peace was possible because human beings have a conscience and a heart, cf. The Speeches Pope John Paul Prepared for His Pastoral Visit to Sarajevo (8 September 1994), Islamochristiana 21 (1995), p.147. Islam also played a special part in other internal ecumenical contexts, cf. Vatican City: 1 July 2004, **Common Declaration Signed together with Ecumenical Patriarch Bartholomew I of Constantinople** (Extract), Pro Dialogo 118 (2005), p.25, where the list of challenges for the present time includes, "to restore to the European Continent the awareness of its Christian roots; to build true dialogue with Islam, since indifference and reciprocal ignorance can only give rise to diffidence and even hatred; to nourish an awareness of the sacred nature of human life; to work to ensure that science does not deny the divine spark that every human being receives with the gift of life." However, the assessment of the concrete presence of the Muslim participants at Assisi was voiced by Thomas Michel SJ, himself a long-standing staff member of the Pontifical Council for Interreligious Dialogue and, as such, responsible for dialogue with Islam. Along the same lines, Fitzgerald, Michael L[ouis] / Borelli, John (eds.), Interfaith Dialogue, A Catholic View, London/Maryknoll 2006, p.85 refers to a "rather timid Muslim presence" at this first meeting. John Paul II himself, in his conversation on Islam with the journalist Vittorio Messori, especially emphasised the second prayer gathering in 1993 for peace in Bosnia, cf.

Johannes Paul II, p.121. It is evident that the confidence of the Muslims was not automatic and had first to be won. In contrast, the list of the Muslims present at the second encounter was impressively long with more than 30 persons including such prominent names as Smail Balic and Dalil Boubakeur. An anthropological reason is also given for the whole event. "Since all human beings have been created by God, and all are members of the one human family, we have a duty to come to the aid of all." (Religious Leaders Gather anew in Assisi (Italy) to Fast and Pray for Peace in Balkans and Europe (9–10 January 1993), Islamochristiana 19 (1993), p.251, cf. p.252). Here, a prayer of the Pope for peace from a specifically Christian-Muslim context must be quoted because it explicitly includes human dignity and human rights as derived from God the Creator (Qunaytra, Syria: 7 May 2001, **Prayer for Peace**, Pro Dialogo 107 (2001), p.173): "Inspire them <i.e. the civil leaders of this region> to work generously for the common good, to respect the inalienable dignity of every person and the fundamental rights which have their origin in the image and likeness of the Creator impressed upon each and every human being." In connection with Assisi 2001 cf. the impressively long list of Muslim participants among those welcomed (Assisi: 24 January 2002, **Greetings on the Occasion of the Day of Prayer for Peace**, Pro Dialogo 109 (2002), p.44); the main emphasis because of the situation was on Christians and Muslims and 30 Muslims really came, more than from any of the other religions invited, cf. DAY OF PRAYER FOR PEACE IN ASSISI (24 January 2002), Pro Dialogo 109 (2002), p.135, and the address in Assisi: 24 January 2002, **Address on the Occasion of the Day of Prayer of Peace**, Pro Dialogo 109 (2002), p. 45–48, which contains wise statements on the connection between human dignity and terrorism (p.46): "*Justice*, first of all, because there can be no true peace without respect for the dignity of persons and peoples, respect for the rights and duties of each person and respect for an equal distribution of benefits and burdens between individuals and in society as a whole. It can never be forgotten that situations of oppression and exclusion are often at the source of violence and terrorism." There is also a Commitment to Peace which the Pope himself submits as the ten commandments of Assisi for peace to the heads of state and of government and in which human dignity and human rights play an important role, e.g Concluding Ceremony, Pro Dialogo 109 (2002), p.153: "4. *We commit ourselves* to defending the right of everyone to live a decent life in accordance with their own cultural identity, and to form freely a family of their own." The French text with an introduction by the Pope, which again emphasizes the one human family and inalienable human rights, is found in Cité du Vatican: 24 Février 2002, **Aux Chefs d'Etat ou de Gouvernement**, Pro Dialogo 110 (2002), p. 171–172. There is also an interesting greeting to a follow up meeting organized by the Community of Sant'Egidio which includes the statement (Castel Gandolfo: 3 September 2004, To the Reverend Brother Cardinal WALTER KASPER, President of the Pontifical Council for Promoting Christian Unity, Pro Dialogo 118 (2005), p.29): "Dialogue releases the courage for a new spiritual humanism, because it requires trust in men and women." Mention must also be made of Vatican City: 17 January 2004, **On the Occasion of the Concert of Reconciliation**, Pro Dialogo 115 (2004), a concert sponsored by the Pontifical Councils on the issue of reconciliation between Jews, Christians and Muslims, where John Paul II in his address, which naturally refers to the music selected, underlines on p.31 that, despite all their differences, the resurrection of the dead is a uniting factor.

81 Vatican City: 1 January 2002, **Message for the Celebration of the World Day of Peace** (Extract), Pro Dialogo 109 (2002), p.37; at the same time, this is a good example of the changes which Lucie Pruvost observes in John Paul II, namely a departure from triumphalist language while otherwise continuing to present the same position, cf. Pruvost, Lucie, From Tolerance to Spiritual Emulation, Islamochristiana 6 (1980), p. 6–7. The problems of religion and terrorism are also discussed in Cité du Vatican: 10 janvier 2002, **Discours du Saint-Père aux Membres du Corps Diplomatique accrédité près le Saint-Siège** (Extrait), p.40, whereas Vatican City: 6 December 2001, **To the New Ambassador of Bangladesh to the Holy See** (Extract), Pro Dialogo 109 (2002), p.20, underlines in a similar way the common values which originate in human nature, although referring back to the address of 2001; Vatican City: 11 April 2002, **To the New Ambassador of Yugoslavia to the Holy See** (Extract), Pro Dialogo 110 (2002), p.176, although here the ideas about the importance of forgiveness are taken up from the address of 2002. A concentration on values is also found in the reference back to the address of 2001, Cité du Vatican: 12 décembre 2003, **A l'Ambassadeur du Qatar près le Saint-Siège** (Extrait), p.21, although there it is emphasised that the truest and most characteristic features of humankind are expressed in these values. The content of an earlier address was more clearly directed to Islam, cf. Vatican City: 1 January 1999, **Message of His Holiness Pope John Paul II for the Celebration of the World Day of Peace** (Excerpt), Pro Dialogo 102 (1999), p.281. "Religious freedom therefore constitutes the very heart of human rights. Its inviolability is such that individuals must be recognized as having the right even to change their religion, if their conscience so demands. (...) The Universal Declaration of Human Rights recognizes that the right to religious freedom includes the right to manifest personal beliefs, whether individually or with others, in public or in private. In spite of this, there still exist today places where the right to gather for worship is either not recognized or is limited to the members of one religion alone." Although this speech is frequently quoted, even in reference to threats to human rights (Vatican City: 11 January 1999, **Holy Father's Address to the Diplomatic Corps:** *The Time Has Come to Ensure that Everywhere in the World Effective Freedom of Religion Is Guaranteed.* (Excerpt), Pro Dialogo 102 (1999), p.284), it is never so pointed. But it is interesting that human dignity is described as a transcendent value which can always be recognised as such by those who seriously seek the truth (Vatican City: 16 December 1999, **To the New Ambassador of the Islamic Republic of Pakistan to the Holy See** (Extract), Pro Dialogo 103 (2000), p. 57).Still more explicit than the address of 1999 is Conseil Pontifical pour le Dialogue Interreligieux where there is reference to the severe punishments applied to such things, for example, although Islam is not mentioned explicitly, and in general to how the identification of religious and civil rights not only stifles religious liberty but can also limit other human rights or may even repeal them completely (cf. p. 513–514, n.699–700). One can even sense a hint of the typical Islamic dhimma (protective) status (p.514, n.701). "Pour éliminer les effets de l'intolérance, il ne suffit pas de 'protéger' les minorités ethniques ou religieuses, en les réduisant à la catégorie de mineurs civils ou d'individus sous tutelle de l'État." Several of these messages are quoted repeatedly and the frequency of these quotations in decisive contexts generally increases – perhaps because peace was becoming an increasingly urgent issue. Another reason could be that the limited selection of issues mentioned above is also representative, since the corresponding speeches of the last five years – always in a Christian-Muslim context – were addressed almost exclusively to diplomats among whom the Islamic world was well represented – a very clear parallel to the addresses to diplomats of predominantly Muslim countries which will be discussed later. Moreover, John Paul II himself usually referred to the most recent address, an indication that these addresses related to specific situations, although naturally certain subjects recurred. Here is another selection of the most important references to anthropology which have not yet been mentioned: Conseil Pontifical pour le Dialogue Interreligieux, p.448, n.613 (1988, positive effects of religion on human coexistence), p.526, n.713 (1991, very positive Image of human beings), p.555, n.751 (great importance of peace in holy scriptures and religions); the quotation leads on to the wish: "Que la conscience de chacun soit pleinement respectée, afin que l'image de Dieu en chacun puisse resplendir et produire d'abondants fruits de justice de *(sic!)* paix et d'amour." This address is quoted frequently: ibid. p.580, n.783, p.588, n.787, p.604, n.806*, p.688, n.881*.

82 Conseil Pontifical pour le Dialogue Interreligieux, p.487, n.663, cf. p.486, n.662, p. 487–488, n.664.

83 cf. Pope John Paul's Address to the Participants in this Colloquium during the Audience of Friday 26 June 1992:, Islamochristiana 18 (1992), p. 320–321, also in Address of H.H. Pope John Paul II to Participants in the Colloquium "Women in Society According to Islam and Christianity", in: Women in Society According to Islam and Christianity, Acts of a Muslim-Christian Colloquium Organized Jointly by the Pontifical Council for Interreligious Dialogue (Vatican City) and the Royal Academy for Islamic Civilization Research Al Albait Foundation (Amman), Vatican City, cf. also p. 123–124. On the question of women and fundamental anthropological considerations, it is also worth mentioning A Letter Sent by the Holy Father to Archbishop Vinko Puljic of Sarajevo "Change Violence into Acceptance" (2 February 1993), p.212: "Even in such a tragic situation they must be helped to distinguish between the act of deplorable violence which they have suffered from men who have lost all reason and conscience, and the reality of these new human beings who have been given life. As the image of God, these new creatures should be respected and loved no differently than any other member of the human family. In every case it should be emphasized most clearly that since the unborn child is in no way responsible for the disgraceful acts accomplished, he or she is *innocent* and therefore cannot be treated as the *aggressor*."

84 Conseil Pontifical pour le Dialogue Interreligieux, p.282, n.402, cf. p. 281–282, n.400–401, p.283, n.402, p. 314–315, n.442–443, p.371, n.514; the statements to participants in an interreligious colloquium on questions of marriage and the family also fit into this picture, p. 613–614, n.813*; His Holiness Pope John Paul II, To Cardinal Francis Arinze President Pontifical Council for Interreligious Dialogue, in: Religion and the Use of the Earth's Resources, Acts of a Muslim-Christian Colloquium Organized Jointly by the Pontifical Council for Interreligious Dialogue (Vatican City) and the Royal Academy for Islamic Civilization Research Al Albait Foundation (Amman), Rome 1996, p.179. Another occasion on which John Paul II addressed Jews and Muslims was his visit in Jerusalem. But no specific Abrahamic anthropology was developed apart, perhaps, from the fact that the golden rule was more or less the same for all three. But it is striking how much the religious leaders are spoken to on the same level with regard to their tasks (Jerusalem: 23 March 2000, **To the Christian, Jewish and Muslim Leaders**, Pro Dialogo 104/105 (2000), p. 175–177).

85 cf. Conseil Pontifical pour le Dialogue Interreligieux, p. 462–463, n.633–634, p. 491–492, n.670, Vatican City: 29 May 1998, **Pope's Discourse to the Members of the Delegation from Al-Azhar**, Pro Dialogo 99 (1998), p.281, cf. also Pope John Paul's Address to the Delegation of the World Islamic League during the Audience of Thursday 28 January 1993, Islamochristiana 19 (1993), p.316, where the common convictions about human dignity are cited as an argument against the misuse of religion.

86 Conseil Pontifical pour le Dialogue Interreligieux, p.477, n.653, cf. n.651, cf. also a general message to Lebanon, p. 301–304, n.425–428, the direct consequences in: Une délégation parlementaire intercommunautaire du Liban en visite à Rome, Islamochristiana 11 (1985), p. 226–227, and also a later speech during the first visit of a Lebanese ambassador, Vatican City: 2 April 2004, **To the New Ambassador of Lebanon** (Extract), Pro Dialogo 118 (2005), p.13f.

87 cf. Conseil Pontifical pour le Dialogue Interreligieux, p. 519–520, n.706–708. It is also interesting that, when referring to the deaths already caused by the conflict, John Paul II speaks about the same belief "qu'ils sont confiés au jugement miséricordieux de Dieu" (p.519, n.706);

on p.727, n.919*, it is marginally a matter of the conflict between Israel and Palestine; in the telegram of condolence on the death of Hassan II of Morocco, in addition to the personal encounter in Casablanca, there is also reference to his commitment for Jerusalem, cf. Cité du Vatican: 25 juillet 1999, **Télégramme du Pape à l'occasion du décès de Sa Majesté, Hassan II, Roi du Maroc**, Pro Dialogo 102 (1999), p.312, and in the speech on the occasion of the accreditation of a new Tunisian ambassador whose country has a tradition of tolerance, a lot of attention is given to the people and the situation in the Near East which is marked and deformed by violence of every kind, see Cité du Vatican: 27 mai 2004, **Au nouvel Ambassadeur de Tunisie près le Saint-Siège** (Extrait), Pro Dialogo 118 (2005), p.16.

88 The speech during the visit to the Umayyad Mosque in Damascus, for example, deals more with the concrete occasion and contains only one section which could be said to have anthropological relevance and that refers – significantly – to young people again (Damascus Syria 6 May 2001, **Discourse during the Visit to the Umayyad Mosque**, p.171): "It is crucial for the young to be taught the ways of respect and understanding, so that they will not be led to misuse religion itself to promote or justify hatred and violence. Violence destroys the image of the Creator in his creatures, and should never be considered as the fruit of religious conviction." But the pilgrimage in the anniversary year obviously had consequences for the dialogue locally, as John Paul II said to the new Jordanian ambassador: Cité du Vatican: 17 Mai 2002, **Au nouvel Ambassadeur de la Jordanie près le Saint-Siège** (Extrait), p.184.

89 Conseil Pontifical pour le Dialogue Interreligieux, p.586, n.787, for the following text cf. also p.587, n.787, for Ghana cf. p.250, n.353, for Senegalcf. p. 551–552, n.744–745, p.553, n.748, for Guinea cf. p.556, n.752, p.558, n.755. A later address in Nigeria emphasised Cardinal Francis Arinze, the president of the Pontifical Council for Interreligious Dialogue, on the one hand, in his role as a living link between the Roman Catholic Church and the country of Nigeria, and, on the other, it strongly underlined the value of human life and the family for Muslims and Christians. It also dealt with religious liberty, religion and violence, as well as corruption which ran counter to God's will for the human family; cf. Abuja – Nigeria: 22 March 1998, **Meeting with Muslim Leaders at the Apostolic Nunciature on the Occasion of the Pope's Second Pastoral Visit to Nigeria to Beatify Fr Cyprian Michael Iwene Tansi**, Pro Dialogo 99 (1998), p. 276–278.

90 Conseil Pontifical pour le Dialogue Interreligieux, p.276/277, n.392, cf. p.276, n.390. In addition, on p.276, n.393, there is reference to a common faith in God as the source of all rights and values of humanity. On several occasions also in a more general speech the Muslims are addressed specifically, e.g. Astana International Airport, Kazakhstan: 22 September 2001, **Response to the Welcome of the President**, Pro Dialogo 108 (2001), p.306. In general, John Paul II makes a great effort in each case to deal with the particular situation positively and to find points of contact; e.g. Astana, Kazakhstan: 25 September 2001, **On the Occasion of the Parting Ceremony**, Pro Dialogo 108 (2001), p.317, where he reformulates the motto of his visit (making it specially anthropological) with a reference to the Muslim national poet: "[C]an a human being fail to love?"

91 cf. Conseil Pontifical pour le Dialogue Interreligieux, p. 322–324, n.451–452 (Cameroon), p.247, n.350 (Kenya), p.257, n.363 (Philippines), cf. on the latter also p.260, n.367. "C'est seulement dans le cadre de la religion et des promesses de foi qu'elle partage que l'on peut vraiment parler de respect mutuel, d'ouverture et de collaboration entre les chrétiens et les musulmans."In Pakistan, John Paul II refers to creation (p.257, n.362) but previously (p. 256–257, n.361) to common concerns. The formulation used in Sudan is literally (The Holy Father's Address to the President of the Sudan, Islamochristiana 19 (1993), p.281): "Man is a rational being endowed with intelligence and will, and therefore he is capable of finding just solutions to situations of conflict, no matter how long they have been going on and no matter how intricate the motives which caused them. Efforts to restore harmony depend on the parties involved being willing and determined to implement the conditions required for peace. But where constructive action does not follow declarations of principle, violence can become uncontrollable."

92 Conseil Pontifical pour le Dialogue Interreligieux, p.585, n.786, concerning the fundamentals cf. p.433, n 596 (USA), p. 315–316, n.444 (Belgium), very basic in line with NA 3 p.498, n.680 (Malta), p.255, n.360 (France) deals with the unresolved problems between Christians and Muslims. Mention must also be made of Argentina, p.424, n.582, and one statement deserves special emphasis (p.316, n.244): "Il ne nous est pas donné de former une communauté unique; c'est là une épreuve qui nous est imposée." As far as the address in Paris is concerned, Fitzgerald/Borelli, p.120, underlines that the Pope addresses the Muslims as brothers (in the faith in the one God): "This may not seem significant until we remember that traditionally the term 'brother' was reserved for fellow Christians. The World Council of Churches, in its documents, prefers to speak about 'neighbours of other faiths'. The use of the term 'brother' by the popes can be seen as a sign of openness and friendship." In summary, John Paul II can be quoted again from Le nouvel ambassadeur de Tunisie près le Saint-Siège est reçu par le Pape, Islamochristiana 15 (1989), p.236: "[I]l m'arrive souvent de tenir, notamment au cours de mes voyages apostoliques, à savoir que le respect de toute personne humaine et de ses droits inaliénables découle de sa création à l'image de Dieu et de son destin transcendant".

93 Conseil Pontifical pour le Dialogue Interreligieux, p. 712, n. 907*

94 Vatican City: 27 May 2004, **To the New Ambassador of Yemen to the Holy See** (Extract), Pro Dialogo 118 (2005), p.22. It is especially a matter of peace in: TUNISIE *Le nouvel ambassadeur près le Saint-Siège est reçu par le Pape (19 novembre 1994)*, Islamochristiana 21 (1995), p.212: "Pour les Livres sacrés des différentes religions, l'aspiration à la paix et sa réalisation occupent une place considérable dans la vie de l'homme et dans ses rapports avec Dieu. On peut même dire qu'une vie religieuse authentiquement vécue produit des fruits de fraternité et de paix, car il est dans la nature de la religion de favoriser des relations toujours plus solidaires entre les hommes, précisément par des liens toujours plus étroits avec la divinité."

95 Extraits du discours du Saint-Père au Corps diplomatique (13 janvier 1990), Islamochristiana 16 (1990), p.294, cf. also p.293. In the following year he took this up again although no longer so fundamentally, cf. The Holy Father's Address to the Diplomatic Corps Accredited to the Holy See (12 January 1991), Islamochristiana 17 (1991), p.277. And another year later he referred very critically to the Gulf War but also to the situation of Christians in countries with a Muslim majority in the Near East and in Africa. "Il est des pays, par exemple, où la religion musulmane est majoritaire et où les chrétiens, aujourd'hui encore, n'ont même pas la possibilité d'avoir un seul lieu de culte à leur disposition. Dans d'autres cas, il ne leur est pas possible de participer à la vie politique du pays comme des citoyens à part entière. Dans d'autres cas encore, on leur conseille tout simplement de partir." (Le discours du Saint-Père aux membres du Corps Diplomatique accrédité près le Saint-Siège (11 janvier 1992), Islamochristiana 18 (1992), p.318, cf. p.317) John Paul II gives a very good explanation of Pancasila in INDONESIA *The New Ambassador to the Holy See Is Received by the Pope (19 June 1995)*, Islamochristiana 21 (1995), p.170: "the national philosophy which calls for belief in God, national unity, social justice, profound respect for human life, dignity and rights, and which insists on that freedom by which citizens determine their destiny as a people. Essential among these principles are religious freedom and interreligious tolerance".

96 Conseil Pontifical pour le Dialogue Interreligieux, p.448, n.613, concerning clearly positive situations cf. p.546, n.737 (Tunisia), p.735, n 926* (Morocco), p.698, n. 891* (Mali), Vatican City: 25 May 2000, **To the Ambassador of Kuwait to the Holy See**, Pro Dialogo 104/105 (2000), p.182, Cité du Vatican: 3 mai 2002, **Au nouvel Ambassadeur du Maroc près le Saint-Siège** (Extrait), Pro Dialogo 110 (2002), p. 182–183, Cité du Vatican: 12 décembre 2003, **A l'Ambassadeur du Qatar près le Saint-Siège** (Extrait), p.21, referring to the rights of minorities cf. Conseil Pontifical pour le Dialogue Interreligieux, p. 740, n. 930* (Turkey), Cité du Vatican: 17 mai 2002, **Au nouvel Ambassadeur du Soudan près le Saint-Siège** (Extrait), Pro Dialogo 110 (2002), p. 185–186, concerning reference to the rights of Christians cf. Conseil Pontifical pour le Dialogue Interreligieux, p.294, n.415, p.536, n.724 (both Algeria), p.720, n.913* (Iran) and, before a somewhat different audience, Vatican City: 18 October 2001, **Message on the Occasion of the 88th World Day of Migrants and Refugees, 2002**, Pro Dialogo 108 (2001), p.322 (and vice versa cf. Vatican City: 20 November 2003, **To the Participants in the V World Congress of the Pastoral Care of Migrants and Refugees** (Extract), Pro Dialogo 115 (2004), p.18), concerning the rights of minorities in general and of Christians in particular cf. Vatican City: 15 May 2003, **To the New Ambassador of the Islamic Republic of Pakistan to the Holy See** (Extract), Pro Dialogo 114 (2003), p.333. "Not least of these <fundamental human> rights are: unprejudiced access to the employment market, full participation in democratic civic life, and freedom of authentic religious practice. (....) For this reason I have said on numerous occasions that corruption, whether it be on the part of politicians, judiciary officials or administrators and bureaucrats (...), is a scourge which affronts the inviolable dignity of every human person and which paralyzes a nation's social economic and cultural advancement."Concerning general references to human rights cf. Conseil Pontifical pour le Dialogue Interreligieux, p.447, n.611 (Nigeria), p.549, n.741 (Senegal), p.640, n.834* (Turkey), p.649, n.841*, p. 700–701, n.895* (Pakistan), Cité du Vatican: 17 mai 2002, **Au nouvel Ambassadeur du Bélarus près le Saint-Siège** (Extrait), Pro Dialogo 110 (2002), p.188, for the positive effects of human rights cf. also Conseil Pontifical pour le Dialogue Interreligieux, p. 687–688, n.881* (Egypt), p.716, n.910* (Syria), including peace in general, as in Cité du Vatican: 18 mai 2001, **Au nouvel Ambassadeur de Tunisie près le Saint-Siège** (Extrait), Pro Dialogo 107 (2001), p.188 (with emphasis on peace education and here addressed to a female ambassador), Cité du Vatican: 6 décembre 2001, **Au nouvel Ambassadeur du Mali près le Saint-Siège** (Extrait), Pro Dialogo 109 (2002), p.28 (again with strong emphasis on education and hence on the family), and in the Near East in particular (Vatican City: 19 February 1999, **Greetings of the Holy Father to a Delegation from the "International Forum Bethlehem 2000"**, Pro Dialogo 102 (1999), p.291). "In particular, we must be confident that it is possible to build peace in the Middle East. The promise of peace made at Bethlehem will become a reality when the dignity and the rights of human beings made in the image of God (cf. Gen. 1:26) are acknowledged and respected"; cf. on this also the statement to the Lebanese bishops that Lebanon had shown it was possible to respect every person's right to religious liberty (Conseil Pontifical pour le Dialogue Interreligieux, p. 723, n.914*). In 2000John Paul II emphasised as well in connection with human rights there that it was particularly urgent "de permettre à chacun d'avoir des conditions de vie décentes et dignes (Cité du Vatican: 26 Octobre 2000, **Au nouvel Ambassadeur du Liban près le Saint Siège** (Extrait), Pro Dialogo 106 (2001), 16). There is a similar connection in the address to the ambassador of Gambia (Vatican City: 18 May 2001, **To the New Ambassador of Gambia to the Holy See** (Excerpt), Pro Dialogo 107 (2001), p.185). "It is in fact a sad commentary that at the dawn of this new millenium serious forms of social and economic injustice, and of political domination, are still affecting entire peoples and nations in different parts of the world, on your own continent of Africa and elsewhere. There is growing indignation on the part of countless men and women whose fundamental rights continue to be trampled upon and held in contempt." Addressing the Iraqi ambassador, John Paul II took up the issue of religious liberty in 2001 as well but the emphasis was on the consequences of the embargo and – in line with *Gaudium et Spes* – on the fact that nothing and no one has the right to reduce people to what they can do or produce (Vatican City: 28 April 2001, **To the New Ambassador of Iraq to the Holy See**, Pro Dialogo 107 (2001), p. 165–166).Over against the Indonesian ambassador, in whose country there were also various conflicts, he stated (Vatican City: 12 June 2000, **To the New Ambassador of the Republic of Indonesia**, Pro Dialogo 104/105 (2000), p.191), "Authentic democracy is based on recognition of the inalienable dignity of every human person, from which human rights and duties flow. Failure to respect this dignity leads to the various and often tragic forms of discrimination, exploitation, social unrest and national and international conflict with which the world is unfortunately too familiar. Only when the dignity of the person is safeguarded can there be genuine development and lasting peace." John Paul II expressed a very fundamental view when speaking to the ambassador of Iran (Conseil Pontifical pour le Dialogue Interreligieux, p.525, n.713). "C'est la défense et la promotion de la dignité humaine que le Saint-Siège poursuit par sa présence au sein de la communauté internationale et par ses relations diplomatiques bilatérales avec de nombreux pays. Cette activité qui n'a pas d'autre but que d'être au service du bien de la famille humaine, est caractérisée par un intérêt prédominant pour les aspects éthiques, moraux et humanitaires des relations entre les peuples du monde"; and similarly Cité du Vatican: 6 december (*sic!*) 2001, **Aux nouveaux Ambassadeurs** (Extrait), Pro Dialogo 109 (2002), p.29. This was also made very clear in: Jerusalem: 25 March 2000, **To the Consul General of Jerusalem**, Pro Dialogo 104/105 (2000), p.179 and in Vatican City: 14 December 2000, **To the New Ambassador of Eritrea to the Holy See** (Excerpt), Pro Dialogo 106 (2001), p.31. "Deeply concerned about the social dimension of human life, the Church contributes to the political order by teaching the inalienable dignity of the human person." The line followed here can be described today as one of the most fundamental in Catholic doctrine in general (Cité du Vatican: 12 novembre 2001, **Aux membres de l' "International Catholic Migration Commission" et de la Fondation"Migrantes"** (extrait), Pro Dialogo 109 (2002), p.15). "L'âme de votre oeuvre est une vision de la dignité humaine fondée sur la vérité de la personne humaine créée à l'image de Dieu (cf. Gn 1,26), une vérité qui illumine toute la doctrine sociale de l'Eglise. De cette version découle des droits inaliénables, qu'aucune puissance humaine ne peut accorder ou dénier, car il s'agit de droits qui ont leur source en Dieu. Il s'agit d'une vision profondément religieuse, qui est partagée non seulement par d'autres chrétiens, mais aussi par de nombreux disciples des autres grandes religions du monde." This naturally applies particularly to the weakest members of a society, for example to the children, as e. g. Cité du Vatican: 27 mai 2004, **Au nouvel Ambassadeur de Mali près le Saint**-Siège (Extrait), Pro Dialogo 118 (2005), p.19. There is also a connection with universal values and norms, cf. Vatican City: 18 May 2001, **To the New Ambassador of Gambia to the Holy See**, p.185. "The Catholic Church will always be a staunch and tireless defender of universal, unchanging moral norms, a role she exercises for no other purpose than to serve men's true freedom. (…) There exist, then, fundamental moral rules of social life entailing specific demands with which both public authorities and private citizens must comply, and which the international community too is required to respect. This underlying morality must guide all aspects of social and political life." And in the address to the Yugoslav ambassador there is reference to "those universal values which are rooted in the nature of the person and ultimately in God" (Vatican City: 11 April 2002, **To the New Ambassador of Yugoslavia to the Holy See** (Extract), p.176), which means, above all, solidarity, peace, life and education. At the New Year reception for diplomats in 1996, there is a relatively clear reference to Saudi Arabia in connection with the subject of religious liberty (Conseil Pontifical pour le Dialogue Interreligieux, p.658, n.851*). "D'autres cependant continuent à pratiquer une discrimination à l'égard des juifs, des chrétiens et d'autres familles religieuses, allant jusqu'à refuser le droit de se réunir en privé pour prier. On ne le dira jamais assez: il s'agit là d'une violation intolérable et injustifiable non seulement de toutes les normes internationales en vigueur, mais de la liberté humaine la plus fondamentale, celle de manifester sa foi, qui est pour l'être humain sa raison de vivre", and still more clearly on the same occasion in 1999 (Vatican City: 11 January 1999, **Holy Father's Address to the Diplomatic Corps:** *The Time Has Come to Ensure that Everywhere in the World Effective Freedom of Religion Is Guaranteed*". (Excerpt), p.284/285). "In other regions, *where Islam is the majority religion*, one still has to deplore the grave forms of discrimination of which the followers of other religions are victims. There is even one country where Christian worship is totally forbidden and where possession of a Bible is /285 a crime punishable by law"; somewhat more generally in 2001 (Cité du Vatican: 13 janvier 2001, **Au Corps Diplomatique accrédité près le Saint Siège à l'occasion de l'échange des voeux du Nouvel An** (Extrait), Pro Dialogo 106 (2001), p.52). "Je voudrais ici vous redire et redire par votre intermédiaire aux gouvernements qui vous ont accrédités auprès du Saint-Siège, *la détermination de l'Église catholique à défendre l'homme, sa dignité, ses droits et sa dimension transcendante*. (...) Le drame vécu par la communauté chrétienne en Indonésie ou les discriminations patentes dont sont victimes aujourd'hui encore d'autres communautés de croyants, chrétiens ou non, dans certains pays d'obédience marxiste ou islamiste, appellent à une vigilance et à une solidarité sans faille." These New Year addresses at receptions for the diplomatic corps are naturally very wide ranging; a particularly good example is Cité du Vatican: 12 janvier 2004, **Aux membres du Corps diplomatique accrédités près le Saint-Siège, à l'occasion de l'échange des voeux du nouvel an,** Pro Dialogo 115 (2004), p. 26–31: from the conflicts in Iraq and Palestine to a fundamental discussion on the identity and character of Europe, naturally also including human rights. However, when addressing the Turkish ambassador (Conseil Pontifical pour le Dialogue Interreligieux, p.559, n.757) and on the occasion of a special concert (ibid. p.531, n.718) there are references to love for one's brothers in the Turkish mystic Yunus Emre. The later address to one of his successors deals in greater detail with the specific situation of secular Turkey but is able even here typically to bridge the gap from creation to human rights (Vatican City: 7 December 2001, **To the New Ambassador of Turkey to the Holy See** (Extract), Pro Dialogo 109 (2002), p.34). "For a secular State, the challenge is to be genuinely open to transcendence: that is, to base itself upon a vision of the human person created in the image of God and possessed therefore of inalienable and universal rights. There are in fact certain rights which are universal because they are rooted in the nature of the human person rather than in the particularities of any culture." An interesting reason for the priority of religious liberty is found in Pope's Address to the Diplomatic Corps Accredited to the Holy See, Islamochristiana 15 (1989), p.252. "The religious aspect, in fact, has two specific dimensions which show its originality in relation to the other activities of the spirit, notably those of conscience, thought or conviction. On the one hand, faith recognizes the reality of the Transcendence which gives meaning to the whole of existence and which is the basis of the values which behaviour takes as its guidelines. On the other hand, religious commitment implies membership of a community of persons. Religious freedom goes hand in hand with the freedom of the community of believers to live according to the teachings of its Founder." There is a very interesting link between the maintenance of diplomatic relations with the Holy See and religious liberty, as can be found in Le nouvel ambassadeur de Turquie près le St-Siège est reçu par le Pape, Islamochristiana 14 (1988), p. 308. "Tout Etat, et plus encore lors-qu'il a pris l'initiative de nouer des relations diplomatiques avec la Siège Apostolique de Rome, se distingue hautement en montrant une claire attitude d'équité à l'égard des croyants qui ont légitimement fait le choix de leur religion." This is well supplemented by an extract from the speech at the accreditation of a new Syrian ambassador, SYRIA *The New Ambassador to the Holy See Received by the Pope*, Islamochristiana 15 (1989), p.234/235. "The diplomatic relations which the Holy See maintains with numerous nations, diverse in culture and in their role on the international scene, have a special nature, as you know. Their principal inspiration is the promotion of the basic ideals which protect and enhance the human person, assuring respect for his or her dignity, striving against difficulties to promote a civilization of toler-/p. 235 ance, mutual help and fraternal love." A variant on the usual priority of religious liberty on the list of human rights is Vatican City: 16 December 1999, **To the New Ambassador of the Islamic Republic of Pakistan to the Holy See** (Extract), p.57 which states, "the fundamental and inalienable rights of the human person, the most basic of which are the right to life, the right to freedom (including freedom of thought, conscience and religion) and the right to participate fully in society. From these basic rights which are essential to the well-being of individuals and societies." Vatican City: 27 May 2004, **To the New Ambassador of Sri Lanka to the Holy See** (Extract), Pro Dialogo 118 (2005), p.20 links this emphasis on religious liberty with an equally clear rejection of any form of proselytism. What is quite new (and perhaps, as already mentioned above, connected with the most recent developments in the European Union) is the emphasis on the separation of religion and state as important in connection with freedom of conscience. "[L]a claire distinction entre la sphère civile et la sphère religieuse permet à chacun de ces secteurs d'exercer ses propres responsabilités (...) dans le respect mutuel et dans la liberté de conscience." (Cité du Vatican: 21 février 2004, **Au nouvel Ambassadeur de Turquie près le Saint-Siège** (Extrait), Pro Dialogo 118 (2005), p.11, where it is again mentioned (in line with *Pacem in terris* of John XXIII) that peace cannot be separated from human dignity and human rights and is based on the four pillars of truth, justice, love and freedom.) Similarly in Cité du Vatican: 15 novembre 2004, **Au nouvel Ambassadeur de la République d'Irak près le Saint-Siège** (Extrait), Pro Dialogo 118 (2005), p.34, where there is additional emphasis on the authority of the law as a condition for preserving human rights. And the final New Year address by John Paul II to a diplomatic corps, which could in a sense be described as a comprehensive statement on human beings and their existence (starting before birth and via the family, social justice [hunger!], peace and finally to freedom), again strongly underlines the separation of religion from the state and its benefits for both sides and moreover adds the fact that the human rights to freedom are especially individual rights – an aspect which is very important in relation to Islam because the latter is inclined to emphasise the rights of the community and to subordinate those of the individual and of minorities to them, precisely because the separation between religion and the state is seen as a rule as non-Islamic and even as a threat to Islam (Cité du Vatican: 10 janvier 2005), **Au Corps diplomatique accredité près le Saint-Siège** (Extrait), Pro Dialogo 118 (2005), p. 44–47). During the long interview which John Paul II gave to the journalist Vittorio Messori, and which has been published as a book, it was not a matter of an address to diplomats, indeed not of any document for an official occasion like the others quoted here, but the readers in mind must certainly also be seen as very international and the final words of the chapter on Islam are quite plain and clear with reference to religious liberty. "In den Ländern, wo *fundamentalistische Strömungen* an die Macht kommen, werden die Menschenrechte und das Prinzip der religiösen Freiheit leider sehr einseitig ausgelegt: Die Religionsfreiheit wird als Freiheit verstanden, allen Einwohnern die 'wahre Religion' aufzuerlegen. Die Lage der Christen ist in diesen Ländern nicht selten sogar als bedrohlich zu bezeichnen. Solcherart fundamentalistische Einstellungen gestalten die gegenseitigen Kontakte außerordentlich schwierig. Dennoch bleibt die Kirche unverändert offen für den Dialog und die Zusammenarbeit." (Johannes Paul II, p.122) A similar line is followed by BANGLADESH *The New Ambassador to the Holy See Is Received by the Pope (10 January 1992)*, Islamochristiana 18 (1992), p.245. "Without good interreligious relations there is a danger that religion could be degraded into a weapon of hostility".

97 Conseil Pontifical pour le Dialogue Interreligieux, p.526, n.713, and earlier p.525, cf. Vatican City: 7 September 2000, **To the New Ambassador of the Arab Republic of Egypt to the Holy See** (Excerpt), Pro Dialogo 106 (2001), p.7: "the gift of the Law which God wrote long ago on tablets of stone and which he continues to write in every age on the human heart" and also the very late address by John Paul II to a group of Spanish bishops is which he castigates secularism as an attack on religious liberty from another angle and in another form (Vatican City: 24 January 2005, **To the Bishops from Spain (1st Group) on the Occasion of their "Ad Limina" Visit (Extract)**, Pro Dialogo 118 (2005), p.49). "This ideology leads gradually, more or less consciously, to the restriction of religious freedom to the point that it advocates contempt for, or ignorance of, the religious environment, relegating faith to the private sphere and opposing its public expression."There is also a lot of content in another address on the occasion of the accreditation of a new ambassador from Bangladesh. Again the starting point is God the creator and from there he takes a direct step to human solidarity in clear terms. "It <i.e. human solidarity> is not a matter of dispensing favours but of recognizing the basic human right to a just share of resources" (Vatican City: 6 December 2001, **To the New Ambassador of Bangladesh to the Holy See** (Extract), p.21). With regard to values, he refers back to the address to the United Nations but what can more or less be considered obvious is the desire of all nations for peace. In general, he considers directly or indirectly that there are major agreements between Christians and Muslims (ibid.). "They <i.e. Muslims and Christians> agree further that the Creator has also revealed a way of life, based upon what you rightly call 'fundamental human values and norms' which have their origin in God himself. (...) For Islam and Christianity, human life is a sacred and inviolable reality, since it has its origin and destiny in God himself. Therefore, it is never possible to invoke peace and despise life, a contradiction found all too often within human societies and human hearts." Here reference can also be made to two interesting summaries of papal statements which, although of an earlier date, in fact point the way things will go. In connection with the addresses intended for Catholics there is: Kayitakibga, Médard, Chrétiens et Musulmans aujourd'hui face à face, BSNC 62 (1985), p. 192–204. Here the great optimism of the Pope is evident which naturally provokes the criticism of self-appointed realists who are accustomed to emphasising the intolerance and aggressiveness of Islam. In reaction to this, the author states (p.204), "Le Pape est mieux placé que quiconque pour savoir tout ce que les catholiques souffrent dans certains pays à cause de l'Islam. Il se place dans son rôle de pasteur qui, à temps et à contre temps, annonce l'Evangile". The counterpart to this, Michel, Pope John Paul II's Teaching about Islam in His Addresses to Muslims, additionally emphasises how important prayer for and with Muslims was for John Paul II and that, irrespective of whom he is speaking to, he always makes the same points: "He does not have one message for Muslims and a contradictory teaching for Christians." (p.184) Anotherinteresting reference is St. Egidio Community, **Rome – Italy: The Visit of the Grand Mufti of Syria to Rome** (December 1985), BSNC 62 (1986), p.218. There it is mentioned that, at the end of the conversation, John Paul II said to the guest: "I also read a passage of the Qu'ran every evening."

98 Vatican City: 27 May 2004, **To the New Ambassadors (Surinam, Sri Lanka, Mali, Yemen, Zambia, Nigeria and Tunisia) to the Holy See** (Extract), Pro Dialogo 118 (2005), p.15. The New Ambassador of Pakistan Received by the Pope, Islamochristiana 11 (1985), p. 232–233, is a good example that people certainly register what the Pope says on such diplomatic occasions as the traditional New Year receptions and, on the other hand, a further, early example of how John Paul II stood up particularly for religious liberty and for the distinction between the political and the religious realm, also comparable to his commitment to human rights in: Bangladesh, The New Ambassador to the Holy See Received by John Paul II, Islamochristiana 12(1986), p.196. Sometimes this commitment also took the form of demanding legal equality for all irrespective of their religion, as in: Iraq, New Ambassador to the Holy See Presents Credentials. Islamochristiana 12 (1986), p.209.

99 Conseil Pontifical pour le Dialogue Interreligieux, p.717, n.911* (Sudan where there had been disputes with bloodshed between Christians and Muslims over a long period), on Pakistan cf. p.701, n.895*, on Jordan cf. p.674, n.868*, but also International Airport, Amman, Jordan: 20 March 2000, Pro Dialogo 104/105 (2000), p.169, with special praise for the Royal Interfaith Institute and for the monotheistic religions in general. "The three historical monotheistic religions count peace, goodness and respect for the human person among their highest values" (cf. for example, Sua Altezza Reale Principe di Giordania El Hassan bin Talal / Elkann, Alain, Essere Musulmano, Milano 2001), and also Wadi Al-Kharror, Jordan: 21 March 2000, **To the People of Jordan**, Pro Dialogo 104/105 (2000), p.170, with a special greeting for Prince Mohammed.

100 Cité du Vatican: 13 janvier 2003, Au Corps Diplomatique accrédité près le Saint Siège à l'occasion de l'échange des Voeux du Nouvel An (Extrait), p.19, cf. also: Cité du Vatican: 7 octobre 2004, Aux Membres de la Commission théologique pontificale internationale (Extrait), Pro Dialogo 118 (2005), p.30. "La conviction de l'Église a toujours été que Dieu a donné à l'homme la capacité de parvenir par la lumière de sa raison à la connaissance de vérités fondamentales sur sa vie et sur son destin, et de manière concrète, sur les normes d'une façon juste d'agir. Souligner devant nos contemporains cette possibilité est d'une grande importance pour le dialogue avec tous les hommes de bonne volonté et pour la coexistence aux niveaux les plus divers sur une base éthique commune." At this point, which constitutes the conclusion of the remarks on the magisterium of John Paul II, a final reference may be permitted to the chapter "Mohammed?" in the book mentioned several times "Die Schwelle der Hoffnung überschreiten". In the context of positive references back to Council documents and his own experience, John Paul II still emphasised very strongly that Islam, in contrast to Christianity, was not a religion of salvation and that this had consequences: "Daher ist nicht nur die Theologie, sondern auch die Anthropologie des Islam sehr weit entfernt von der christlichen." (Johannes Paul II, p.120) It is important to mention this because otherwise, in the "official" statements, it is the convergences and not the differences which predominate. Concerning the role not only of individual speeches but of John Paul II's work in general cf. Akasheh, Khaled, Considerations on Forty Years of Religious Dialogue with Muslims (A Report), Pro Dialogo 116/117 (2004), p.198 f, who underlines the Pope's journeys, the two gatherings for prayer in Assisi, of which less account has been taken here because they were multi-religious, and finally the Pope's rejection of the Gulf War of 2003, which was not able to prevent the war as such but did not consider it a religious war. There are also two interesting Muslim evaluations of the statements by John Paul II, namely by Ayoub, p. 169–184, and Abu-Rabi, Ibrahim M., John Paul II and Islam, in: Sharwin, Byron L. / Kasimow, Harold (eds.), John Paul II and Interreligious Dialogue, Maryknoll 1999, p. 185–204. These two presentations point simultaneously to major agreements and major differences. The documents and speeches cited are the same to a large extent and, at the end, also the book mentioned above with its views which both consider to be the true, personal opinion of the Pope. This makes it clear that previous opinions about what dialogue should be, where it should start, how it should be conducted and where it should lead have a strong influence on how negatively or positively the Pope's commitment is seen. Whereas Abu-Rabi consistently attempts to assign the various statements, which were certainly different in nature, to the various contexts and in the process shows great understanding for John Paul II as the head of the Roman Catholic Church trying nevertheless to approach the Muslims positively, one can clearly sense that, for Ayoub (as also for many Muslims), dialogue implies recognition of an equal way to salvation (or preferably even a better way). For this approach, John Paul II is naturally still far too Catholic, especially when speaking to Catholics in Muslim countries. These two contributions in the same book could almost serve as prime examples of how wide the spectrum of classification and evaluation can be depending on the standpoint adopted. The first part of the book provides a good survey of the statements by John Paul II not only on Christian-Muslim dialogue but also on interreligious dialogue and its various areas in general, a survey which is naturally far less comprehensive that the one presented here but more easily accessible for anyone who simply wishes to get a basic general overview (because it is arranged by subjects and not chronologically).

101 Conseil Pontifical pour le Dialogue Interreligieux, p.769, n.876, for further arguments cf. p. 768–772, n.874–879, p.810, n.955, p. 815–816, n.968–970, explicitly on Muslims cf. ibid. p.757, n.839. There is also an interesting list of obstacles to dialogue p. 818–819, n.975–976 (starting with the fact that dialogue is not easy even on the interpersonal level), as well as the statements on ecumenical cooperation in interreligious dialogue p. 838–839, n.1016 and the remarks on ecumenical and interreligious cooperation in the realm of the mass media p. 773–777, n.880–893. The juridical provisions for mixed marriages are of similar interest more for practical than for fundamentally anthropological reasons cf. ibid. p. 854–856, n.1033 and p. 858–861, n.1039.

102 Kongregation für die Glaubenslehre, Erklärung *Dominus Iesus*, Über die Einzigkeit und Heilsuniversalität Jesu Christi und der Kirche, edited by Sekretariat der Deutschen Bischofskonferenz, Verlautbarungen des Apostolischen Stuhls 148, Bonn 2000, p.31, cf. p.26f, in: Konrad-Adenauer-Stiftung e.V. (Hrsg.), Vatikan und Ökumene, Die Debatte über die vatikanische Erklärung "Dominus Iesus", Sankt Augustin 2000. The documentation contains the wording of the declaration and also a survey of the reactions and commentaries on it, including an interesting statement by Cardinal Franz König that this is a document "by theologians for theologians" (König: Streit um Vatikandokument ein "Sprachproblem", http://www.kna.de/kna/di enste/bsptxt/7tdw.h), in fact disagreeing with Bishop Karl Lehmann, the chairperson of the German bishops' conference, as far as Christian-Muslim dialogue and efforts at evangelisation are concerned, and an interesting question from his side. "So muss man sich fragen, wie weit die Glaubenskongregation in Rom bei der Formulierung des Dokuments mit dem Rat für das Gespräch mit den nichtchristlichen Religionen oder dem päpstlichen Rat für die Förderung der Einheit der Christen mit Bischof Walter Kasper zusammengearbeitet hat" (Höher, Sabine / Siebnmorgen, Peter, Wem gehört Jesus, Bischof Lehmann?, Welt am Sonntag (10 September 2000)), or a reference to the opposite view of Perry Schmidt-Leukel (Geyer, Christian, Ringparabel, Frankfurter Allgemeine Zeitung (6 September 2000)).

103 Vatican Council for the Pastoral Care of Migrants and Itinerant People, Istruction Erga Migrantes Caritas Christi (The Love of Christ towards Migrants), Vatican City 2004, p.44, cf. also p.45. There are also interesting remarks on migrants and religion in general, p. 41–42, which emphasise, e.g., that Christians should provide help in preserving a transcendent view of human beings. The comments on interreligious dialogue as such on p. 45–46 are interesting for the line they follow (precisely not agreement and "peace" at any cost), whereas p. 42–43, directly before the statements about Muslims, mention four fundamentally important points: social institutions should be shared but places of worship should not in order to avoid misunderstandings. In Catholic schools, where so desired, account should be taken of other regulations (e.g. provisions concerning food) and no compulsion should be exercised to participate in religious activities, although religious education and the general approach should not be abandoned. There is a general dissuasion of interreligious marriage although to different degrees depending on the religion (unfortunately this is not spelled out further but the statements quoted in the document on the following pages provide food for thought). The last point is that of reciprocity, especially with regard to the attitude to minorities and to justice in legal and religious affairs, indeed mutual respect in general. "Reciprocity is also an attitude of heart and spirit that enables us to live together everywhere with equal rights and duties." (p.43) The aspect of reciprocity also becomes very evident when, in the context of a general audience, John Paul II congratulates the Muslims on the inauguration of the mosque in Rome. "This event is an eloquent sign of the religious freedom recognized here for every believer. And it is significant that in Rome, the centre of Christianity and the See of Peter's Successor, Muslims should have their own place of worship with full respect for their freedom of conscience. On a significant occasion like this, it is unfortunately necessary to point out that in some Islamic countries similar signs of the recognition of religious freedom are lacking. And yet the world, on the threshold of the third millenium, is waiting for these signs!" (The Inauguration of the Mosque in Rome (21 June 1995), Islamochristiana 21 (1995), p.177)

II The Secretariat's Publications

The Secretariat for Non-Christians first became known through the publication of various documents, which naturally also concerned Islam[1]. This was a conscious preliminary step towards dialogue and, as such, deserves specific assessment, that is to say, a separate chapter. With regard to the presentation of these documents, this chapter will follow the movement from inside to outside. This refers above all to the circle for which these texts were intended. At the outset, also chronologically speaking, are the *Notes sur l'organisation et les règles d'action du Secrétariat*. This dossier was already compiled during the Second Vatican Council and served as an internal guideline for the new Secretariat for Non-Christians. As such, it was never published, but was only intended for internal use. The next circle is formed by the publications of the *Commission pour les Relations Religieuses avec les Musulmans*. This commission was instituted in 1974 by Pope Paul VI to support the Secretariat. It is composed of consultors from all five continents appointed for a period of five years, who were soon assigned the task of working on specific topics that were considered important at the time. These deliberations were actually published, but in completely different ways. If publication by the commission itself was not worthwhile, they were published in the internal bulletin, in other cases as a book or rather as a kind of brochure, clearly belonging to the category of "grey" literature. These publications come at second place here, because they are more or less clearly addressed to an internal readership of church decision-makers and not to the general public. Since the themes covered are frequently very topical even today (for example, violence, politics, or freedom of religion) and are treated with great care from an international perspective, they also take up the lion's share of this presentation. The introductions to the Christian-Muslim dialogue are addressed to an extensive Catholic audience, first of all the French *Orientations pour un dialogue entre chrétiens et musulmans*, which were very quickly sold out, and then ten years later, after a great deal of practical experience in dialogue, the revised version in English *Guidelines for Dialogue between Christians and Muslims*. The book *Religions – Thèmes fondamentaux pour une connaissance dialogique* points in a similar direction, but only partly deals with Islam. This book, a reaction to suggestions from outside, was supposed to serve as a scientific key to statements on dialogue and on individual religions and is particularly interesting for our topic because it has a decidedly anthropological introduction. The booklet *Religion in the World* is also the result of suggestions from outside and contains only one chapter on Islam. It is distinctive because organisations operating internationally had expressed their desire for information about the world religions in a more popular style. Thus it is aimed at readers outside the church. On the other hand, the booklet *Chiesa e islam* was written at the beginning of the pontificate of John Paul II entirely in Arabic, obviously in order to inform Muslims about the pioneering texts on Islam which had been produced up to then within the Catholic Church. *Meeting in Friendship* (which is cited here in the extended and updated version of 2002) was also directed towards Muslims, being simply an anthology of all the messages of greeting for Ramadan which had so far been despatched. At the end of this chapter a questionnaire is to be found concerning humans, or rather the answers from different religions; this was not published as a separate book, but rather as one issue of the internal bulletin and was intended to serve as a kind of Christian further education. Two of the answers did indeed come from a Muslim background and are naturally of great interest for anthropological considerations.

1 The beginnings of the Secretariat's work

The publications concerning how the Secretariat began its work are unfortunately not very informative.[2] As a later secretary, Pietro Rossano, reported in retrospect, reference was made the conciliar documents and papal statements which already existed, whereby he recalls that this ranged as far as the encyclical *Redemptor Hominis* by John Paul II, stating that conversion would only deepen personal identity and destroy nothing. That alone shows that this cannot be a really authentic voice from the very beginnings.

According to Rossano's account, the Secretariat sees itself as the outward sign of the Church's interest in people who belong to other religious traditions. This becomes concrete in three areas: outwardly in friendly relations, communication and dialogue; inwardly in the task of arousing interest for this topic within the Church, to communicate knowledge[3] and to encourage communication and initiate dialogues. This point is important in so far as it respects the authority of the local churches[4] and also seeks to promote ecumenical cooperation. The last area is the issue of inculturation. This started in practice with the selection of consultors and a first meeting with them when they were in Rome for the third period of the Council. From this one may conclude that they were bishops. It is obvious that there was great enthusiasm, but also great uncertainty. There was no clear understanding of dialogue itself and especially its aim (which might possibly be unacceptable for non-Christians). One had the Council's new anthropology and ecclesiology, as well as the clear will of Pope Paul VI, but no more. In this situation, Cardinal Marella proposed to the Council Fathers that one should make contacts at the human level. Internally, people were very cautious, especially on the part of the Congregation *de Propaganda Fidei*. They wanted action to be taken where their own task could not be fulfilled either legally or factually – another clear indication that dialogue was hardly defined or delineated, quite the contrary. What do Christians and non-Christians have in common, what can Christians learn from others? This is where universal human brotherhood comes into play as the basis of dialogue, the common purpose of loving and serving the creator, alongside natural law, traces of a primitive revelation, and the grace of God which he does not deny to people of good will.[5] It was not necessary to appoint members, because the bishops of the missionary countries are, so to speak, automatically members. The shining example is the Pope, meaning at that time Paul VI, who emphasises the common spiritual and ethical values and calls for more cooperation. Although other demands evidently came from the circle of consultors[6], there was no willingness to organise an interfaith conference quickly to cover the topics of justice and truth, freedom and peace, remaining open on the one hand to the circumstances and especially to the bishops' questions and on the other hand extending the ideological basis, that is to say, clarifying the position of non-Christians before God on the basis of the Bible, the Church Fathers and the Council. In 1967, there was one person specifically responsible for Islam[7], Father Cuoq, who also undertook journeys to Muslim countries, and from that year on there were also annual messages of greeting for Ramadan. The purpose was clearly to make the Church known to non-Christians.

There were no real meetings with non-Christians, or only some initial approaches. That changed under the presidency of Cardinal Pignedoli. He urged the bishops' conferences to appoint dialogue commissions or at least dialogue commissioners, and he worked out his own dialogue programme: the reflections and studies were be continued (although in fact the publications ceased), but not only the knowledge about non-Christians should increase, but also the friendly relations with them and the hospitality towards them. The focus of the dialogue should be local. Ecumenical cooperation was very good and extensive, but there was an imbalance concerning the dialogue with non-Christians with regard to the representativeness of religions and their relation to contemporary issues. Both were missing, which meant that Catholic Church was convinced that it had to play a leading role in dialogue. At this time dialogue was also outlined more precisely: the participants should be aware of their identity and firmly anchored in it. Comprehensive and honest respect for the other partners and their religious identity was required. Reciprocity was also necessary, and one had to be aware of the common elements, meaning the partial solidarity in the spiritual search[8] and in the religious values, and one also had to be convinced (as a Christian) that the partner did not achieve human and religious perfection and the gospel was the answer to his search. Patience was needed for small steps (warming to one another, mutual understanding, finding common ground, sharing one's own convictions) and one had to be aware of the difference between the gospel itself and its cultural context. One should not forget that the gospel itself is dialogical because Christ was so himself. Last not least, dialogue means not just talking about religious affairs, but about things people are interested in, but in a religious light. That made it easier to move on to concrete cooperation, for example in favour of righteousness, morality and peace. Thus non-Christians would grow in their spiritual values, and Christians would get to know their limits, but also the opportunities of their faith. Summed up in one sentence, dialogue is "a meeting with non-Christians motivated by love and by a spirit of service, sustained by a sense of respect and solidarity, with the purpose both of listening to the other, understanding him in his spiritual journey, in his hopes and in his problems, and of helping him to know, appreciate and desire the message of Christ and to wish to share it in some way, and at the same time of broadening our own understanding of the message of Christ, and our way of accepting this message existentially so as to be able to express it in a better way."[9] Thus dialogue is a very personal matter, a matter concerning people who have convictions which they live up to. For Christians, and here it becomes anthropological, dialogue is the fruit of an anthropology centred upon humans as the image of God and the object of his love. People need a personal relationship with God and their fellow human beings in order to be completely human. This anthropology is described like this: "It is an anthropology in which the poles of man and revelation, reason and faith, religion and gospel, eros and agape are not resolved in antithesis nor in indifference or isolation, nor in equivalence and equality but in a profound and reciprocal relationship that finds its archetype in the perfective relationship existing between creation and redemption."[10] Or to combine both in a summary from another lecture by the same author: "Dialogue is born from the conviction of the value of the subject as being essentially in relationship to God and to the other, as the depository of values and of experiences which can enrich me, as the object of a history of salvation and of a general revelation, as being gifted with an inalienable freedom which prevents me from considering him simply as the 'object of mission'."[11] The relationship between creation and salvation is thus of decisive importance for the dialogue. Because humans are by nature hearers of the word (according to Rahner), this dialogue makes sense. This basic structure, and in general his nature and also specific culture, constitutes the special value of each person. The main problem is the tension between identity and openness, between firmness and flexibility, between a universalistic longing and the divine significance of the other. The Christian should be aware that he does not know everything about humanity's path to God and God's ways of dealing with humans. Here he is not perfect and can perfect himself with regard to the knowledge and application of his faith, of which he does not know all facets and values. Scripture and tradition had always spoken of an economy of wisdom and salvation outside Israel and the Church. However, one also dissociated oneself clearly from a pluralistic theology of religion.

Since this essay was written at a later time, it already addresses the first phenomena of disillusionment. The Christian-Muslim dialogue is particularly difficult on account of four things: the legal character of Islam, which means that the *status confessionis* comes up immediately in social questions as well; the rejection of the historical-critical method; the conviction that one cannot learn anything from the Christians anyway, because one knows them and is better; and because of the lack of distinction between the spiritual and the temporal, between religion and politics. In addition, while for Christians dialogue includes the acceptance of others and their rights, this is not the case for Islam. Is it possible to demand reciprocity, or does that contradict the gospel? How should one respond to the mission of Islam and other religions in Christian areas?[12] In addition to these questions, the possibility of deepening bilateral commissions with followers of a different religion was also raised.

What is hardly touched upon in this document, which covers many points, is the importance of the new approach in anthropology at the Second Vatican Council. It was only this which allowed a new movement towards people of other religions, namely dialogue. Without the new emphasis that all human beings already have a connection with God, that they have values that they share with Christians, and that they also yearn for more, dialogue would be meaningless, and Christians in particular could not harbour any expectations from it. On the basis of the new anthropology, however, dialogue is also an area of theological learning for Christians. In the interest of a seamless connection, natural law and natural theology are emphasised, which only have to be positively reinforced by the gospel. It is also important that the dialogue must and should not concern itself primarily with religious topics, but that it is more about casting a religious light on the issues that are important to people. The further development of the guidelines of the Council and the Pope by the Secretariat therefore bears the mark of humankind and thus of anthropology.

2 Publications of the Commission pour les Relations Religieuses avec les Musulmans

A series of publications by the Commission pour les Relations Religieuses avec les Musulmans (C.C.R.M.) is particularly important for thoroughly preparing and completing the task that had been set. This commission was called into existence by Pope Paul VI on 22 October 1974 together with the Commission for Religious Relations with Jews and is naturally part of the then Secretariat for the Non-Christians, later the Pontifical Council for Interreligious Dialogue. This is already evident in the personnel structures: president and secretary of the Secretariat are president and vice-president of the Commission, while the secretary is the head of the Islam department. There is also a group of seven, later eight, consultors, each appointed for a period of five years and representing all five continents if possible, as well as the advisers available to the institution as a whole, including Islam specialists, of course. The consultors to the Commission showed a little more presence in public when they were assigned to work on certain topics.[13] On the one hand, these were studies on current trends in Islam, resulting in a whole series of publications which will be discussed in detail below. On the other hand, they concerned approaches to harmony and conflict. These were published in a number of internal bulletins and will be examined more closely in the next section. They differ completely in their literary character, as the editors were well aware, so that it is hard to recognise a single common denominator, even though they deal with the same topic and are in all cases important. Each of them should therefore have an effect, in particular with regard to the way in which they speak of humans and their existence.

2.1 Harmony and conflict

2.1.1 Truth and violence in Islam

Starting point was a text study by Michel Legarde on the relationship between truth and violence in Islam – which significantly was the longest contribution. There was no equivalent article shedding light on the Christian side of the topic, which is regrettable, because one might gain the impression that the use of violence in the name of truth is an exclusively Muslim problem.[14] But the existing essay does at least give a very profound and differentiated account of the Islamic side of the problem, thus certainly making an important contribution to a discussion which has been widened by the events that have taken place since its publication.

2.1.1.1 Muslim understanding of the Quran: a lengthy road from peace to violence

Philosophically speaking, the initial question gave room for various answers: violence and truth have nothing to do with each other; violence could weaken the truth; violence could serve the truth; and the truth could legitimise and exert violence. In dealing with this question, Legarde examined the Quran, two Hadith collections, certain Quranic commentaries, some essays on the Holy War, and several works by modern authors. The crucial point is that this intellectual heritage has helped to shape the mentality of the Muslim community on this issue. According to Legarde, the starting point was, of course, the Quran with all its contradictions, but these are necessary to do justice to human reality. Accordingly, the statements run the gamut from violence to tolerance, so that the Muslim attitude on this issue was obviously not exactly defined at the time of the Quran. When considering the Muslim attitude today, which sees in the Quran the immutable and eternal divine truth, it is only important that the Muslim tradition itself assumes that the more pacifist traditions are predominantly Meccan and that there was no clear mandate for *jihad* before Medina. Nonetheless, the issue should not be narrowed down to this. Awareness of a completely new relationship with God evoked an attitude of dissociation from those who did not want that relationship. Over time, and in stages, this attitude became violent.

First stage: eschatological and inner-worldly separation
The Quranic terms for this attitude all express separation, distance and denial.[15] This separation refers first of all eschatologically to the Day of Judgment, which is also called the Day of Separation. These separations also lead to concrete actions in the present: believers should not make friends or alliances with unbelievers and should avoid doing business with them. Even more, Muslims should clearly desist from doing obedience to unbelievers. In post-Quran times, these separations spawned the idea that Muslims lived in the House of Peace and all others in the House of War. At least asylum and hospitality continued to be valid explicitly.

Second stage: Non-Muslims are strictly speaking "religiously outdated"
In the second phase, the Jewish and Christian communities as religious predecessors were condemned to non-existence in the spiritual sense: they are superseded. Islam is the true religion, the only one considered to be a religion in God's eyes. This community will remain in eternity and thus shares analogously in the attribute of God's own subsistence. But there is also an inner-worldly place for those who are different.

Third stage: the Quranic ideal of the rule of the faithful as God's rule
Logically, the separation signifies exclusivism, which in turn promotes the rule of the stronger over the weaker. The Quranic text conjures up the ideal of believers' rule. The Muslims are superior to all others, because God is with them. According to the classical Quran commentaries this is also meant in a physical sense, for example most clearly in Sura 47:35, where the Muslims are forbidden to make peace if they are in a stronger position. God is the supreme judge who allots attributes of his sovereignty to the true believers. This is also associated with a time of patience, for in a temporary position of weakness one should support what the unbelievers say.

Fourth stage: subjection by force
But this period of patience passes, and when the time is favourable, the believer must also show that he is right by and through force. After that, when the normal system of subjection has been established, he can also offer samples of his generosity. The background is that God's generosity is not for the benefit of the unbelievers, but on the contrary will only increase their sins. Therefore the believers must stand firm over against the unbelievers, even if the latter are stronger (see for example the Suras 9:73,123; 48:29; 66:9). But one should treat the People of the Book in the best possible way, and above all not argue with them. However, later legal comments place restrictions on too general demands for politeness. The most famous and most cited passage for Islamic tolerance is Sura 2:256, which states that there is no compulsion in religion – and the way of truth is distinct from error. This is interpreted by the commentators to mean that the truth is compelling in itself and needs no external compulsion, which leads Legarde to comment: "Ce qui fait que le 'libéralisme' religieux coranique se trouve sensiblement tempéré."[16]

Fifth stage: not only severity in relationships, but physical violence
It is only a short step from severity in social relationships to various forms of physical violence, and the Quran clears the way for it, whereby it should be kept in mind that this is not always groundless – it is perfectly legitimate to defend oneself. Legarde paraphrases the well-known expression *jihad* as a permanent mobilisation of the believer's body and possessions against the infidels. This is exactly what characterises the true believer; there is no substitute for it, and it is absolutely necessary to obtain God's protection and reward. The physical and moral pressure on the unbelievers described above can lead to armed conflict if the enemy should divert the faithful from the right path and from prayer in the mosque, if he should persecute them, drive them out of their homes and separate them from their children. Muslims should continue

to fight as long as the enemy does so. If he retreats, makes peace proposals and lays down his arms, then peace should be made. In Sura 9:5,29, which would justify arbitrary violence against unbelievers, the majority of commentators agree that these verses are fully valid, but there are doubts as to whether they may not be subject to conditions. The armed conflict can expand into large-scale organised military campaigns. With the Quran, this becomes one of the means of attacking an unbelieving and hypocritical opponent who will not rest, and it is apparently also a means of strengthening one's own community.

Quran: a wealth of material for a Muslim feeling of superiority
Legarde's systematisation of the Quranic statements is very commendable and more objective than it may appear in my short summary (a good example is to be found in his remarks on jihad, *p.292). It is of great worth that he not only takes trouble to look at the texts themselves, but also to ask how they are understood in the traditional interpretations. But it is already clear that the revealed text provides a wealth of material for a Muslim sense of superiority over all others which, despite noteworthy moderations (especially in the context of the time they were recorded), definitely encourages social or military consequences. From a different perspective (which Legarde does not specifically identify) one could also maintain that visions which were originally eschatological are increasingly transported into the real world, into society and politics. It would be really interesting if there had been a comparable development within the Christian tradition on the scriptural level alone.*

2.1.1.2 Sunna: Preference for less tolerant traditions

The second level examined by Legarde is that of the *ahadith*, whereby he only considered the Two Sahihs, Bukhari and Muslim, which are generally reckoned to be authentic and therefore most widely used in teaching and in general life. For this reason they were in a position to shape the general mentality and to create something like a common opinion on many issues, including *jihad* and violence. But what is quoted today and by whom? In general, those traditions that emphasise the steadfastness of Muslims are esteemed by both classical and modern authors to a measure that in no way corresponds to their real importance. In other cases as well, traditions are often cited that are less tolerant, such as the *hadith qudsi*, which states that the Earth serves as a mosque for the Muslims, which unbelievers render unclean, and that this mosque must be cleansed of all impurity. Practical consequences also arise from the popular quotation of further traditions, namely that two religions may not exist together on the Arabian Peninsula and that adherents of two different religions may not inherit from each other.[17]

Muslim: Homage to Muhammad's heroism
However, if one looks at their statements on *jihad*, the two hadith collections are oriented in quite different directions. In Muslim, there is only one of 64 "books" which is dedicated to the subject, and 32 of the 51 chapters in this book deal with Muhammad's historical military campaigns. Some of the practical actions of that time were later taken as precedents – whether it was forging temporary alliances with non-Muslims, or breaking these alliances and driving them away, depending on the situation. Legarde judges these texts to be a tribute to Muhammad's historical heroism, rather than an encouragement to war and violence.

Bukhari: treatise on religious chivalry
The case is different with Bukhari: 75 chapters in his book on *jihad* deal with reality, whereas 106 chapters are concerned with the ideal image of the new human, the good Muslim, whom he sketches out and endows with what would be known in Western terminology as chivalrous traits. It is not about fighting for the sake of battle, war and terror, not even about a religious idea, but rather about this chivalrous ideal. The martyr is, so to speak, the culmination of this ideal of honour, glory, reward, victory and paradise. To this end, he must have a taste for battle and be able to incite others to do so by setting an example of courage and steadfastness. The only motive is the desire to uphold the Word of God. But there is nonetheless room to go into questions of raiment or the lady in the background! Questions of the everyday life of warriors are also covered, as well as their relationship to those under their protection, the *dhimmis*. Bukhari devotes 21 chapters to religious aspects. In Legarde's assessment, Bukhari is also not an apology of violence, but a treatise on religious chivalry. He also sees aspects which would lead to a milder relationship with unbelievers.

2.1.1.3 Abrogation greatly restricts tolerance
These sources are naturally only the starting point. It is the way they are handled that is important, as Legarde summarises it: the position of the Quran in questions of violence and tolerance, as in many other fields, is quite well-balanced, indeed almost neutral. It was the intellectual tradition that deprived Islam of this original innocence and came to interpret it very one-sidedly. It started with the famous question of abrogation and substitution. This was the principal solution for the obvious problem of contradictions in the text of the Quran, which were no longer acceptable since the dogma of the absolute incomparability of the Quran

had very soon been accepted by the entire Muslim community. The text of the Quran itself provides the basis with Sura 2:106, which states that God can replace one verse with something better or similar, which, of course, can only refer to commandments or prohibitions. No less than 81 classical authors have written a treatise on the topic of abrogation. Sura 9:5,29 quite explicitly permits a mission with the sword, at least against the polytheists. According to three of these authors, this verse overrides 124 other verses of the Quran that are more conciliatory or impose some conditions on the fight against the infidels. Sura 2:191, the first one to call for battle against non-aggressors, abrogates the verse which immediately precedes it and prohibits exactly that. All kinds of active or passive tolerance towards non-Muslims are abolished, whereby other Muslims believe that the commentators tended to widen abrogation unnecessarily, although the opposite should be the case. For desisting from action and trusting in the sole judgment of God was also part of the abrogated behaviour. Even the statement in Sura 109:6, which assigns to everyone his own religion, is no longer valid. Likewise, certain conditions have been abrogated which in fact marked the transition from resistance to struggle, such as the observance of alliances and the enemy's peace proposals. Legarde quotes another example: the famous Sura 2:256, which speaks of the fact that there shall be no compulsion in religion. According to the Mutazilites, this clear and open verse could not be abrogated. However, in Sunni opinion it only refers to the Peoples of the Book and the Magi. But even when it came to unbelievers who had converted to Judaism or Christianity, the Sunnis were of divergent opinions. And finally there were those who simply believe that this verse is also abrogated by Sura 9:5,29, so that one should even fight the People of the Book.[18] *It is clear that there is a tendency towards limitation of tolerance, but disagreement about the extent in which this is to be practised.*

A strong god and his rights: theological path to intolerance
In other respects, too, the Quranic commentaries are tendentious and restrictive, and introduce violence into texts where none is to be found: they build up an extensive system of reward and punishment, creating moral restraint and peer pressure. On the positive side, success, well-being and victory are promised in this world and paradise in the afterlife. However, the range of punishment is emphasised much more greatly than that of reward. It takes the form of numerous trials, illness, death, the Last Judgment and hellfire. Islamic theology exacerbates the problem by its argumentation. With regard to the image of God, Muslim theology – as opposed to Jewish or Christian theology – places emphasis on God's strength. Only a mighty God, as he appears in the Quran, can convincingly be called God. Believers need a strong God who is really able to intervene, and

only such a (Quranic) God is the true one. This strength also has an effect on the Muslim community, which is the best community because it prescribes good and prohibits evil on a secure basis, namely by combat. Because the commitment to warfare is more pronounced in Islam, the Muslim community is obviously the best of all. The well-known Hanbali theologian and jurist Ibn Taymiyyah criticises the Christians *expressis verbis* for rejecting armed combat. He claims that *jihad* is the prerogative of Islam, and therefore the proof of its superiority over the other religions. More than a few modern authors adopt this kind of reasoning. If one accepts the idea of a strong God and a strong religion, it is relatively easy to understand why Muslim theology generally tended to solve problems radically, favouring the stronger, namely God and Islam. A theological connection is established between the unity of God and the martial efforts of *jihad*. This connection is logical in itself: the unity and uniqueness of God ultimately leaves no room for anyone or anything else. So it makes no sense to go into discussions about human rights, because only the rights of God are really important. From there, it is only a small step to declaring *jihad* in order to establish the rights of God and gain respect for them. Legarde states that one can trace this basic problem into various theological questions, such as the question of the creation of human deeds. Formally, human freedom is guaranteed, but what is most important is that God is in no way limited in his freedom, not even by any morality. Thus, in the opinion of the majority (which is contested by the Mutazilites with regard to God's righteousness), God is also the creator of evil, disloyalty and unbelief. God would not be the almighty sovereign if he were not able to command us to do things which are absolutely beyond our strength. If all of this is put into practice, it leads to the notorious problem of applying Islamic law to non-Muslims, which is at once very old and yet very topical and which many Muslim authors tend to affirm. The liberal ones limit it in such a way that non-Muslims in this world are not affected by Islamic law, but that their fate in the next world depends on it. Overall, therefore, the development of intellectual systematisation in post-Quranic times clearly tends toward intolerance. The Quran itself repeatedly appeals for leniency, especially in mission, which should refrain from all kinds of violence. Fanaticism would be directly contrary to the will of God. But in the reflection on Quranic passages, only sporadic and isolated moderation is to be found. Thus the original observation may be made again here: the statements are in stark contradiction to one another, advocating severity on the one hand, leniency on the other. Some exegetes manage to find a middle ground by pointing out the need for righteousness: on the one hand, God has given the command to fight the unbelievers. Their unbelief does not entitle them to be pitied, but on the other hand they should not be treated unfairly, for example by imputing to them evil deeds that they have not committed at all. Legarde's detailed textual analyses[19] thus result in an open picture on the question of religion and violence, but with a clear tendency; violence is more pronounced the further one moves away from the original texts.

2.1.1.4 Traditional legal literature: tendency to limitation of violence

Another interesting question that Legarde also investigates is how violence in religion is handled in practice. Consideration of the subject from a purely theoretical point of view, with its underlying intellectual stringency, clearly tends towards the simple victory of violence over tolerance. It is therefore all the more important to look at the real coexistence, because non-Muslims were indeed living within the Islamic community. Hence, a large amount of literature on legal practice was produced, dealing with the handling of conflict and violence (and also providing fertile ground for the treatises on *jihad*). First of all, Legarde looks into al-Ghazali's approach, who proposes a rather simplistic solution: since persuasion alone will only have the power to win over a minority (albeit a good one), one should rather prefer the violent solution. Even if that means losing the best, one would gain larger numbers, and by steadfastness and example make them true believers. (However, there are grounds for doubt as to whether the desired results would really materialise. The procedure is certainly drastic: al-Ghazali was actually a theologian and not a politician or a soldier, so he could develop ideas without paying too much attention to reality.) In contrast, jurists have written treatises on Islam which do not eradicate the element of violence, but treat it with proportion and realism. Out of the abundance of material, Legarde selects al-Mawardi, Ibn Taymiyyah and al-Andalusi, each of which has his own standpoint which he denotes respectively as integration, limitation and relativisation of the obligation to violence. Al-Mawardi, for example, does not see *jihad* as a separate topic, but only deals with it in the context of public and administrative law. Rejection of the call to Islam is an absolute prerequisite to justify an assault. Women and children should not be attacked at all. He even recommends pardon for prisoners of war who might lawfully have been beheaded or enslaved. Ibn Taymiyyah – although a representative of Hanbalism, which is not particularly liberal – goes even further by restricting personal commitment to *jihad* to a situation of defence. He is also convinced that Islam is threatened less by external enemies than by internal foes, that is to say, by Sufis and Shiites. He reckons Jews and Christians to the internal enemies, because they are dissidents from Islam. For this reason they should be put under serious pressure. Among those who, according to Ibn Taymiyyah, may not be killed are priests, elderly, blind and disabled people. In any case, to fight does not necessarily mean to kill. Al-Andalusi is even more extreme. Officially, his treatise was concerned with *jihad*, but it contains practically no religious content. To explain this, Legarde agrees with the generally accepted interpretation that the religious terminology was simply the only one available, which is why it is still used today. The chapter ends with a short consideration of the legal literature on *dhimma*. Here also the laying down of an unequal status leads to a limitation of violence: it does at least establish a certain reciprocity in rights and duties.[20]

2.1.1.5 Violence in the present – five different forms of Muslim apologetics
After this, Legarde jumps straight into the present, which is marked by an emphasis on human rights, by non-violence, democracy and freedom in all forms. He postulates that Muslims consistently have a clearly apologetic tendency, trying to assert or to prove that Islam has always best represented these values, for example as the most tolerant religion. However, there are very different ways in which this claim is made, and Legarde refers to five trends in contemporary Islam that vary greatly in their argumentation: reformist Islam, revolutionary Islam, fundamentalist Islam, modern or rather modernising Islam and utopian Islam.

Reformist Islam: Refusal to recognise its own violence
He uses the term reformist Islam to refer specifically above all to the school of Mohammed Abduh and a Quran commentary from his inner circle, although similar ideas are also presented by other Muslim authors. The first argumentative step tries to show that *jihad* had always been defensive historically, that Islam had never been spread by the sword, and that only in Islam true peace and tolerance exist. Contrary statements found in the Quran only referred to the people who wanted to lead the believers away from their religion. Where verses in the Quran appear to link conversion with fighting, then this should be understood politically in the sense of the necessity of subjugation to legitimate political rule. In the next step, the argumentation then somewhat prematurely makes the desired state equivalent to reality: Sura 2:256 forbids forcible compulsion to faith, and therefore that cannot exist. In general, therefore, the greater *jihad*, meaning personal or collective asceticism, is preferable to the lesser *jihad*, meaning warfare (whereby Ibn Taymiyyah actually does not acknowledge the hadith which sees the greater *jihad* in spiritual struggle). And even the lesser *jihad* is regarded more as martyrdom than as violence against others. In a further step, violence is claimed to be no longer the hallmark of Islam, but of every group of people looking to preserve their own integrity. In his preaching, Muhammad had to face resistance in the society of his day, and it was inevitable that he might slip up and make a remark to the effect that the end justifies

the means. In a last step, their own violence is finally pushed to one side, either by pointing out that the Peoples of the Book were even more violent, or by casting doubt on or simply disowning all Muslim authors who affirmed the legitimacy of offensive *jihad* or physical violence (and they are neither seldom nor insignificant). Responsibility for violent behaviour is denied by casting it onto others or simply closing one's eyes to one's own past. The third element in this context is rejection of the system of abrogation in favour of individual solutions.[21]

It may be remarked that here Legarde indeed describes a system whose various elements are frequently to be found, especially among Muslims living in the West. Legarde brilliantly describes and analyses a relatively common pattern of arguments that is often employed in Christian-Muslim dialogues as soon as they touch upon the issue of violence in Islam.

Revolutionary Islam: forcibly creating the new world order
However, as Legarde points out, there is also a tendency in the other direction, namely towards offensive *jihad* as represented, for example, by the leading thinkers of the Muslim Brotherhood. It is quite evident that Islam has fought violently throughout its history, but not to force people to Islam, but rather to anticipate and repel aggression and to ensure the safety and freedom of the faithful. Anticipation and prevention of any possible hostile surprise attack, that is to say, offensive *jihad*, is legitimate, not just the defensive variant. The previous interpretation is explicitly rejected. Jihad was total revolution against any kind of human absolute sovereignty, whether by governments or laws. One could only offer peace and reconciliation to the Peoples of the Book after they had been humiliated and dominated. Zionism, Freemasonry, Orientalism, and mission should only be combatted with the sword. According to Legarde, this attitude continues to be found in many school textbooks, which to this day depict violence as something which is perfectly normal, should the need arise. They see conspiracies and plots everywhere: Christians and Jews, who are classified as the enemies of Islam, are trying to pull the weak onto their side. At the same time, the argument always contains an element of liberation, in the sense that one is freed from the human yoke and is subject to God alone. Thus it is a worldwide revolution. The goal of Islam was to establish freedom of belief and a certain kind of humanism, and this could only be achieved by the establishment of a strong order, namely the Islamic one. Its establishment, maintenance and expansion need physical force, that is, offensive *jihad*. Since Islam requires public order, strength and struggle are necessary. This is a normal and natural reality, without which there can be no vibrant and dominant Islam. The struggle is large-

ly spiritual, moral and psychological, but also involves physical and military strength.²²

In my judgement, the assessment becomes very clear at this point that people are to be liberated from human rule by the establishment of the divine, Islamic order. Since this order is also and not least of a secular nature, very earthly violence can also come into play. There is an undeniable logic in this argumentation, and even its view on humanity is one that is fundamental and repeatedly to be found.

Fundamentalist Islam: sects employ violence, Islam implements divine law
Nonetheless, there is also an even more theological argumentation for the *jihad* which Legarde calls fundamentalist; he invokes as the representative of this view Mawdudi, who also wrote a book on the subject. Mawdudi's argument is based on the theological comparison of Islam with all other religions. According to Islamic belief, Islam is the ultimate religion, the only one of divine origin, while all other religions are sects, human inventions. Consequently, the Islamic community as well is not a purely human creation with a purely human constitution like all other states. On the contrary, the *umma* is also the ultimate community, created directly by God as the best of all communities. Therefore, it is removed from all extremes and can in its entirety never be mistaken. Its divinely revealed basis and unique source of legitimacy is Islamic law. Consequently, the recourse of sects to armed force is simply arbitrary and illegitimate warfare (*harb*), because it lacks divine justification. Islam alone can practice *jihad*, indeed, it has even made it a major practice of devotion to God and regards it as the pearl of its crown. Islam alone has received the commandment in the Quran: "Dans ce cas, et dans ce cas seulement, l'utilisation de la force n'est plus violence, mais simple exercice du droit divin de juste coercition. "²³

The similarities to the previous position are evident and are also named by Legarde. It only remains to add that in this way the divine right of punishment (of which Christianity can also speak) passes into the hands of the Islamic community, that is to say, of humans. This is certainly a very special understanding of human "vicariousness" and one which poses even more urgently the question as to the exercise of control and the danger of arbitrariness and abuse.

"Modern" Islam: the unity of God is the only guarantee of democracy
Another interesting variation which is represented by the Muslim Brotherhood sees the justification for the *jihad* not in its utility for the benefit of the Islamic community or in a divine commandment, but rather in the unity of God. This line of reasoning is very interesting, but not necessarily easy to understand; the unity

of God is the only democratic principle on which one can form the foundation of a politically free society. Thanks to God's unity, no person is dependent on another, but on God alone. In the case of Jews and Christians, on the other hand, the law proceeds from authorities and priests, creating a dependency on these authorities and therefore preventing people from realising any form of democracy. God intended humans and beliefs to be different, so that there would be room for people's freedom. However, humankind showed a natural tendency towards a totalitarianism that negates these differences. Every strong civilisation awakens the belief that only its own values can lead to humanity. The task of *jihad* is to establish and protect the coexistence of nations, cultures and religions, and thus to protect differences and the freedom of choice. That is why *jihad* in the service of the unity of God is the foundation of every true democracy.

Utopian Islam: option for nonviolence
The fifth trend is described by Legarde as modernising with a utopian tendency, because he thinks that it cannot find general acceptance, not even a majority, but he sees crosslinks to non-violent movements throughout the Muslim world. This applies first of all to the question of the abrogation of tolerant verses in the Quran by militant ones. This abrogation can only be valid once, because an ideal remains an ideal. Furthermore, the problem of violence is an inseparable part of the Islamic moral sphere. But if one resorts to violence without making a distinction between combatants and non-combatants, then violence becomes unacceptable. Muslims should fight for justice, which means that human life and the other elements of divine creation are sacred. Therefore, the new form of combat is that of non-violent action. Islam has a potential of passive resistance and disobedience to unjust powers in the sense that one should obey God alone.

Classical Islam: sophistic reasoning cloaks violence
All in all, as Legarde himself notes, many of the opinions remain classical and unchanged, but there are also a number which differ quite considerably. He concludes from his research that a great deal of the argumentation is based on two sophisms: that the Quranic revelation, and in consequence Islamic law and theocracy, are perfectly divine, so that their consistent application to human society is the only way to free humankind from themselves and from tyranny. This disguises the most fundamental and inhuman despotism that exists. The second sophistry consists in the claim that the freedom of non-Muslims is scrupulously guaranteed within the Islamic framework, as long as they comply with Islamic laws and regulations. Legarde comments that this is equivalent to saying that a prisoner may be considered completely free, so long as he stays inside his cell.

2.1.1.6 A new image of the Islamic community as a solution?
In his concluding remarks, Legarde brings the idea of *umma wasat* into play. While modern Islam considers itself to be the world champion in tolerance, the actual deliberations of Muslim theologians regarding the perfection of the Islamic community lead to a decidedly immodest conclusion: the community is so good that it can impose its criteria on all. Legarde, on the other hand, takes the Arabic term *wasat* literally as the middle way in the sense of mediation, reconciliation, self-restraint and non-violence, and also as the just way in the sense of righteousness based on concord. It is therefore a community of peace, a mediator on different levels, for example in the religious sense between Jews and Christians. The Islamic community is ultimately a community of tolerance.[24]

Thus a very critical article finishes with a very reconciliatory vision, whereas many centuries of variously justified violent reality predominate strongly over this vision of peace and tolerance, which scarcely extends to more than one page at the end of the long overview. The question arises as to whether this vision can really overcome the sophistic arguments mentioned above, or whether, in accordance with the Islamic version, humankind must not be freed from itself by greater force. Even though Legarde's assessment is now somewhat dated, the current majority opinions still lie on the latter side.

2.1.2 The attitude of the Franciscans – a contrast to the Muslims
The second paper in this series deals with the experiences of Franciscans in interfaith coexistence. The reports are of a wide variety and too personal to be reproduced here meaningfully. Only the retrospective reference to the history of the Order and to the history of the Church as a whole is of general interest. The Second Vatican Council was decisive in this respect. It was regarded as a kind of commission for this ministry, whereby the roots are far older, going back to the Bible itself. There the Franciscans recognise an understanding of church that is far removed from an exclusive club in which everything is regulated, but rather resembles a tent in which nomads receive strangers as if they were God. This requires "divine" behaviour by Christians: giving freely, as they have themselves received freely; loving first, accepting others, who are different from themselves; and even rejecting the very notion (and this is interesting, given the relationship between power and truth in Islam) of using force to defend the truth, the glory of God and the Church. In addition, Franciscans should, like Jesus, sing the praise of non-Christians. Holding a religion in high esteem, although one actually rejects it, is difficult. It is easier to adopt the attitude of

an exclusive club and to demand that everyone else should behave and think like oneself. As a Franciscan, one should prepare people to encounter others without any complexes and free from aggression, not to oppose them. As far as the Second Vaticanum is concerned, the Franciscans refer mainly to statements from *Gaudium et Spes*, *Lumen Gentium* and *Nostra Aetate*. *Gaudium et Spes* emphasised that even non-Christians could arrive at a Christian life, albeit in a different way. In addition, the Church recognised the good in non-Christians and stated in *Lumen Gentium* that these values were to be promoted. Moreover, according to *Nostra Aetate*, we could not truly call on God if we refused to treat any person (from other religions) in a brotherly way. These people could even bring us a ray of truth – a concept which was difficult to convey either in the Church or in Islam, as could be seen after more than a quarter of a century. God, however, would wish us to recognise Christ, our brother, in every human being and thus bear witness to the truth.

2.1.2.1 Building on the heritage of St. Francis

Of course, the specific Franciscan heritage is also mentioned, starting with Francis himself, who visited the Sultan of Egypt in the midst of the crusades, and the *Regula Non Bullata*, which also allows Franciscans to do so, even if this means having to expose themselves to enemies. But as Franciscans, they have also surrendered to Jesus Christ with their body. One way to fulfil this special mission is to be subject to every human being, for the sake of God, and to confess one's Christian faith. The official resumption of this tradition took place on the 800th birthday of Francis of Assisi in October 1982. According to the Rule of 1221, the brothers should not criticise others; but they should be gentle, peaceful, and unassuming, courteous and humble. In the course of history, not all brothers lived this ideal, so the letter also asks forgiveness from those that had been hurt. This course is also followed in the General Constitutions of 1985, in which, for example, "the friars are to be especially concerned to go humbly and devoutly among the nations of Islam", or in a document from the 1991 General Chapter that intends to stimulate and support the ministry of the friars in lands with a Muslim majority.

2.1.2.2 Encounters with Muslims – a necessity for Christians

The text by Gwenolé Jeusset then goes into detail, as is again quite typical for the Franciscans, on the best opportunities for a Christian-Muslim encounter. In his assessment, the so-called ministry of reconciliation is more problematic, which ranges from the struggle for justice for immigrants to interreligious dialogue. Finally, he mentions the creation of places of presence and prayer in the Muslim territory. He does not see a fraternal relationship with the Muslims as the fruit of a spiritual opening, but as a Christian requirement, a necessity. Once again, the role model of Francis of Assisi comes into play. Muslims can be seen in three ways: naively, malevolently, or realistically. In the light of the Second Vatican Council, one should view Muslims with a positive prejudice.[25]

This article is extremely interesting because it shows that the supposedly revolutionary new perspective of the Second Vatican Council with regard to Muslims and others was not so new after all, but that Francis of Assisi, within the Catholic Church, had already pointed in this direction three quarters of a millennium earlier, but had unfortunately remained unheard for a long time.

2.1.3 Between harmony and conflict: clarity as the goal of dialogue

The third contribution of this collection, on the other hand, is somewhat out of place because it does not consider an aspect of the Christian-Muslim comparison (which is really the exclusive topic of this work), but rather contrasts the monotheistic religions with those of the Far East. Here the contrast is between harmony and conflict, at least in common judgement and in the title of the article, even if the author Roest Crollius calls this very judgement into question[26]. He regards it not so much as a judgement of the religions, but rather as a description of the different mentalities in West and East Asia which have developed in thousands of years. The former tend towards exclusivity, whereas the latter are characterised more by inclusivity. Western Asia is more concerned with the clarity of distinction, even if it means conflict, while East Asia is more interested in reconciling differences and harmonising them. The interesting thesis of the author is that both harmony and conflict are necessary in order to achieve a real dialogue, since either attitude on its own can only culminate in a monologue. Another interesting observation is that harmony is not only to be found in the religious traditions about the origin of human history, but also appears as a promise of a future reality. Otherwise, however, the focus is largely theological and not anthropological, thus going beyond the scope of our question. An emphasis on the endlessness of truth involves a contemplative attitude, an approach through intuition, community and even identification. On the other hand, where more attention is paid to the correct conception of the truth, the consequence is a discursive approach and thus the emphasis is laid on the intellect. These attitudes also determine attitudes in dialogue: whilst one side presses towards an inclusive solution, the other attaches great importance to a clear recognition of the truth and the courage to defend it. Sociologically, this means a clear distinction between those who have the truth and those who do not. Against this background, and this corresponds with the first article in this series, it can be seen how the Holy War comes about, and this applies to all three monotheistic religions. (The fact that the religions of the East have also waged wars is disregarded.) The author does not come down in favour of one position or the other, but shows that they both harbour dangers in their respective isolation: while the West Asian attitude tends to underestimate the majesty of truth, the East Asian one is inclined to underestimate the human mind. The different ways of thinking simply do not start from the same preconditions, and in dialogue one can only seek to understand and express the similarity in difference. To this end, a certain common background of experience is indispensable and also existent, for both sides emphasise that the inner journey presupposes self-denial and that it is only through death that one can gain life. Any form of egoism, even in the guise of spirituality, would lose and pervert this spiritual path. The differences in mentality are also transcended by the experience of modernity. The use of modern technologies apparently leads to certain processes of mental assimilation, whereby the author sees the solution in a creative implementation of present-day possibilities. One last point concerns what Christians commonly designate as structures of sin or injustice. They necessitated solidarity with and for the poor in order to alter those very structures. Otherwise, academic efforts would only serve to cover up injustice and violence. The author also sees these three areas as suitable topics for the dialogue, whereby Roest Crollius is principally concerned with clarification. That is for him the goal of the joint path of dialogue and should not be compromised by misconceived sympathy or fear of failure to reach agreement.

2.2 A reference work on Islam – Information for decision-makers

According to the introductions of the following publications, the Islam studies of the Commission were to provide exact and objective information about the Muslim world. These seven dossiers, drawn up by the aforementioned consultors between November 1992 and June 1993, form a kind of comprehensive reference book providing an overview of historical trends and the current state of affairs. This reference work covers the areas of *West Africa, Southern Asia, Western Europe, the Middle East* and the *international Muslim organisations* (International Organisation of Muslim Brothers, Popular Islamic Congress – Khartoum, World Muslim Conference, Muslim World League, Organisation of Islamic Congress, World Council for Islamic Da'wah, International Islamic Federation of Students), *Maghreb* as well as a worldwide summary of the *number of Muslims* in each country and a single volume on *Algerian Islam*.[27] As can be seen from the introductions, these dossiers were sent to an unspecified circle of recipients and thus clearly belong to the internal publications and the field of "grey literature". They apparently served principally as information for a circle of decision-makers and were not intended for wide circulation, although they do appear in a list of publications mentioned in 2004 (see note 1 and below). From an anthropological point of view, one must here emphasise above all the meticulous way in which information about the other party has been collected, which indicates that statements and decisions were indeed issued and made against a background of sound information. In addition, these dossiers give a good impression of the diversity of Muslims and of Islam as it has evolved and continues to evolve in widely differing places in the world and over the course of many centuries. It is neither possible nor appropriate to reproduce this here, due to the abundance of extremely compact and detailed information, but the worth of such a reference work, compiled by competent experts with relevant (and also clearly detectable) prior knowledge, can hardly be overestimated.

2.2.1 Special case Algeria

The single volume on Algeria forms a special case, which is why it is described here in a specific section. It was presumably given this form because of the extreme tension in Algeria (which was also particularly critical for the Catholic Church, one need only recall the murders of its official representatives) and the opportunity given by an individual volume to go into greater analytical depth. The description of the separate currents within Algerian Islam also sometimes deals with whether and how the topic of humankind is treated. This is crucial for the figure of the great Algerian thinker Malek Bennabi (1905–1973), who can hardly be attributed to any particular tradition in other ways, but maybe has the nearest affinity to the Islamic humanist Mohammed Iqbal. According to Bennabi, Islam must rediscover its original dynamics of producing an "homme intégral"[28] whose actions are always in accordance with his ideal and needs and who always fulfils his dual role in society as both actor and witness. Such a person can then no longer be colonised, representing, as it were, a response to the social trauma of the colonial era. However, parallel to this true Islamic tradition, according to which human beings must be transformed, he sees the experience of a Descartes. Thus he is very much in favour of the autonomy of human reason. On the other hand, the movement of the "juste milieu"[29], located between the Islamists who identify Islam with sovereignty and the laicists who only wish to see Islam as a religion, looks for the balance between religion and the world, between the spiritual and the temporal, and therefore between faith and reason, divine revelation and human responsibility. According to the understanding of this movement, it is the latter which the Quran emphasises, not simply prescribing one political, economic or legal system. Religion must be reconciled, not mingled, with reason, as was the aim of the Muslim philosophers. Another variant of this issue is offered by the so-called modernists, who are especially interested in human rights and there in particular for the equality of women and men. The background is that modernity is seen as the bearer of universal values produced during the history of humanity. With regard to Islam, it is not the universal categories of modernity in themselves which are problematic, but their formulation in Western culture. And with regard to traditional Islam, it is not so much the contents which are problematic, but rather the methods. In Muslim scholasticism, reason is not denied but reined in, thus falling victim to literal definitions and formalism. Thus, the reformists do not demand a return to the sources (Quran and Sunnah), but a return to humans as rational beings. A good example of this is to be found in the question of human rights. A typical Islamic declaration of human rights would not only introduce Islamic law as an interpretational factor, but would also justify human rights by quotations from the Quran and the Sunnah. This kind of argument is no longer found in the reformist milieu. One can therefore speak of the search for an enlightened Islam that does not break with its spiritual heritage, but adapts it to the new historical conditions from which political and personal liberties have emerged. The public debate on the concepts of individual autonomy would be falsified if one were to cite divine authority as an argument. It is precisely this idea of personal rights which is at the centre of the debate on modernity. The core of this notion (various political rights, democracy) is seen as a universal achievement that extends beyond the bounds of the Occident. It is only relatively recently that religious freedom has been debated in public. Clearly, the Algerian modernists demand the right to believe or not to believe, and to freely choose and practice the religion of one's choice (in contrast to the traditional conclusion that apostasy from Islam is considered as capital treason). They do not use Sura 2:256 as an argument, but rather reason and natural law.[30]

Here it becomes very clear how much depends on how humankind, with its possibilities and limitations, is seen from the Muslim side, and that it is not (or no longer) possible to speak of a unitary point of view. On the contrary, very different opinions are struggling against one another, unfortunately not only with the weapons of the spirit and the word. The dossier naturally restricts itself to a stringent description of the situation, as befits its mandate, without suggesting any consequences, but it is nonetheless interesting to follow up this kind of Islamic-anthro-

pological argumentation and to see whether and how intensively it reappears in other contexts.

2.3 Joint prayer – with limitations

One of the topics later considered by the Commission was the question of common prayer with Muslims, based on a study conducted by the Pontifical Council for Inter-religious Dialogue. Joseph Stamer summarises the most important elements in an article in Pro Dialogo. Since prayer is a more theological topic, it is neither possible nor necessary to repeat all these considerations here, but only those that concern the human plane. First of all, one should always pay attention as closely as possible to the respective context before any consideration or decision. For example, there is a tendency for violent conflicts to be reflected in prayer meetings for peace, starting with Assisi in 1986. In a quite practical sense, it appears that Muslims rarely take the initiative in such joint prayers. There are also definitely differences in the understanding of prayer – Muslim prayer is much less personal, its focus on Mecca gives it a strong community dimension. The author sees Christian prayer, on the other hand, as deeply focused on the Eucharist. As for popular piety on both sides, there is still a wide range of unresearched similarities. It should also be noted that the first Muslims had no hesitations in performing their ritual prayer in places of Christian worship, ignoring statues, icons, etc., following Sura 22:40b – which, however, was restricted by later Quran commentators. Conversely, the Christians also prayed in the "Prophet's Mosque" in Medina. Moreover, the Muslims are commanded to intercede for the unbelievers, following the example of Abraham. Nonetheless, there is no mention of joint prayer in Muslim sources. Stamer concludes that official and liturgical joint prayer is not possible. It would damage Christian identity. In addition, one should not underestimate the danger of confusion and syncretism, especially for children and poorly instructed Christians. Nevertheless, he emphasises that all true prayer comes not from Christian or Muslim initiative, but from the Spirit of God. It was also clear that joint prayer endows all dialogue initiatives with a spiritual dimension and focused the dialogue on what is essential, on God and on the common search for his will. As a result, the survey mentioned provides for officially organised prayer meetings on the occasion of national or international celebrations, or at meetings of Christians and Muslims. Why should one not place the worries that concern both sides into the hands of the one who is himself peace? Prayer for one another should come before prayer with one another, however. Furthermore, it was necessary to consider the situations in advance, in order to establish some cautionary pastoral rules.

For example, when deciding on textual readings, one might draw on the Psalms, which represented a kind of common heritage. Songs and music were the elements which might cause the most friction. It was imperative to prevent regular joint prayer of Christians and Muslims from acting as a substitute for other prayer practice, as had sometimes occurred in schools or among young people. *Stamer's contribution is ultimately an invitation to more spirituality, as it is demanded by both sides.* Following Paul VI (*Ecclesiam Suam*), Stamer concludes, "Il y faut la clarté et la douceur, il y faut la confiance et la prudence."[31] Despite all reservations, the question posed by the document – "Prier avec les musulmans?"[32] – is answered with a limited yes. *Both the yes and the limitations can serve as guidance in practice.*

2.4 Politics (and Law): Insights into Islamic Anthropology

The next major publication by the Commission, *Religion et politique: Un thème pour le dialogue islamo-chrétien*, is an anthology containing the results of its work between 1995 and 1999. This concentrated above all on the question of politics, which had previously emerged as an extremely important and long-standing issue within Christian-Muslim dialogue. The focus was principally on countries with a Muslim minority, alongside the special case of the more or less laicist Turkey, but the inner parallels with Algeria (see above) are abundantly clear. It all comes down to the question of authority: what is the authority wielded by God, and what authority is wielded by humans as individuals or as a collective (a people) in matters of legislation and formation of society or in personal life? May a Muslim accept an authority parallel to the divine commandments; are humans entitled to certain discretionary powers, and if so, in which areas and under what circumstances; or is the life of Muslims (and thus, according to Muslim understanding, of human beings as such) defective if they do not subordinate their authority completely to divine authority? These questions and their answers are seldom formulated so directly in an anthropological sense, but rather in a political or, to be more precise, legal context. But behind the detailed answers, which are a veritable treasure trove of all kinds of developments in the past and present, a basic theological and anthropological concept is to be found: the relation observed between a person and the divine authority determines what that person is allowed to do in practice, and what not. The degree of freedom that is considered compatible with the divine authority certainly varies considerably, especially but not only in the modern age (a further parallel to the Algerian case), and is occasionally marked by a political-legal pragmatism which is a far cry from conventional theological anthropology, at least in the eyes of a Christian. The anthology itself is roughly divided into three sections, the first of which is more fundamental in character, the second contains relevant articles from the *Journal of Muslim Minority Affairs* (abbreviated JMMA or JIMMA) and the third and last concentrates on various local situations. *Once again, we are confronted not so much by a reference work, but rather by a work with precise and highly concentrated information which saves the reader a lot of other reading, assuming that the literature and other information described would even have been accessible at all.*

2.4.1 The fundamental political issues in Islam
2.4.1.1 The big controversy: how much sovereignty is allowed to humans?

The publication is introduced by placing, or rather intertwining, Quranic concepts of authority with the current reality of the Muslim minority living in North America. The situation for this minority is such that in the area of public institutions they had to adapt to the values of others, whether academic, social or cultural. This ranges from questions of clothing to those of biomedical ethics, and causes problems in the question of education, among others. On the theoretical level, the question immediately arises as to where a Muslim's responsibilities now lie, and whether different authorities even fit into a Muslim framework. After all, Islam regulates all aspects of human life, and politics exists to create the appropriate framework and conditions. Against this background, certain passages from the Quran are quoted (in particular Sura 4:59 and 5: 44–48) together with their interpretations with the intention of determining how they define authority, whom a Muslim must obey and in which cases. Particularly in political discussion, it is fundamental to Islam (Sura 5: 44–48) that authority is based on the fact that sovereignty belongs to God alone and that he alone is the lawmaker. If, in a political system, humankind (the people) is the sovereign, then this is diametrically opposed to the basic concept of Islam. Furthermore, the Islamic side often defines this by declaring that Islamic law must be the basis of the law system, whereby it would be equally possible for the Torah or the Gospel to assume this role. It is amazing how often the Hanbalite Ibn Taymiyyah is named in connection with the dispute over interpretation. His ideas have a great influence on the Muslim reformers, even on those who are quite radical. He is the originator of the claim that Islam is inherently theocratic. For example, he is referred to in a pamphlet that is reckoned to be the ideological manifesto of Anwar al-Sadat's murderer: if a Muslim ruler owes his position to laws created by unbelievers, his rulership amounts to apostasy from Islam and he himself becomes an unbeliever. And if an Islamic state cannot be established without warfare, then the believer is obliged to fight this war. Traditionally, this passage was considered strictly historical and was either not applied to later Muslim rulers, or else in a completely different way (fear of God, not of humans, as a leadership quality), and it was also always used as evidence that Islam was the compromise between the revenge justice of Judaism and the exaggerated spirituality of Christianity. All in all, there are once again many different Islamic concepts of authority. The key question is how to deal with competing claims to authority. The Quran only provides the ideological context, but this is clearly contradictory to the models of democratic secularism that prevail today. But there is also the approach which was indicated above, namely that Islamic law must also embrace human experience made historically and in the present. So finally the authoress has to acknowledge two findings: Muslim writers on the subject still generally assume that Muslims are in the majority, and there is no consistent opinion.[33]

2.4.1.2 How far can the Islamic model of society be spiritualised?

The next article deals with the divine rule and its establishment from the perspective of South Asia. More specifically, it is about a dispute between the influential Maulânâ Sayyid Abû'l A'la Mawdûdî, who provided the framework for the modern discussion of such concepts as Islamic revolution, Islamic state, Islamic ideology, and his former follower Maulânâ Wahîduddîn Khân, a very fundamental dispute that was never really brought to a conclusion. The controversy concerns where the key to the understanding of religion (dîn) in the Islamic sense is to be found, and what inevitable consequences follow for the Muslim community. For Mawdûdî this key lies in the order, in the system of life (nizâm), and all other values such as faith, submission (islâm), fear of God and good values are only the moral basis for it. For Wahîduddîn, on the other hand, this attitude is mistaken and a breach of tradition, because it declares one part and one consequence to be the whole. But it is religion as a relationship between God and his servant which forms the whole picture, and this comprises much more than mere obedience to God. The values mentioned are only an expression of this relationship, which grows ever deeper. In this context Wahîduddîn makes use of many drastic anthropological comparisons: it would be equivalent to understanding a human simply as a social being without taking account of his soul, or to taking a photograph of a person from behind, thus ignoring his face, and finally to making the leader of a prayer meeting a leader for the entire world. The difference is clear and momentous: should religion focus on God and the last things, on a relationship to God with heart and mind, which automatically affects social relationships and leads to people's subordination of their free will to God and obedience to God's commandments, or should religion be a new system of life brought about by a revolutionary system change such as has never actually taken place in Islamic history, as Wahîduddîn maintains? For him too, religion is a divine order, but never purely external. It is the completion of human nature. Therefore one could not see the order (*nizâm*) in such an undifferentiated way as Mawdûdî. There would be something essential and additional which had to do with the things of external life and did not come into effect generally. Actually, both are at least in agreement that Muslims *ought not* to use force to achieve their goals, but while Wahîduddîn is very strict on this point, emphasising that in the implementation of matters of faith and worship violence may only be tolerated in relation to Muslims and that compulsion in religion is otherwise strictly forbidden by the Quran, Mawdûdî employs a completely different line of argument. In his opinion, to reject Islamic law amounts to a violation of the law and a revolt, and true believers were obliged to overcome such rebellious resistance. Good advice was usually insufficient, violence and war were required in order to get into government. *Thus, religion becomes practically synonymous with a totalitarian state with God at the head, in which there is room neither for human self-determination, nor for the rule of one human over another.*

The two differing opinions can now be applied to a wide variety of passages in the Quran, each with completely different results. If one were to go along with Mawdûdî, according to Wahîduddîn, socio-political action would gain supremacy over the inner, spiritual dimension of Islam, resulting in more ideology than spiritual depth. The struggle for superiority and power would be all that matters, and the consequence would be an extremist attitude, unwilling to contribute actively to peaceful and just coexistence in pluralistic and secular societies. On the other hand, Wahîduddîn's concept is also not as rounded and smooth as it seems, and credit is due to the author of the article for pointing out three major flaws. One of them is that Muhammad himself would surely have understood Sura 9:32f. as a summons to wage war in order to assert the superiority of Islam, to eradicate polytheism and to subdue the religions of the peoples of the book – not to mention the liquidation of political enemies for the good of the Muslim community. And it was Muhammad who was the spiritual and practical model for Muslims. Another flaw lies in a lack of clarity in Wahîduddîn's own argumentation. He also speaks of the need to fight for religion in a confrontation between religion and non-religion, even if this is not the real task and the main work. However, it is an open question which situation is meant and whether, for example, non-religion is to be understood as secularism. The last ambiguity refers to an issue that is hardly broached, namely the question of human rights and equality in the opposite situation, when Muslims are in the majority and the non-Muslims are minorities. What he implies here indicates that in such a case violence may be justified for the external enforcement of good and prevention of evil.³⁴

In this case, too, it is evident that Islam has a relatively concrete idea of what it means in detail for society that God is the absolute Lord and that humans are subordinate and obedient to him; this concept can hardly be spiritualised, certainly not for a Muslim minority, and even less so for a Muslim majority.

2.4.1.3 Pluralism and Laity – many open questions

The last of the more fundamental articles deals with the anthology *Pluralisme et laïcité, Chrétiens et musulmans proposent*, which, however, does not seem to offer very much on the question of anthropology. But in the section on Islam there is an observation that sounds almost like a commentary on Mawdûdî's ideas quoted above. The problem of modern Islam was that it turns the pyramid upside down, that is to say, calling for a reformation that did not start at the grassroots, like Luther, but at the head, the top, with politicians and religious leaders – and that could not work. But a system with Islam as state religion might very well be conceivable if there were pluralism of inner-Islamic interpretations and acceptance for others, meaning a certain form of secularism. However, nothing is said as to how that might look in practice, and above all how it could be implemented, and that is a great pity. There is also no comment on the frequently quoted claim that the modern state has no continuity with other political forms, but is a church in a different form.³⁵ *If this assertion were to be confirmed, it would directly affect the above-mentioned question of authority and loyalty with which Muslims are confronted. But there is no hint in this direction, let alone a discussion of this question, so that this article is significantly inferior to the first two.*

2.4.2 Perspectives from the Journal of Muslim Minority Affairs

The following section looks at Muslim minorities from their own perspective, specifically in Europe (excluding Russia) and West Africa. It is also interesting to note that both articles emphasise the relatively great neutrality of the *Journal of the (Institute of) Muslim Minority Affairs* on which they are based, which is demonstrated by the fact that non-Muslim authors also write in it, but also in terms of content. This neutrality is attributed to the late former publisher Syed Zayn al-Abidin, who was strongly of the opinion that Muslims could live comfortably as a minority as long as certain minimum conditions existed.

2.4.2.1 Muslim minorities in Europe: plea for a European Islam

In the above-mentioned sense, according to Michael Fitzgerald in his article, the magazine would support a European Islam, which is also in line with the overall view of the articles on Europe he had studied: there were hardly any purely political demands, but it was rather a question of satisfactory living conditions with regard to work, education, religious and cultural expression. When looking more closely at individual countries, the immigration situation and the resulting discrimination, e.g. poor housing conditions, was especially noticeable in Scandinavia and Germany.

Muslims in Germany – a special case in many ways

Overall, Germany is covered relatively extensively because, at least for the period covered by the articles, the labour market situation for Muslims was worse there than in France, Belgium or the Netherlands. Moreover, Muslims in Germany were only guest workers, which led to inadequate housing, wage discrimination, and failure to take account of their special needs in education (teaching of German, the mother tongue and Islam). Hence the demands for government intervention to protect against exploitation and discrimination and to provide adequate language, writing and vocational training, because the general problem of Muslims in Germany was their poor knowledge of German and of the German legal system. It is interesting that it was only in connection with Germany that the situation of Muslim women was specifically addressed. They were not allowed to take part in things such as sports together with men, so that their main leisure activities were television, videos and reading magazines. Arranged marriages and the threats of violent punishment in the case of sexual transgressions led to a high proportion of women with mental disorders, or caused these women to seek refuge in German institutions. It is also emphasised that Germany is a secular, but not a secularised, state and well-suited for the diaspora – it was even possible that Muslims could enjoy more personal and political rights there than in Muslim countries.

Muslims in England – ultimately not accepted?

As far as England is concerned, there are clear parallels with North America: the principal concern of the Muslims is upbringing in an environment that upholds moral values and respects religion, so that young Muslims can develop a positive self-image. More specifically, the state system should offer education in Islam and also in the pupils' native language, and there should be gender-segregated schools. Since the Education Reform Act of 1988, which stipulates that school assemblies and religious education programmes should be primarily Christian in order to reflect the surrounding culture, Muslims see that the pressure towards assimilation has increased. Their response to this consists partly in making use of their right to refuse participation, for example in sex education lessons or music and dancing classes. Overall, English society classifies Muslim peculiarities regarding clothing, food, worship or burial as deviant behaviour, and this can lead to racist reactions, against which there is no specific law in England. This experience is in line with the overall impression of Muslims, who think that Muslims are never really welcome in Europe, no matter how they try.

Muslims in the Netherlands – integration into the system of pillarisation
This could also be related to the fact that they are not prepared to understand the system of a particular country and therefore develop separatist tendencies, as is the case in the Netherlands, for example. Islam is integrated there into the so-called pillarisation system, and the Muslims get help as a minority, but not because they are Muslims – which is what they would like, not to mention even greater financial support for the building of mosques or for Islamic religious education, even in Christian schools, or the introduction of Islamic public holidays. On the other hand, the Dutch are very critical of the mosque schools and suspect that they are breeding grounds of fundamentalism.

Muslims in South Western Europe – no peculiarities
In France, it is also a question of the Islamic identity of Muslim children, but due to the system there the demands are of a different kind and tend to focus more on prayer rooms and mosques. It is very interesting that the Muslims declare that France should stop seeing itself as a Judeo-Christian society. Portugal and Spain also fit very well into the overall picture. The Islamic community of Lisbon has even undertaken to promote friendly relations with other religious groups. Muslims in Spain have an average to fairly high cultural background and are divided in their attitude to Spanish society: some Sufis consider it to be ungodly (*kufr*), others do not. However, cooperation with the Spanish government should lead generally to the recognition of Muslims' civil rights with regard to education, marriage, taxes and military service.

Muslims in Greece – a Christian "Islamic" world
The situation in Greece, on the other hand, is in some ways reminiscent of the traditional Islamic world, but with reversed roles. Officially, the Turkish minority has Greek citizenship and minority protection, but the state interferes in many ways in religious matters, e.g. in the building of mosques. The Greek Muslims demand access to the media, protection of their historic buildings and lifting of travel restrictions. The group of Bulgarian-speaking Muslims are also faced

with particular difficulties, because they are a minority scattered across several countries, and such minorities are generally seen as a threat to national unity.

Muslim minorities in Eastern Europe – first insights into a difficult situation
There is a special situation in Eastern Europe, because at the time of the investigations quoted the influence of communism was still very strong, and neutral information was not always available. Only in Poland was the situation for Muslims really good. With regard to Bulgaria, there was controversy as to the information that was available, e.g. the authorities vehemently denied that people had been forced to change their names. However, it was possible to demonstrate that things had been getting worse for Muslims continually from 1946 onwards: their schools were closed, and atheist education included in the state curriculum. As from 1958, massive anti-Muslim propaganda began: circumcision and Muslim funerals, Ramadan, pilgrimage and Islamic festivals were banned, and marriages conducted under Islamic law were no longer recognised. But what provoked the most resistance among Muslims was the fact that official letters were not recognised if they had been signed with a Muslim name – as was later admitted. Now Muslims want their old names back and freedom to practise their religion. Former Yugoslavia forms an even greater special case, even though there could naturally be no mention here of the civil war, although that later demonstrated the explosive nature of these issues in great clarity. Two parts of the country are affected, Croatia and Bosnia. The Croatian Muslims are especially remarkable, because they adopted a number of Christian customs and practices: they drank alcohol, ate pork and allowed their women to marry non-Muslim men. The Bosnian Muslims were glad to have been granted a separate nationality in the Yugoslav system, although further demands from their side met with violent counter-reactions. Their problem was that Bosnia had been dominated by the Serbs since 1918. Smail Balić, who was himself of Bosnian origin, has described the Islam of his homeland as liberal. They upheld basic human values and advocated, in place of Islamic law, family values that they considered Islamic: humility, a respectful attitude between men and women, obedience and respect by children towards their parents. These Muslims did not want an Islamic state (on the contrary, the Muslims had once been leaders here in terms of democracy, modernity, pluralism and secularisation), but the Serbs just suspected them of such ambitions. The sad result was: "In Bosnia there is an Islam belonging geographically, historically and culturally to Europe, yet the rest of Europe has difficulty in accepting it."[36]

2.4.2.2 Muslim minorities in West Africa – some difficult questions

From the historical point of view alone, the situation of the Muslim minorities in West Africa is quite different. The author himself believes that the journal is not quite objective, but sees a certain glorification of the past compared to the period of colonisation, even though large-scale Islamisation only took place in this region during the colonial period. But this does not fit in with the basic Islamic pattern, in which religious freedom is directly opposed to Islamic order. Here there is an echo of the common Islamic interpretation, which claims that the colonial period had a destabilising effect both on traditional religion and on Islam, and that the indigenous population in particular had not made proper use of the religious freedom brought by the colonisers, so that Islam had been at a disadvantage compared to the Christian missions. In addition, some countries (Benin, Burkina Faso) are entirely missing, while Nigeria monopolised and also completely politicised the debate from 1988 on. Thus, the question of sociological problems and the coexistence of various communities is completely ignored. Moreover, the differences between French-speaking and English-speaking countries are hardly mentioned. The question is left completely open, why in the former countries the elites and most religious leaders favour the secular nature of political institutions and try to give them a positive direction, even when Muslims are numerically in the majority. Nevertheless, and perhaps for that very reason, the articles under review offer some interesting insights into our subject as far as the religious background of the Muslims and their consequent reactions are concerned.

Nigeria: the influence of Islamic ideal culture
On a closer look, the case of Nigeria with its dispute over the introduction of Islamic law is not so special. The main features of the debate also emerge in other countries in the region. The debate itself is not new; what is new is that nowadays virtually all the countries of the world are somehow affected by these questions, since Muslim minorities now exist practically everywhere. Taking a closer look at the case of Nigeria, the first striking fact is that the economy has meantime become almost the most important factor in current politics. Most of the time, political leaders are judged on the success of their economic projects. And in this area, an Islamic ideal always affects the convictions of Muslims, even if they live in a pluralistic society or in a political system in which they do not feel sufficiently recognised. For Nigeria, this ideal finds relatively clear expression in the ideas and work of Usman dan Fodio (1752–1817). His goal was a more just economic order, because a person is also responsible to God in his business affairs. Economic fraud, extortion and corruption are to be regarded as forms of idolatry, i.e. economic justice is directly linked to the proclamation

of the unity of God. In addition, economic activity has to contribute to the strengthening of the Muslim community, which is why he advocated sedentarism, urbanisation and tradesmanship and rejected nomadism and begging. (A Muslim government should first of all remove everything that is bad in religious and temporal affairs and take care of the current well-being of the people.) Looking at the situation in West Africa, this ideal seems remote – but at the same time it is a highly topical issue. People are demanding answers to problems such as the fight against corruption, serious control over the use of public funds, distribution of land – and they recall the Islamic ideals of the past. Especially in Nigeria, the widespread opinion is that Islamic law forms the entire basis for legitimacy and justification of the state. For that matter, the worst phase of Nigerian history had been colonisation: it had not been possible to fulfil the obligation of holy war, the Christians had not been treated as protected citizens, the Quran punishments had been abolished or replaced and the alms tax repealed in favour of secular taxes. The last remains of Islamic law after British rule had then been swept away in the judicial reform of 1968. Today, Nigeria was stuck in the midst of a majority/minority problem that is not atypical of the region, without being able to define who belongs to which group. Specifically, the Muslims are actually strongly represented in the executive and in the legislature, but they still demand even more, which is probably more the result of an emotional feeling than an objective assessment of the situation. Apart from that, Nigeria often behaves as a Muslim country in external contacts, which it finds difficult to do internally.

Who is a minority anyway?
This is a good example of the fact that in this part of the world it is not easy even to define a minority. A minority cannot only be defined on a simply numerical basis, but also from the socio-economic viewpoint, for example. To be even more specific: one could ask about the women alone, or about the clergy, or the religious education of the youth. There are so many conceivable situations and combinations; it might be that minorities have a good standing in every respect, or else that they are not making sufficient use of their economic strength or their influence in society (for example, because they disagree among themselves). Art and culture can also serve as an integration factor for minorities. Especially in West Africa, Islam, pre-Islamic customs and spiritual traditions often mingle with one another, so that it is not clear where the minority lies, with Islam or with traditional religion. Particularly with regard to land ownership and economic traditions, there are still many pre-Islamic regulations that are still in force or have re-emerged. The articles reviewed here contain practically no information about the situation of women in the region.

Muslim conflicts with modernity
But regardless of specifics, with the modern state the colonial rulers brought a social model into the region that is completely alien to Islam, and this raises problems. But here, too, there are notable differences between various secular states: wherever secularism imposes stronger limitations, there is a greater likelihood that the population's participation in development efforts will be blocked in the name of Islam. But some problems are of a completely fundamental nature: since the birth of Islam, the Muslim state has been obliged to organise schools and pay for their upkeep. But what will become of that obligation in a pluralistic or even Islamophobic system? It would be theoretically possible to form a counterbalance to the Christian schools, for example, but this is in fact seldom the case, both qualitatively and quantitatively – the Muslim schools are often somewhat behind the times and refuse on principle to enter into any compromise with modernity. What is missing is a clear conception of what Muslim life should look like today; instead, all that Islamic scholars of the various traditions often have in common is just extreme aversion to everything that comes from the West. Otherwise, they often argue among themselves. The best way to describe the Muslim minorities in West Africa is to say that they are minorities between emigration (*hijra*) and struggle (*jihad*), which of course immediately leads to the important question of whether pragmatism can only be a temporary solution for Islam. In the face of a strong system that is not well-disposed towards Muslims, every Muslim minority is tempted to emigrate, internally or externally, or even more so to practise active resistance. This was the attitude adopted by Muslims in the colonial era in particular with regard to Western-style education, and it is still prevalent when it comes to development by a secular government or administration, although there are also quite different examples. What is more typical is a constant wavering between the desire for a pure, authentic Muslim state and permanent compromises with a much deeper African reality. Repeatedly and solely in the name of the Quran, religious leaders and politicians compromised by their collusion with modernity are the victims of violent attacks, but the real reasons may be ethnic resentment and socio-political disappointment. It cannot be said with certainty what is true. It is easy to blame the disunity of Muslims, with all the above-mentioned consequences, on politics and its motto *Divide et impera*. Here, too, a Muslim ideal exists, which ought to be reinstated, and its motto is "Faith and Brotherhood" – or even that all Muslims are one. The basic dilemma is indeed whether Muslims can cooperate at all with political institutions whose laws override Islamic law. Even religious freedom is not sufficient, if Muslim life is fenced in and regulated by an administration that calls itself Christian; that also appears intolerable. This fundamental denial of plurality also provides the reasons for most of the political demands

of Muslims in West Africa. But they are not typical for Muslim minorities, they are rather typical for the different varieties of political Islam in general, and they are often brought forward in the name of religious freedom and democracy. The catalogue of demands is correspondingly extensive and wide-ranging: Islam should be part of the state education, and at the least the state should finance Islamic instruction. The whole of public life should be organised in such a way that it supports the Muslims' faith rather than being a constant temptation for them (like the freedoms of the West). The state must also actively promote Muslim practice by space in the media, mosque construction, work-free Islamic holidays and organisation of pilgrimages. Muslims should be judged according to Islamic law, and there should be Islamic courts for that purpose. Finally, there is insistence on Muslim political parties.[37]

This shows clearly that the Muslim demands in West Africa are significantly greater and more extensive than in Europe or North America, even if they overlap partially. One has to take into account that such demands are not made by all Muslims there, but altogether they tend towards a markedly Islamic society, tailor-made for the needs and basic understanding of a person of very traditional Islamic faith. Whether these demands are directly stated or not – this ideal and vision is in any case deeply rooted, which means that it is constantly present and virulent, because it can never be completely reconciled with any other model, e.g. the Christian one. As the article also states, one must reckon with a debate on these issues wherever Muslim minorities exist, and that is now almost everywhere the case. That was presumably the reason why an anthology on this subject was considered necessary at all.

2.4.3 Islam in regions

The third and final section of the book contains a number of regional studies which have certain repetitions in content and authorship. The studies concern Europe, India, Turkey, East and West Africa. It is natural that for the first and last mentioned the authors also resort to the material that they had evaluated for the two previous literature reports. The two reports on Turkey are completely new and extremely detailed, whereby the first report in particular has a very comprehensive historical review. Unfortunately, they cannot be reproduced here in such detail, since this presentation concentrates on anthropological aspects, but they can be highly recommended as a source of information.

2.4.3.1 The practical situation in Europe

The report by Michael Fitzgerald on Christians and Muslims in Europe begins by emphasising the importance of taking note of the factual religious plurality in Europe and supports this with the historical reference to the lasting cultural presence of Muslims in Europe, even after they were no longer present in large parts of the continent. He also points out that Western Europe's attitude towards Muslims in Bosnia and Albania would not fail to have an influence on future relationships between Christians and Muslims. Socially, the European Muslims belonged mostly to the working class, especially in insecure jobs, and unemployment was particularly high among young people. In addition, there were also Muslims who had set up their own businesses, shops, cafés or restaurants, and there were Muslim farmers in the former Yugoslavia. The legal situation of Islam in Europe was as heterogeneous as the legal situation of the churches in Europe. Legal recognition affected in particular education, taxation, and hospital and prison counselling. In some countries, it was still in process on account of the internal disagreement and lack of representation in the respective Muslim institutions. Islam in Europe was predominantly the result of a wave of immigration, and for both sides it was the external situation which had priority, e.g. economic and social concerns such as housing and language training. It was only after existence had become more permanent that mosques (and other components of ritual religious life) became important. The field of education was and remains especially important. Discrimination existed as well as racism, but they were directed against immigrants in general, not against the Muslims. It has already been mentioned that the Muslims themselves experience this differently. In addition to the above-mentioned designation of their religious needs as deviations, they feel that they are judged according to what is happening in Algeria, Afghanistan, Iran, Pakistan, or Saudi Arabia, and not by their own behaviour. They want to adhere to Islamic family law themselves – and this is denied to them with reference to Islamic criminal law. Mosques and Islamic centres were generally suspected of being fundamentalist. Moreover, Muslims are afraid of Christian mission, especially when churches call for missionary action.

Can there be a European Islam?
It is clear and logical that the transition from temporary to permanent residence has made it important for Muslims to ensure that they can live their religion as a whole. But this raises two crucial questions: can Europe accept the existence of religious communities within its boundaries that refuse to see religion as a private matter and make demands not only in matters of cult but also in the social sphere? And conversely, can the Muslim communities adapt to their minority situation? The discussion of these two parallel questions goes back and forth. The

nineteenth century, more specifically the Romantic Age, had produced national cultures in Europe, but they allowed regional or religious subcultures to exist as long as they respected the dominant culture. This could also show the direction for Islam, but voices are raised that claim Islam cannot live in a minority situation. On the other hand, one could cite South Africa and India as examples of the other position (the founder of the Institute of Muslim Minorities' Affairs personally recommended a partial application of Islamic law under such circumstances, meaning only personal law). A contrary opinion is held by those Muslims who say that capitalism has also failed and that Europe is in a stage of moral decay. In addition, there are actually young Muslims who want to marry a Christian woman in order to bring her to Islam. Even if she remained a Christian, the children would be Muslims. Large families would ensure the growth of the Muslim community, which would one day be in the majority. The question of the future path of Islam in Europe can thus not be so easily answered from the Muslim position.

There remains the question of openness to dialogue. With particular reference to practical experiences, Fitzgerald replies that dialogue with partners from other religions may well be easier than the dialogue between Christians and Muslims. In dialogue, the Church respected everyone's freedom, had pointed out deficits in behaviour towards Muslim immigrants at an early stage, and its attitude was truly free of ulterior motives; that could not be emphasised too strongly. The irenic and naïve attitude of Christians in the past is a problem that no longer holds, says Fitzgerald, but has rather been superseded by anxiety and the fear of losing their own cultural identity. One of the issues raised In this context was the question of reciprocity and the allegation that Muslims had been given good conditions in Europe, while Christians were suffering in the Muslim countries, for example under restrictions on the freedom of assembly, hindrances to spreading the Christian message, and prosecution of conversion to Christianity. It was to be noted, however, that there were major differences between the Muslim communities and that it was not so easy to hold anyone accountable, since there was no connection with treaties between states. In this situation, Christians demanded that Muslims reinterpret their tradition in the light of the demands of modern life and with an openness of heart and mind. From the outset, the relationship between state and religion had been seen differently in the two religions, and it might therefore be better to seek reconciliation rather than reciprocity, for example by defending the religious freedom of Muslims in Europe in the hope that the Muslim states would guarantee Christians comparable freedom. The article concludes on a positive note, but also with a series of questions: "Sur le plan purement humain nous pouvons parler d'une certaine unité. Ne voulez-vous pas réfléchir avec nous sur la façon dont vit l'humanité? Que pouvons-nous faire ensemble pour améliorer nos relations? Quels sont nos rêves pour le prochain millénaire? Quels projets entreprendre?"[38]

2.4.3.2 Turkey

Next there are two articles about Turkey written by one and the same author, Xavier Jacob. The first deals in great historical detail with the relationship between Islam and politics in the country, starting with the transition from the caliphate to the Republic down to the present day. The second is a much shorter description of the current situation of Islam in Turkey, whereby naturally some of the content is repeated from the first article. In this context it is the general picture which is of interest, not so much the details specifically concerning anthropology.

The phase of secularisation

Here two phases can be identified: first of all that of transition from the caliphate to the nation state, which describes how the influence of Islam, no longer a state religion, was continuously on the wane and was relegated from public to private life. In the newly created state, all sovereignty emanates from the people, in contradiction to classical Muslim doctrine that sovereignty is due to God alone. In 1926, the first statue of Atatürk was put up in Turkey, although Muslim tradition prohibits any representation of a human being, especially in the form of a statue. It is also interesting that women's suffrage was already introduced in Turkey on 5 December 1934 – long before all European women had been given this right.

The democratic road to the Islamic state of Turkey

The development was reversed in 1945 when the multi-party system was introduced in order to meet the requirements of the United Nations. The elections proved that the above-mentioned reforms had not reached the general public, which led to a more liberal religious policy and an erosion of secularism. Additional arguments for the turnaround were constantly repeated, stating that the unity of the nation had to be backed up by unity of religion (this can take on racist and militant characteristics to the effect that a true Turk must be a Muslim) and that the abolition of official religious instruction had opened up a space for sects and extremists – although in the meantime the extremist movements have actually grown and consolidated. In the end, this development possibly lent more importance to religion than in the Ottoman Empire, because at that time there were no elections; the caliph appointed people to various positions, and usually they submitted to him and his wishes. On the one hand, a retrogression set in (since 1950 the Islamic institutions have steadily increased; for example, an average of a thousand or more new mosques have been founded per year, and the disputes flared up again regarding beards and especially headscarves – as a reactionary symbol – and about the position of women in general). On the other hand, the numerous different movements share the opinion that a secular state cannot teach authentic Islam, and should it do so nonetheless, it intends to use Islam for its own purposes. Some accept this, because they are anxious about the secularity of the state or their own religious freedom (Alevis), while others do so because they are anxious about true Islam and its complete freedom of action (Islamists). *De facto*, the religious instruction often ends in an inner dichotomy: on the one hand there is Muslim, often Islamist religious education, on the other hand a school system that is training a secular youth. It is also interesting to note that Turkey has been a full member of the Organisation of the Islamic Conference since May 1976, i.e. is officially an Islamic state in the field of foreign policy.

The question of Turkey's future
After 1980 and after long preparation, the administration was infiltrated by Islamists, in particular the Ministry of Education and the Ministry of Internal Affairs and thus the police (which gave itself an outward image as demonstratively European and modern, emphatically advocating individual freedom, even fighting in the name of freedom and democracy). From this point there were official calls for the introduction of Islamic law, since God's laws are more important than the laws of human beings. Within a population which is 99.5 percent Muslim, the Islamists are the strongest party with 21 percent. The private financial support for fundamentalist activities of various kinds is substantial and particularly well organised by the formation of large companies, financial groups and banks. According to the army (see below), there are about 20 Islamist holdings, and the Islamist opposition in parliament does its very best to block all measures intended to strengthen the secularity. In addition, it can be observed that politicians, when in power, are permissive towards Islam, but when they are in opposition they criticise the rulers for the same permissiveness. The Islamists were only unable to infiltrate the Foreign Ministry and above all the army, which carries out a purge every year. All in all, it comes down to the fact that, in the past, politicians tried to put Islam at their service, but at present, politics seems rather to serve Islam. Up to now, the state still has official control over law and religious freedom. But the suspicion was seen as not entirely unfounded that this republican system was to be overthrown, and that even the very new, demonstrative advances towards the Christians, especially to the Catholic Church and the Pope, might well be rather public gestures than honest intentions. The adaptation of Islam to the 21st century remains a burning issue in Turkey as well, but it is difficult for various reasons: not only are there more traditionalists than re-

formists (as can be seen in the dispute as to whether there can be salvation for non-Muslims), but there is always great perplexity and disagreement when seeking for concrete solutions to specific issues. And once again, the crucial point of controversy is whether people conform to God's commandments or whether they adapt faith and religion to their lives.[39]

The background to the political development in Turkey is again the question of how and in which Islam the divine and the human realm are distinguished from one another, and of the extent to which humans can exercise freedom of choice.

2.4.3.3 Questions of the past in India

According to widespread opinion, the Indian Muslims share the problem of reconciliation of the Muslim faith with modernity, but also in conjunction with their minority status. More specifically, in the past, the issues of political power and social organisation, which is central to Islam, have always been answered in an "either/or" fashion, being used to dealing with autocratic politics, not with democracy. The question of power sharing is therefore a completely new challenge. As is well-known, a nomocratic ideal for a universal social order had already emerged in Medina, but in fact Islamic law was not simply implemented word for word in administration and politics, but served more (or less) as a symbol for the lasting divine guidance in the life of the Muslim community. Basically, it is sufficient to be in possession of Islamic law, and its full implementation is not necessary. Especially after the Mongol invasion, it was more important that the Muslims were united by religion and law than by politics and the constitution, and a pious ruler was indeed free in administrative, financial and military affairs, and even in the exercise of law going beyond Islamic law. In India, these questions were raised in a completely new way under British colonial rule: can Muslims be loyal subjects of the British crown? Can they serve a non-Muslim government? (It is interesting that in this context the term "house of war" is regarded as a political analysis rather than an actual declaration of war.) The main answer was that cooperation would be possible as long as Islamic values were not violated and the religious and cultural character of Muslims was preserved, that is to say: one side did not interfere in the religious education of Muslims, while the other side did not interfere in politics. The Muslims developed various new responses to this situation: looking into the principles of Islamic law, or else political apartheid in the sense of *imperium in imperio*, meaning establishing an autonomous Muslim society, even in opposition to Muslims with a modern education, a kind of nomo-democracy leaving education in the hands of Muslims in order to remain culturally separate. However it remains unclear how this could be compatible with the financial and economic system of a modern state.

Pragmatic: in India first and foremost a Muslim only in matters of faith
The article begins at the situation described above with a systematic evaluation of a survey completed in 1988 among Muslim scholars to assess and tackle the new situation. Four different kinds of response are recognisable: one is pragmatic, one modernist, one is fundamentalist and mission-orientated, and the fourth seeks to deal with the legal situation constructively. The pragmatic approach, which tends to the political left wing, postulates that this situation is not new at all and solves the problem in a very Indian Muslim manner (see above): according to this solution, a Muslim in India is first and foremost a Muslim only in matters of faith, but in all other matters primarily an Indian. Muslims elsewhere would see this differently, but the growing industrialisation made other opinions impossible. One variation of this position is to see the solution in an Islamic party that finds Islamic solutions.

Modernist – faithfulness to the religious spirit is sufficient
The modernist approach, on the other hand, trusts in reinterpretation in the light of change, with the side effect of minimising the necessity of holding to the consensus of religious scholars. The basic problem is seen in the fact that Islamic law is a combination of law and religion, whereby the former needs to change constantly and adapt to the situation, as did Muhammad, while the latter is essentially unchangeable, since morals are lasting. Therefore, one had to define religion and law within the twentieth-century concept of thinking, distinguish between them in Islam, and interpret Islam in the light of modern science, metaphysics and theology. With regard to the principal question in this entire context, the conviction behind this thinking is that God and his will could not be truly sovereign in practice in a modern state. In the modern world, however, it was sufficient to be faithful to metaphysics and to the religious spirit. Ali Asghar Engineer offers a good example here, who stated that the five pillars of Islam are sufficient. The problem with this approach is to know whose consensus is decisive, and how it would be possible to reach such a valid consensus.

Fundamentalist mission-orientated: theocracy, theo-nomocracy, theo-democracy?
The fundamentalist missionary approach, on the other hand, regards the Islamic political system as indispensable for Islam, so that theocracy or theo-democracy is an essential precondition for full Islamic life. This represents an ideological and practical contradiction to popular sovereignty, democracy and secularism. The solution offered here is a kind of theo-democracy founded on the Islamic

concept of the governorship of humankind. In practice, this raises a number of contradictions: The Muslims should neither willingly cooperate with the other system, nor participate in it, but only use it; on the other hand, they should not violate its civil laws, but bring about a peaceful system change. There is no clear answer to the question of how this is to be achieved. On the contrary, it is alleged that the majority has manipulated the political system in such a way that democracy always results in the rule of the majority society... . In any case, the framework would have to be changed before considering the question of Islamic law and modernity. Sometimes, a connection with mission is made by maintaining that only the early missionary activities of the Muslim community (Mecca and the beginnings in Medina) could serve as a model.

Legally constructive – but with many limitations
It is notable that all those who favour the legally constructive approach plead for significant alterations to traditional Islamic law. In concrete terms, the situation of Muslims in India was such that, given their numbers (between 100 and 150 million) and their history (800 years of successful government of the country), they ought by rights to be asked by their fellow citizens to take a leading role. The Muslims should stand out morally and thus attain the leadership in the end. Their fellow citizens would almost inevitably come to them, begging them to take the lead in various areas. God had set the Muslims apart by right guidance, and they should remain so, upholding their own social and cultural system and Muslim personal rights. It was unthinkable to make concessions in this area, which would amount to a rejection of Islam. The bourgeois, republican and religious law had to ensure that no individual non-Muslims or a non-Muslim state should exercise control over Islamic law. For the community, the law of the land should apply, but Muslim personal rights should be valid in private affairs. With all these limitations, however, it is questionable whether unity could still exist in the fields of economy, education and culture. This approach is closest to the Salafist school. But even according to Hanafi definition, India could not be a house of war, because it is not completely dominated by unbelievers. That would be the case if Muslims did not have access to government and did not enjoy the freedoms they have. Thus, no republican country with a non-Muslim majority could be a house of war today. Nevertheless, the problem always arises as to how Muslims should be elected representatives of a state that is not theocratically or theo-nomocratically organised, but sees itself as the source of and supreme authority over the law. The ultimate goal is always the establishment of the rule of the right religion, and a distinction is made between accepting secularism as an ideology, meaning as a matter of faith, and accepting it pragmatically as an inescapable fact, as a method for organising and governing

a highly pluralistic state. Not infrequently, a movement in this world would have to adopt such an attitude in order to retain the possibility of achieving its full goals and rights. There were typical examples of this both in the Quran and in Muhammad's actions, and everything was legitimate which aimed to serve Islam and humanity. To summarise, it is emphasised once again that this survey had been completed before certain concrete questions and problems emerged in the Indian context which would have been of interest and importance in this respect. All in all, however, one could conclude that major advances in Islamic legal and theological thinking would be required in order to enable a truly Islamic justification for participation in the life of the democratic and secular republic of India[40]. *Looking at the other contexts, it must be said that this is not merely the case in India, but more generally for the whole relationship of Islam to democracy and secularism.*

The Catholic position in Kenya – truly realistic?
The next article describes the situation in Kenya, taken to represent all East Africa, as it were, whereby great attention is paid to the position of the Christian churches, above all the Catholic Church. Ecumenism was rather rare among missionary churches and limited to cases where religious politics threatened Christian life, e.g. also in an Islam environment. The general tendency among mission churches was to the tune of: leave us the (religious, spiritual, and moral) education, medical work, and freedom to proclaim our religion, and we'll leave politics to you. Thus the influential fields of politics and economy remained with the secular powers. Changes only took place after the so-called second independence, that is, the introduction of the multiparty system, whereby in the whole of East Africa religiously based parties were never allowed, even though the Muslims would have liked them. The position of the Catholic Church in Kenya was as follows: there is a fundamental difference between party politics and politics as a social structure. The latter is closely linked to God's plan for creation, that people should live in fellowship. Politics in general as well as party politics should remain within the bounds of morality and be directed towards the well-being of the general public. These conditions are mostly lacking in the region, where human rights violations and corruption are commonplace. Nor can politics be autonomous (even the emperor belongs to God, to use biblical terms). Authority comes from God, but is bound by laws, and can only thus lay claim to obedience and loyalty. The state cannot interfere in the rights of God, the church, of individuals, the family and the parents (in that order). In turn, the church's mission relates to all aspects of human life; it is, so to speak, the soul of human society. Among the tasks to which the church is dedicated are human development and the promotion of justice, love and peace. It also has a mission to prepare lay peo-

ple for life in society and to give them a Christian view of politics and economics. The lay world is more concerned with politics, economics and sociology, while the church has to protect the values of the kingdom of God, that is to say, truth, life, justice, love and peace. Current key issues include human rights, justice and peace, together with abuse of power and corruption. Measured against this, neither the Protestants nor the Muslims in Kenya have such a clear and profiled position. A sensitive issue for Muslims is the topic of Islam and the slave trade, for example. But it is generally true that the religious leaders represent a moral authority that should not be underestimated and form, so to speak, the conscience of society. Impartiality and universality are important characteristics both of Christianity and of Islam, and they are rooted in faith in God the creator and in the moral principle that all are equal. This is a powerful tool against tribal, racial and ethnic divisions. In this sense, the two faith communities should have a common mission. Echoing the second Vatican Council, there is reference to the sacrament of unity, peace and reconciliation between peoples. This is an astonishing statement, extended as it is to Christianity and Islam in general. However, it is also mentioned that up to now the opposite is mostly the case: ecumenical cooperation could be strengthened (for example, in education programmes, in the area of human rights, justice and peace, the fight against corruption), but it should also be mentioned that in the last ten to fifteen years the conflicts between Christians and Muslims had rather increased.[41]

Thus, this article offers a detailed and anthropologically sound Catholic view of politics, clearly based on ideas of the Second Vatican Council, but in particular an almost spectacular view of what Christianity and Islam might be able to contribute to humanity on the basis of their common anthropology. Unfortunately, the social and religious realities are not so simple and cannot be reduced to these fundamentals, as the overall perspective proves.

2.4.3.5 West Africa
This becomes even clearer when looking at the next and last article, which concerns the situation in West Africa. It offers a good summary of the historical background and the different countries, which naturally cannot be reproduced in detail here.

The phase of secularisation
The overall development is highly volatile and has certain parallels to Turkey. Here, too, the secular phase dates back to the colonial era and has survived it, whereby increasing political freedom (multiparty system) is leading to a gradual

erosion of this secularism in favour of traditional Islamic concepts and also radical Islamistic tendencies. The initial paradox, as already mentioned, is that under half a century of non-Muslim colonial rule Islam has grown in numbers more than ever before in this region. Today, many countries in West Africa have a Muslim majority, but all of them have the principle of secularity enshrined in the constitutions, with the exception of Mauritania. In earlier centuries, African Muslims had already developed a great ability to live under non-Muslim governments without too many problems and difficulties, and their religious leaders and scholars were well capable of justifying this situation – as a Muslim, one should simply not really love the others and not accept their religion. Politically, Islam had been paralysed by the colonial rulers, and in newly emergent independent states the population was accustomed to secular institutions which spawned the new elite. At that time, there was no tension between the creation of modern secular states and the strong Muslim proportion within the various populations. On the contrary, Islam was pushed even more strongly into the private sphere than under colonial rule. However, secular does not mean the same as secularistic. The prevailing conditions are not similar to those in France, but the state and its representatives regard religion and the participation of religious communities in national life as positive and important, without giving preference to one community at the expense of the others. Dialogue and balance are part of African nature, one might say. Tolerance is seen as a positive assessment of others and of their diversity, as recognition of the contribution of all to the common good and as respect for one's fellow human beings. Regarding religious education and civil and criminal law in general, the state is neutral and much has been left open *de facto*.

Processes of re-Islamisation
Subsequently, this openness was mostly filled islamically. Chad is cited here as an example of the inner emigration, which is mainly practised in contrast to the French school. Islam and Arab culture have become factors that strengthen the Muslims' own personality on the one hand and also serve to express opposition to the occupying forces, to foreigners and their culture. In West Africa, the encounter between Islam and a monolithic power generally leads to the politicisation of Islam and to a gradual transition within the secularist concept towards the establishment of an Islamic order via authority. Since 1969, with the oil crisis and the Arab-African solidarity against Israel, the "Islamisation" of these states has set in. No matter on which side of the political spectrum, support of the masses is basically the most important goal, and to that aim it is impossible to ignore the religious leaders. Currently, the situation of secularism in West Africa is such that it accords religious communities an important place in society.

In the general social crisis, Islam openly presents itself as an alternative solution, a comprehensive system encompassing people's spiritual dimension, their life and success in society, and the exercise of economic and political power under the law of God. Religion becomes the only true guarantor of development. With the help of Arab Muslim countries and international Islamic organisations, Islam has been able to exploit the newly created political freedoms (especially free speech) to present itself as a solution to the crisis. Often, this propaganda is anti-Western: the West, often confused with Christianity, is the reason for Africa's current misery. It is not a long way to a call for a new holy war. Suras and hadiths against the unbelievers are quoted extensively. Thanks to the new freedom of speech, certain Muslim movements are in a position to demand the Islamisation of the nation's life and structures, even in the name of the democratic majority principle. Specific demands include the separation of the sexes in public buildings, transport and markets, special conditions for Ramadan, an Islamic banking system, the construction of mosques, organisation of pilgrimages and religious public holidays. Juvenile delinquency and drug abuse are named to justify the demand for Islamic education. Chad even calls for the introduction of Islamic corporal punishment. According to the author, a detailed study of the financing of political parties in West Africa would also bring surprising results, because nothing can be achieved without resorting to the large commercial resources, and they are fundamentally in the hands of convinced Muslims. The traders might not all want an Islamic state, but there is a real danger that the multiparty system might be derailed and secularism called radically into question. The latter was always presented as an anti-value, imported from the West and synonymous with atheism. In Chad, for example, a conference on human rights was broken up by radical young Muslims. This also showed the pressure exerted by a radical Islam in some of the countries of the region. The governments were afraid of a violent Islam according to the Algerian or Sudanese pattern, which was why mollifying gestures were often made towards the Muslims. For the non-Muslims, in turn, this led to the feeling that they were already living in a Muslim state, strengthening the impression that Islam and political power were inseparable. This had a greater effect than direct demands for an Islamic state.[42]

One thing becomes especially clear in this collection on Islam and politics. The situation is complicated and has many facets, both historical and geographical. However, one clear tendency in geography and particularly in history demonstrates the attitude of Islam towards the spectrum of political conditions for the existence of humankind. This is for Muslims by no means unimportant or completely interchangeable, but it is a question of how humans themselves are seen to be subor-

dinate to God and therefore "put in order" by God. The extent to which this tendency manifests itself, or is even reinforced by those who explicitly do not want it, and the individual forms it has taken, is taking and will take in future – these issues form a vast, quite overwhelming field.

2.5 Experiences with dialogue on the eve of the third millennium

2.5.1 Christian experiences with dialogue – hope lies with the youth

On the eve of the third millennium, the *Commission pour les Relations Religieuses avec les Musulmans* posed the question as to how the Church had conducted Christian-Muslim dialogue since the Second Vatican Council and what kind of dialogue Christians and Muslims would like to see in the third millennium. For this purpose, they drew up two symmetrical questionnaires for Christians and Muslims and sent them out to the friends and acquaintances of the Commission's consultors. The responses were collected in the secretariat of the CRRM; there were about 20 from the Christian and 15 from the Muslim side. They did not issue a separate book about it, but at least a short evaluation was published in Pro Dialogo[43], which is now summarised here. From the Christian side (containing responses from Asia, Africa, North America and the UK) a wide range of different experiences with Muslims was described. On the positive side, it was reported that a great many human values had been discovered which Christians could share with Muslims. On the negative side, they specified the ignorance and superiority complex of the Occident, whilst in the Orient, on the other hand, the religions were forced to live together, although no relationship was desired. Overall, however, the authoress believes that the real-life relationship would gradually lead to theological reflection and a re-consideration in the area of evangelisation. Individual statements indicate that there has been a movement from mistrust to trust, bringing people closer, especially in the area of the youth, who are most likely to recognise the need for Christian-Muslim relations, and also in the case of the better educated, who are more open to the international dimension. However, there are also voices that do not yet recognise any major changes, and are partly fearful of Christian proselytism. Overall, however, the young Muslims are considered to be more open. As for the concrete dialogue, even critical observers admit that this is easiest on a personal level. Muslim hospitality is judged to be generous and honest. There is dialogue at all levels: in everyday life, in spiritual experience, social action, and sometimes even in theological dialogue, whereby the most frequent exchanges occur in connection with social and human issues. Here the support of the Church was seen to come principally from the Council and the Pope, while the responses from the

Orient did not see any support for dialogue coming from the hierarchy, nor was there any formal preparation for dialogue. Overall, therefore, it must be acknowledged that the attempts to promote dialogue can also be frustrating, at least from the Christian point of view.

2.5.2 Muslim experiences with dialogue – carry on despite all hindrances

The equivalent Muslim responses come from Asia, Africa and Europe. They emphasise the unifying human dimension: Christians are people like themselves, and this is a major reason for mutual understanding. The topics are life and sharing. There are also those who are interested in the others' faith. Here as well, dialogue is described as discouraging – but at the same time stimulating! In other respects, too, the parallels are striking: there had been a progression from suspicion to understanding and finally to respect. The Muslim side also stated specifically that these processes were sometimes compromised by eagerness to convert. Muslims consider that theological dialogue belongs to Europe, while coexistence in daily life is found in Asia. The religious leaders on the Muslim side gave scant encouragement to dialogue, but they at least had a positive attitude to it, while the rank and file were sceptical. The Muslims also criticised the media for caricaturing Islam. Conversely, it was inter-religious solidarity for a just world which fostered dialogue. The dialogue should be continued, because it served as an effective weapon against violence and hatred. Dialogue obviously has a positive effect, especially on the people involved in it. Muslims are considered to be relatively indifferent to dialogue (for Muslims it is sufficient if the message is passed on and the people of the book are invited; on the other hand, for Muslims it is not enough simply to talk to one another, which in turn can act as a block for the Christians), but that should not hinder the dialogue. One ought rather to ask about the causes of indifference and hesitancy. For the future, the Muslim side identifies various obstacles to dialogue: the claim to possess the truth, suspicion towards those who have begun a dialogue from a position of strength, the believers' mutual hegemonic intentions, the mythologised unity of the ummah, which makes the partners appear to be a monolithic block, failure to prepare thoroughly, and the different expectations connected with the dialogue – as well as the Christians' refusal to recognise Muhammad as the Messenger of Allah. Nevertheless, the motto remains: carry on until the end of days! And specifically: stop demonising others, do not give fundamentalism a chance, start by educating children, work together for the minorities and the poor, do not be afraid to diminish yourself for the benefit of others, do not be deterred when certain people accuse you of betraying your own community, make the results of the dialogue widely known, emphasise the spiritual dimensions of dialogue. Other proposals included: an internal Muslim dialogue parallel to the dialogue existing between Christians, a rethinking of Muslim doctrine in the face of new challenges, an emphatic declaration by the Church that there are no ulterior motives behind the dialogue, a clarification of the concepts of dialogue and preaching, which still comprise a source of confusion and incomprehension.

It turns out that both sides consider the personal level to be of the utmost importance in dialogue, both for its furtherance and for its hindrance, whereby the former outweighs the latter. Thus, the general opinion is that in the new millennium there are great opportunities for the dialogue that began with the Second Vatican Council. One should not give up – and maybe pay attention to some suggestions.

2.6 Religious liberty as a topical theme

But the Commission itself also thought about how it was to proceed in the next five-year term. One option that was envisaged in the anniversary year was a rereading of the history of Christian-Muslim dialogue for the healing of memories. However, the decision was then made in favour of the issue of religious liberty. The structure of the publication was defined in advance: first there was to be a theoretical section on religious liberty in Islam, at the international level and in the US, which was to be considered as a special case; then a second section would deal with concrete efforts to control religious freedom at an international level and in three different national contexts – Pakistan with its Muslim majority, Nigeria with its unclear majority, and France with a Christian majority. It was also not ruled out that the Muslims would be asked to work on the project as well.[44]

But the publication *Religious Liberty: A Theme for Christian-Muslim Dialogue* departed from this concept in the first section. After a general juristic introduction, the theoretical part is oriented more towards the positions of those participating in the dialogue, i. e. the Catholic or Muslim view. This makes sense, since the introduction points out that the issue of religious liberty is interesting for dialogue, because Christians and Muslims have a different understanding of freedom, so that the practical application has a different form both at the level of the individual and of society.[45]

2.6.1 The theoretical foundation of religious liberty
2.6.1.1 Religious liberty – the absolute human right

The first article deals with the development of the concept of religious freedom – which was, indeed, the first human right of all to be recognised (long before the UN Convention on Human Rights: for example, 1598 Edict of Nantes, 1648 Westphalian Peace, 1689 Toleration Act, England). Article 18 of the Universal Declaration of Human Rights of 1948 refers to the right to religious freedom, namely freedom to change and to practise religion, whether alone or in community. Thus, in the opinion of the article's author, religious freedom is not only binding for the member states of the United Nations, but for all states as international customary law. Under Article 18, the framework for these rights is the personal right of every person to freedom of expression, conscience and religion. This status as a fundamental personal right means that religious freedom is an indestructible and inviolable right with respect to any external force. As a result, every human being ranks as a *sanctum sanctorum* and religious freedom as the only freedom that really deserves to be termed "absolute" and "holy". Charles Malik, the Lebanese philosopher who played a huge role in drafting and passing the convention, recognised at this point that every nation was obliged to recognise the diversity of views on the "last things".[46] It should also be stressed that it is not about a right to be or to do wrong, but a right that belongs to innate human dignity. This, in turn, owes more to the European and American tradition of human dignity than to the more individualistic Anglo-American tradition. This is also evident in the fact that the Declaration of Human Rights begins and ends by mentioning community, which is also consistent with Catholic social doctrine. Human dignity is the seedbed for the principles of social order. Thus, religious liberty is of a social and civil nature, which means that it must be anchored in the constitution. The dignity of human beings, in turn, results from their intellect and free will, leading to the privilege of personal responsibility. Therefore, it is not only protection from outside pressure which is required, but also psychological freedom. Even if a person failed to assume this responsibility, he or she would still be entitled to inviolability, which is not dependent on subjective disposition, but on human nature itself. The right to freedom of religion is connected with a number of other rights, the observance of which is not self-evident in all countries.

The crucial question: changing religion and equal rights
The actual cause of controversy was the question of the right to change one's religion. This topic had been tabled at the express request of Charles Malik, who as a Lebanese had witnessed how converts had fled from religious persecution to his homeland as a safe haven. The essence of freedom is the possibility of

change. To deny people this possibility is to enslave them. During the voting, the most persistent opposition to the right to change religion came from Saudi Arabia, whose representatives argued that missionaries throughout history had been the precursors of political intervention. In the end, Pakistan argued that the right to freedom of religion was fully compatible with Islam, since Islam was also a missionary religion. Finally it must be said that the right to religious freedom is limited by the exercise of other rights and freedoms, by the requirements of morality and public order, and for the good of the democratic state. In the Declaration of Human Rights, there are two other articles alongside the right to religious freedom which are relevant in this regard: Article 16 on marriage and Article 26 on education. The former states that all people have the right to marry without any limitation due to religion. They are entitled to equal rights as to marriage and its dissolution. This last regulation led to protest by several representatives of states with a Muslim majority. It was alleged that Western standards and family relationships were simply being written into the human rights declaration. Finally, Pakistan accepted the text and pointed out that equal rights were not necessarily synonymous with identical rights – a position also held by Eleanor Roosevelt, the American delegate. Article 26 stipulates that education shall be directed to the full development of the human personality and to the strengthening of respect for human rights and fundamental freedoms. It shall promote understanding, tolerance and friendship among all nations. Parents have a prior right to choose the kind of education that shall be given to their children. Both of these terms, namely tolerance (which is otherwise only mentioned in the preamble to the Declaration[47]) and the emphasis on the rights of the parents, were incorporated into the Declaration of Human Rights on the basis of experience with the Nazi regime and its educational methods; in the case of the parents' position it was thanks to the involvement of the Dutch priest Leo Beaufort, supported by the later Pope John XXIII.

Sturdier than morals – a treaty for freedom of religion
Another important document on religious freedom is the *International Covenant on Civil and Political Rights* from 1966. Its particular significance lies in the fact that it is a treaty that actually binds the member states legally. (There are a number of states that have not signed, including Bahrain, Brunei, Indonesia, Malaysia, Maldives, Mauritania, Oman, Pakistan, Qatar, Saudi Arabia and the United Arab Emirates.) The wording is very moderate, speaking only of the freedom "to have or to adopt a religion or belief of his choice"[48] – there is no reference to a change of religion. That was a concession to some Islamic states that had expressed their reluctance. However, the UN Human Rights Committee, which is responsible for interpreting the document and monitoring its application,

states clearly that this includes the right to replace one's current religion with another. In contrast to the right to freedom of worship, this right is absolute and cannot be restricted in any way. It is one of the rights that can be denied to no one. The text of the document makes it clear that the right to freedom of religion cannot be limited on account of national security interests. Another important section of the document states that any advocacy of national, racial or religious hatred that constitutes incitement to discrimination, hostility or violence shall be prohibited by law. In addition, all religious minorities are granted the right to enjoy their own culture, to profess and practise their own religion, or to use their own language. In addition, the covenant declares the right to freedom of expression, to peaceful assembly and association.

1981 – Focus on religious communities
In 1981, the UN Sub-Commission on Prevention of Discrimination and the Protection of Minorities adopted a document dealing exclusively with religious freedom and the prevention of intolerance or discrimination on religious grounds. The document is, however, not yet binding, even though the text seems to imply this. First of all, a worldwide study on the subject was drawn up, whose (Indian) author attributed the concept of freedom of expression and faith to the religions, which in almost all cases had elements of tolerance. This study was a milestone for the UN. The declaration of 1981 expressly includes the right to have a religion or belief of one's own choice. However, there is no explicit reference to the freedom to adopt a religion of one's own choice – the group of Islamic countries had been unwilling to accept this and had made approval of the entire declaration dependent on the exclusion of this formulation. For all others, this weakened the declaration text and also represented a step backwards from the previous two statements. This resulted in the addition of an additional article stating explicitly that none of the wording of the document should be interpreted as limiting or nullifying these two statements. It was thus clear that the right to conversion had not been abolished or restricted. It should also be mentioned that discrimination – unlike intolerance – is clearly defined by law. The declaration strengthened the rights of parents as regards religious education. However, religious education was also to promote respect for religious freedom. It is more difficult, if not impossible, to define which religious practices impair children's mental health and full development and should therefore be prohibited. Maybe the most decisive point in the 1981 Declaration, however, is that it addresses for the first time the requirements of religious communities and makes mention of the freedoms that they need. These include, for example, the right to create and maintain appropriate charitable and humanitarian facilities, the freedom to produce, acquire and use the items and materials necessary for reli-

gious rites, the freedom to write, publish and produce books in this field, the right to appeal for and receive financial and other assistance from private individuals and institutions, the right to observe holidays, and the right to build and maintain relationships with individuals and communities in religious matters. All in all, it can be said that the declaration has certain legal implications, that it expects compliance and that it represents, as it were, international customary law. However, the list of rights was not yet complete and the formulation generally unsatisfactory.

Catholic commitment to human rights and religious liberty
The overall analysis of human rights documents emphasises once again that there had never been an intention to produce a completely uniform practice. The UN Secretary-General at that time, Kofi Annan, also emphasised that no single model of human rights could be used by all states as a blueprint. But conversely, that does not mean that human rights are relative. The first international conference of the Arab Human Rights Movement (Casablanca 1999) rejected any attempt to use civilisational or religious specificity to contest the universality of human rights. Pope John XXIII also recognised the moral importance of human rights: their formulations are reflected in his encyclical *Pacem in Terris*. Pope John Paul II also made it abundantly clear that the universality and indivisibility of human rights did not exclude legitimate cultural and political differences in the exercise of individual rights, provided that the standards set by the Declaration for all humankind were respected. John Paul II also saw an apostolate of human rights for the Church in the task of promoting by means of dialogue the right to religious liberty as a universal principle of moral order, an order rooted in the innate human dignity. Therefore, the Church advocates a binding human rights convention with monitoring mechanisms. Since no such convention has so far come into force, there is no international monitoring organisation for religious freedom, although it is urgently needed.[49]

2.6.1.2 Catholic definition of religious liberty: freedom of conscience
The second article deals specifically with the official Catholic view of religious liberty and how it has developed historically. The author is convinced that from a Catholic point of view the concept of religious liberty has always been present, but constantly in a process of maturation and also undergoing changes of ideas, culture and legal sources. In order to define religious freedom, one would first of all have to define what did not belong to religious freedom, what did not constitute its very essence and value. Freedom of religion did not mean that people are entitled to be a law unto themselves, which would make

humans completely independent of God. Freedom of religion could also not mean that all religions were equal. From the Catholic point of view, freedom of religion is the right of every person to confess his faith according to the dictates of his conscience. This was, as it were, a paean to the freedom of the creature, part of God's plan to which human conscience remains bound. In other words, freedom of religion means the right of every human to develop a relationship with God according to conscience, protected from any form of external compulsion. From the Catholic point of view, autonomy does not refer to religion as such, but to all those who aim to curtail the religious feeling of a person in some way. A person is, according to Catholicism, not free from the obligations that follow from religion, but people's freedom is violated if they are prevented from following their conscience in religious matters. The relationship between God and humanity is therefore actually a legal one, and the content and justification of religious freedom are a kind of circular reasoning. One of the strongest motivations for the Church to promote religious liberty is the Great Commission. Every human being is entitled to these rights and also benefits from them. Outstanding among all these human rights is the right to come to the knowledge of the truth and to follow it of one's own free will. Freedom of religion is freedom of conscience, which should not be restricted in any way. To be more precise, as *Dignitatis Humanae* states, this means that coercion may never be used in spreading the gospel. Only completely voluntary consent respects human dignity and freedom. Freedom of religion, understood as freedom from external coercion, becomes, alongside the call to faith, the second part of the very nature of the act of faith: this is the only way in which reasonable and free submission and devotion can be conceivable. This is a doctrinal, consequently necessary argument. The right to religious freedom must be granted to every human being, primarily by public, political authority. Formulated on an individual basis, everyone is called to form a personal relationship with God in accordance with the law God himself has established. It is the duty of humanity to know this law. Respect for religious freedom can even become a way leading to more and more conformity to God's will and plan. The priority of conscience is absolute: even if it happened to be objectively wrong – with a clear conscience, as it were – it had to be followed. All in all, religious intolerance is extremely harmful to humankind.

Christ's death and Christian martyrs – a plea for religious freedom
Freedom of religion also has preconditions, according to the author, referring first and foremost to the Bible and the Magisterium. First of all, one may say that nowadays people's right to conduct research in the religious field – to a certain degree – is generally recognised. To argue from a more specifically Christian point of view, Jesus Christ brought the relationship between people and God onto

a very direct and personal level, based and lived on the foundation of truth and love. Freedom of action guarantees complete ability to act responsibly. Personal recognition that freedom is necessary for faith leaves one no alternative to the desire that the social dimension be organised in a way that supports the exercise of religious freedom. In practice, this means that the social environment should not exert any pressure to belong to a religion against one's will, nor should it prohibit by force the confession of one's chosen faith. Faith is only possible if exercised freely. Love, such as Christianity commands, can only be lived freely. Christ, too, showed great respect for personal freedom. Thus, if God calls people freely, then their response is also free. Nothing should stand in the way of freedom and thus of human responsibility. It is well known that if people were not prepared to listen, the disciples were to go away. Religious freedom therefore means that there is no possibility of using force; it means freedom to choose. But an important question concerns the boundary between preaching and proselytism. The history of Christianity shows that after the persecution and death of the apostles Christians must necessarily have come to advocate genuine religious freedom. But public opinion was convinced that their decision not to recognise the sanctity of the Emperor constituted a real threat to the social order. The death of the Christian martyrs, like that of Christ, appears to be the supreme act of freedom of thought, freedom of conscience and freedom of religion. If a Christian were to deny Jesus Christ, he would no longer be worthy to be called a person, because he would have denied his own conscience. Rather than take up arms and defend his own life, he accepts martyrdom and respects his neighbour, believing him capable of conversion. There is written evidence for this in the writings of Laetantius, who says that religion is a matter of will, which cannot be imposed by force, and indeed that there is nothing apart from religion which is solely dependent on the will. Hilarius von Poitier argues in the same vein when he refuses to resort to violence or political interference in the field of religion. Pope Nicholas I (858–867) also vehemently rejected torture, because it severely compromised freedom of conscience.

Past mistakes
But it was not always that easy. Perhaps the best example of this is the Crusades: even the most disinterested among those who initiated them and carried them out were convinced that they were defending what we would now call human rights, namely free access to the holy sites for Christians and lessening the pressure of Islam. In practice, however, the Crusades were highly ambivalent, an extreme combination of religion and politics – and failed almost totally to achieve the original goals. Other Christians preferred to follow the Sermon on the Mount. Raymond Lull, for example, insisted that conversion had to be a matter of choice,

because God respected human freedom. Calvin, on the other hand, wrote that God demands that all humanity be set aside when it came to fighting for his honour. The author states that the Christian conscience suffered a setback in the thirteenth century. There were a few exceptions, such as Las Casas or the papal bull *Sublimis Deus* (1537) by Pope Paul III, which states that the Indians may not be robbed of their liberty and personal possessions. But an outstanding negative example was the persecution of the Spanish Jews.

The popes of the twentieth century – increased commitment to human rights
Then the author turns to Modernism, focussing on the twentieth century, and examines papal utterances on religious freedom, beginning with Leo XIII. According to him, people have the right to fulfil their duties to God in complete freedom, which is why society is obliged to respect this freedom of conscience. Nonetheless, there remains an unanswered residue question as to how one should deal with actions related to untrue religions, and whether in this case, too, a right to such actions may be justified by conscience. Leo XIII did not explicitly comment on this, but the author understands hiem to be implying that for the sake of the common good public authority should allow all believers to confess their religion freely, even if they had no right to do so for the reasons given above. The encyclical *Mit brennender Sorge* ("With burning concern") issued by Pius XI dealt with the conditions of the Catholic Church in the German Reich under Nazi rule. It states that the believer has the inalienable right to confess his faith and to practise it in the way he deems fit. All the laws which suppressed or encumbered the confession and practice of such faith were contrary to natural law. In the case of Pius XI as well, it may be concluded from the context that a sincere conscience is decisive, whether or not it is right or wrong. Pius XII declared that whoever desired peace in society had to respect human fundamental rights, namely the right to an intellectual and moral life, especially the right to religious education, and the right to private and public worship and charitable works for religious reasons. The context indicates that these are rights granted to human beings as such, not merely to Catholics. In his encyclical *Pacem in terris*, John XXIII also addresses the human right to worship God according to the precepts of a sincere conscience, that is, the right to public and private worship. Specifically, this is a right to which all human beings are entitled as persons. From the very outset of his pontificate, Paul VI was involved in the issue of religious liberty because of the Second Vatican Council. He remarked with pain and displeasure that in certain countries religious freedom, like other basic human rights, were outweighed by political, racial or anti-religious intolerance. There was still so much injustice against honesty and the free confession of religious beliefs. Here again, the right to religious freedom was not limited to

Catholics – that would mean that real rights were only granted to those who were, in their own opinion, in full accordance with the order established by God. In that case, non-Catholics would not be able to be completely free in the field of religion, and their religious freedom could not be legally guaranteed by positive laws. But Christianity did not want to fight against people, but rather in their favour, and wanted to defend what was sacred to them and not to be suppressed: the pursuit of God and the right to express this in forms of worship. John Paul II himself struggled together with the entire Church to ensure that religious freedom, as desired by God and inscribed into human nature, be actually exercised. Personal dignity and the nature of the search for God made it imperative that the state apparatus did not interfere. However, religious freedom should not be misinterpreted as a right to error. Religious freedom should be recognised in the legal system and be made a civil right. The exercise of human rights remains linked to the historical and cultural traditions of a religious community in a particular nation, which means that special recognition from the state for such a religious community does not represent discrimination against others. The Church is aware, however, that the right to religious freedom is violated by many states, going even so far as to make it a punishable crime to give or receive catechesis. In principle, the Church recognises the legitimate autonomy of the democratic order, does not favour any particular constitutional solution and does not weigh up differing political programmes against each other, other than in connection with their religious and moral implications. Overall, the Church was capable of defending freedom of expression for its moral judgment whenever necessary. Or to summarise:

> "For John Paul II, a correct vision of religious liberty implies for the Church the freedom of expression, of teaching, of evangelisation; as well as the liberty to worship in public, the liberty to organise itself and have its own internal rules, the freedom of choice, of education, to name and transfer its own ministers, the freedom to construct religious buildings and to acquire and possess goods adequate for its activity. Finally, it implies the liberty of association for ends not only religious, but also educational, cultural and charitable."[50]

Freedom of religion, presented as uninterrupted doctrine of the church
It can thus be seen, says the author, that it is first of all a matter of the existence of an objective moral order, and then of different relationships within this order. It is a matter of obligation and the inviolability of fundamental rights, starting with the right to freedom of religion. It is also a matter of the public dimension of religion. For this reason the authorities are called upon to ensure that the religious freedom of individuals and communities is supported in practice. According to the author, this had been the uninterrupted doctrine of the church. The truth of the gospel had acted as leaven in society, as far as freedom was con-

cerned. The church had the duty to promote and protect this freedom of the believers. For the Church, religious freedom is based on rational principles as a manifestation of the personal dignity of human beings, a value which is valid in every regime, in every society, and in every environment.

This concluding summary leaves a rather strange impression when it claims that human rights had continuously been part of the Church's teaching. This doctrine had at least not been resolutely advocated at all times, as is even proven by the brief historical outline. Did the Catholic Church sometimes simply neglect and deny an important doctrine? Or was it the case that this doctrine only came to be formed as a result of massive external indications and at a relatively late stage, in spite of signals within its own tradition which had pointed in this direction but had obviously been fairly easily ignored or reinterpreted?

Looking at Islam, but first into one's own history
The next article concerns the Catholic opinion on religious freedom in modern Islamic thought, starting with the closing considerations of the previous article, namely the long and rocky road taken by the Catholic Church towards the recognition of human rights and religious freedom by the Second Vatican Council. Christian Troll believes that many of the questions faced by the Catholic Church along this path are very close to those faced by Islam today. First of all, he refers to the statements in the Council document: no one should be forced to act against conscience, whether in the private or public sphere, alone or with others – within reasonable limits. The justification for this is the dignity of humans, who are equipped with reason and free will as well as with personal responsibility. This dignity can be recognised from the revealed Word of God and from reason and should be realised in civil law. People can only fulfil their duty in the manner appropriate to their nature if they enjoy psychological peace and freedom from external coercion. The right to religious freedom is based on human nature itself, not on a purely personal conviction. Therefore, even those who do not fulfil their obligation to seek the truth enjoy the right to non-intervention. Proper public order must be respected. Freedom from coercion in religion is not just for individuals, but also for the community, because community is required both by human social nature as well as by religion. Therefore, states should not enact any legal or administrative measures which prevent such communities from selecting, training, appointing, and installing their own priests, from communicating with religious authorities and communities in other parts of the world, from erecting buildings for religious purposes, or acquiring and using appropriate possessions. Religious communities had to refrain from any action that reeks of unseemly pressure or improper incitement, especially in connection

with poor and uneducated people. Furthermore, it is part of the freedom of religion that religions should not be prevented from demonstrating the special worth of their teaching in relation to the proper order of society and for all human activity. Every civil authority has the innate obligation to protect and promote inviolable human rights. This is achieved by just legislation and in general by creating favourable conditions. When special constitutional recognition is accorded to one religion with regard to demographic conditions, the right of all citizens and religious communities to religious freedom must at the same time be guaranteed. Moreover, the equality of citizens before the law should never be openly or covertly hindered for religious reasons. Finally, religious freedom must therefore be given adequate legal protection throughout the world, in order that peaceful and harmonious relationships be formed and deepened.

The Muslim starting position – a move towards humanity?
As for the majority position of traditional Islamic doctrine on human rights, two issues are absolutely binding: people have a fundamental obligation to accept Islam. Muslims, however, must remain faithful to this belief and law at all costs, for they have bound themselves for all time – not only to God, but also to his community, the umma. The unity and well-being of the community are closely linked to the faith and faithfulness of Muslims. But when it comes to proving that apostasy deserves the death penalty, the two hadiths usually cited as evidence are more than doubtful, even if the scholars of the different schools of law agree on this question. Traditional modern legislation has doubts in such cases where the punishment is based on one single authority. Therefore, it is rather the apostates' opposition to believers – their antagonism and attempts to dissuade them from the faith – which is given as the reason for depriving them of their rights. But there are also discussions as to what exactly constitutes apostasy and how women should be treated in this context. It is only recently that larger Muslim organisations have taken a different view on religious freedom and other human rights, at least in their official statements. Troll cites as an example the Islamic Charter for Germany, issued by the Central Council of Muslims in Germany. Here (in §11) religious freedom is welcomed and the right to change or to reject religion recognised. This is justified by the fact that the Quran forbids any exercise of violence and any compulsion in matters of religion. Anthropological reflections have acquired a new meaning, which is an element of the general Western influence. More and more Muslims are participating in a way of life and thought centred around humankind and characterised by a consciousness of subjectivity and individuality that was alien to pre-modern Islamic culture.[51] But the contrast between theocentrism and anthropocentrism is actually inappropriate to describe the changes that have affected religious awareness

in the Islamic environment. Humanity and God can stand in the centre of the worldview simultaneously. Here Troll explicitly agrees with Rotraud Wielandt, whose research on the subject he generally refers to. She denies that there has been a shift from theo- to anthropocentrism in modern Islam. But the new status of anthropological reflection in Islamic thought is a fact, and results indeed from the influence of modern culture which was first developed in Europe. The concepts of human dignity and personal freedoms, along with other values (and also non-values), have gained influence in the Islamic world. Freedom, with its concrete political and legal implications, has become increasingly attractive to Muslims. But so far, they have not been able to change the attitude and teachings of any of the great traditional institutions of Muslim learning, nor of any noteworthy Muslim movement. Freedom in the sense of the Arab *hurriyya* was the opposite of slavery, the equivalent of nobility and generosity, and described the state of a mystic. The word is not used in the context of free will and predestination, which is why freedom as understood by modern Europeans was a term foreign to the Muslims. As far as early modernity goes, that is to say, until about the middle of the second half of the nineteenth century, a concept of innate freedom for every person existed among the Young Ottomans and among the followers of Ahmad Lutfi as-Sayyid in Egypt, the intellectual father of a whole generation of liberal intellectuals and politicians. In recent decades, the most noteworthy Islamic theological theories were not forthcoming from professional Islamic theologians or from the ideologues of Islamism, who are mainly scientists, engineers and doctors, but rather from Muslim scholars of religion and Islam who are familiar with modern ones critical historical and philosophical methods. Such Muslim voices have to reckon with opposition from the religious establishment, and occasionally with drastic repressive measures on the part of the state establishment, which hopes to become popular in those traditional Muslim circles wielding power over the masses, especially wherever a strong Sunni or Shiite majority exists.

Quranic testimony: God himself creates human dignity
Where are the sources of Islamic argumentation on human rights? To answer this question, the article names four passages from the Quran: Suras 17:70, 2:30, 33:72 and 7:172. Sura 17:70 says that God has honoured the children of Adam. The modern Arabic term for human dignity refers to this root. The Quranic idea is that God has endued humankind with honour that he wishes to see respected under all circumstances. God has given humankind a position above all other creatures. Humans have received from God the ability to have dominion over the world. In addition, God has issued many commandments aiming to protect humans from being dishonoured. According to conventional interpretation,

Sura 2:30 states that God has placed human beings as his representatives (*khalifa*) on earth. This is similar to the Jewish and Christian concept of humankind as the image of God. A further interpretation says that it is the gift of reason that enables humans to be deputies of God on earth. Sura 33:72 says that God entrusted heaven and earth to humankind, because all others had rejected it. This is the most important and most cited passage concerning the justification of human dignity. The fact that the world was entrusted to humans by God was only the result of humans' personal free choice. There are also more recent interpretations according to which this trust means freedom in the sense of moral responsibility. This exegesis is often found in Arabic theological writings from 1970s onward. Sura 7:172 says that people bore witness to God as their Lord. The Moroccan philosopher Lahbabi believes it is the basic definition of humankind that they should be witnesses. What all these Quranic passages have in common is that human dignity was created by God himself. Therefore it can neither be earned nor lost. Human dignity is not bestowed on the basis of religion, nor of honourable behaviour, but exists in the very being of humanity.

Problematic cases: slaves, women, corporal punishment
On the other hand, in the Quran there are two specific stumbling blocks in this respect: the question of slaves and of women (whereby one might add the topic of corporal punishment as well). The Quran takes slavery for granted, but does not declare it to be binding; on the contrary, it frequently recommends that slaves be set free. This attitude towards slavery is generally accepted, with the exception of certain scholars in Saudi Arabia, Pakistan and African countries such as Mauritania. But the same distinction between social preconditions and norms of permanent validity is not accepted by the more conservative theologians when it comes to the position of women. Modern-minded Muslims regard traditional discrimination against women as a violation of human dignity and attempt to harmonise the Quranic statements about women with modern concepts of human dignity.

The spiritual struggle for the freedom of humankind
Contemporary Muslim philosophers and theologians do not generally make a connection between the grounds of human dignity and those of human freedom (and the various rights derived from them). Those who are more conservative assert that these civil liberties were commonly accepted in Islam from the beginning. At best, this freedom is derived from the obligation of human beings to make use of the reason which God has given them. But a different approach sees freedom as that which has been entrusted to people, which means that human dignity is to be found in their freedom. Wielandt differentiates between

three different ways of connecting human dignity with freedom: the first assumes that God is free, which is why deputising for God can only mean that humans are also truly free. The second one sees creativity as the crucial point in this deputisation, and the third sees in it a certain degree of freedom and the capability for independent action. Lahbabi's argues in this context that the prerequisite of witness truly represents human freedom: affirmation can only be a personal witness if the possibility of denial exists. The theme of freedom is also important in other contexts. Many Muslim authors are convinced that God requires only voluntary obedience. Should people be forced into obedience, that would not only be worthless morally and unworthy of humanity, but even less worthy of God. But in this case it is impossible to assume at the same time that a person's will and actions are ultimately predetermined by God. From the tenth century on, belief in predestination was standard in Muslim orthodoxy, but from the end of the nineteenth century there was once again a broad tendency to believe in the freedom of the human will. To a large extent, present-day Muslim thinking favours this free will, because many of the thinkers refer directly to the Quran, which sides with human responsibility – for which free will is indeed the prerequisite. Contemporary Muslims emphasise that God wants nothing but the consciously free, active and responsible submission of humans. Conservative authors understand the Quran in such a way that humans have the right and the duty here on earth to do all that is required of them in accordance with the degree of insight granted them by God. But how can human freedom be reconciled with the unconditional obligation of people to obey the will of God? Or is freedom bounded by religious law as the clear expression of the will of God, as taught in classical and traditional Islam? The conservative position claims that it is necessary to say "yes", at least for those who are already Muslims. Therefore, there can be no right to freedom of action in contravention of sharia law. Theologians and philosophers from the fundamentalist position, on the other hand, are in agreement with modern freedom ethics in so far as God only wishes obedience to be voluntary. In their opinion, the revealed norms can only be implemented on the basis of free personal commitment, not by external coercion. The fulfilment of such norms should be a moral action, therefore by necessity free. But how do civil liberties and religious obligations relate to one another? Which are the duties of the individual and which are those of the state? What is the purpose of the legal system and the state that supports it? As far as the relationship between civil liberties and religious duties is concerned, conservative and fundamentalist theologians give precedence to duties before rights – one can only demand one's rights after fulfilling one's duties. Thus, there is no freedom as fundamental as the obligation to fulfil God's will. According to this view, it is ultimately not an act of freedom to say "yes", but rather something that needs to be done, even under forcible pressure, at least by those who are already Muslims. Therefore, there are no civil liberties in regard to religious duties. Civil liberties can only follow from conformity to sharia law and the revealed principles of social ethics. Only in this way can a person receive what he or she deserves according to God's will, including freedom, if one may still call it that. Every human being only gets the degree of freedom that God has granted them. Freedom, then, is one of the desirable consequences of fidelity to God's commandments. Freedom is not the way of fulfilling religious duties intended by God. Today's Muslim theologians and philosophers, who are seriously convinced that God only wants voluntary obedience, have a completely different view on the connection between civil liberties and religious duties. Revealed norms should only be fulfilled by self-commitment, not by external pressure or compulsion. Thus, the confession to Islam is not automatically identical with commitment to sharia law. In this way, religious law would belong to the realm of secular rule over the body and not to the realm of spiritual rule over the soul, that is, to religion in the strict sense.

What is the role of the state?
On the grounds of the nature of religiosity, a religious society is not only in a position to reconcile the freedom of belief and their diversity with democracy, but it even requires them. Power is not helpful, but faith needs security and freedom in order to thrive. When a religious government does not protect the security and freedom of belief or the further development of religious understanding, it forfeits its religious legitimacy. In the context of medieval political thought, it was assumed that individual obligations were completely in accord with the actions of the state. Only later was it questioned whether the state is indeed justified in forcing an individual to follow the will of God in the form of sharia, or whether it is unlawful to limit personal responsibility and freedom. The ideal Islamic state is organised on the supposition that the population is identical with the religious community. According to Islamist thinking, the responsibility for the fulfilment of God's will lies principally with the state. The state has the duty to bring individual Muslims back onto the path of true faith and subordination to the sharia, if necessary by force. However, Muslim intellectuals in particular consider it evident that the norms of belief are primarily a matter of personal conscience and that it is not the state which puts religion into practice. Since 1970, this has been the topic of repeated discussions: what should be created first, an Islamic society or an Islamic state? Those who argue for the former say that Islam can no longer be imposed politically, but only on the basis of education and winning acceptance. Even the others have meanwhile developed a keen awareness for the value of freedom as a basis for religious commitment and

democratic ways of promoting Islamic values. There are certainly Muslims who demand an Islamic state, but are convinced that it can only be established by peacefully winning over a majority of the population for this idea. But here, too, the state's aim is to subordinate the community to the rule of God. In the meantime, however, there is reference to the fact that God has endowed humankind with reason and will, that is to say, the concept of humans as God's representatives, responsible in the field of the legal system and obliged to re-formulate the legal regulations continuously. The goals of revelation are to be transformed into the intentions of the people.

Religious freedom and Islamic law – the key problem is unresolved
Vice versa, the claim that personal freedom is only allowed within the limits set by the provisions of sharia amounts to an objectifying and unhistorical understanding of revealed truth. The Quranic statements on people's role as representatives and their responsibility are almost invariably interpreted as the foundation for the doctrine of human dignity and freedom, whereas no one saw it in that way 200 years ago. The problem of time is also clear from the example of Sura 12:40, which indicates that God alone stands in judgement. According to the Pakistani scholar Riffat Hassan, this is the most important Islamic bulwark for the defence of freedom, not least freedom of religion. Historically, this verse was cited to justify both total oppression and complete freedom of thought and religion. It becomes clear that a hermeneutic is required that takes the social and historical conditions of every human understanding of revelation seriously and thus eliminates all arguments justifying totalitarian claims by rigid interpretation of sacred texts. Names such as Farid Esack stand for this approach. On the other hand, it is argued that the normative authority of religion would be lost if such modern hermeneutical considerations were allowed. The alternative position maintains that the extent of revelation is limited by the possibilities of dialogue with human reason. This means that the truth of revelation is always linked to a living historical context. Therefore, when interpreting statements of revelation from a certain period of time, the first question concerns what the text was saying for its own time, and after that, what it has to say in the context of present considerations. Human historicity is a reality, even among Muslim thinkers. For example, the opinion exists that sharia could not simply be equated with the will of God. It was only a means of fulfilling the will of God, which aims to bring well-being for the earth and justice between the people. (This approach goes back to the Islamic legal principle of general well-being.) Thus, Muslims were not to be alienated from their own religion. Laws should also be congruent with the legal consciousness of their subjects, at least to a certain degree. The Quran itself had already embraced this principle. For example, in the cases of

slavery and polygamy, the will of God went far beyond what was really demanded at the time of the Quranic revelation. One should therefore ask about the social conditions which were ultimately intended by God for us. Troll closes his reflections on this topic by referring to Wielandt's concluding reminder that other circles within the Muslim world will only be influenced by such ideas if the premodernist concept of rulership, which repeatedly mobilises religious loyalties, is modified in countries with a Muslim majority; in other words, this could only happen if Islam's role as an ideal for ideologies of unification and harmonisation were reduced. But both authors are of the opinion that the inner logic of Islam does not stand in the way of accepting the concepts of human dignity and freedom. Troll concludes with the sentence:

> "On the other hand, should the concept of religious freedom in its full and proper sense one day prevail in Muslim societies, then the traditional structures of the religious Islamic institution would be radically transformed in ways difficult to predict with any degree of precision now."[52]

Thematically, this article certainly forms the core of the whole publication, and this conclusion could hardly be more equivocal: an Islam exists which is compatible with human rights, but it is currently only supported by a small modern minority, who are not even living in countries with a Muslim majority. Such majorities, and the Islam practised there, would undergo massive change if religious freedom should really come to prevail there. It is up to each individual to come to a realistic assessment of such probabilities. It may be that movements for democracy have set this process in motion, but it is by no means a straightforward and direct path. Up to now, the democratic majority has tended to move back to an even more traditional Islam.

2.6.2 International religious freedom in practice
2.6.2.1 Pakistan: a sad road to intolerance

In contrast, the report on religious freedom in Pakistan offers insights both into the past and the present. It recounts a somewhat sorry story: Muhammad Ali Jinnah, the founder of Pakistan, was a fervent champion of religious freedom, so that Pakistan was not founded on Islam, but on Western-secular nationalism. Thus the members of minorities are equal citizens with constitutionally equal rights and equal responsibility. Jinnah's speeches prove that he dreamed of a democratic and secular Pakistan and not a theocratic state. In the Lahore resolution of 1940, practically a constitution, Islam is not even mentioned. But Jinnah died in the autumn of 1948, and in the following spring Islam was declared to be the basis of the new state. According to the 1973 constitution, Islam is the state religion. Under Zia ul Haq Pakistan even became a rigid Islamic state: sharia courts were established, *hudud* punishments introduced, and non-Muslims reduced to classic second-class citizens. The religious freedom granted by Islam with its famous "no compulsion in religion" clearly does not extend to Muslims, who do not have the right to change their religion. This would result in a death sentence passed by the religious leaders. In Pakistan it is even difficult to fight for religious freedom and human rights at all. When the constitution states that religious freedom is subordinate to law, order and morality, then that refers to Islamic law. In plain language, this means that nothing may be said which criticises Islam or the Prophet Muhammad and that laws must be in conformity with Islam, so that in some cases Quranic laws are also applied to non-Muslims. Minorities may profess and practise their faith, but they are not allowed to preach it, and religious education may only take place at home and in private schools. A great deal may be attributed to the strong influence of extremist religious and political parties, which in certain provinces enforced the introduction of strict Islamic laws against the will of the central government, for example the blasphemy law, which had not led to a (possible) capital punishment at the time the report was written, but only to life imprisonment, and had also been abused to settle up with Christians for private reasons. Christians and Hindus have the feeling that they have very little religious freedom or none at all. The National Assembly is religiously dominated, leading to extremism and ultimately to terrorism. Religious discrimination and persecution of religious communities is based on state laws. The explanation offered is that a weak state sought refuge in religion, as could be seen for example in the rapid increase of mosque schools, whose numbers rose twenty times over in little more than 20 years (until 2000).

The educational system of Pakistan is marked by Islamic ideology. Each subject was taught according to Islamic guidelines; the minorities were not mentioned in the history of the state, although their role had been partly decisive. One theory postulates that it was because the role of the minorities in nation-building, education, health and defence had been ignored that there was too little religious freedom and tolerance. Religious tolerance or harmony was not on the school curriculum. Islam was simply taught as the true religion, and other religions looked down upon. There was a great need for revision of the curriculum in order to remove this disparagement of other religions. The direct comparison is made to the education practised by the Taliban in Afghanistan with its religious intolerance, fanaticism and discrimination against women, minorities and Islamic sects. The intolerance of the mosque students towards Ahmadis, Hindus and Christians (in descending order) can be indisputably proven by statistics. Such opinions in a more moderate form are also commonplace, so that one may well say that the majority is prejudiced against the minority and not prepared to treat them as equal citizens. Those most likely to favour this are the students at the English elite schools, while the mosque students are really disturbingly intolerant and militant and are also growing in numbers, because the lower class masses are joining them in droves. The mosque schools are the largest non-governmental organisation in Pakistan, forming their own world based on self-help. The report urges the government to deal with this and renew the curricula, since there is no education on religious freedom. That was the only way for Islam to prove itself as a religion of tolerance and peace, protecting human rights and raising its voice against discrimination. In retrospect, the worst situation was shortly after 11 September 2001, when the people took to the streets in protest against their own government, sympathising with the Muslims in Afghanistan, and a massacre was committed in a church. That may have been a turning point, because then the minorities were officially declared equal citizens. The issue of discrimination and the violation of human rights was taken up; discussions and events on the subject took place, and the revision of the curricula was initiated in order to include human rights issues in them. The report expressly acknowledges this, but also makes further suggestions. The current regulations and the national constitution should be brought into line with the UN Charter on Human Rights, and Pakistan should sign the relevant human rights conventions. A culture of peace should be promoted. Pakistan's educational system should be revised in order to engender religious tolerance and mutual acceptance without any kind of discrimination. The role of minorities in nation-building should be given its proper place. Interfaith dialogue should also have its place, and all media should be used for education. Above all, however, all discriminatory laws curtailing the rights of minorities should be revised.[53]

So while this report is both sobering and frightening, it does not abandon the hope that a new generation might be educated to become more compassionate.

2.6.2.2 Nigeria – the historical background to a conflict

The report on religious freedom in Nigeria also begins in the past and describes how Islam entered the country some 700 years before Christianity. Nonetheless, the growth of Christianity has led to the situation that the population is divided almost equally between the two religions, which makes Nigeria the largest Christian-Muslim nation of the world, and at the same time the one with more religious violence than in any other African country. Historically, a theocracy existed up to the British colonial era, and the sharia was in force and also used to oppress the population. The British colonial power certainly cooperated with the Muslim princes in opposing strongly radical Islam (*Mahdi* movement) and it introduced three types of courts, a British one, a sharia court and also a "customary court"[54]. Decisions taken by the latter were however only adopted if they were not inhumane and in violation of natural justice, for which there were special provisions. (Thus executions with the sword, cutting hands off, and banishment were prohibited.) This did not solve the problems, but British policy had made religion an instrument of conflict. The issues of sharia law and human rights are still virulent in modern Nigeria.

Nigeria's problems, and sharia as an illusory solution
One crucial problem, for example, is the fact that eleven of the nineteen northern states have introduced sharia law, which has led to excesses: non-Muslims were also summoned by force to these courts; the courts were themselves corrupt and even passed judgment on people who did not belong to their jurisdiction. In the North, religion represents the sole only for the fears of non-Muslims. During the wrangling over the constitution, parts of the Muslim population saw the sharia as a way of tackling the problem of injustice, of which there was plenty. The North had dominated the country, even under the military government, and there was massive corruption. Productivity was decreasing, and all efforts were concentrated on oil, which had never really been used for the benefit of the people. On the other hand, religious as well as ethnic identities were re-invented, as it were, and exploited in the struggle for rights and privileges. In the end, military rule intensified ethnic and religious tensions and destroyed the foundations of the nation even more than colonial rule. Some critics even speak of a "balkanisation" of the country. The report clearly blames the military for the situation and for failing to provide a solution so far, and believes that democracy will lead to greater consensus in the course of time. In fact, the constitution stipulates that no state within the Federation shall adopt any religion as state religion, but the Muslims oppose this. This would allegedly encourage the government to ignore religion and to strive for secularism. But the government did not take a clear stand on the guidelines for dealing with religion. The ensuing, extremely fierce debates, accompanied by violence, destruction and death, revolved around such issues as the observance of Sunday as a public holiday, relations with the Vatican, clothing regulations for school children, pilgrimages, building plots for churches, sites for mosques, and the return of schools to non-state institutions. For example, when the government simply changed its status in the Organisation of the Islamic Conference from "observer" to "member", it aroused increased suspicion among non-Muslims, and the Christians were con-

vinced that they could not trust the government to treat both religions fairly. Also, the Advisory Council for Religious Affairs, a body comprising the leaders of the religious communities, was unable to function on the grounds of mutual distrust. Archbishop John Onaiyekan did not regard this as a religious issue, but rather as an expression of deep-seated injuries. The conflicts had also not arisen for purely religious reasons. Onaiyekan considers that they stem from rivalries over access to limited resources, feelings of exclusion and unfair treatment. This opinion is shared by the author of the report. The elites treated the people unjustly, and despite the marking with labels of religion, region or ethnicity the people's situation did not necessarily improve. In other words, despite Muslim rule, ordinary Muslims felt powerless. In his report, the author regards the ethnic groups as the biggest problem in Nigeria, whereby Islam and Christianity transcend the ethnic borders, but had enough problems of their own and were not able to solve the problem of religion as such. He shows one example of a governor who went ahead with the propagation of sharia legislation and thus exerted pressure on his colleagues to do likewise, something for which they were completely unprepared. Over and above this, great tensions and divisions even existed between Muslims, and many issues were disputed among Muslim scholars and teachers. There remains the important question as to whether sharia may be classified as unjust, or whether it contains the means of self-correction for the victims of injustice – the sentences tend to be acclaimed as proof that sharia really works.

So does sharia indeed represent a threat to freedom and justice, or is it an expression of the principles of pluralism in a multi-religious state? According to the report, sharia was actually seen in the political process as a way for Muslims to assert themselves after losing their political power to a Christian president from the South. But in practice it turned out to be a setback, a blatant violation of Nigeria's secular nature in the words of the Catholic Bishops' Conference, which found an echo among most non-Muslims. The rights of innocent and law-abiding citizens were being trampled upon, and they could not go to court on account of well-founded fear for life and possessions of their own person and the whole family. Normal Muslims, on the other hand, saw sharia as the easiest way to counter the corruption, injustice, and social degeneration that had taken a hold on society. Only a small group of Muslim intellectuals considered it to be a completely pointless exercise, because it was not possible to impose a sharia state on a post-colonial state modelled on Western democracy. Instead, one should question what place Islamic law had in a modern democracy – for example, in Nigeria only one of four schools of law had ever been practised; and apart from that, cases involving such matters as the pregnancy with a "sleeping child" (a spectacular case) were simply ridiculous. The Muslims, who were supposed to benefit from this regulation, agreed unanimously in private conversations that it had not paid off. They had expected relief with the help of the sharia, but only suffered more in the end because they themselves became the target – women in particular were an easy target for those who wanted to prove that sharia works.

Nigeria – a spark of hope for the future
The issues of the future of religious freedom in a plural state and the role and position of religion in a democracy are thus important for Nigeria's development. At the end of his report, the author points out, among other things, that ordinary Muslims and the Muslim elite had completely different expectations regarding the introduction of sharia law. The elite was aiming to gain political advantages, while ordinary people had hoped for an end to corruption and moral degeneration and, above all, for a strengthening of *zakat* as a kind of social security that had been completely neglected by the Muslim elite in northern Nigeria. The governors' expectations had not been fulfilled, because things did not turn out as easily as they had imagined. The sharia courts were not so bad from the point of view that they offered justice to many people for a small price – while the state courts remained corrupt. All in all, the entire discussion on sharia throws more light on the deficits of the Nigerian state and the corruption that has brought it to its knees. But even the anxieties of many non-Muslims have not been fulfilled. The author mentions those dialogues that had already taken place and had gone well, and expresses the hope that they would break down the walls of mistrust in the course of time. He also considers it important for non-Muslims to emphasise that they do not oppose sharia, because they are in favour of those weaknesses which the proponents of sharia want to fight.[55]
To sum up, one may say that in this case, too, the current situation is not exactly positive, but that democracy and dialogue are seen as a hope for the future.

2.6.2.3 France – the general situation of Muslims
The last report comes from Jean Marie Gaudeul of France and refers immediately to the fact that France is so strictly secular that no census on religion is allowed. One is therefore dependent on opinion polls, which show that up to a quarter of those people who should by origin be Muslims actually say that they are non-Muslim or merely of Muslim descent. It is astonishing that Christians and Muslims in France are so alike when it comes to strengths and weaknesses of religious affiliation. Jean-Marie Gaudeul draws a conclusion which is hardly less astonishing: "Religious freedom leads to a flittering away of attitudes and options."[56] But this actually does not belong to the description of the situation,

which starts with a presentation of the original legal situation, drawing on a contribution by Dalil Boubakeur, the rector of the Grand Mosque of Paris (from 1921, a legacy of France's colonial past). Freedom of conscience and the practice of religion are allowed in France provided that they do not disturb public order. Religious celebrations are public, but salaries for priests are not paid, except for those in pastoral care at state institutions. This French laicism is based on the constitution of 1958, but it goes back to the French Revolution. In the field of education, this means that only lay people may teach in the state schools. National sovereignty means that the state is independent of any authority that is not recognised by the entire nation. But equality can also lead to inequalities that should not exist. One such case affects the funding of the religious communities – the Muslim population in France is poor, and the Muslim communities do not have any state-owned buildings, where the state would be responsible for the upkeep, so they do not have the resources for their work. On the other hand, says Boubakeur, the distinction between the House of Islam and the House of War no longer obtains. According to Sura 4:60, it is necessary to obey those who rule over a place. During the colonial period, the *ulema* in Algeria called for the introduction of secular law, and the French administration, which was supposedly laicistic, exercised full authority over religion and appointed all the clergy. Most of the Muslims in France largely come from repatriation and family reunification for immigrant workers, which was permitted by law in 1975. In the meantime, Islam is the second largest religion in France with two million French citizens.

Most problems concern representativeness
In this situation, the Ministry of the Interior insisted on the foundation of a religious association in accordance with a law dating back to 1905, in order to regulate religious activities such as mosque construction, Islamic teaching, and the activities of the imams. Muslims are recognised by the state *de facto*, but they do not have anything resembling the status of a legal corporate body; for each problem that crops up, a makeshift solution has to be found. Because of its laicity, the state is unable to organise a representative corporate body, but may only decide whom it will invite for talks. In the ensuing political negotiations, Muslims managed to have the reference to the right to a change of religion deleted in the declaration of their compliance with the French constitution. (Gaudeul tries to justify this by arguing that otherwise the level of civil liberties might have been confused with the level of dogmatics.) The problem is that the four organisations and six mosques delegated to talks with the state are not representative, because only five to ten percent of Muslims in France practise their religion. What remains is not so much a national representation of Muslims, but rather a kind of national representation of those Muslims who attend worship. For this reason

the Minister of the Interior also tried to negotiate with quite different branches of Islam in an attempt to get a broader representation of those who were not connected to any particular mosque. Terrorism and its hotbed, fundamentalism, are a constant temptation to politicians to intervene actively, rejecting certain representatives or seeking out others in order to choose the supposedly appropriate partner. It was particularly the young Muslim movements which opposed this selection process, arguing that it gave predominance to old believers, who were frequently illiterate and controlled by the consulates of their home countries. In addition, proposals were made for the erection of mosques, which would bring Muslim communities into the same situation as today's Catholic communities. Another difficulty is that the secular state prevents it from granting imams a legal status. In official parlance they are "workers" and only have the status of visitor; they often do not speak French. If they were to teach the language and culture of their home countries, they might perhaps be integrated into the state schools, but these subjects are increasingly being seen by the schools as marginal and even as a barrier to integration. There is also a possible problem in distinguishing between the culture of the homeland and religious preaching. Even though French Muslims are trying harder to escape the influence of foreign countries or potentially fundamentalist organisations, they are as yet unable to stand on their own two feet, but require help from Muslim countries.

And what was the dispute about the headscarf all about?
Further conflicts resulted from the demands of a small minority for girls to wear headscarves, for the observance of Muslim dietary laws and festivals, the modification of curricula in sports, science and sex education as well as the introduction of Islamic education in schools, although no religious education takes place at state schools. The legal situation was that schoolchildren were not obliged to conceal the signs of their religious affiliation as long as they did not disturb the public order, were not too conspicuous and did not serve the purpose of proselytising. Refusal to take part in certain subjects is not possible and would lead to expulsion from the school. The problem with the headscarf is that it is also an expression of a social system in which men and women do not have equal social status such as the French constitution upholds. Feminist Muslim associations wanted the school authorities to take an even stricter stance and demonstrated on the streets. Following these debates, a commission was set up to investigate whether there were more such cases in which secular principles and women's equality were being violated in the suburbs, schools and clinics. When it transpired that the secular nature of these institutions, and especially the status of women, had indeed been undermined, there was a public outcry, and in March 2004 a new law was introduced banning all visible religious signs in pri-

mary and secondary schools, in particular the headscarf. As a result, two French journalists were taken hostage in Iraq and threatened with execution, should the law not be repealed. A small number of girls had asked to be admitted to Catholic schools because of the headscarf. While most Muslim children had attended state schools ten years before, more and more were now going to private Catholic schools (and comprised in some cases more than 50 percent of the pupils), but they did not even have enough places for their own students and would certainly no more tolerate a fundamentalist attitude than the state would. For financial reasons, Muslims would be unable to run their own schools. Overall, however, Gaudeul does not believe that the headscarf necessarily signifies a denial of integration; he would often rather see it as a "flip through a period of return"[57]. By this he understands the feeling that one has to show one's difference at some point, if one conforms in all other ways. Furthermore, as far as schools were concerned, a project existed to introduce religious studies as a subject. The French state also employed and paid chaplains in those institutions in which the inhabitants had no freedom of movement. Here the Muslims were looking for equal treatment, or there was maybe a mimicry effect. The big problem, however, was that such counselling services did not really exist in Islam, and also differed substantially from normal Islamic practice. Thus, there was no one who could give guarantees to the state for such Muslim clergy, and it happened straightaway that those who came to such positions either did nothing or else asked their Christian colleagues for instruction. Another critical point was ritually slaughtered meat. As long as Muslims were only staying temporarily in France, the legal opinion was that they could buy meat from ordinary butchers. But then, similar to the case of the headscarf, pure food was rediscovered as a new ritual distinction over against the non-Muslims. The outcome, however, was massive hygiene problems with slaughtered meat, or shortfalls in the supply of living animals at the time of the Islamic Feast of the Sacrifice. In the question of polygamy, marriages that have taken place abroad are recognised, but only the first wife has the status of a legitimate spouse in France. Regarding public holidays, the general public is not prepared to make changes, but Muslims, like Jews, can be granted special leave on three religious holidays, i.e. the Breaking of the Fast, the Feast of the Sacrifice and the birthday of the prophet Muhammad.

Developments in France: relationships and coexistence
As for the role of the Catholic Church, it had initially helped Muslims as foreigners, then as underprivileged social partners, frequently lending them rooms for their religious services in the face of the problems mentioned above. In 1974, the S.R.I was founded, the Secretariat for Relations with Islam. Meanwhile, the two religions were on an increasingly equal footing, thinking together about

the role and visibility of religion in tomorrow's pluralistic society – laicity is supposed to be given a more neutral face. In the meantime they conduct a kind of extended dialogue, opportunities for constant exchange. Gaudeul judges that the population wants the religions to manifest themselves, but to speak with one voice in facing the great problems of coexistence. This was, in turn, not unproblematic. Public authorities could call on the religions when they themselves were no longer able to guarantee public order, and one community alone could lose the right to speak in its own name when it saw those values endangered which it considered important. Nevertheless, there was no alternative to such meetings, because the public and the media no longer listened to one party; in order to be heard, one had to speak in unison. In this context, Gaudeul also addresses more recent events. 9/11/2001 led to the question of taking on responsibility for peace. What place did religion have in society and in the exercise of power? What was the influence of religion in such confrontations and what should it be? Gaudeul considered the spontaneous reactions to be inadequate. The question was quickly posed as to a possible pretext for the theories of the Islamists. For example, people referred to the injustices which led hosts of individuals to despair completely and be prepared for any kind of sacrifice or acts of violence. More recently, however, one had come to think more about those mechanisms of religious thought that could lead believers, especially the young ones, to turn to intolerance and terrorism. Tariq Ramadan, for example, said that it was time for self-criticism, after blame had always been laid on the others' doorstep. The West was not responsible for all the suffering of the Arabs. The question remains as to how future generations may be educated in a true culture of peace. Moreover, the Palestinian conflict cannot be overstated, says Gaudeul. This conflict had caused serious riots in France; synagogues had been attacked as well as Jewish schools, cemeteries and individuals. The counter-demonstrations had also led to violence. The Catholic Bishops' Conference had been working to improve relations with the Jews, but this had apparently led to a negative reaction on the part of those who wanted to talk about the socio-political dimensions of a just peace in the Middle East. Meanwhile there were Jews who appealed against things that would make Israel lose its soul, and also young Muslims who vigorously opposed violence in their own ranks. Dialogue was urgently necessary in the life of society, and the problems of living together had to be given priority over theological questions. Gaudeul sums up by saying that Islam in France is no longer an immigrant religion, but a religion of French citizens or of foreigners in the integration process. There was theoretically complete freedom of religion, but in practice there were repeated clashes between public order in France and lifestyles and cultural options deriving from other countries. Only a few decades before, Islam had offered opportunities for cultural friction between the French and the Muslims, but now it was increasingly a question of faith and universal ethical values, which took place entirely within the context of French culture. Potential moments of tension are often a symptom of growing integration. The biggest problems are at the level of practical religious practice. In anticipation of the centenary of the concept of *laïcité* in France, Gaudeul concludes that laicity is now less a question of ideological conflict than of an opportunity for a positive exchange between all those involved.[58] *Thus this last situation report is quite different in character from the first two, but it also ends on a positive note with respect to the future.*

2.6.3 Monitoring religious freedom
2.6.3.1 Who can determine the boundaries of a religion today?

The situation reports almost inevitably raise another important issue: who is actually monitoring the observance of religious freedom around the world, and how? One may say first of all that human rights are one of the most prominent areas being handled by non-governmental organisations – and the largest and most powerful of these organisations exercise considerable influence. Recently, according to the report by Jane Dammen McAuliffe, they have discovered their particular interest in religious freedom. This is now a very complex topic: there is both negative and positive religious freedom, that is to say, not just freedom from persecution or freedom to worship and preach, but also freedom to defy or reject religion. So it is a question of the extent to which a certain tradition is willing to permit freedom in thought and action within its own self-understanding. Who decides what is internal and what is external; what is permitted to those who do not share traditional premises? Does a person of different religious persuasion suffer restrictions or even aggressive attacks? There have been major changes over the past 50 years – demographically, but also in questions of authority and authenticity. The context is global religious pluralism, but also the postmodern concern about identity. Postmodernism is characterised by an increased sensitivity to differences and the conviction that knowledge is historically and socially anchored, meaning that social contacts shape and filter human perception. There is no longer an objectivist, that is to say, absolute, universal and timeless truth. Objectivity, neutrality and impartiality were once considered authoritative, but they have since been unmasked as subjective positions. Any culture, together with the image it creates, has developed historically and is not an object that may be definitively interpreted. Every human being, even, has different identities – the individual is both complex and variable. Many individual factors shape and filter the perception of reality, resulting in "self-consciously situated scholarship"[59]. Nowadays, religious pluralism and its implications for freedom of religion are of close personal relevance to almost everyone. Here McAuliffe cites a sociologist to the effect that faith today requires inner knowledge and must be constantly renewed and brought into line with life experience. She also cites American sociologists who say that a plurality of religious possibilities increases the amount of religion consumed – although this has not yet been empirically proven. But it always comes down to the old key question of who is speaking on behalf of a religion, and a further complication is that religious rhetoric is also used for political purposes, thus leading to more and more hybrid identities: insiders and outsiders at the same time. Then reference is made to the relationship between knowledge and colonial power. Much of our science had become the accomplice of nineteenth-century imperialism, so that knowledge did not serve the quest for new discoveries, but rather supervision. To be more precise, those who were doing research came from a culture shaped by religion; they discovered parallels to Christianity which were non-existent, and created the other religions in their own image. Even the secular concept of religion since the Enlightenment was Christian. So we are multiple identities, our knowledge is a construction, a kind of cultural conversion has taken place – how are we supposed to assess authenticity and authority and weigh them up against each other? Instead of such an individual affair, Muslim-nationalist movements now want to establish "one forceful and authoritative representation"[60], says McAuliffe, for example by actions such as the destruction of the Buddha statues of Bamiyan by the Taliban. The perpetrators themselves feel that they are Muslims and declare that their actions conform to Islamic values, which was actually true in this case. The situation is similar when acts of violence against people are committed in the name of Islam. In both cases both approval and rejection are expressed by Muslims or in the name of Islam. This very fundamental and detailed introduction is meant to show that freedom of religion and its supervision is not as easy as it may seem at first glance.

2.6.3.2 American committees monitoring religious liberty

This introduction forms the backdrop to a survey of past efforts to monitor religious freedom. They were first made by religious organisations and their human rights activists who identified religious persecution in all parts of the world. More specifically, it was predominantly evangelical groups who investigated the persecution of Christians – and this then spread to the field of conservative politics. The question now was how best to combat religious persecution and discrimination. Might the situation of the victims be further aggravated by drawing attention in this way? It was decided to abide by the standards that were recognised by each nation concerned. The US Congress was then to draft an annual report that would lead to government action. In addition, a two-party commission was created as well, which should also deliver an annual report. Thus the affair is tackled from two different angles, and it is interesting to see how the two reports relate to each other. In the Department of State, an Office on International Religious Freedom is established, and an Ambassador at Large is entrusted with the task of presenting an annual report on the status of religious freedom in 194 countries. (When this post once remained vacant for a year and a half, the question immediately arose as to whether the new government was no longer interested in this issue.) The Commission on Religious Freedom is smaller, both in terms of size and in its field of work. Its first report focused on the three countries of Sudan, China and Russia, where there were particularly blatant cases of violations of religious freedom. They also criticised the State Department office for its policy. The State Department report, on the other hand, defined categories and types for the various forms and contexts in which individual religious freedom is violated. It focuses on political types and ideologies, but also on history, culture and tradition in each country and on whether religious persecution is actively pursued by governments and rulers. At the turn of the millennium, not many Muslim countries were mentioned – only Afghanistan, Iran and Iraq. It turns out that religious freedom is particularly denied to religions that do not have official recognition. Certain religions are seen as a potential or current threat to national security: the Baha'i in Iran, internal discrimination and persecution within Islam, and the oppression of the Shiite majority in Iraq. There is also the area of disregard, when laws on religious freedom exist but are not adequately enforced. For example, in Egypt the process of gaining permission for the construction of a church building still resembles an obstacle course, and in the Moluccas 2,500 Christians were killed in sectarian conflicts. Then there is the area of legislation and politics, which applies to Israel, for example, where Arab citizens (who are mostly Muslims) do not have equal access to education, housing and the labour market. Finally, according to the US State Department report, there is still false stigmatisation, which they think affects Jehovah's Witnesses or Scientology. It is also interesting to see which sources have been used in compiling the report – the Office of Country Reports and Asylum Affairs, reports from US embassies, non-governmental organisations, religious organisations and individuals. These also include, for example, a meeting of the United States Conference of Catholic Bishops on international politics, or a visit to the Vatican to discuss issues of religious freedom. The second report already mentions improvements by country. This shows that in the current situation of instantaneous and global communication, where governments no longer control information, public opinion has the power to effect change. As far as Christian-Muslim relations are concerned, Egypt is cited as an example, where more Copts have come to occupy important party positions, and Christians in general have more representation in public and political life. In 1999, there was a decree by Mubarak that all places of worship should be subject to the same building regulations, which has clearly facilitated the church repairs. The government's response to sectarian violence against Christians has also improved. In France, Islam has become a state-recognised religion and it is expected that this will also lead to state funding for the construction of mosques. In Greece, the construction of a mosque with an Islamic cultural centre was permitted for the first time in the modern age, and it is planned to remove the religious affiliation from passports. In Kuwait, the Vatican has been granted a permanent representation, which is also seen as a sign of more tolerance for Christians. Qatar has given permission to build a first Christian church, although previously only Wahhabi Islam was allowed to be practised in the country. Here, too, it is clear that the criticism in the other report has led to reactions.

All in all, McAuliffe considers the State Department report valuable because it is truly comprehensive and global, it heightens attention in many areas of government for the demands of religious freedom, it makes public denouncements and leads to sanctions and other actions. (The 2002 report, for example, notes that progress has been made on the situation of Christians in Indonesia, Turkey, Egypt, Kuwait and the United Arab Emirates, as well as significant progress with regard to religious freedom in Afghanistan.)

2.6.3.3 The question of balance in representation

The second report, drafted by the United States Commission on International Religious Freedom (USCIRF), draws on the reports of religious groups and non-governmental organisations and is useful in order to criticise the methodology of the other report, for example in the evaluation of sources. When the cases are simply named one after the other, it may well be that the most important cases are overlooked. The respective context is naturally also important, i.e. how long and extensively a religion has been present in a particular area, or whether the state also interferes with other rights and freedoms. But even terminology has implications, or the question of whether religions are perceived as monolithic. In effect, international norms and treaties are needed, because there is a pressing problem of religious freedom all over the world. Of course, it is difficult to reconcile a moral initiative with foreign policy, and the accusation followed promptly that the definition of religious freedom was based on a Western, American understanding of religion. Of course, it is always dangerous to define a hierarchy of human rights. However, it is helpful if the discussion on religious freedom is not given a bias, for example if is stated that in countries such as Sudan, China and

Russia persecution affects both Christians and Muslims. In Russia, for example, there are constant reprisals against Protestants, Catholics and Muslims in the population as well as against missionaries from abroad. It must be generally taken into account that in matters of religious freedom the emphasis always lies on situations of restriction and oppression.

2.6.3.4 Other monitoring institutions and their particularities

The State Department's report is still the most detailed, but there are other organisations that produce similar reports, such as the International Helsinki Federation for Human Rights, which starts with the Helsinki Final Act and deals with the OSCE member states. Its report is closest in type to that of the US State Department, be it only because it is appears annually. One focus of this report is religious intolerance and all matters of religious persecution and religious freedom in general. For example, the European Union had organised a conference on legal and practical challenges in the field of religious freedom, in which issues of the recognition of religious communities were dealt with together with the question of how to recognise legitimate restrictions on the activities of religious communities, for example on the grounds of public safety, order, health or morality. The discussion also turned to proselytism and blasphemy, because they affect the right of others to religious freedom. In addition, there is a further religiously based organisation, the Institute for Global Engagement (IGE), which was founded by the first Ambassador for questions of Religious Freedom, Robert Seiple, to offer additional criticism of the State Department's report from a non-governmental position. This criticism is evidenced by the fact that countries such as Saudi Arabia and Turkmenistan are added to the list of negatively rated countries (which raises the question as to whether the US State Department had left them off the list for political reasons). Then there is also a Canadian organisation created by the Oslo coalition on the occasion of the 50th anniversary of the Universal Declaration on Human Rights and funded by the Norwegian government. Their reports are sent to 25 states, which send their responses back. The main topics are religious extremism, discrimination against vulnerable groups such as women and minorities, and abuses of blasphemy bans. They also produce texts on issues such as the registration of religious organisations, the rights of parents and children, new religious movements, proselytism, tolerance and understanding through education, interfaith dialogue, joint religious initiatives on common issues, and conflict resolution initiatives. All these publications are described as being very helpful. Finally, there is another organisation that was founded in 1941 by Eleanor Roosevelt and today also criticises the reports from the State Department. In particular, it has a very helpful list of the key criteria for its own reports. It is the United States and the United Nations which have the most comprehensive monitoring abilities, and in principle they are generally in agreement as to which countries violate religious freedom most severely. However, there are differences in tone, for example with regard to Saudi Arabia. When making further comparisons between the various reports, it is striking that the UN report also presents the communication with the respective states and their responses, that the two American reports classify the violations into various categories, and that it is only these two reports and those of the Helsinki Commission that are published annually. The UN report repeatedly reproaches the State Department because its report only occasionally led to actions, and because these actions bore no relation to the seriousness of the infringements or to the worsening of conditions. In order to make the similarities and differences clearer, the article contains a survey of the various reports using selected case studies.

2.6.3.5 Saudi Arabia with or without political considerations

As one example, the UN report mentions two relatively spectacular cases in Saudi Arabia – one person was arrested on suspicion of witchcraft, and a singer was sentenced to death for having compared his suffering to that of Muhammad. This case was not to be found in the State Department's report, but it did state that there was no religious freedom in Saudi Arabia, as the public exercise of non-Muslim religions was forbidden. There had been a whole series of cases in 2002 in which mainly Christians had been prevented from practising their religion. The report explicitly emphasises that it made no comment on the efforts non-Muslims had to make in order to organise their religious practice whilst concealing this from authorities, neighbours and employers, nor on the fact that there was no viable way to complain officially about abuses and that there was no independent supervision of government action. The report would also not say anything about the informal pressure intended to dissuade migrant workers from making such complaints, or about the educational system that reinforced some of the prevailing intolerant attitudes towards non-Muslims and the Shiite minority. Overall, however, the situation was similar to the previous year. The UN report criticises this inadequate description of the extent to which the religious practice of non-Muslims and Muslim minorities is restricted. For two years, the United States had been requested to declare Saudi Arabia to be a country warranting special interest (a sort of standard term in cases of serious violations of religious freedom), but nothing of the kind had taken place. In 2004, Saudi Arabia was then classified in the same category as Eritrea and

Iran. This is particularly noteworthy because the description otherwise remained practically identical.

2.6.3.6 Turkey – which minorities are actually recognised?

The next example chosen is Turkey. The UN report mentions that here the conflicts concern places of worship. A directive of August 17, 2001, placed severe restrictions on the places where Protestants, Bahai, Jehovah's Witnesses and others are permitted to worship. Turkey also confirmed this, but emphasised that the administration was not intending to close such places which already existed, and had in two cases even allowed empty church buildings to be used again. The US State Department's report mentions some restrictions on religious groups and expressions of religion in government offices and state institutions, for example in universities. There was also occasional harassment, such as imprisonment for alleged proselytism or for unauthorised meetings. In a number of cases, Christians were not allowed to proclaim their faith or worship in private homes, but the majority of restrictions concerned Muslims. Some people were even said to have lost their position in the civil service because they were suspected of being involved in anti-state or even Islamist activities. In the eyes of the National Security Council, fundamentalism is a threat to public safety. In particular, advocacy of Islamic law is seen as a threat to the democratic secular republic. Thus, Fethullah Gülen is charged with attempting to infiltrate the military, and no explanation is given as to the anti-terrorism law behind this. But positive developments are also mentioned: the Ministry of Religion has ordered that women should also be allowed to participate in daily prayers, at Friday prayers and funerals. Men should not use the Quran as a pretext for domestic violence, and the advancement of women did not violate the spirit of the Quran. The Civil Code of 2001 gave men and women equal status, so that now women may also decide in marriage matters. On September 11, 2001, the Ministry of Religion issued a declaration to be read out at Friday prayers in all mosques, stating that there is no Islamic justification for any form of terrorism. The secular majority opposes the interference of even moderate Islam in politics. It is also mentioned that Islamist newspapers regularly publish anti-Semitic material. The Helsinki report, on the other hand, takes a rather different view of the situation in Turkey, giving the country a poor record on human rights in 2001, especially with regard to national minorities. It takes particular exception to the definition of minorities, which only accepts those minorities named in the Treaty of Lausanne in 1923, meaning Jews as well as Greek and Armenian Orthodox Christians. Thus the government was officially unable to fulfil the Copenhagen EU criteria on the protection of minorities, since most minorities, such as the Alawites, the Assyrian Catholic Christians, the Kurds and the Roma were not recognised. Therefore, it would have been helpful if the American report had first defined the term minority.

2.6.3.7 Iraq – how political is the question of human rights?

The third country taken as an example is Iraq. For nine years, the country had rejected requests from the United Nations for inspection visits, so that information could only be gathered from interviews with newly emigrated Iraqis, from opposition movements with contacts to the country, from interviews from other sources and from published reports. The reports from the UN and the US both agree regarding the situation of the Shiites in the country, where the US observes a brutal campaign against the Shiite majority. In assessing the situation of Christians, the reports come to somewhat different conclusions. The American report saw an attempt to undermine the identity of the Christian minorities (Assyrians and Chaldeans) and the Yazidis. The UN report, on the other hand, spoke of a process of Arabisation, but said that Christians were otherwise generally free to practise their religion without hindrance. Although the US had access to the UN report, this fact was not mentioned. This raises the possibility of selective use of information in this case. For example, there was no mention of the fact that a UN rapporteur had meanwhile been allowed to enter the country, nor of his findings. This gives rise to the suspicion that reports from the US State Department are connected with the aims of American politics.

2.6.3.8 Germany – recommendation for corporate body status for Muslims

As an example of a country outside the Muslim world, Germany was chosen, which had been visited by a special rapporteur of the United Nations from 17–27 September 1997 as the only European country. The UN rapporteur recommended that Islam should be given corporate status in Germany. The US report mentioned the discussions about Islamic religious instruction in schools, about imams in the army and the construction of mosques, but said nothing about the aftermath of 9/11, in spite of raising the topic of developments in society, such as increasing criticism of the Israeli government and the complaints by Jewish groups that the print media were pro-Palestinian. The Helsinki report noted that the anti-terror laws restricted the groups entitled to seek asylum in Germany. People who supported terrorist organisations or used violence for political ends would be excluded. The American report did not mention this – maybe because the situation is too similar to that of their own country?

2.6.3.9 The struggle against terrorism and the question of who is a terrorist

This leads on to a separate section dedicated to the specific situation of the struggle against terrorism. The UN report warns against overlooking violations of religious freedom when a state supports the fight against terrorism. At the moment, the United States only has one overriding goal – the protection of its own citizens, of national security, and democracy around the world. This was at least partly achieved by combatting terrorism and its supporters. This would offer a unique opportunity to use relations with countries such as Pakistan, Uzbekistan and Afghanistan in order to encourage their governments to make the necessary improvements for the protection of religious freedom. But one of the biggest problems in the fight against terrorism was that no precise definition existed for terms such as terrorism, extremism in the name of religion, radical or militant religious-extremist groups. That was already affecting countries like Pakistan or Turkey. Such definitions were particularly indispensable when religious rhetoric came into play while working internationally to fight terrorism. The US was asked to clarify the connections between actions to combat terrorism and the violation of human rights.

2.6.3.10 Combatting terrorism and the American perspective on religious freedom

The next State Department report made it unmistakably clear that for the US a policy in favour of religious freedom is a means towards fighting terrorism. 11 September 2001 had shown that people could and would use religion for terrible things, manipulating and destroying other people as mere instruments. That was not new, but after the Cold War some scientists had predicted that religious differences would probably cause greater conflict between cultures – a very clear allusion to Huntington's thesis of the clash of civilisations. So if the commitment to religious freedom were to be successful, then it would be one of the most effective and persistent antidotes not only to religious discrimination and persecution, but also to violence in the name of religion and the potential clash of civilisations. Citizens saw religious freedom as a social good, and religious persecution and violence had no justification in their eyes. They tolerated religious differences, and many saw the practice of religion as constitutive for human freedom and dignity. The American Report on Religious Freedom even quotes President Bush here, who states that in the best case religion produces productive and charitable citizens and a stable society. To deny someone religious freedom was to deny him humanity. The right to ask the most basic questions about the origin, nature, value and meaning of human life, and then to live accordingly, was absolutely central to freedom. The Bureau of International Religious Freedom, as the department is officially called, was intended to promote religious freedom worldwide. The ambassador in charge was one of the chief advisers to the President and the Secretary of State with regard to international religious freedom. But the work of this bureau also served to promote further fundamental interests of the United States, such as the protection of other important human rights, the promotion of democracies and the fight against terrorism. The annual report was supposed to provide the basic facts on the status of religious freedom worldwide and thus become the main source of American policy in this area. The previous reports had been criticised by the governments they had admonished, but welcomed by human rights organisations as a worldwide reference work on religious persecution. Beyond this self-image, however, questions were raised about the objectivity and effectiveness of this report. For example, Pakistan is reported to have taken steps against religious extremism and militancy (whatever that is), but with mixed results. Approximately 2,000 members of suspected groups were arrested, but there were ongoing rumours that it was the party leaders – of all people – who had had protection from government circles and had been able to go underground in time. Such actions are then reported under the heading of "restrictions of religious freedom", and it is difficult to say what the actual position of the United States may be. Applause for such activities would be paradoxical at best; here again, there is no proper definition of what is actually extremist or fundamentalist. Only rarely is the question asked as to whether this is not in reality a political agenda.

2.6.3.11 Reporting on religious freedom – weaknesses and opportunities for improvement

At present, according to McAuliffe's report, there is not a single monitoring body for religious freedom without American membership, and the question is whether Americans are really prepared to sanction religious persecution. The Turkish military, for example, dismisses practising Muslim soldiers on the grounds that they are fundamentalists and thus less loyal to a secular democratic state. The American report makes no more than a reference to this. The United States is also criticised for emphasising individual human rights whilst ignoring the issue of cultural norms, and for opposing an international criminal court. A further argument against the United States is that, after 9/11, Arab-Muslim minorities in the country frequently suffered discrimination. The UN report mentions the terrorist screening used by some US airlines, as well as negative stereotypes of Islam in the media, hate crimes and also discrimination at places of workplace. The complaints received by the United Nations are mostly about discrimination on ethnic or religious grounds. Supposedly, arbitrary arrests have been made, people detained for too long, offered no legal counsel and not permitted to make contact with anyone. The United States, however, denied having acted on religious grounds. The UN report notes a tension between the fight against terrorism and democratic processes. It would be necessary to strengthen the credibility of the system for the protection of human rights. Of all the reports that McAuliffe was able to investigate, it was only this UN report which gave a definition of extremism, warning against it and demanding that it be fought with a minimum of common rules of conduct. The overall implications McAuliffe draws from her comprehensive and very detailed study are that governmental organisations may not be adequate to handle interreligious problems, since self-interest in politics is tending to increase. Conversely, non-governmental organisations have a problem because they lack the power to enforce their proposals. McAuliffe sees an ever-increasing need for religious bodies to allow others to worship freely, while it is even more urgent that they permit different interpretations of the dominant religion in a particular country. She quotes as examples Greece, Islamic laws in various Muslim countries and evangelical movements in the US. Many of the violations of religious freedom are the result of cultural conditions or religious laws in constitutions. For example, under classical Islamic law, non-Muslims are (at best) second-class citizens. The situation in Europe and the US may be seen as a mirror image: Muslims do not have the same religious freedom as their Christian and Jewish fellow citizens, and are not *de jure*, but certainly *de facto* second-class citizens. Even in the West, there is no clearly defined consensus on what religious freedom means and what it does not. In the United States, for example, there is no internal control body for religious freedom, apart from the state courts and the fact that religious freedom is enshrined in the Bill of Rights. On the other hand, there is a European Court of Human Rights that actively examines the situation of (mostly non-Christian) migrants in Europe. The EU Constitution should also grant religious freedom. Monitoring of religious freedom even on a religious basis is seen by McAuliffe as a corrective for a political monitoring (which is alien to many countries). Where there is no affiliation to a government, there are also no constraints of diplomacy or foreign policy. On a positive note, she says that a new stage had already been reached by the production of documents and demands for improvements.[61] That is undoubtedly correct.

2.6.4 Religious freedom from a religious point of view
2.6.4.1 What has changed for humans?
The last contribution by Jean Marie Gaudeul seeks to take a look at the issue of religious liberty from a religious, but at the same time more personal, perspec-

tive, in order to find out what is actually happening in the conscience of a believer who demands freedom of choice such as did not seem necessary a few generations ago. Some tended to say that this was simply a pretext, and that people actually wanted to free themselves from their obligations to God. Some even understood religious freedom as liberation from their entire religious heritage. There were conservatives In Christianity and Islam who denounced a certain representation of human rights as anti-religious. But at the same time, there was a new awareness among Christians and Muslims around the world that comprehended religious freedom as the key to a better service to God. There were still areas which only received world news in a muted form, as it were, but changes turned out to be all the more violent the longer they had been postponed. Developments were simply accelerated, among them the new ways in which religious, political and ideological messages were received. In the last century, a kind of concentration-camp universe had existed in many parts of the world, in which laws could be dictated to the conscience of the inhabitants. This, together with the seductive power of advertising and the development of globalisation, had made the danger of conditioning obvious even to those who saw less clearly. Experiences and values were often incompatible. Everything had become relative, even holy scriptures, because other holy scriptures had existed at the same time or earlier. Thus they were perceived as literally relative, that is to say, either comparable to one another or not. Even the strongest believers could no longer trust in a message without consciously choosing it. A protected society was artificially excluded from the rest of the world by bloody violence. But that could only lead to unbelievable suffering and a tragic backwardness, both from a cultural and material point of view, before the whole system would implode under the pressure of life. In an environment of this kind, the desire for religious freedom had grown instinctively and inwardly, and this had been recognised by the Second Vatican Council. The historical roots had lain in Europe and North America, fuelled by the desire for freedom from absolute monarchies and also from the misuse of religious institutions. It came to a political movement which at the same time rejected religion. This simultaneity paralysed political thought on this point for a long time: was it possible to accept freedom which rejected submission to God? It was not until the Church once again became the target of persecution, as it had in the beginning, that it rediscovered the fact that freedom of religion may be claimed in order to be more faithful to God. However, a mere idea is not necessarily put into practice. The implications of religious freedom, especially among Christians and Muslims, must be deduced from the behaviour and the hopes of those who live by this freedom of religion. Many non-theological factors have an influence on the self-image of the believers and their relationship to religion: urbanisation, the resettlement of large parts of the populations, globalisation. It is increasingly up to individuals themselves to decide about their faith and their actions; it is no longer the case that society and institutions make such decisions for them in advance. Gaudeul quotes specific Muslim and Christian examples. According to their conviction, everything else would be immature faith, and thus not the will of God. The emphasis lies on inner conviction, which involves a certain privatisation of religious affiliation. This can also result in a decision to follow a different faith, or none at all. De-Islamisation can, for example, also be observed as a real fact. The believers are leaving slowly, on tiptoe, so to speak, their heart is just no longer in their faith, even if there are still various factors holding them within a civil religion – just as it was in ancient times. That is why there is suddenly so much interest in conversion and apostasy, says Gaudeul.

2.6.4.2 Tensions within Islam

Turning a blind eye to such freedom to believe better, however, would lead to misunderstandings and disastrous stubbornness. Muslims themselves say that the transition to a tolerant and pluralistic society cannot take place without pain and tragedy. The greatest tensions existed at the universities, and the typically Islamic student was produced by those who are indeed no longer Islamic. It was unavoidable that minorities with complexes were created, which were either seeking proud isolation or liberation by force. Gaudeul, however, sees this as a general phenomenon in a religiously authoritative milieu.

Then he once again provides examples from both religions for the democratisation of religious institutions. When lay people gained maturity, as it were, and set the tone, then religious institutions would discover anew that they were to serve the faith, not the other way round, and that religion was made for humans, not humans for religion. That could also be an opportunity for Islam, in that God would be honoured in humankind and old values revived, such as faith, goodness, greatness, solidarity, respect and tolerance. Competence in religion comes from experience of life in the modern world, whose doubts, questions and problems are perceived by the grass root believers but often remain unknown to religious scholars. But the message has to be appropriate to their situation. Even Muslims have the feeling that the Quran's ethical vision and metaphysical origin have been lost. In the wake of Islamisation, there had indeed been a revival of many laws and institutions of medieval Islam, but also an increasing ethical deterioration. It was no longer possible to continue according to the traditional patterns. Religious and cultural pluralism introduce new ways to deal with the differences between people. In many areas, it is no longer possible to build one's life according to traditional religious rules dating from a distant past, and no nation or individual can go into total isolation in order to find answers to these questions.

2.6.4.3 Passing on the faith – not like it used to be
Religions have to accept the obvious truth that their realms, whose borders had so long been immovable, are now crumbling both inwardly and outwardly. Their only option is to be prepared to accept change. Each faith must ask itself where its place and task is to be found in the new world order and must seek factors of renewal within its own traditions. Religious communities would now begin to speak a second language of moral concern reaching beyond their own circle. But rules or values regulating behavioural details and religious rituals cannot form the basis for something new. On the other hand, a certain convergence in ethical values is emerging between all religions. For example, one Muslim says that he prefers general freedom and justice to an Islam which does not respect these values. For him, the yardstick for authenticity of the faith and its loyalty to the original message consists in respect for religious freedom, and he quotes a fourteenth century theologian in this matter: God's law and religion is to be found wherever there are signs of truth and justice. All else is not existential. Gaudeul judges that the truth of itself is able to win over intelligentsia as followers, and many Christians would think like Gamaliel in the Acts of the Apostles, having more faith in God's truth than in condemnations and fatwas. But all this calls religious leaders into question, and also the way in which they served and defended the message. (Whereby it is doubtful whether the content of the message can actually remain completely unchanged if it is served in a different way.) It is precisely the religious professionals who are irritated by this. For centuries, the various religions have all followed a very similar path: they demanded conversion, and then rapidly created a way of promulgating the faith through the families, in the belief that faith would be passed on from parents to children by upbringing. Thus Islam was organised around the patriarchal family. In this way, people belonged to a religion in the same way that they belonged to a family or a nation. There was political pressure, and personal decisions were prohibited or prevented. To change one's religion amounted to betrayal. Down to the present day, one can find concrete examples of parents who want to pass on certain values which are no longer important to the children. The leading authorities of all religions should urgently address the fact that from one generation to the next they are losing a considerable number of believers who are only counted as Catholics or Muslims on paper, says Gaudeul. This leads to the question of how faith really is passed on, and what kind of upbringing is needed, as adolescents moving on to adult life become increasingly aware of their freedom. In any

case, the risk inherent in the path chosen up to now was that the individual would direct his freedom against those around him and against the faith that they were trying to convey. Gaudeul implies that there is an ample number of people in the religious milieu who have had enough of religion and do not want to hear about it anymore. It is true that a child needs rules, a language and culture, family and social customs. Certain facts of religious tradition belong to all these. Early education comes from authority, but that cannot simply be imposed. Excessively exercised, it merely leads to rebellion.

2.6.4.4 Believers in the modern world – how can this work?

Many parents had recognised this and would themselves agree that a child's wishes and decisions can be expressed at a very early age, whether or not it wants to believe or to practise this belief. Up to what age would one be able to compel them – and risk turning them into atheists? Gaudeul points out that young people are constantly bombarded by other proposals coming from their friends or the media. If they have not learned to choose freely, they will fall for all kinds of suggestions. The real problem for religious teachers is to help young people to develop their capability to make an adult decision in the midst of this tumult, as did the first converts. The role of the family differs from previous times and is perhaps infinitely more important for this reason. According to Gaudeul, if it is true that nowadays there is a crisis of the family, it may arise from the fact that many people are unable to resist conditioning that undermines the values of conjugal love and marital fidelity. Gaudeul prophesies that in future only those who have been prepared by their education to be nonconformists will become or remain believers. Only those with adequate inner freedom would be able to withstand the influence of a world whose techniques of indoctrination are steadily becoming more sophisticated and efficient. In Islam, this attitude may be most clearly observed among the Sufis, who maintain that the transition to adulthood should enable every person to make a conscious choice; even those born as Muslims should confess the faith of their own free will, as if it were for the first time. Thus it is not a question of passing on the faith, but rather of recommending it. The well-known religious coercion, be it brutal or subtle, does not lead to the freewill belief required by the word of God. This raises the question of new forms of socialisation. In Christianity as in Islam, there is an almost universal, general protest against the clergy or *ulama*. For example, the imams are criticised for being too weak in the spiritual realm, and since spirituality seems to be the most popular topic, young Muslims in France are critical and ultimately dissatisfied. A similar phenomenon exists in quite traditional Muslim countries. According to a Christian testimony cited by

Gaudeul, conversations are important because they serve as guideposts for future life. A fellowship founded on common faith in Jesus meant compassion, love, and sharing everything, in good times and bad. It was not the building which was important, for one could pray in all places.

2.6.4.5 A believing community is good, but it has to be convincing

According to Gaudeul, the desire to belong to a believing community certainly exists, but unity could not simply be enforced authoritatively or by social pressure. If people are no longer so gullible, one is obviously tempted to manipulate a group in a different way. It is particularly dangerous when this temptation consists in welding a group together by hatred against another group, which is contrary to the positive content of the faith. The positive contents are essential; despite growing cultural equality, people have the feeling that they are more and more alone in the face of great existential decisions and less and less prepared for the trials imposed by life. Birth and death, sickness and suffering require answers, and these no longer result simply from social conformity. Fashion and advertising are ever more intrusive, while family, upbringing, schools, churches and mosques can ultimately only offer options and values. At critical times, people have to make their decisions alone. Whenever some claim is made, the individual can always hear other voices in the background, says Gaudeul, other suggestions, other revelations, other sacred scriptures. Today people are sceptical of doctrines, ideologies, and proclamations, but at the same time they yearn for testimonies, confidence-inspiring messages, and anything else that can show them how more or less well-known personalities make decisions that affect their lives. Gaudeul derives from this the conclusion that in all major religions doctrine must be evident in testimony, otherwise it will not work. Even Paul VI had said that his contemporaries preferred to listen to witnesses rather than to teachers, and if they listened to teachers, then only because they were witnesses.

2.6.4.6 Objectivity and subjectivity of religion – a shift in emphasis

However, religious authorities are accustomed to delivering statements in the name of a timeless, eternal, and absolute Christianity or Islam. Yet the religious leaders who make these statements are perfectly ordinary people. And thus, as even an Iranian academic has expressed it, one has to distinguish between religion itself and religious interpretation. The former, for example a revealed text, remains constant, while our interpretations of this text vary over the course of time. Thus we only receive interpretations of religion: Islam consists of a series of interpretations of Islam, and Christianity is a series of interpretations of Christianity. And because interpretations are always historical, the element of historicity always belongs to them of necessity, even though all these interpretations have to be justified. Gaudeul mentions that the risk of presenting a human interpretation of the message as a divine message must be avoided, for it would be idolatry, strictly speaking. Objective truth can only be obtained together with all the subjectivities accompanying it. The privatisation of the faith is indeed a very real phenomenon. One can observe a kind of religious zapping, so that decisions are often made according to the strength of the emotions that are aroused. This leads to the birth of a new type of religious person, one who is more spiritual than ritualistic, more mystical than dogmatic, and also more uncertain. This is taking place at different rates in the various continents and countries, but is nonetheless a general phenomenon. The old-fashioned unanimity has given way to an à la carte-religion or religiosity, in which the ingredients of the faith are re-composed according to the needs. Both effects are to be found in history: there were times in which ties broke up, and times when people came together around specific messages and institutions. But all that would not change the fact that the religions were obliged to find a new way to proclaim their message. To serve humankind does not mean to enslave them. Their longing for more religious freedom should not simply be understood as refusal to believe in God or to submit to him.

2.6.4.7 To whom God has given freedom, from him should it not be taken by humans

And finally, God himself took the risk of addressing every person, in order to receive a personal response. Thus the religious specialists on both sides are servants, who may sometimes be approached for their advice or knowledge, but whose interference is increasingly perceived as an indiscretion. It is a matter of God's extraordinary freedom, who leads each person according to his will (a hidden quotation from the Quran). And in concrete terms, it concerns a new spirituality, or maybe rather a new anthropology, because while the influence of religious authority in guiding conscience is diminishing, its advice in the modern debates is being increasingly sought in order to remind us of the meaning of human existence in face of frequently confusing developments. Christianity says that the only sacrifice that God demands from his creation is love, and that consciousness and freedom are constitutive to the nature of humankind. God does not manipulate us by attracting our automatic attention. The Most High, the Almighty, takes the freedom of humans seriously. Humans are respected, also in their ability to say no. Gaudeul asks at this point whether it is on this account that angels express their fear in Sura 2:30 as Adam is created. God proposes to us that we obey his commandments – so how can we attempt to enslave those whom God himself wished to address as partners, as his representatives? God seeks the heart of every single person. I have to reply to him at the last judgment. It is impossible, even unthinkable, that I should prevent someone from giving his own free answer. Any attempt at manipulation here is abhorrent and a tragic misunderstanding of the divine mystery. In fact, religious leaders and teachers are tempted to apply more subtle methods of conditioning or guidance. But believers are becoming increasingly aware of this. And the fact remains that a forbidden fruit is more seductive than one which is allowed. God himself causes the immense yearning for freedom to spring up in human hearts, and freedom which exists should not be suppressed by family upbringing and religious preaching. Love does indeed have its own demands and its own logic. In conclusion, Gaudeul emphasises that all these considerations were not new, but originated from the basic revealed texts and had only gained a new force among thousands of believers. They might also contribute to a new sense of kinship between Christians and Muslims in the face of people's struggle against oppression of any kind. In any case, a page in the history of the world and of religions would now be irrevocably turned. One might get the impression that many people were tossing to and fro in a maelstrom of beliefs, and that the privatisation of faith was in danger of leading to a rejection of revelation. But precisely on account of the reality of the dangers, God would be faithful in his guidance. His final thought is, "[T]he one who wants to be free, the better to serve and to love God will find what he seeks"[62].

Thus, right at the end of a host of detailed and specific expositions, the clear picture appears that God himself has given people – both Christians and Muslims – the freedom to seek God of their own free will. Actually quite logical and simple, if only it were not so often difficult in practice.

3 Guides to dialogue

3.1 Orientations pour un dialogue entre chrétiens et musulmans

The first publication on the subject which was intended by the Secretariat to reach a wider audience was the *Orientations*, points of orientation for the Christian-Muslim dialogue. It was designated as a first step from the very outset, translated into several languages, and intended to encourage dialogue. That appears to have succeeded, for the booklet was quickly sold out, and before the appearance of the completely revised next edition, which will be reviewed in the next section, a great deal of experience in practical dialogue was gathered. The *Orientations* played a pioneering role and quickly became "historical", so that subsequently they are not even explicitly mentioned. But in view of what was initially only stated on an abstract or internal level, they certainly represent an important step towards the realisation of dialogue with Islam and above all with Muslims. The idea was (and is) to speak about people and about life, which can connect people more strongly than religious texts, and ultimately to build a human community in which everyone feels better understood, respected and loved.[63]

3.1.1 A long road from knowledge and respect to openness and sympathy

This is a clear anthropological focus, which is continued in varying emphasis and forms throughout the following chapters, whereby the priority lies at first on the fundamental attitude within dialogue. This was seen to be more important than the actual topic of the dialogue. It begins with the simple recognition that one cannot engage in a dialogue with religious systems, but only with people who are embedded in a religious horizon of experience, but are more than just objects of study. (Especially with regard to Islam, the love for the object of study tended to be much greater than the love of real Muslims, even among experts.) It concerns the existential level of someone dealing with concrete problems in their life. That is why new problems are more interesting for a dialogue than the old ones, because one is talking about present-day or future people. But the relationship between one person and another automatically presupposes a third aspect, namely the relationship to God underlying every dialogue. It is particularly in dialogue that one is constantly challenged to bear witness to the primary source of one's being and to formulate it anew. But the text also states quite specifically: Christians should take the first step without asking whether this is sensible from a human point of view. The attitude of the heart was decisive, welcoming and truly accepting the partner in the dialogue – precisely because hospitality in the Muslim world enjoys such a high and religiously significant prestige. To know a person meant to know a friend and to recognise their good qualities and hopes. This required a change in mentality for Christians, indeed a conversion which would take a long time. This included well-founded knowledge of the other person's religion. That would go beyond the acquisition of expert knowledge and would have to include the expectation of being able to learn something personally from the dialogue partner. The authors also emphasised that sympathy based on friendship and respect is only possible if one believes that a Muslim can approach God in prayer. Muslims took the faith aspect for granted, but they were quick to suspect proselytism, which is why it might be appropriate to suspend religious dialogue for a while without concealing one's Christianity.

3.1.1.1 The basis: creation and shared values

At a very early stage, it is evident here how the clear hierarchy of truths, as represented by the Second Vatican Council, is connected with the image of a common road to the truth, to God. The goal of the dialogue which has to take place is formulated quite anthropologically in this way: "l'acceptation pacifique et joyeuse de l'autre tel qu'il est et (...) l'enrichissement mutuel pour une meilleure affirmation de la Vérité et du Bien dans l'humanité, tels que Dieu les a révélés dans sa création."[64] *The basic attitude is therefore very clear and is partially explained in detail, but the formulation of the fundamentals is not very precise, extending to no more than a general reference to creation and common values; this is fairly typical, as may easily be established by taking a look at the basic texts.*

3.1.2 Humans in community – a perspective expressed with pride and conviction

The same may be broadly said of the following chapter which deals with Islamic values, mainly in order to prepare a future Christian dialogue partner for these conceptions. A Christian who does not share these values should nonetheless beware of committing a faux pas by offending a Muslim counterpart by a thoughtless remark which might be considered derogatory. Sometimes common values are visible (at least partially), sometimes the reader gets a glimpse of the Muslim perception of himself and humankind. This is the case at the very beginning, which deals with Islam as a community, with the unity of the Islamic world, as reflected in its strongly uniform view of humankind, its fate and relationship to God. The Muslim was only fully aware of himself as a member of the Muslim community, in which strength, peace and the richness of human dignity were to be found. In the depths of the heart the Muslim was proud of this affiliation. This conviction made it possible to explain much of the individual and collective behaviour of Muslims and Islamic nations. Under the influence of nineteenth-century modernists and modern societies, large numbers of Muslims had come to acknowledge that every honest and upright person conscientiously living according to a faith other than Islam is a believer. Finally, there is reference to the fundamental theological and anthropological problem of the relationship between God's omnipotence and the freedom of humans. By emphasising God's unique power to act, one makes his righteousness questionable, for how can a person be held accountable for something for which he is not responsible? Conversely, if personal freedom and responsibility is emphasised, then God's omnipotence is called into question. It was held that at present the emphasis in the Islamic world lay on the freedom and responsibility of humans, so that there was potentially a chance for Christian-Muslim dialogue to overcome this fundamental contradiction without detriment to the mystery. One last important area in this context is that of the prophets and messengers. In this regard it was important to emphasise the person of Abraham, whose attitude of faith and obedience to the paradoxical will of God is revered by Christians and Muslims alike. One decisive aspect is that the Muslim tradition down to Mohammed Abduh speaks of three ages of humankind: the childhood with Moses and the Torah, which issues commandments just as a teacher would command children; the period of youth, in which Jesus and the gospel spoke to the heart and appealed to the feelings; and finally adulthood with Muhammad and the Quran, meaning the age of Islam, the religion that is consistent with reason and written in the heart of every person born as a Muslim.[65]

This is indeed the first passage which clearly describes the basic features of a Muslim conception of humankind and especially the conviction, pride and love with which they are advocated. However, there is no further mention of the fact that is precisely this conception of humankind and the conviction behind it which might also lead to problems in concrete encounters.

3.1.2.1 One Muslim is not like another – the delicate question of group formation

The *Orientations* also deal with the diversity of Muslim dialogue partners, starting with the dominant character of Arab Islam and the clear overall tendency towards uniformity and universalism. All kinds of differences, including those over religion, were seen as family matters that did not concern non-Muslims. This attitude is also shared by the *Orientations*. The great majority of the Sunnis had no direct contact to Islamic sects and were unaware of their existence (this was at least the case in 1969). Wherever they had become conscious of the schisms, they found them painfully dishonourable for Muslims, so that it is recommended that Christians should not regard this as grounds for satisfaction. In addition, one should take care not to invite representatives of such groups to dialogues or to take sides at all in such questions. Anyone who invoked the Quran and the Prophet Muhammad was a Muslim from the Christian point of view. Particular attention is paid to the attitude of different groups towards modernity. In general, the Western world and culture functioned as role models, and this is also seen as an advance on the path of dialogue: through the convergence of cultures, people would also converge on a spiritual level. The *Orientations* distinguish between de-Islamised Muslims with Western culture, traditional Arab scholars, Muslims from the poorer classes and those employed in the modern working world. The latter were interested in the progress of their society, whereby they were confronted with resistance and problems of a moral, religious, sociological or psychological nature. This is where a chance is seen for a true dialogue that would also go into basic spiritual values such as the dignity of the individual, freedom, and concern for the common good. Summing up, it is said that truly atheist, areligious Muslims are rare. Therefore one should always see a Muslim as a believer who would be able to sympathise with Christian believers and engage in dialogue with them about the fundamental values of human life and society. In addition, it is pointed out (apparently for a good reason) that one should not categorise people, but rather approach them as human beings, not as representatives of any particular group.[66]

For a sensitive reader, the entire text represents a walk on a narrow path. On one side there is the summit – a grandiose new encounter with people of Muslim faith, which has apparently been meticulously prepared and should be implemented in exactly this way. On the other side a chasm yawns, consisting of the mere fulfilment of external requirements, because the necessary inward conversion can ultimately not be worked, but only entreated.

3.1.3 Christians struggle with themselves in dialogue

The following chapter confirms this in a striking way. To start with, it states that dialogue is rather a fight with ourselves "qu'un affrontement de l'autre (le mot affrontement étant pris dans le sens de la philosophie du dialogue)"[67], but in conclusion it says that dialogue is not only a cry from the heart, but above all an act of intelligence. A basic tension clearly becomes apparent here: what does it mean to be a person taking part in dialogue, how can one accept such a role, or refuse it? This problem is not directly covered, but just appears on the margins of quite practical questions as a kind of backcloth or generalising resumé. Indeed, many important matters are mentioned almost casually, for example the Council statement advocating a departure from past enmity between Christians and Muslims in favour of a commitment to social justice, moral values, peace and freedom in the interest of all people.[68] This had been the most frequently cited statement in the Arab press and had aroused a deep interest among Muslims, who saw in it the indication of a new era. It is in fact a question of mistakes made in the past and the best ways of avoiding them in the future. This begins with a recognition of the deep-seated bitterness among Muslims, who recognise a historical thread leading from the Crusades through colonialism down to the present day, continuing in the issues regarding the state of Israel and the involvement of the West in this situation (where the Church should seek humane and Christian orientation, taking sides with those who suffer the most and showing sympathy beyond mere words). This persuasion extends to the Muslims' deepest conviction that even if Christians valued and loved Muslims as individuals, they did not value and love the Muslim community. Therefore, Christians should be particularly interested in the religious and social problems of Muslim peoples and, if possible, give preference to the Muslim solutions. Moreover, it was not enough merely to love the Muslims, one also had to come to respect Islam.

3.1.3.1 Christian prejudices: from fatalism to fanaticism

The question of mutual prejudice also belongs to this complex. There is a long list of preconceptions in relation to Islam: fatalism, legalism, religion of fear, moral laxity, fanaticism, inflexibility. In particular, the question of the so-called fatalism in Islam is directly related to the image of human beings. The primary question here is the relationship which has already been mentioned between God's omnipotence and humankind's responsibility. Emphasis is laid on the predominant role played for hundreds of years by the Asharites, who placed great value on the omnipotence of God, and how the religious brotherhoods had ultimately transformed such submissive piety into passive resignation, possibly as the last refuge of the poor and socially vulnerable. However, one should not overlook the even earlier school of the Mutazilites, who even designate people as the creator of their own deeds (human freedom being a Quranic expression) and are now once again enjoying increasing popularity. A lot of attention is given to the question of Islamic fanaticism. It is pointed out that it is widely believed in the West that Islam spreads its faith with the sword and violently represses all resistance, which causes irritation to Muslims in their turn. A Muslim would reply by recalling Islam's tolerance towards the peoples of the book, who are accorded guest status in Muslim countries and nowadays treated as equal citizens, and he would finish by repudiating the accusations with reference to historical proof of Christian fanaticism. Thus the result is a stalemate from a purely human point of view. There are some anthropologically significant aspects: all Muslims are proud of their faith and of belonging to the community of the Prophet. It is therefore logical that they want to share this status with others. The laws of God and of humankind as defined by the Quran should be respected throughout the world. As for the so-called "holy wars" waged by both sides, armed conflict was considered to be no longer an expression of today's mentality. Humanity had truly made progress by recognising that acts of violence were no longer tolerable, especially when committed in the name of religion. In order to prove this opinion, reference was made to the majority of Christians and Muslim reformers who acknowledged that the struggle for the establishment of just and honourable peace was of equal, or even greater worth. This opened up a wide field for dialogue.

3.1.3.2 Difficulties on the Muslim side

Taking a look at the Muslim side, the main theme is their possible openness to dialogue, which appears to have been aroused on the one hand by the Second Vatican Council, principally among the elites who would like to talk to Christians about social, moral, cultural and even spiritual issues. But there are also other reactions, even mistrust, and it is necessary to recognise the reasons for them. The interest in mutual understanding forms the basis of every dialogue, but many obstacles to this are to be found on the Muslim side, because the Muslim believes that with the help of the Quran he already knows Christianity well, indeed even better than it knows itself. This is why many Muslims show little interest in Christian values, even if they encounter them on a daily basis and appreciate them. On the one hand, the Quran calls on Muslims to engage in dialogue with Jews and Christians, but on the other it is clear that there is nothing for Muslims to discuss. From the Christian point of view, two things appear to be evident: the dogmatic formulations which the Quran attacks can only refer to a heterodox Christianity, but in their description and Muslim interpretation they amount to a denial of the Trinity and the incarnation. It would appear to be a hopeless task to prove the opposite, and the Muslims seem to be impervious to it. However, Muslims were always joyfully surprised to discover that Christians also confess the One God. There is a clear recommendation that Christians ought to use language that can be truly understood by their Muslim counterparts, whilst demonstrating other aspects more intensely in their lives to make them clear to their Muslim friends. The authors of the *Orientations* acknowledge that this might be interpreted as mere tactic, but they disagree. Finally, they deal with the fact that Muslims are quick to apply their own situation to the Church, assuming that there is the same connection between spiritual and worldly, even political affairs that they know from their own tradition. It would be more or less as difficult for a Muslim to perceive the distinctions between Christianity, the Church, Christendom, and the Western world as it would be for a Christian to understand exactly what it means for Islam to be both a religion and a community at the same time, and to see the consequential special connection between spiritual and temporal issues. Over and above this, for the reasons mentioned above, it is hard for Muslims to grasp the freedom of Christians in matters concerning the secular order. This means that they reproach the Church for the actions of individual members of Christendom, even though they have long been forgotten by Christians themselves. It would be a good thing to adopt the position of the Second Vatican Council and explain the nature of the Church as a sacrament of unity, whose prophetic task is to call all people of good will to unity, to the advancement of moral values and to the service of God. For a significant number of Muslims, the Church was already a moral guarantor, and they appreciated its statements and activities in favour of humanitarian aid, brotherhood, and the call for peace and justice, even if they tended to understand and instrumentalise purely religious statements in a political sense. Dialogue did not mean agreement, not even a common language, but respect for the other in his otherness and understanding of what the other person was saying in his own terminology. A cornerstone of this was that the Muslim might believe that Christians were politically, or indeed fundamentally, disinterested.[69] *Once again, it becomes very clear that dialogue, particularly between Muslims and Christians, is in danger of closely linking high and noble goals, fundamental ideas and concepts with realities which sometimes seem to be quite banal and all too human.*

3.1.4 Anthropological key statements
3.1.4.1 On the interpretation of *Gaudium et Spes:* movement in society
The most important chapter of the entire presentation from an anthropological point of view is the chapter which follows, the second to last. It deals with perspectives for the Islamic-Christian dialogue and is notable for its frequent references to Council texts and encyclicals, which immediately indicates its fundamental character. In the very first sentences it says that every human action presupposes a certain conception of humanity and its relations to society and to God, and that therefore every dialogue will one day inevitably arrive at the fundamental questions of the conditions of human existence, as *Nostra Aetate* had formulated. And straightaway it moves on to *Gaudium et Spes* and the newly evolving order of things. The problems are thereby seen to be grouped around two crystallisation points: the progress of the human person and the establishment of a more brotherly society. These were the issues of the time which ultimately faced us all. First of all, it is stated that humans are by nature free beings and oriented toward a goal that transcends themselves; thus they should not be reduced to a structure or ideology, but had to be principle, subject and goal of all these institutions themselves. This basic doctrine offers a broad common basis for an exchange between Christians and Muslims – whereby there is no mention of those Christian and Muslim traditions which do not exactly under-

stand human freedom in this way. Three points are selected out of the wide-ranging programme which would be possible: the role of the person in modern society, the family, and the relationships between cultures. With regard to the first point, opinions on persons, society, and the relationship between them differ according to the various milieus, above all concerning the degree of freedom that can be allocated within the structures of the respective socialisation. One should refrain from delving into the everyday imperfections, as did *Gaudium et Spes*. Here Christians and Muslims might well collaborate on many points to allow everything to triumph which respected the creator's honour in the face of humanity. The family is seen as the primary stronghold of the highest values, of life and love, and also as the principal instructor of conscience and faith. The family faced many different problems, including the immense question of feminism and the advancement of women. The Muslim family differed clearly from the western one, being much more patriarchal, whereby one should beware of underestimating or destroying the values of the patriarchal family (solidarity, help beyond the immediate family, family spirit) in favour of the freedom of persons and couples; there were dangers and errors within our individualism – which was increasing in any event. On the other hand, Muslims were looking for a development of the traditional family and had real interest in the Christian family. Unfortunately, their real-life example did not always testify to Christian love, but exhibited distortions due to a civilisation tending towards hedonism.

3.1.4.2 On the interpretation of *Populorum Progressio:* the importance of cultural dialogue

The encounter between traditional and modern culture provides a reason for quoting from the encyclical *Populorum Progressio*, which states that a civilisation contains elements of universal humanism, but cannot be introduced elsewhere without adaptation. Especially with regard to the Muslims, one should not underestimate their cultural level and interest in their own culture, and the Islamic societies were still determined by a humanism from which we could learn. The importance of cultural dialogue is strongly emphasised, since it is here that a new conception of humanity and society is formed for future generations. As for the second area of more societal issues, two problems are highlighted as being particularly topical: economic and social development, and the peace question. In order to formulate concrete goals, they make recourse to *Populorum Progressio* once again. The goal of development is in all cases the service of the whole human being, including material needs and the requirements of intellectual, moral, spiritual and religious life. No dialogue about humans or society is possible without reference to something absolute, neither for Christians nor for

Muslims, and for Muslims this is even one of the basic structures of their faith. The ultimate reason and goal of dialogue therefore do not consist in the discussion and solution of common practical problems, but in the common faith in God and the joint search for him as the true meaning and goal of human existence.[70] *Here it becomes clear that theology embraces anthropology, both for Christians and for Muslims.*

3.1.5 Christian-Muslim spirituality – humankind "attains" its rightful place

As if it were the finishing line crossed at the end of a race, the last section deals with the spirituality of the Christian in dialogue. Without explicitly mentioning it, the authors once again take up an idea from the Second Vatican Council, namely that of God's saving dialogue with humankind, in which we participate. It is emphasised that Christians discover this divine activity in the cultures and religions of others (and some Church Fathers may be helpful in this respect) and should adopt it in their own faith. The Muslim formulation regarding the ecumenism of the peoples of the book is very consciously employed here. Five aspects of concrete religious experience are investigated, and they are specifically selected to underline what we have in common, not what divides us. The Christians knew of the divine grace which desired to gather the whole of humanity in one body, and similarly the Muslims wished for the unity of all believers. Indeed, they were even more sensitive than the Christians for the importance of unity realised among people by belief in a single God. The aspects examined are: the believer's relationship with God, the Word of God, the role of the prophets, the importance of fellowship, and prayer. As far as the relationship of God to human beings is concerned, the Muslims see themselves as completely dependent on him and completely subject to him. This idea is not strange to Christians, but they also affirm that divine grace and love bring people into close fellowship with God. The Muslim calls himself a servant of God, as do Christians, but they also see themselves accepted as his children, thanks to God's generosity. Many Christians, however, tend to forget that they are creatures, according to the order of creation, and thus totally dependent on God, that they are fundamentally and first and foremost servants, unworthy servants. Conversely, the Muslim does not refer to himself as a son, as it seems impossible for him to rid that word of all carnal connotations, but he sees himself as a chosen servant and senses, as it were, a divine fondness for humanity. So while the Christian believes everything the Muslim says, the text maintains that the Muslim moves instinctively toward the realities expressed by the Christian. In other words, the Christian lives in the Kingdom of God both with its "already" and also with its "not yet", while the Muslim lives in the here and now. These considerations should lead Christians to re-examine the expressions of their spirituality in order to bring together the best of both sides, as it were, under the primacy of God's transcendence. This would be an ambitious spirituality that not only rejected shocking expressions giving the impression of a too human familiarity with God and a displaced religious infantilism, but also put humans in the place they had attained as dependent, incomplete servants in the face of God's greatness.

3.1.5.1 Prophets – apex of human and moral progress

A Christian spirituality that was consistent with Muslim thought would have to pay respect to every divine word (whereby it almost seems that Jesus' respect for the law of Moses is here equated with respect for the Quran) and also to study the "signs" of creation, human history and personal experience. But both sides were in danger of placing the divine word into the confines of closed, manmade systems. Special emphasis is given to the function of the prophets as moral and spiritual leaders for humankind, which is ignorant, undecided, and weak in the search for God. The prophets brought God closer to humanity by proclaiming his judgement, his coming, and his grace. In the perspective of world history they represented the summit on humanity's path to moral and religious progress. When asking what message for the history of humanity God wished to deliver through one prophet or another, it is necessary to show respect for one another's rich heritage.

3.1.5.2 The challenges of spiritual fellowship

On the other hand, the subsection on community makes very clear statements. The solidarity of the Islamic community attracts more spectacular attention than that of the Christians. This solidarity is evaluated as an authentic human value and is an image of the human fellowship to which our (Christian) faith leads us. The Church also strives to unite the human community into one body. Nonetheless, there are naturally differences: the Muslims tend to limit fellowship to those of their own faith, while Christians, according to Acts 10:35, do not exclude other believers, as long as they live in righteousness and the fear of God. Muslims express this fellowship through the sharia, which tends to make cultures uniform; but Christians emphasise the harmony of cultures, seeing as their common heritage all that is good and true in them (this is once again a notion to be found almost word for word in the Second Vatican Council). Muslim spirituality tends to subordinate spiritual freedom to the requirements of unanimity, while Christians focus on freedom, which tends to weaken solidarity. A balance did exist, but was difficult to uphold and might differ depending on time and place. In summary, one could say that the Muslim community aimed to promote the spiritual and temporal well-being of the society of the faithful, while the Church community was oriented towards the well-being of individuals. The scandal of Christian schisms had been a stumbling block to Muslims right from the start. We should nurture the spirit of unity among Christians, sacrificing our egoisms and prejudices: "Comment pourrions-nous parler d'union aussi longtemps qu'un profond esprit oecuménique ne nous habite pas? Il y a encore beaucoup à faire sur ce point entre chrétiens."[71] In another necessary step, we should no longer regard the non-Christian as our adversary, but should see in him a brother who is seeking just like us. When viewing him and his culture, we needed an attitude that was increasingly loving – neither dominating nor indifferent, and certainly not contemptuous.

3.1.5.3 The right attitude in prayer – no easy question

The last section deals with prayer. Prayer was important to Christians and Muslims, and they would both call themselves servants and worshippers of the one, true and living God. Islamic prayer in particular had to take place in the attitude of the repentant sinner, and a Muslim did not simply pray as an individual but also as a member of the Islamic community. He believed that this way of praying was the only one that was worthy in God's eyes, and should be practised by all true believers. Muslims cannot accept that we call ourselves children of God although we are sinful humans (Sura 5:20). They regard this assertion as huge arrogance, which becomes almost intolerable as they get to know us better. The *Orientations* classify this as an appeal to our conscience. Once again, it is a question of God's gift, which is incomprehensible for humans, and of the fact that we should lead our lives on behalf of all of our human friends, in order to bring them closer to their calling, too.

3.1.6 No anthropology, but the discovery of the brother

All that has been said hitherto remains, in principle, in line with what the Second Vatican Council had said and what the internal guidelines had extracted from it in a practical, but more general way. But when taking the early date into consideration, some individual statements seem very bold, and the authors were probably well aware of that (the *Orientations* were mainly developed by Joseph Cuoq and Louis Gardet, as the preface mentions), because the epilogue mentions the allegation of idealism. It concerns Islam as a faith, and therefore as a way to people's fulfilment in God. If Christians, whether clergy or lay people, were not to achieve this level, then the whole Christian-Muslim dialogue which was being sought would make no sense. At the end the authors state very pointedly that they are not talking about the Muslim as an enemy in past or present fights, or as a rival in our projects, but about the man of faith, our brother.[72]

And so the Orientations *is not about elaborating a Christian or Muslim anthropology, but rather about proposing the need for and realisation of such a discovery. The comments and comparisons made in the text, which have been presented here and are also quite convincing, must be seen in this light lest they be misunderstood or criticised as incomplete. Under this condition, the text* Orientations *remains remarkable and well worth reading.*

3.2 Guidelines for dialogue between Christians and Muslims

3.2.1 Introduction: a common anthropology of Christians and Muslims

A good ten years later, the newly formed Islamic Commission proposed in its very first decision to issue a new, completely revised edition of this book on the basis of the reflected, practical experience of dialogue. Once again, this is a document "in progress" that does not claim to be definitive, but the fact that noticeable progress has been made is correspondingly stressed in the forewords. But the foreword by Bishop Rossano, who had been decisive in establishing and shaping the dialogue work, goes a great step further. He says that mutual knowledge and relationship has to correspond to the love for one another in order to be veracious and honest, and that, in turn, is the prerequisite for dialogue and communication. Although he does see problems between Christians and Muslims in the use of terminology (basic concepts such as human rights were understood differently), he nonetheless considers it possible for them to share their commitment to people and the world, and he justifies this *expressis verbis* with the anthropology that is common to Christians and Muslims – more clearly than is later stated in the actual *Guidelines:*

> "From this faith in God springs an understanding of the human person which is virtually the same: created by God, 'servant of God' (...), the crowning element of the universe, steward of God's gifts, subject to the law of good and evil, called to attain God as ultimate end."[73]

3.2.2 Clearly visible: dialogue experience made so far

A comparison of the structure of the two volumes reveals a number of interesting aspects. The partners in the dialogue move into the foreground, and there are more of them. For the first time it is not only Muslims of all shades who are in focus, but also the other Christian denominations taking part in this dialogue (see the list of dialogues at the end of the *Guidelines*). Both aspects may well be attributed to the experience made in dialogue; it is not necessary to start with one's own dialogue position, if one is meanwhile confident enough with it to put the partner first. This is also confirmed by the fact that the main subject is no longer the Christian attitude in dialogue, but rather how and where it may take place, which implies two background levels, the "purely" human one and the meeting as believers. The result is that the chapter on the values of Islam reverts to third place, but has been greatly extended – presumably on the basis of the experiences that have been made. This is also clearly noticeable in the wording for the fourth chapter. This is no longer about ways to prepare for and adjust to dialogue, but about how to deal with the current obstacles. The very general perspectives of dialogue have been replaced by areas of cooperation, and the topic of Christian spirituality in dialogue by fields of religious agreement which are still possible. The introduction also still speaks of the unity of humanity as God's ultimate goal, but at the same time poses the (almost Muslim) question of the positive aspects of diversity in the divine perspective, stating that dialogue is for those people who shared the same values and expected just this.[74]

3.2.3 Islam and Muslims – what changes have taken place meanwhile?

On a closer look, there are even more changes and developments, which to a great extent bear the clear stamp of practical dialogue experience. In many ways, the edges have been taken off the image of the partners in dialogue. First of all, there is a reference to the very remarkable historical fact that Islam indeed portrays itself as a friend of the Christians, but from the very moment when the first Christians became protégés of Islam, a long series of confrontations began at all levels, which only gave way to a more positive view of Islam in the recent past – whereby the Orthodox Christians are tellingly not mentioned in this context. All in all, it is stated that the differing perceptions of Muslims on the part of Orthodox, Reformation and Catholic Christians are likely to continue for a considerable time, but that they tend to complement one another in their views. One enlightening remark is that those who are particularly sensitive to justice and human rights may well have difficulty in penetrating the nature of a Muslim's religious experience. This corresponds to the fact that the human rights issue has become an increasingly urgent problem in the past years, as evidenced by the statements of the Pope. Another interesting remark is that it depends on the temperament of the individual whether he sees and emphasises more the similarities or the differences in dialogue. This raises the question of whether this is merely supposed to be a description of past experience or whether further conclusions may be drawn from this regarding the image of human beings. The common service to humanity (as an anthropological basis for dialogue) is generally seen to be secondary to the absolute nature of God as theological basis. But this may be rated as a fundamental theological decision. As to the view of the Muslims, only small but significant differences are to be found in comparison to the first version. The question of whom one should invite, and whom not, seems to be on the right track. It is interesting (and perhaps a consequence of dialogue experiences in various parts of the world) that the Druze and the followers of the Aga Khan, for example, are still reckoned to the Muslim community, and only the Ahmadiyya are definitely outside. What is missing is the reference to modern culture as a unifying factor – it was now obviously the time in which Muslims were seeking their own national or Islamic model. This return to their own religion and culture also left clear traces in the Muslims' subdivisions and designations: modernity plays a smaller part, while the group of fundamentalists has arisen anew. Moreover, there is no more mention of de-Islamised Muslims. These are no longer named first, but rather – maybe because they have often proved to be the first partners in dialogue in practice – the working-class Muslims with a very traditional Islam in faith and practice. They are very numerous, but are also described as "silent Muslims"[75]. When they are migrant workers, they suffer from their uprooting and lack of social integration, but the religious values of their homeland remain firmly rooted within them. Thus they are particularly responsive to values such as faith, prayer, work, gratitude, hospitality, generosity, patience in suffering and acceptance of death. However, their new experiences had made them increasingly sensitive to values such as human dignity, freedom, equality and brotherhood, as well as for the message of the beatitudes. Behind those with a religious (or traditionally Arabic) education, who claim to possess the teaching authority for the masses of uneducated believers, there follows the group of so-called modernists, which comprises government officials, intellectuals and technicians. They were only apparently alienated from Islam, but had a sharpened Islamic consciousness. However, they felt more keenly than others the cultural gap between classical Islam and today's world, and therefore tended to separate the temporal from the spiritual. The focus of this group lay on ethics and a specific kind of humanism, so that social issues such as human rights, birth control and welfare of the family, and the struggle against underdevelopment would easily form the terrain for an encounter.

The last group, the new one, are designated fundamentalists. They were very self-confident and therefore difficult interlocutors for Christians. Altogether, the modern world is seen as a common challenge; it enables, for example, the advocacy of human dignity and in this way, that is to say with the topic of humanity, presents an important horizon for dialogue alongside God and the world.

3.2.4 Places and paths of dialogue
3.2.4.1 Positive developments

The next chapter, which, according to the title, deals with the places and paths of dialogue and is mainly aimed at Christians, like its predecessor in the first version, begins straightaway with an anthropologically interesting definition of this dialogue. It is characterised as "the way of living which rejects solitude, shows concern for other persons and believes that others are part of what makes up being a person."[76] One does not need to be a theologian to participate in this dialogue; it is a dialogue of professional, economic and political values, which takes place in family relationships, in daily work and often also in the common fight against underdevelopment. There is talk of the religious humanism of Christians and Muslims, which could be enriching. What is new in this elaboration is the view that dialogue is dependent on four prerequisites: accepting one another in mutual understanding, living side by side, sharing and participating, and also joint ventures and risks. This welcome in the spirit of Abraham amounted to more than courtesy and also more than tradition. If both sides affirmed their absolute honesty and their absolute goodwill beforehand, then – and this is new – they could never suspect one another of proselytism – or of trying to impose their dialogue on the other side, even if they had too hastily claimed to be in possession of the truth. Positive expectations towards the dialogue ensured that the dialogue partner never became an object. The concept of participation led to this realm of sharing the most modest but crucial aspects of life, helping one another in the great issues of the world, of humankind and God – a trinity which is mentioned again quite frequently and in which humankind has a one-third share. Particular emphasis is given to the commitment to each other with the human goal "that the others are required less and less to suffer the humiliation of being poorly known, poorly understood and poorly received."[77] As far as the last of the four requirements is concerned, the emphasis is strongly on the free play of forces, that is, on human initiatives and on their dynamics. All of this, however, is insignificant compared to the statement which Rossano presumably took up, namely that Christians and Muslims share the conviction "that human beings are not the result of chance or of determinism, but that they are part of a marvelous *(sic!)* plan of the living and eternal God, who wills *(sic!)* finally to *(sic!)* bring together into community all whom he has created."[78] In this sense, both Christians and Muslims were witnesses to God who above all desired to serve him and gain his favour. They both paid attention to the claims that God had upon them – the sentence even continues by saying that the closer they got to each other, the more they would become aware of his presence with them. However, it is also said that they would need to see consistency between the normative expression of their religion and their personal experience – what this might lead to, is left open.

3.2.4.2 Tensions and difficulties

In a certain contrast to what was said at the beginning of the chapter, it is further stated that the encounter in dialogue presupposes a difficult spiritual development, which consists, for example, in relinquishing prejudice and opening the heart to others. This true spiritual freedom only came from turning to God, e.g. away from the evil of not esteeming the religious experience of others. It amounts to a spirit of humility and truth and a dialogue that is not merely formal or a show. In this context, it also addresses the pillars of Islam as religious values. In general, Christians and Muslims should surpass one another in the good works of faith, that is, in the promotion of life, justice and peace, and in the defence of human rights. The open or partial contradictions contained in what has already been said are presumably clear to the authors themselves. An encounter which is so intense on the religious and human level implies a certain contradiction, because there are differences in the understanding of people's calling – knowing God as Father through Jesus Christ, or living in complete submission to God according to the Islamic faith – and this is connected in both religions with a universalist claim. At this point the headline states "Undertake the impossible, but accept the provisional"[79], because here is a contradiction that can only be resolved in God, who, in words almost reminiscent of Muslim thought, will one day let them know why they were different. In other words, at the moment both resemble travellers who do not know exactly where God is leading them. This in turn – and here comes a turn of phrase that was not yet to be found in the previous version, but is consistent with the remaining development – presupposes complete respect for the decision of a person of a certain religious tradition, even if this decision involves changing one's religion in order to respond better to what conscience perceives as the call of God. As already mentioned, this is in accordance with Council statements that see religious freedom rooted in human dignity.

3.2.4.3 A new starting point

Ten years of additional experience have certainly led to this result: Christians have now taken the first steps which were called for at that time, so that the starting position is now a little different. The basic attitude of openness and love is unchanged, regarding the other partner as he would like to be and to become, and expecting something positive from him. But one thing has in the meantime become clearer; as long as one only considers the beginning and end with God one can assume a significant common ground in the view of humanity. At the same time, great differences prevail in detail when it comes to the way in which people should act according to God's will between these points, during this earthly life. This tension continues to be the crux of dialogue. On the one hand, dialogue between human beings is completely normal in the world, but the tension remains because, on the other hand, it is just when faith comes into play that this dialogue really requires a turnaround. For while it is easy to see others as human beings, it is difficult to see others as people of faith without immediately associating them with negative qualities. *The less simple experiences have obviously made it clear which are the basic common prerequisites in attitude or action in order to conduct a dialogue (the four points mentioned above) and that dialogue is often more resistant than one had dared to hope, for example in the suspicion of proselytising; all in all, an impressive and predominantly positive interim balance.*

3.2.5 Muslim values from a Christian perspective
3.2.5.1 Keyword and principal value: submission to God

The next chapter is about the other partner's values – and here, too, it is noticeable that the emphasis is laid on persons, not on the religion. It starts with an insight which is possibly banal, but nonetheless important: ideals do not necessarily coincide with everyday reality. This chapter as well is intended primarily for Christians, that is to say, it focuses on those values that are most easily accessible to them and which, in the experience gained so far, have proved to be accessible to both partners and thus binding. This is reflected not only in the increased material, but also in a completely new, more pleasing and illuminating presentation. The new key word that has moved up to the top position is submission to God, whether as an individual or as a community. Since monotheism belongs to the human nature created by God, it is offered to all people as a model of a perfect religion. It is a question of restoring people to the place where they fulfil their original covenant with the creator. The believers' dignity, standing and responsibility lies in their total submission to God. This means devotion, trust and obedience, most certainly in an active and responsible sense. It corresponds to a person's permanent dependence on God. Reference is made to the linguistic relationship between "believer" and "security" in Arabic. It is clear that this does not mean resignation and the evasion of responsibility, but that it is rather a shield against pride and presumption on the part of humanity. Muslims call themselves servants of God because they are convinced that people cannot claim to be God's partners or associates. All in all, this leads to the very Islamic virtue of steadfastness, which is also seen as a Christian attitude by recourse to Jesus.

3.2.5.2 Subordinate values: book, prophet and community
Subordinate to this steadfastness, which is a direct result of submission, are the values that constitute the ways to achieve it, namely contemplation of a book, imitation of a prophet, and support of a community. Thus a certain order is already created within the values. (The first version, in contrast, still placed the focus on the aspect of community.) In these subordinate areas, especially in the first two, the anthropological aspects are more concealed and indirect. There is mention of a need to meditate on those signs that speak of God, humanity and the world, and of the "rules of sound religious judgment for 'inspired books'"[80], but unfortunately these rules are not defined, nor is it said where they come from. Are they taken from the Christian tradition, or do they belong to the realm of human reason, or perhaps both? As for the realm of the prophets, the Islamic side regards the history of salvation as humankind's constant need to worship one god, submit to his law, and follow his prophet or messenger. In the eyes of Muslims, the formation of the corpus of Hadith is seen as a sign of the growing importance of the prophetic model. It is followed only later by the Muslim community, as the last element, both temporally and logically. This is not paralysing nostalgia, but rather a continuously renewed, utopian vision, demonstrating an Islamic version for the organisation of society, development of the economy, and the promotion of flourishing humanism among the faithful. A more specific present-day effect is the consolation and fraternal support felt by a Muslim at Friday prayers in the mosque and especially during the pilgrimage, but also in concrete action. The community guarantees security (everything that belongs to a Muslim is sacred to another Muslim: blood, possessions and honour) and even infallibility. *Summa summarum*, both Christians and Muslims might be designated servants and witnesses of the transcendent, eternal God. On the other hand, the idea that humanity was only created to worship God is Islamic. The five pillars of Islam belong to this conception, as well as Islamic law; Muslims were only able to find their inner balance as human beings if they followed these divine precepts. Here (Catholic) Christians think more in the categories of permanent and universally valid law. The common ground, however, is that people feel a dramatic conflict between good and evil in their personal lives, whereby the further interpretation concerning satanic powers which have been in-

volved in human disobedience since Adam's sin probably belong to the Christian position despite cautious formulation. In conclusion, there is another new element: asceticism and mysticism are identified as important human expressions of the search for God.

3.2.5.3 Points of agreement – too good to be quite true
Once again, Abraham comes into play as a model to all humanity of faith and submission. Overall, his significance has definitely increased while others, such as the Islamic idea of the three ages of humankind, have been left out. Alongside the Islamic community, the conflict between human freedom and responsibility and God's omnipotence has considerably less weight and pointedness in this chapter. *Had it indeed been possible, as the* Orientations *had signified at this point, to overcome this contradiction without harming the mystery? In this paper, the portrayal of humanity's active submission is so harmonious and so positive that one might well think that there is no problem here and never had been, at least from the Muslim side; that is just as unlikely as that this question could already have been conclusively discussed. Thus, as has been said, the emphasis had truly and clearly been placed on the connecting values, which means that the realistic next chapter forms a sobering, but necessary, counterbalance.*

3.2.6 Obstacles in dialogue: allegations in all directions
This chapter on the current obstacles in the dialogue is at least much shorter than the corresponding chapter of the first version. Here, too, the influence of experience gained in the meantime is already revealed in the title: it is not a question of attitudes and prejudices, but of concrete problems. But first of all, the shadows of the past are in the foreground in the form of a long list of legitimate accusations from both sides. It is evident that the believers must constantly counteract the pressure deriving from their own background, which has shaped them before they could shape it. Christians should be particularly aware of the fact that Muslims believe they have been unjustly humiliated for the past few centuries, both politically and culturally, and they should ponder the causes and effects. Christians should point out that the influence of economics, ideology and politics was at least as strong as that of religion, if not more so. In this way, it might be possible to come to a common reinterpretation of past events, in which Muslims, through self-criticism, could come to a more balanced view of the relative responsibility of both religions for the tragic events of the past. All those involved should examine whether they are truly blameless, especially with regard to the separation of religious values from injustice. It was indeed

true that religious people often did not recognise the requirements of freedom and the secular state's autonomy, thus exercising power that had nothing to do with faith or charity. All in all, it would be good if believers were to try to draw lessons from the past, in all humility, but also without complexes.

3.2.6.1 From fatalism to fanaticism: still the typical prejudices against Islam

What follows is a short and concise list of typical Christian prejudices, incorporating many elements from the first version. Regarding the allegation of legalism, it is (once again) stated that all religions are at risk, since they all contain a moral code that cannot avoid regulations and prohibitions; nonetheless, it is pointed out that the inward attitude is important in Islam. A new and also very anthropological aspect is the statement that Islam takes especially good care of the inner need for security felt by religious consciousness. Respect for parents is also given a very positive appreciation. A long passage deals with the accusation of fanaticism. Muslims consider this particularly annoying and also make accusations in the opposite direction, as already described in the first version. On the other hand, this is a particularly persistent stereotype among Christians, presumably because there are examples of social pressure and communal solidarity in certain parts of the Islamic world which non-Muslims interpret as latent intolerance. Conversely, for Muslims the Islamic community is synonymous with peace, justice and brotherhood. That which one might understand as fanaticism is usually the result of an all-encompassing view, an almost totalitarian understanding of the relationship between religion and state – whereby it is only a small step from totalitarian to fanatical, even though that is not stated explicitly. It is at least pointed out that, in their eagerness to enable God's law to gain the upper hand throughout the world and thus ensure (in a second step) that human duties and rights (in that order) are respected everywhere, Muslims are allowed to resort to means that are not prohibited in Islam, such as the calculated application of the alms tax. Today, however, the urgent problems were religious-cultural pluralism and the new relationships between political power and religious communities. The question of the "holy war" belongs to this complex. Here there is a very interesting passage stating that all historians now agree that there is no place for such practices, and that humanity is not prepared to accept the use of force in the name of religion. This leaves a wide scope for speculation as to which groups do not yet agree on this question, and it represents a clearly retrograde step in comparison to the first version, presumably due to the fact that appropriate changes and experiences have taken place in the meantime. Now reference is made to the real, great battle, which is the spiritual struggle against all forms of injustice, hatred and warfare, all of which begin in the human heart with selfishness, pride and violence. That is why this expression is applied in many countries to the struggle of the entire population against economic and cultural underdevelopment, that is to say, to the struggle for justice, love and brotherhood – even if this occasionally demands a recourse to violence, a fact which is not commented on any further. The great struggle is also identified with the efforts of all people of good will to respect and apply the Universal Declaration of Human Rights – unfortunately, no details are given as to the type and number of such advocates. Then the question of change is taken up together with the reproach to the Muslims that they could never be more than just consumers in the modern world. Here the text refers to the mechanisms, particularly those of Muslim reformers, which enable adaptation to local society. Finally, there is a brief reference to the Christian accusation that Islam is a religion of fear, while Muslims often describe it as a religion of love. One should not make the mistake of equating the nature of a religion with a particular historical manifestation of that religion.

3.2.6.2 The Muslim view of Christianity: unrecognisable to Christians

The next section looks the other way round, seeing Christianity and Christians through the eyes of Muslims. It is made quite clear that this is not based on purely personal opinions, and that experience shows that Christians are confused, even shocked, and do not recognise themselves, even if the allegations that texts have been falsified have now often been modified to poor interpretation. The Christians did not see themselves as a people of Scripture, but rather as the people of a person. Regarding the core issues, the opinion remains that fundamental progress is not to be expected, and the most that one could demand from the Muslims was more respect for what Christians consider to be central to their faith and life and the proof of the sublime position of humankind. A rapprochement on the theological level still appears to be impossible, but more respect would at least signify rapprochement on the anthropological level. Finally, and this is of particular interest in our context, the question of the Church as a secular power is addressed, as seen in the eyes of Muslims. First of all, all believers recognise the need for a social framework in which faith is preserved and transmitted, and in which certain religious and moral values are conveyed through ritual practices and human relationships. Muslims admired the organisational structure of the Catholic Church and the effectiveness of Christian schools, hospitals and social work, and appreciated the moral influence of encyclical and pastoral letters. They were also aware of the activities of the World Council of Churches. The distinction between Church, Christianity and Christendom, which had already been mentioned in the *Orientations*, remains a problem for Muslims. Their own background leads them to classify traditionally Christian countries as Christian nations and the Church as the religious expression of the ruling power, so that they then looked for Christian politics, which had at one time actually existed. This, in turn, explains why Muslims view much of what was negative in the past as the official position of the Church. For that reason Muslims nowadays believed that the Church should interfere much more often and directly in the affairs of Christian countries in order to dictate (in a very Muslim fashion) what is right and to forbid what is wrong. The *Guidelines*' conclusion from all this is that Christians and Muslims should come together in order to rethink the role religion should play in the formation of a human society, with special regard to the risk of intolerance when a state is based on religious principles – for which there were many tragic examples in history. An important sub-question is, for example, what foundations Muslims can recognise for the pluralistic societies in which they live today. The last entry in this list is the accusation directed against Christians that they have been unfaithful to the message of Jesus (first of all dogmatically with regard to early Jewish Christianity, but also having either under- or overstated it in the practical implementation and having strayed away from the direct contact between the believer and God – a reproach which could also come from Protestants). On the other hand, there is admiration for Christians who have tried to practise the ideals of the gospel seriously in their lives. Certain past and present Muslim scholars have also made statements resembling those on the Christian side, considering that righteous non-Muslims have access to salvation. The overall assessment and the resulting conclusion shows, however, that one is probably still confronted with a long and difficult path to an unknown outcome:

"It is evident that Muslims find it difficult to accept the Christianity of today as authentic and to believe in the complete sincerity of those who practice it. History seems to corroborate their doubts. The only way they can be freed of such feelings is to see Christians who practice truly the ways of the Gospel, in obedience to the message of Jesus. To the degree that witnesses of the Gospel are open in spirit and not exclusive there will be progress toward understanding and mutual acceptance."[81]

3.2.6.3 Practical issues are the true difficulties: from intermarriage to mission

What still remains is the new field of enduring practical difficulties[82], where increased knowledge and the elimination of prejudices have not necessarily led to progress. Among the very practical examples are the dietary rules, but mixed marriages are even more problematic. In particular, the legal regulations for the Christian woman in a Muslim marriage often led to bitterness and hostility. The counter-proposal recommends first of all more information, followed relatively abruptly by the concluding remark that mixed marriages could and should represent a particularly favourable environment for dialogue. The third point which is brought up is the obligation of the apostolate, or in other words the question of mission. There were indeed practices that stood in disconcerting opposition to the ideals that had been agreed upon in dialogue. Above all, it appeared to be the case (note the extremely discreet formulation) that it was difficult for Muslim societies to recognise the principle of religious freedom, which includes the right to change religions, as stated in the Universal Declaration of Human Rights. Conversely, Christians found it problematic when the right of Muslim minorities to religious autonomy was supposed to extend to the areas of law, culture and politics. A compromise would be necessary for a pluralistic society, but it was not even easy to achieve real mutual understanding at this point. *Here, too, the path in front of Christians and Muslims is long and the outcome uncertain, but at least it is recognised and clearly named for the first time.*

3.2.7 Cooperation: deepening spiritual conversation by common service?

The chapter which follows is about cooperation. It is still characterised by a relatively large number of quotations from ecclesiastical documents, as had been the corresponding chapter of the previous version, but here, too, the influence of practical experience in dialogue hitherto is noticeable. Everything is a tick more concrete, a little less fundamental. It does start with a quotation from *Gaudium et Spes*, but this leads right into modern life and the modern world with all its contradictions and the breakdown of all kinds of relationships. These problems were experienced by all alike, so everyone should work together to overcome them, whereby believers should do so on the basis of their faith and their love for humanity. Only then is reference made to the questions of the deepest meaning of life, which had already been quoted by the Council and which remain acute, because progress has not eliminated suffering. A new aspect in comparison to the first version, which presumably reflects developments during the decade between 1970 and 1980, is the issue of creation and its completion, which are dealt with as a sub-topic. God himself had sought humanity's cooperation in this field, and as a sign of this high calling humankind had been made, so to speak, a reflection of the world in miniature. A combination of Christian and Muslim thinking seems to form the background to this idea and formulation – humankind as God's representative in the world (Muslim) and as his image and crown of creation (Christian). Specifically, it is about a relationship of respect, obedience and flexibility between humanity and nature instead of pollution, violence and subjugation. Mass consumption had to be restrained and technology given a human face. By contrast, the next two areas of focus, service to people and to society, are already well-known. All believers should actively promote the inalienable human rights worldwide – some of them, including Muslims, were already doing so. And while serving others, Christians and Muslims could also press ahead with their dialogue concerning the spiritual values that unite them and share their view of the human greatness that inspires them – an interesting thought, to deepen the spiritual exchange by joint service, which was maybe inspired by experiences already mentioned in the first internal publication, and possibly even by later experiences. The best testimony to respect for human dignity, however, would always be the service to the poorest and the oppressed, to orphans, disabled and mentally sick persons, lepers, to those marginalised by society, to the elderly and the dying. The dignity of atheists is said to be fully recognised, "even though their denial of God implies certain serious possibilities for undermining human dignity"[83]. This formulation certainly plays on the fact that Islam has massive problems in this regard. All in all, however, cooperation in this area is well conceivable, for justice and mercy are also Muslim ideals, for example (Sura 4:36) towards orphans, the needy, neighbours, travelling companions, migrants and slaves. This section ends with a bold statement:

"All those, then, who go back to Abraham in their religious traditions can together join with all who hold to the Universal Declaration of Human Rights, calling attention to the fact that the people who should be given first priority in the defense of those rights are the ones who have been long deprived of them and whose cries of distress continually rise to God."[84]

3.2.7.1 The image of humankind and justification for cooperation
It is noticeable that this topic has been kept in, but it has changed considerably. It focuses much more closely on Christian and particularly Muslim statements on the position of humans, and thus also on the justification(s) that can exist for a joint commitment in this area. It is also clear that this has not been conclusively discussed, even if there is an explicit emphasis on points of agreement here. This is particularly clear in the sentence quoted above on supporting human rights. One of the most controversial topics is how they are to be formulated specifically. This can be seen in the *Guidelines*, where, in the description of the Muslim position, the rights of God are mentioned first, secondly the duties of humankind, and only in the third and last place of the human rights. By and large, the individual topics have been repeated, even if there are some alterations. In the first version, when referring to human advancement, the focus was on family and cul-

ture, but now it is seen as a task for society. The even more fundamental question of service to humanity gives the highest priority to the right to life as such and, most emphatically, the commitment to the poorest and the most disadvantaged. Just as was the case above with the newly included environment question, this may well mean a general shift in priorities, at least at first sight, which is not limited to Christians and Muslims, but encompasses all people worldwide.

3.2.7.2 The area of brotherly society: also family and culture

At the beginning of the next section, an inclusive society is named as the prerequisite for all this, one which respects the values that promote human dignity, in a word, a society of brotherhood. This idea is certainly echoed in the Islamic world, but not even the Muslims themselves agree on the details and ways to achieve it. For purely practical purposes the following are then assigned to this theme: marriage and the family, culture (i.e. the two new topics), economic and social life, politics, the unity of peoples and peace between nations. The social sphere has thus been extended and the personal sphere restricted, since marriage and the family as well as culture are now seen more from the point of view of society than of the individual. This may also be due to the influence of the Muslim side, for which both are clearly not a private matter, but a deciding factor for the community. If that is so, it would signify once again that the practical experience of dialogue with Muslims made in the meantime has shown increasing influence. This is distinctly noticeable on consideration of the very concrete and practical subtopics: freedom in the choice of a marriage partner (which can by no means be taken for granted in the Muslim world), financial independence for young families, adequate housing, the wife's responsibilities, children's rights, responsible fatherhood and motherhood, marital harmony, presence of the elderly in the household, relations between the generations. Christians and Muslims often had the same ideals of life and love, and could therefore act jointly, whether against abortion and child murder, or in sex education, marriage preparation, centres for mothers and children, groups for family and upbringing, or for support for couples in separation. A special role could be played by the partners from Christian-Muslim mixed marriages. As far as culture is concerned, people's right to culture is once again emphasised.

3.2.7.3 Righteousness in business and politics

The issues of culture, justice and politics are again connected directly to *Gaudium et Spes*. There is a clear statement on the inhumanity of all wars, including terrorism in particular. Everyone has the duty to help their fellow human beings to refrain from violence and to resolve their conflicts on a humane basis instead. As the frequent references show, these are not new ideas, even if specifically more pointed than in the first version, especially on human rights, but this is a development that has also been recognised in other parts of the paper.

3.2.7.4 God's gift as the foundation of shared commitment

The last section of this chapter returns to the basics again. The dialogue partners are seen as representatives of experiences made by all humans. Whether Christians or Muslims, both should strive for the divine ideal, because life as such (and every kind of possession) is the gift of God. Therefore they should also reject any form of discrimination and defend freedom in the interests of responsibility to their own conscience, to fellow human beings and to God. For freedom is itself a gift of God, just as peace and unity in diversity are God's gifts. Every common commitment is rooted in God's gift and the conviction shared by Christians and Muslims alike that it is God's gift. Thus both sides only have to act according to what God has already granted as a gift, to put it in almost Protestant terms. In this way, a chapter that had started by naming problems and questions actually ends on a doubly positive note, by showing what God has bestowed, and thus what is available to both parties, despite all the tasks comprised.[85] *The fact that the authors dared to write in that way may also be understood as a big step along the path that has been chosen.*

3.2.8 Religious agreement – a further step on the chosen path

In the last major section, too, the title already indicates a clear development on the practical level, even if the content arrangement has remained essentially the same. But the perspective is no longer directed towards Christian spirituality, but it is about possible religious agreement, which is already a very far-reaching idea. Starting from the previous chapter, ideals and values are taken as a common ground for service to humanity, but they have to be supplemented by the decidedly divine dimension. Specifically, this means the possibility of speaking about religious experience, which in turn is seen as personal response to the divine initiatives towards humankind. The believers would have to be ready to go beyond the sociological limits set by their traditions and environment. Again, the movement from "inert" to "dynamic" can be seen, which had already been emphasised in the first version. The aim is to understand the Word of God within the broad range of human experience. Referring to John Paul II, the routes may be different, but the goal is the same, namely the deepest aspiration of the human spirit, meaning the quest for God, the meaning of human life and the full dimension of its humanity. There is a creator and fulfiller of all creatures, and in dialogue this deep desire for unity will become particularly visible over and over again. In the light of the accusation that religions have been the cause of division and war, believers who hold Abraham as their model (once more, the emphasis on his function as a role model!) should prove the contrary by translating their faith in the One God into deeds of human solidarity and friendship. As in the first version, albeit in a somewhat more positive way, the connection is made to inner-Christian ecumenism: in both cases – quoting John Paul II again – loyalty, perseverance, tenacity, humility and courage are required. This is surely an extraordinary combination, placing high demands on people. As far as the Muslims are concerned, it is emphasised that their hearts would be touched by the living example of people who obviously lived under the guidance of God.

3.2.8.1 Consensus in common life – ultimately brothers or rivals?
After these more general statements, the text moves on to possible elements of agreement, as applied – as was already fundamental in the first version – to the level of common life here and now. The subtitles dealing with God, the Word of God, the prophets, communities and prayer are practically identical with the previous version, but they are extended to the area of holiness, a point which had in the meantime been a topic of Christian-Muslim dialogue. In the section on God, two aspects are above all relevant to the field of anthropology. First of all, it is the *imitatio Dei*, and secondly the emphasis on the fact that Christians opposed every form of deification long before Islam. This is a much more self-confident attitude than was evident in the first version. As for the Word of God, it is generally seen as a revelation *inter alia* of many truths about the hidden destiny of humankind. While for Muslims the Quran was also a law for humankind, Christians believed that the Word had become flesh as a perfect human being. However, both shared an attitude of respect for the Word of God and its practical observance as well as the aim of growth of the human spirit; in addition, they agreed that the Word of God can only accomplish its purpose if it is respoken as a human word. Here, too, a greater self-confidence can be detected in comparison to the first version with regard to the fact that the divine word always has a human dimension which must be taken into account. In dialogue, Christians and Muslims could share their insights into the human acceptance of the divine Word in their respective traditions. The prophets are seen as a mercy of God for human beings, bringing them among other things truths about themselves, for people needed spiritual guides who were both like them and unlike them – another new and interesting anthropological state-

ment uttered in passing. The classification as the apex of the moral and religious progress of humanity, however, is no longer mentioned. The problem of "sisters and brothers" or "rivals and adversaries", which had been addressed in the first version, is formulated more weakly here in the form of a question against the background that we were all searching for God and called upon to practise Abrahamic hospitality – note that Abraham is once again cited as an example. The basic comparison of the two communities with all its implications has been deleted.

3.2.8.2 Equality and inequality in prayer and mysticism

Prayer is generally seen as a submission of the body and readiness of the soul, as identification with the will of God. Both Christians and Muslims organised their day around regular prayer. Through this, and through the cycle of festivals, all events of life, from birth to death, became the focus of prayer and thanksgiving. The reader, however, already wonders whether this image of Christians is not depicted through rose-coloured Muslim spectacles, as it were – at least with respect to large numbers of them and the daily practice of prayer. As far as the concrete attitude and self-conception are concerned, the great differences are then named after all – a Christian does not feel and see himself in prayer simply as a humble servant, to whom submissiveness is the most important attitude. The specific issue of praying together is also addressed, but principally with regard to the difficulties. Direct participation in the prayers of the other faith should be avoided, as well as insistent invitations and premature assumptions of similarity. This could be interpreted as syncretism or as disguised proselytisation. Respect for one another's faith, on the other hand, means avoiding any suggestion of direct or indirect incorporation. Under the heading of holiness, there is again a reminder of the *imitatio Dei* and of the fact that such examples of holiness have always been very attractive to Muslims. The path of human sanctification, mysticism, was the same everywhere:

> "Every effort of self-purification on the way toward God requires the same disciplines, such as examination of conscience, control of desires, restraint of the tongue, detachment from the world, surrender of all earthly possessions, renunciation of all forms of prestige, abandonment of all pride and ambition. The first steps in a mystical approach to God are marked by the renewing grace of conversion, the strength coming from confidence in God, thanksgiving, godly fear and hope, poverty and detachment. Finally, through meditation upon the oneness of God and surrender to his providence, the way leads to a state of ardent desire, loving familiarity and tranquil nearness."[86]

3.2.9 The new Guidelines – practical and "more Abrahamic"

This is very broadly worded, while the Conclusions[87] make it very clear that the spiritual level of dialogue is indeed important to Catholic Christians, but that Muslims in their religious practice aimed towards the level of everyday life of society. (Altogether there were four necessary levels of human communication: dialogue of the heart, i.e. sharing as brothers and sisters, dialogue of life, meaning the promotion of human values, dialogue of speech, thus also speaking of humankind, and dialogue of silence as the beginning and end of all dialogue.) Here, too, the practical experience becomes palpable, also in the fact that dialogue as the principal dimension of the life of believers still has to be defended inwardly, even though it now encompasses almost all countries and all areas. The basic prerequisites are the mutual recognition of one another's values – here again the model of Abraham is cited, who really plays an important role in these *Guidelines* – and human cooperation. Distrust and prejudice were often instinctive and could only be eliminated by information and practiced love. Conversely, one also had to be aware of limits in dialogue in order to avoid setbacks in relations with one another. *All in all, the* Guidelines *are exactly what they claim to be: a further step on the path of Christian-Muslim dialogue, a step which draws attention to the practical experiences that have been made in the meantime, but retains the basic direction and is open towards the future.*

4 Religions – Thèmes fondamentaux pour une connaissance dialogique

4.1 Introduction through the "homo religiosus"

To judge by the official internal list of publications on the topic of Islam (mentioned in Note 1, produced for the 40th anniversary), then the first official and scientific work was a copious publication about the different religions, including Islam. Its origins go back to a number of individual booklets which had been compiled by the extended staff of the Secretariat (together with advisers and correspondents) and then put together as a compendium representing a key to statements on the dialogue with single religions, especially in the anthropological sense: "[D]e telles indications pratico-essentielles pourront difficilement être assimilées et avoir une efficacité sans une préparation mentale adéquate et sans une étude systématique et comparée des essentielles composantes de l'expérience religieuse de l'humanité."[88] In this sense, the *Orientations* which we have previously analysed appear almost to have anticipated this process, which is possibly one reason why they do not (any more) appear in the official list of publications. The new book, which is more of a religious-scientific nature, also begins with a general study of humankind and religion, before diversifying into individual questions and individual religions. This study is about the nature and behaviour of the *"homo religiosus"*[89], who is at the basis of all religions. According to the Preliminaria, this religious nature of humans can be demonstrated by historical and psychological analysis, and despite all differences there is a fundamental analogy in the forms in which it expresses itself. The book as a whole is mainly addressed to a (predominantly younger) Catholic readership, which is to be prepared for dialogue in the spirit of the Second Vatican Council. Although the introductory chapter includes a detailed bibliography on the subject of humankind, there is no scientific debate of any significance, not even footnotes, and the points of reference named are no longer particularly up to date from a modern perspective. *In this sense, it would be preferable if the book were not just to be reprinted, but re-issued as a revised edition, even though this would certainly be hard to do, since Piero Rossano, who was responsible for the anthropological introduction and had done a great deal of theological groundwork*[90], *is no longer alive. Still, when speaking, for example, of internal processes of transformation and the religious crises they initiated, one would certainly no longer refer to exegetical liberalism with respect to Islam – that has no influence on the present-day Islamic landscape and was probably never much more than a projection from the Christian side. And one can also only have very limited expectations towards the adaptation of sacral totalitarianism to a new, technical-industrial civilisation.*

4.1.1 Confrontation with religious criticism

The anthropological chapter begins with a fundamental discussion on religion. Alongside the specifically theological criticism of religion (as opposed to belief, which reminds one of Karl Barth, although he is neither explicitly mentioned in the text, nor in the bibliography), it also refers to the fundamental criticism of religion, which denies that religion has any objective foundation. Theological criticism leads to the necessity of exactly defining what is religious or sacred, because that is not understood in the same way by everybody. With regard to Christian (or more exactly, Catholic) anthropology, it is stated that the sacred dimension in humans, as far as it exists at all, consists "dans sa possibilité profonde de se reporter à Dieu"[91]. The question of fundamental criticism of religion is not raised at all; instead, the religious dimension in the human soul is practically accepted as a presupposition. When the text maintains that some sciences dealing with human beings "voient dans l'expression religieuse une manifestation de tendances spécifiques, particulières et profondes de l'être humain"[92], then one would be grateful, irrespective of the need for brief explanations, for some examples, or indeed substantiation, of what is meant. On the other hand, the reference to the Council is abundantly clear, especially to *Nostra Aetate*, which had seen in religions, or rather in their innermost being, "rien moins que le point de rencontre profond entre les hommes, susceptible de réaliser l'unité et la fraternité de la grande famille humaine"[93], whereby Christianity, with its reference to God as creator, Saviour and Providence, was especially open to all people.

4.1.2 The religious dimension of humanity – the relation to the absolute

The next subsection is altogether more substantial and more argumentative. It begins with the statement that the religious person forms the basis of religions, as was being increasingly acknowledged by psychology, anthropology and sociology, but sadly no references are provided. The religious philosophy of Romanticism in the nineteenth century had been the first to concern itself systematically with this aspect of human consciousness. The fact that this was a relatively late development within intellectual history and a clear tendency against enlightenment is not discussed. It is about discovering a religious space within humans, an ultimate horizon, which is nonetheless not a purely human structure, not transcendental. The following sentence is typical: "Mais assurément, c'est existentiellement une réalité."[94] On the basis of both history and the philosophy of religion, one could speak of religion as "disposition spécifique et mode autonome de la conscience humaine"[95]. This is stated without further argument, although the reasoning is at least partially explained later. Humans were oriented towards a different reality and found the present reality unsatisfactory. They posed the question of their origin, the causes and meaning of their existence and actions. They were oriented towards an absolute good and sought abundance, something complementary to themselves, and therefore had a fundamental tendency to integration. Examples from philosophy are on the one hand Kierkegaard, on the other Nietzsche, who does not see humanity as a goal, as an end, but as a transition, a bridge. There is an obvious reference to *Nostra Aetate*, which describes man as someone seeking an answer to the secrets of his existence. However, Rossano makes a connection between Greek philosophy and the Christian tradition of defining religiosity as a psychological echo of an ontological relationship manifested on the level of conscience. There is only very brief treatment of objections (for example, Marx, Freud) as having only as much value as their philosophical basis, without any explanation or justification of the obviously differing assessments of various philosophical approaches. It is "la constation d'un dynamisme religieux inscrit dans le coeur de l'homme"[96] which is decisive or, in a somewhat different formulation, a "disposition pour la religion inhérente à l'homme"⁹⁷, that is to say, the existence of this dimension, not its possible interpretations, When it comes to what characteristically distinguishes this religious disposition from others, for example, from a purely ethical one, Karl Barth is now cited explicitly, who saw in religions an expression of the ultimate, extreme possibilities of humans. This is modified to: "[L]a religion est la dimension de sa profondité dans toutes ses fonctions"⁹⁸; once again. one explicitly avoids understanding religion as (just) one particular realm of people's spiritual capabilites, but it is indeed " la tension *dernière* de l 'être humain vers son horizon *dernier*"⁹⁹. This last horizon was also the origin of a person's need for unity in his entire life, oriented towards the supremacy of the highest justifications – a very neutral expression. This was visible in in all religions, and particularly strongly in Islam, which refused to be seen as a religion in the purely traditional sense, but was rather a complete lifestyle, culture, political order, social system and philosophy at the same time. The relationship to the Absolute was always necessarily also an ethical relationship: "[L]a recherche de l'Absolu coincide essentiellement avec la recherche du bien et la fuite du mal"¹⁰⁰. God's love never existed without justice and charity, and religious progress could only ever be achieved in two areas: morality or mysticism. Another important statement is that the mere existence of religious potential in human beings does not necessarily lead to its adequate. let alone perfect implementation – doubts and crises were absolutely normal and human, both for individuals and for groups. It is also mentioned that religious life needs freedom and that religiosity can express itself in different, even areligious ways, whereby the expressions used by the monotheistic religions are relatively similar to one another.

4.1.3 Religious life as experience of polarity

The following subsection on religious experience is even more descriptive. If the religious disposition of humanity consists of his relationship to one who is the first, the highest and absolute, then religious experience is to be defined as the totality of his feelings and inner reactions in the context of these relationships. To a certain extent, this religious experience was common to all people of all cultures, even if interpreted differently. As seen from the Christian (Catholic) side, it was a divine gift that transformed and elevated the human religious disposition, whereby humans contributed to it by their will. On the Islamic side, submission to divine law was considered the basis of spirituality. Overall, religious experience is "un phénomène non immédiatement perceptible dans ses causes, mais hautement signicatif dans ses manifestations"¹⁰¹. In general, however, religious experience is considered to be the cause of happiness, and this happiness is a measure of the genuineness and authenticity of religion. Culture, age, sex, occupation and also class have a diversifying influence on the religious experience, but some basic lines are general, such as the distinction between sacred and profane. Here Rudolf Otto is quoted and the typical reaction of a person who comes into contact with the sacred, "sent immédiatement l'inégalité du fait de ses limites et de son imperfection ontologique et morale"[102], and reacts with uneasiness and fear This situation led to submissiveness or trusting devotion, whose strength determined in turn the depth and vitality of religious life. The love of God was also a common phenomenon running through all religions and obviously necessarily preceded in all religions by an experience of dissatisfaction and unrest. In any case, the goal is always a state of security, liberation and peace. The long list of possible elements of religious experience fills several pages and cannot be repeated here, but some contrasting pairs should be mentioned which are obviously particularly significant for religious experience because of their contrariness: death and life, whereby life is always in the foreground, combat and peace, whereby the combat always leads to peace (for example spiritual combat, but also the holy war in Islam), movement (faith as pilgrimage) and rest (as goal), zeal and order, sin and grace, as well as the whole complex of transgression, punishment and salvation, individual/society, down here (earth)/up there, fellowship/contemplative loneliness, love of God/ love of one's neighbour (in Islam, for example, as a pilgrim to Mecca said, 50 percent for this world and 50 percent for the other), but also fear/love, spontaneity/regulation, inwardness/outwardness, contemplation/action. This polarity seems to be a basic element of human religiosity, the last proof of which is the tension between apparently almost magical happiness in this life and a purely otherworldly conception of salvation, which can hardly be resolved alone in the latter: "[C]'est la preuve que la poralisation dont il est parlé plus haut, est inscrit dans la disposition religieuse même de l'homme."[103]

Whether or not this can and must be taken as a proof is an open question, but as already mentioned, the strength and emphasis of this article lies on the level of description rather than argument, and as such it is certainly good and important, especially for a reading public that is presumably not versed in religious studies and is unfamiliar with the great diversity of phenomena.

4.1.4 Constant elements in human religiosity – prayers, rites and more
The last subchapter moves, as it were, into the next wider circle and deals with forms of religious expression and structures. It stays on the same track, maintaining that strict methodological procedures will lead to the recognition of per-

sistent elements enabling a real knowledge of the *homo religiosus* at the bottom of all religions:

> "[É]tant donné qu'il existe, comme on l'a vu, une analogie, ou même une coincidence fondamentale d'attitude et d'expériences religieuses dans l 'humanité, évidemment il y a lieu de s'attendre aussi à des analogies et affinités entre les formes d'expressions de la religion."[104]

The first point to be mentioned here is prayer addressed to a Thou, which shows, "à quel point le rapport personnel est connaturel a l'homme dans son dynamisme religieux."[105] In Christianity, as in Islam, prayer is even a religious duty, and mysticism is very similar in both religions anyway. It should also be mentioned that prayer is always community prayer. Another point of similarity are rites: initiation rites, rites of transition, rites based on the cycles of life and nature. Following this, religious symbols are named next (and myths are also to be included in this context), and also religious philosophy and theology and certain concise formulations, for example in Islam the creed and the five pillars, in Christianity the Ten Commandments and the triad of faith, hope and love. Furthermore, there is reference to mediators such as prophets and founders of religion, for example, who are also reckoned to belong to the common structures of humanity's religious sense.

Altogether, this article on religious anthropology wavers constantly between different positions: it touches critically upon philosophical models, and towards the end upon some that are based more on religious studies. There is a call for strict methodology, but no more than that, and in the end the authors arrive at their own model, which practically presupposes an inherent religious dimension of humankind and corresponding religious experience. Two forms of deviation appear to be regarded as fundamentally possible – either the genius and experience of the peoples, or a departure from the original nature of humanity. This model is not really given any justification, not even theologically, since the article claims to be religious-scientific, but over and above that is is apparently not justifiable:

> "L'unité croît dans la mesure où l'on descend du plan des manifestations historiques et phénoménologiques au plan ontologique de la nature humaine, à une profondeur devant laquelle la recherche expérimentale semble devoir s'arrêter: le domaine de la disposition religieuse inscrite dans l'être humain, le royaume des interrogations capitales que tout homme se pose, le lieu secret du rapport de chacun avec l'Ultime et le définitif."[106]

Regarding Christianity and Islam, it only remains to remark that at this point the problems are almost negligible anyway, since both religions are very close.

4.2 Humankind's striving for salvation – the basis of interreligious dialogue

The following chapters deal with individual aspects of religions, first and foremost the search for salvation. This decision is justified once more with fundamental anthropological aspects, in the very first sentence: "Si l'attitude religieuse de l'homme se caractérise comme une tendance, un mouvement dynamique vers un terme ultime qui est sa paix, on doit reconnaître que la recherche du salut et l'engagement qu'elle comporte s'inscrivent dans le dynamisme même de l'esprit religieux, bien plus, ils paraissent en être l'aspect fondamental et universel."[107] The behaviour of people in this respect might be described as follows: on the one hand there is a form of existence from which one wants to escape, on the other hand an ideal goal to aim for, as well as the ways and means (and also persons) that facilitate or enable the transition from one to the other. This would even apply to secular contexts; an equivalent hope for unity and peace also existed there, together with the typical belief of the modern age that work and security were the solution to all problems of humanity. Whilst these opinions differed in their concrete forms, the soil in which they grew was largely similar, if not to say identical, so that every human being can understand and empathise with his brother's disquiet. This deep-seated harmony was also recognisable for researchers. The similarity of the matrix also provided the reason why encounter and dialogue between people of different religions could take place at all. It is only the human search for something transcending the horizon of their existence which makes spiritual dialogue possible. This dialogue brings people closer to each other and also closer to this their goal – at least, this is the hope.

4.2.1 Salvation in Christianity – also for people outside the Church, but not without it

As already mentioned, the following texts describe the appropriate situation in individual religions, including Christianity. According to this account, salvation is so central to Christianity that the whole history of humanity is presented as a history of salvation. The crucial points are Jesus Christ's incarnation (emphasis of the Eastern Church), suffering and death (emphasis of the Western Church) and resurrection. What is more interesting in our context, however, is to see how the position of humans is depicted and how, for example, the concept of original sin, which is so controversial among Muslims, is carefully and skilfully avoided. Following biblical traditions, humanity is portrayed as proud, wayward and unfaithful. In accordance with Paul, sin and death are called "puissances ennemies enracinées dans l'humanité dès les origines et qui l'entraînent vers la perdition et la condamnation"[108]. As John had taught, Christ brought freedom from "l'impiété et les convoitises de ce monde, la servitude du péché et l'impureté, l'aliénation et la ruine, l'impuissance, la désagrégation et la mort, un état sans espérance."[109] The terms for salvation all described "mystérieuses réalités"[110] which could be accessed by faith and baptism. It is also important that in this salvation the present and the future are interlocked: salvation is already realised in part, but full realisation is yet to come, so that it is also possible to lose salvation. It is certainly more difficult to specify what can be an aid to salvation, because there had sometimes been heated discussions about the coexistence of freedom and grace, of divine predestination and free will, of faith and sacrament. Here there is once again a very skilful expression: "Ainsi on arrive au salut grâce à un dynamisme spirituel et externale, sacramentel et éthique, individuel et communautaire."[111] However, the essentials are listening to the Word of God, worship and the sacraments and, following Vatican II, the Church as the universal sacrament of salvation; nonetheless, it is seen to be possible that people who do not know the Church may partake of salvation, depending on their life situation, not without the Church, but beyond visible affiliation. At the same time, the council statements on the relationship of salvation to humanity's major ideals – freedom, justice, truth and peace – are emphasised. Here again a picture of humanity comes into play which shows its attachment to a nostalgia that has shaped it from the very beginning – precisely the same picture which had been sketched out in the long introductory chapter on humankind.

4.2.2 Salvation in Islam – optimistic view of humanity

Significantly, the description of Islam's understanding of salvation begins with the sentence: "L'Islam se veut une religion *totale*, dans un triple sens: religion de tous les hommes, religion de tout l'homme, religion des 'deux vies'"[112] – the latter referring to life in this world and the next. Islam sees itself not only as the last and most perfect religion, but as the reinstatement of the natural religion of humans, who had entered into a monotheistic covenant with God ("le pacte prééternel de monothéisme"[113]) before their creation or birth. This is anthropologically significant; it forms, so to speak, the opening and the conclusion of this contribution and is repeated in the text under the heading of Muslim salvation history. According to a hadith, every human being is born a Muslim, and it was only his parents who made him or her a Jew or a Christian. Thus, a person is at birth good and believing, a fundamentally optimistic view of human nature. This in turn could explain the strict rejection of any mediator of salvation, which is so characteristic for Islam. It is consistent with this picture that Adam and his wife obviously fell under the influence of Satan when they were disobedient and expelled from Paradise, and that for this reason all people are subjected to trials, hostilities, and death: "Il y a donc là le récit de la faute personnelle d'Adam *(peccatum originans)* et même de ses conséquences physiques pour l'humanité."[114] But there can be no question of original sin, *peccatum originale*. In addition, Islam addresses the whole person in all aspects of their life. In principle, everything is religious, and there is no distinction from the secular area, so that state and religion are inseparable. This meant that Islam had to be the state religion, shaping all areas of family, social and political life. Further, in its self-understanding Islam ensures people's welfare in this life and the next life in a balanced manner; no one is neglected for the benefit of another, and body and soul are supplied with what they need. This was very often used as an argument for the supremacy of Islam, particularly against Christianity – Jesus' reply to the rich young ruler would be met with complete incomprehension.

As far as the details are concerned, the author focuses heavily on the development within the Quran between the Meccan and Medinan suras. The balance mentioned above clearly belongs to the latter, while in the Meccan period there is an emphasis on social justice and the Judgment of God, which in many ways has strongly biblical traits. It is clear that faith needs works, first and foremost social justice and aid for the poor, which justify God's verdict. Altogether, in the Quran faith appears to consist of works. It is only at the end of the Meccan period that the idea of predestination emerges. Faith and unbelief are free actions by humanity and, at the same time or even beforehand, the gift of God: "Ainsi le Coran comme la Bible, suivant le genre littéraire sémite, affirme simultanéament deux vérités apparement contradictoires, mais en réalité complémentaires: Dieu est tout-puissant et l'homme est libre et responsible."[115] With the decisive Medinan period, however, the entire focus shifted away from social justice to temporal victory as a sign of divine favour and the veracity of the Quranic revelation, and thus also towards holy war, which becomes a good work par excellence. Thus, a defeat is perceived as a religious scandal, and in Orthodox Islam there is no mysticism of loss and suffering. Salvation is reserved for Muslim believers, although some verses in the Quran include Jews and Christians, for example,, but these are generally considered to have been anulled by others. However, in Islamic intellectual history there have been movements that assume that a non-Muslim who does not know Islam or only has a misconception of it, will enter Paradise if he follows the light of his conscience or his reason (which discerns good and evil and demonstrates the truths of the faith) as well as the teachings of the prophets (in the case of Christians, Jesus Christ). In the reform movements of the twentieth century this even extends to the statement that it is not necessary for salvation to believe in the mission of Muhammad. But one common theme is valid: "Dieu est à l'origine et à la fin de la destinée de l'homme".[116]

Simply by contrasting the positions of that time (Piero Rossano had worked on the religious-scientific studies and Christian theological contributions, while those on Islamic studies came from different sources, all within the advisory staff of the Secretariat) it becomes clear that many similarities exist with regard to human beings as creatures, but there are great differences of opinion on the measure of their need for salvation, whereby Islam is clearly (more) optimistic about people's natural possibilities. In addition, there is a differing assessment of what is specifically related to religion in the field of human life. Here, Christianity makes stricter distinctions between various areas, while Islam gives greater weight to worldly existence and regulates it accordingly, leaving practically no leeway for individual, secular activity.

4.3 Humankind – open to a relationship with God

Significantly, deeper differences in anthropology also appear in the next section of this volume, which is officially devoted to theology and only deals with anthropology indirectly and of necessity, for one can only speak of God in his relation to humans. The introduction concerning religious studies, which is again written by Piero Rossano, makes it clear that all religions share a relationship structure – humanity desires to make contact with the ultimate and supreme reality. God is the highest value for humankind. This leads to faith and a gamut of attitudes and feelings between fear and love, whereby this relationship often expresses itself as a search for security and harmonious union with the cosmos. Despite all differences between religions and concepts of God, the common anthropological baseline is again emphasised: "Du point de vue historique, soit la religion, soit son centre, Dieu, apparaissent comme un phénomène polyethnique et polymorphe et répondent à la quête de l'Ultime et de l'Absolu inscrite au fin fond de tout homme."[117] In addition, it is common to all religions that for them God is always far greater than anything that people could say about him. Biblical religions, that is to say, all that relate to the revelation of the Old Testament, had a special position. For them, there are elements causing both separation and connection between creator and creature. The world and humankind bore traces of God and his greatness. Both Christianity (here again, although not explicitly, Catholic dogma is meant) and Islam recognised a natural theology and religiosity. But altogether there are three ways of seeing people: either as part of divine being, as insignificant and mortal (materialistic or existential), or in such a way "que l'homme soit conçu capable d'absolu et ouvert à une communication infinie"[118] – which is clearly the attitude advocated here and leading to the statement that all religions were calling people to this communion with God, without which the true nature as a human being could not be realised.

4.3.1 Islam: humanity cannot really love God

The article on the Muslim position first emphasises the traditional Muslim view according to which Muhammad was supposed to proclaim God as the creator (as in the oldest sura in the Quran) who both cares for humans and exercises absolute sovereignty over them. Humankind is created only to worship God. God is at the same time infinitely close to people and totally inaccessible to them – complementary contradictions, as it were. This essay emphasises the special importance of the covenant God made with humankind before the creation of the body. At birth, every human being bears the sign of this covenant like a seal on his heart and has no excuse for not knowing God at all. The only question that remains is whether or not human reason needs to be supported by revelation for this. According to conventional opinion, it is by revelation that reason is made aware of this command. The prime example for the revival of this pact, for the sudden recollection that God is behind the order and harmony of the universe, is Abraham. Muslim theology and philosophy has dealt extensively with the question of what human reason can know about God, whether of its own accord or through written revelation, but some problems remain, e. g. the relationship between God's omnipotence and humankind's free actions. The Mutazilites emphasised human freedom, while the Asharites, for example, "nieront la réalité intrinsèque, ontologique, du libre-arbitre humain, tout en veillant à en sauvegarder phénoménologiquement l'apparence (psychologique) d'effectuation."[119] This problem re-emerges when the attempt is made to harmonise human freedom with total devotion to God (Islam). On the one hand, it is the only possible attitude, on the other hand, this attitude should not suppress human initiative, quite apart from the Islamic view of humanity as God's representative on earth. However, this problem and its "solution" are not altogether alien to Christianity, as is expressly stated: "Nous noterons enfin une précision, qui n'est pas sans avoir sa répondance en théologie chrétienne: l'action mauvaise est imputable à l' homme seul; cependant que l'action droite, qui mène à Dieu, et qui engage aussi la volonté, ne s'accomplit que par la Miséricorde du Très Haut."[120] Finally, it is interesting that Orthodox Islam rejects the idea that people could love God, since love is seen either as an act of will, and the finite human will cannot have an infinite object, or else that love at least presupposes a proportional conformity, which is not possible between creator and creature. Faith in God charges humans to love one another, but in relation to God there can only be love for his law, his commandment and blessings, "mais non de Dieu Lui-même et en Lui-même."[121] *It is clear at the very first glance that this represents a diametrical contrast to the Christian view of humans and their relationship to God.*

4.3.2 Christian anthropology: only Christ is the true human being

On the subject of anthropology, the Christian article is almost purely Christological. All theology dealt with the meaning of God for humankind, and even more with the significance of us humans for God. It speaks of the history of encounters between God and humankind. God himself takes the initiative for such encounters, and he adapts himself to humans and their mentality or milieu, especially in Jesus Christ. Once again, the striving to overcome evil and achieve universal salvation is depicted as characteristic for human beings. Altogether, it is relationship which defines humans, both relations to their fellows and also to the environment. Contradictions characterise their existence: good – evil, death – life, egoism – love, history – timelessness, earth – heaven, humanity – God. And it is precisely because Christ is really God that he can also be the kind of person that humans should be. This is expressly declared to be an important truth of Christian anthropology and forms a distinctive feature throughout the entire presentation. Whoever wants to truly comprehend human beings and their making in the image of God, as the Bible describes it – a sign of the frequently invoked depth of human nature – must behold Christ, the second, the true Adam. He alone is absolute love which sacrifices its life. Humankind as such can only become human in the image of God by seeking refuge in Christ.[122]

For this description, it certainly plays an important role that it is really referring to theology, and that Christian theology cannot speak of God without speaking of Christ. By a certain logic it also true that theology cannot speak of human beings without speaking of Christ. But one may also question whether this must of necessity always be expressed in terms of a so pronounced, so exclusive, Christological and Trinitarian theology. This exposition is certainly not easily going to convince (or even intended to convince) a non-Christian on the basis of the humanity that we all share.

4.4 Good and evil

The last section of the book on religions and their basic themes deals with good and evil – as the fundamental world pattern, but also as the cornerstones of concrete morality and ethics. The introductory chapter is broadly based on religious studies and only goes into more detail about Judaism, Christianity and Islam at the end. In Judaism and Christianity, this also includes human motivation for obedience or disobedience to the divine laws. The Old Testament defines evil as unfaithfulness to God and as unlawful acts. The historical influence of Genesis 3 is also of interest for Judaism and Christianity. This was presumably originally intended as an explanation for the living conditions of humankind (otherwise only to be found in the Old Testament wisdom literature), but other aspects emerge in late Judaism and particularly in Christianity. These include the seductibility and weakness of women, but also the contrast between Adam and Christ for the history of salvation, whereby salvation was the starting point for the Pauline-Christian doctrine of evil and original sin, and not vice versa. Admittedly, the anthropological conclusions that have been criticised benefited from the fact that both themes had been combined in one episode: the disobedience of the forefathers and the fact that the human being had (received) a bad heart and lust was simply there. In addition, the uniqueness and continuity of both salvation history and humanity led to the development of the doctrine of the transmission of original sin, taking advantage of the obvious fact that one becomes human by being born. To counteract dualistic interpretations, Augustine defined evil, for example, as a lack of the goodness to which one was obliged.[123]

It is striking that this very interesting and, as already remarked, fundamental report hardly deals with Islam at all. Apart from that, the account from a Christian point of view which follows later completely ignores these most fascinating anthropological thoughts and arguments, as will be seen. Instead, long passages concentrate in an almost pragmatic fashion on ethics. That is highly regrettable from the point of view of this investigation.

4.4.1 Good and evil, God and humans – the Islamic concept

The article on the Muslim position is concerned with what behaviour towards God and other people the average Sunni Muslim considers to be prescribed. In Islam, good and evil are the object of the fundamental choice offered by God to human freedom, whereby it is important to look closely at what that means and what it does not. It is important, first of all, that the norm of distinction between good and evil comes from revelation, namely the Quran, which lays down divine rights and human rights. Up to that point all Muslims would agree. The profound differences lie in the issues of the imputability of human action and the ultimate legitimation of morality. Here the article looks into the limitation mentioned above, which claims that human action is highly dependent on

God's creative action. God does not only sustain creation, He never ceases to create. By continuous, fresh acts of creation, each creature becomes a new being at every moment of its existence. Of itself the creature is nothing. Thus the creatures' deeds in themselves are ineffective without the power that God creates in them. God does establish a kind of habitual relationship between people's deeds and their effects, but he could interrupt it at his discretion. Therefore, it is God himself who creates the free action of humankind, also in the area of morality, which immediately raises the question of people's own responsibility. A person feels, so to speak in his soul, that he is free to choose. This leads to statements about human freedom which sound quite illogical: "Bien qu'elle n'ait aucune efficacité sur les actes qu'elle choisit, la liberté humaine existe."[124] In fact, at the moment when a human acts, God creates in him the ability to do so. On the basis of this ability, people perform their deeds, and God rightly gives them the credit for them. Therefore, it is precisely when acting freely that humans are the object either of divine assistance or of God-forsakenness, for God also creates the capability for disobedience. God is not obliged to hold his servant to the right path. As for the evildoer, only God can want this evil and create it in him. Ultimately, it is the – positive or negative – predetermination by God which is crucial, and all free actions of humans are included in it. Good and evil are thus God's work, and he attributes responsibility to the person to whom he has previously given the capability for it. Thus evil comes down to people, not God. One can safely call this a "conception negative du libre arbitre humain."[125]: the human appears as a being forced into the form of a free being. However, God is decisive, not the nature of anything in itself: humanity only becomes the subject of rights and duties through divine command, whereby God also has the freedom to command what is unjust or impossible. However, disobedience to divine law is classified as a blow to God's gracious contract with the Adamitic race. But the denial of God's oneness is even worse than disobedience. It is the rebellion that renounces this contract, the only sin endangering eternal salvation, for which the faith imbued by God in the human heart is normally sufficient – human beings' first duty, but also their first good deed.

Over and above faith, duties are submission to God's will (*Islam*) and the so-called five pillars of Islam, at least with the will to obey in each case. It is the principal duty of the community, in which every Muslim has a share according to present understanding, to command what is good, meaning once again the five pillars of Islam, but also other regulations. All in all, this refers to the divine law and forms the foundation of Muslim morality, both in the personal and in the social field. However, from a rational point of view all this is neither good nor bad, and therefore human deeds cannot be essentially qualified as good or bad. Reward or retribution are not necessarily justified by obedience or disobedience, but rather depend on God's free choice, whereby a reward appears to be pure grace and retribution pure justice. Again, the factual situation remains strangely ambivalent: "Bien que purement gratuits de la part de Dieu, récompenses et châtiments sont une réalité avec laquelle l'homme doit compter."[126] Punishment can be avoided, for example, by expunging one's own failings by repentance, which effectively means stopping this wrongdoing, being determined never to commit it again, and, if the wrongdoing had affected other people, to make amends or be exempted from doing so by the victim. Delaying this repentance already counts as a new misdeed. But many people also believe that other acts can offset failures, such as reciting the Creed a hundred thousand times, going on a pilgrimage to Mecca, or participating in the Holy War. In any case, there is usually no connection between human repentance and divine forgiveness; only in the special case of the denial of God's oneness is people's repentance a precondition for God's forgiveness.

4.4.2 Inwardly torn apart – humanity from a Christian point of view

The depiction of good and evil from a Christian perspective begins with a recourse to *Nostra Aetate*. The author makes it clear from the outset that he would like to go beyond purely rational basic principles, i.e. general natural ethics, and that his position is based on Holy Scripture and the Magisterium. This decidedly Catholic position is also clearly noticeable at various points, while other Christian denominations (some of which are expressly mentioned, others only recognisable by reading between the lines) adopt different positions. But first of all, the basic Christian message, which is universally valid, states that human history is a history of sinful humanity that has to be freed from sin. Morality is God's personal injunction to every human being to true godliness – and it is by free response to this love that humans attain their value; a rejection of that love can only be tragic. There is frequent reference to this human freedom and how it comes about (which is a highly controversial issue among Christians), but it remains vague or, to put it more positively, mysterious: Christ, acting through the Church and the Holy Spirit, "éclaire nos intelligences et stimule et vivifie nos volontés en une synergie mystérieuse où Dieu intervient et l'homme déploie toutes ses libres initiatives."[127] Altogether, God never denies his light to those who do what is in their power. For a more detailed description of the dramatic reality of sin, which is part of the drama of human existence, the author refers back to Augustine. Original sin appears as a sin of disobedience, whereby the decisive event takes place inwardly: the will to be self-sufficient, the attempt to resemble God in one's own strength. This total autonomy destroys the relationship between people and God, which was not only a relationship of dependence, but also of friendship, so that God appears to be people's rival instead. This has far-reaching consequences: a person experiences an inner conflict, as Paul describes it. Sin is a factor of disintegration, within the person himself, in his relationship with God and also with fellow human beings, who are suddenly unable to understand each other. One can also say that sin is, above all else, refusal to love. If one continues along this line of thinking, one arrives at the fact that God's commandments cannot simply be fulfilled, but must first be accepted, something that takes place in a person's innermost self, in the conscience and the heart. God is already there, has placed within the heart this desire to search for him, and speaks in a language that once again respects human freedom in a mysterious way. It goes even further: "C'est en présence de Dieu, mieux en climat divin, que l'homme agit, en dialogue avec lui, et en co-opérateur et de sa création et de sa Rédemption."[128] Christian morality is not a human creation, it is God's injunction and people's response and active cooperation. God's transcendence is preserved. Despite all similarity, there is still more dissimilarity between the creature and the creator, and it is God who, in the history of salvation, repeatedly tries to tear people away from sin. On the horizontal level, this ultimate divine love liberates people to love one another, even as far as to love their enemies. Charity is the only absolute Christian value. This charity entails a repulsion from evil and a firm security in goodness.

4.4.3 Concrete morality in Christian understanding

When it comes to a more detailed specification of Christian morality, the article cites the Ten Commandments and the Sermon on the Mount. Every human being does indeed have the opportunity to hear the voice of God, according to the interpretation of Romans 2, for God had written the essence of the Ten Commandments eternally into their hearts, but he wanted to become more explicit. The Bible even condemns a person who fails to do something demanded by his conscience. On a closer look, however, it appears that the inner voice mentioned above does not carry very far: with rationality one may discover within oneself the fundamental law of doing good and avoiding evil, but a more precise definition is dependent on the background conditions, that is to say, it depends on family, society, or the historical period in which one lives. Elsewhere it is even stated that God breaks into people's lives, and that since then it has not really been possible to find any meaning except in God: "Toute autre hypothèse de recherche ou pire encore, toute autre tentative de construire un monde sans Dieu ou contre Dieu serait non seulement chimérique et irrationnelle, mais criminelle et fatale."[129] This is reminiscent of the argumentation regarding the Fall, but the idea of God breaking in is rather irritating. More specific is the finding that absolute love demands an absolute answer. This is designated typically Christian. A more interesting aspect in contrast to Islam is given by three important biblical points regarding morality. These include a clear stand against personal revenge, an equally clear plea for reconciliation as the ultimate aim, and also: "Nul ne peut mobiliser Dieu même sous le prétexte de croisades."[130] But that did not mean peace at any price. At state level, self-defence was justified, but that did not include the right to endless violence, for example, to the destruction of entire cities (*Gaudium et Spes* is quoted here, and it sounds as though there are echoes of the experiences in the Second World War). The article concludes by stating that Christ effects all this in a Christian through the sacraments: "À l'homme interpelé par Dieu, dans le Fils, la route est indiquée et la force est donnée, non d'une manière motrice externe mais bien au plus profond de lui-même en vertu d'un processus vital, qui part de Dieu, invite, attire et transforme et qui, en un mot, dynamise tout l'être humain qu'il saisit, tout en respectant sa liberté."[131] *This brings us back once again to the question of the relationship between divine action and human action or freedom, which finally remain side by side without a precise clarification of their relationship. This happens on a different basis and above all in a very different manner from the previous, very philosophical and then almost paradoxical representation of Sunni Islam, but it becomes clear that we are confronted by a fundamental question of theological anthropology for which there is no simple answer.*

4.4.4 Orthopraxis is easier than common dogmas

Finally, a short article once again ties up the discussion about good and evil in the religions and, in a sense, even the book's entire results, assuming that it is a universal human experience to distinguish between good and evil and to place them in contrast to one another, whether by instinct or by conscious reflection, and that all religions claim for themselves the right and the authority to teach morals and to make their followers shun evil and seek good. Particularly in the realm of Judaism, Christianity and Islam, the opposition of good and evil is bound to deeds and thus to people's freedom and responsibility. The resulting ethics were heteronomous: perfection was to be achieved by obedience and submission to a transcendent creator God who had revealed himself. There were even general lines along which religions recognised what was good or bad. The Golden Rule is mentioned as an example of this. In the Christian (to be more precise: Catholic) tradition, this rule is seen as a synthesis between natural law and revealed law. According to Augustine and Thomas Aquinas, its inclusion in the gospel was intended by God to complete and fulfil what people had tended towards by natural instinct since their beginnings. In Islam, there is a hadith which states that one should wish for one's brother what one wishes for oneself, and this hadith, according to the commentators, refers both to a Muslim and to an unbelieving brother, thus expressing the universality of love. This love, however, expresses itself in the desire and will that the unbelieving brother comes to know and accept the Islamic faith. This shows very nicely that the Golden Rule is on the one hand very important and fruitful for the peaceful coexistence of the human family, but that one should not on the other hand overlook and underestimate the nuances. One thing is certainly clear:

> "La convergence sur une orthopraxis est apparemment bien plus facile à l'humanité que ne l'est l'acceptation de doctrines communes (orthodoxie)."[132]

This conclusion at the close of a theoretical comparison will be more than confirmed when it comes to the practical dialogue with Muslims.

5 Religions in the World – a publication for international organisations

This booklet is not actually mentioned in the list of titles concerning Islam, but it is particularly interesting because, like the book in the last section, it was published in response to strong external demand. In this case it was organisations working internationally that expressed an interest in information about the world's religions on a more popular level, which would naturally include Islam.[133]

The chapter on Islam is written by Maurice Borrmans, a proven specialist who combines expertise with a very easy-to-read style. The very first words make it clear that his starting point is the real person to be encountered, the Muslim, who may or may not be well versed in the religion. The chapter's approach and orientation are anthropological in a very positive sense. He begins by explaining that every Muslim was striving to be submissive to God, thus according him the honour to which he is entitled as Lord and creator. At the same time, a Muslim would strive to obey God in all things, for God's will was to him a law of life and salvation. In addition, he understood himself to be part of a worldwide community of believers, which was the best possible community for humankind, according to the Quran. These first few sentences sketch out briefly and precisely the frame which is crucial for Muslims and their thinking, feeling and acting. The chapter goes on to explain how unity with this will of God means conformity to the Quran as revealed, not (merely) inspired scripture. The Sunnah is portrayed in a similarly vivid way: nothing in Islam can be decided or revised without reference to the Quran and Sunnah. This tradition often took the form of proverbs and maxims with catechetical content, so that they could be understood by children or simpler minds. The section about angels demonstrates the meaning of humanity in Islam especially well. The role of Satan (*Iblis*) is particularly interesting. He refused to obey God's command to prostrate himself before Adam, and has since then sworn irreconcilable hatred of the human race and tried to alienate humankind from God. The Islamic faith in the Last Judgement is also described in great detail. The Quran and tradition specify: the examination in the grave, the resurrection at the end of time, the great gathering of the risen, the judgement over them with severe retribution and the intervention of God's grace. It was clear to the majority of Muslims that faith alone could save them, so that a believer who has sinned would not stay in Hell for ever, but only until his sins had been expiated. Only the unbelievers, who had worshipped other gods, would have to stay in Hell for eternity. Paradise is conceived as an abundance of created goods that satisfy all the physical and spiritual needs of the believer. God will be pleased with them, but that does not mean that they would always behold him. That would be just a short additional mercy. It is also interesting to see how this brief text presents the problem of predestination in contrast to retribution according to works. In a further, most human portrayal, Muslims are shown to be inclined, on the basis of the experience of traditions and the practice of piety, to praise the mysterious will of God and thus to abandon themselves to predestination, following Abraham's example, even though the reform movements laid great emphasis on freedom and responsibility (which in turn is an interesting comment when it comes to assessing and classifying Muslim statements on this subject). All in all, Muslims, like Christians, were aware of the dialectic of human freedom and divine election, and also of the difficult discussion about the relationship between faith and works. The concrete practice of faith is also described in great detail, in particular the requirements of ritual purity, but also the effects of Islamic law in the area of everyday life. Traditional provisions of family law are described and evaluated at great length, as well as the relatively broad provisions (from a Catholic point of view) concerning contraception and sexuality in general, the nonetheless strict attitude to abortion, the authorisation of polygamy and a divorce practice which is clearly too extreme in the eyes of the author, who also considers that the children are too exclusively assigned to the father and that the wife is placed in a rather subordinate position in various areas. He sees the mystical currents as counterbalance to a certain juridification. Finally, he also briefly discusses the traditional and modern groupings in Islam and the relations with Christians, which are in principle friendly. Here he successfully describes the concerns of the individual groups and their effects, including those of the reformists who want to return to the purity of the sources, of the modernists who want to reconcile faith, reason and science, and of the modern "irreconcilables", as he calls the Muslim Brothers and Wahhabis, who recognise only the Quran as a constitution. He closes these short, but very precise and understandable explanations with another extremely human comment:

"Following the intellectual class, the liberal career, the world of commerce or of labour (industrial or agricultural) to which each one pertains, all these ways of being a Mussulman take on further characteristics which add one thousand human richnesses to one and the same fundamental religious experience which has been described above."[134]

6 Chiesa e islam – early John Paul II and his focus on Islam

The next publication (from 1981) mentioned in the list is once again of a completely different type. It is a relatively slim volume, but comprises a total of five languages. The title, *Chiesa e islam*, and the table of contents are in Italian (and both are to be found in Arabic in the same volume), but none of the other contributions is in that language. Apart from the extract taken from *Nostra Aetate 3*, these are speeches made by John Paul II to Muslims during the first three years of his pontificate, as one of the headlines in the book states, but that is not quite correct, because it starts with a talk given to the Catholics in Ankara. All the texts have been translated into Arabic, and with the exception of the (German) greeting to the Muslims in Germany, they also appear in English and French. These few facts alone lead to a number of fairly certain conclusions. There is no explanatory foreword to the texts, and this most probably indicates that they should speak for themselves and serve, as the title says, as fundamental and pioneering Church texts, not so much internally as externally. So this little publication is principally addressed to Muslims, as are most of its content and the one language which is used throughout. It is astonishing that after only three years of the pontificate of John Paul II (and even before his groundbreaking speech in Casablanca) such a booklet appeared, showing that at this early stage of his long pontificate he was already determined to make moves towards the Muslims as one of his major priorities.

In retrospect, it is noticeable that many, but not all of the texts were also included in the anthology compiled by Francesco Gioia for the then Pontifical Council for Inter-religious Dialogue, and that there were changes, some of which were not insignificant. At first, it seems logical that *Nostra Aetate 3* was included. However, even if one pays attention to the frequent references to this text in the first speeches, it appears that the focus lies more on the high regard for Muslims and their religious heritage and less on what one could do jointly. This only comes to the fore in the later speeches, alongside problems that also exist. If one also considers that the speeches were held in Turkey, Ghana and Kenya, in France and Germany, in Pakistan and in the Philippines, then a representative picture emerges which is remarkable, given the short and early time period. This concerns not only the inner development, but also the geographical surroundings and the starting point, but the picture is consistent in its basic features with the fundamentals described in the first chapter. The address to the President of Pakistan did not find its way into the later collection, possibly because its content is largely identical with that of the speech to the people of Pakistan. Likewise, John Paul II's greeting to Muslims in Germany was not included in the collection, maybe because it seemed too unimportant, but more likely because there was no Italian, English or French version available. This is really a pity, because it contains in a very compact form many thoughts that had not so far been formulated so directly, but are very important and interesting. This includes the idea that a person who goes to a foreign land with his faith (even though he has only moved there to find work) is one of the numberless pilgrims who have set out, ever since Abraham, to seek and find God. In addition, the public exercise of prayer in Islam is recognised as exemplary for Christians. But there is also a warning not to allow faith to be misused for human or political interests.[135]

This demonstrates once more exact knowledge of the situation and the ability to handle it concisely but soundly. This is the generally positive characteristic of the entire booklet.

7 Meeting in Friendship – Messages to Muslims for the End of Ramadan

This is the title of a collection of messages at the end of Ramadan in the years 1967 to 2002 (with the inexplicable exception of 1970, when there was none; on the other hand, there are two for the year 2000, in which the end of the month of fasting almost coincided with Christmas – one referring to the usual occasion, and the other dealing with the 2000th anniversary of Christmas). An earlier collection bearing the same title was quickly sold out, necessitating an updated reprint, which appeared in English and French. The messages concerned are translated into many different languages, especially those of the Muslim world, and printed as single sheets, which were then solemnly handed over in a folder by the Papal Nuncio, for example. Initially, the Bureau Chief for Islam was responsible for these messages, and then, from 1973, the President of the Secretariat for the Non-Christians, later the Pontifical Council for Inter-religious Dialogue, himself. It came to an exception in 1991, when Ramadan ended shortly after the end of the Gulf War, and Pope John Paul II used this occasion to turn to the Muslims himself. An excerpt from this message is also on the back cover of the anthology. In the table of contents as well as on the individual sheets the messages are given titles or topics, but inside the book they are only designated by the year, and a short key statement from the message is printed in larger type in a separate box on the page. The Pope's own Ramadan message is emphasised by differently coloured paper and the papal coat of arms in colour. Thus the booklet as a whole makes an appealing, easily accessible impression, so that important contents are conveyed even when leafing through quickly. The reader is invited to take a closer look, which is probably also the purpose of the collection. It is another matter to what extent this actually happens. In some cases, the later messages make reference to reactions to earlier ones, in a similar way to the illustrated book on the history of the dialogue. On the other hand, a rather more internal report (see above) lays emphasis on an exceptional, presumably well-intended and politically motivated procedure: after the Nuncio had handed over the message, it was also read out to the press, which had received an unusual invitation, and thus gained appropriate public attention. This would be in line with an assessment by Tarek Mitri, a Lebanese who was at that time Coordinator for Interreligious Relations and Dialogue at the World Council of Churches. He told me that what was really important about these messages was not their content, but the gesture itself, the fact that the Papal Nuncio handed them over personally in an official ceremony. After this presentation, the folder would normally not be opened at all, nor the message read, since – as Mitri sees it – no one was interested in a commentary on an item of Catholic dogma, a statement by the Pope. It is certainly impossible in this context to resolve the contradiction between these statements or to clarify their respective validity, but in any case at least one important human goal has been achieved by these messages: human empathy, in this case sharing other people's joy at a feast. Besides, the degree of attention actually given to the content of these messages may well vary from one country to another. It certainly also makes a great difference whether they are dealt with at the official level or, as obviously is increasingly the case, within the private sphere of good neighbourly relations and visits between Christians and Muslims. Nevertheless the question still remains, whether there is sufficient regard for the interests of the Muslim addressee, or whether the emphasis is not primarily on self-presentation.[136]

7.1 The beginning under Father Joseph Cuoq (1967–1972) – burdened by the past

It may be that a closer look at the individual messages, their anthropological content, but also their approach to the Muslim counterpart can help to bring more clarity. It makes good sense to focus on the individuals responsible for the messages. The first was Father Joseph Cuoq, who was in charge of the Department of Islam at that time. He signed the messages from 1967 to 1972. On the whole, they are still very strongly marked by the novelty of the situation, by the fresh start after a great deal of mutual contempt and all the suffering it evoked. The salutation as "friends", or even "brothers", is new, as well as the notion that a concept such as pilgrimage may refer both to Christians and Muslims. As a result, the wider family of humankind soon comes into focus, the way they could receive help by a brotherly cooperation of Christians and Muslims in the face of concrete problems such as war, hunger or unemployment, and the kind of responsibility they both would have. One ought to work jointly for humankind, and thus gain new confidence in them. This appeal is made again and again, sometimes distinctly, sometimes more subtly, but it must also be emphasised that this talk of friendship is not a mere formality, but meant quite seriously without ulterior motives, and that one should first of all just start to meet one another more often as friends. There is a clear awareness of the broad horizon of people and their problems on the one hand, as well as of religions and their solutions on the other, confronted as they are by humanity's modern idols such as money, power and amusement, but it is all very new. The messages also try to respond to the current situation in the various parts of the Muslim world and thus to address the partner more directly, but albeit with a clear understanding of the great difficulties, since one can only offer words for the solution of concrete problems. The double roots of fraternity are seen in faith in God and the knowledge of having been created by God, and these find expression in these good wishes at the end of Ramadan. To quote from the message of 1972, the last one written by Father Cuoq, and the only one which is so decidedly anthropological, "Many of our friends, both Muslims and Christians, have encouraged us to renew our wishes every year. For this is indeed an opportunity for us to encounter one another simply as believers, setting aside all cultural, social, national or racial barriers, and to contact one another in what is most basic to our common humanity, where fraternity and solidarity have their roots. We all long to communicate with each other at that hidden level of our being where we discover that we are all the creation of the same God, made out of the same clay."[137] *Perhaps one could, or should, have added the reference to the Muslim idea that humans were created from a clot of blood, as a sign that we have knowledge of other people's thoughts, which in the realm of creation are broadly similar, quite the same as our own.*

7.2 Cardinal Sergio Pignedoli (1973–1979) – against the backdrop of real encounters

The messages have a slightly different tone between 1973 and 1979, when Cardinal Sergio Pignedoli was responsible for them. Certainly, there is a great deal of continuity, for example the emphasis on the lack of ulterior motive, the appeal to God as the basic foundation of friendship and cooperation between Christians and Muslims, as well as the call for more cooperation between the two in the interests of all humanity. But this now takes place against the backdrop of actual encounters, which are sometimes described in detail. There is not only an appeal to the spirit of fraternity. According to the Cardinal, it has already become perceptible in charity, the will to forgive, common commitment to the strengthening of human values, and in the peace that is desired by all creatures and called for both by Christianity and Islam. It sometimes looks as though Islam has rubbed off on to the texts, when they say that both religions now had the role of communicating the will of God to people, and of reminding them of the religious and moral values that would lead to mutual respect, among other things. It is also noticeable that there is now more often specific mention of children and adolescents who, as representatives of the future, should learn to live together fraternally from an early age in order to build a new world of love and understanding, free from discrimination. Sometimes this is all even underpinned by an anthropological substructure, whether by alluding to the same earth and sky which all people shared, or by making statements expressed as rhetorical questions, as in the 1977 message: did not both religions hold humans in highest esteem? The Bible spoke of values such as intelligence and freedom, which people bear within them because of their creation in the image of God, and did the Quran not speak of humanity as God's representative on earth? But people's present-day living conditions did not correspond to God's wonderful plan for humans. Was humanity not threatened by the tyranny of technocracy and political systems of oppression? The former reduced them to cogs in the machine, forcing them to obey without understanding, to abandon their freedom, their dignity and some of their highest principles. The latter also curtailed their freedom, in particular freedom of expression, making them robots of an ideology. Precisely because the Feast of Breaking the Fast was also an invitation to do good, it could be a reminder of what it means for humankind to be God's representative on earth, and thus guarantee them the dignity and respect that God had already bestowed upon them. The earlier message of 1974 also strongly emphasised that human greatness, as well as the brotherhood of man, is only recognisable in their encounter with God,. At the same time, this realisation is propagated as a applying to all peoples, for they all proclaimed that someone who built a civilisation without God would do so against their brother humans, so that it could not exist in righteousness. The relationship between humanity and God was, so to speak, an inheritance of all peoples, and this recalls to some extent the thoughts and formulations of the basic document.[138]

Conversely, the God-given greatness of humanity cannot be established by empirical reality, but is only recognisable in God and by faith. Christians and Muslims had a lot of work to do in the empirical reality, and the texts always call upon them more or less clearly to tackle it.

7.3 Archbishop Jean Jadot (1980–1983) – materialism as a common opponent?

The messages of the next President (actually Pro-President), Archbishop Jean Jadot, repeat this call even more urgently, and it is also noticeable that in all four of his messages he refers to Pope John Paul II. Otherwise, the common faith is once again named as the sure foundation of human dignity, fraternity and freedom, as well as divine wisdom as the best guarantor of frequently violated human dignity. It is striking that materialism twice appears as a common counter-image, from which people and their values should be protected; only in this way could God be truly honoured as creator, as the Muslims did and as was expected by many people, particularly those who were left alone by our world.[139]

7.4 Cardinal Francis Arinze and the great influence of Pope John Paul II (2000)

This common vision for those who are most vulnerable is also something that appears over and over again in the messages by Cardinal Francis Arinze, frequently in the context of common opposition to a materialistic world view, which is seen as a particular threat to the younger generations. Especially those marginalised by ill-treatment, unemployment, drugs and permissiveness should be given priority. Attention is drawn to a great number of specific fields of action: defence of humanity, above all minority rights, peace and disarmament, cooperation in combatting hunger, aid to refugees and serious efforts to help them return to their homelands. It was true that fasting helped to develop a sense of community, to bring the highly privileged and the especially needy together, to restore priority for spiritual values and detachment from material goods, but also to restore solidarity and submission to the creator. Humankind is seen quite positively, there is even talk of trust in humanity, but only combined with dependence on God. Pope John Paul II is once again cited extensively, sometimes in longer text passages when it comes to justifying designation of Muslims as brothers, or as a source of topics when he had emphasised certain points within the Catholic Church. There is frequent reference to special topics such as the Year of Youth or the Year of Peace – the latter was taken up by the Pope regularly. A great deal of space is devoted to the first Prayer Meeting in Assisi, seeing prayer as the rhythmic accompaniment of people's life and work, joys and sorrows. Even more explicit and interesting from an anthropological point of view is a quotation from a prayer spoken at the famous meeting with Muslim youth in Morocco, in which God is expressly referred to as the one who gave us the inner law according to which we should live. This is in accordance with another pointed statement which already appears in the second Ramadan message by Cardinal Francis Arinze: "We are firmly convinced that seeking to liberate man by releasing him from his submission to God is to take a false path."[140] This is, as it were, an anthropological handshake with Muslims. The Ramadan message written personally by Pope John Paul II in 1991, shortly after the Gulf War, is primarily a sign of solidarity with the Muslims, dealing with the values held in common by Christians and Muslims, which they can offer to people as an alternative to the attractions of power, wealth, and material pleasures. Although this was John Paul II's only personal Ramadan message, it remained relevant, being quoted in four of the messages by Cardinal Francis Arinze.

7.5 Cardinal Francis Arinze (1984 – 2001): all about life worthy of humans

The way human beings are depicted in these messages remains basically the same. They often speak of the one human family, and the obligation to seek concrete solutions for a truly peaceful world, preferably in cooperation. Humans move, so to speak, within the frame given by God as creator, Sustainer and ultimate goal, and this involves duties not only to God, but also to their neighbours and to human society in general. But how do these duties look in a pluralistic society? That is indeed an open question, and it is raised because in a community of this kind religion has to show that it contributes to a fraternal and peaceful society and not just to tensions. Internally, the argument is simple and clear: if

God is one and creator of all, then it follows that the human family is also one. We share a common history and common hopes for the future. We believe that God's will has sovereignty over all humanity, and know that it is God's will that every person be treated with respect. This is all very evident, but the messages do not have room to go into the presumably more difficult details, also because they are designed to be no more than a greeting and a congratulation. At the most, more than a few questions and topics are marked for further clarification. Issues that are centred around people are also addressed, such as the question of the family. Moral life and family values are rated as very important topics. But protection of the environment is also an issue, once again for the reason that we all have one creator, inhabit the same planet and should behave towards our neighbours in a fashion corresponding to the grace of God we have received. In addition, it is expressly stated that man is God's representative in his creation (a clear allusion to Islamic thought) and must therefore play a responsible role in the world according to God's will. Moreover, the destruction of the environment is a threat to humanity, especially in some countries where hedonism and consumption are rampant. Fasting was a suitable help in cultivating virtues that ought to be exercised more often: restraint, moderation, discipline, and the spirit of sacrifice. Education in environmental responsibility and respect for life were also of indirect importance for peace. Genetics is a further issue raised, because it affects human nature directly. Thus, many new and interesting subjects are brought up, but the question repeatedly raised is that of cooperation, whereby the old, ever-present topics of humanity are named such as the right to life, the dignity of the human person and human rights, but also social justice, peace and freedom as values which are essential for a dignified life giving glory to God. This is not elaborated anthropology, but the reference to the creator is clear. Those tasks that could and should be tackled jointly are well recognised and strongly emphasised, together with the relevant agreements in ethics, particularly in commitment to the weakest (which is unsurprising in the context of Ramadan). Only once in the many messages of greeting for which Cardinal Francis Arinze was responsible does he go into the problem of human temptation and sin. The problem of original sin is not addressed, the formulations remain strangely vague, not going beyond the area of temptation, inclination to pride, hardness of heart and duplicity, or the tendency to be self-centred and to forget one's creator – thus the inclination to unbelief and injustice, as the Quran puts it. However, all the messages contain references to the consequences of such behaviour, which form the background for the calls to joint action. The broad range of topics already mentioned above can also be put down to the fact that they cover a period from 1984 to 2001 – by far the lion's share of all the messages produced up to now.[141]

7.6 Archbishop Michael Fitzgerald (2002–2005) – many unusual events

The most recent messages were written by Archbishop Michael Fitzgerald, at that time the new President of the Pontifical Council for Inter-religious Dialogue. With one exception (2003), they are strongly marked by specific events, first of all the aftermath of 11 September 2001, for the 2002 Ramadan message deals extensively with the theme of conflict. It names not just concrete grounds, but also anthropological causes, for example selfishness and the immoderate pursuit of power, domination and wealth at the expense of others, and pleads for peace education. In 2003 the message follows the same track, being based more closely than any other on the document of a Pope, in this case (occasioned by the 40th anniversary) on *Pacem in Terris* by John XXIII. Peace was supported by the pillars of truth, justice, love and freedom. This gives rise to further interpretations: truth means that humans should acknowledge that they are not their own master; justice implies respect for the dignity and rights of every human being; love starts with the recognition that we are all members of the one human family; and freedom as an essential characteristic of humankind is the prerequisite for all of this. In comparison, the message of 2004 appears very neutral, although it also mentions John Paul II. It deals with the rights of children (life, family wherever possible, nutrition, clothing, protection, education, health care wherever necessary) and the many threats with which they are confronted. It is emphasised that both Christians and Muslims see in children a blessing from God and have often joined in defending the family and the rights of the most vulnerable. It was necessary to reflect on the dignity of every individual, as God himself intends it. By contrast, the message of 2005 is, of course, marked by the death of John Paul II and an appreciation of his manifold contributions to Christian-Muslim dialogue, as well as by the early statements by Benedict XVI which indicated continuity in this field.[142]

What is common to all the Ramadan messages is the repeated attempt to address the specific situation, whether on the basis of papal declarations, of the level of development of encounters with the Muslims or of topical political situations. Also, the call for common action for humanity, especially for the weakest members of society, is always in the foreground, against the background of the framework that God has given to all human beings as creator, Sustainer and Judge. That is, in turn, a clear theological-anthropological statement.

8 Questionnaire on Mankind: too difficult to be made into a book

This is not a book, but it was supposed to have become one, and is particularly interesting in this respect: one of the projects initiated by the Secretariat for the Non-Christians was to send a questionnaire to qualified personalities of the various great religions, asking about the beliefs about the attitude of their faith towards people, their nature and actions, their rights and duties and their problems at the present time, based on the source texts and the recognised doctrine of the respective religion. This amounted almost to a written dialogue and was intended to serve as the basis for the advancement of the actual dialogue, which up to then had barely taken place, and certainly not in the form of thematic conferences. According to Rossano's internal report, Secretary at that time, the extent of the answers received was insufficient to produce a book. Moreover, they indicated clearly the fundamental problems these religions had with requests like this – ranging from the lack of a dogmatic authority to complete isolation from issues raised by modern culture. In retrospect, the Western and Christian influence in the formulation of the questions was recognisable. Instead of a book, Volume 32 (1976) of the internal bulletin was dedicated to a selection of these responses, two of which come from Muslims, Mohammed Aziz Lahbabi and Rashid Ahmad Jullundry. The former is considered to be practically the inventor of the concept of personalism in the Arabic-Muslim area; he studied at the Sorbonne in Paris before returning to North Africa. The latter worked at the University in Islamabad, Pakistan, in the field of Modern Languages/Arabic. A simple comparison of these two answers demonstrates the problem indicated above, that it is virtually impossible to speak of representativeness when the questions touch on modern issues.[143] Lahbabi is well aware of this issue and starts out, even before actually answering any of the questions, by emphasising that currently no dogmatic authority exists that is recognised by all Muslims, and that his answers can therefore be no more than his personal opinions in accordance with the spirit of Islam.

8.1 Humanity's calling: Servant of God – but how?

As for the question of humanity's meaning and purpose, both refer to Sura 51:56, according to which God only created humanity to serve him (translated as *adorer* or *worship*). But while Lahbabi relatively quickly interprets this as meaning that people should acknowledge the unity and uniqueness of God and conform to him, i.e. to his law, Jullundry adopts a much broader line of argument. This starts with the divine origin of humans, created in God's image, and destined to be God's representatives on earth and altogether a manifestation of God's honour, beauty and perfection. But then the human soul split up into two parts, one animal part and the other angelic, whereby the specifically divine qualities, such as life, knowledge, power, hearing, sight and language, indeed everything that is godly and perfect, are only constituents of the latter part of the soul. If human beings should fall away into the purely animal part, meaning materialism, and lose their connection to God, then they no longer had the right to be called human, because they have then missed their true calling, the relationship with God. Thus, in order to realise their purpose, humans had to fight against a part of themselves, whereas in Lahbabi's paper the same process seemed rather to be a completely natural outflow of people's love for God. For him, It is just as simple to find the cause of human misfortune and failure: if a person abandons the task given to them by God (righteousness towards humans and protection of all creation), then inner ruin and also external misfortune cannot be avoided.[144]

Here it is already clear that, despite agreement in the broad outlines, great differences also exist in important nuances, for example how they see the distinctly negative sides of humans, how they judge whether or not there is a simple and direct path between people's failure to meet the mark and the tangible consequences, or in the reflection on modern categories such as materialism.

8.2 Peace – difficult in many respects

Both authors give a great deal of consideration to the topic of peace. For Jullundry, three things are primarily constitutive, so to speak: that humans are always looking for peace, even if appearances give the lie to this, that inner and outer peace can only be achieved if the broken relationship between humanity and God is restored, and that no real peace can exist without righteousness. He goes on to elaborate on this in two fundamental thoughts and two points on a more practical level. The basis for the Islamic concept of peace is first of all the idea, which also is based on Islam, that all human beings and peoples are brothers, God's family, irrespective of all else that distinguishes them, explicitly including religion. In the name of this brotherhood, every system is rejected that brings people into competition with one other and exploits individuals or groups, including slavery – although it is not mentioned that neither the Quran nor Islam have actually abolished the latter. It is stated that this fundamental idea of brotherhood has been undermined in the name of a false concept of religious dogma, leading to the rule of one party over others, but no concrete examples are given. In a further step, righteousness is recognised as an absolute value, parallel to the guiding idea of human brotherhood. It is not made clear where the dividing line is to be found, or in what way righteousness is "more" than brotherhood, if that is the case. In conclusion, freedom of religion and even freedom to preach, so long as it is not propaganda against other religions, is presented as a matter of course, but here (again) there is no reference to the fact that Islam hardly complies with this, at least in the eyes of non-Muslims. *All things considered, this representation remains unsatisfactory and leaves the impression of a self-reflection which has been, or wants to be, spruced up.*

Lahbabi, on the other hand, chooses a different, not to say simpler approach. For him peace is, from an Islamic point of view, a task given by God to humanity, just as he sees Islam in general as a collection of forms of behaviour and thinking with regard to other people. (In Jullundry's opinion, religion was, on the contrary, not a collection of doctrines and theological beliefs, but a living spirit of truth and personal devotion to God). Peace has different levels of meaning, ranging from quietude to safety and peace of mind, that is to say, the total absence of warfare, both inwardly and externally. But peace of mind involves a constant struggle, which offers Lahbabi the opportunity for a long digression about the impossibility of translating the expression "Holy War". At this stage, his explanations become apologetic. He classifies war as something that permanently destroys all positive values such as justice, charity and freedom, whereby he also goes into the question of when it is justified to fight against exploitation, chauvinism, etc., in order to defend peace and humanity's positive material or moral achievements. This amounts to a reversal of Jullundry's train of thought: it is only at the very end that he states explicitly that the option for peace is also the option for humankind, the sacred masterpiece of divine creation in soul (direct emanation of the divine breath) and body. Here he also speaks briefly, without going into details, about people's dignity and personal values.[145] *While Jullundry's presentation left the impression of being too self-contained, unproblematic and thus polished up, Lahbabi's presentation lacks the anthropological-philosophical argumentative stringency one would have expected from him, possibly on account of the questionnaire context. Only at the very end, and extremely briefly, does he comes to consider why the issue of peace is so important to people. Once again it appears that Jullundry adopts a more western approach (in the anthropological sense) in comparison to Lahbabi's astonishingly traditional line of argument.*

8.3 Social inequality – for Islam an evil

In contrast, Lahbabi is very precise and consistent on the question of the growing inequality among humans: according to Islam, people were created by one God and therefore share the same nature. The only inequality between people that the Quran accepted was inequality in piety, and every other inequality was an evil. Social inequalities had nothing to do with differences in piety, but were a consequence and expression of wrongdoing and lack of morality. On the contrary, poverty endangered people's physical, social and moral existence and ultimately their belief in God. To combat material poverty was therefore an urgent duty for every Muslim, as shown by the Islamic institution of the "alms tax". Jullundry goes in a similar direction, offering no anthropological justification at this point, but rather a socio-political intensification. He also emphasises that Islam takes from the rich to give to the poor, and that according to Islamic understanding the state should protect those who were unable to protect themselves and should prevent the emergence of classes. Here again he brings into play the idea that people should free themselves generally from bondage to material goods by using them for God's purposes. However, his criticism is clearly directed against the currently prevailing capitalist system as the root cause of social problems, and even before that he rejects democracies in the name of Islam, because they did not ensure righteousness everywhere, corresponding to Islam's basic concern.[146] *Here again it is clear that Jullundry's argument is much more pointed towards the current situation than Lahbabi's. Thus he lies closer to what was expected, to a Christian model as it were, than Lahbabi, although fundamentally more critical.*

8.4 Family issues: the controversial problem of birth control

As far as the question of birth control, marriage and divorce is concerned, there are astounding discrepancies or questions which are dealt with by one author, but not by the other. Regarding population growth and birth control, Lahbabi clearly explains that this is frequently discussed in Islamic circles, but that there is no agreement between the "modernists" who are in favour and the "conservatives" who reject control measures as a lack of respect for life created by God, although they otherwise have no problems with modern biology. Islam rejected celibacy, even regarding it as a disgrace, and saw divorce, which was not to be confused with repudiation, as a necessary evil. At least on the last point, Jullundry would agree. Unlike Lahbabi, he claims that it is recommended not to marry if there was no desire for marriage and one could live a life of absolute chastity. As for the question of birth control, he states that a mother should breastfeed her child for two years and not have another child during that time. However, modern forms of birth control were not allowed, since they led to sexual anarchy and destroyed chastity. Abortion was absolutely forbidden as a form of birth control, because the Quran had already condemned killing children for fear of poverty. Only if the mother's life were in danger could an abortion be considered legitimate.[147] *Here it becomes very clear that very modern issues could lead to a strong divergence or difference of opinion – at least on a methodological level – whereby the more traditional side obviously tends to see itself as the representative of undivided Islam.*

8.5 Industrialisation and the dispute about secularisation

There is much more agreement on how to evaluate industrialisation – namely, fundamentally positive with limitations regarding its humanity. In the eyes of Islam, human beings are masters of the universe. So long as industrialisation helps people to satisfy their needs, it is only positive. Lahbabi even goes so far as to say that celebrating humanity's creativity is equivalent to praising the creator. But the limit is clearly reached at the point where such a development becomes inhumane and exploitative, degrading people to objects. Once more, Jullundry is more explicit and clear at this point. He emphasises that sovereignty over and enjoyment of the world are bound to the divine will, and definitely renounces the present-day industrialisation that has made people slaves of the machine and exploited and colonised parts of society and entire nations. On the other hand, the Quran also rejected conscious poverty and abstention from the world as the spiritual path and goal of humanity. The question arises as to whether this much more pointed attitude may also be related to the context of the Indian subcontinent, which was a much more worthwhile object of exploitation and colonisation than the much poorer North Africa. In a certain sense, this seems to be confirmed by looking at the next question about secularisation, which is directly related. Lahbabi simply regards this as irrelevant to Islam, while Jullundry refers explicitly to the Indian situation and makes fundamental and quite positive remarks on this subject. To be sure, he starts by stressing that Islam is a system of morals based on revelation, and that a Muslim must follow this morality both in private and in public life; God is the measure of all things, not humans, and here Islamic law plays a role. But at the same time he notes that this law – unlike Islam – is strongly subject to change in time and space. Moreover, revelation and intellect, which is also a divine gift, could not get into each other's way, and everyday worldly affairs had always been a matter of intellect and consultation since the times of the Prophet. In addition, in all its history Islam had never developed an organised church or a system divided between spiritual and temporal power. The state had no right to impose its faith on the people, whereby the founder of Hanbalism and his personal fate are named as a prime example. Conversely, secularism was the product of certain circumstances in which religious organisations were unable "to provide proper guidance for people in their temporal affairs"[148] – a thoroughly Islamic view of things. The basic principles of secularism are in his view freedom of thought, meaning that certain areas of human life such as opinion, social forms and customs were not determined by religion, as well as lack of interest in a life after death, and finally truth independent of revelation. In India, there had been a positive development of secularism (from a negative to a positive attitude towards religion). The common ground of different religious traditions was accepted, and all people saw themselves as heirs of this foundation based on eternal reality – incidentally a very Hindu-oriented idea. Thus, Jullundry concludes that there is more concurrence between Islam and secularism than difference, for example the principles of freedom of thought, the rule of intellect over temporal matters, and the impartiality of the state towards its citizens as long as it is not anti-religious or immoral, whereby the divine guidance for human reason is once again brought into play. On the whole, however, this new secularism was perhaps the only seriously feasible path for multi-religious societies – and had indeed already been compared by Muslim scholars of the Indian subcontinent to the Treaty of Medina, which the Prophet had concluded with the non-Muslims. *What Jullundry's answer clearly expresses at this point, is how there is indeed an influence of space and time in a certain area and therefore a range of possible answers in terms of Islam. It would certainly be interesting to ask (as was not done here) when, where and by whom deviations are seen to be legitimate.*

8.6 Human freedom – differing priorities

Lahbabi gives a very fundamental response to the question of human freedom: people are born free, and this freedom is also necessary, because otherwise they would be neither morally nor socially responsible for their deeds. The responsibility as such is the same for all, because it is a personal dimension. Differences only exist in the degree of responsibility, which varies according to the concrete possibilities. However, it is clear that no one with normal opportunities is entitled to live at the expense of another, and it is also clear that there may be no subordination to a creature that sins against the creator. Jullundry, on the other hand, places much more value on virtue, without which there can for him be no real freedom, also with recourse to the Quranic statement already quoted that differences between people exist only in the degree of their piety. Islam also made no distinction between men and women, in the sense that both could reach the truth and lead an ideal life. However, women had their own rights and obligations, and free interaction between men and women for purposes of amusement was not allowed. Power was a grace bestowed by God to people, and Muslims were even encouraged to seek power in order to establish justice among human beings, and to create an atmosphere in which a Muslim can develop his personality on the basis of moral values. A similar principle applies to wealth. Ibn Taymiyya is even quoted with the words, "[i]f religion is separated from power or power from religion, peoples' affairs are ruined."[149] *To a certain extent, these responses show how differently the two authors have been influenced, in this case in a philosophical-religious way, in spite of belonging to the same religion. Such differences are also reflected in the answers to the immediately following question about the ideal image and the path leading to it.*

Lahbabi again underscores the inner connection between freedom and responsibility: social commitment and the struggle against injustice and the corruption of morals as well as against all bad wishes. Jullundry, on the other hand, is very traditional and oriented towards concrete statements from the Islamic tradition: according to him, the prophets are the perfect humans, then come those who sacrifice their possessions and their lives for the sake of the truth, in other words the martyrs. The consequence is that to live, work and endure suffering for the truth is the ideal image of human life. The basic characteristics of a perfect person were described in the Quran in this way: faith, good deeds, earnestness (as opposed to hypocrisy), combatting the evil within (e. g. jealousy, hatred, lies, lust, addiction) and without, steadfastness and modesty, as well as determination, modest appearance and dignified behaviour.

8.7 Individuals and transcendental collectivism in the Islamic community

When it comes to the next question about the relationship between the individual and the community, it is surprisingly Jullundry, and not Lahbabi, who brings the question of personal responsibility into play. Both, however, point clearly to the concrete experiences of the past in order to explain contemporary Muslim reactions to this question. Jullundry's answer is almost more fundamental: to be moral in being and in action, people must be free. Both concrete individualism (in the form of capitalism and colonialism) and concrete collectivism (which developed into totalitarianism) had curtailed this freedom and thus people's moral possibilities, although Islam certainly favoured a healthy form of collectivism, see the Islamic community. For this reason the state had the right to restrict the activities of individuals if they ran contrary to the interest of the general public, but did not extend to the suspension of an individual's responsibility for his actions. It is a kind of transcendent collectivism, whereby Islamic society is not an end in itself, and Islam certainly supports a democratic collectivism working for people's material well-being. These are all very basic, but also fairly nonspecific considerations presented here by Jullundry, while Lahbabi is astonishingly definite, stating straightaway that Islam is part of the Third World as an earthly city. Just like Jullundry, he had experienced individualism of a liberalistic nature in the shape of past imperialist regimes and the present-day neo-colonial forces. He believes that this led Muslim thinkers to opt for a kind of Muslim socialism marked by the Islamisation of the banks and the nationalisation of certain sectors. These were both supported by texts from Islamic tradition, since Islam does not only forbid interest, but also monopolies on primary products such as fire (energy), water, common pastureland also, at least in part of the tradition, on salt (as food and source of wealth), as well as prescribing the alms tax and charitable acts.[150]

8.8 All people are brothers – or not entirely?

Then the questionnaire moves on to a very different level, namely the question of the attitude towards other people (and thus also the use of force) and towards other religions (which is not really directly related to anthropology). As far as the very first question is concerned, Jullundry's answer is extremely short: Muslim relations with other people were based on peace; war was only allowed temporarily, if Muslims had become victims of aggression. In addition, Muslims were supposed to hold out a helping hand to those in need. Even on a religious level, Islam was not hostile to anyone, for it did not see the truth as the monopoly of a particular race or group of people, but recognised all the prophets. Lahbabi is first of all more fundamental on an anthropological level: according to Islam, other people are brothers, since all humans are Adam's descendants, so the relationships between them can only be fraternal relations based on equality. The Quran also explicitly states that believers (not just Muslims, because according to Islamic understanding faith is an innate gift inherent in human nature) should be as brothers to one another. Nonetheless, he lists a fairly broad range of exceptions: violence is justified, even demanded, when someone tries to destroy the truth, i.e. the belief in God, and if someone's person, dignity, possessions or homeland are attacked. Basically, however, the person is sacrosanct, unless he puts other people in danger. In general, Lahbabi classifies violence against the weak as criminal, but revolt/violence by the oppressed as just. For Islam, everything inhumane was also directed against God and should therefore be countered by holy violence, a just struggle. Regarding people of other religions, he actually quotes Sura 5:78 and 82, which emphasise that the Christians and some of their priests and monks are particularly friendly towards Muslims, whereas the Jews are described as also being devout, but especially hostile to the Muslims.[151]

8.9 Life after death: a certainty

Finally, the last question is about life after death. Here there is clear agreement. People have the choice between two paths; by faith/pure intentions and good deeds, they can earn the joy of the next world/fellowship with God, for after death they will have to answer for their life in this world. It is interesting that Lahbabi concludes with personal statements and wishes, from which I would like to quote one passage:

> "Constat: fiasco lugubre des politiques en cours, ces temps-ci, par ce monde là. Où chercher la sauvegarde? Peut-être les religions et les idéologies humanistes, en engageant leur foi ardente dans une sincère collaboration, arriveront-elles à redonner da têtes bien faites à ce monde décapité. Tous les souhaits de succès à toute initiative dans ce sens!"[152]

9 Not anthropology is problematic, but its concrete effects

On the whole, it is very clear how wide a range of themes in the field of Islam is covered by the publications of the Pontifical Council for Interreligious Dialogue and how much they vary (which is to be seen as necessary and positive), whether in fundamental matters or individual problems, in strictly theological anthropology or in efforts to perceive and respond to the needs of the counterpart. It is precisely the latter which is a constant and never-ending process of spiritual learning, and ultimately it is only the Muslim himself who can say how far he feels spoken to, whereby not all Muslims can and must be of the same opinion. In any case, the honest attempt is clearly discernible, as is also the specifically Catholic train of thought coined by the popes, especially in the person of John Paul II, whose efforts in the promotion of Christian-Muslim dialogue brought great progress. It is also abundantly clear how strongly the similarities in the understanding of humanity (God as creator and goal of humankind) are emphasised when the Muslims are addressed directly, especially in contrast to a purely materialistic world view. This is right and proper in the perspective of real cooperation for the benefit of real people, which is the expressly declared goal and was also received and appreciated in this sense. But this makes it easy to ignore differences on the level of human anthropology that are otherwise quite obvious. The more the texts are intended as internal information, the more clearly they address these differences. Even if both religions see humans as God's creation, it makes a big difference whether they are made in God's image or whether they are God's representatives. At a closer look, the freedom and innate dignity of humans, regardless of correct behaviour towards the creator, are indeed greater if they are God's image rather than God's representatives. Although both should not act independently of God and his commandments for humanity and the world, but the creative freedom accorded to people as God's image by biblical-Christian tradition is indeed great, even taking all the similarities in fundamental orientation to the hereafter and in specific ethical issues. For human beings as God's representatives, as the Quranic-Islamic tradition sees it, the highest priority is the greatest possible implementation of the very practical, comprehensive and detailed will of God and to do so with all one's strength. And it is precisely at this point, where it is not so much a question of anthropology as such, but rather of its (necessary) impact on society and the state, that the great difficulties emerge repeatedly without any apparent prospect of solution. To be sure, one should not overemphasise difficulties when working together to achieve something important, but it is also misleading and, in a sense, dangerous to simply sweep them under the carpet while at the same time emphasising the common battle against materialism and secularism. In particular, there are differences between Christianity and Islam in their attitude to the modern secular world, and it is not possible to wage a joint battle as easily as it sometimes appears when speaking to Muslims.[153] It is certainly difficult to find just the right expressions and the correct tone of voice, but it is only possible to build up good collaboration on a solid foundation, so that it is therefore especially sad that the book about humans seen from the perspective of other religions, as it had actually been planned, did not at least materialise for Islam.

Notes

1 cf. Pontifical Council for Interreligious Dialogue (ed.), Recognize the Spiritual Bonds which Unite Us, 16 Years of Christian-Muslim Dialogue, Vatican City 1994, p. 48–54, the chapter "PCID and Dialogue", which gives a brief overview of the work of the Secretariat, structured according to the respective presidents (with a subchapter on regional dialogues), whereby the final section "Letters for Id al-Fitr" occupies almost as much space as the entire history of the Secretariat or Pontifical Council. If one takes into account that the years under the direction of Cardinal Marella (1965–1973), i.e. almost the whole first decade of the existence of the Secretariat, were dedicated only to this written work, then this fact alone reveals a very clear emphasis, a conscious decision which was also opposed by some members of the advisory board (to be found in Rossano, Pietro, The Secretariat for Non-Christian Religions from the Beginning to Present Day: History, Ideas, Problems, BSNC 41/42 (1979), p.92f., in which he reports on the meeting of advisers on 28–30 April 1965) and which distinguishes the Vatican Secretariat for Non-Christians from the corresponding structures of the World Council of Churches, where such basic documents were in any event products of actual dialogue, cf. Sperber, Jutta, Christians and Muslims, The Dialogue Activities of the World Council of Churches and their Theological Foundation, Berlin/New York 2000 (Theological Dissertation Augustana University Neuendettelsau 1996), p. 73–74. From a historical and ecumenical point of view, this publication by the Secretariat has a comprehensive but also popular character, marked by a mixture of informative continuous text, interspersed with original quotations and many pictures. The latter are, in a sense, the principal feature of this booklet, the part of the documentation that is nowhere to be found in such an impressive abundance of detail or precision. The images certainly speak a language of their own and are of great significance. They strongly remind us that much of what is dealt with here in this study was also, and possibly often, originally a tangible, even violent human encounter. The level of the images is also an important anthropological level and dimension, but one that would go beyond the scope of this work, which is why the texts from this volume are used here to supplement other sources wherever relevant, but are not taken as a (sub-)topic. Significantly, the book is also not mentioned in a list of publications expressly addressed to professionals, such as the plenary assembly of the Council, although it is still up-to-date and is gladly passed on, cf. Akasheh, Khaled, Considerations on Forty Years of Religious Dialogue with Muslims (A Report), BPCDIR 116/117 (2004), p. 203–204. The idea had evidently arisen at the plenary assembly, as first remarked in Michel, Thomas, P.C.I.D. Dialogue with Muslims since the Last Plenary, BPCDIR 82 (1993), p.45: "It was felt that a timely project might be the preparation for a dialogue between the Catholic Church and Muslims in the past 15–20 years." In the final report on the discussion F[itzgerald], M[ichael] L[ouis], Final Discussion of the Plenary Assembly, BPCDIR 82 (1993), p.89, it is described in this way: "It would be useful to have a book on the dialogue which has been carried out by the Catholic Church with Muslims during the last fifteen years. This could be an illustrated volume, with some of the most significant texts." It is a different, yet similar case, with the volume Fitzgerald, Michael L[ouis] / Borelli, John (edd.), Interfaith Dialogue, A Catholic View, London/Maryknoll 2006. Even though Archbishop Kevin McDonald emphasises in his preface (p.vii): "What Archbishop Michael offers is an authoritative voice in an area that is crucially important but sensitive and delicate," the volume is nevertheless no official publication of the Secretariat and is therefore here not taken into account in its entirety, but only insofar as it is relevant in certain places as a supplement and confirmation. It is a very comprehensive overview, not overloaded, easy to understand and very balanced, and can only be highly recommended for reading. Particularly helpful is the appendix The PCID: A Vatican Structure for Dialogue (p. 239–241), which also deals briefly and substantially with the work of the Commission for Religious Relations with Muslims. It should also be mentioned here that there are other important publications by the Secretariat, but they are not tailored towards Islam and have therefore not found a place here despite their general importance, for example, the document on the relationship of dialogue and mission (German title: Die Haltung der Kirche gegenüber den Anhängern anderer Religionen), which is to be found in BSNC 56 (1984) in a total of six languages, including Arabic, presumably with regard to the fact that mission is one of the most provocative issues for Islam, although that has – from the Islamic point of view – theological, not really anthropological reasons.

2 There is a document "Notes sur l'organisation et les règles d'action du Secrétariat" in the library of the Pontifical Council for Inter-religious Dialogue which is not publicly accessible, but which is expressly marked as intended for internal use. Thus, in the following description, I can only refer to Rossano's account from 1979 mentioned above (p. 88–109), which was originally also written for an internal event, but later published. In addition, reference should be made to the obituary for the first secretary, P.Pierre Humbertclaude, and thus a recollection of the first ten years of the Secretariat, cf. Pedretti, G., A la mémoire du P.Pierre Humbertclaude S.M. (1899–1984), BSNC 55 (1984), p. 114–115. A slightly different picture is painted by the later secretary and president, Archbishop Michael Fitzgerald, in Fitzgerald, M[ichael] L[ouis], The Secretariat for Non-Christians Is Ten Years Old, Islamochristiana 1 (1975), p.91: "1974 saw a change in policy, perhaps inspired by the new President, Cardinal Pignedoli, but also determined by the development of new structures of dialogue. In the last ten years 25 Commissions for Dialogue have arisen in all the Episcopal Conferences of the world. Rather than hold a general assembly of its consultors the Secretariat decided to go out into the field." On p.90 he writes on the beginnings: "It was felt that, before engaging directly in dialogue with Muslims, there was a need to help Christians to reflect on the documents of the Vatican Council and upon their implications. A new spirit was to be created which could have practical consequences, for instance in the way in which Islam is described in books written by Christians. Thus, while encouraging contacts at local level, the Secretariat did not feel called to organize official meetings between Christians and Muslims. It preferred to concentrate more on sensitizing Christians to the need for dialogue and the ways in which it could be prepared." The basic method of obtaining confirmation, as it were, at the grassroots level, would be dealt with later, even if the results were no longer so tangible, cf. Arinze, Francis A., Prospects of Evangelization, with Reference to the Areas of the non Christian Religions, Twenty Years after Vatican II, BSNC 59 (1985), p.138–140. Apart from that, it is interesting that he speaks of great difficulties on the road to interreligious dialogue (that alone is significant!).

3 Later, there were also very detailed considerations as to who should be prepared for the dialogue, how and by whom, especially with regard to talks between Christians and Muslims, for example in a lecture at the conference of the Secretariat with consultors and professionals from 26 to 28 September 1971 in Paris: Gelot J[oseph], Éducation des chrétiens en vue du dialogue, BSNC 20 (1972), p. 63–68.

4 The idea behind this emphasis on the bishops is obviously that perfect discipline alone could guarantee the unity of all in the Church, which is especially necessary in relation to the non-Christians. One may well understand this position, for example with reference to the Quranic polemic against Christians, which is not specifically mentioned. Nonetheless, internal discipline is a long way from resolving the problem of the different denominations.

5 The influence both of general Catholic ideas of natural law as well as concrete concepts of theology of religion is perceptible. For example, Pietro Rossano said: "some consider the religions (...) as the natural expression of the genius of the peoples (Daniélou)" (Exposure from the Catholic Point of View, BSNC 17 (1971), p.105). Again and again the model of various stages is visible, which sees the Catholic Church and its teaching, as it were, at the peak of truth and perfection which are achievable and therefore desirable for humanity. This also represents the inner logic for missionary impetus.

6 The actual role of the consultants was quite different at different phases of the work, as is explained in various other places, but the task as such and the contacts were always taken seriously, for example Machado, Felix [Anthony], Bangalore – India: Report: Asian Consultors and/or Secretaries of the National Episcopal Dialogue Commissions from Asia, 5–8 July 1998., BPCDIR 99 (1998), p.345/6: "It is the practice of the PCID to meet with its Consultors on the continental level, at least once in their five-year mandate, to reflect on the theme of /p.346 interreligious dialogue, taking into account the specific concerns relevant to each particular situation." An even more fundamental and weighty assessment is to be found at Fitzgerald, The Secretariat for Non-Christians Is Ten Years Old, p.91: "The Secretariat has always held regular meetings of its consultors. These have enabled study to be made of fundamental points concerning dialogue, and also have provided the Secretariat with information on the way dialogue is progressing in various parts of the world. The consultors also made practical recommendations to the Secretariat. For instance, at the Paris meeting of September 1971, the section on Islam recommended that greater efforts be made to make Christians aware of the need for dialogue. It expressed the wish that each episcopal conference would appoint a bishop to take charge of dialogue. It insisted that care should be taken in catechisms to explain relations with Muslims in a positive manner. It hoped that more people would be prepared for a dialogue with Muslims and that there would be greater collaboration in this line with other Churches and Christian confessions." A similar line is taken by Fitzgerald, Michael L[ouis], Report on the Activities of the PCID: November 1998–October 2001, BPCDIR 109 (2002), p.73, which points out that efforts were also made to maintain a certain representation regarding religions, but also with respect to geography and the various positions, ranging from the more academic to the very practical, and that one made sure that some of the consultors were always close at hand, meaning available in Rome. On p. 92–95 he also goes into detail on the publications, even describing the structure of the internal bulletin: excerpts from papal speeches and statements, articles that were usually written by staff and consultors and containing substance, without being too specialized, among them as a rule all the lectures held at meetings with the consultors, reports from various dialogues worldwide and finally bibliographic information. In a later text, Fitzgerald expresses his thanks to the Pontifical Gregorian University for the good cooperation and advice given by the lecturers in the field of interfaith dialogue: Fitzgerald, Michael L[ouis], The Pontifical Gregorian University and Interreligious Dialogue, BPCDIR 107 (2001), p.247.

7 There is enlightening information about dialogue with Muslims, above all in the Maghreb, in Gelot, J[oseph], Éducation des chrétiens en vue du dialogue, BSNC 20 (1972), p. 63–68. The lecture was held during the conference of the Secretariat with consultors and other specialists in Paris (26–28 September1971). It is clear, and also clearly stated, that for the so-called professional religious dialogue with Muslims at least a considerable part of a lifetime is required. Then the lecture deals very critically with the role of church institutions in former colonies, which had not adapted to an environment that was now almost purely Muslim, but had rather offered their services in the same way as before, whether – implicitly – for a smaller circle with respect to worship, or – explicitly – for a new Muslim clientele, in the case of education and schools, for example. To take account of Islam or not was practically left to the individual's own efforts. The author sees it as particularly important to be familiar both with the religiously relevant high Arabic language and with the local dialect, for which he even presents a relatively precise syllabus. In spite of all this, it was very important not to lose one's Christian identity: "Plus qu'on n'est parfois porté à le penser, les musulmans veulent avoir affaire à des gens qui craient à qu'ils sont." (p.68) With regard to the continuing training, a workshop on interreligious dialogue and conflict resolution should be mentioned that was organised by the Pontifical Council for Inter-religious Dialogue from 24–31 July 1999 for participants from various African countries in Nairobi. This dealt with Islam and with Christian-Muslim relations in Africa, and in quite practical terms with the possibilities for mediation between Christians and Muslims in the Sierra Leone conflict by the Interreligious Dialogue Council. For the purposes of broad impact, the participants were supposed to send the reports of this workshop to their dioceses, but at the same time it was emphasised that inter-religious dialogue and inculturation should be included in the training at the seminaries – in this respect it would seem that in more than 30 years not enough has been done in places where it would be particularly necessary. It is also interesting to note the remark on the need to take the (presumably common?) African culture into consideration during dialogue with Muslims, giving rise to the conclusion that this could, in a certain sense, be even more characteristic than religious affiliation, for which assumption there are further indications in documents about Africa mentioned here, but also in others; cf. Nairobi, Kenya: Interreligious Dialogue Workshop Session, 24–31 July 1999, BPCDIR 102 (1999), p.346–348. The report from another workshop makes it clear that the multicultural and multi-religious situation in Africa is not simple either, which means that the Pontifical Council for Inter-religious Dialogue continues to be called upon to provide aid for education and training. As far as Islam is concerned, dialogue on life and cooperation should also be promoted. It is also interesting to observe that Islam in Africa imitates the Christian mission with its schools, hospitals and other welfare institutions, while in areas where it is in the majority (Chad and Senegal are mentioned as examples) it is sometimes afraid of the spread of Christianity, as described in Isizoh, Chidi Denis, Yaoundé, Cameroon: Workshop on the "Challenges of Interreligious Dialogue in Sub-Saharan Africa", 20–25 March 2001, BPCDIR 107 (2001), p.254f. A project on the contribution of Africa to the religious heritage of humanity also points in this direction, a joint undertaking by the Pontifical Council for Inter-religious Dialogue and the World Council of Churches. Muslim experiences were also included, albeit rather incidentally, and it was explicitly stated that such workshops and dialogues would be useful elsewhere as well, which is certainly true if one reflects on an anthropological level alone on facts and issues such as these, with the necessary modifications: "Each person is in the 'we are' kinship of biological, anthropological, sociological, psychological and historical realities (...) What does it mean to be a human being in spiritual terms of humanity, in biological and anthropological terms, a citizen of any country or nationality?" (Mbiti, John, Enugu, Nigeria: The Contributions of Africa to the Religious Heritage of the World;. 8–13 January 2001, BPCDIR 107 (2001), p.257, cf. also p.255f.259) – even ignoring such specific issues as African slaves in some countries of the Arab world, which challenge one to ask questions and to take concrete action. The preface of the book by Michael Louis Fitzgerald and John Borelli, Interfaith Dialogue, A Catholic View, London/Maryknoll 2006, brings the recollections of the two protagonists on their respective lives in the field of interreligious dialogue (p. 1–12 and p. 13–24) and offers two interesting and absolutely different examples of the way something turned out in practice which otherwise was described more generally and theoretically. Education and dialogue are also seen as a way to combat fear for one's own identity and of fundamentalism, so Machado, A Summary of Reports by the Members: Interreligious Dialogue Promoted by the Church, p.127.

8 The importance of prayer (in combination with cross-religious work for the well-being of people and for peace and harmony in society) is mentioned later in a prominent place, just after 9/11, according to Machado, Felix A[nthony], A Summary of Reports by the Members: Interreligious Dialogue Promoted by the Church, Pro Dialog 109 (2002), p.126f.

9 Rossano, p.102.

10 ibid., p.103.

11 Rossano, Piero, Exposition from the Catholic Point of View, p.104 (on the occasion of the meeting of the Joint Working Group, 7–12 June 1971). Nowhere else is there such a concise, exact, yet comprehensive summary of the anthropology underlying the dialogue with people of other faiths on the part of the Catholic Church. By contrast, all other statements can only represent substantiations and clarifications with respect to a specific situation. A concise and helpful summary of such substantiations can be found in Duval, Léon-Etienne, Brève reflexion sur le dialogue, BPCDIR 72 (1989), p. 319–322. They are also important for the Christian-Muslim dialogue, cf. Sabanegh, E[douard] S[ami] Martin, Les Chrétiens d'Europe face aux travailleurs musulmans immigrés, BSNC 57 (1984), p.312: "Qu'il suffise de signaler que dans les années à venir, il faudrait attacher plus d'importance à la recherche anthropologique afin d'asseoir les bases d'un vrai dialogue pour l'ensemble de l'Église." However, at least as regards internal assessments, these things are in practice more restricted to "political" issues (Muslims with fundamentalist tendencies, the frustrating situation in countries with a Muslim majority), says Machado, Felix [Anthony], Interreligious Dialogue in the Various Regions of the World, BPCDIR 92 (1996), p.268 f. This may be coincidence to a certain degree and extent, but it also demonstrates a fundamental tendency or threat for Christian-Muslim dialogue in particular.

12 Deliberations on Islam had already taken place at the Paris meeting of the Secretariat with consultors and experts (27–29 September 1971), according to Forward, BSNC 18 (1971), p.135. The Islam Section recommended bringing the benefits of dialogue more strongly to the attention of the local hierarchies, i.e. the bishops' conferences. Special concerns – apart from increased financial and human resources – are the training of the people (so, more generally, Fitzgerald, Report on the Activities of the PCID: November 1998-October 2001, p.75 f.), the various local catechisms and their positive portrayal of Islam, the practical commitment to justice and peace, the situation in India and the concern about unauthorised statements and actions, especially regarding the Middle East conflict: "What it has particularly in mind is the making of certain declarations in favour of Israel which offend Arab Opinion." (Recommendations of the Four Section (sic!) Addressed to the Secretariat, BSNC 18 (1971), p.214, cf. also p.213.215) On the subject of Islam, Cardinal Arinze explained (Prospects of Evangelization, with Reference to the Areas of the non-Christian Religions, Twenty Years after Vatican II, p.133) that religious and political leaders were mainly unwilling to conduct dialogue, as were high-ranking individuals also, that religious freedom for Christians in Islamic countries was far from assured, even though many individual Muslims were friendly and moderate, and that many Muslims made no distinction between religion and politics. On the other hand, there were also Christians who considered dialogue with Muslims a waste of time. In addition, particularly in Asia, the two communities were highly prejudiced against one another: the Muslims accused the Christians of being westernised and allied to the West in all kinds of ways, whilst the Christians for their part accused Muslims of not taking them seriously as equal citizens on a personal and structural level (denial of a religiously pluralistic state, Islamic law). But in general he remarks that "after the dialogue euphoria following on the Second Vatican Council it is no surprise that these difficulties show up themselves more and more." (p.131) As far as the concrete situation in individual countries is concerned, whether in a general sense or regarding dialogue, the reports in the internal bulletin are more interesting, especially the reports presented at the plenary session of the Secretariat and published in BSNC 57 (1984), such as Michel, Tom, Islam in Asia Today, p. 318–327, or maybe not quite so outstanding, Kayitakibga, Médard, Chrétiens et musulmans en Afrique sub-Saharienne, p. 328–338.

13 In his editorial, BPCDIR 81 (1992), p.265, Michael Fitzgerald states that the consultors had been used "as a 'think tank', a reflection group on study certain themes." For the further developments, cf. Akasheh, Khaled, Bureau pour l'Islam Rapport d'activités: novembre 1998-octobre 2001, BPCDIR 109 (2002), p.98.
14 cf. ibid. The essay itself is Legarde, Michel, Violence et verité, BPCDIR 81 (1992), p. 282–327.
15 ibid., p. 284–287 provides details regarding terms and references for this first stage of the relationship of truth and violence in the Quranic tradition. The argumentation on the other stages (p. 287–294) is also based on key concepts from the Quranic tradition. The references to the eschatological connotations in the Quranic texts mentioned by Legarde are mine.
16 ibid., p.291.
17 cf. ibid., p.294f., as well as note 18, which names not only the evaluation, but also the individual positions evaluated. Here one learns that both traditions are particularly popular among the Muslim Brothers, who emphasise that Islam believes and cannot diminish, and that Islam commands and cannot be commanded.
18 For the comments on *jihad* cf. p. 294–299, on abrogation cf. p. 300–305, whereby p.304 identifies the groups to which the abrogations refer, especially the unbelievers. Legarde also deals with the fact that there can be mission with the sword: "Les prophètes ont pratiqué tour à tour 'la mission par la conviction' (...) et la mission par l'épée; par la suite, les savants musulmans ont calqué leur mission sur celle des prophètes; quant aux princes musulmans, ils se sont cantonnés dans la mission par l'épée."
19 On the theology cf. ibid., p. 306–309, whereby note 36 p.307 is especially interesting. Legarde's concluding remarks are to be found on p.309f.
20 On these practical considerations cf. ibid., p. 310–315.
21 cf. ibid., p. 315–318.
22 cf. ibid., p. 318–321.
23 ibid., p.321, cf. also p.322.
24 cf. ibid., p. 322–327.
25 According to Jeusset, Gwenolé, Conflits et harmonie, BPCDIR 81 (1992), p. 328–335.357–359. The pages in between with the examples themselves are also well worth reading.
26 According to Roest Crollius, Ary A., Harmony and Conflict, BPCDIR 81 (1992), p. 360–377. It is also interesting that Roest Crollius does not associate fundamentalism and syncretism directly with religion, but with cultural psychology. He classifies them both as two different ways of coping fairly effortlessly with the uncomfortable phenomenon of pluralism of ideas and values. Both had, each in their own way, a superficial interpretation of texts and facts. In both cases they drew a clear dividing line between those who were inside and those outside. And both of them also showed a tendency to produce leader figures: "The search for security calls forth leader figures, either prophets of fundamentalism or gurus of syncretism, the latter not less sure of themselves than the former." (p.376)
27 In detail these are: Stamer, Josef, Les tendances actuelles de l'islam en Afrique de l'Ouest, Dossiers de la C. R. R. M. Courants et mouvements dans l'islam contemporain 1, Rom 1992, Troll, Christian, Islam in South Asia: Facts, Movements and Trends, C. R. R. M. Reports Recent Trends and Movements in Islam 2, Rom 1992, Platti, Emilio, Tendances dans l'islam d'aujourd'hui: Europe Occidentale, Dossiers de la C. R. R. M. Courants et mouvements dans l'islam contemporain 3, Rom 1992, Donahue, John, Islam in the Middle East, C. R. R. M. Reports Recent Trends and Movements in Islam 4, Rom 1992, Michel, Thomas, 1991 Estimated Muslim Population of Various Countries, C. R. R. M. Reports Recent Trends and Movements in Islam 5, Rom 1992, Pruvost, Lucie, L'islam dans les cinq pays du Maghreb arabe, Dossiers de la C. R. R. M. Courants et

mouvements dans l'islam contemporain 6, Rom 1993, Pruvost, Lucie, Tendances et courants dans l'islam algérien, Dossiers de la C. R. R. M. Courants et mouvements dans l'islam contemporain 7, Rom 1993. The meaning and purpose of this series is explained in each case at the beginning of a dossier, even before p.1. These dossiers are not included in the list of publications on the Christian-Muslim dialogue mentioned in note 1, as they are obviously of inofficial or semi-official character. (However, the corresponding publications are recommended, for example, by Francis Arinze to the Synod of the African Bishops' Conference, so Arinze, Francis, In Promotion of Interreligious Dialogue in Africa, BPCDIR 87 (1994), p.210.) But in another place Khaled Akasheh describes this commission, which exists since 1974, as a "'think tank' for the study of questions arising from relations with Muslims" (Report on the Activities of the PCID: Relations with Muslims, BPCDIR 101 (1999), p.218). This phrasing hat already been used by Michael Louis Fitzgerald, so Editorial, BPCDIR 81 (1992), p.256. This issue of the Bulletin consisted mainly of three essays which form, as it were, the small-scale counterpart to the separate publications here and do not deal with current trends in the Islamic world, but with the theme of harmony and conflict.

28 Pruvost, Tendances et courants dans l'islam algérien, p.9, cf. also p.7–8.10

29 cf. the chapter ibid., p. 15–18, which deals with this movement.

30 cf. ibid., p.18–24. Significantly, the French term for enlightened Islam is "islam des lumières" (p.21). This type of reasoning is certainly closer to the Catholic standpoint, as a comment by Cardinal Arinze at a Christian-Muslim meeting brilliantly reflects: "Une personne raisonnable accepte que le pluralisme religieux est une réalité. Nous n'obtiendrons pas l'unité religeuse par la force: physique, psychologique, économique, politique, sociale ou quelle qu'elle soit. Cette unité forcée ne serait pas digne de la personne humaine. La religion doit etre proposée et non pas imposée. ... La violence utilisée pour défendre une religion est mauvaise. Dans un certain sens, le violent suppose que Dieu est faible et qu'il a besoin d'être défendu par l'extrémiste qui porte l'épée et le pistolet. Ceci est inacceptable. C'est une insulte a Dieu. (...) Le principe de la liberté religieuse doit etre accepté par tous les croyants. Chacun doit etre libre de toute contrainte dans le domaine de la conscience et de la religion, de sorte qu'une personne, en tant qu'individu ou membre d'un groupe, puisse adorer Dieu selon ses convictions. Ceci inclut le droit de changer de religion", quoted according to Baltimore, Secrétariat pour les Relations avec l'Islam: Lettre n° 51 (Nov. 1995). The theme of human rights in Islam is treated fundamentally for example by Honecker, Martin, Religion und Politik: Zur Geltung der Menschenrechte in Christentum und Islam, in: Busse, Heribert/Honecker, Martin (publ.), Gottes- und Weltverständnis in Islam und Christentum, EZW-Texte 123, Stuttgart 1993, p. 15–27.

31 Stamer, Joseph, Prier avec les musulmans?, BPCDIR 96 (1997), p.370, on the previous statements cf. p.357–369.

32 ibid., p.357 (italics removed)

33 On the general classification cf. Commission pour les rapports religieux pour les musulmans (éd.), Religion et politique: Un thème pour le dialogue islamo-chrétien, Cité du Vatican 1999, p.9 and Akasheh, Khaled, Bureau pour l'Islam Rapport d'activités novembre 1998–octobre 2001, BPCDIR 109 (2002), p.98 f (he provides information on the composition and working structures of the Commission, for example that it consists of eight consultors who have a mandate for five years and should if possible come from five continents and who obviously meet annually; plans also existed for work on other topics, such as the re-reading of the history of the Christian-Muslim dialogue to promote the healing of memories in the light of the Great Jubilee of 2000 or documentations on dialogue intended to bring a closer understanding of Christianity to Muslims. The new Commission, which has been in office since 2000, has decided to work on the topic of religious freedom; the results are to be published in English and French and possibly also in Arabic, whereby in May 2001 at least six exposés were finished and the general structure of the book could be discussed. The first part is supposed to give a theoretical approach to the topic in Islam, on an international level and in the USA, which forms a special case. The second part was to deal with concrete efforts to monitor religious freedom, first at the international level, and then in three countries with very different majority situations, namely Pakistan, Nigeria and France. There was a possibility that Muslims should be asked to work on the project at a later phase.), otherwise cf. McAuliffe, Jane Dammen, Rendering Allegiance to the Word: Qur'ânic Concepts and Contemporary North American Concerns, in: Commission pour les rapports religieux pour les musulmans (éd.), Religion et politique: Un thème pour le dialogue islamo-chrétien, Cité du Vatican 1999, p.14–16.23–25.28–36. The following statement is of interest, not for the policy, but rather for the style of the dialogue, p.27: "But despite assessing the activity of debate negatively, the Qur'ân nevertheless engages in it constantly. This frequently forensic character constitutes a large part of its rhetorical power and it shapes the context and contours of subsequent intellectual and cross-cultural engagement."

34 according to Troll, Christian W., Divine Rule and Its Establishment on Earth: A Contemporary South Asian Debate, in: Commission pour les rapports religieux pour les musulmans (éd.), Religion et politique: Un thème pour le dialogue islamo-chrétien, Cité du Vatican 1999, p. 37–52.

35 cf. Blockhausen, Denise, Pluralisme et laicité, in: Commission pour les rapports religieux pour les musulmans (éd.), Religion et politique: Un thème pour le dialogue islamo-chrétien, Cité du Vatican 1999, p. 54–57.59.

36 Fitzgerald, Michael [Louis], Muslims in Europe, in: Commission pour les rapports religieux pour les musulmans (éd.), Religion et politique: Un thème pour le dialogue islamo-chrétien, Cité du Vatican 1999, p.77, also according to p. 63–76.78. Over and above such reflections (and of course the papal involvement in this area), it can be seen that in Bosnia there was at least an indirect engagement, idem, Sarajewo: A Christian-Muslim Colloquium, 13 November 1999, Pro Dialogo 103 (2000), p. 83–84. As for the Netherlands, the article by Machado, Felix A[nthony], A Summary of Reports by the Members: Interreligious Dialogue Promoted by the Church, Pro Dialogo 109 (2002), p.129, shows that that the Dutch bishops, with regard to the Church's own schools, were certainly considering how the students could learn more about (and also more respect for) other religions – in the hopes that this would motivate them to learn more about their own faith. With a view to the general situation of non-Christians and in particular of Muslims in Europe, a conference of experts had already been held in the mid-seventies, in cooperation with a representative of the WCC, from 13–14 March 1974 in Luxembourg. Here too the distinction was made between the situation in Yugoslavia (which still existed at the time), where Muslims, as former Christians, saw themselves consciously opposed to Christians, and the situation of migrant workers in the rest of Western Europe. The bodies concerned with them tended not to perceive the latter as religious people with their own values, but only took account of their economic and political situation. It is also interesting that at that time one expected that the economic situation would lead to a decline in immigration, but that nevertheless a certain number would stay for good. It was strongly recommended that an independent Islam Commission be formed in the Secretariat and that the issue be generally followed up at all possible levels, preferably ecumenically and with Muslims, cf. Canivez, Jean, Luxembourg: Les Non-Chrétiens en Europe – Brève relation sur la rencontre au Luxembourg (13–14 Mars 1974), BSNC 26 (1974), p. 145–147.

37 according to Stamer, Josef, Présentation des minorités islamiques et de leurs revendications politiques dans l'Afrique de l'Ouest à travers le "Journal of the Institute of Muslim Minorities' Affairs" (JIMMA), dans: Commission pour les rapports religieux pour les musulmans (éd.), Religion et politique: Un thème pour le dialogue islamo-chrétien, Cité du Vatican 1999, p.79 – 83.85 – 98.

38 Fitzgerald, Michael L[ouis], Chrétiens et musulmans en Europe, dans: Commission pour les rapports religieux pour les musulmans (éd.), Religion et politique: Un thème pour le dialogue islamo-chrétien, Cité du Vatican 1999, p.117, cf. also p. 103 – 116.

39 according to Jacob, Xavier, Islam et politique (Turquie), dans: Commission pour les rapports religieux pour les musulmans (éd.), Religion et politique: Un thème pour le dialogue islamo-chrétien, Cité du Vatican 1999, p. 120 – 127.129 – 130.132.136.138.141 – 144 and idem, Situation de l'islam en Turquie, dans: Commission pour les rapports religieux pour les musulmans (éd.), Religion et politique: Un thème pour le dialogue islamo-chrétien, Cité du Vatican 1999, p.145.148 – 159. In the first of these articles Jacob quotes (p.133) a professor for constitutional law at the Istanbul University, Ismet Giritli, as follows: *"C'est le politicien ou le parti qui se montrera le plus permissif et meilleur musulman qui aura le plus de bulletins de votes dans son panier."*

40 according to Troll, Christian W., Islam and Pluralism in India, in: Commission pour les rapports religieux pour les musulmans (éd.), Religion et politique: Un thème pour le dialogue islamo-chrétien, Cité du Vatican 1999, p. 160 – 165.167 – 177.

41 cf. Nnyombi, Richard, Christianity and Islam in the Political Life of East Africa: The Example of Kenya, in: Commission pour les rapports religieux pour les musulmans (éd.), Religion et politique: Un thème pour le dialogue islamo-chrétien, Cité du Vatican 1999, p. 180 – 181.183 – 185.192.194 – 195.

42 according to Stamer, Josef, Religion et politique en Afrique de l'Ouest: Des voix musulmanes, dans: Commission pour les rapports religieux pour les musulmans (éd.), Religion et politique: Un thème pour le dialogue islamo-chrétien, Cité du Vatican 1999, p. 198 – 204.206.210.212 – 215.217 – 218. One detail which is interesting in our context is the passage which describes the end of the human rights conference p.217: "Le Séminaire s'est terminé dans la confusion sur une vague déclaration soulignant la convergence entre Islam et Droits de l'Homme." That is not exactly encouraging.

43 cf. Blockhausen, Denise, Les relations islamo-chrétiennes depuis Vatican II et les perspectives de collaborations islamo-chrétienne (sic!) pour le troisième milléniare, Pro Dialogo 102 (1999), p. 332 – 340.

44 cf. Akasheh, Bureau pour l'Islam Rapport d'activités: novembre 1998-octobre 2001, p.98f., idem, Considerations on Forty Years of Religious Dialogue with Muslims (A Report), p.197 says that the Commission became a sub-structure of the Pontifical Council, preparing documents to be made available to nunciatures and local churches. He enumerates: "Contemporary trends and movements in Islam / Religion and Violence / Praying with Muslims / Reactions to Nostra Aetate / Religion and politics / Religious freedom". This shows that these documents were not to be accessible to a wider public *per se*.

45 cf. Introduction, in: Pontifical Council for Interreligious Dialogue / Commission for Religious Relations with Muslims (ed.), Religious Liberty: A Theme for Christian-Muslim Dialogue, Vatican City 2006, p.7

46 cf. El-Hage, Youssef Kamal, Religious Liberty: A Universal Human Right, in: Pontifical Council for Interreligious Dialogue / Commission for Religious Relations with Muslims (ed.), Religious Liberty: A Theme for Christian-Muslim Dialogue, Vatican City 2006, p.12

47 The preamble defines tolerance very widely, understanding it as respect, acceptance and appreciation of the different ways of being human. Tolerance is harmony in difference and makes peace possible. Tolerance could never be used to restrict people's rights. Tolerance was not to be understood as a concession or condescension, but tolerance means responsibility for upholding

human rights, pluralism (including that of culture), democracy and the rule of law. Social injustice must not be tolerated, but the right had to be recognised that people are entitled to live in peace just as they are. No one should impose his views on another, cf. ibid., p.20, for the other statements cf. p. 13–19.
48 ibid., p.21, for the rest of the document cf. p.20.22–25
49 cf. ibid., p. 25–31
50 Buonomo, Vincenzo, The Catholic Vision of Religious Liberty, in: Pontifical Council for Interreligious Dialogue / Commission for Religious Relations with Muslims (ed.), Religious Liberty: A Theme for Christian-Muslim Dialogue, Vatican City 2006, p.52, for the rest of the presentation cf. p. 33–51.53f.
51 51 cf. Troll, Christian W., Religious Freedom in Modern Islamic Thought: A Catholic Perspective, in: Pontifical Council for Interreligious Dialogue / Commission for Religious Relations with Muslims (ed.), Religious Liberty: A Theme for Christian-Muslim Dialogue, Vatican City 2006, p. 55–60, whereby the remarks as from p.60 refer to Wielandt, Rotraut, Menschenwürde und Freiheit in der Reflexion zeitgenössischer muslimischer Denker, in: Schwartländer, Johannes (Hrsg.), Freiheit der Religion, Mainz 1993, p. 179–209, a study which Troll describes on p.62 as "masterly".
52 ibid., p.77, for the previous positions cf. p. 60–76
53 according to Channan, James, State of Religious Freedom in Pakistan, in: Pontifical Council for Interreligious Dialogue / Commission for Religious Relations with Muslims (ed.), Religious Liberty: A Theme for Christian-Muslim Dialogue, Vatican City 2006, p. 83–101
54 Kukah, Matthew Hassan, Religious Liberty in a Plural Society: The Nigerian Experience, in: Pontifical Council for Interreligious Dialogue / Commission for Religious Relations with Muslims (ed.), Religious Liberty: A Theme for Christian-Muslim Dialogue, Vatican City 2006, p.107, cf. also p.105f.
55 cf. ibid. p. 108–124
56 Gaudeul, Jean Marie, Islam in a Minority Situation: The Situation in France, in: Pontifical Council for Interreligious Dialogue / Commission for Religious Relations with Muslims (ed.), Religious Liberty: A Theme for Christian-Muslim Dialogue, Vatican City 2006, p.126, cf. also p.125
57 ibid., p.139, cf. also p. 127–138
58 cf. ibid., p. 140–147
59 McAuliffe, Jane Dammen, Monitoring for Religious Freedom: A New International Mandate, in: Pontifical Council for Interreligious Dialogue / Commission for Religious Relations with Muslims (ed.), Religious Liberty: A Theme for Christian-Muslim Dialogue, Vatican City 2006, p.153, cf. also p.151f. (in the original the quotation marks were obviously used here to indicate the coining of a new expression.)
60 ibid., p.156, cf. also p.154f.
61 cf. ibid., p. 157–183
62 Gaudeul, Jean-Marie, Free to Serve God Better, in: Pontifical Council for Interreligious Dialogue / Commission for Religious Relations with Muslims (ed.), Religious Liberty: A Theme for Christian-Muslim Dialogue, Vatican City 2006, p.212, on the previous explanations cf. p. 187–211
63 for an assessment cf. note 1, otherwise Secrétariat pour les non chrétiens (éd.), Orientations pour un dialogue entre chrétiens et musulmans, 2ième éd., Rom 1969, p.7.9, Marella, Paul, Présentation, dans: Secrétariat pour les non chrétiens (éd.), Orientations pour un dialogue entre chrétiens et musulmans, 2ième éd., Rom 1969, p.5, as well as (in retrospect) Rossano, Pietro, Foreword, in: Pontificial Council for Interreligious Dialogue / Borrmans, Maurice, Guidelines

for Dialogue between Christians and Muslims, Interreligious Documents 1, New York / Mahwah 1990, p.1 and Jadot, Jean, Preface to the 1981 Edition, in: Pontifical Council for Interreligious Dialogue / Borrmans, Maurice, Guidelines for Dialogue between Christians and Muslims, Interreligious Documents 1, New York / Mahwah 1990, p.7. A review praising the English translation is provided by Fitzgerald, Michael L[ouis], Maurice Borrmanns, Guidelines for Dialogue between Christians and Muslims. Translated from the French by R. Marston Speight. (Interreligious Documents I) New York / Mahwah, N.J., Paulist Press, 1990, pp.vi-132., Bulletin Pontificium Consilium pro Dialogo inter Religiones 76 (1991), p.149. In a contribution to the Joint Working Group in Stuttgart (7–12.6.1971) Rossano had already mentioned that Protestant organisations in particular had also shown great interest in this work, cf. Exposition from the Catholic Point of View, p.103. In "The Secretariat for Non-Christian Religions from the Beginnings to the Present Day: History, Ideas, Problems", p.94. he again emphasises the importance of this publication. The publication is highly praised at the first joint dialogue with the Jordanian Al Albait Foundation, cf. El Assad, Nassir El-Din, Inaugural Speech, in Religious Education and Modern Society, Acts of a Muslim-Christian Colloquium Organized Jointly by the Pontifical Council for Interreligious Dialogue (Vatican City) and the Royal Academy for Islamic Civilization Research Al Albait Foundation (Amman), p.10. The reaction was altogether very positive, cf. Greco, J[oseph], The Activity of the Consultors, BSNC 12 (1969), p.206. Charbel Gravrand, in his article L'Afrique au Secrétariat, Pro Dialogue 72 (1989), classifies the series of guidelines generally under anthropology, cf. p.358f. This article provides information on the development of the Secretariat which goes far beyond African concerns. As for Islam, he mentions, for example, that in September 1964 it did not yet belong to the Secretariat's responsibilities (p.351), but was only assigned to it in 1965 (p.353), and that after the creation of an Africa department the field of Islam in Africa was also assigned to that (p.361). For an even more detailed appreciation cf. Zago, Marcello, Les documents du conseil Pontifical pour le dialogue interreligieux, Pro Dialogo 72 (1989), p. 362–367; Zago risks a look into the future (p. 374–376) regarding the production and type of possible documents, whereby he speaks of the fact that relations with other religions are still in their infancy. It is interesting to read this "wish list" more than a decade and a half later: many things have been achieved, especially in terms of documented dialogues between Christians and Muslims, but what has been implemented so far is rather a modest version of his visions, and many themes still remain, as do the claims to representativeness and reciprocity. So there were still ample tasks for a "dicastère d'animation" (p.374), as he characterised the Pontifical Council for Interreligious Dialogue. Here one should also mention Pontifical Council for Interreligious Dialogue (ed.), Journeying together, The Catholic Church in Dialogue with the Religious Traditions of the World, Città del Vaticano 1999. This relatively recent publication deals with a number of religions and only has a small section on Islam (p. 75–83), but does at least also include a section with statements by John Paul II on Christians and Muslims (p. 84–89). The original publication is the work of two French Islamologists, Joseph Cuoq and Louis Gardet. The former was virtually a founding member of the Secretariat for Non-Christians and responsible from March 1965 for the area of Islam. It was under his direction that the tradition of messages at the end of Ramadan was started, which are mentioned below in detail. These two facts show what was important to him: to create a climate of appreciation and friendship on both sides. The effect of this indirect activity among his own ranks was important to him. For the same purpose he set up a network of correspondents. Here one can find exactly the same attitude that was held almost universally, namely to act indirectly through the churches on the ground, and therefore not to attach great importance to one's own dialogue conferences (according to Fitzgerald, Michael L[ouis], Joseph Cuoq (1917–1986), BSNC 63 (1986), p.339–341).

64 Secrétariat pour les non chrétiens, Orientations pour un dialogue entre chrétiens et musulmans, p.25, cf. also p. 11–24.26–30
65 cf. ibid., p. 34–56
66 cf. ibid., p. 59–74
67 ibid., p.75
68 ibid., p.75/76 quotes *Nostra Aetate* in the French version: "Si, au cours des siècles, y est-il dit notamment, de nombreuses dissensions et inimitiés se sont manifestées entre les chrétiens et les musulmans, le Concile les exhorte tous à oublier le passé et à s'efforcer sincèrement à la compréhension mutuelle, ainsi qu'à protéger et à promouvoir /p.76 ensemble, pour tous les hommes, la justice sociale, les valeurs morales, la paix et la liberté".
69 according to ibid. p. 77–109
70 according to ibid., p. 111–124
71 ibid., p.140, cf. also p. 125–139
72 according to ibid., p. 147–152, cf. also Marella, p.5
73 Rossano, Foreword, p.2, cf. also p.1.3, Jadot, p.7, Arinze, Francis, Preface to the English Translation, in: Pontifical Council for Interreligious Dialogue / Borrmans, Maurice, Guidelines for Dialogue between Christians and Muslims, Interreligious Documents 1, New York / Mahwah 1990, p.5; p.12 of this volume is the exact history of these guidelines, giving the names of all involved. The original is in French like the totally revised first edition; on the translations into other languages (the English one was surprisingly late) cf. the book review Fitzgerald, M[ichael] L[ouis], Maurice Borrmans, Orientamenti per un dialogo tra cristiani e musulmani, Pro Dialogo 70 (1989), p.127. Fitzgerald, Michael L[ouis], Twenty-five Years of Dialogue, The Pontifical Council for Inter-Religious Dialogue, Islamochristiana 15 (1989), p.110 also mentions that at that time, in 1975, it was likewise decided to do a study on the theological status of Islam, which, as he puts it, "never materialized in any official document." Like the first edition, the revised version received a lot of praise from the Muslim side for its objectivity, cf. El-Assad, p.10. Here we do not deal specifically with the study Mission and Dialogue, because it does not just concern the Christian-Muslim dialogue that is the subject of this paper. But it should be mentioned, because for one thing it lay close to the heart of the third president of the secretariat, Archbishop Jadot, who emphasised that dialogue should not remain a matter for specialists, but become a concern for the entire Christian community. This was also evident in the way the document materialised – it was not prescribed from above, but was the product of a worldwide discussion by those affected. And the categories which we now take as a matter of course – a dialogue of life, ideas, action and religious experience – did not emerge from theoretical consideration, but out of the practical experience of dialogue worldwide, according to Thomas Michel, Growing towards "dialogue in community "a Reflection on Archbishop Jadot's Tenure at the Secretariat for Non Christians, p.69f.
74 Secrétariat pour les non chrétiens, Orientations pour un dialogue entre chrétiens et musulmans, p. 159–161 in comparison to Pontifical Council for Interreligious Dialogue / Borrmans, Maurice, Guidelines for Dialogue between Christians and Muslims, Interreligious Documents 1, New York / Mahwah 1990, p.iii-vi.9–10.115–120. For further details on the first version, reference is made to the review directly preceding the text and its comments.
75 ibid., p.23, cf. also p. 13–24(previous facts).24–26(following contents)
76 ibid., p.28
77 ibid., p.33, cf. also p. 29–32.43
78 ibid., p.37, cf. also p. 33–36. As far as the problems in dialogue are concerned, it should be noted that the two main authors, Maurice Borrmans and Ary Roest Crollius, were both also par-

ticipants in the very first dialogue in Tripoli, which certainly did not go off smoothly. Against this background, the sentence on p.34 has a quite different ring to it: "It is even good to experience times of suspicion and of frustration which will oblige the interlocutors further to clarify to themselves the reasons for their encounter and the motives of their cooperation". Regarding the quite opposite question of how close non-Christians are to the kingdom of God, it is pointed out that theologians had always maintained that even a non-Christian of goodwill could be saved, but there were great differences as to the exact modalities. In general one could say that this goodwill/faith consists in a life devoted to an absolute, proved by a clear conscience and a moral life.

79 ibid., p.40 (emphases removed), cf. also p. 37–39.42 for the complete context
80 ibid., p.49, cf. also p. 45–48 and for the rest of the section p. 50–66
81 ibid., p.83, cf. also p. 68–82
82 according to p. 84–87
83 ibid., p.92, cf. also p. 88–91
84 cf. ibid., p.93
85 cf. ibid., p.88.93–99
86 ibid., p.110, cf. also p. 100–109
87 cf. ibid., p.112f.
88 Marella, Paul, Présentation, dans: Secretariatus pro non Christianis (éd.), Religions, Thèmes fondamentaux pour une connaissance dialogique, Rome 1970, p.6. This volume exists in several languages, whereby only the editions in French and Italian (Segretariato per i non cristiani (ed.), Religioni, Temi fondamentali per una conoscenza dialogica, 4. ed., Fossano 1987) were available to me. Given that the later edition is a simple reprint, which was not even provided with a new preface, I have decided to make the quotations in French, which is more widespread in Germany. For the interesting background to this publication, please refer to Rossano, P[iero], Presentation of the Volume "Religions" Published by the Secretariat, BSNC 16 (1971), p. 36–40.
89 Rossano, P[iero], L'homme et la religion, dans: Secretariatus pro non Christianis (éd.), Religions, Thèmes fondamentaux pour une connaissance dialogique, Rome 1970, p.9, cf. also p. 10–92, sowie Marella, p.6. This first chapter had already been published as a supplement to the internal bulletin as early as 1968 under the title Man and Religion, cf. Fitzgerald, The Secretariat for Non-Christians Is Ten Years Old, p, 93, note 15. The homo religiosus is a central theme for the Secretariat, as is also demonstrated by Geneva: Visit of the President and the Secretary of the Secretariat for Non-Christians to the Ecumenical Council of Churches (15 January 1974), BSNC 26 (1974), p.137f., and in even more detail in a lecture held by Piero Rossano at the return visit on the topic of "Our Programme and Method", published in BSNC 26 (1974), in which he says p.142: "The first characteristic of our method is that we meet our non-christian brethren in the capacity of religious persons endowed with religious values, and we join them in a dialogue because we believe alike in a Reality which transcends this world and our senses. This common belief has a power of unity and give (sic!) a basis of common sharing and mutual exchange and collaboration." His second lecture on the same occasion, "Contents of Dialogue with Non-Christians", also published in BSNC 26 (1974), goes into more depth theologically p.143/144: "This option of the 'homo religiosus' has been done on the base of a definite theological evaluation of religiosity and a positive estimation of the fundamental religious experience, originated in the creation of man as 'Imago Dei'. As created by God, in and for Christ, man is basically in quest of God (…) and oriented to God, even though he is wounded by sin. His quest of an Absolute is in fact the expression of the 'menschliche Kreatürlichkeit und fraglichkeit (sic!)'. / p.144 (…) This doctrine has been codified in the Vatican II and represents the basis and the inspiration

of our activity." This is made possible, as is clearly stated here, because the image of God is seen to have been damaged, but not fundamentally impaired by sin, as is the case in Protestant theology, for example. Rossano, The Secretariat for Non-Christian Religions from the Beginnings to the Present Day: History, Ideas, Problems, p.95 makes it quite clear that the decision to base the work of the Secretariat on the concept of homo religiosus had already been taken under and by the first President, Cardinal Marella, and represents one of his great achievements. Somewhat later (p.103) Rossano writes on the connection between this anthropology and dialogue as the task of the Secretariat: "[D]ialogue for the Christian is the fruit of an anthropology in which the person stands at the center as an image of God and as the object of his love. Man is seen to be so structured ontologically as to have a vertical and horizontal relationship which relates him deeply to God (homo religiosus) and to the other (Mitmensch). Indeed, man is considered not to attain his fullness without a personal relationship to the Thou of God and of neighbour (sic!)." He had already mentioned p.100 that this approach, which had been adopted by the Second Vatican Council, goes back to the dialogic thinking of Martin Buber. Further evidence for this is Fitzgerald, The Secretariat for Non-Christians Is Ten Years Old, p.88, where he writes on Cardinal Marellas plans and remarks: "An effort would be made to acquire an objective knowledge of different spiritualities and of the different ways the human mind expresses its approach to God." He dedicates virtually the entire next page of his article to corresponding statements by Rossano.

90 cf. Rossano, L'homme et la religion, p.22 on Islam, p.51 and 60 more on the general stage of research, regarding the intention of the volume cf. Rossano, Exposition from the Catholic Point of View, p.103f., where he clearly states that this work was aimed at theological students. In "The Secretariat for Non-Christian Religions from the Beginnings to the Present Day: History, Ideas, Problems", p.94 he goes on to say that the idea was proposed by relevant academic institutes.

91 Rossano, L'homme et la religion, p.18, cf. also p. 16–17.26–27
92 ibid., p.20, cf. also p.19.21
93 ibid., p.25, cf. also p.24
94 ibid., p.30, cf. also p.29
95 ibid. p.30
96 ibid., p.36, cf. also p. 30–35
97 ibid., p.36
98 ibid., p.37
99 ibid.
100 ibid., p.41, cf. also p. 38–40
101 ibid., p.50
102 ibid., p.55, cf. also p. 51–54
103 ibid., p.67, cf. also p. 56–66
104 ibid., p.72
105 ibid., p.73
106 ibid., p.89, cf. also p. 73–88.90–92
107 Rossano, P[iero], Introduction, dans: Secretariatus pro non Christianis (éd.), Religions, Thèmes fondamentaux pour une connaissance dialogique, Rome 1970, p.95, for the rest cf. p. 96–99.
108 Rossano, P[iero], Le salut dans le christianisme, dans: Secretariatus pro non Christianis (éd.), Religions, Thèmes fondamentaux pour une connaissance dialogique, Rome 1970, p.105, cf. also p. 101–102
109 ibid., p.108, emphases removed

110 ibid., p.106

111 ibid., p.109, the further explanations taken from p. 110–114. To conclude, it is stated p.114 in a good Catholic sense as was only to be expected: "Le vrai chrétien a conscience de toute ces réalités et travaille à son salut en fréquentant les sacrements et en écoutant la Parole de Dieu" – with the explicit inclusion of trust in St. Mary's mediation and the intercession of the saints.

112 Caspar, R[obert], La recherche du salut dans l'islam, dans: Secretariatus pro non Christianis (éd.), Religions, Thèmes fondamentaux pour une connaissance dialogique, Rome 1970, p.115

113 ibid., p.121, cf. also p. 115–116 and p.137, where it says: "Ainsi, être musulman, c'est suivre la loi de la nature, et ne pas l'être est une anomalie, une sorte de parjure."

114 ibid., p.121

115 ibid., p.122, cf. also p.123. In Hadith and theology the tendency is: "Tout est déjà 'écrit' lorsque l'embryon est encore au sein de sa mère" (p.127).

116 ibid., p.136, cf. also p. 123–125.127–135.138

117 Rossano, P[iero], Préliminaire, dans: Secretariatus pro non Christianis (éd.), Religions, Thèmes fondamentaux pour une connaissance dialogique, Rome 1970, p.228, cf. also p.227.229

118 ibid., p.235, cf. also p.229.232–234

119 Gardet, L[ouis], En islam: Dieu et le croyant en Dieu, dans: Secretariatus pro non Christianis (éd.), Religions, Thèmes fondamentaux pour une connaissance dialogique, Rome 1970, p.354, cf. also p. 344.347–348.350–352. An interesting contribution to the subject which should be mentioned here because it was presented at an internal conference of the Secretariat with consultors and specialists from 27–29.11.1971 in Paris, is Arnaldez, R. Les valeurs de l'Islam, BSNC 19 (1972), p. 21–29. Apart from exact descriptions of Sufi understanding of God and humans, he also states that the absolute monotheism in Islamic understanding and the constant divine creation of all things leads in practice to the fact that not only does natural law not exist, and with it human rights as natural law, and that human freedom of action is only present in the human conscience, but that every kind of inner-worldly causality of human action is completely ruled out. For example, the acquisition of wealth has nothing to do with external circumstances or the work and skill of the individual, but is seen to be a pure gift of God to the person concerned. If he accepts it as such, then this wealth is legitimate, but if a person ascribes it to himself, then it becomes an illegitimate appropriation, which he will have to answer for. Certainly there are points of contact with Christian spirituality, but a Christian view is more differentiated and generous in terms of personal freedom as well as of creation and its orders.

120 120 ibid., p.360, cf. also p.357.359

121 ibid., p.363, cf. also p.362

122 according to Festorazzi, F[ranco], Le Dieu vivant dans la révélaton chrétienne, dans: Secretariatus pro non Christianis (éd.), Religions, Thèmes fondamentaux pour une connaissance dialogique, Rome 1970, p.365.368–370.372–373.387.389.395.400. For the numerous statements on christological anthropology cf. p. 377–379.386.391.393–394 and finally p.396: "Pour que l'homme devienne l'image du Père il doit recourir au Christ, qui en est l'image idéale."

123 according to Bianchi, U[go], Le concept du bien et du mal dans les religions, dans: Secretariatus pro non Christianis (éd.), Religions, Thèmes fondamentaux pour une connaissance dialogique, Rome 1970, p.428.431–438

124 Gelot, J[oseph], Le bien et le mal en Islam, dans: Secretariatus pro non Christianis (éd.), Religions, Thèmes fondamentaux pour une connaissance dialogique, Rome 1970, p.529, cf. also p.552 and p. 525–528. On his own admission (p.523, note 1), the text is strongly based on an article by Louis Gardet dealing with the relationship between God and human fate.

125 ibid., p.530, cf. also p.529

126 ibid., p.546, on the previous explanations cf. p. 529–536.538–545. Particular attention should be drawn to the modern presentation of Muslim morality (p.541, note 56); it practically consists of a long quotation from a Muslim statement on the same topic. The subsequent explanations refer to p.547.551–552.

127 Greco, J[oseph], Le bien et le mal, Perspectives chrétiennes, dans: Secretariatus pro non Christianis (éd.), Religions, Thèmes fondamentaux pour une connaissance dialogique, Rome 1970, p.557, cf. also p.555f. Subsequent comments on this topic are not much clearer, for example (p.571): "La réponse de l'homme sera toujours libre. (...) Si l'être humain est évidemment conditionné sous plus d'un aspect, il reste toujours le maître, s'il le veut, de ce domaine intime d'un ordre de grandeur infini où il domine tout conditionnement. Or c'est cette enclave divine, la plus noble qui soit, qui en l'homme crie ontologiquement qu'il est de la race de Dieu".

128 ibid., p.565, cf. also p. 558–562.564 In the context of the definition of primeval or original sin p.579 is also interesting with its indirect suggestion that the human aspiration to be God could even have been created by God "dont l'amour n'a de cesse qu'il n'ait réalisé ce paradoxe de rendre participants de la nature divine les enfants des hommes, appelés à devenir authentiques fils de Dieu!" On the other hand it is stated quite clearly (p.580): "Le refus libre, donc conscient et volontaire, de reconnaître cette origine divine et la tentative d'arrachement de cette origine divine et la tentative d'arrachement de cette dépendance foncière en un geste d'auto-suffisance, voilà qui est à la racine de tout péché. Mais toute trahison de l'amour de Dieu est du coup trahison de l'amour de l'homme."

129 ibid., p.579, cf. also p. 566–573.575–578

130 ibid., p.584, cf. also p.577

131 ibid., p.586/587, cf. also p. 584–585.588; p. 591–596 contain a reproduction of the Didache.

132 Rossano, P[iero], Le bien et le mal dans les religions: vue rétrospective, dans: Secretariatus pro non Christianis (éd.), Religions, Thèmes fondamentaux pour une connaissance dialogique, Rome 1970, p.601, cf. also p. 597–600.602f.

133 cf. Pignedoli, Sergio, Preface, in: Secretariatus pro non Christianis (ed.), Religions in the World, Città del Vaticano cf. also p.7

134 Borrmans, M[aurice], Islam, in: Secretariatus pro non Christianis (ed.), Religions in the World, Città del Vaticano cf. also, p.117, cf. also p.85f.88–91.95.98–101.105f.110–116

135 Since it has already been explained that Segretariato per i non cristiani (ed.), Chiesa e islam, s.l. 1981 appeared in several languages (cf. also bibliography), I will only quote from now on from the English version of the articles, as far as they are available. It starts on p. 5–7 with an excerpt from *Nostra Aetate* in French, English and Arabic. There follows, fully under the impression of this Council declaration, the address to the Catholic community of Ankara (Extract from Discourse of Pope John Paul II to the Catholic Community in the Chapel of the Italian Embassy in Ankara, in: Segretariato per i non cristiani (ed.), Chiesa e islam, s.l. 1981, p. 12–15), then the very short and accordingly insubstantial address to the Muslims in Ghana (Accra. Greeting to Leaders of the Muslim Community in Ghana, in: Segretariato per i non cristiani (ed.), Chiesa e islam, s.l. 1981, p. 19–20), and, in reverse chronology as it were, as may be seen in comparison with Conseil Pontifical pour le Dialogue Interreligieux (éd.), Le dialogue interreligieux dans l'enseignement officiel de l'église catholique (1963–1997), Documents rassemblés par Francesco Gioia, s.l. 1998, p.246f, n. 350 und p.249f, n. 353, the speech to the Kenyan Muslims, which refers once again very clearly to *Nostra Aetate* (Nairobi: Greeting to Leaders of the Muslim Community in Kenya, in: Segretariato per i non cristiani (ed.), Chiesa e islam, s.l. 1981, p. 25–26). This is followed by the addresses in Paris (Paris: Message to Representatives of the Moslem Community in France, in: Segretariato per i non cristiani (ed.), Chiesa e islam, s.l. 1981, p. 33–34) and Mainz (Gruß des Papstes an die Muslime in Deutschland, in: Segretariato per i non cristiani (ed.), Chiesa e islam, s.l. 1981, p.37) as well as two from Pakistan: Extract from Message to President of Pakistan, in: Segretariato per i non cristiani (ed.), Chiesa e islam, s.l. 1981p.39–40, exactly the one that was no longer included in the later collection, and Extract from Address to the President and the Civil and Religious Authorities Present at the Airport before Leaving Pakistan, in: Segretariato per i non cristiani (ed.), Chiesa e islam, s.l. 1981, p. 44–45, in which *Nostra Aetate* is quoted once again. However, the text was shortened at the beginning by two sections, as shown by comparison with Conseil Pontifical pour le Dialogue Interreligieux. p.256, n. 361, which proves that this speech was held at the beginning of the visit, not at the end. It states namely: "Commençant aujourd'hui un nouveau voyage pastoral, je suis heureux de ce que le Pakistan soit le premier arrêt sur la route." The last contribution is the speech to the Philippine Muslims, which deals very strongly with the political situation (Discourse to the Representatives of the Muslim Community (Davao Airport – Philippines), in: Segretariato per i non cristiani (ed.), Chiesa e islam, s.l. 1981, p. 50–53). The importance of this theme for John Paul II, which was already visible at an early stage, is assessed in Fitzgerald, Twenty-five Years of Dialogue, p.114: "That there was sufficient material for a booklet, after only three years of pontificate, shows that the Pope was living up to the commitment he had made to the participants at the 1979 plenary."

136 cf Pontifical Council for Interreligious Dialogue (ed.), Meeting in Friendship, Messages to Muslims for the End of Ramadan (1967–2002), Vatican City 2003, in particular the Presentation by Felix A[nthony] Machado, p. 5–7, additionally p.62 (1997), p.73 (2001), p.76 (2002), Pontifical Council for Interreligious Dialogue, Recognize the Spritual Bonds which Unite Us, p. 52–54 (on the original background cf. Ramadan 1969, BSNC 13 (1970), p.52f), Akasheh, Considerations on Forty Years of Religious Dialogue with Muslims (A Report), p.199f, idem, Bureau pour l'Islam Rapport d'activités: novembre 1998-octobre 2001, p.104 (in which he makes a number of suggestions for improvement to prevent the recipients getting two copies, or that the local bishops might add their own text to adapt the message to the specific context), Jacob, Situation de l'islam en Turquie, p.153f. Fitzgerald, The Secretariat for Non-Christians Is Ten Years Old, p.93f. already mentions these messages, which were also being transmitted by Vatican Radio, and reports p.94 note 18 on a very interesting reaction by the late King Faisal of Saudi Arabia, who appreciates this message precisely because Islam, as he emphasises, is the religion of truth, the religion of freedom, the religion of cooperation and the religion of toleration. It would be hard to better this as an official acknowledgment of a gesture while simultaneously ignoring the content. Jacob, Xavier, L'islam turc et le dialogue, Islamochristiana 15 (1989), p.237 expresses a similar thought when he mentions that the national media drew attention to the Pope's travels, but not to what he said – not even in Turkey itself. He supposes that they are only really interested in a common front against materialism and communism. An example for the positive reception of these messages is Isizoh, Chidi Denis, Report of Meetings in Ghana: 18–30 January 1998., Pro Dialogo 99 (1998), p.345. At this point one should also draw attention to the initiator of these messages, Joseph Cuoq, cf. Lanfry, Jacques, A la mémoire du P.Joseph Cuoq, premier responsable pour l'Islam (1964–1974), Pro Dialogo 72 (1989), p.389. The article is also an excursion into the history of how the Vatican and the Catholics did (and did not) take note of Islam and Muslims. During this period the reactions were also collected, cf. Salama, Andraos, Le Message du Sécretariat aux Musulmans pour la fin du Ramadam 1407/1987 et ses échos dans le monde, BSNC 66 (1987), p. 307–314. The message for Ramadan in 1987 refers to the multi-religious peace prayer of Assisi. The replies reproduced here are divided into groups. First of all (p.309) there are some of the most eminent persons and positions, then as the first official subdivision the embassies (p.310f.) and in second place the universities and religious institutions (p.311f.). Many high-ranking Muslim personalities and institutions write very, very politely and cordially, expressing their grateful thanks – but also vaguely. In fact, it is only the statements from the field of university and religious institutions which really refer to the content. Maybe one could or should give a two-fold answer to the question of how much attention is paid to the messages. In the academic and religious realm they do indeed take a close look, while in the political sphere it is rather the gesture that counts, the attention, appreciation, dialogue and cooperation as such. This context is also covered in Michel, Thomas, 25 Years of Letters to Muslims for Id Al-Fitr, Bulletin Pontificium Consilium pro Dialogo inter Religiones 84 (1993), p. 300–302. Michel is prepared to answer the criticism that Muslims are possibly indifferent to these messages. He distinguishes between the official reactions and the personal letters of thanks, whereby the latter come in several cases from very important personalities. These proved that the messages had been read very carefully, so that even tiny typing errors had been detected and corrected. This even reached to offers from the Muslim side to proofread the messages of greeting, which appeared in three languages: English, French and Arabic. Thus it is evident that the messages were being read attentively, although the nature of the attention varies. It is also important to point out that the practice of sending messages of greeting on the occasion of festivals has now been extended to Hindus and Buddhists, and that Sikhs have alos expressed a desire to receive such a message, which may also say something about the way these messages are perceived, cf. Machado, A[nthony], Summary of Reports by the Members: Interreligious Dialogue Promoted by the Church, p.128. Regarding the significance of the greetings ritual, one should refer to Channan, James, Pakistan – Karachi: 9th General Assembly of the World Muslim Congress (30 March – 2 April 1988, Karachi), Bulletin Pontificium Consilium pro Dialogo inter Religiones 70 (1989), p. 120–122. In conclusion, reference should also be made to Akasheh, Considerations on Forty Years of Religious Dialogue with Muslims (A Report), p.200, who argues that the messages were crucial in helping Muslims develop positive feelings for Pope John Paul II. He also refers to the peace prayers of Assisi, but also to the Pope's opposition to terrorism and its causes, especially in the West. In a different report Akasheh also mentions that Turkey's supreme Islamic authority has begun to send messages of greeting to bishops at major Christian festivals, idem, Some Reflections on Islam and Christian-Muslim Relations in Certain Countries, Pro Dialogo 116/117 (2004), p.255.

137 Cuoq, J[oseph] M., 1972, in: Pontifical Council for Interreligious Dialogue (ed.), Meeting in Friendship, Messages to Muslims for the End of Ramadan (1967–2002), Vatican City 2003, p.16, cf. also p. 8–9 (1967), p. 10–11 (1968), p. 12–13 (1969), p. 14–15 (1971)

138 cf. Pignedoli, Sergio, in: Pontifical Council for Interreligious Dialogue (ed.), Meeting in Friendship, Messages to Muslims for the End of Ramadan (1967–2002), Vatican City 2003, more precisely the messages of 1973 (p.18), 1974 (p. 19–20), 1975 (p. 21–22), 1976 (p. 23–25), 1977 (p. 26–28), 1978 (p. 29–30) and 1979 (p.31)

139 cf. Jadot, Jean, in: Pontifical Council for Interreligious Dialogue (ed.), Meeting in Friendship, Messages to Muslims for the End of Ramadan (1967–2002), Vatican City 2003, more precisely the messages of 1980 (p.32), 1981 (p.33), 1982 (p.34) and 1983 (p.35), all of which are very short, but make reference to John Paul II.

140 Arinze, Francis, 1985, in: Pontifical Council for Interreligious Dialogue (ed.), Meeting in Friendship, Messages to Muslims for the End of Ramadan (1967–2002), Vatican City 2003, p.38, cf. also p.39, p. 36–37 (1984), p. 40–41 (1986), p. 42–43 (1987), p.44 (1988), p.46 (1989), p. 47–48 (1990)

141 cf. Arinze, Francis, in: Pontifical Council for Interreligious Dialogue (ed.), Meeting in Friendship, Messages to Muslims for the End of Ramadan (1967–2002), Vatican City 2003, p. 52–53 (1992), p. 54–55 (1993), p. 56–57 (1994), p. 58–59 (1995), p.60 (1996), p. 62–63 (1997), p. 64–65 (1998), p. 66–67 (1999), p. 68–69 (2000), p. 70–71 (2000), p. 73–75 (2001). The papal Ramadan message is on p. 49–51 and begins under the papal coat of arms with the title "To my beloved Muslim Brothers and Sisters", signed (handwritten) Joannes Paulus PM II.
142 cf. Fitzgerald, Michael L[ouis], 2002, in: Pontifical Council for Interreligious Dialogue (ed.), Meeting in Friendship, Messages to Muslims to the End of Ramadan (1967–2002), Vatican City 2003, p.76 f. as well as the unpaged reprint Pontifical Council for Interreligious Dialogue / H.E. Archbishop Michael L[ouis] Fitzgerald, Constructing Peace Today, Message for the End of Ramadan, 'Id al-Fitr 1424 A.H. / 2003 A.D., Vatican City s.a. Thankfully, the two most recent messages were sent to me directly by e-mail.
143 cf. Rossano, The Secretariat for Non-Christian Religions from the Beginnings to the Present Day: History, Ideas, Problems, p.97 f., as well as Preface, BSNC 32 (1976), p.105 f., where it is mentioned that it had not (yet) benn possible to print all the answers received – as Rossano might have said, one may speculate on the true reasons and the criteria of selection. On Lahbabi and his representativeness cf. also note 37 of the chapter on the magisterium.
144 cf. Lahbabi, Mohamed Aziz, Islam, BSNC 32 (1976), p. 155–158, Jullundry, Rashid Ahmad, Islam, BSNC 32 (1976), p. 186–189
145 cf. Jullundry, p. 189–195, Lahbabi, Mohamed Aziz, L'Islam et la paix, BSNC 32 (1976), p. 175–185; this line of argumentation also includes the interesting statement in Fitzgerald/Borelli, p.217, that in Islam forgiveness has to do principally with the relationship between God and humans, i.e. that forgiveness is primarily something divine; that is why the idea of a vicarious suffering and sacrifice for sin is an inconceivable attack on God's divinity. The most important concern in the human realm is the restoration of justice, which does not rule out forgiveness and the request for forgiveness between people.
146 cf. Lahbabi, Islam, p. 158–160, Jullundry, p.195 f.
147 cf. Lahbabi, Islam, p. 160–162, Jullundry, p.196 f.
148 Jullundry, p.201. cf. also p. 197–200.202 as well as Lahbabi, Islam, p. 162–164
149 Ibn Taymîyya, al-Siyâsa al-shar'iyya, p.177, quoted by Jullundry, p.204, on human freedom cf. also ibid., p.203 and Lahbabi, Islam, p. 165–167, on the ideal image of humans cf. ibid., p.167 and Jullundry, p.204 f.
150 cf. Jullundry, p.205 f. and Lahbabi, Islam, p.167 f.
151 cf. Jullundry, p.206 f., Lahbabi, Islam, p. 169–171
152 Lahbabi, Islam, p.173, cf. also p.171 f. and Jullundry, p.208
153 This danger exists not only for Christians, but perhaps even more on the Muslim side, as Rossano noted in 1971 (Exposure from the Catholic Point of View, p. 107–108): "It seems that in some circles, particularly in some Islamic circles, that the theme of secularization is the favourite ground for dialogue; in any case we may note that many exponents of the non-Christian religions look to us to see how the Catholic Church is behaving in the face of this phenomenon." The theology of religions is not to be discussed in this treatise on anthropology, but a small remark is permissible. In a speech on the occasion of a book presentation on this subject by Jacques Dupuis, Michael Fitzgerald makes the cautious remark that religious pluralism is a fact and that the practice of interreligious dialogue and the cultivation of positive relationships with people of other religions may be described as *locus theologiae*. The statements of Christians in this connection did not have to be acceptable for Muslims or members of any other religions. (It is

already a great step that other religions can shed some light on the Christian truths concerning salvation, as has already been stated by the Catholic magisterium.) – Fitzgerald, Michael L[ouis], Toward a Christian Theology of Religious Pluralism by Fr Jacques Dupuis, Pro Dialogue 108 (2001), p. 335–338.

III The Christian-Muslim Dialogues

1 The beginnings and basic structures

1.1 Rome 1972 – an unintentional dialogue

As can be seen from the structure of this discussion, the first main interest of the then Secretariat for Non-Christians was to clarify the basic issues related to its own position and work and to assist the Catholic Church worldwide in its encounter with other religions and their representatives. During this initial period under Cardinal Paolo Marella as President, the Secretariat was therefore evident in public almost exclusively through its publications; other contacts were rare and there was as yet no conception of organising dialogues of its own. What was later described as the first attempt along those lines was the second general assembly of the advisers from 3–6 October 1972, which was to deal with how other religions understood themselves, and to which two non-Christians were also invited, a Buddhist who was then unable to attend for health reasons and Prof. Muhammad Talbi from Tunisia as a representative of Islam. Under the presidency of Cardinal Sergio Pignedoli beginning in 1973 a major change came about. Publications receded notably into the background and there were far more direct contacts of many different kinds instead, exploratory conferences (in March 1974 in Luxembourg for Europe and in June 1974 at Bamako for West Africa) and also fully fledged dialogue conferences. Cardinal Pignedoli turned the Secretariat into an arm of the Holy Father which was extended to followers of other religions.[1] As previously in its publications, the Secretariat also benefited from the expertise of advisers and, particularly in connection with Islam, from the Pontifical Institute for Arabic and Islamic Studies.

The first Christian-Muslim dialogue – in a sense unintentional and hence also almost unnoticed – came about in the context of one of the annual meetings with the advisers, namely during the meeting already mentioned in Rome (3–6 October 1972). On the second day, there was an interesting confrontation between Caspar as the expert for Islam and the Muslim Professor Talbi on the question of Islam and Christianity in their relationship to modernity. The report indicates that others also participated actively in this dialogue but unfortunately it does not go into the details of the discussion and only reflects Talbi's main argument (and his own experience), namely that there is a discrepancy between the traditional and simple formulations of the faith among older people, for example, and a carefully considered faith in a modern context. This discrepancy could not simply be bridged by religious reformism and the great danger involved was the alienation of young people through indifference, although the Quran really contained all the necessary requirements for developing modern religious thinking which could support faith. But the discussion in general had tended to move away from the concrete historical religions and their problems and dealt more with religion as such and human religiosity. So contemporary culture led representatives of religions to turn their sights to anthropology and the nature of human beings. That also meant that the religions devoted their resources increasingly to human beings and their social advancement. Peace and peaceful coexistence were one of the major, long-term but certainly positive challenges for religions in this connection. Among its short-term aims, the Secretariat should take up issues which were simultaneously of religious and human interest, such as human rights from a religious point of view. As far as Islam itself was concerned, Islam in Asia became a focal point of interest.[2]

1.2 Grottaferrata 1975: a key role for the dialogue with Islam

1.2.1 Analyses of situations in various countries

A still greater key role with reference to the paths followed and still to be followed in the dialogue was played by the conference of advisers and experts organised by the Secretariat in Grottaferrata near Rome from 12–15 October 1975. It was based on a questionnaire which served to identify the contemporary situation of dialogue in various fields including Islam, and to decide on how to proceed. Particularly with regard to the counterpart to Islam, the programmatic limitation to *homo religiosus* and the reticence about the practical implementation of new behaviour congruent with the faith were seen as problematic, because in Islam the connection between religion and the worldly order is considerably more direct. Among the various reports compiled in an issue of the Secretariat's own bulletin, however, only very few touch on this fundamental question; the vast majority constitute a kind of momentary view of the dialogue situation in different areas and countries. Indonesia, the Philippines, Pakistan, the Arab world with special attention to Egypt, North Africa, especially Tunisia and Libya, are represented, and also France. Of course, the real dialogue situation is as different in practice as are these countries and their history (which naturally makes these reports a treasure trove of interesting details), but the net impression is that Christian-Muslim dialogue was still mainly at an initial stage and/or was seen as a matter for specialists, especially for the White Fathers, and not as a concern for congregations and churches as a whole. The reason for this is often that the local Christian churches (unlike the Vatican as a central organisation) are perceived and viewed – if at all – as foreign (western) bodies in these countries, whether or not they really are with regard to their history, or whether only their different structures or even just the eyes of the Muslims make them appear so. Of course, this is particularly extreme in the Arab world which, as a rule, simply equates Arab with Muslim and believes there should be no such thing as Christian Arabs, and these are sometimes not even allowed to teach their own mother tongue because that is the language of the Quran. No other report is as critical of Muslims as this one which was actually written by the Secretariat's officer for Islam, Abou Mokh, himself who also came from that region. Not all the examples are as extreme as that of Saudi Arabia, but he is critical of the schizophrenic situation in general where Islam believes, on the one hand, that it has the right to engage in propaganda for itself in the whole world among nations and individuals and builds mosques and cultural centres throughout the Christian world, but, on the other hand, denies Christians these rights in the Muslim world and prohibits any form of Christian propaganda. The Arab Muslims especially were not prepared to learn anything from others, expected nothing of Christians or of dialogue and were sufficient unto themselves because their religion was the most perfect and their ideology the best. Islam alone could save the world, restore its balance and give it peace and justice. Such exclusivist tendencies could also be observed on a quite practical level. The Arab world was spending fabulous sums for the maintenance and defence of Islam, but only for that. "Le monde arabe musulman est exclusif dans sa charité ainsi que dans son comportement. Dans leurs hôpitaux, leurs écoles et leurs oeuvres, les musulmans se sentent uniquement concernés par les musulmans. Jamais les fonds de charité musulmans ne sont partagés par les adeptes des autres religions".[3] And against this background the question arose whether, as had been considered for Pakistan, it really made sense to hand over the Christian schools and hospitals to the Muslims so that they did not constitute a stumbling block for approaching Muslims.

This raises the question acutely for the reader whether, under such conditions, there can ever be a dialogue on an equal footing, indeed a dialogue worthy of the name at all, and what it should and could then look like. As the example of Pakistan demonstrates, even the inclusion of religious propaganda of non-Muslim religions as well in the constitution apparently is of no use in a predominantly Muslim context. And a study of Muslim documents from Egypt dealing with Christianity shows that, although the tone of the presentations has become more friendly, nothing has changed in the content at all. Since unfortunately the discussion has not been reported, it is not possible to ascertain whether such questions were raised at all and, if they were, how they were answered.

1.2.2 Grottaferrata and afterwards: many different questions and issues

There were only two documents which asked more fundamental questions, the one generally along the lines of whether it was possible to concern oneself only with religious dialogue, as the Secretariat had done thus far, and to refer the practical questions to other institutions of the Curia, whereas other organisations, such as the WCC or the WCRP, were dealing above all with the contemporary, practical questions of humankind and tending to lose sight of the religious issues. The speaker considered both approaches one-sided and erroneous because both areas were so closely inter-related that they could not be separated in a meaningful way and it was evident that religion and/or ideology were the key to behaviour. His concrete example in relation to Islam was popular fatalism; if someone did not believe that human life could be improved before a person died, because God had determined everything in advance, any attempt to improve the earthly life of people would appear fairly useless to them unless it were combined with a religious dialogue about the relation between human beings and God. This is, indeed, a very important thought. As regards the specific queries and suggestions about dialogue with Muslims which affect more than a single region, they really have little to do with the anthropological level which is being studied here and are more theological in nature, unless they are looked at from the point of view of all that causes distrust or unease among Muslims. That includes the question of political power (a mutual concern), the question of evangelisation/ conversion and dialogue, as well as Islam being considered just one of the non-Christian religions whereas Judaism was granted a special place. *It is clear that these questions do have an effect at the human level but especially the last two are purely theological questions and, in addition, are among the most difficult for a theology of religions.*

The actual dialogues of the Secretariat itself made clearer and entered more deeply into the questions which were reflected in the statements from Grottaferrata, also e. g. with regard to the understanding of dialogue as a whole. While on the Catholic side dialogue meant, for example, the acceptance of others and of their rights, in comparison to the Muslims it was observed that the same was not true in reverse. This again raised the question of the necessity of mutuality or made it more acute – a question which was (and is) not so easy to answer unequivocally from a Christian standpoint either. Thus the dialogues themselves opened up a new, broad area precisely (also) from an anthropological point of view (particularly the human spiritual values as expressed in various religions were to form the basis for dialogue) and that will now be described. But it was significant that – not only among the Muslims – justice, peace and human development were very high on the list of priorities of the dialogue partners, namely all non-theological but anthropological or simply human issues and problems which also constitute a challenge for joint action – another matter which had already been mentioned at Grottaferrata.[4]

1.3 Challenges of the dialogue and how it was presented

The presentation here will be based on the individual Muslim partners in the dialogue and not on individual sub-themes because the differences between individual Muslim organisations and countries with which contact existed[5] are sometimes quite considerable, and for this reason the dialogues and their results also turned out very differently, and it is interesting to see along which lines certain relationships developed over the years and exactly how common points and differences came to light in the process. One single dialogue partner from one religion is simply not sufficient because it is impossible to cover the whole spectrum of a religion in that way. However, it is generally not easy to find the right partner(s). If the desire, as became increasingly important as time went on, is to eliminate the very roots of religious fundamentalism and terrorism, it is essential also to have a dialogue with hard-liners despite all the difficulties involved or, when political conditions make official dialogue impossible, to resort to informal dialogue and, in the process, always to make sure that the dialogues are not being exploited by one's counterpart and one is not compromising one's own cause. These are naturally all insights which are sometimes very costly although this is rarely stated clearly. In addition, a summary of this kind must not forget that one of the main tasks of the Secretariat was to encourage dialogues at regional and local level, to support them in every way possible but without actually organising them. Regional dialogues, which will be discussed in greater detail later, often have the advantage of exercising a positive influence on a previously difficult situation of dialogue and of providing hopeful stimuli for the future. Whether as a preparation for larger dialogue conferences of its own (as the sections of this chapter make clear) or as preparation for local dialogues, the visits by representatives of the Secretariat were of great significance as is illustrated, for example, by a report on a journey by Monsignor Rossano dealing with the initial contacts made in the countries of Iraq, Pakistan and Bangladesh (6 February – 4 March 1977). Such visitors were considered somehow as ambassadors of the Pope who, for his part, enjoyed great respect because he embodied the spiritual element in this world. So a visit of this kind could frequently become a catalyst for both sides to approach one another, especially for the local Christian churches which often have the problem of being weak and not firmly rooted in their context, something which makes dialogue very difficult if not impossible. The aim is by means of dialogue to bring the message of the Gospel into the human setting and the human discussion and for this it is necessary to be very familiar with both sides. Faith must take a personal form, especially in reference to young people for whom forms and institutions mean little. But the freedom to express one's faith is also very important, including the social forms of this expression; this is seen as directly connected with dialogue and the Gospel. The tendencies which came to light in the countries visited can certainly also be found elsewhere, apart from the fact that the spirituality of Asian Islam is more open to saints and mediator figures than is the case in the originally Arabic Islam. Materialism and secularism are also seen as a danger on the Islamic side; this is a front on which they would be happy to cooperate with Christians. They are concerned about the young people because they have difficulties, very similar to those of the Christians, in trying to express and transmit the faith in today's world. Suggestions for dialogue went along the lines of culture, humanism, the metaphysical bases for social behaviour and living together, so these were all more anthropological subjects against a theological background, defined here as juridical, ethical and philosophical cooperation on a theistic basis. The very practical issue of Jerusalem and Palestine was also always present but became less and less important the further away one was.[6]

1.4 After some time: wishes addressed to the Muslims and small steps backwards

The first general assembly of the Secretariat from 24–27 April 1979 with all the staff, members, advisors and selected ecumenical guests was still, when looked at attentively, strongly influenced by the initial, not very easy experiences made and which had caused this dialogue to be considered difficult. The representation of so-called official Islamic bodies was said to be a continuing problem. It also had to be accepted that many Muslims followed the pattern laid down in the Quran of leading non-Muslims to Islam by means of the best dialogues. Finally, it should not and would not be forgotten that, for the Muslims, dialogue always had a more or less strong political slant. On the other hand, when the Muslims tried to respond to the challenges of the modern world, one could also find many attempts to do so by re-establishing Islamic law, and this gave special encouragement to so-called Islamic fundamentalism irrespective of whether of the Gaddafi, Muslim brotherhood or Saudi Wahabism variety. The wish on the Islamic side was for Islam to be seen in a less ideological or, more precisely, juridical-political way, a view of Islam which considered it more as a way to God. So far, this shift in emphasis towards underlining (what Christians consider to be) the really religious dimension of Islamic experience can be observed mainly among Muslim minorities in non-Islamic countries.[7] So these experiences fit very well into the overall picture with its necessarily different aspects. As far as dialogue events themselves were concerned, under the immediate successor to Cardinal Pignedoli, Monsignor Jean Jadot, again considerably less attention was being paid to dialogue meetings organised by the Secretariat itself, and its members were being encouraged instead to take part in dialogues organised by other organisations, the reason also being given that convincing dialogue simply did not take place at the highest level but rather in the many dioceses around the world. This is an approach which was followed consistently from then on: a threefold division of the work into Islamic studies, which were undertaken mainly by the specialist advisors of the Commission for Religious Relations with Muslims (see also Publications), the role of the organisers of dialogue at the local level which was considered particularly important and, finally, its own involvement in dialogue which fell far behind the other two and was to be reserved essentially for the international level. To go into all the other levels in detail would exceed the limits of this investigation, but it should not be forgotten – and this was repeatedly recalled by those responsible – that perhaps the most important dialogue is the one which appears least glamorous to the world outside because it takes place at the local level where people meet quite directly and in a lasting way. The insights and results which come out of this in the course of years of work should not be underestimated and certainly deserve to be made known to a broader public[8] although this is naturally difficult to implement in practice. Here an attempt will be made to outline the developments where there were several and, at best, a whole series of dialogues with specific partners. Those that must be mentioned above all are Libya and the World Islamic Call Society, the Royal Academy for Islamic Civilisation Research (Al Albait Foundation), Amman, as well as the contacts with Tehran. When a series of dialogue meetings took place, the Catholic side endeavoured[9], on the one hand, to ensure a degree of continuity among the participants and, on the other, to involve leading experts from the Catholic side, some of whom then met Muslim academics for the first time – an advantage for both sides which was certainly intended.

2 Contacts with Saudi Arabia – working session on human rights

As early as in April 1974, Cardinal Pignedoli visited Saudi Arabia and was also received there in audience by the king at that time, King Faisal. This helped to prepare for another event which resulted from an idea of Sean MacBride, the

Under Secretary General of the United Nations, and which took place after quite a number of postponements from 24–27 October 1974 when the Saudi minister of justice came to Europe with a delegation of Muslim legal experts to discuss human rights with European lawyers. In the process, he paid a visit to the Vatican which was the climax of the journey and there were some joint working sessions, mainly on the issue of human rights, in which the Pontifical Commission Justitia et Pax also participated. Although this was more of a visit by nature – also including a private audience with the Pope – than a dialogue conference, it is nevertheless worth mentioning for this early stage.[10] After all, it dealt with a very important issue related to the consideration given to human beings and one which is disputed between the two religions.

3 Contacts with the World Islamic Call Society, Tripoli, Libya

3.1 The very first dialogue: Tripoli 1976

3.1.1 The invitation to a dialogue with Tripoli – numerous contradictory factors

The very first contacts with Libya and with Muslims in general in the context of an official dialogue after the Second Vatican Council did not come about through World Islamic Call Society based there but from 1–6 February 1976 at the invitation of the head of state, General Muammar al-Gaddafi personally, to the capital of his country. This dialogue was extremely noteworthy in various respects and it is still difficult to reconstruct because the decisive documents are not yet officially available. So this evaluation is also subject to certain limitations. With regard to the events concerning the preparations for the meeting and during the meeting itself, the evaluation is based above all on the very detailed description written for Islamochristiana by one of the main protagonists, Father Maurice Borrmans, which contains a wealth of detail. To supplement this, reference was also made to the official report and the English translation of the concluding declaration which were published in that form in the Secretariat's own bulletin, provide closer insight into the preparation of the meeting and make it easier to interpret the official Arabic final declaration. Unfortunately, this meeting is known less for its interesting overall development and more because of the éclat over the final declaration and how it was produced. In an all too human way, that continues to overshadow both the memory and the follow up to this day.

But to begin with, above all, something about the developments which preceded it. Officially there were two delegations, one Christian and one Muslim, comprising finally 12 (15) and 14 (18) participants (there is some lack of clarity about the numbers comparing the two reports), but there were about 500 observers from 62 countries of Africa, Asia, Europe and America, of whom about 100 were journalists – a large amount of attention for a religious meeting and certainly not always an advantage for its course and its evaluation. (Comment: "What had been expected to be a quiet exchange among experts turned into a festival of dialogue."[11]) The initiative for this came from Libya, from the Socialist Arab Union. This probably had to do with the fact that shortly before, as a result of a different initiative, there had been Christian-Muslim dialogues in Broumana, Cordoba and Tunis and there was a wish not to lag behind or even to do better. The Vatican in the form of the Secretariat for Non-Christians was the ideal partner for this. And the Secretariat, for its part, did not want to turn down the invitation. The whole event was organised in only four months which is a very short time if one considers the preparatory periods for later events with Vatican participation and also the time generally required for preparing much less impressive events. In this case, what happened was that in May and July 1975 a Libyan delegation led by the minister of foreign affairs had visited Cardinal Pignedoli, initially as a private, polite visit and then officially in the Secretariat. During the second visit, the conversation had turned to a Christian-Muslim dialogue. Cardinal Pignedoli expressed his general readiness for this, wherever it might take place, on condition that it dealt with a religious and not a political subject. His suggestion was the subject of youth. The Libyan foreign minister suggested a general Christian-Muslim dialogue with Tripoli as the site and December 1975 as the date. He then produced an invitation for the Holy See to attend the anniversary of the revolution (1 September) and urged that a preparatory committee be appointed to organise the dialogue. A delegation from the Vatican State Secretariat participated in the celebrations of the national day in Tripoli on 1 September 1975. The secretary of the Commission on Islam, Father Abou Mokh, was a member of this delegation and took advantage of the opportunity to discuss the details of the arrangements for the dialogue conference with the Libyan minister for foreign affairs and other representatives. From that moment onwards, the practical preparations were in the hands of Father Abou Mokh, on the one side, and of a representative of the Libyan embassy in Rome, on the other. It was also the former who, when the first difficulties arose, flew to Tripoli again and also travelled to the conference itself two days ahead of the Vatican delegation. But, prior to this, on 1 October a rather late joint preparatory meeting gathered in Tripoli with the whole Vatican preparatory commission to which Msgr. Rossano, the secretary of the Seretariat for Non-Christians, and Father Borrmans of the PISAI also belonged; there the purely religious nature of the encounter was again confirmed (this was again stated officially at a joint press conference in Rome on 19 January 1976), and the issues as well as the dates were decided. At that point, two delegations of twelve members each were envisaged as well as about 20 silent observers; not only were the numbers ignored later, as has already been stated; there were also repeated departures from the complete observer status during the conference. The Christian side wanted to put together an ecumenical delegation (thinking of a Coptic bishop, a representative of the Greek Orthodox Church and a representative of the World Council of Churches). The Libyan side was to prepare the invitations immediately and also to ensure that the addresses were made available in writing one month in advance. The names of the speakers were to be fixed by the end of October. At its own request, the Libyan government wanted to cover all the costs. The only thing that happened immediately was that the number of observers was increased very quickly and in a very one-sided way by the Libyans who invited both Christians and also mainly Muslims. Since the Secretariat for Non-Christians itself had issued only a few invitations (twelve participants and the same number of observers) and, in addition, the Coptic Patriarch of Alexandria and the World Council of Churches had declined the invitation, the Muslim participants and observers present were considerably more representative than those on the Christian side. Those who were obviously not represented were the famous Muslim universities, especially the Al-Azhar, and Saudi Arabia. Among the Christian observers, the Eastern Orthodox churches were strongly represented. Their presence was very lively and, above all, markedly anti-Jewish. The Christian (actually Catholic) delegation, which had expected a much more modest and more academic framework (cf. above), was clearly very much taken aback. In addition, the promised prior exchange of papers had not worked out (only one was distributed shortly beforehand after much insistence), indeed, before the conference it was not even known who the Muslim speakers would be so it was not possible to gauge the tenor of the meeting.[12] At that point, some people even suspected that it would no longer be possible for it to be an academic conference and it would remain on a more social, popular level.

3.1.2 The framework of the conference – what was it to deal with?
The Vatican delegation consisted of the president of the Secretariat, Cardinal Pignedoli, the secretary, Monsignor Rossano, the officer for Islam, Father Abou Mokh, a Syrian, and several advisers to the Secretariat, three West African bishops, two priests from Arab churches and an Indian layman, member of the Vatican Commission Justitia et Pax, so it considered itself to be representative. The Muslim delegation was led by the Libyan minister of education and was certainly representative as far as standing and countries of origin were concerned but in

no way officially represented any Muslim institutions. The official welcome by the minister on behalf of the Arab Republic of Libya did not only reflect how the host country saw itself and its revolution but also, in a sense, set the direction for the dialogue conference: to honour Islam and to work for the liberation of humankind from all forms of injustice and slavery. Underpinned by numerous quotations from the Quran (including Sura 49:13 which emphasises the creation of humanity from one man and one woman and the formation of communities and tribes), the minister emphasised that the dialogue between Muslims and other people was as old as Islam itself and also the importance of trust and openness. In his reply, Cardinal Pignedoli underlined two common values which justified this dialogue, the worship of God, the creator and lord of the universe, who had made his will known to humankind and taught them the ways of justice and peace and, derived from that, the fraternity of all human beings which obliged us to be agents of mercy and service. The gospel really equated the commandment of love for one's neighbour with the commandment to love God. But the Cardinal also referred to Sura 5:48 according to which God could have made all of humankind into one single community if God has so desired. Overall, he emphasised the importance of respect and love for others and also of the desire to be a witness of this universal brotherhood to others and the importance of the willingness to serve.

3.1.3 Contrasting religion and ideology: (very) bad prospects for ideology

The first real working day was devoted to the issue of religion and ideology (a suggestion of the Libyans) with addresses by Prof. Utba from the University of Tripoli and Anthony Chullikal of the Commission Justitia et Pax. Prof. Utba immediately approached the problem anthropologically but making the question a criterion for deciding whether ideologies were able to respond to all human needs, be they individual or collective, material or spiritual, and naturally in all places at all times. He went straight on to deny this because ideologies would never be able to offer a comprehensive view of human beings, societies and the universe in a framework which allowed every living being its place and its relationships to the others. Religion alone could provide this; it alone was the source of order for life and for society, and naturally Islam in particular which was best suited to regulating human life, human groups and the international community. Behind that lies Islamic law which also governs morality and inter-personal relationships and, in a comprehensive view, makes human beings God's representative. He spoke particularly about relationships with one's fellow human beings and how they were regulated by the Islamic system simultaneously with strictness and smoothness. "[L]e système islamique se caractérise par son fondamentalisme (...), sa globalité, sa souplesse et son réalisme. Il prend en considération les vérités de la vie, les caractères des gens. Les principes ont fait leurs preuves. Le système islamique forme pour l'humanité une civilisation permanente, une révolution continuelle, un message immortel... L'Islam, dernier des messages célestes (...), est un message général et mondial, valable pour tous les temps et tous les lieux, et donc pour tous les hommes "[13].

The Christian speaker also discussed the inability of ideologies to respond to all the problems of humankind for human individuals. Religion, on the other hand, as the search for God, encompassed all aspects of life. It was the only protection for the inalienable freedom of human beings who were persons not because they were social but social because they were first of all persons. He came to the very Christian or rather Catholic conclusion, "[L]'unité de la personne et la multiplicité de ses relations internes et externes temporelles et spirituelles, individuelles et sociales, sont tissées ensemble et subordonnées à la vocation spirituelle intrinsèque, respectant pleinement la possibilité humaine de communion avec Dieu. Dans son économie divine, Dieu guide les efforts concertés de l'humanité dans la construction d'un royaume humain sur terre: nous participons donc avec Lui de façon optimiste à la création d'un monde meilleur. La foi chrétienne laisse donc la place à un pluralisme légitime. Nous ne sommes pas constraints par un seul 'formulaire'. La création continuelle est à la base de notre espoir"[14]. In other later and simpler words: Christians find guidance in their religion, a kind of conscience, but it does not give them ready-made solutions once and for all.

Both of these addresses were probably hard for the average listener on the other side to understand because each was too strongly marked by its own religion, linguistically and also in its basic conceptions.

Although Msgr. Rossano believed he could observe that both Christians and Muslims placed great trust in human reason (to what extent Protestant Christians would see this in the same way as Catholics is also a question; here there are some basic differences in the fundamental view of human beings although the practical outcome is then very similar when compared with the Muslim side), but the Muslims present considered even the temporal ordering of earthly polity to be a dangerous autonomy of reason and were very critical of the principal of "laïcité" and a secular state. The question was also raised whether a Christian was not too often torn between the spiritual demands of his religion and the confusion and disorder he experienced when faced with the possibilities of political choice in a society which was not guided by religion. This is certainly a very honest, profound and typical question of Muslims addressing Christians. Unfortunately, the answer apparently was lost in the excitement caused by the appearance of the head of state, Gaddafi. This made the discussion turn to other, more political questions. Gaddafi himself also spoke vehemently against a secular state and secular legislation. The constitution of a state should not depend on the whims of governments and on party wrangling, and secular legislation was wrong because it had lost its natural sources, namely religion and customs. This touched on a problem which belongs to the fundamental differences between Islam and Christianity and which logically could also not be resolved by this conference.

3.1.4 Features common to both – and how they were swamped by Muslim polemics

The next day was devoted to the features common to Islam and Christianity, also in the practical realm (a subject suggested by the Christian side), and it was marked by the addresses of Father Maurice Borrmans of the PISAI and Professor Ismail Faruqi of Temple University, Philadelphia. The former spoke (in Arabic) against the background of the Council documents, indicating, for example, that both religions sought and saw God as the creator, the one who loves humankind and is its fulfilment, who is merciful and shows mercy and raises people from the dead. In addition, Borrmans referred to the common ways of responding as humans to the divine call, namely faith, worship and the recognition of divine commandments as agreement from the human side – in a common image: adopting divine ways of behaving. Commitment to human beings on both sides also had profound theological roots: because we believe in the living God, we stand up for respect for life wherever it is threatened. Because we believe in a God who is just, we combat every form of discrimination whether its motivation be sexual, racist, cultural, religious or national. Because we believe in a God who is free to create and to renew, we defend the rights to freedom everywhere. Because God is peace and calls people to himself, we endeavour to create an international, fraternal community. Christians were conscious of human dignity, were concerned about human community and convinced of the value of human activities. Borrmans invited the Muslims to a dialogue on values in the context of common service for life, justice, freedom and brotherliness.

The lecture by Prof. Faruqi, on the other hand, concentrated mainly on an original religion from which Judaism was also derived but which he explicitly called Arab and not Semitic (just as he also described Abraham as an Arab and considered the three monotheistic religions to be Arab); he saw this religion as constituted by the following: a dichotomy of the real between creator and creature, a communication of the creator with the creature through revelation, the gift to the creature of an ability by means of which the creature can implement the plan of the creator in the world and, finally, the realisation of moral values by the creature thanks to additional power and the use of reason, and hence also responsibility and sanctioning. However, the second part of the address consisted mainly of polemics against Zionism, Christian mission and Oriental studies. Although he voiced a warning about attributing the shortcomings of Christians to Christianity as such, and generally admitted the task of mission – one had to respect that every person made an effort to enable his neighbour to share in exalted wisdom and religious truth –, he still saw Christian missions as a sin against human freedom because, especially among Muslims, they always took place in a necessary, secret collusion with colonialism, according to his conviction. And he also demanded at the end that reason rediscover its limits, although here he was thinking of the human sciences which had to lead to a healthy understanding, moral behaviour, a balanced family and the complete rejection of racism and materialism.

This was certainly not, as was also observed on the Muslim side, the positive contribution desired with a view to future cooperation but more of a step back to useless accusations in which the discussion then mainly got bogged down and which Prof. Faruqi did not abandon even to the bitter end, despite all the theoretical and practical positive arguments against them.

3.1.5 Faith and justice – a very Muslim theme

The next day the subject was faith and justice – also a subject suggested by the Libyans. The addresses were given by Prof. Al-Ghuwayl, a lawyer, and Father Camps from the University of Nimwegen. The whole of the Muslim contribution was an excellent presentation on Islamic legal discernment in the area of interpersonal relationships or, more precisely, of economics, and it ranged from the Quran via traditional interpretations to its implementation by Libyan socialism. It was possible to construct a complete social system on the principles of Islam. The starting point was the unity of faith, thought and action or, more exactly, the central certainty that human beings are God's slaves, a slavery that liberates them from other, alienating slaveries. The brotherhood of all humanity was also derived from this, as well as the common right to benefit from the whole of creation. On the basis of numerous passages from the Quran, the speaker then explained that the non-Islamic cultures were unable to protect human beings, whereas Islamic culture gave them the task of thanking God by their work without ever being unjust to themselves or to their brothers. This task situated human beings far away from the economic ways of materialism, whether practical or historical, such as had been adopted by the capitalists and other atheists.

As God's representatives on earth, human beings were aware that they benefitted from universal submission because God had made everything subject to them, minerals, plants and animals. The consequence was a surplus of resources on this earth, enough for all people to live on in a humane way. So it was sufficient to obey God's law, the Islamic law. In the second section, the speaker then went into some detail with regard to possessions and payment for work. Possessions were seen as subject to God's plan and entrusted to human beings. Profits and benefits were to be considered gracious gifts from God which had to be distributed according to social justice; hence Islam also had a conception of common property which comprised everything directly created by God, and of private property which the community allowed the individual to have because it was not needed by the community or because certain things could only be done by private initiative. This private property was subject to the alms tax which benefited the poor (as the speaker explained in detail) and also served to purify private wealth and thus led to a degree of social security. Therefore work, faith and science worked together for development in all fields and guaranteed right action. By this, the speaker understood that access to property was only the right to use it and that the poor had their share of good things from the creator thanks to a just distribution of wealth.

Camps, for his part, adopted a quite different approach in his address and began by discussing the question of social justice in non-monotheist societies, namely India with its caste system and its caste-less people, Buddhism and its efforts to escape from the world, and also the "primitive" religions which hardly knew social justice apart from the solidarity within a clan and found it hard to recognise. Against this background, Camps drew the conclusion that social justice had something to do with faith in (the one) God. Human beings were a unity and one should never separate their material from their spiritual life. The following spiritual values in the monotheist religions promoted the development of justice: faith in a personal God who had made each human being as a person; positive evaluation of creation; linear conception of history which was progressing towards a future kingdom of peace, justice, unity and love according to God's plan; awareness of the social nature of human beings: justice was brought about with the followers of one's own religion and with all people. Naturally, there was reference to the social prophets of Israel, as well as the Quran and, above all, Luke 4:14–22 which was a call to eradicate poverty, liberate prisoners, heal the sick and the blind and to help those who were crushed by the difficulties of life. The lecture had to accept the accusation of being one-sided: the anthropology of the non-monotheist religions was not the only cause of social injustices.

That is certainly true although it was probably not the aim of the lecture to list all the causes of social injustice but, on the contrary, to show the special common relationship to justice which the monotheist religions had. However, the discussion as a whole made clear that there were fundamental difficulties of understanding, especially among the Muslims, to be able to grasp Christian freedom in this respect and to see it positively – quite apart from the fact that many unbelievers also struggle for social justice. The various statements by the Muslims repeatedly made it clear that, according to their conviction, it was a shortcoming that the gospel only made suggestions and did not contain any precise divine law that solved all life's practical problems. The fact that Christianity rejected a theocracy was a deficit in the sense that it did not completely embrace the spiritual and the worldly within itself, nor organise both the one and the other. However, there was a view on the Muslim side as well that it was not religion that would solve the problem of under-development but rather science and technology. But the immediate response was that social questions belonged in the political realm and that quite clearly depended on Islamic law. At this point, there was an exhortation – from the Muslim side – to remember that the preparatory commission had promised the Vatican that political issues would not be discussed and that it was surely desirable that everyone respect this promise. Ths touched on a very important point, particularly with a view to the final declaration, and it became clear – and from a Muslim mouth – that there had really been clear instructions from the Vatican and equally clear agreements concerning politics on which all had agreed.

But it is also clear that it is much more difficult for Islam and for Muslims, because of their comprehensive, fundamental approach, to exclude any realm as non-religious and hence neutral, even if one does not speak immediately about an area like politics which is so important for human coexistence and therefore religiously ordered. The very comparison of the subjects desired by the Christians or the Muslims speaks volumes. The subjects requested on the Christian side were more on the theological level whereas those on the Muslim side all also had a practical, legal, social, political slant – and all combined because the exclusion of these questions would amount to an amputation of Islam since, in the eyes of its followers, that is precisely its main content, indeed, demonstrates its superiority to ideologies and other religions. So, in a sense, the difficulties were (and are) bound to arise.

3.1.6 Prejudices and misunderstandings: subjects desired by the Christians and a festival for the Muslims

The last major subject (again a Christian suggestion) was the question how one could overcome divisive prejudices and misunderstandings. The addresses on this were given by a parish priest from Algeria, Father Lanfry, and a philosopher from Algeria who was living in Paris, Muhammad al-Aychubi. There were few statements that were directly anthropological unless one counts the impressive human effect produced by Father Lanfry's request for forgiveness. From his list of proposals, one could also include the condemnation of every form of proselytism which failed to recognise the demands of religious liberty, and the absence of any kind of religious or racial discrimination in Christian aid to the "Third World". But he also mentioned various points which, in his opinion, were obstacles to progress in Christian-Muslim dialogue including, for example, the misunderstanding of dialogue as a holy alliance against the forces of materialism and communism (which had been said on the Muslim side several times), the Muslim equation of the actions of individual Christians with the actions of their (secular or atheist) states in which they were free citizens, and finally the fact that both sides decided in favour of unchangeable doctrines which made coexistence difficult in societies which were not pluralist (as systems). He himself appealed for the acceptance of pluralism. *At this point, at the latest, the dialogue reached the all too well known problem, albeit in a somewhat different form, and the question is whether the patience demanded will be sufficient on its own to solve this problem.*

The other address consisted in practice only of the classical accusations against the Christians (and the Oriental scholars) which Faruqi had presented in a similar way before, with special emphasis on the crusades, and of a series of basically exclusively political demands under the overall heading that Christians should in future refuse to cooperate with an oppressive and unjust international order. What this implied was a condemnation of Zionism, even before the just distribution of the material and technological wealth of the West, and refraining from the papal demand of a special status for Jerusalem and its holy sites.

Fortunately the discussion was more peaceful than one might have expected after that. The reason was, on the one hand, that the Christian delegation considered it (understandably) useless to respond to these attacks and also made this known. But also because another Muslim statement (Ibrahim) summarised the problems in a rather more dignified way as, firstly, of a dogmatic nature, secondly of a combative nature (although he was generous enough also, and in first place, to mention the combative spread of Islam and not merely colonialism and Israel), and thirdly of a cultural nature because there were also differences and superiorities in this realm. More attention was devoted to the Christian address but here the response given pointed out, for example, that Christians in Muslim countries naturally enjoyed all the rights of dhimmi (protected persons). With regard to Christian missionary activity in the charitable and humanitarian field, it was praised and declared useless in the same breath, indeed even described as offensive in Muslim countries. Here again a very fundamental problem came to light. The statement by Msgr. Sangaré, the archbishop of Bamako, the capital of Mali, was also very interesting and revealing; three things seemed very important to him: not to import the quarrels and conflicts between Christians and Muslims of the Mediterranean region into Black Africa, to avoid simplified slogans such as that Christianity was the religion of the white people and Islam the religion of the Blacks, and finally to respect all minorities everywhere.[15]

3.1.7 The final declaration: chaos and a Near East political scandal

This then was the initial situation against the background of which a final declaration was to be drawn up. Borrmans described it in retrospect as follows, and this fits the individual observations made thus far. "[I]l s'y affirmait un *'climat arabo-musulman'* qui allait crescendo et où le politique et le religieux ne se connaissaient plus de frontières"[16] – certainly also encouraged by the hosts by means of one-sided or clearly erroneous reporting. In the course of the meeting, Cardinal Pignedoli had often intervened to counteract the obsession with political problems but apparently he was finally unsuccessful. It seemed probable and fairly evident that it would not be possible to agree on all points for a final declaration, indeed, that fundamental differences and conflicts of interest existed. Both sides had worked independently on the draft of a final statement in the meantime and the suggestion was also made (because the time for producing a joint final declaration was short) that perhaps the two statements should simply be juxtaposed. But then a common final declaration did come about somehow or other. The first thing that must be said about it is that its only official version is really the Arabic one; in other words, that is the only one which was in any way countersigned by the Catholic participants although never as a whole, neither by the head of the delegation nor by anyone of the four participants who had worked on it in practice, but only in individual parts by these four individual participants. The French and English versions can strictly speaking only be considered semi-official because they were produced without any assistance from the Christian side. However, it was these which then hit the headlines worldwide and, even though somewhat revised (the French version), adopted by Osservatore Romano and thus in a sense made official. It must also be pointed out that, for its production, the basis was only the Muslim drafts and no use was made of the excellent Christian preparatory draft[17] which clearly dealt with the happenings during the seminar and, at the same time, was more succinct, whereas the Muslim (and later "common") draft must be seen more as a collection of individual paragraphs of different nature and content. Moreover, at the time when it was officially read out, it was not available as a whole to any of the Christian participants, and the leader of the Vatican delegation, Cardinal Pignedoli, was not even familiar with the content because the translations were only produced later. "L'improvisation et la confusion continuaient donc à être de règle"[18], to put it cautiously and with maximum politeness in relation to the hosts. It is clear that there was no time for all of this, and indeed nothing worked out well on that day as it normally should have done, so the rather cryptic formulation "La rédaction des Conclusions se développa en des formes imprévues qui sont difficilement imaginables pour qui n'y a pas participé"[19] appears to be the best description possible.

Thus, at the end, the final declaration consisted of 24 paragraphs including one (§ 5) on human dignity and its mobilisation against any form of injustice, one (§ 6) on the rejection of any form of racial discrimination and one (§ 8) on religious liberty, all three of which are theoretically important for an anthropological examination,[20] and also a paragraph (§ 23) on the establishment of a mixed commission for the continuation of this dialogue, but all of this remained theory on paper because there was also a § 20 on the distinction between Judaism and Zionism and a § 21 which affirmed the national rights of the Palestinian people and a common interest in the protection of Jerusalem, and from that moment onwards there was practically no talk of anything else, not about the other points in the declaration and also not about the happenings during the conference itself. That the Muslim side wanted these points in the declaration and that the Vatican did not want to say anything on these points from the very beginning is evident from the descriptions of the preparation and course of the meeting; so the open question is how exactly these two paragraphs slipped in nevertheless. Unfortunately there was also no possibility of clarifying the matter straight away because the press conference planned never actually took place. Finally, far into the night, the two delegations agreed that these two paragraphs, which certainly discredited Zionism as an aggressive, racist movement without any link with Palestine and, on the other hand, affirmed the right of the Palestinians to return and Jerusalem as an Arab city, should be considered a Muslim statement and should be submitted as such to the Vatican secretariat responsible for political issues – which naturally promptly rejected them. But feelings had already been aroused, especially on the Jewish side (because there had just been a declaration of the United Nations along these lines) but also elsewhere, and it was not so easy to calm them down again, even though it would certainly be an injustice to this dialogue to see it only in the light of these two paragraphs.

3.1.8 Lesson from Tripoli: dialogue is not that easy

But unfortunately this is all too human and it (also) led the Catholic side to reflect on the difficulties of dialogue and the requisites for its preparation – and some Muslims as well. On the Catholic side it was expressed as follows: "[I]t must be acknowledged that dialogue with the Moslems is difficult: there is a discrepancy of history, of culture, of social evolution, of religious mentality which only now we fully realize. The dialogue state of mind has still to be born in them, and for the Christian part the vocabulary of dialogue and sufficient theological categories have to be developed in order to be able to think Islam. The very commonest words such as dialogue, prayer, monotheism, revelation, will of God, have different meaning for them and for us. (…) [T]he most concrete result of the days in Tripoli and the most valid methodology for dialogue seem, for the moment, to be meeting and personal welcome: to feel and call each other brothers in the name of the same God and to offer to the Muslims the example of humility, disinterestedness and brotherhood in an interpersonal relationship of friendship in which the only thing sought is the good of the other."[21] Almost 30 years later, the person then responsible for Christian-Muslim dialogue still put it in the following way: "While appreciating Libya's generous hospitality, this initiative created a negative opinion with regard to dialogue among the Christians, due to what was perceived as political exploitation of the dialogue."[22] But the meeting also had another practical success. In December 1977, the Catholic Church in Benghazi was opened. On the Muslim side, at least, one of the spokespersons for this dialogue later made the observation that dialogues at which there were hosts and invited guests had no chance of success.[23] If one considers the later development and the dialogues, whether in Libya or elsewhere, which were the result of the cooperation between the Vatican and various different Muslim structures, it would appear that this insight has been generally accepted. The further back in time it lay, it was also possible to get a better view of the positive aspects of this dialogue and to emphasise them. For example, UNESCO in Paris marked its third anniversary with a celebration at which the then moderator of the Vatican delegation, Msgr. Rossano, made a speech which strongly underlined the points common to both sides, especially that neither in Islam nor in Christianity could human beings be perceived as such or in their relationship to the world without reference to God. That distinguished both from atheist systems and naturally had many practical consequences. The stability of a legal order and personal and social rights and duties were seen as depending on their religious basis. And, the other way round, human beings were certainly seen in the mirror of the universal divine law and thus as intelligent and endowed with reason, and this was expressed in their activities in the world. Perhaps, according to Rossano, absolute respect for the conscience of the others could even solve the problem that both religions are essentially missionary. But it would certainly be a great day in the history of humankind when both religions really discovered they were brothers.[24]

3.2 A new beginning from 1989 onwards: also many anthropological elements

After a long break in relation to Libya, the Pontifical Council for Inter-religious Dialogue again had regular contact with the World Islamic Call Society, Tripoli, Libya, in the form of annual short meetings in a small group of officials, starting on 18–19 March 1989 in Tripoli with an introductory visit by Cardinal Arinze, and continuing on 14–15 February 1990 in Rome where it was a matter of the understanding of mission or *da'wah*. Naturally, anthropological elements also came in here, for example at the meeting in 1990 on service in the area of promoting human development and liberation. The Muslim side continued to lay emphasis on knowledge of God, the creator. There was agreement that, inter alia, an attitude of respect for human dignity and freedom was necessary and also a humble sense of service to humankind. Moreover, a whole series of things were mentioned that were to be considered wrong or at least problematic: compulsion, enticement, manipulation, mockery. In addition, there were difficulties connected with modern society, e. g. the influence of materialism on the present generation, but also migration, especially in the form of refugees from war, dictatorial regimes, and economic and social need. That gave rise, among other things, also to the problem of the religious rights of minorities. There were international conventions dealing with this but they needed to be implemented. The question of mutuality was also raised, as well as the expectations Christians and Muslims had of one another with regard to religious liberty. On this, there had been improvements in the past few years, also thanks to dialogue and to the growing contacts between Christians and Muslims, even though there were still difficulties in certain places.[25]

3.3 Malta 1990: the importance of tolerance

Continuing these meetings, the organisations held a joint symposium from 22–23 November 1990 in Paolo, Malta, on "Co-existence between Religions: Reality and Horizons", in other words, the question of tolerance. The opening addresses by the two presidents underlined the importance of tolerance for human civilisation and human co-existence on the small and the larger scale. Cardinal Francis Arinze, in particular, went straight to the heart of the matter and cited the relevant documents of Vatican II to explain that the Catholic Church did not consider tolerance to depend on outward circumstances but as based on the dignity of human beings. It was a matter of a fundamental human right, namely the right to obey one's conscience. This applied both to the level of the individual, where it implied a positive, interested and cooperative attitude to others, and also to the level of communities, because both human nature and the nature of religion necessarily led to such communities. The welcome speech from the Muslim side, on the other hand, emphasised above all that they had gone beyond the stage of getting to know one another and reached the stage of cooperation, and it made very positive mention of certain papal statements.[26] *Considering the difficult start to these relationships (cf. above), this can certainly be viewed as great progress despite all the necessary reservations.*

3.3.1 Christians in the cosmology of Islam

The first round of addresses dealt with the Christian view of Islam and the Muslim view of Christianity and then moved on to present the practical conceptions of tolerance. These initial addresses naturally provided less information on the view of human beings because each dealt with the other religion. The Catholic address, in particular, followed this line very strictly and, in practice, consisted of a summary of the statements of the Council and of *Ecclesiam Suam*, supplemented by a quotation from the Casablanca speech of John Paul II. What was decisive was the common view of God as creator and judge. Islam was also praised, for example, for its tolerance even though there were some problems there, and also the cultural achievements of Muslims to whom Christians owed a lot, and finally also their spiritual experience and the general fact that they sought God with honest and open hearts. On the other hand, in the presentation from the Islamic side, it was much less a matter of Christianity as such and how it saw itself and more a question of how Islam in general sees humankind and hence also its religions or religion as such. Islam was a call for cosmic harmony and thus also to harmony within humankind. In its very name, Islam was peaceful submission to the laws which led to security and comfortable peace, which was another way of describing the element of harmony, the harmony between human beings and the world. Only when people submitted their lives to these fundamental laws (and naturally first understood them), could they enjoy well-being. The rest of creation obeyed these laws anyway; only human beings had the possibility of obeying or disobeying, but it was expected that they would voluntarily harmonise with the whole universe which praises God. Therefore the main problem people had was the loss of a sense for communion with nature. The departure from these laws was seen by the speaker, Ibrahim Ghwel, as the hour of birth of human existence. People needed guidance, which no doubt was to be understood in connection with the Islamic concept of right guidance, in order to become a part of the harmonious universe. Within this broad framework, humankind was seen as a family whose diversity should be understood as enrichment. God had created a man and from him his wife, and by means of marriage and precreation all the skin colours, races and languages had come into being. Human beings and still more the prophets were therefore all brothers and their diversity a sign of God's sacred creative power like the diversity of nature in general. This again was an expression of the great trust in objective, scientific research which would never be able to come to any other conclusion. According to Islam, no segment of humankind was superior to any other, but each tribe and nation had its own land and its own message appropriate to its needs (and here Islam united within itself what was essential in each message and abandoned what was outdated in the process). However, religion remained unchanged by this: "Religion, in other words, is a conception of human history based on unity and monotheism, and the diversification in human life is but an indication of a cosmic law that regulates every diversity. Thus, this is diversity in unity or unity through diversification."[27] Not all of these prophets were revealed in the Quran which is why a believer had to take seriously every "struggler to the sakes of humanity"[28] in history because he might be one of these unnamed prophets. *Here it is evident that the Islamic view of humankind is indeed very closely connected with religion and Christians and Christianity are quite simply part of a much larger overall design.*

3.3.2 Tolerance – a structural problematic case
3.3.2.1 Muslim tolerance – really so simple and sure?
In this connection, in Islam the unity of the human origin, human dignity, respect for rights, dialogue and right exhortation, but also the defence of places of worship all play a part. Overall, the address by Muhammad El-Saadi was full of quotations from the Quran and Sunna, proofs that in Islam religious tolerance is firmly rooted among the duties of a Muslim and thus safeguarded

against fanatical tendencies or changes in conditions and circumstances. As had already been stated, the unity of humankind by virtue of its origin was rooted in the Quran, so no nationality, race, skin colour and also no belief should dominate any other. Similarly, according to the Quran, God had endowed all human beings with a special dignity above all other creatures and this deserved respect independently of the faith of the individual. Again, the unity of the divine messages to the various prophets was emphasised and that Muslims should respect all of these. Naturally, the quotation from the Quran saying that there must be no compulsion in religion was mentioned, as well as the other, also quite well known quotation that, if God had so desired, all people would have one faith. And so the conclusion was that God created people free and gave them the freedom to choose their faith, and therefore there are different faith convictions. To make people responsible for this was thus not a human task; it belonged to God alone at the final judgement. An additional Quranic quotation was included which was supposed to prove that all places of worship should be respected, maintained and defended by Muslims, just like the mosques. The so-called Holy War was instituted to protect religious liberty and the sanctity of places of worship. In addition, the Quran commanded cooperation and justice for all and finally the free exercise of religion. "Every human being has his own religion which he may preach with wisdom, gentle exhortation, dialogue, logic and proper argument within a framework of politeness and mutual respect and without inflicting harm on anyone or infringing upon the rights of anyone."[29] Thus, for the speaker, assuming their practical implementation, tolerance, co-existence, peace and social justice were sufficiently safeguarded.

The little problems about this Islamic tolerance, which is certainly recognised on the Christian side, such as the way of dealing with polytheists, the dhimma *status and the real legal and practical treatment of church construction, the lack of religious liberty for Muslims paralleled by the prohibition of mission, quite apart from the absolute limitations in Saudi Arabia, were not mentioned in the slightest and there was just a hint that it was and is not always easy. For connoisseurs of the subject this is naturally unsatisfactory and leaves one with the impression that entering deeper into the subject is at least still necessary, and that this at the most only constituted a first round to counteract fundamental reservations.*

3.3.2.2 Catholic Church and tolerance: admittedly there were problems ...
The Catholic paper by Thomas Michel cited, on the one hand, the exhortation of the Apostle Paul that Christians should make their "tolerance" known to everyone (his interpretation of Phil. 4:5), and also Mt. 5:45 according to which God

causes his sun to shine on the just and the unjust, as well as the general issue of human dignity and love of one's neighbour, before moving on very quickly to deal in great detail with the many statements by Pope John Paul II on this question in general and in direct relation to Muslims. The aim of this contribution was not to give a comprehensive picture of the issue of tolerance but to indicate how the Catholic Church was currently working for a peaceful society under the pope of that time. It was certainly recognised and admitted, in practice, that a monotheist faith did not always constitute a reliable foundation for the dignity, brotherhood and freedom of persons, as John Paul II had formulated it, and this was traced back more or less directly to a narrow, if not egoistic, understanding of religion. What mattered was to recognise the otherness of others positively and with appreciation without dominating them, and to bring more tolerance and trust into religious culture and thus into culture in general. Trust, in particular, was not a universal element in human relationships but rather something that needed to be built up slowly, patiently and sometimes painfully.[30]

One is immediately struck by how different these two contributions are despite the fact, or precisely because, they both make a major effort to quote the highest available authorities on each side for the tolerance of their own religion. In Islam, these are the Quran and Sunna (both of which assume situations in which a certain degree of Muslim tolerance was possible and established), whereas for Catholic Christians the authority of the magisterium resides in the person who holds the highest office at a particular time and he was very much committed to this issue, as has already been sufficiently demonstrated. Scriptural evidence to parallel that of Islam is rather more difficult to find because the persecuted Christian community of biblical times was far removed from a situation in which it would have made sense to call for tolerance of people of other faiths. Here, the problem lies less in the many and certainly great exceptions from a principle of tolerance, as in Islam, and more in the way people understood their own tradition and handled it as soon as they were more or less the unchallenged masters of the situation. This was referred to at least briefly as well as the conditions that have changed making pluralist societies the rule today in many respects – and that this should be accepted positively as beneficial diversity (which almost sounds Islamic) and not merely grudgingly as a necessity. In this sense, the Catholic contribution was also a consciously/purposefully positive view, at least by recognising the fact that there had also been problems, which is perhaps easier than admitting that there are problems, and that they are also of a structural nature and not simple, albeit very long-lasting, erroneous developments.

3.3.3 Tolerance and its practical implementation
3.3.3.1 On the Christian side: a plea for the separation of religion from the state

The third round then dealt with practical implementation. The address from the Christian side was given by Maurice Borrmans. He began by emphasising how markedly the point of departure had changed compared with earlier times. Nowadays, almost everywhere people of very different religious backgrounds were living side by side and they had to be treated equally, so the old style of society which could assume a unity of religion, among other things, even if it assigned clearly defined standpoints to followers of certain other religions (there was express mention here of the "*dhimma*" status), was no longer conceivable in the modern world precisely because it did not respect the rights of all. He also pointed out that, according to the information they themselves provided, about one third of the Muslims world-wide were not familiar with a socio-political system marked by Islam and nor would they ever be. It was clear that the challenges differed. For Christians, the question was whether they really felt at ease and free to develop in a country that was trying to implement Islamic ideals and law. And, on the other hand, whether Muslims could also accept as a community to live their Islam as a personal religion in a pluralist society of secular inspiration where the state left the responsibility for religious practice and its social organisation to its citizens. Could a common view of human rights serve as a common charter for such new, pluralist states? Was it not necessary to find some kind of common programme with values and behaviour which, on the one hand, took account of modernity and technology and, on the other, respected the different religious and political loyalties of the citizens? Here, he also cited John Paul II's Casablanca speech according to which God had created human beings with equal dignity but with different gifts and talents, and he pointed out that the Pope had spoken repeatedly about the rights and duties of minorities. The real question was whether the declarations on human rights could not form this common basis which would also allow very different citizens to live together in peace, "especially if their interpretations could eventually, little by little, be made to converge by means of an exacting and sincere dialogue".[31]

The speaker identified certain agreements with the Indonesian concept of "*pancasila*" precisely with regard to equality and justice for all, on the political level as well, and with the Maliki principle *masalih mursala* (the five essential interests which had to be protected for everyone and which came into play if there was no religious text applicable for solving a particular case), namely religion, physical integrity, family descent, material possessions and intellectual abilities. After these more fundamental reflections, the address then turned to more concrete issues and problems. Here the speaker emphasised several times that the Christians had, in a sense, made an advance payment as far as their approaching Islam was concerned and he asked what had already been done on the other side or what was conceivable at all. There was special emphasis on representation in the cultural field, religious instruction (mutual, correct presentation in school books was explicitly included), no kind of discrimination in the charitable realm and respect for the religion of those in need. It would be desirable to have a common commitment of Christians and Muslims in ecological movements, because finally everyone knew that respect for nature was a form of intelligent obedience to God, the creator, and his creation. But the basic prerequisite was that society organise pluralism at all levels and offer the possibility for people to decide according to their conscience, e.g. as far as food was concerned or clothing (whether a woman wanted to wear a veil or not), festivals (both religions having state holidays) or matters of faith, namely freedom of worship with the possibility of having churches and mosques, priests and imams, and explicitly also conversion in either direction without fearing any reprisals under civil law. As far as dialogue was concerned, he quoted the archbishop of Algiers, Msgr. Teissier, who had said that the principle of equal value was important and also that Christian-Muslim studies should be undertaken by persons who had the same level of university education.

But all these concrete points related back to the one basic issue that, in view of the plurality and complexity of modern state structures, the separation between religion and the state was an absolute necessity. In history, a distinction had always been made between the two (emperor and pope, caliph and sultan). This distinction, coupled with human rights, led to a secular but not secularised state, one which did not only tolerate but also respected and valued religion. Reference was also made here to the results of a study conducted by the *Groupe de Recherches Islamo-Chrétien* which came to the same conclusion. At the same time, that was what so-called Western societies were trying to practice, and many Muslim immigrants confirmed that they were respected as individuals and as a community and they had received help. Finally, it was a question of an attitude of openness to other people and also to God, hence the final question, "Does not genuine pluralism finally open one to the mystery of God who has created us different but complementary for His glory?"[32] *This was certainly a bold plan but unfortunately there was no equivalent on the Muslim side.*

3.3.3.2 The Muslim side: separation between religion and politics is not an issue

The Muslim contribution began by placing strong emphasis on the fact that any practical application of sacred texts was human and thus subject to discussion.

So it was absolutely necessary to make a distinction between Islam and Muslims and between Christianity and Christians. Then Wajhi Kawtharani underlined how much Muslims sympathised with the religious liberation movements in Eastern Europe and were happy that the God who had been killed and relegated to the collective sub-consciousness was returning to Europe. Starting from here, the address dealt with the issue of injustice in the world in general and especially against Muslims but went no further (although the speaker officially denied limiting his remarks to such a list). As long as that was the case, it was also not possible to demand tolerance of these people. The speaker did not deal at all with the issue of the separation between religion and politics, as suggested, so to speak, by the Christian address; it was clearly not his concern or problem; on the contrary, "Perhaps a more useful method of dialogue is that which enables us to acquire better knowledge of our problems and the issues of our life and existence be they economic, cultural or political, or the rights of peoples and individuals. This methodology is not a call for the politicization of religion or for a decisive separation between religion and politics. For religion is concerned with justice, freedom and equality. (...) Religion is the only refuge for peoples when they are oppressed and the sanctuary for humanity when it is tortured. Islam and Christianity are equal partners in playing this humanitarian role in this fearful time of ours."[33]

It is precisely at this point that the fundamental differences began to appear which were so characteristic of the first dialogue in Tripoli and had made it politically so sensitive although not so intended on the Catholic side. The differences of approach really are so great that it is difficult to mediate between them, and the one theme becomes two quite different themes which give the impression of speaking past rather than with one another. Indeed, it is far from clear to what extent the concern of each party can really be understood by the other and, above all, whether it can be taken up positively, or whether the collisions are not more fundamental and insoluble in nature. It can certainly be almost as difficult to define concretely and especially to agree what (divine) justice in a situation is as to come to an agreement on a definition of monotheism.

3.3.4 Religious tolerance in Libya: very much dependent on Islamic law

The address on religious tolerance in Libya given by Fr. Sylvester Magro as the official representative of the Catholic Church in Libya occupied a special place to some extent. Naturally his address was mainly very concrete but it also contained some statements of a very basic character. He started from the assumption that human beings had a God-given need to live in brotherhood, peace and freedom as was appropriate for the children of God. In addition to this, the decisive factor in this area was religion and religious tolerance in particular. Nowadays, most religious leaders were also promoters of justice and peace between people although they could always do more for a new world order free of war and divisions. His practical experience was that people were also prepared to be educated for religious tolerance. The religiosity of the Christians was accepted and welcomed by the Muslim Libyans and even frequently encouraged, at least as long as it was a matter of persons. But this tolerance did not extend to religious writings. The latter were usually confiscated and censured initially by the customs authorities, for example, and only released weeks later.[34]

This is a very interesting observation which can perhaps be explained by the fact that Islam traditionally assigns Jews and Christians a clear place in a Muslim society and allows them religious liberty within their community, but everything that does or might relate to the outside or to Muslims is prohibited, and the concrete holy scriptures of Jews and Christians are considered falsified in comparison with the Quran. This would again be a convincing example that religious liberty is no problem as long as it is in harmony with the traditional conceptions of Islamic law but that it becomes a problem as soon as there are tensions and overlap.

The joint final statement attempted to reflect the line of argument of the meeting in broad outline; but it did not reach any clear common resolutions or declarations of intent although there was discussion of practical implementation and it was resolved to continue the work in groups of experts.

3.4 Tripoli 1993 – religion and mass media: a joint supervisory committee

From *3–6 October 1993*, a conference was held in Tripoli on *Religion and mass media*. There were three rounds of short, specific addresses on the positive role of the media in the formation of real people, but also on the destructive use of the media in relation to other people's religious attitudes and how religious institutions could correct this. From an anthropological point of view, naturally the first point is especially important, namely how the media can contribute to educating a real and also religious person, how they can lay the foundation for virtue, love and goodness and also bring people to reject all that humiliates people; and how the media can also encourage respect for people irrespective of religion or race and promote a spirit of tolerance when presenting religious doctrines. The importance of cooperating with the media was emphasised, as well as the production of suitable and attractive programmes to promote common human values, for example a sense of honour and family ties to counterbalance materialism, racial discrimination and strife. The media workers on both sides in particular, needed to cooperate to encourage positive initiatives and correct false presentations. Minorities should also be given access to the necessary media. There was immediate agreement on forming a joint Christian-Muslim committee to keep watch on how religion was presented in the media, to take action against unworthy and distorted presentations and to praise and recommend those that encouraged inter-religious understanding and respect. In addition, the importance of educational institutions was again underlined when it was a matter of overcoming prejudice and extremism.[35]

3.5 Continuation in Vienna in 1994: still more common commitment

The theme of *Media and Religion* was taken further by a small scale meeting of about a dozen participants at the Catholic Press Club in Vienna from *7–8 October 1994* where five Muslims and six Christians again analysed the current situation of positive and negative examples (without addresses being given on these) and reflected on how a positive influence could be exercised. In addition to what had been decided in Tripoli, it was agreed that suitable material should be made available to news agencies, and they should appear in public together if false representations had been made. They should also continue jointly to keep watch on the situation and generally form a network of experts on Christian and Islamic concerns. In addition, they could sponsor seminars for journalists dealing, for example, with the reasons for and types of misunderstandings, distortions and prejudices in this field and with their effects on Christian-Muslim relations.[36] *More than ten years later, it is even clearer than it was then that the subject is fundamentally relevant and positive actions are required.*

3.6 Rome 1997: the subject of mission raises many questions

The conference themes overlap or merge repeatedly in other realms as well. From *27–30 April 1997* in Rome the question of how *mission or da'wah* were to be understood was taken up again, explicitly including concrete practice and, above all, the prospects for the next century. Ten participants from each side, from countries and regions such as Libya, Sudan, Nigeria, Jordan, Iraq, India, Europe, North and South America, discussed this and on one day a public session was also held in the PISAI. This conference in 1997 did not reach any common conclusions; the open exchange had only contributed to clarifying the positions on each side. It was noted that there were agreements and differences when comparing the two concepts and also when comparing the practice in the nineteenth and twentieth centuries, for example the relation between *da'wah* and reform during that period, or the way missionaries and missionary congregations in Africa saw themselves and their religious view of the situation. It was interesting here that the letters and diaries of the missionaries used as sources for this purpose definitely showed a contrast between missionary and colonial strategies and attitudes. What was common to both concepts was that they addressed their central message in each case to all nations and all people. In addition, this was naturally an opportunity for a critical and self-critical appraisal of the practice on each side. At least it was emphasised that, with a view to the future which was the subject of a third round and signs of which were also visible at a number of points, special attention should be paid to respecting human dignity. This included respect for the other's religion when speaking or writing about it and also freedom of conscience as a part of religious liberty; both parties agreed, in line with the celebrated Quranic formulation, that there should be no compulsion in religion. Nor should economic emergencies be exploited; on the contrary, Christians and Muslims should combat injustice and exploitation together and no longer engage in *mission/da'wah* in competition with one another but in a spirit of cooperation and as a service to humankind. A further interesting idea on the Muslim side was that it was necessary to establish a link between the two views of history. But there was also a strong sense of Muslim self-confidence. A Muslim speaker from Panama underlined the future strength of Islam because it could provide direction and peace in the lives of individuals and societies which were afflicted by the illness of moral degradation resulting from the secular freedom which drove people to godlessness. The final session of the conference took place in the Pontifical Institute of Islamic and Arabic Studies and led to a lengthy discussion of the key questions that had arisen: the concept and practice of dialogue and mission, the question of diakonia and mission and the possible misuse of *diakonia* with a view to later conversions, the question of a possible common response of Christians and Muslims to the spread of secular (meaning godless) ethics and, in that connection, ways of relating reason and right guidance or revelation to one another. Of course, there was also discussion about religious liberty, including free choice and changing one's religion (cf. the results mentioned above), but also about it being a possible aim of *da'wah* to strengthen the system of *sharia*. In the discussion it was also stated that the right understanding and the right use of religious language was important in every inter-religious dialogue. However, the most important point in this dialogue was that Christians and Muslims really had to study their respective religions in a context of respectful encounter, shared concerns and as much cooperation as possible.[37]

All this cannot completely remove the suspicion that the fronts had hardened in the course of the years and the dialogues had not made any real progress, at least not on this question.

3.7 Tripoli 2002: dialogues between evaluation and concepts of the future

However, there were also dialogues on other subjects. On 22 February 2001 there was a meeting in Rome to prepare a dialogue at which it was also agreed to draw up a joint document on the Islamic-Christian dialogue or possibly to work out a joint charter for dialogue. A further preparatory meeting took place from 28 June – 1 July 2001 in Libya; the dialogue itself was to be held from 4–9 March 2002 in Tripoli on the subject "A culture of dialogue in an era of globalisation", although the exact date was postponed a bit to 16–18 March. The participants came from 15 different countries and the organisation by the World Islamic Call Society was explicitly praised for its excellence. There were various lectures, naturally directly on the subject of the conference but also on other issues, such as an evaluation of the previous Christian-Muslim dialogues, an address on sympathetic or strategic dialogue, on the requisites for and means of dialogue and – surprisingly because it was a very theological question – on religious pluralism and the challenges of exclusivity, inclusivity and globalisation from an Islamic point of view. In addition, two studies were presented with responses, namely on Islam between globalisation and universality and on Christian-Muslim dialogue in an age of globalisation. Of course, there was also discussion of the addresses and the report lists some of the points that were raised. Firstly, there was a need to evaluate the process of Christian-Muslim dialogue thus far and to map out a route for the future – a need to which this study may perhaps also contribute. Dialogue itself was seen as particularly urgent because, in view of the outbreaks of cruelty, injustice, violence and oppression, an answer was required to the claim about a conflict of cultures. What the participants demanded of dialogue was precise knowledge of the position of the others, honesty and a religious approach. In addition, dialogue required mutual acceptance precisely of otherness, naturally including freedom of belief and religious practice. Globalisation, the other pole of the conference theme, was seen mainly in its economic sense but with many cultural, political and social aspects and implications. Globalisation had both positive and negative sides. Christians and Muslims wanted to strengthen the positive aspects together and to counteract the negative ones, including defending human dignity. It was also not to be forgotten in the whole discussion that it was a matter of a message of truth, justice and love and hence the main focus had to be on religious values. Dialogue did not remain limited to intellectual discussion; it also included positive relationships in everyday life and the will to work together in the service of humankind. As a result of globalisation there was a new need to cooperate in helping the poor, the weak and the needy, in promoting development, striving for justice and working for peace in the world. For the sake of dialogue, it was necessary to pay attention to the influence of the media and to youth who needed a programme of education for dialogue. Both organisations were called upon to undertake joint programmes as a response to the challenges of globalisation. The concrete wish for the meetings already taking place was greater continuity. And the exchange of information had to be developed further, for example in the context of visits or on the occasion of specific events where advice could be helpful. Where there were tensions between Christians and Muslims, the two organisations could cooperate to reduce these tensions. As far as the recommendation to create a culture of dialogue was concerned, the idea was perhaps to organise an encounter between young Christians and Muslims and to do so initially in an African country. The requests expressed during the meeting led immediately to the concrete result of appointing a Coordinating Committee which was to exchange information, plan colloquia at regular intervals and coordinate other activities. Regular yearly meetings were envisaged.[38]

Here, it is again striking that the subjects and structures of the dialogues with different organisations overlapped with increasing frequency in the course of time or were even almost identical.

4 Muslims in Europe – familiar demands in a difficult situation

The meeting which took place from *19–21 November 1976* in Vienna was by nature and indeed in name more of a meeting than a dialogue. The Secretariat for Non-Christians invited the bishops' conferences of Western Europe, the World Council of Churches and a few leading Muslims living in Europe, half a dozen in all, to three days of study and reflection. The initial outcome was a kind of stock taking of the number and origin of the Muslims in Europe (nearly two thirds of the total of about nine million are immigrants, a good third of them from what was formerly Yugoslavia). The overall framework world-wide was a revival of Islam, flanked on the one side by a will for renewal but, on the other, also by some degree of aggressiveness. The situation of the Muslim migrant workers in Europe was seen fairly negatively. The secular society into which they had come offered them no legal guarantee for a secure, humane and religious life. Their human dignity and cultural and religious identity were considered to be constantly threatened. Hence it was urgent to sensitise the Christian public to the inalienable rights of immigrants, namely their right to work, to security, to fair wages and to culture. In addition, the Christians needed to learn more about Islam and its positive values, for example submission to God, along the lines of the relevant declarations of the Second Vatican Council. It was also possible to refer back and point to other declarations, bilateral commitments and initiatives. The Muslims present especially emphasised that they expected a lot of the church and desired serious cooperation to solve the problems of their brothers in the faith. It was recognised that all believers faced some kind of common challenge from the new questions of the technological and secularised world. The openness of the Catholic Church in relation to Islam since the Second Vatican Council was explicitly recognised and the good relations between Catholics and Muslims especially in Yugoslavia were underlined. This resulted in special confidence in the Catholic Church, particularly in view of the painful situation that their brothers in the faith were considered second class citizens in Europe if not even new slaves. On this basis they voiced the following wishes: to be allowed to keep their faith and practice it freely, and also to have suitable schools with Quranic teaching which was adapted to the demands of the modern world. In addition, they wanted to be recognised as a religious community under public law so that, like the other religions in Europe, they had the right to religious programmes in the mass media of radio and television, the right to Islamic instruction for Muslim children (wherever religious education is provided) and the recognition of all their human rights like other citizens, although this is not spelled out further. They also spoke about Christians in Muslim contexts, specifically Christian immigrants, especially technicians, in Muslim countries, and that the latter tended to experience how Muslims were being treated in Europe. Muslims were very sensitive about issues of hospitality and about everything that could harm their honour and dignity, but this was not spelled out further although it would have been very interesting and important, especially from an anthropological point of view.[39]

Different sensitivities and how they are rooted in different religions in each case would certainly be an important social issue as such, today perhaps even more so than at that time.

5 Dialogue on the question of spirituality: the thirst for God

Spirituality was a special subject. It did not belong to the dialogue with institutions but was still a concern – on the Catholic side, naturally, but in the course of the years repeatedly also for individual Muslims because no religion can be reduced to a law or an ideology. The amount of dialogue in this area was therefore limited and in a class of its own both organisationally and otherwise. In contrast, as far as the realm of anthropology was concerned, the contributions were extremely productive because the question who human beings are or should become in their relation to God and to other beings was naturally raised here in all clarity and with a depth and to an extent which was hardly achieved elsewhere, because here it was a question of the "created thirst for God"[40], as stated in the welcome to the first such dialogue. This took place from 6–7 May 1985 in Rome on the premises of the Pontifical Institute for Arabic and Islamic Studies (PISAI). The Muslims invited (with one exception) were only from the Indian sub-continent, namely from India itself, Pakistan and Bangladesh. The idea for this event had come up in December 1983 in India on the occasion of a seminar on Islam for the bishops there. The seminar had been combined with visits to Islamic institutions, buildings and pilgrimage sites. This had given the participants the idea that it could and should be possible the other way round to offer Muslims a similar experience of Christianity and naturally Rome was the ideal place for this. All the practical organisation was done by the Secretariat for Non-Christians, namely its officer for Islam, and by Christian Troll S.J., one of the advisers to the Secretariat and, at that time, working in the Vidyajyoti Institute in Delhi. The funding came from various Christian and Muslim sources but the main sponsors were Missio Aachen and the Center for World Thanksgiving in Texas. The official theme was holiness in Christianity and Islam. Initially, it was a question of the concepts and examples of holiness with responses from partners of the other religion in each case. On the second day there was even a criticism of each other's holiness and a panel discussion on it, as well as various small contributions from the Islamic side. It was rounded off by sightseeing in Rome itself (Early Christianity and Renaissance) and by two excursions, one in the footsteps of Saint Francis to Assisi and La Verna and the other to the origins of the Benedictine order in Subiaco. At the end, the participants of the colloquium were received in private audience by Pope John Paul II.

5.1 Holiness from a Muslim point of view

5.1.1 Holiness in Islam

First of all, something about the addresses in so far as they are of interest for our study. The start was made by Prof. Ziaul Hasan Faruqi from Delhi, a specialist on the history of Islam and of Sufism in India, and the latter was especially obvious. He approached the question of the conception of holiness in Islam in a very un-Islamic way initially, namely by referring to Rudolf Otto and Blaise Pascal. His concern was the non-rational level which, when related to human beings, can be evoked or awakened but not taught. It was a matter of the beginning and end of the things which remained hidden from human beings despite their reflective capabilities and for which they were dependent on God, the creator, in order not to draw false or fatal conclusions in their ignorance. These insights which, in a sense, were to be found among Christian philosophers as well, were also found in Islam – and they were then spelled out in their Islamic form against this background. This was certainly a clever move which made or was to make it easier for outsiders of whatever kind to approach and accept the contents which then followed and which might otherwise have been rather strange or alien. The distinction between visible and invisible was also made in the Quran, and in the visible world there were sufficient divine signs to enable people to realise that there must be a creator behind them, although there were always people who closed their eyes to these insights and persisted in unbelief. However, the invisible world was accessible to every creature only through intuitive experience and that too only by the grace of God. There followed a very detailed section on angels, their nature, their standing and function according to the Quran and to tradition, confirmed, as it were, by the religious philosophy in the introduction mentioned above and explicitly put into words. "The pure rationalist would say that all this is vague, indefinable and cannot be proved: hence unacceptable. But here there is a different feeling, a different faculty that comes into play. It is intuition that comes from a deeper level of human nature, a non-rational level which represents an inner urge for contact with the celestial world."[41] The angels served in some way as a link between God and human beings. They punished wicked people and caused noble souls to repent, refrain from evil and act rightly. The purpose of their creation was through them to influence individual persons and humanity as a whole, similar to the way a person's mind influenced their body. When a human being thought of God, that was already an emanation of God's shadow which this highest gathering had passed on. Those who spiritually occupied the highest ranks among human beings such as prophets and saints were accepted into this circle after death.

5.1.2 The special standing of human beings according to Islam – full of tension

It was only after these explanations, the details of which filled more than three pages, that Faruqi came to humans themselves who, according to the Quran, have immense rational and spiritual capacities. God had made human beings out of clay and then breathed his breath into them, giving them ability to hear, to see, to feel and to understand. Human beings were God's representatives (vicars) on earth and as such God taught them the names of all things (Sura 2:30/33), namely the inner nature and characteristics of all things including feelings. The devil (Iblis), however, only saw the clay, the lower side of human beings, not their higher side which God's spirit had imparted to them. For this reason, the devil refused to bow before human beings as God demanded of all the angels and he was punished for this, but the punishment had been postponed until the day of the last judgement. He took advantage of this delay to lead people astray. But he could do no harm to God's servants who were purified by God's grace. As a rule, Sura 7:172 is interpreted to mean that there was a bond between God and human beings at the time when they had another kind of existence which was known only to God and of which only God was aware, a kind of divine existence. Everything else, knowledge but also feelings, force of will, understanding and the ability to distinguish, were then added. That was what made human beings unique compared with the animals, who had only five senses, and even compared with the angels. Human existence as a whole was a mirror of God's properties. Ghazali even went so far as to state that God had created Adam in his own image. So the perfect human being was the one whose life was increasingly an expression of the divine properties. But, in view of the challenges of the forces of evil, human beings had to struggle (that was their task) to purify themselves completely and subject themselves to God in order to be able to enter into contact with the invisible. This pointed to a real drama. Although creation as a whole was good and human beings were created in the best state, as it were, they were also impatient, arrogant, ungrateful and easily inclined to all evil because they made the wrong use of the limited free will which they possessed. But they sensed by nature a proximity to the divine soul and desired to come closer to it. They were in constant tension between the mutually exclusive elements of their nature, between the material and the divine. Their existence in this world was one of alienation. That was described by Faruqi in amazing terms: "Surely, man enjoyed originally a state of bliss and felicity. But he transgressed and was deprived of that bliss. He, however, belongs to God and should return to Him. His worldly existence is, therefore, a state of separation, estrangement and alienation from his true self which is nothing other than something of God's spirit. Life, as lived on the material plane, is not at all satisfying for him. Inextricably bound to physical phenomena, it is only futile, absurd and full of sorrow. His soul, therefore, yearns for its original abode. It is a perennial restlessness."[42] The Sufi traditions, in particular, were very evocative in this respect. They explained the phenomenon by the principle of the mutual attraction of all things that were of the same nature. Concretely, they saw the love of God in its double meaning which constituted the unifying bond between human beings and the divine. In other words, whatever a human being was seeking, the soul of the object sought yearned for him. All steps taken by human beings to bring them out of the dilemma described and closer to the divine world could be called holy. There were three stages of the soul involved. At the lowest level it was inclined to evil which would lead it to perdition. And naturally the devil made evil look so attractive that human beings were rapidly disposed to believe wrongly that worldly life was a lasting matter. At the next stage, the human soul struggled hard to gain control of the lower side of human nature and sought refuge in God as the protector of the believers. At the last stage, the soul had laid aside all lower earthly lusts, in a sense cast the world to the devil like a piece of bread to a dog, heard the divine call, shared to some degree in the properties of the holy gathering and could therefore itself be described as holy.

5.1.3 Holiness in Islam: not without quite concrete deeds

Then the argument became really concrete, i.e. it was a matter of the Islamic ideal of righteousness and spirituality, and so the first sentence mentioned that God did not only desire the complete submission of human self-confidence but the total submission of the whole of life to his will, as laid down in the Quran and Sunna. "*Such believers, designated in the Qur'ân as 'friends of God', may be termed as 'holy' and their piety and righteousness, and the moral and spiritual quality of their character as 'holiness' in them.*"[43] Again there was emphasis on the nature of the struggle although this might not be obvious from outside. It was a question of dying before death. Worship could not simply be restricted to prayer; it had to embrace the whole of life as also stated in Sura 6:161f. Another key term was the fear of God which (as in the Jewish-Christian tradition as well) was seen as the beginning of wisdom. In practice it implied restraint: you keep watch on your tongue, your hand, your heart to keep it from evil and do so from within, not as a result of external pressure or as an outward show. The purification of the will was at least closely related to love. According to Ghazali, the fear of God enabled people to give up dubious things and everything which went beyond what was absolutely necessary for living, namely just to build the house in which you live yourself, not to pile up food, to view the world in the knowledge that you will have to leave it behind you and not to spend a minute without thinking of God. The Quran itself repeatedly explained that there was only one humankind, everybody was equal before God and the person who was most respected in God's eyes was the most righteous person (Sura 22:37) because it was not flesh and blood which could reach God but only spirituality. And again, the fear of God naturally meant rightly following God's commands and prohibitions which implied, of course, in addition to rejecting things such as worldly temptations and corruption, that Islamic law, in particular, had its part to play. Good deeds, e. g. compliance with the five pillars, were rewarded by perfect peace and repose for those who performed them, but again it was the mental attitude and the intention behind the formal fulfilment which counted. In a righteous life, love of God and of one's neighbour were the highest ideals and the final goal – which certainly also sounds very Christian. In Sura 2:177 this was then spelled out quite concretely with reference e. g. to orphans, the poor, wayfarers and slaves. The love of God should be expressed by love and mercy towards one of God's creatures – even if it was a sparrow. On this there was a traditional Muslim hadith which sounded much like Matthew 25. Of course, perseverance was one of the key terms. "Love is a great killer and love of God requires a completely exclusive devotion to Him, renunciating all that is other than God. The lover (the sincere believer) is also tested by God in his sorrow, loss and suffering; and if he faithfully maintains a cheerful attitude of resignation and understanding in his sufferings and misfortunes, he finds God very close to him."[44] To resist one's passions and natural instincts required a large amount of patience; it was a permanent battle against one's own lower self. Ghazali wrote that there were three stages for a man: firstly, he abandons his lower self, then he renounces the world and is content with his fate whatever it may be and, in the last stage, the stage of love, he is happy about everything that God, his Lord, does for him. According to another holy man, it was easier for a believer to leave this world for the world beyond than to renounce all worldly relationships for God's sake. If the poor were righteous and satisfied, according to Muhammad they were the most blessed of all people and, if they even gave away something of the little they had, they were the most devout of all. Their patience and forbearance would bring them nearer to God on the day of judgement; hence love was the key to paradise for the poor, just as love for the poor, the orphans and one's neighbours was generally among the axioms for a good life. Divine characteristics were mirrored in human life when all were treated equally, just as the sun, the clouds and the earth were the same for everyone. The Quran described such persons as Friends of God and they were known and recognised less by their miraculous deeds than by their seriousness and steadfastness in the faith, by obedience to God and by friendliness not only to their fellow creatures but to all living beings, by their prayer which came from within that God would show mercy to all people, by a pure heart (no enmity, no revenge), by refraining from all greed and all worldly attractions, by refraining from slander and calumny even to the extent of taking care not even to see the weaknesses of others. It was possible to divide the saints into four categories according to their individual characteristics but that was not so important for the image as a whole.

The response by Paul Jackson concentrated on three questions from an anthropological point of view: whether one could really speak about killing the self or whether the ideal was not rather to control it, that the *"pirs"*, the spiritual leaders, had not been mentioned although they played a decisive role in Indian Sufism, and finally, the most important question, that of the use of the rich, historical heritage in order to produce holy Muslim men and women in the present.

5.1.4 Among the Muslim examples: Muhammad was far the best
That question was answered in some sense in the following address which dealt with practical examples of holiness for Muslims. First and foremost, and this can hardly be emphasised enough, the Prophet Muhammad was cited here because the Quran itself describes him as such and recommends imitating his behaviour. All Muslim holiness attempts to follow Muhammad's example without ever being able to compare with it. And, indeed, it is not possible to live as a Muslim without following Muhammad's guidance even in normal aspects of everyday life which were and are scrupulously defined, examined and followed. Naturally, as a side effect, that has major consequences not only for individuals who are striving for holiness but also for the sense of community and the practical, sociological expression of the religious community as such. Nor is it merely a question of striving for personal holiness; it is a matter of Islamic law and hence of the religion itself. Here, one cannot mention all the details because really no detail was ignored, even the most intimate. *But it is still striking (cf. below) to see how much the addresses from the two different sides had parallel structures, that the primary, dominant example was Muhammad or Jesus and that everything else was derived from that, albeit with different reasoning and in different forms.*

On Muhammad and his role it was stated, "The Qur'ân has reinforced the position of the Prophet in Islam and has declared: 'Do whatever the Prophet orders you to do, and abstain from what he forbids you from'. Love of the Prophet has been declared to be an essential and inseparable part of the love of God. The objective test of this love has been declared to be the earnest desire to emulate, imitate and follow the Prophet in all the spheres of life. Muslims through the ages have looked upon the Prophet as an *Insân-i-Kamil* (The Perfect Man) and have aspired to imitate him."⁴⁵ The "saints" then named were the founder of the Indian Chishti order, namely a Sufi order, and four in the line of his disciples covering the period from 1141 to 1356 in Northern India about whose lives and workings there are records. What is common to them all and chacteristic of this order is its firm distance from the Muslim rulers, great practical concern for needy people because, according to a traditional saying of the prophet, God is found by their side and, as a practical consequence, usually the great poverty of the saint and his family. *One cannot ignore certain parallels with the Christian saints listed below; they even go so far as forgiving a physical attack.*

5.2 Holiness from a Christian point of view

5.2.1 Incarnation as a key term

The corresponding Christian address on holiness was given by Father John Carroll Futrell who chose a descriptive, phenomenological approach such as had become customary in spiritual theology. He began with an anthropological observation: all religious systems ultimately understood holiness as unity with the Holy One. Christianity in particular saw the absolute as something that broke into history, with its climax in the incarnation in Jesus Christ, by means of which we were invited to share God's life now and for ever. Therefore incarnation was the key word also for Christian holiness. In the biblical understanding, the whole of humankind was encompassed from the very beginning to the end of history in a breath-taking, dynamic movement of divine love which then called us to respond, to open ourselves up to this life and to live this life of love ourselves and thus change the world. So Christian holiness was expressed necessarily in inter-personal relationships and in "transforming holy action on the world"⁴⁶, currently meaning promoting social justice, for example. But Christian holiness also presupposed penitence and conversion, opening oneself up to the saving love of the crucified Christ. Practical Christian holiness expressed union with God, with God's love, in relation to others even including love of one's enemies. For this, one only and above all needed to consider and imitate the earthly life of Jesus that went beyond a mere conscientious decision between good and evil to a spiritual knowledge of what God wanted in a particular situation, because one could feel God's presence, the presence of the Divine Spirit in a specific decision precisely because one lived in the Holy Spirit. This also required an orderly life of prayer and, the speaker insisted, spiritual guidance for which there were various schools and techniques that had developed and continued to develop in the course of time. Additional help came from the community and the sacraments, especially the Eucharist. In addition, he referred to the traditions of asceticism and mysticism. Finally, the perfection of Christian holiness could only be brought about by death through which one entered into eternal union with God.

The "response" to that came from Dr. Riffat Hassan from the USA and did not really deal directly with the address or, at least, only to the extent that he pointed out that, in Western countries, Islam was immediately recognised as having a social, cultural and political nature, but that people had great difficulties in identifying and recognising the religious dimension of Islam.

5.2.2 Christian examples of holiness

In addition to the basic address, there was an address on examples of holiness for Christians given by Sister Suzanne Le Gal which, as she wrote, had been selected according to very personal criteria. Nevertheless, some structures can be identified beginning, as one would expect, with Christ himself as an example of holiness. The speaker emphasised, on the one hand, Jesus' complete submission to the will of God and, on the other, his special love for those whom society relegated to the margins: women, heretics and foreigners. Jesus had also wanted to teach his followers this attitude of universal brotherhood. Brotherly love was expressed especially in compassion and more concretely in forgiveness (even as far as death on the cross). Every human being whom Christians saw as a saint had lived according to this example. She referred specifically to Francis of Assisi, one of the best known saints, to Anwarite, a young nun and martyr from Zaire, to Giorgio La Pira, for many years the mayor of Florence, and to Father Maximilian Kolbe who became a martyr in Auschwitz. The striking thing they all had in common, in addition to following the example of Jesus in their concrete circumstances – and here frequently his poverty was underlined – was the inner and then also outer freedom they had gained, a freedom for extraordinary deeds even to the extent of giving their own lives. It was also striking that they were probably all unmarried (the description of the life of Giorgio La Pira also points in this direction although it is not stated outright).

Unfortunately, the response by Dr. Riffat Hassan did not refer to this at all, nor to the similarities with or differences from the Muslim pattern, but dealt, on the contrary, with the question raised following the counterpart address, namely who could be considered examples in Islam and who determined the norms of holiness since it had no official canonisations. She confirmed that popular piety determined the norms and that it obviously set less store by the exemplary character of holiness than by the divine presence in the form of miracles, blessing and the possibility of acting as advocates. Dogmatically, this is a form of the immanence of the divine which Islam otherwise firmly rejects but which perhaps corresponds to a human need to have God closer in a more tangible way. On the question whether a special blessing could be inherited, she as a woman pointed out that, strangely enough, the saints had predominantly been men[47]. In general, therefore, a number of interesting views and questions arose which could provide sufficient material for further discussion.

5.3 The mutual criticism

5.3.1 From the Christian side: different emphases

But, to begin with, there was criticism from each side in the form of addresses of the concept of holiness on the other side. The Christian address given by Jean-Marie Gaudeul took account straight away in its introduction of the difficulty of the concept because it really demanded too much and could easily lead to persons evaluating others simply according to their own conceptions. Especially when dealing with the subject of holiness, the only possible criterion was God; holiness was God's gift which human beings could always only receive in their creaturely weakness and imperfection. So Gaudeul stated that he did not want to engage in (negative) criticism but to distinguish (positively) between the views. Experience showed that people were critical and reacted with particular sensitivity at points which were their own weak points, whether as individuals or as religious communities. He therefore attempted to give a Christian description of Islamic holiness and concentrated on three points. The first was the unity and uniqueness of God which Islam emphasised so strongly and which – at the level of holiness and inner life – excluded transitory fame, ambition, pride, striving for earthly profit, power and pleasure and which was really only interested in God (and not even in paradise). The second was contained already in the word "Islam" – devotion, submission to/under God and his will – and could certainly be described as one of the main characteristics of Islamic holiness, although there was naturally also a similar basic attitude among Christian saints despite some characteristic differences. Christians assumed that God had a quite personal plan for each individual and it was their task to discover and follow it in normal, everyday larger and smaller decisions. On the Islamic side, such things were the prerogative of the prophets, and everyday life was sufficiently governed anyway by the provisions of Islamic law – which could naturally easily result in important things being considered unimportant because the classification of Islamic law evaluated them as merely indifferent. Thirdly and finally, it was a matter of the transformation and purification of the person, of the self, outwardly (by obedience and following rules) and inwardly (by coming closer to the nature and characteristics of God). This was again something common to both religions, with some degree of Islamic emphasis on the former and a similar Christian emphasis on the latter. But he rejected other types of 'holiness' linked with official positions, inheritance, miracles or scholarship as being too far removed from its original meaning. "History shows us, in Islam and in Christianity, that holy functions have not always been entrusted to holy people, and that even miracles can be performed by people who are far from holiness. True holiness, the type of holiness which we have been talking about yesterday and today, seems to bring us to a point where we are reminded of our common condition of humble servants called by God to know and do his will. How could we turn holiness into an honorific title?"[48] *This indeed displays more self-criticism than criticism of Islam.*

Question from the Christians: holiness or only apparent holiness?
The second part of his address dealing with the threats to holiness from a Christian point of view also assumed that this was not just a concern of one of the religious communities but finally, basically, concerned them both because, in relation to the Creator, there was always the possibility of trust or distrust, love or indifference, obedience or rebellion, God-centredness or self-centredness, and the structure of the problems was the same despite all the differences. The first was the problem of equilibrium also between good characteristics such as prayer and action. Gaudeul's very pointed question here was, "Is it not one of our problems, at the present time, that we see many of our brethren, in Islam or in Christianity, live that sort of 'unbalanced' faith with the resulting conflict between people who choose *militancy* as a priority and those who choose *mysticism* as their first goal?"[49] The second was the question about change in the other religion. Both Islam and Christianity had tried in the past to create societies in which divine law was followed. Many people on both sides liked to forget today how imperfect those societies were and glorified them as a golden age to which one should return. Instead, he suggested, it would be better to reflect on the effects which such attempts would have on the holiness of such a society and on the holiness of those who wanted to create it. In fact, many people set such great store by reforming society that they completely (happily) forgot their own conversion in the process. He again he spoke very directly (and did not just restrict himself to the example quoted): "My fear is that this mobilization of masses of believers against the sins of other people, whoever they may be, may lead to more sinfulness, more violence, and more suffering, because zeal for God will be channelled exclusively towards political or legal reforms or revolutions, and the attention of most believers will be diverted from that other *Jihâd* which is the struggle of each person against his, or her, own sinfulness."[50] And it was known that stronger compulsion in religion had only produced more apparent

holiness – hence probably also the famous Quranic statement that there should be no compulsion in religion. In any case, and that was his third point, man-made holiness was an illusion which made God very small. Fourthly, one could also set one's own militancy or one's own religious community in the place of God and thus practice idolatry. And fifthly and lastly, and this he described as a very Christian view, holiness in Islam was perhaps too much identified with certain specific forms of behaviour which were nowadays no longer possible for everybody everywhere. Did this make holiness impossible even for people who really wanted it, or were many of these external forms of behaviour ("a certain dress, pious language, solemn face, and so on"[51]) perhaps not real holiness at all?

It is striking how the most intimate questions of the human attitude to God are linked with the most external questions of society and politics. That is what constitutes the relevance of this particular contribution and its queries.

5.3.2 From the Muslim side: holiness must be and remain sanctification

The contribution from the Muslim side was significantly not called criticism but an "appreciation" of Christian holiness and attempted to approach this holiness from inside via the biblical witness which identified it as the state of a "thing" which had been set apart for God and to serve God. By means of the subdivision into and connection between moral and cultic holiness, the speaker came to the way the Christian church understood itself as a communion of saints, of those redeemed from sin, and from there to the importance of the sacraments for holiness in Christianity, especially baptism, the sign of sanctification for him, and the Eucharist, the ritual of sanctification in his view, but also of penitence, confession and forgiveness in a sense as a sacrament for restoring holiness after sin. With all reticence as far as simplification is concerned but nevertheless as an idea to be considered, he observed a theological development of the understanding of sacraments in the Western churches away from this sanctification and towards fixed elements of delimitation which created group identities. Here, the mystical theology of the East with its oriental character had also preserved the force of sanctification better, at least in the eyes of the Muslims who were also "Eastern". In addition, Ayoub was inclined to equate the "Christian" Holy Spirit with the "Muslim" right guidance of God, and to place great emphasis on right action in connection with holiness where holy was understood as good and wholesome. In the modern age of science and technology, the feeling for this had been lost in comparison with earlier times which had been denounced as superstitious. This very science and technology needed to be sanctified by

God's right guidance. The expressions used in this connection were not exactly prudish: "The three greatest evils of the human soul are self-centeredness – individual, national, international – arrogance and greed. The sin of regarding man as measure of all things is as old as humanity itself. Since the renaissance this sin has taken on forms of ideology – nationalism, Marxism and humanism. This /p. 97 self-centeredness leaves no room for God in our lives."[52] In conclusion, he again underlined that the Arabic word for "holy" in the Quran was used in a positive sense for God alone. A "holy" life would be described as "rightly guided" and "holy" things were simply "good". Nor were there any saints; a *walī*, a friend of God, was more a man of God in the Old Testament sense than a saint as understood by the Christian churches. Really all people and things were holy by nature; they only needed to regain this status of holiness in which they were created by God by endeavouring to lead a holy life in the sight of God. And this meant that life itself could be sanctified by good works, thoughts and relationships. Indeed, in the service of God, life as a whole must acquire the character of a sacrament (the corresponding actions are sanctifying and the relevant requirements provide the necessary discipline for holiness) which was why there were no sacraments as such in Islam.

Here, Islam's positive image of human beings becomes very clear, namely that human beings only need divine prescriptions in order to achieve a status which is pleasing to God and thus holy – but these prescriptions are really necessary and must be followed. Despite all the high esteem for human beings and human abilities, nothing can succeed without this divine right guidance, hence also the assessment of the Renaissance and humanism as de facto godless ideologies. Despite all the knowledge and grasp of Christian content and despite all the similarities with Christian formulations, particularly in this contribution compared with all the previous ones, there is very clear evidence here (as was also stated) of the distance from and criticism of the West and the churches of the Western tradition – a provocative stimulus for reflection that was probably at least partly deliberate but which one might not have expected to be so direct in connection with a subject like holiness.

6 Relationships with Turkey – strong academic cooperation

The relationships with Turkey were predominantly of an academic nature, apart from the fact that, long before the Second Vatican Council, namely from 1935 to 1945, one of the forerunners of inter-religious dialogue was working in Istanbul, namely Angelo Giuseppe Roncalli, later Pope John XXIII. On the occasion of his beatification, there were celebrations not only in Rome (with a high-ranking delegation from Turkey) but also in Istanbul. Especially between the University of Ankara and the Gregorian University, one of the pontifical universities in Rome, a lively academic interchange developed in the form of an exchange of lecturers but also in joint seminars. The significance of this academic cooperation should not be underestimated, especially as far as its duration and almost natural character is concerned. (Since October 2000, Jesuits have also been resident officially in Ankara where they teach at the faculty and receive students and scholars, while the Dominicans have a similar programme in Istanbul.) But this can naturally not be dealt with fully in this study of another pontifical body since there has been only occasional overlap of staff and organisation. One example was the seminar jointly organised with the Pontifical Council for Interreligious Dialogue on problems of religious and theological education in the modern age which was held from 10–13 May 1989 in Rome under the official title "Communiquer les valeurs religieuses à la jeunesse aujourd'hui" with the cooperation of the deans of seven other theological faculties from various Turkish universities (Gaziantep, Erzerum, Bursa, Konya, Izmir, Kayseri, Marmara University Istanbul). The other Roman universities and institutes were represented by professors. To some extent, the addresses dealt with clearly anthropological issues such as the dual nature of human beings according to the Quran or the necessity of religious convictions in the education of young people to safeguard human rights. The discussions which followed emphasised, among other things, the need for more profound anthropological studies in the area of theological education[53]. An extensive programme of academic visits and a papal audience rounded off this high-ranking event. Later on there were also relationships with the Office for Religious Affairs, namely a joint declaration of intent by both sides (April 2002). It began by observing that, for quite banal reasons, there were increasing numbers of contacts between followers of different religions: travel had become easier, communication media transcended national and religious frontiers and, of course, people were more mobile for economic reasons or as a result of conflicts. In this situation, it was recognised that religious differences could be exploited to sow enmity between nations although, when rightly understood, they could be enriching, just as religions could also have a positive significance in society. Basing themselves on the existing good relations between the Republic of Turkey and the Holy See, the two sides committed themselves to eradicating misunderstandings and prejudices in religious affairs, to upholding the freedom of religion, belief and the conscience, to promoting training programmes for the knowledge of other religions as well as inter-religious dialogue, the latter especially at the university level wherever religion was taught, and finally to keeping watch on the implementation of this joint declaration of intent by means of regular encounters between representatives from both sides. On the Catholic side, it was considered that the document significantly facilitated the contact between academic institutions.[54] This was an approach which had already proved valuable as far as the existing contacts in the university context were concerned and also in connection with institutional contacts in general, considering the contacts with the Azhar University, for example.

7 Cooperation with the Al Albait Foundation

7.1 Beginning with religious education

The Pontifical Council for Interreligious Dialogue had been cooperating for many years with the Royal Academy for Islamic Civilisation Research in Amman (which itself was dependent on the Al Albait Foundation). The patron of this foundation is His Royal Highness, Prince Al-Hassan bin Talal of Jordan, for a long time the official heir to the throne of his country. Among all the dialogue activities of the Pontifical Council for Interreligious Dialogue, this series is of special importance because it is the only one that has been comprehensively documented and published. As early as February 1982, there had been conversations between Msgr. Jadot, Msgr. Rossano, Dr. Sabanageh and the then Crown Prince about how Christian-Muslim dialogue could be promoted. In November 1984, Cardinal Francis Arinze, who had then taken over the leadership of the Secretariat for Non-Christians, as it then was, and Dr. Sabanageh had been guests at the first dialogue conference organised by the foundation in Windsor (with the Anglican Church as its partner) and then also at the second in Amman in September 1985. Within this general framework and certainly referring back to the Council's declaration *Nostra Aetate*, which had called upon Christians and Muslims to overcome the conflicts of the past and to find a way to cooperate, *a first joint conference* took place from *6 – 8 December 1989* in Rome (following the series of dialogues with the Anglicans and that with the Orthodox Church); it dealt with *religious education in modern society*, especially at universities and colleges. In his welcome speech, Cardinal Arinze emphasised that religious education began long before school, namely in the setting of parents and relatives, and it should really accompany a person throughout their life. Naturally, it was particularly important for young people who needed to learn to distinguish what was positive and valuable and how they could live a life in harmony with the will of God and worth living. The other party also underlined from the start that neither dialogue nor this subject were particularly alien to Islam, quite the opposite. Dialogue, irrespective of the specific issue, was necessary in order to develop trust. Understanding was necessary, and here the two introductions to Islam which the Secretariat had issued were mentioned with appreciation, in order to make cooperation possible for society and for life to continue. But differences were also necessary in order to produce and develop ideas in the search for the truth.[55] After these introductions, the main subject was tackled and the first sub-theme of the conference examined religious education in a pluralistic world, significantly with the key terms identity and openness.

7.1.2 Religious education between identity and openness
7.1.2.1 Natural knowledge of God in Islam
The Muslim contribution began by examining the expression "pluralist world" in greater detail between the historical development of the Christian, Muslim and other blocs and the new challenge of the revolution in the media, as well as the factual proximity of others. Although, according to Abdul Aziz Kamel, secularism was not really an Islamic phenomenon, he called for the inclusion of such people in the reflections. The traditional behaviour of the religions alone was no longer adequate for this new situation. The aim in mind was clearly peaceful coexistence in a context of tolerance and mutual respect, something which had historically been quite rare. He then described Islamic identity as two circles revolving around the centre of monotheism, the prophetic circle with Muhammad as the seal of the prophets and – of greater interest in our context – the broadest possible circle of *fitra* – normally the natural knowledge of God which he called intuition. In any case, it was a reference back to creation and its unchangeable, divine order which also included human beings. Creation was, as it were, God's open book and, according to Islamic logic which, as he emphasised several times, knows no dichotomy between religion and science, progress in education was always also a way of coming closer to God. On this basis, the combination of identity and openness was also absolutely no problem. His remarks on concrete Islamic rules also led to the statement that, in Islam, the earth was compared with a large mosque. And even in Islamic architecture, in the typical forms of Islamic decoration and in Islamic calligraphy, he identified reflections of the harmonious link between heaven and earth, the balance between the individual and the community, between law and beauty, the great unity which human beings needed. With regard to the practical situation of non-Muslims in an Islamic society, naturally the classical *dhimmi* status was explained, emphasising that Muslims could only not have such people as friends whom they had fought for religious reasons and who had at least helped to expel them. Otherwise, freedom of cult and protection were not only possible but required, against the back-

ground of all people being part of the vast human family which God had created and endowed with special dignity, which God cared for well and indeed to which he had subordinated the whole of creation. This framework was particularly important in order not to cause tensions due to a limited view, nor to upset relationships with other approaches to faith. In connection with openness in practice, he firstly referred to the Mexico declaration on culture which understands education in a very broad sense as a process of counselling, of exchanging ideas and experiences and an appreciation of the values and customs of others (within which basic human rights naturally also have their place). He attributed special importance to the equal value of all forms of education also mentioned there but which were not a reality for a long time because of the eurocentrism of science. In connection with religion and religious science he explicitly identified with the approach of Mircea Eliade who describes humans as religious beings (here there is an indication of agreement with the approach which Piero Rossano had also presented as fundamental for the Secretariat for Non-Christians and its work) and even recognised borrowed religious elements which had simply taken on a secular garb among those who had abandoned religion. Kamel examined holy places and holy times (which were often interconnected) and then explained that every religion had its own myth of the universe (in Islam it was the universe as the school of faith) and of the origin of life, as well as a special role for particular elements of the universe such as water or mountains, for example. His last point was the key role of one or more persons however it may appear or be viewed in practice. In addition, all religions had in common the existence of an ethical law with some concordant provisions and some special ones. The speaker saw these remarks as a model of how Islam could unite mentalities in a flexible framework of identity and pluralism.

At the end, he offered some assistance, as he saw it, for the realm of religious coexistence. This included, among other things, the need to accept identity and pluralism because they belonged to the most important characteristics of our time. Every religion and culture should be presented as it saw itself, and one should not try to force one's own understanding on it as often happened in relation to weaker nations in international cultural centres and in publications. In the religious realm, in particular, this was very crucial and complex. Followers of different religions were able to cooperate in theory and in practice in many areas, especially with regard to the ethical law already mentioned. He also wanted to make a special effort in relation to secular people who believed that religious thinking contradicted science or was opposed to it. As he had emphasised several times, that did not correspond to the Islamic conception of science which also included divine inspiration and traditional science. At the same time, he pleaded for more comparative religious studies. When describing one another, one should not make the other person's religion despicable. One should contribute to understanding by identifying cultural particularities. The practical links between religion and work and production should be seen more clearly; that would strengthen both the human and the religious element simultaneously. The relations between religious "leaders" should be strengthened in every way (significantly there is reference only to men here) so that this was also reflected in their writings and hence also among their recipients. Against the background of the fact that all the followers of one major religion formed a majority in one place but a minority in another place, all those who wished their minorities to have a peaceful life should also be concerned about the other minorities, with an emphasis not on strength or weakness but on human dignity. And here he concluded with praise for the countries in which the presentation of Islam in school textbooks had already been corrected.[56]

The presentation as a whole had the tendency to form individual parts which are not too closely connected, but the conception of a global concentration on peaceful coexistence between the religions is very important and the suggestions deserve further reflection, both theoretically and with a view to their practical implementation.

7.1.2.2 In the Christian response: also emphasis on ethics

The official response to this came from Rodulfo Galenzoga, a Philippine priest from the problem area of Mindanao, who spoke about his own culture shock including an identity crisis during his studies in Europe, and how he then experienced his identity in the central Christian truth that God loves humankind including himself. He was a child of God and a brother of all other people irrespective of their race, religion and culture. That made him certain of his identity, and the encounter with other religions and cultures only served to reinforce this. "In God's mysterious ways, our oneness and identity is in the many."[57] Only from the context of this love was it possible to voice other, critical truths. In this connection, he also made a few remarks on the emphasis which Kamel had placed on comparative religious studies. Their strength lay in making clear the parallels and the basic pattern of human beings as religious beings in search of God. The danger was that religions might no longer be accepted as forms of revelation but treated as if they were merely products of culture and reason, means of curing human uncertainty. So he wanted initially to place the emphasis on ethics as a starting point for dialogue, understanding and cooperation in order e. g. to work on common tasks such as greater justice, eradicating poverty and a new relationship with the earth. This task of creating a new peaceful world was so holy that one should never dehumanise oneself in the process.

7.1.2.3 The Christian position: identity and openness – no problem as God's image

The corresponding Catholic address by Michel Sabbah, the Latin patriarch of Jerusalem, was considerably more compelling that the address by Kamel, especially in its anthropological argumentation, but it did not contain such very practical, concrete conclusions. He also began discussing the subject with some brief reflections on pluralism. Pluralism had always been a reality which was necessarily connected with the human race even to the extent of different conceptions of truth. But the modern information society with its communication media and greater ease of travel had literally removed all borders, apart from the fact that plurality was also found within individual societies in the form of different religions, mentalities and minorities. Every citizen and every believer had to learn to deal with all these different influences without aggression and also without spiritual alienation. So the question was how religious education tackled this, and that led, in analogy to the Muslim contribution, firstly to the question of how identity and openness were understood before one could be clarify what exactly was the task of religious education. Identity was defined first and foremost by the freedom, self-determination and independence which God had given to every person. Each had a unique dignity which every other individual and every social organisation had to respect, whether it be the state, church, school or even the family; indeed, these should even help individuals to make this image a reality as far as possible. Here Sabah quoted the French philosopher Maritain who defined identity by reference to totality and autonomy. An individual was a complete self, not part of a whole. He/she could comprise the whole human and non-human universe within him/herself through knowledge and love. He/she was even superior to the whole world because he/she was spirit and, as such, God's temple (the latter is a Pauline conception) which was created for eternal life. So the primary identity of human beings was their immediate relationship with God, the Absolute, expressed in the forms of existence, truth, pre-eminence and beauty. Humans were not first and foremost material beings. They were not part of a complex ruled by different forces and influences. They were influenced by these forces but simultaneously superior to them; they were their master and the identity of human beings was mainly defined by this superiority. That was the basis for self-determination, freedom and dignity. That was the fundamental and general identity of each individual which was then coupled with the specific identity resulting from the links between the individual and a particular society, family, nationality, state, religion and milieu. Among these relationships, some were of a fundamental nature such as family, nationality and religion; others depended on the latter such as mentality, behaviour, customs and traditions. It was the right and duty of each person to preserve their identity in these two dimensions while being exposed to and dealing with various influences. One task of education was to respect this identity, indeed to promote both of its dimensions. But in this process the human being had also to remain open, open for the milieu which supported and nourished him/her spiritually and materially, and for the universe as a whole, because the whole universe was God's creation which God had created for humankind's benefit and over which God had given humankind dominion. All human beings were brothers and sisters. Humans had to maintain this relationship because it was so willed by God, the Creator; indeed, that was the only way for humans to grow. So openness meant above all openness for the way in which God viewed his creation and acceptance of the divine plan for this creation. "Fundamental personal identity which is based on totality and autonomy, since its foundation is the image of God in man, is for the same reason based on openness which sees the image of God in every human person. For the human being who is an independent whole meets all God's creatures in God their Creator. Consequently any closing off of oneself from another goes against the authentic definition of identity."[58]

It was necessary to distinguish between this openness, which took shape in growth through interaction with everything that was good in human beings (one senses an echo of the wording of Vatican II), and a disintegration or loss of self-determination and of spiritual freedom in which influences overwhelmed a person and could deprive them of their identity. Openness did not mean ending the links which gave people roots in a particular setting (family, nationality, religion) which was seen as God's gift. But nor did openness mean fleeing from the equilibrium that was provided by the framework of tradition and custom. A negative influence, especially the danger of scepticism and its transmission, was considered to be equally widespread and ungodly. Education had to counteract this by developing a personality which was aware of its own identity, had its own values and could distinguish between right and wrong when it encountered something new. "The role of education is to develop and establish a liberty capable of being open, of throwing itself into an adventure which will lead it to spiritual enrichment, not to suicide."[59] Education in general, and here he quoted the Greek philosopher Pindar, consisted of helping people to be human. In this, religious education had a religious approach depending on the grace and direction God had given people in their specific milieu and context. At this point, very much like on the Muslim side, God and the world, humans (humaneness) and God, are ultimately seen in harmony. "For sound education is that which enables a person to interact with God, and to rise to the level of divine logic. A religion which is built on the foundation of a sound humanism is one which brings man nearer to God and to His creation at the same time."[60] Human dignity was derived from the dignity of God himself through creation and redemption. Education had to awaken an awareness of this God-given dignity so that human beings could preserve and develop it. The educator also had to be aware of this dignity and act concretely with respect and responsibility. Human beings were composed of spirit and matter and education had to maintain this balance. Body, intellect and soul had to be able to grow. Humans existed in relation to God and other persons; both relationships needed to grow and be inter-connected. The criterion for this was the dual commandment of love where the second half implied that one should love one's neighbour as God loved him/her even if he/she possibly had a different understanding of truth (God). First and foremost, human beings possessed the dignity granted them by God. Therefore they were objects of love, whether their attitudes were right or wrong. Irrespective of the circumstances, love came first and all other commandments and ways of acting were derived from it. In this sense, education also meant education for loving or, more precisely, to the view which God had of one's neighbour and of one's self. The goal of education was God's perfection so God himself was the example. This was not strange because we were God's image and should always look upwards to God, also for the sake of our own salvation. The risk and possibility of errors had been taken into account, even by God himself, as Patriarch Sabbah formulated it in true Catholic terms. "God created man capable of being in His image, and He called him to realise this image in himself. He endowed him with the necessary natural powers. Then, through the incarnation and the redemption he put within his reach an increase of grace and power as a means to doing good. (...) [M]an who is the object of trust is a sinner and capable of error. He has a tendency towards evil, just as he is capable of doing good. His nature from the beginning has been wounded by original sin, but he has been redeemed, and the capacitiy of doing good has been capable of rising like the Lord Jesus Christ (...). For this reason he must be given trust."[61]

To summarise: if God had placed such confidence in human beings, the educators should do the same, apart from the fact that education was based on freedom and self-determination and without these no education was possible. The principles described could also easily be applied to differences in the religious realm, although a clear distinction had to be made between the attitude to truth as such – where there could be no compromise – and the attitude to persons or to mere expressions of truth. And no one had the right to challenge or violate the dignity of another person for the sake of the truth. God himself was the model. God created human beings free and independent like himself and God respected this freedom and independence and did not compel them to accept the truth or to behave in the right way. Sabbah explicitly called for respect for the loyalty of pupils to their religion and nationality (the combination raises the question whether perhaps it was influenced by Sabbah's situation as a Palestinian Christian) and to accept what was good and make it known. It was necessary to sow a love of the truth, love of one's neighbour and freedom all together in the hearts of the faithful. The faithful had to be liberated from fear of and aversion to people of other faiths. There was a strong emphasis on seeking the good everywhere (cf. Vatican II) in faithfulness to the truth and to oneself and also in faithfulness to the brotherhood of humankind and in the process, in his view, setting an ideal example. "There is a historical model for a successful cultural mix. This is Greek culture which both Islamic and Christian cultures handed on, having been enriched by it, without either losing their own identity."[62]

Creation and the brotherhood of all people – but with different points of emphasis
This was a real anthropological foundation for education in general, for religious education and for religious education in a broadly and religiously pluralistic world. The way in which the doctrine of being in the image of God is interpreted is clearly influenced by Catholicism; effects of Vatican II are also evident, as already mentioned, and the emphasis on the milieu (nationality, etc.) as God given is reminiscent of a number of statements in the first, internal document of the Secretariat (cf. above). The very positive evaluation of Greek culture can also be described as a typically Catholic tradition, although it is naturally used here as a bridge to Islam as well. In contrast to the Muslim presentation, it is striking that the idea of creation, of the brotherhood of humankind and of the world as a whole as a reflection of God can certainly be described as common. What is lacking on the Muslim side is the idea of humans being in God's image, which was interpreted mainly as freedom and independence given by God to human beings and human nature in the context of creation. There is greater emphasis there on the harmonious inclusion of human beings in creation and society as a whole, something which is at most secondary in the Christian presentation. The Christian emphasis on love as the key to all action is also striking, whereas on the Muslim side there is greater recourse to legal provisions and regulations. Despite all the similarities, however, it is also clear that there are major differences of emphasis in the view of human beings and hence of society: the importance of individuals, their freedom and love, on the Christian side, and emphasis on a harmonious, ideal society in which everyone and everything has its orderly place, on the Islamic side – and both based on God's creation. The strongest link is that, because of this cre-

ation, both see human beings not only as fellow creatures but also as brothers and sisters.

7.1.2.4 Islamic commentary – more than detailed

It is significant that the Muslim response to this address also identified ideas which were completely in line with Islamic principles and rooted in Islamic law, and Al-Abbadi then also went into this in such detail that his "response" almost amounted to a second address. He saw agreement in the distinction between a general, fundamental and a special identity. The former related to belonging to humankind and the latter to belonging to a nation, a state, a religion or a society. With regard to someone's personal identity this meant that it had two dimensions, namely one that was global because the person was part of the global human identity and the other of autonomy in relation to other persons. There was only one human race but every individual had his/her own character and particularities. Openness was welcome but loss of identity had to be avoided. Then Al-Abbadi provided an insight into why this conference theme had been chosen at all. Neglect of or even hostility towards the spiritual and moral dimensions of life threatened human society with disintegration and therefore more value had to be attached today to religious teaching and education. Moreover, in view of the pluralistic nature of human societies, two principles had to be maintained: the safeguarding of the identity of followers of a particular religion and, at the same time, the safeguarding of openness, cooperation and bridge building between the followers of all religions, although he also spoke against comparative religious science from a secular point of view because it treated religions only as social phenomena and not as divine inspiration. That was certainly not easy but possible and necessary and in all realms, and it was essential to include the eradication of distortions and criticism, especially relating to Islam of course. "This requirement is particularly urgent as far as Islam is concerned, since knowledge of Islam in non-Muslim countries is, more often than not, flawed by lack of objectivity and by reliance upon hostile sources which are unworthy of trust."[63]

Finally, he presented the view of Islam on the subject and this turned his response into an address which was really superfluous because a Muslim address on this had already been given. He emphasised that God had sent the Prophet Muhammad for the whole of humanity. Islam was opposed to any kind of watering down, loss of identity, blind imitation or borrowing of elements from one religion to incorporate them into another, and that was why, as had already become clear, he particularly agreed with the parts of Sabbah's address which aimed at the preservation of specific identities, concluding with the sentence,

"The truth is that focusing on the specific identity moulded by religion into man's soul will enable religion to achieve its aims in human society."[64] In the Islamic view, human beings were really inclined to do good although they also had to endeavour to resist and combat evil. Here, Al-Abbadi gave another definition of human identity which also explained why Islam calls all prophets and their successors Muslims. "In Islam, man's distinctive identity is based on his submission to God alone, on his adherence to His precepts and on his surrendering himself to His laws... This is precisely the basis of the terms 'Islam' and 'Muslims' in the Arabic language."[65] Religious pluralism, national differences and the division of human beings into peoples and tribes were recognised by Islam, although the provisions in the Islamic tradition which he mentioned concretely, especially the *"dhimma"* status, really only apply to Jews and Christians. And one can at least partly doubt if there is full religious liberty for the latter or a social status according to the principles of justice, compassion and tolerance. Respect for contracts and agreements formed a wider circle together with cooperation in organising practical affairs. Since Quranic statements emphasised that God had created humans from a man and a woman and made them into peoples and nations (Sura 49:13), Islam had always been very much against any attempt to deprive people of their rights and dignity. Naturally, Islam had also demanded that people acquire knowledge and information continually and by all means; here he cited a hadith, for example, according to which Muhammad had set prisoners free in exchange for each of them teaching ten Muslim children to read and write. In line with Sura 17:70, which refers to the special honour and exalted position of Adam, Islam insisted on human dignity and the human rights to life, liberty and dignity. In the Islamic view, humans consisted of soul and spirit (here the body was not mentioned) and they should treat their nature in a balanced way. Therefore Islamic education endeavoured to cover all aspects of human life and to produce individuals who maintained ethical and moral principles even if these went against their interests and wishes. Moreover, the basic principles of Islamic law had enabled Muslims to integrate the whole human inheritance of science and knowledge into Muslim civilisation and in the process still to preserve the specific Islamic identity.

Overall, this contribution does provide an insight into Islamic anthropology and a wealth of relevant quotations including some which are normally cited less frequently, but precisely those sections constituted duplication for the conference as such and here for the reader.

7.1.3 Youth, natural science and faith
7.1.3.1 The Christian view: integration of evolution into creation and of knowledge into faith

Some sections of the two addresses which discussed the question of science, faith and the concerns of youth also dealt in a very basic way with issues of anthropology. The Christian address was given by Sister Rosaleen Sheridan. At the very beginning, she emphasised that science and faith were not in contradiction, as many people believed; on the contrary, science could lead to a better understanding of creation and of the place of human beings in the natural order. The details of her approach were strongly influenced by Teilhard de Chardin whom she also quoted explicitly later on. She saw God as the creator of the evolution process which had produced the natural beauty and harmony that led young people automatically to ask about the cause. The Creator God was in some way the love and harmony which were reflected in his creation, in perfection. God wanted to express his love to persons and therefore created human beings so that they could hear and feel the love of God and freely accept or refuse this offer because love was a gift that was given freely. According to this definition of human beings, freedom and love were perfectly united. "God wished to become identified with the created object of his love so he had to create a being that was capable of this unique union, a being that desired that congenial love. Man is called into being so that love might transform him and it is precisely this readiness or attitude that constitutes the central core of what is most authentic and innermost in man. It is based on man's experience of a self who desires in freedom to be known and loved. (...) Man's life or spirit is a continual reaching out towards God. Man continually transcends everything in the pursuit and realization of his own becoming – forever on the road and journey towards God – man's very condition is part of creation which can reach beyond itself, and being transformed into God himself be a revelation in time of the very essence of God."[66] God's incarnation in Christ was also seen in this context; as the crown of creation, Christ had always been present in the created world and in each human being. He was, as it were, the evolutionary breakthrough for all people of all cultures in history because God's offer of love and total human receptivity were united in him. In addition, because of this role of love, human beings were social beings by nature. Revelation was understood basically starting with creation and with the spirit which dwells in human beings. "God is involved where people are. He is present in their growth, in their aspirations in *(sic!)* their movement away from destructiveness to community awareness, and in the extension of their responsibility to include the whole human family. (...) God's message, uttered in Christ, reveals the hidden dynamism present in human life everywhere and at *all* times. It is the message that explains to *all* men what is going on their lives and what their mission is, enabling them to interpret the 'inside' talk that takes place when they listen and respond to the gratuitous gift of love that is present at their deepest centre. Since man is never totally in possession of himself, never fully what he could be, he must deal with situation after situation, experience after experi-/p. 52 ence in the understanding of self on his journey of becoming. Neither God, nor religion, nor revelation can be adequately communicated from outside as a message.[67] But religious traditions and ways of believing could be a help here for young people. In line with Vatican II, this scheme could also easily include the other religions with their striving for perfection, truth and beauty and their concern for basic human questions and experiences such as death and resurrection, suffering, love and compassion and life after death.

Against this background, she then proceeded to deal with the subject itself. Here she emphasised that young people in particular were looking for the meaning of their lives and for human values, were hungry for truth, wanted to know reality, and the fate of those who were less fortunate, e.g. in the "Third World", was a serious and urgent concern for them today. They wanted to be able to identify themselves with a group or a concern and at the same time to be something special. The degree of freedom in a person's life was an indication of the maturity of a person's thinking and it was especially liberating to grasp that every person wanted to be loved and was lovable. The view of present-day secular society was very critical with its value system based on consumerism, egoism, the desire for profit and the avoidance of pain. This hedonism was a hidden form of idolatry and modern people were alienated from themselves with all the consequences, e.g. that an alienated person cannot love his/her neighbour or God. As far as knowledge itself was concerned, there were various forms such as science, experience, myths, intuition and faith. Some things had to be understood to be believed and some had to be believed to be understood. As far as the totality of reality and God were concerned, they exceeded what an individual person could investigate. Science dealt with the "how" of creation while faith raised the question "who" and "for what". "The human is free yet God is in total control of history; we humans are trustees of God's creation guided by God's laws. Our aim is to affirm man's total dependence on the universe on *(sic!)* people, and in particular on God. (...) For the Christian it is God not persons who is lord of the universe and man's value is due to the fact that he is created in God's image as God's personal representative to continue developing the world according to God's guidelines and values."[68] Logically following all these arguments Rosaleen Sheridan saw self-knowledge as the root of religious knowledge and she also saw the learning of young people mainly as experiential learning about how God speaks through the events and people in one's real life. It was even necessary for the child to be kept alive in every person so that the sense of awe, openness and for gifts and one's uniqueness as a child of God could be preserved because this implied freedom and joy. In conclusion, she saw the grateful relation with the creator as the point common with other religions.

7.1.3.2 Muslim criticism: negative image of human beings alienates Christians from their faith

The address was subjected to strong criticism in the official response given by Zainul Abedin. His first objection was that he could not describe the dichotomy between science and religion as traditional because, in the Islamic world, literally "the ages of religious awareness had also been the times of rationalism, scientific development and innovation. Religion, understood in its essential connotation is a system of truth which when sincerely held and believed has efficacy to bring about desirable changes in human affairs. The first 8–9 centuries in the history of Islam bear testimony to this observation."[69] He saw the responsibility for this dichotomy mainly in the mediaeval conception of human beings developed by the Church Fathers. In the first 15 centuries of Christianity, Christians had been led to believe that the human body with all its earthly connections and implications was profane and despicable. In this connection he quoted Augustine who said a lot about the consequences of the fall, namely sin and death, from which he deduced that Augustine attributed no positive value at all to humans and their affairs. But the passage which he quoted from the document of Pope Innocent on the wretchedness of the human condition was considerably stronger. "Formed of dust, dirt and ashes and what is lesser, of the most contemptible seed; conceived in sensuality, in the flame of desire, in the stench of lust and, what is worse, in the stain of sin, born to care, pain, worry, and what is more miserable, to death... man does all that is shameful, common and vain... He shall become a heap of putrefaction, riddled with pestilence which shall stick eternally."[70] Apparently, the church authorities had not only considered human nature to be evil; they also saw everything human, their actions, wishes and even the bearing of children as bad and sinful. In his view, this had two implications. It was a crime against God to have a family and bring up one's children. If human beings attended to humans, nature and earthly things, that was evil. The latter had led to the division between science and religion and the former to alienating people from their faith. It was not possible to speak about eternal love in the shadow of eternal sin and no one was able really to hear talk about love if their whole being was condemned as totally worthless. If one wanted to make what Sister Rosaleen had said a reality, it was necessary to set the traditional teachings of the Church Fathers aside completely.

His last point of criticism was that the address – in contrast to its title – did not strictly deal with young people, whom he defined as people between 16 and 20 whom urban society did not yet consider totally grown up. These young people were certainly not alienated by machines but, on the contrary, fascinated by technology. They were sceptical but only about adult values and norms as young people had always been. He agreed that they were looking for love and security but he did not believe that an appeal to love one's neighbour would help here because he, also in reference to other studies, tended more to the opinion that this love was not possible in an urban, mobile, competitive society with its small flats and its concentration on gain. This adult society needed to be changed in politics, the economy, pedagogy, architecture and a number of other things. Islam, on the other hand, placed its strongest emphasis on the best that parents could give their children being a good education and that a Muslim was never frustrated. The former gave the children security while the latter eliminated alienation. If everyone followed this approach, anxieties and, above all, frustration would simply not arise among young people.

It is a great pity that the further discussion has not been recorded because this response offered plenty of fuel since it really did lay its finger on the shortcomings of the presentation which simply followed the extremely modern approach of Teilhard de Chardin too closely without saying a word about the traditional images of human beings and the world. It must naturally also be stated that the response, in a typically Muslim manner, concentrated on the concept of original sin in a form which would indeed no longer be found in the modern world and, in the process, completely lost sight of the powerful "nevertheless" of God's love and grace (which incidentally was not the case in the previous Muslim response). No Christian would claim that it is a crime against God to have a family and bring up one's children, and therefore the alienation of Christians from their faith for this reason can also only be considered a completely unfounded idea. On the other hand, he undermined his own arguments by demanding a complete restructuring of society, on the one side, because a mere call for love of one's neighbour could not provide young people with the love and security they wanted and needed, but he only talked, on the other side, about a good education and even claimed that a Muslim could not be frustrated and hence alienated: "Frustration does not happen to a Muslim"[71], and seemed to believe that such Islamic doctrinal statements alone could prevent anxieties and fears among young people. At best, that is incomplete because five lines simply cannot be sufficient to provide a reasoned solution to such a problem and, at the worst, it is simply totally blind to reality. It can only be underlined that it is a great pity that we have no knowledge of the ensuing discussion because this was indeed a clear expression of one of the typical Islamic reproaches against Christianity, namely that it has a much too negative image of humans and the world and concentrates too much on the world to come, does not keep the right balance between this world and the next and between the different elements in human nature, all of which Islam does as was repeatedly stated in a quite neutral way in other Muslim contributions.

7.1.3.3 The Muslim conviction: faith in God solves all the problems of young people

The Muslim speaker on the theme, Wahbah El-Zuhaili, logically assumed still more strongly that faith and science were not in contradiction but rather two essential truths of life which reinforced one another and led to God, the One, the Creator, whereby faith clearly had the upper hand and, where necessary, a corrective function in relation to science. One of the effects of science was modern technology which had in turn had effects on human beings. They were more closely bound to life, its appearances and temptations and, above all, quite passionately to the luxurious style of living which modern machines had imposed on them, as El-Zuhaili put it. Faith was an inborn element of human nature and nature revealed to humans that they were weak and incapable and needed God, the Creator. All healthy, reasonable, scientific knowledge, in whatever field, and especially when it was a matter of the mystery of life and of the human mind and reason, recognised its own limits and demonstrated that God is the Creator and the Almighty; here the absolute unity of God was almost more strongly emphasised than God as the Creator of the universe. Faith and the oneness of God also led to unity and equality among the nations. The tendency to have faith had been rooted in human beings since their creation. Faith in God brought the human soul satisfaction and gave people the strength they needed to understand life and its problems. Providence gave the human soul calm and security, provided the things that were beyond human capabilities were left to God. El-Zuhaili also emphasised that Islam had always called on people to learn more and increase their knowledge. The spread of Islam was indeed based on science; pre-Islamic paganism was equated with ignorance whereas the age of civilisation and enlightenment had begun with Islam. Islam had adopted many things in this connection from the Greeks, the Indians and the Romans and developed them further. Here, a clear dividing line was drawn against the mediaeval Christian church compared with which Islam was presented as the religion of science, research, knowledge, reason and education. And the Islamic attitude had never changed, neither in the Middle Ages nor in the nineteenth century, the age of the industrial revolution in the West. For young people, in particular, faith in God had only positive consequences, the address concluded rather abruptly.

"Obedience and surrender to God keep them spiritually healthy and happy and ease pain and anxiety which result from passionate attachment to the world of material. Faith and obedience to God must be combined by the belief in His prophets, Holy books and metaphysics as told in the Holy Qur'an. Sound faith directs man to conclude that Islam is the last of the revealed religions. It also leads to a complete acknowledgement of al-Sunna and to an ideological commitment to Islam as a comprehensive way of life which results in self-esteem. Thus, Islam is deemed a cause worth fighting for. Once man believes in the monotheism of God (especially the youth), his soul is set free and made independent. He refuses to be dominated by other human beings and fears no one but God, the Almighty. As a result, a young generation emerges enjoying distinctive and positive characteristics such as chastity, courage, nobility and honesty. This young generation is fully aware of the fact that God is the only Provider."[72]

His analysis of the situation of youth in the West was disastrous and, in total contrast with that of the previous speaker, focussed on the indifference and lack of responsibility among young people, their false approaches and wrong behaviour. Young people in the West generally lacked faith and so they were spiritually empty, dissatisfied and restless. Young people in the socialist countries were being deprived of human rights, human dignity, freedom, equality and democracy. Young people in the Arab countries and in the "Third World" had no political rights, were threatened by the West culturally and ideologically and suffered from the terrible destructive triad of poverty, ignorance and illness. His recipe, at least, is as clear as it could be. "Youth reformation cannot be achieved without faith in God, the Almighty. Faith in God leads to the prosperity of religious brotherhood, equality and order. Once they believe in one God, and live as one nation under religious brotherhood, they become a generation capable of fulfilling objectives and dispelling pains out of their country. As a result, there would be no contradictions between the reality of the underdeveloped society and the requirements of true faith. I am inclined to believe that the problems of youth can be cured only if we adopt an education based on religious belief, respect of the Muslim family system and concept of sacrifice. Relating youths to life and society, cultivating in them independence, respect of work, and wish to be productive, helping them to get rid of Western intellectual invasion and making opportunities of work more available, can also be extremely helpful."

He then made concrete suggestions about this upbringing by parents at home and education in colleges and universities with strong reference to the Muslim tradition. Parents had to lay special emphasis on prayer because it brought people closer to the Creator and to good deeds. In addition, the tradition set great store by the dignity and pride of people, especially their dignity which was acquired by religious belonging and not by striving for money. As far as further education was concerned, they should support and encourage it, and a good relationship between the younger and the older generation and a solid link with reality rather than living in an imaginary world were foundations of a good education. Apart from this, El-Zuhaili quoted mainly from a declaration of the Association of Arab Universities, e.g. that teachers should set an example, that writing and reading should be promoted, especially the reading of Islamic literature and works of Islamic law. Attention should be paid to the choice of friends; they should be educated and well brought up. Overall, much attention is devoted to dangerous influences. "[T]he educators (...) /p. 68 (...) also should warn them against losing their identity and should teach them how to be independent and not to be easily influenced by the cultures, traditions and behaviours which are alien to our good tradition."[73]

It appears that here, in order to create a link with the subject of the first lectures, identity was being much more strongly emphasised than openness because it seems to be very much in danger, despite the show of sovereignty in relation to science, and in the practical context of young people perhaps not every reality so automatically led to Islam as was repeatedly being so emphatically claimed. It seems that a high degree of guidance in the right, "healthy" direction was considered necessary. This contrasts with the Christian address which, although it presented similar lines of thought about the presence of God in creation and in human beings, finally placed much more confidence in this presence in education than did this Muslim address, at least, although it was based on Muslim authorities from Muhammad to the present.

7.1.3.4 Christian response: emphasis on the theology of creation again

The Christian speaker who was responsible for the response, (again) succumbed to the temptation to give a counter-address from the Christian point of view, in a sense providing duplication instead of mainly responding to the lecture itself. The anthropological interest in his address was e.g. his emphasis that human beings should take precedence over science and technology because they were both the subject and the goal of all scientific results. In addition, he at least dealt briefly with the problems of the past (e.g. the case of Galileo Galilei) and presented them as a misinterpretation of Holy Scripture; when the Bible and theology spoke about creation, they did not intend to make any analytical, horizontal statements like the natural sciences but rather vertical statements about the relation between the created realities and the other, un-created reality – God. Then it almost sounded Muslim when he underlined that there were many indications that the whole universe could be summarised in a law, a kind of world formula, that pointed to the oneness of God who had also set the stamp of his oneness on creation. Everything – science, nations and peoples – moved in that direction. Another aspect which was certainly connected was that geographical borders had become practically insignificant today and that the fates of the nations with regard to the economy, security, stability and peace depended on one another. There had been some progress but also progress towards potential enslavement and even the extermination of humankind, e.g. in the form of conditioning by the mass media, by the degradation of the environment and by nuclear deterrents. But this was less an effect of science itself than of ideological, political and economic factors which determined progress. This insecurity particularly affected the younger generations who breathed in this atmosphere from birth and were now exploiting life with all its possibilities as if they were hunted in a desperate attempt to forget the threatening future. The youth of today needed a new ideal in order to channel their energies and find answers to these problems. A minority had already changed direction towards a different future marked by love of one's neighbour, peaceful coexistence and great respect for nature, which brought the speaker back to the theme of unity, not to be confused with uniformity. Again, the key factor was God as the basis of universal brotherliness. If we looked to him, we could see all people as candidates for this brotherliness irrespective of race, nation, ideology or religion. The Christian faith taught that human beings were created in God's image and as his counterpart and would therefore only find their fulfilment in their relation to God. This relationship gave them the strength to love other people and build up an atmosphere of respect, appreciation and trust. The history of humankind was a long and strenuous rediscovery of the reality that God, the Creator, was the Father of his creatures and we, men and women of the whole world, were his children and all members of the same human family of brothers and sisters.[74]

The theology of creation is clearly the central issue when dealing with the way of looking at others and at coexistence, irrespective of the details of the problem.

7.1.4 Religious education at universities
7.1.4.1 Muslim view: only the vicarious responsibility of human beings makes religious education meaningful

The last round of addresses dealt with religious education at colleges and universities. Although the subject was intended to relate to the present and future, the Muslim contribution by Abdul Aziz El-Khayyat was a general review which should be of interest especially for those who would like to know what happened in Islamic educational institutions of all kinds in the past and the present and which approaches they adopted. His entry into the subject was naturally of a more fundamental nature, namely that religious education in the narrower sense is defined in Islam as connected with Islamic jurisprudence. But, in the broader sense, medicine, chemistry, biology, astronomy, philosophy, pedagogy, engineering, mathematics, languages, etc. are also religious sciences for Islam because Islam calls upon people to study and to learn these sciences, and indeed life-long learning is a duty. However, it is sufficient if certain people carry out this duty of studying the sciences on behalf of everyone. So everything was taught in the classical Islamic universities during the classical period and no distinction was made between religion and other sciences. The separation between religious and secular sciences came about only under the influence of the West, as well as a clerical approach to religious doctrine which finally resulted in its being detached from social developments. From that time onwards, the classical Islamic universities only taught Islamic law. It was not until Islam was newly discovered as a religion which offered progress, a style of life and a system of statehood, that these developments were partially reversed. In the meantime, universities such as the Al-Azhar were again offering sciences like medicine, economics or art, and Islamic education had also become very important in the other faculties of the universities. The general aims of an Islamic education from the very beginning were listed as: good manners (which play a predominant part time and again in a striking way), ideal values, belief in the almighty God and the religions of revelation, love and good treatment of all people, the rules and teachings of their own Islamic religion, and an introduction to Islamic culture which enabled children to develop a reasonable way of living. Religious education on a higher level "aims at creating a good Islamic person who stands on his own merits, mentally, spiritually and socially, and who has faith in God"[75]. In this broad survey, it is interesting that the author knew of no example of a co-educative school. In connection with types of schools, he pointed out that, in the Islamic view, puberty as the transition from childhood to maturity was also the established age for the beginning of religious duties, indeed the age at which a practical, independent life began in which one was able to work and became intellectually productive and mature. Therefore there were also conceptions and ideas about not wasting this time totally with university education but rather linking working and learning more closely. Alongside the orderly and institutionally supervised (usually by the ministry of education) school and university education, there was also what he called "irregular" religious education in mosques, Islamic cultural centres and Quran schools, not to mention the sermons and the at least two teaching units given each Friday in every mosque of the Islamic world. Then there was also semi-regular religious education in centres and institutions for the further education of preachers and imams.

Significantly enough, the anthropological basis for this wealth of detail, which even included the curricula of very different institutions, was only provided at the end, in a sense as a possible future prospect, but nevertheless in detail and also emphatically. It this connection, he said that the reason for religious education to take place was that human beings were God's representatives on earth, chosen to bear responsibility for caring for the earth. God had honoured human beings by giving them reason, intellectual abilities and an inclination to learning. God had chosen people as his vicars because he had created them in the best way so they were always God's best and most beloved creatures. He had placed everything in heaven and on earth at the service of human beings. In return, God required people to fill and enrich the earth with love, faith and good works. Reward and punishment in this world and the next had been created in order to ensure that human beings would also carry out these tasks. The prerequisite for persons to be able to care for the earth and to correct people were a firm spirit, the possibility to choose, knowledge of good and evil, God's right guidance and teaching through the prophets, and hence naturally a religious education that honoured and promoted people in soul and body. Religious education helped to correct people's behaviour and to make them good members of society. Everything that persons did to please God and improve their behaviour in society was seen as an act of worship for which they were created by God and in which all people were equal, as in their rights and duties, irrespective of skin colour, race or language, because God had created them all. Islamic education was based on moderation and tolerant treatment of others and was concerned about the realistic aspects of a person's life. It showed people a comprehensive picture of the world and of life. In one sentence, "Modern Islamic religious education derives its essence from faith in God, Islamic creed, and the fact that Islam provides for this creed and legislation to guide the humankind regardless of race, colour, and religion in order for everyone to live peacefully under the loving and tolerant teachings of Islam"[76].

It is regrettable that this comes at the end and not at the beginning so the question is not asked to what extent Islamic religious education as it really is also lives up to this very far-reaching demand.

7.1.4.2 Christian criticism: the little illogicalities of Islamic education yesterday and today

The official response regretted that no definition had been given of religious education and that it was assumed that it was known. The speaker would also have liked to have more information on certain points, for example that the older Islamic tradition had made no distinction between the religious and other sciences, and that Western influence had forced a clerical conception on the teaching of religion. He was also interested in the details of the various reform efforts, the shortening of the length of studies and the reasons why several religious schools were not recognised by the state authorities, as well as how the speaker really judged the irregular form of religious instruction, supposedly that which took place outside of schools, colleges and universities.[77] *However interesting these questions may be for concrete practice, they are naturally very specialised, on the other hand.*

7.1.4.3 Christian position: education for consciously Christian and hence human values

The Christian speaker and Jesuit, Vincent Duminuco, chose a quite different structure for his presentation. He began by dealing briefly with the very old tradition and significance of teaching in the Christian Catholic context and then went on to describe the major social changes between 1940 and 1980, which had brought about a tremendous upsurge of Catholic universities on the outside while there was simultaneously internal erosion right across the board as a result of increasing specialisation in society and hence also in the universities, but also because of the loss of religious socialisation which had been taken for granted in family and society and was evident among the students. What Catholics had taken for granted by way of content and behaviour had been eroded in societies which were increasingly marked by urbanisation, industrialisation, affluence, education, the media, secularisation and a pluralist system of values. Catechesis also had to change in this situation. The relation between religion and all other aspects of life was no longer as organic as in stable, rural societies where faith provided the basic assumptions for living and helped to form a coherent world view. In such cultures, the expectations were relatively similar and general with regard to the different aspects of human life and human affairs, namely concerning the type and duration of marriage and family life, ethics in business and profession, social customs and what constituted a good life; in brief, in relation to the nature of human beings, their place in the universe and their destiny. What had happened could be described as a paradigm shift leading to fragmentation of what was now a secular society and of life into partial realms, one of which was certainly still religion but it had nothing more to do, for example, with economics and business. This model could better be called mechanical than organic and often led to the disintegration of persons and society, with a feeling of alienation, loss of a sense of responsibility and with extreme injustice. The developments of the twentieth century had shown that a person could indeed be educated but morally bankrupt and that therefore education did not automatically lead to greater humaneness and Christian living. Consequently, it was childish to believe that every education produced virtues, and it was becoming increasingly obvious that anyone who wished to exercise a moral influence on society needed to ensure that education took place in an organic, moral framework.

These considerations formed the background for a future vision for Catholic universities on a theological, anthropological basis. The Catholic world view was clearer today than in the past, especially where it was challenged by the value system of contemporary society. The so-called "value free" education had proved to be a myth; today not even the natural sciences were considered neutral. A value was something valuable, something which gave one grounds for living and possibly even for dying. Values gave life meaning, made it meaningful. They gave a person identity, a face, a name, a character. Without values a person was at sea. Values were central to one's own life, to everyone's life; indeed Duminuco went so far as to say that they determined the quality of life. Moreover, every academic discipline in the area of the human and social sciences had to admit that the values which it conveyed were based on an assumption as their starting point, namely an assumption about the ideal human being. Here, he believed, the conception of justice in the Gospels could become effective. And in Catholic universities important human problems and concerns should be investigated and any partial or distorted view of humanity should be generally rejected, whereas other educational institutions often lost sight of a concentration on human beings because of specialisation. Faith should no longer only be understood and lived in a vertical, spiritual way but also with concrete reference to concerns such as justice, care for the poor in the world, attention to questions of the economy, human rights, ecology, etc. Everything else was not only inappropriate for the Catholic faith but also ineffective in a secular climate. In Catholic education as a whole, it had always been a question of the intellectual development of each individual to the full extent of his/her God given talents and in a holistic sense, which also included self-discipline and initiative, integrity and precision and, on the other hand, excluded superficial, negative views of the individual.

There remained the difficult question of practical implementation. Here, Duminuco laid great weight on interdisciplinary work, not only because no one science alone was sufficient to explain the whole of creation but also because the human problems in their complexity on the threshold of the twenty-first century simply required cooperation, for example the questions of genetics and medical technology, of the beginning and end of human life, of human rights, poverty, homelessness and many others. The constant progress of our human possibilities faced us with very difficult moral questions which could not be solved by one discipline because it was a matter not only of technical but of human values. The leaders of tomorrow needed to be confronted with the questions of the consequences of progress and the moral implications of seemingly financial decisions. Really, such an overall approach should be taken for granted for every university and a Catholic university was not even conceivable without this combination of various forms of knowledge with human values and theology, nor without this comprehensive view and realisation of human beings. Academic inheritance and tradition also set store by human dignity and a good life in an inclusive sense, e.g. academic freedom, the demand for excellence of schools as well as students, including moral responsibility and sensitivity, by treating religious experiences and questions as central to human culture and human life. In this connection, a Catholic university would attach special value to the realm of the contemplative capacity for God and the world which belonged directly to the centre of human existence. The global approach was also typically Catholic and moreover simply necessary in view of many modern challenges in the fields of ecology, economics and genetics. In the realm of religious education, the privacy of the students and their personal attitudes of faith and values had to be respected and, indeed, in a whole series of Catholic universities, professors of other faiths were offering courses in their own religion. But religious education should not just be a matter for the head; it also had to provide room for spiritual and diaconal practice, the former again also with consideration for non-Catholics among the students whose number had increased enormously. Service of one's neighbour for purely altruistic reasons was also seen as a great benefit which needed no justification of its own, but a Catholic university should also be concerned to recognise the suffering Christ in one's neighbour and to provide a possibility for reflecting on the experiences of serving the poor, elderly, handicapped, sick and marginalised in order to make the students defenders of Christian values, who could certainly also question economic and social systems which caused or tolerated suffering and need that they would want to mitigate, and thus become agents of social reconstruction or allow their own decisions about their status and their style of life to be influenced. Many things were already underway, also attracting the highest attention, but the real work had to be done locally on the campus.[78]

This contribution by Duminuco has a clear structure related to the situation and to practice, exposing important foundations and connecting them with the social situation and its challenges, quite in line with the Second Vatican Council, as he himself mentioned. The difference between the Muslim and the Christian contributions in their descriptions of the situation is striking. Whereas the latter described an increase in secularism and the crumbling of a closed Catholic world (concretely especially in North America and Southern and Eastern Asia), the former described more of a reduction in secularism which was presented as Western influence in the Arabic Islamic world and a return to the interconnected approach of Islam. This certainly also helps to explain a number of differences, especially of tone, in the presentations of the two sides.

7.1.4.4 Muslim view: a confirmation of its own reservations about secularisation

It is also significant that the official Muslim response sensed a complaint in the address about secularism having destroyed the organic relationship between religion and the other aspects of human life although consciously it was only intended as an observation. This, as it were, hit the nail on the head for Muslims. The speaker also added that probably all secularists and even many religious intellectuals would surely contradict the presentation in the sense that religious education of the old kind had veiled secular education for centuries, and this had led to intellectual confusion and then also to many psychological disturbances. He also stated that many highly qualified Christian theologians had proclaimed loudly that secularisation was a liberation and had its roots in the biblical faith, that it constituted the coming of age of Christians, and that many people in the meantime saw an almost organic link between secularisation, humanisation and Christianisation. Muslim educators had repeatedly heard from their Christian colleagues that secularisation was humanisation and so it was interesting to observe that, in the so-called Christian West, there were growing doubts about the value of secular education, especially in connection with humanisation. This led him to say a few words on the subject from an Islamic perspective. In the confrontation of the Muslim world with Western ideas about knowledge and education, it had immediately been felt that something was lacking in these modern conceptions which was considered important on the Islamic side. For example, there was a lack of emphasis on human beings perfecting themselves which did not only include intellectual but also ethical, religious progress. On the contrary, such ideas even seemed to undermine this progress also when it was praised as being value-free. Against this background, the conception of the "Islamic university" had begun to develop. Against the backdrop of the general criticism just described, the idea of de-westernising and thus de-secularising knowledge had come up and was being hotly debated by some Muslim intellectuals at that moment. The myth of value-free education and the interest in the real human problems were in any case important issues for any religious education. In addition, the stress on justice was very encouraging because the Quran identified a connection between the denial of religion, the rejection of orphans and the neglect of feeding the helpless, and it had made clear from the very beginning that no one could be a believer in the full sense if they did not at least try to create a just social, political and economic order. With all respect, however, he doubted whether the desired programme of reuniting all forms and areas of knowledge again was feasible, however necessary it might be. Finally, he also regretted that many students from a pluralist, secular background were not able to make much sense of the traditional formulations of dogmas, although he expressed confidence that the Catholic tradition was rich enough to provide other explanations.[79]

Overall, this was further confirmation of where the different emphases lie and how that can also influence one's listening, but that it is very helpful as a whole to have a counterpart with another point of view, and that this system of official responses to an address makes still more clear where the decisive points lie in general and in reference to a particular religion as a counterpart. It can offer confirmation and thus highlight common views but it can also make clear in a critical way that at certain points things are not as simple as one had perhaps believed oneself.

7.1.5 Education for the future: tension between personal identity and common values

Both the Christian and the Muslim participants (a total of 33 specialists and young people from 18 countries) agreed at the end that God had created human beings with dignity, freedom, rights and duties. For religious education in the modern world, that meant concretely that it should not only emphasise the development of a religious identity, however important that might be for the well-being of individuals (and beyond), but also contribute to a better understanding of common values. Differences should not be ignored in the process but recognised in order to create mutual respect for the attitudes of faith and values of the others on both sides. Teachers of religion should encourage young people to acquire all the different types of knowledge – scientific, human, religious and ethical – and help them also to view this knowledge in the light of their religious convictions. Young people wanted to know what was good or harmful in modern developments, particularly also in reference to religion and a meaningful life as believers. It was especially important to emphasise religious identity for school children, and religious values and the religious character of educational institutions at the level of higher education where the different i.e. secularised social conditions were particularly evident and to which a response had to be given by reformulating the religious nature and values of educational institutions. The participants also agreed to include among the documents the papal address from the private audience in which John Paul II had underlined that religious education was not merely talking about God but accompanying young people in their search for God so that their wish to know him and carry out his will could be deepened.[80] At the end of the meeting, an invitation was immediately extended to a next meeting in Amman on the theme "Rights and education of children in Islam and Christianity".

7.2 Amman 1990 on children's rights: from the unborn to school children

7.2.1 The rights of the unborn

This conference on *"The Rights and Education of Children in Islam and Christianity"* was held from *13–15 December 1990* in Amman. The theme was divided into three blocks: the rights of the unborn child, rights and education of the pre-school child and finally rights and education of children in school (as far as government and society were concerned). In addition, there were three opening speeches by the directors of the two institutions and by Crown Prince Hassan of Jordan in person. The latter believed that the underlying theme of the meeting was the duties which adults had towards children. When would people finally start thinking about the role of children in the (much discussed) problems such as poverty, the environment or (civil) war? Cardinal Arinze in his words of greeting referred back to the dialogue about the family (or more correctly about education) and stated that the attitude to children both born and unborn said a lot about a society. By neglecting and harming children, a society signed its own verdict, and for him the large number of abortions was clearly a focal point.[81]

7.2.1.1 Catholic moral doctrine: life and dignity from conception onwards
This was naturally an obvious transition to the first main subject of the meeting, the rights of the unborn child, on which Cardinal Arinze himself gave the main Christian address. It was a matter of the relation between modern medicine and religious teaching and that humans were persons from the very beginning and

had to be treated as such. From the Christian point of view, it was also a matter of the absoluteness of the will of God, but in order to be heard in the modern world it was also necessary to have good arguments in the sense of a humane basis for a humane society. According to the Catholic understanding (he immediately remarked that not all Christians were agreed on this), the Christians' moral doctrine was based on principles which were derived in a special way from divine revelation but also from right reason applied to natural law. As far as the specific question was concerned, a person was a unity of body and soul and human life was a gift from God. It began with the union of the egg cell with the sperm cell and God alone had the right to determine its beginning and end. But human beings had the right to life and dignity from conception to their natural death. According to Cardinal Arinze, the Catholic Church had always been against abortion and now saw itself justified by modern biology. The next problem in this connection was *in vitro* fertilisation because the unwanted fertilised egg cells were destroyed and, according to Catholic teaching, this was the equivalent of aborting human persons. The last problem was pre-natal diagnostics, at least to the extent to which it was carried out with the intention of conducting an abortion if the results were displeasing. Precisely in the case of abortion, the problem had now become that the state legislation of many countries had legalised it, but this was far from being a moral justification as human beings had no power at all to make laws against God, our Creator. Even if the embryo were to be declared a non-person by law, in the eyes of God it remained a human being with the rights of every human person, and for this reason canon law also automatically excommunicated anyone who performed an abortion. On the contrary, the embryo deserved special protection, i.e. good nourishment from its mother and good hygienic conditions. Therapeutic measures on an embryo were possible within appropriate limits but not experiments, for the reasons stated above. *In vitro* fertilisation cannot be reconciled with human dignity in the Catholic view, especially not if someone else's sperm is used and the sanctity of marriage is thus violated. But the main reason was that the process from marital intercourse to the education of the child was seen as a unity which was violated by *in vitro* fertilisation and also by surrogate motherhood and hence, despite theological discussion, the Catholic Church rejected artificial insemination even for a married couple because all of these procedures deprived the act of procreation of its dignity. Since human life is the life of a person, it cannot be a technological product. This, in a sense, was the present medical state of the challenges although new ones could be expected and so such conferences were very important in order to make the ethical principles known.

7.2.1.2 Muslim response: the subtleties of Islamic law

The Muslim response was longer than the original address because it began by repeating many of its statements and it regretted that the address had only presented the Catholic point of view. As an Islamic starting point, it began by emphasising that no contradiction is allowed between religious facts and scientific discoveries, indeed, scientific progress further confirmed religious facts. In practice, Islamic law even underlined the study of contemporary medical questions in the light of its religious prescriptions since Islamic law was intended to govern all aspects of humankind. A moral framework guarded scientific research from errors and from harming human life. Concretely, according to Sura 42:49, human life was a gift, blessing and creation of God where the human being was created in the womb with a harmonious nature consisting of body, spirit and reason. The real beginning of the life of the foetus was when God breathed the spirit into it after 120 days. (If a birth occurred earlier, survival, for example, was not considered possible; in the case of a still birth, it was only after that period that an official confirmation of the death was required.) This distinction was traditionally made in Islamic law; only after that time could blood money be demanded for a foetus, for example, (one twentieth of the usual blood money) and after that point an abortion was absolutely prohibited. Abortion before that time was viewed in various ways, partly as generally allowed and partly as allowed when the foetus as such was not yet recognisable, while most lawyers identified the beginning of life with the implantation of the foetus and declared abortion prohibited after that. In addition, abortion facilitated promiscuity and was therefore a plague on society and, moreover, foetuses were then more easily degraded to stores of spare parts. Whenever El-Abbadi spoke of morality, human values, protection of human life, of the family and sanctity of marriage, which he believed were accepted by the so-called heavenly religions, he was primarily concerned about quite concrete laws and provisions of Islamic law in general or of certain specific schools of law. The life and dignity of the foetus were clearly protected in Islam and scientific experiments on the foetus, including an illegitimately aborted one, were prohibited, so abortion to this end was not worthwhile. Experiments on a still birth or on a premature birth incapable of surviving (cf. above) were however possible provided they did not disfigure the foetus or violate its dignity and respected the Islamic provision that the dead should be buried as quickly as possible. The same provisions applied to organ extraction. As far as *in vitro* fertilisation was concerned, there were no reservations on the Islamic side provided it took place in the context of an existing marriage, i.e. without a surrogate mother and without someone else's sperm, if the couple absolutely needed children. However, there were various warnings not to fertilise more eggs than could really be implanted because they received life and were prepared for life. There was a final criticism that the Catholic presentation had mentioned only the right to life, dignity and health, whereas the unborn child in Islam had far more rights and could e.g. inherit or receive endowments, and also had the right to be the descendant of its father.[82]

It is therefore clear that, in the Islamic context (and this was not totally unknown in mediaeval Catholicism), a pregnancy was and sometimes still is divided into a phase before and a phase after it can be ascertained by traditional methods, which made (and sometimes still makes) a difference in how it was viewed and the protection required. From the moment when the existence of a pregnancy was quite sure, the unborn child had always been protected, but as a result of medical progress that moment now comes earlier.

7.2.1.3 Islam and the rights of the unborn child: a wealth of detailed regulations

Many of the points already mentioned then came up again in the official Muslim address but in greater detail, for example the question of God's breathing in the spirit which referred back to Sura 23:12–14 and to Sura 86:5–7 according to which hearing, seeing, perceiving, movement and feelings were traced to the spirit. The period of 120 days related to a traditional saying of the Prophet Muhammad according to which the creation of the human being in its mother's womb took 40 days, then for 40 days it was a little lump like a leech and finally a lump for 40 days. After that, God sent an angel who breathed the spirit into him/her. According to Sura 42:49, God decided whether it would be a girl or a boy. What was decisive was not the sex but the righteousness of the children – and a righteous son was certainly considered the greatest thing one could wish for. The speaker's references to the Islamic tradition totally determined his remarks in every respect, even including such sentences as "So we must not confuse established Quranic facts with modern scientific theories, however advanced scien- / p.35 tific research may be"[83] and a chapter entitled "The Fetus's Right to Defining the Pregnancy Period which Decides Its Lineage"[84]. Behind this lay the statement in Sura 13:8 that pregnancy and breast feeding lasted 30 months, although the minimum duration for pregnancy was given as six months. No maximum duration was specified which was the reason why a number of Muslim scholars pleaded for not laying one down.

In this way, a whole number of very special rights of the embryo could be found in Islam, beginning with the choice of the mother where special value was attached to her fertility, but also to her good name and her spirituality. In this connection, there were also the prohibitions of marriage with very close relatives which were supposed to prevent inherited diseases, and the right to dissolve a marriage in the case of sickness and handicap, especially a mental handicap of the mother. The child was to enjoy optimum conditions for its development from the very beginning. To this end, even before marital intercourse, it was important to pray that God would keep the devil away from a potential child, and this continued with further prayers for the child and better behaviour of the parents in order to set it a good example. A Muslim family was generally seen as so religious and good that the mother automatically transmitted this happiness to the child. In addition, because of the cessation of menstruation during pregnancy, she was able continually to offer the obligatory prayers and was thus in constant contact with God and also calmed and relieved by prayer. And, through the mother, the foetus also offered the obligatory prayers although it did not yet have the duty to do so. One should also read the Quran to the unborn child, just as it was customary to pronounce the call to prayer to the ear of the newborn child in order to confirm the inborn nature of the person and to anticipate the whisperings of the devil. The worst crime against an unborn child, on the other hand, was if the mother ate prohibited meat; that condemned the child to hell. In general, every kind of excess was prohibited because it was harmful for mother and child. On the other hand, a pregnant or breast feeding woman did not have to fast, nor to worry about her maintenance, because work was always connected with annoyance which could be harmful for the child; cf. the general, emphatic exhortation by Muhammad not to get annoyed. Even in the case of disputes and divorce, her provision was guaranteed. And during this time her husband should give her special assistance with the work so that work that was too heavy would not cause a miscarriage – for this, too, Muhammad's example could be quoted. By way of modern phenomena, the speaker referred here to the damage to unborn children from Aids and the abuse of alcohol which Islam prevented because it prohibited sexual intercourse outside of marriage and the consumption of alcohol.

Among other provisions of Islamic law, there was mention of the postponing of inheritance issues until the birth of a child so that the child can then receive the share appropriate to its position and gender. Only in the very last part of the address was there reference to the protection of the life of the unborn child, and it became very clear that there were different degrees of the crime of infanticide depending on the various stages described above right through to the child that had been born. The concrete evidence of its being a crime was that the person causing a miscarriage always had to pay blood money – as a rule the amount of blood money required for a slave or for a newborn girl, namely five per cent of the otherwise normal rate, and the miscarriage might not only have been caused by a concrete action but also verbally, for example. A pregnant woman could also not be executed or subjected to physical punishment in order not to harm the innocent child. On the other hand, a couple had the right to an abortion at the early stage mentioned and by common consent. This right did not exist in the case of an extra-marital pregnancy because there was no legitimate father who could give his consent. In that case, the divine right to life of the unborn child was seen as unlimited, also explicitly so that one crime would not be covered up by another. A woman was fundamentally prohibited from taking anything that would end her pregnancy; but if she did so and thus caused a miscarriage, it was her duty to set a slave free and to pay the father of the child compensation amounting to the compensation for a slave.

In summary, it becomes very clear that, as is so often the case, the conceptual approach is much more juridical and leads to endless details, but it is also evident that the protection of unborn life is not as absolute as on the Catholic side. Abortion is possible on specific conditions and even where it is prohibited the punishment is much less severe. Unborn life also has a price. That makes it valuable but simultaneously in some sense a commercial object, and these remarks made it very clear that, despite all the dignity God grants human beings, in Islam or rather in Islamic law the price varies considerably depending on gender, age and status.

7.2.1.4 Nothing new in Christian education

The official Christian reaction did not deal with this question at all, however. It only regretted that nothing had been said about the issues of modern medicine and tried to present the Catholic view on that with its three basic criteria. From the beginning of its formation the foetus was a human being. As such, it had a right to life and freedom from bodily harm from the moment of its conception until death. The foetus had the right to be born in a context of marriage and a family. These criteria and the comments made on them naturally agreed with what Cardinal Arinze had already said in his address earlier on[85]. Overall, one cannot avoid the impression that, despite all the agreement between the Christian and the Muslim position, there were still fundamental differences in the ways of arguing – more philosophical and theological or more juridical and theological.

7.2.2 The rights of the pre-school child
7.2.2.1 With infinite precision: Islamic education of the pre-school child

This was also evident (and was stated here really for the first time) in the next round of argumentation which dealt with the rights of the pre-school child. The Islamic contribution on the subject which, as was to be expected, went into infinite detail, was more than double the length of the Christian one and, as frequently happens when it is a question of concrete views on particular questions, it was a brief compendium of Islamic law on the subject. In the address by Professor Abdul Aziz Khayyat, there was a glimpse of practical reality only once right at the beginning when he observed – taking up the ideas of the Crown Prince – that the rights of children were the duties of society, or more specifically of parents, educators, providers or the state, which were well fulfilled as a rule in progressive states but worst, by comparison, in Oriental, African and Muslim countries. However, amidst the wealth of the detailed, modern and good provisions of Islam, this sad fact was completely lost from view and the question of the reason for this blatant imbalance was not even asked, let alone answered. On the contrary, the speaker complained that many people who had worked on the question had failed to recognise the positive influence of the Quran, the Sunna and the Islamic juridical tradition. In the Islamic tradition, there was a lot of literature on education, first and foremost al-Ghazali whom he quoted in detail. The latter described children as blank pages on which everything had yet to be written. The temporal and eternal fate of the child depended on the parents' education. Parents should protect their children by teaching them good manners, keeping them away from bad company, not letting them get accustomed to an easy life and not teaching them a predilection for jewellery and luxury goods. Otherwise the children would spend their lives pursuing these things and that would be their ruin. It was also very important to al-Ghazali that the nursing mother only ate the food allowed because, after consuming prohibited foods, her milk contained no blessing (similar to what had been said earlier about the pregnant mother). In addition, the child should be active in order not to succumb to lethargy. Other signs of lethargy to which children should not get accustomed were crossing one leg over the other, resting their chin on their hand and supporting their head with their arms. They were to be instructed to sit properly and not to chatter because that was uncouth and a sign of descending from bad parents. They should also be taught not to take oaths, not to made stupid, obscene or spiteful speeches, not to swear and also to keep away from those who did such things. Parents were to teach children these things before entrusting them to a teacher.

The rights of the infant according to Islam
After this interesting excursus, Khayyat looked at the status of the child in Islam which was fundamentally very high because Islam rejected celibacy and monasticism and called in various ways for fertility in order to populate the world and develop it further. In addition to wealth, sons were counted among the pleasures of this world. As far as the child was concerned in the phase from birth to starting school, this period was divided into the first two years best known in Islam as the "breast feeding period" (for that length of time, children could and should be breast fed according to Sura 2:223), and the time from the beginning of the third to the end of their fifth year, namely early childhood or the kindergarten age. During these two phases the child had the right to be educated and brought up by parents, relatives and the state. Prior to this, its right to a father who carefully selected its mother was again emphasised, or to a mother who rejected an unsuitable father which was why Islam had given her the right to choose her partner herself. Indeed, the three rights of a son connected with his father were summarised: to choose the mother with care, to give him a good name and to teach him the Quran. But naturally far more rights of an infant can be derived from tradition (the English terminology gives an idea of the specific expressions in Islamic law related to the degree of obligation: desirable, recommended or a must), beginning with pronouncing the call to prayer in its right ear and the prayer itself in its left ear immediately after its birth so that the child heard God's word first and was comforted and entertained by it. In addition, the child had the right to be given a beautiful name either on the day of his/her birth or seven days later. Ugly names, or such as expressed servitude to something or someone other than God, were to be avoided. Circumcision for men was a rule (for women the tradition described it as a sign of decency). The child should be breast fed by its mother for up to two years; this became a must if the child did not accept any other milk – provided that the mother had milk. Even in the case of a divorce, the father could not deprive the mother of this right, and if a nurse were necessary she had to come into the house of the mother. The idea behind this was that the child should get the best and the milk and care of its own mother were the best for it. It was also recommended to chew a date (or something else sweet) and to rub some of it around the mouth of the newborn, and on the seventh day after its birth to kill a sheep or even two for special merit. It was desirable also to shave the head of the newborn on the seventh day. All of these provisions gave the child health and strength and connected its education with the unchangeable rules of Islam. On the other hand, the speaker considered other traditional ways of acting as harmful and therefore not to be followed. Then came the second phase, a phase still within the family but also in a crèche or kindergarten.

The six basic characteristics of Islamic education
As far as education was concerned, the following six basic characteristics had to be present in any Islamic education. Firstly, faith education which strengthened the link with God and the faith of the child. This took place, on the one hand, by the transmission of knowledge particularly of what is allowed or prohibited and, on the other, by being gradually introduced to prayer and fasting. According to Muhammad, children should be introduced to three things: to love for their prophet, love for his household and to the recitation of the Quran. Secondly, moral education following the lines of the Islamic regulations which led to desirable characteristics and deeds such as genuineness, friendliness, justice, a sense of duty, courage, self-control and also a sense of shame, whereas harmful behaviour such as envy, lying, calumny, etc. had to be avoided. According to Muhammad, it was a form of respect for children to educate them well. Thirdly, physical education, namely suitable food, accommodation and clothing, again all following the Islamic provisions even including sport. Fourthly, mental education and here it was emphasised that in Islam faith and knowledge were closely linked. Fifthly, gradual social education so that the child could lead a more active social life when it started school. The child should sense that it belonged to a family and this family was part of society and acted accordingly. The child should help at home, in the crèche and in the kindergarten, behave well and have good relationships. It should treat older people with consideration and respect and also start going to the mosque very early on to participate in worship, hear the recitation of the Quran and the preaching and be spiritually satisfied. And, last but not least, practical education because Islam had always connected learning with acting and faith with deeds.

The new rights of the child according to Islam
So that was the general educational framework in which the child was to grow up, but it also had quite concrete rights – nine of them. The first right was the right to maintenance by its parents, especially by the father, even if the parents were divorced and, if the parents were no longer living, by the grandparents. Secondly, children had a right to the same treatment as their brothers and sisters, i.e. no child should have preferential treatment e.g. with presents. Thirdly, they had the right to compassion and genuine love; that was a special problem for orphaned children, which was why Islam repeatedly emphasised care and compassion for orphans. Fourthly, children had a right to playing and fun. Muhammad had also played with his grandchildren and Islam required that children learned to swim, to shoot and to ride. In addition, it was good for children to get used to useful games in the home and garden, for example, picking fruit or caring for animals. Fifthly, children continued to have a right to care; up to a cer-

tain age, this was provided mainly by the mother, even in the case of a divorce, provided she did not marry another man. Which female member of the family then had to fill the gap had been laid down in every detail. Sixthly, children had the right to paternal guardianship, i.e. the father had to stand up vehemently for the interests of his child in whatever area. Paternal guardianship could only be exercised by the father or the grandfather. The guardianship of other persons, which could become necessary and was then a (seventh) right of the child, is distinguished in the Arabic Islamic context from paternal guardianship both linguistically and legally. This second form of guardianship could also be exercised by the mother, the father's brother or other suitable persons. The eighth right of the child was that of legitimate descent from its father; according to Sura 33:4f, in Islam adoption does not exist. This descent was normally proved by birth within marriage; otherwise it was the concern of the father to prove that the child was his. A foundling had the same right as any other free Muslim and every normal child to care and education and recognition as a witness, only that for him/her the state accepted responsibility or handed him/her over to a foster father. Finally, and the order is certainly significant and strange from a Christian point of view, the ninth and last right of the child was the right to life, i.e. it had to be protected from aggression, hunger and other forms of dying and must not be sold into slavery or bondage. This meant positively that the child had the right to food and drink, clothing, medical care and life together with its family and relatives, as well as to a safeguard against hunger. In Islam, this protection was the duty of the family, the state and the institutions responsible for children.

At the end, the speaker again came back to the subject of education which, in a sense, surrounded the core of actionable rights, had less legal safeguards (recommended, to have precedence and meritorious are the terms used here) but was no less important even so. In conclusion, there were four more points that seemed important to Prof. Khayyat. Firstly, a good example. Here, parents should point the way in word and deed and try to hide any bad characteristics they might have, not to lie and not to make any false promises. Because the child was the mirror of the family, especially of the mother, and the parents could not give a child anything better than a good education, as Muhammad had stated; and later tradition had a saying that anyone who only developed his/her possessions and not also the character of his/her children would lose both in the end. The second point was good company, because, as Muhammad had said, a person normally imitated his/her friends and should therefore choose them well. Special attention had therefore to be given to the education of the eldest child who was then the example for the younger brothers and sisters. And the parents or the family should look for well brought up children of the same age from good families so that their children did not learn obscene language or bad habits from bad company. The third was friendliness to children in contrast not to discipline but to violence because – as Islamic scholars had said – violence in education created an inclination to violence and destroyed what we would today call social competences. So it could be said that Islamic education wanted firstly to create good people and then good citizens. At this age (in contrast to later, according to a hadith), it was better only to threaten children or at most give them a slap on the hand. Finally, children should be taught the characteristics which Islam demanded, namely spirituality and fear of God in public and privately, manliness (hence not getting accustomed to comfort and luxury, not fulfilling all wishes), respect, care and concern, obedience and serving one's parents, learning the Fatihah and other Suras by heart, being accustomed to truth, patience, responsibility, courage and various everyday Islamic rituals at home, in society and in the mosque.

7.2.2.2 The Christian inquiry about the real situation

The wealth of detail, as has already been mentioned, was overwhelming and therefore the task of giving a Christian response to this address was not easy but Nete Rasmussen coped with it in a masterly way. On the one hand, she made clear that sacred texts, whether Bible or Quran, did not speak about our rights as human beings but about our duties to God and to human beings. It was only from the latter that rights could be derived, for example those of children. Among the large variety of rights mentioned by the previous speaker, she underlined as the most important (and also the most obvious) the right to maintenance of the financial kind. She then pointed out that both the Bible and the Quran saw children as future adults in their roles in family and society, just as they saw women as wives and mothers and not just as, and with their value as, children (or women) as such. The latter was a development in the West and, even there, only in the last few centuries since the renaissance, and this was reflected in the UN Convention on the rights of children. Similarly, neither the Bible nor the Quran distinguished between education and ethics, not even linguistically as the West did today. For the former, education was and is morality, above all, to make a person into a good and valuable member of society. Today, in contrast, the idea was fairly widespread that one could lead a good life both as a child and as an adult without bothering too much about knowing ethics or acting ethically. Finally, Nete Rasmussen took up the point which had been made right at the beginning of the previous address but had then disappeared from view, namely practice. The question was not what the ideal situation should be according to the Quran and the Sunna, which was how Muslims usually answered every question, but what were the practical ideas and how were they being implemented in various countries, contexts and social groups. She herself described this as being perhaps a typically Western approach, to look first at the situation as it is and then to move on to the definition of how it should be, but in that way she really did raise an important point which related not only to this address but to many other addresses and many dialogues in general.[86]

The danger is frequently that a Muslim ideal situation is compared with a Christian factual situation and this leads to an imbalance in the argumentation and sometimes also in the mood. It would really be desirable to agree on a common way of approaching the themes in order – when dealing formally with the same subject – not finally to compare apples with oranges.

7.2.2.3 Christian education approached in a quite practical way

Hence, it was typical, for example, that the Christian speaker approached the subject by reflecting on her own practice as a mother and a Christian, beginning with some fundamental considerations, especially that recognising that a child had rights (as the UN had done) meant that the child was a person, a subject and not an object. This was obvious to believers because for them God was the Creator of all life and humans only transmitted it so life did not belong to them. She also saw a correspondence between rights on the one side and duties on the other, which was particularly evident in the case of human children who depended longer and to a greater extent on the help of others than any other species. This was a challenge to the parents with the support of the community around them. But, since human beings were not just material but also spiritual beings, from the very beginning it was also a matter of developing this second aspect. Practical tests had demonstrated that children needed loving care wherever possible in the form of individual attention, and that they needed regularity with regard to time, place, persons and education in general. The best place for this was the family. From the moment of its conception, the child was a human being that already had a specific genetic heritage. Therefore education was more a matter of bringing something out (as conveyed by the English and Latin term *educatio*) than of putting something in. The aim of education was to bring about a free, responsible adult who was him/herself able to take decisions for his/her life. For this, security and encouragement were necessary. And, however small a child might be, one should never lie to it or make false promises but rather admit that even adults could not do (have) everything. The child had to be respected. Precisely at the very beginning, this affected the course of its life and it would then learn gradually to adapt to other people and circumstances. Its body also had to be respected – this was directed against sexual abuse, in particular.

The child should not be humiliated, mocked or treated as an object in any way. But respect also meant honestly pointing to limitations from the beginning, teaching the child responsibility and respect for others. And finally it meant making the child familiar with its spiritual side, its relationship to God. Authority was necessary but more in the form of firm, calm resoluteness which led a child by the hand than as domination which demanded this or that. Initially, it was a matter of rules for health and hygiene, then of life in community and finally of the first steps in society. A "must" could gradually become a "should" which left room for self-discipline, for personal convictions in life, for one's own moral framework or, in other words, for autonomy which had been quoted as a goal of education. The further a society developed in terms of knowledge and science, the more important even pre-school education became for the future of the child. A child should be able to listen with concentration and, to that end, quiet and sensitivity were necessary as a counterweight to dispersion and superficiality and as a help to finding one's own purpose and not losing one's identity – and finally to find one's way to God. The educator should also teach the child to overcome its fear, whether physical or moral. That was very important because normally the root of fanaticism, violence and aggressiveness was quite simply fear. In addition, it was important to encourage the child to admire things and to teach it idealism and selflessness. Having no other aim than profit deprived people finally of all nobility and resulted in frustration. Imagination and play as well as humour were very important for learning new things, learning to lose and not taking one's self too seriously. In addition, one should let children unravel as much as possible themselves; they also learnt from their mistakes but not from being passive. The child had to learn to share (possessions and knowledge) in order to learn solidarity and that it would lose nothing by sharing but rather gain something else from it. A child should not be too much obsessed with possessing things. Finally, precisely as Christians, children should learn to forgive because they had no right to revenge or punish for an evil deed. Revenge was destructive while forgiveness was constructive. The love children came to know, whether of their parents, of one parent or of God, had a key function as had practical examples; in that way adults could continue their own education and improve themselves. But finally educators should also not be too perfectionist – after all, God was also there. Children should be seen as sons and daughters of God, just as, in reverse, children should look up to adults as if they were God.

7.2.2.4 From the Muslim side: many reproaches for this

The Muslim response began with speculations about the influence of Islam on education and the school system in Europe, but these were not even consistent in themselves, and it then emphasised that, with Eph. 6:4, Christianity was the first religion that called on people to treat children lovingly, while Islam was the first to have recognised the fundamental rights of children. These included the right to be cared for and protected, to learn and be educated, and additional rights in the realm of inheritance law. But the spirit of these rights had not yet taken shape in many Islamic countries. The Quran saw childhood as a phase of human life that began even before conception. Children and money, according to Sura 18:45f, were the joys of life on a lower level, but the lasting joys were good deeds. It was only after these preliminaries that Husni 'Ayesh discussed the address itself which he accused of being too much influenced by modern thinking and modern pedagogical conceptions, so that it had neglected the fact that, for centuries, Muslims and Christians had been reproaching one another for the way in which the others educated their children. He also reproached the speaker for not giving any biblical, especially New Testament, evidence for her views, although he did not commit the error of thinking that one quotation would solve the problem so that it would not be necessary to look at the practice at all. And he accused the speaker of not paying attention to the temporal limits of the theme and of having quite obviously discussed childhood right up to puberty by referring to questions of concentration which were very minor in the case of small children. But the roots of prejudice lay in the earliest years and she had not dealt with this problem at all. Moreover, she seemed to support the UN Convention on children's rights – but did this also apply to children's rights to religious liberty as stated there and how did a child benefit from that? Nor had she said anything about who could defend children from tyrannical treatment by their parents or what she thought about children being beaten (because schools were perhaps the only institutions which still did so although he did not say where). And she had said nothing about the relationship between the sexes (here he even mentioned texts of the Bible), nor about handicapped or retarded children or the differences between children, nor anything about co-education (which could not really be a problem at the pre-school age). Finally, he touched on the problem of adoption versus guardianship and what he saw as the problematic Christian emphasis on love. "Christianity is the religion of love. The emphasis in Islam is on mercy and compassion which protect love from turning into its antithesis in the upbringing of children, e.g. sadism. Islam is the religion of mercy."[87]

At least to Christian ears, this is a most exaggerated hypothesis and an almost malicious claim, whereas his other remarks on the address, with a lot of fundamental consent and agreement, were certainly pertinent. It is also interesting that both ad-

dresses mentioned that parents could or should improve themselves by their automatic function as examples.

7.2.3 Rights and education of the school child
7.2.3.1 The educational situation in the Arab world is a catastrophe

The last round of the discussion, finally, dealt with the rights and education of the school child. The Christian address on the subject came from Egypt and began by underlining that it was the religions, above all, and again Islam in particular, which had promoted the UN Convention on the rights of children and hence logically considered them desirable (with the exception of the statements about adoption). As far as school education was concerned, he started by limiting its real significance with two points. Firstly, by saying that education did not begin at school but in the parental home and that many different factors in the child's context contributed to and often conflicted in its education. Secondly, only the more fortunate half of the children in this world went to school at all, not to mention handicapped children. He then examined the Convention on the Rights of the Child in detail, concentrating on interesting individual points, beginning with the fact that the Convention spoke about all the needs of children and turned them into rights except for the need for faith and religious education. The terms "religious" and "spiritual" had deliberately been avoided in view of the international community of states (and the atheist members among them). Instead, it spoke only of neutral values which needed to be transmitted to the child. The child (all in line with Article 29 of the Convention) had a right to a homogeneous and comprehensive mental, physical and cultural development in which its talents and abilities were to be promoted as far as possible (with special protection for children of minorities according to Article 30), and this in turn naturally implied duties for others. It could also be interpreted as a call to share in bringing up a human being created by God. However, according to this article, one would need to discuss human rights with a primary school child from the very beginning, something which the speaker considered a practical impossibility on the basis of his pedagogical experience. A theoretical discussion on human rights was not possible at that point but they could be practised in the school, for example, by not using beating as a punishment out of respect for the child, but it would also be important to ensure that the parents cooperated in this. Precisely this uniformity between government, family, school and the house of God, and how it could be achieved, was the concern of the speaker and a question for him, because reflections and declarations alone, which never came down to the local level for which they were intended and where they were necessary, were really of no use. Another question was how one could bring children not to be concerned only about themselves but also about people in need (and, for example, to share?) and finally, the most important point, how one could educate children for peace and love of their neighbours and for the implementation of human rights particularly in areas which were marked by violence and religious extremism.

Amin Fahim then dealt in detail with the special efforts in Egypt which were reflected particularly in an extension of the duration of compulsory schooling and a gradual abolition of fees for educational institutions (since 1962 universities had also been free of charge). Theoretically everything was marvellous but in practice reality was far removed from this – and had been for decades. The building of schools and their equipment were still not keeping up with the demand as a result of economic difficulties and the increase in the population. At least half of the Egyptians over ten years of age were illiterate and, as parents, they automatically set less store by school education for their children, especially when their children's help was needed elsewhere. In fact, parents in general saw the purpose of schooling less in the education itself and more in good examination results which finally made well paid jobs possible. The curricula were good but the teaching methods (mere repetition) hopelessly out of date especially in the field of religious education where the formation of the conscience and links with everyday life were so necessary. And the holidays were so long that in the end there were only 120 school days left and hence too little time for learning content, and this again led to a system of additional, expensive private coaching. His own organisation was trying to counteract this but precisely this contribution made clear how steep the path is if one does not only present the theory but has the courage to look seriously at the discrepancy between what should be and what is, as this address did in particular. So the Muslim commentator also came to the conclusion that a child had the right to learn and to a suitable education but that the school education of Egyptian children did not correspond to what the UN Convention understood by an education appropriate to the personality, freedom, abilities and future tasks of the child.

Prof. Omar El-Sheikh went even further to say that this situation was reflected in the whole Arab world where one could assume 50 percent functional illiteracy, 80 percent cultural illiteracy and 99 percent technological illiteracy among the population. Girls were at a disadvantage compared with boys, the rural population compared with town dwellers and the poor compared with the rich. His question was to what extent society could help the state to shoulder this task, also financially. A further question was what exactly was meant by appropriate education for the children according to the UN Convention, which school(s) corresponded to it and what part religious education played in it. And the question of the relation between school education and education outside of schools was important to him and whether the latter could be taken seriously at all.[88]

Thus, both of these contributions contained (self-) critical questions. A child has the right to an appropriate education but how can and should it look like in practice? How can it be implemented under the various types of difficult conditions in the Arab world?

7.2.3.2 On the Muslim side: the problem of an education that promotes identity

The Muslim address on the theme given by Hani Abdul Rahman was not so directly related to practice but provided any number of reflections on the theoretical basis. The first point of departure was that being good was an inherent human characteristic which implied concretely that the human being was very well able to learn from different experiences. Education was a human affair, had human aims which were good for people and had to be carried out by a human institution. In education, the (human) educator related to another receptive person in order to produce a good human citizen possessing information and values as well as habits and skills. Especially with regard to values and trends, a selective process in the educational system was necessary, corresponding to the reference framework of a particular society and its specific modern situation. In a democratic (or similar) society with a clear identity and stability that was no problem, but it could become very difficult and even dangerous in societies that were undergoing rapid social change, had no clear cultural vision and where individuals had plural loyalties. That led to confusion in education which then increased the confusion in society and among individuals, especially when there was an identity crisis in the society in any case and democratic processes were weak. That then caused the division of society into supporters and opponents of change. As far as this process was concerned, what differed was less the problems than the way in which different societies approached these problems, namely the categories of thinking in the society, among the people, which determined the very perception of the problem. Every society dealt with problems and new developments within the framework of its identity and philosophy. Logically, the speaker was concerned mainly about harmony and equilibrium between all the components also in the education of children. Since the Arab world was changing, what existed in practice was the exact opposite. However important money might be for organising an education system, the philosophy behind it was even more important and its consistent implementation and agreement with the specifics of the society in question (which also made the acceptance and implementation easier because usually these values were invisible, meaning not reflected on, and so people clung to them all the more). One's own reflection on these constitutive elements in relation to other cultures and to the components and possibilities of the human mind and the general human framework was therefore indispensable, particularly in view of the technological revolution with its human consequences as presently experienced by the Arab world. This kind of philosophical or wise reflection made a person more human e.g. by removing prejudices or introducing reason into situations where previously impulsiveness and emotions had predominated. Naturally, there was also an immense difference between genuine (thought out) convictions and mere conventions. Often there were caesuras and words and actions did not agree which made it impossible for education to succeed. Nor was it possible to make an eclectic selection of elements from different sources whose philosophy was contradictory. Of course, the speaker was relying very heavily on the long neglected Islamic charter which was supposed to bring about the practical implementation of values such as responsibility, devotion, honesty, trust, resilience and productivity and had been under discussion in Arab Islamic countries for a long time but never put into practice. Nevertheless, in the review that follows of the question of education in history, the perception of Islam is very positive while that of Christianity is extremely negative (probably mainly because of the prohibition of scientific thinking). "In Medieval times darkness engulfed educational concepts in Europe, whereas prosperity and development were characteristic features of the noble educational concepts which prevailed in the Islamic Arab world as a result of the spread of Islam and of adherence to divine guidance."[89]

Yet again: Islam and its legal system
With a big jump, the account then moved directly to the UN Declaration of Human Rights which had recognised a declaration on caring for children that had already been drawn up in Geneva in 1924. This was, in a sense, the basis for the UN Convention on the Rights of the Child that was based on the following principles listed by the speaker: the validity of these rights for all children without distinction and hence protection from racial or religious discrimination, as well as the welfare of children as a primary concern of legislation; concretely their protection from neglect, cruelty and manipulation (children should always be the first to be given protection and aid), as well as everything than was necessary for them to be able to develop mentally, physically, spiritually, ethically and socially in a natural and healthy way. In practice, this meant sufficient food, medical care, a roof over their heads and also entertainment. Moreover, every child had the right from birth to a name and a nationality. Handicapped

children should receive the care and education appropriate to their needs. In order to become a mature person, the child needed love and understanding, namely to grow up with its parents. The child also had a right to an education that was compulsory and free of charge, at least in the initial stages. Children should be brought up on the basis of understanding, tolerance, international friendship, peace and universal brotherhood and in the awareness that their gifts and energies should be used to serve their brothers and sisters.

Then the speaker discussed how individual Arab countries really put this into practice, and only after this introduction did he come to the question what view Islam had of children's education. Islamic legislation, he said, had emphasised the protection of motherhood and childhood in a way that was more comprehensive and went further than modern laws, by which he meant that Islamic laws were derived from the Creator himself who had linked the ethical norms with rewards in this and the next world. Islamic education was always concerned finally with relating the finite to the infinite world. In this connection, Ghazali was quoted again with his statement that a child is like an unpolished precious stone – it had innate inclinations but finally it was its parents who made the decisive and distinctive mark on it. Again, the ideal to work for was equilibrium in all directions. He again emphasised that care of children began in Islam even before marriage with the choice of the right partner. In addition, the life of the child was protected; murdering children was prohibited even in war and abortion was strictly regulated. Pregnant women and children before puberty were not to be punished (and nor were sleeping or mentally ill persons). Moreover, Islam was exemplary in its rejection of any kind of discrimination based on skin colour, race or faith when it was a matter of the resources for living. What was implied was a life in freedom and responsibility. "Man's right to lead a free life has always been protected in Islam. However, this granted freedom has been guided by Quranic controls."[90] So the child had a right to a stable family life and a careful education. The protection of the family again started before its real existence, namely with the necessary consent of the man and the woman to make it a success. For women, the main task was to bear children and care for them, while it was meritorious for a man to provide for mother and children, particularly girls and orphans. The Islamic state had to make the overall conditions and regulations available for this, including the two years for breast feeding – a provision which was often quoted in this connection. As far as learning itself was concerned, in Islam it was a duty for men and women and its non-observance was punishable. Also the Islamic state was the first to offer education for all free of charge, as could still be seen in the religious schools where students received financial aid for their maintenance. A lot of importance was attached to good manners and especially to good language (but also to history and poetry and to swimming, riding and archery for boys). Cruelty and humiliating speech should be avoided in education, however, also among boys, although strictness and beating were still provided for, for example, if ten year old children did not yet offer the compulsory prayers. On the other hand, after fulfilling their duties playing was also prescribed; in any case, Islam was a realistic and therefore perfect religion. Its practical demands – and this agreed with his initial reflections – aimed firstly and repeatedly to establish harmony between the environment, school and technological progress and hence theory and practice as well as the various stages in the development of the child, so they formed a uniform field of learning and would not bring about an identity crisis for the children but rather promote belonging and respect for and responsibility towards the state and society in the past, present and future. But then they also aimed to really give all children the same opportunities, even those growing up in the Sahara, and to avoid negative elements in the education of children which, for the speaker, also included mere repetition. In contrast, he set special value by good examples and everything that encouraged the child's courage and freedom, starting with motherly care and including the child's freedom of opinion, responsible information media right through to special aid for the poor and weak and the destruction of anything that promoted stagnation or a weakness of character. Children had to learn to use their intelligence sensitively to solve problems and against oppression and subjection so that they would not fall victim to preconceived opinions.

7.2.3.3 A practical Christian question back: what happens to the Christian Arab children?

The response to this address was given explicitly from an Arab Christian point of view and began by tackling the question of the Middle Ages or rather of what happened afterwards: why the Arab world stagnated and the West caught it up and overtook. Was that the consequence of a wrong children's education or of ignoring God on the Arab side? Moreover, it should not be forgotten that many Christian Arab scientists had contributed to the periods of Islamic Arab glory. What Antun Naber was able to confirm totally was Ghazali's statement that children were entrusted to their parents, and here it was important to him that it was not society which did that but God himself who – at least according to the Christian conviction – had instituted marriage with the purpose of procreation and the bringing up of children. Therefore it was also almost impossible to replace the family if it neglected its educational duty. Society and the state also had responsibility but more for the overall conditions and not as a substitute for the family and its right to educate. He naturally saw that the church had a very strong role and responsibility. In that, or in Christian ethics, he also saw the solution for the education of Christian Arab children who had been completely left out of the Islamic scheme of the speaker. Indeed, the major shortcoming of the address was that it had not touched at all on the question of Christian religious education for Christian Arab children. They definitely had a right to this as citizens with equal rights, but the way it was dealt with in different countries certainly varied and was often discriminatory, perhaps also as a result of a hidden religious fanaticism. He also reproached the speaker for not discussing the destructive effect of the mass media, particularly with regard to sexual libertinism and violence, whereas the Gospels had already warned that children should not be ruined. In contrast, the school was the ideal place for learning right behaviour, namely understanding, tolerance and friendship, peace and international brotherliness as well as service to one's fellow men. If the government paid enough attention to schools, the ideal society was more or less guaranteed, as well as progress and wellbeing for the Arab nation.[91]

However justified the earlier concerns may have been, this conclusion is surely rather too enthusiastic although, of course, the previous speaker had mentioned that the way to a better society had always been seen historically in children and their education. But it cannot be expected that the ideal society will be brought about tomorrow or the next day either by noble Christian or Islamic educational ideals and, at best, one will come closer to it and, in a worse case, at least a number of negative developments can be stopped. And obviously an ideal education can also not be had so easily and simply.

7.2.4 The best for children: peace, justice, development and health care

The following points proved to be especially important during the discussions following the addresses and were included in the final report. The life of the unborn child had to be respected (one way for doing this was ante-natal care) because life was a gift from God. Therefore, for all applications of progress in the field of genetics and ante-natal technology, a moral framework was required that was based on divine provisions and ethical principles. The primary educators of the child were the parents. For this reason, legislation should allow the mother the time and means to care for the development of her child in its first few years. For education by example and playing, both parents were important which is why the sanctity of marriage was so much stressed. Special attention needed to be paid to disadvantaged and handicapped children with cooperation between parents, state institutions and voluntary organisations. Finally, it was underlined that the parents had the primary right to decide on the education of their children and the state should respect this. All children had a right to religious education and also a right to an education that respected their attitudes of faith and their values. In the interest of harmony and tolerance, however, children should also learn to understand and respect the faith of others. Teachers should do so as well and be aware that no value-free education exists. Therefore the curricula should also promote a spirit of tolerance and cooperation. In general, education should have priority at the national and international level because there were still millions of children who could not attend school. Everything should be eradicated that might threaten the life and wellbeing of children as the weakest members of society. In this respect, peace, justice, development and health care were good and hence also the best way to safeguard the rights and education of children[92]. The third conference, it was decided, should be held in June 1992 in Rome on the woman in society according to Islam and Christianity.

7.3 Rome 1992: women in society

This *third dialogue conference* on *'Women in Society According to Islam and Christianity'* was held *from 24–26 June 1992* in Rome with 30 participants and naturally again in connection with a papal audience. The majority of the speakers were women and it dealt with the fundamental status of women, contemporary problems and challenges and also future prospects and future possibilities.

7.3.1 Opening addresses
7.3.1.1 For the Christians: show me how the women are doing!
In his introductory address, Cardinal Arinze made several references to the previous dialogue conference at which, when discussing the rights of pre-school children, the role of the mother had been specially emphasised because a child, as he put it, comes into the world not quite in its final form. Every person owed their life to a mother. As mothers, women were the givers and nourishers of life, not only physically but also spiritually, religiously and for human culture, because they brought up children and transmitted their identity and their first lessons about God to them. But naturally the role of women in society had to be much broader in scope including the concept of a kind of advanced spiritual motherhood such as that embodied by nuns but also by women teachers, nurses, social workers and the like. Both types of mother taught children respect for older people and people of the same age, interest in and service for the weaker and less fortunate in society, moderation, patience, generosity and faithful-

ness. Here Arinze referred to Isaiah 49:15 where God used the image of the mother to describe his faithful, passionate love for humankind; and indeed the Arabic word for merciful – one of the most frequent attributes of God in Islam – was derived from the Arabic term for the mother's womb. The love and protective care of a mother thus gave the initial impression of what God was like. Also when it was a practical issue of more respect, appreciation, understanding and cooperation between Christians and Muslims, it should not be forgotten how important the role of the mother was in this. If she taught a child that Christians and Muslims were brothers and sisters who for God's sake were to love one another and live together in peace, then the child would not forget that throughout its life. Only after this passionate plea did Arinze also speak about the woman being fully a valuable member of society and having full social rights. She was of equal value to a man and should therefore have the same opportunities in the field of education, choice of profession and other things. If one wanted to know how a society viewed human dignity in general, one only had to look at how the various realms of that society treated women. Examples of discrimination and injustice against women were degrading them to objects of men's desire through prostitution, pornography and advertising, but also in unequal pay for the same work, unjust working conditions, and discrimination related to promotion and special benefits. Even in families there were injustices towards women when preference was given to boys, and when mothers were used as maids and beasts of burden whose services were taken for granted and for which no thanks were due. Both Christianity and Islam considered justice an essential component of life and that human beings were endowed with dignity and inalienable values. They should reflect together on what they could do for women to have an appropriate, dignified role, and here he also brought men directly into the picture, interestingly enough. "Men are called to a conversion whereby aggressiveness is tempered with mercy, where power is mitigated by participation, where pride is restrained by humility, where economic and social burdens can be lifted in a joyful, easy familial environment. It could be a worthwhile subject for a future colloquium for us to examine together the proper role of man in society according to Christianity and Islam!"[93] Unfortunately, that meeting never became a reality.

7.3.1.2 For the Muslims: opportunity to correct misunderstandings
The opening address by the president of the Royal Academy for Islamic Civilisation Research, on the other hand, despite its promising title *'Woman the Mainstay of Human Life'*, had much less personal content and was predominantly of an official nature, even including the mention that immediately after the con-

ference the Joint Committee would be meeting to fix the date and theme for the next meeting. It was only when he came to speak about attacks on Islam connected with the subject of the position of women that he became much more personal and involved. He saw the meeting mainly as an opportunity to correct false conceptions about the wretched position of women in Islam which were in fact far removed from reality. (A controversy had developed between Christian and Muslim scholars on the article in a UN agreement on the abolition of all kinds of discrimination against women and this was obviously still clearly in the background here.) Women were positively the mainstay of human life, its continuation and development. Quantitatively they comprised half of human society. But, above all, the status of women generally and in religion in particular was subject to many distortions and misunderstandings in the mass media and various other circles. "For those who intend to distort the image of religion, and particularly of Islam, find in what they think to be the status of women in Islam an easy means to exaggerate wrong ideas about this religion. In fact what they believe to be the status of women in Islam is derived either from indiviual cases that resulted either from misapplication or misunderstanding, or from the writing of those who are prejudiced against Islam especially the people who are ignorant of the reality of this religion."[94] So it was quite clear what the expectations were on that side.

7.3.2 The status of women
7.3.2.1 Women in early Islam – stronger than the average woman of today?

The first address on the status of women in Islam also clearly endeavoured to meet these requirements, as was evident from its structure. The first part dealt with the theoretical status of women in Islam, namely with the legal provisions concerning women, while the second part examined their practical implementation in Muslim society, significantly enough under the heading "Distortion of the Image of Woman's Status in Islam".[95] It suggested that the bad image of Islam at this point was perhaps not the fault of spiteful outsiders, as the opening address had suggested, but rather that of the Muslims themselves who had consistently followed another practice which caused the theory to be forgotten. But we shall look firstly at this theory as presented by Prof. Dr. Abdul Aziz Khayyat, the vice-president of the Royal Academy for Islamic Civilization Research. Islam, he stated, saw women as human beings like men without distinction, apart from what nature prescribed for the human structure, as he called it. That included the different human functions of the two sexes according to which women are prepared to become pregnant, bear children, breast feed them, bring them up and care for them. To what extent the latter two functions can be described as purely biological is really a justified question but there was no attempt to even begin to ask it; on the contrary, the speaker moved on immediately from this "basis" to a key statement. "Islam's rulings came to decree equality between males and females with regard to rights and duties except where dictated otherwise by their physical nature and their respective responsibilities."[96] He added passages from the Quran here which related to creation and to reward in the world to come. As far as religious, moral and social duties were concerned, men and women were equal and the same applied to education, work, transactions, marriage and the home. In addition, Islam had honoured women from birth and, according to his interpretation of Sura 16:58f, it condemned those who condemned the birth of a daughter. A hadith promised paradise to anyone who neither had a daughter buried alive nor hurt her nor gave preference to his/her sons. As far as wealth was concerned, men were also not given preference; women received their share of the parental inheritance as well. According to the Quran and Sunna, nothing excluded a woman from being a judge and certain Muslim legal experts had even explicitly admitted this, like al-Tabari who only excluded murder cases and the imposition of the *hudûd* punishments. Abu Hanifah also saw things in a similar way; women were so tender and compassionate that they could not pronounce judgements such as execution or whipping. Examples were given from the early history of Islam and similarly in the realm of professional life going as far as pursuing a trade, engaging in commerce, teaching men and women, preaching, being active in politics and participating in a Holy War. There had even been prominent female scholars in many disciplines. Islam had also granted women the right to determine their husbands themselves (although the legal guardian still had to give his assent) and not to be married against their will. Moreover, Islam had honoured women as wives by making them responsible for the household and this then gave the husband other duties, mainly related to the bride price and maintenance. From the Islamic point of view, marriage was not a means of enslaving a wife; on the contrary, she was assured of good treatment even including the protection of her faith if she were not a Muslim but belonged to one of the religions of the Book. Islam granted special honour to mothers, even more than to fathers.

Islam's new honours for women
The speaker then listed a series of nine points which expressed Islam's reverence for women. Firstly, Islam had liberated them from the physical curse with which they were stigmatised in other religions. That meant that, in Islam's understanding, it was not Eve who led Adam to eat of the fruit of the forbidden tree but the devil himself. According to the text of the Quran, Eve was therefore not the temptress but even, on the contrary, too worthy to begin the disobedience. Sec-

ondly, Islam had taken into account that, because of their female nature, women were responsible for the home and children and therefore did not have the duty to support the children financially; that was the duty of the husband who had to support his wife according to his standing. She was the lady of the house and had no duty to do housework unless she did so voluntarily to assist her husband. Thirdly, Islam had granted women the right to be witnesses and not only in matters which exclusively concerned women. The argumentation on the details is enlightening.

> "This right was jointly given to two women so that each of them may remind the other concerning the event or the case they are giving evidence about especially in crimes like murder and criminal attack, because women are usually more sentimental and easily moved than men and therefore they must give joint evidence. Men, on the other hand, are in closer contact with societal events and consequently a man was accorded the right to give evidence by himself (...). (...) / p. 25 One-woman evidence or testimony is accepted in matters related to women like testimony on birth, descent, suckling etc. It is an honor accorded to the woman that she is not exposed to what the man is made to confront in transactions, sales, debts and other dealings leading to law courts and appearing before the judge and being cross-examined by plaintiffs, attorneys and others. Thus woman's testimony is required only in cases of urgency".[97]

The general line had already become clear earlier when it was a question of which cases could potentially be decided by a woman judge, but this went still further in keeping a woman out of society even though, as had been stated above, she was really entitled to participate in it fully. In a sense, it is a special honour which Islam gives women – if one adopts this line of argument – if, as far as possible, they do not exercise the social rights which Islam in fact grants them in order to protect their own sensitivity.

But the speaker was evidently not aware of the contradiction involved here because he went straight on to the fourth special honour which Islam grants women, namely the right of inheritance whether as a widow, daughter, mother, sister or grandmother depending on the precise juridical situation, where daughters only received half of the inheritance of sons because they did not have the duty to maintain the family. The fifth honour of Islam for women was the possibility of polygamy if the first wife was sterile or sick or there were simply too many unmarried women who had no one to support and protect them. This saved women from selling their honour whether as concubines or prostitutes such as could be observed in many European and Eastern countries. Sixthly, Islam had given women independent financial security so they were totally free to take charge of their possessions without needing the permission of a husband, father, brother or son; and here the speaker underlined that this right had only recently been granted to women in the West. The seventh type of honour which Islam granted women was the possibility of divorce, although the speaker began by discussing the divorce right of men in detail and only then moved on to mention that the woman had the honour of being able to have her right to divorce guaranteed in the marriage contract or the right to ask a judge to divorce her from her husband because of a lack of understanding, beating, abuse or compulsion to do something prohibited or to refrain from something compulsory. She had also been given the right to ask her husband to grant her a divorce on payment of compensation in order to escape from a bad married life, although divorce was really prohibited and advantage should only be taken of this possibility if it was necessary. The penultimate aspect of special honour was the veil:

> "permission of adornment, meeting of men for a legitimate deed and prevention of secluded meetings that lead to forbidden immorality. (...) The veil was imposed on women by Islam which allowed them to expose their faces, hand palms and feet-nothing *(sic!)* else. (...) / p. 28 (...) This is enjoined to prevent immorality resulting from woman's nudity and to enhance her modesty and chastity."[98]

This argument also contains a certain contradiction because it is clearly less a matter of honour shown to women, in fact, than of something imposed upon them because otherwise they would not behave morally enough. But the speaker also failed to recognise this contradiction in his own argument and he immediately came to the ninth and last point, the honour of the right to select (or reject) their husbands themselves and in general to be allowed to marry at all, although naturally the consent of the guardian was important in order to safeguard the bonds between the families.

But there are also home-made problems...
The positive statements thus far have already shown that this type of argumentation was not always so easy for an outsider to accept, but the four pages that followed, containing the details of all the deviations from it that were not allowed, demonstrated that the marginalising tendencies concerning women had gone much further.

The speaker described them as the consequences of foreign, especially Persian and Byzantine influences which had come to predominate above all during the periods of Muslim decadence and isolation, namely under the Mamelukes and Ottomans. Then there had been a period when the preachers were concerned to purify Muslim society from corruption. And this has led to a falsification of the traditions related to women and the family – again a logic which is not completely convincing for an outsider, at least. The next point was easier to understand on the consequences of the encounter with the West, especially with the freedom which Western women enjoyed and with their anti-Islamic attitude, although this very classification was tendentious because it implied an intention which was probably not present in that form. This, in turn, had given rise to a quite different reaction, imitation "in the same way / p. 29 as the vanquished imitates the victor and as the weak follows the steps of the strong"[99] with the aim of breaking down the barriers between men and women. They were sitting, according to the image of the speaker, between two stools, outdated tradition on the one side and alien Western conventions on the other. The list of distorted traditions that follows is impressive but it is striking that, precisely in connection with the statement that all women were an evil and that their worst evil was that they were indispensable, no information is given about why they are not reliable. Certain false practices which had found their way in were also criticised including, for example, that women were prohibited from participating in prayers at the mosque.

Would Muslim women also have been so apologetic?
Overall, this was an address with an enormous amount of detail relating to individual regulations and problems of which there are more than enough on this issue but which are connected anthropologically to the one factor mentioned right at the beginning, that in Islam men and women are equal in origin and purpose but that, for the rest, the difference between the sexes, namely the woman as mother, is very much emphasised with its very broadly defined effects. In that role, women were very much honoured and should be much appreciated but the consequence is that this protection tends to undermine their rights to participate in the life of society because that would demand too much of feminine sensitivity. In addition, even though it is sometimes presented as an error, the tendency exists to see women as sexually endangered/dangerous, although they are not considered by the Quran to be responsible for the sin of Adam, and hence they should better be kept away from men however great the necessary distance might be. One is also struck by the efforts to see the regulations on divorce law and polygamy as positive, respectful, indeed, as an honour for women. Here it would have been interesting whether a Muslim woman would have presented this point in the same way or have been more critical – without it being necessary therefore to accuse her immediately of unbridled imitation of the anti-Islamic West. Anyway, it is striking that the two other contributions which were given by Muslim women laid it on much less thickly. Since there were no official responses at this meeting and there is no report on the discussion, the question whether and what contradiction there may have been must unfortunately remain unanswered. But perhaps it would have been good for the subject generally if the women had been left to discuss it

among themselves, even at the risk that, in such a complete freedom from approaches that are automatically influenced by men, some unconventional views might have been expressed, and that on both sides.

7.3.2.2 Also on the Christian side: women of equal value but not equal

At least the Christian counterpart address was given by a woman, Sister Stefania Cantore. She approached the question of women's identity and thus of her own identity on the basis of the biblical witness and adopted a historical approach which started from creation. Human beings had been created in the image of God and as man and woman. Therefore both men and women, whether as individuals or together, represented human beings as God intended them to be. Before and beyond all their differences of gender, race, etc., they were marked by their likeness to God which meant that they could enter into a familiar relationship with God and act according to his example, and this could safeguard the continuity, beauty and perfection of his creation, whether through the existence of other people or through care for nature. To see oneself as God's image meant recognising, praising and loving God as one's creator. This mysterious dignity of likeness to God was something which men and women had in common. They had been created as relational beings for one another, as was clearly shown in Genesis 2. So it was evident as a summary of creation that men and women were created by God with the same dignity in perfection and harmony. But this did not correspond to the present situation from which people, both men and women, were suffering. This could be traced back to their doubting the truth of God's word, breaking the creatures' relationship of trust with the creator, indeed, no longer accepting their status as creatures. Hence men and women now had to fear God, were no longer in harmony with themselves, with others or with nature, could no longer completely open themselves and face one another as equals, and all the other consequences described in Gen. 3. "In the lack of harmony, from which all reality suffers, in which the original man-woman mutual understanding is broken and in which relationships of struggle and dominion infect all humanity, woman often came to be subservient to man. Even if man apparently has the upper hand, imposing himself on woman as her master, in his depths he remains 'alone,' *(sic!)* just as the woman herself also remains alone. Their bond is henceforth falsified and unsatisfying".[100] The only remedy suggested for this was love, which did not see the woman in a subservient, dependent position in relation to the man, nor her possibility or impossibility to become a mother, but which respected and appreciated her for her real value. Examples of such relationships and also of cooperation between men and women in mutual respect did exist but they were exceptions and tokens, as it were. The Wisdom writings also reflected this problem that men and women were apparently incapable of regaining their original relationship of mutuality, even though they always yearned for it with some kind of nostalgia.

But Jesus had restored the dignity of human beings – of every human being – that had been overshadowed by sin and injustices, and he had done so for men and women equally contrary to the traditions of his time. Addressing a woman as "daughter of Abraham", otherwise only used for men and in the plural, was a sign of this, as was the general rejection of divorce as a return to what had really been intended at creation, and also his behaviour over questions of ritual purity (which constituted a special problem for women), and the praise of Mary who, unlike her sister Martha, did not take the place of the housewife but of a disciple, and other examples right through to the fact that the first apostles of the resurrection were women – all of these were signs of this new attitude in which men and women were children of God in God's mind and their standing in society was no longer of importance. Then her last point was Mary, the mother of Jesus, in a sense as the archetype of women, "*the woman*, the original beauty of the feminine person created in the image of God"[101], "the woman attentive to God and to humans"[102] over and beyond the biblical witness. Her conclusion for the new, original role of women was the following.

> "The woman 'seen' by Christ can live fully her likeness with God. No longer subservient, the woman is free to serve, following the Lord Jesus who did not come to be served, but to serve. Woman, after the experience of subordination, is particularly called to 'serve' society, seeking to prevent the resurgence of every kind of relationships of power and dominion among men. Once her equal dignity with man is no longer seen in terms of equality with him (with the consequent impoverishment of the human communi- /p.47 ty), woman can find a point of reference for her proper feminine identity in Mary and in the female disciples of Jesus. In the light of the Gospel, gestures dictated by love, bearing witness that Jesus is alive – all this appears to be some of the most important characteristics of woman. The precise modalities of putting these into practice are, however, neither defined nor definable."[103]

Despite all the differences of detail from the Muslim address, there is surprising agreement. Here, too, the same (original and re-established) dignity of the woman was emphasised, but then it was also underlined that this precisely did not mean equality; on the contrary, and in the breadth of the human community which had to be defended, the woman's role was more that of serving, as was natural and corresponded to her nature, but here the serving element was not introduced by legal regulations but by a newly acquired voluntariness. There are no precise legal provisions but their place is taken by the excessive example of Mary; however the tendency, the basic structure, is very similar, although on the Christian side it is formulated in a much more refined, discrete way – perhaps because formulated by a Christian and perhaps also because formulated by a woman.

7.3.3 Concrete problems of women
7.3.3.1 On the Christian side: from the problems to anthropology

The following two addresses dealt more with the concrete problems for and challenges to women or, as the Christian speaker Eleonora Barbieri Masini put it, with the effects of society on women but also with the reverse effects of women on society, even if these were perhaps not so immediately obvious. Here, she made a distinction between industrialised and developing countries. In the industrialised countries, industry and the state had taken on many functions of the family in the economic and educational realms whereas the role of the family in forming the personality had become more important. Hence, families in industrialised countries were more vulnerable and at the same time confronted with higher expectations than in the past. In contrast, in developing countries there had been hardly any changes in the family context and especially women in these countries still saw their main task in the upbringing of their children. The spread of cell families and extended families, which were sometimes more accurately known as "protective families" (whose function the state often had to adopt otherwise), varied from region to region; the latter form was widespread particularly in Africa and it was not least the women who kept them going. There had been a general rise in the marriage age, in the industrialised countries even to above thirty, which naturally also had effects on family structures. For these and other reasons (greater number of separations and divorces, extra-marital births, migration and/or desertion) in the meantime the head of the family was much more often a woman. The main challenges were the higher average life expectancy, which meant that middle-aged women had to care both for the children and for elderly people, and massive migration movements, evident in the developing countries in a major rural exodus and urbanisation without previous industrialisation (which was the great difference from the industrialised countries). But the large and very large towns (with more than one or more than eight million inhabitants) were a very difficult context for families and especially for women, beginning with the fact that such families had to buy everything. Irrespective of all feminist debates, those were the greatest challenges especially for women in the developing countries according to their own estimate. Participation in the world of labour also offered nothing of the supposedly fascinating experience that had been propagated in the industrialised countries in the sixties and seventies. In the developing countries, women were employed for little money mainly in areas for which they were considered particularly suited because of their nimble fingers, namely in the electronic industry or in tea, coffee and flower plantations. Apart from this, women were more often found in offices and administration, in service or retailing, agriculture and animal husbandry. The transition between housework, informal work (agriculture for personal consumption or for marketing, assistance with book keeping, household services) and formal work had generally become more fluid. Overall, women worked more than men, often without being aware of this, because they were simply used always to be working, and most of the formal, professional work was available at the very age when they could bear children which constituted an additional challenge. As far as the influence of the media was concerned, it was especially strong among Eastern European women with their institutionalised and severe double burden of factory and household, because the media had given the impression that women in the West were rich without needing to work.

Special attention had to be given to ensuring that the developments and challenges of the information society did not bypass women who traditionally had been more involved in practical, everyday life, and that decisions relating to the family, work or participation in the life of society were not taken without them. Whether or not they were caring for children, women had generally undertaken the role of supporting the weaker persons in society, had committed themselves to environmental concerns and been successfully involved in many different self-help groups both in industrialised and developing countries. With regard to the challenges related to migration and integration, it was the women who could create or destroy tolerant communities. Much depended on the humanity and sensitivity of women and their way of dealing with problems as far as education for mutual respect was concerned. Women were only gradually becoming aware of their importance for the present and future of society. Their abilities had always been there and been exercised but they and their bearers had so far been invisible. In addition, their values belonged predominantly to a pre-industrialised society less designed to amass material possessions; this was often a disadvantage for them in an industrialised society but could be very important in view of the new challenges: the ability to deal with complex, constantly changing situations, adaptability, flexibility, self-sacrifice, the ability to do several things at the same time, the exercise of personal and public power predominantly for others, a conception of time which was less fragmented and more related to the natural continuum and gave women a kind of seismograph for the smallest changes and warning signs, a sensitivity for the marginalised (especially among Christian women) and a holistic understanding of development.[104]

Particularly this last section moved on from describing the situation to fundamental issues and did not only provide a summary of the observations and studies done thus far but also, without necessarily saying so, an anthropological description of women. It is surprising that it included almost no reference to religion although the empirical findings seem to confirm much of what had previously been presented as religious, Christian anthropology, particularly in relation to the life of the next generation and to a mainly serving function in general. But this was not discussed because the concentration on practical problems and their potential solutions was given absolute priority.

7.3.3.2 Muslim women in the Arab world – more than a problem

As far as the Muslim address by Prof. Omaymah Dahhan of the University of Jordan was concerned, it was interesting that, in addition to the explosion of science and technology which had enabled women to have other functions than the traditional ones in the home, and the concern for an equal share in the benefits of economic growth, she mentioned the United Nations Universal Declaration of Human Rights of 1948 and the declaration of the United Nations Decade for Women (1976–1985) on the effective and rapid inclusion of women in development. She was concerned about the situation of women in the Arab world because there, despite economic and political differences, the historical, cultural, social and religious conditions were similar throughout the region. There were problems in a total of five areas beginning with the education system in which a large number of girls did not attend school, and this in turn resulted in a large number of female illiterates especially in the poorer rural areas, surprisingly enough even in so rich a country as Saudi Arabia, whereas Jordan and Lebanon had the lowest rate of illiteracy. At the other end of the scale, among those graduating from universities there was a large concentration of women in the subjects which were traditionally assigned to them, namely education, medicine and especially gynaecology, the social sciences and art. In the technical field, however, there was scarcely any work for women in the Arab world and in the trades only in such areas as fashion and hairdressing or secretarial work. With regard to formal work, the employment rate for Muslim women in the Arab world was the lowest in the whole world. This was due, on the one hand, to a large number of children with five to seven children per woman and the burden of work in the home related to this, so women often worked outside the home only when they were very young and, on the other hand, to the poor education or an education that concentrated only on "female" areas (in which the unemployment rate was also higher). That was typical of many societies but especially pronounced in the Arab countries; the same applied to the influence of tradition (the husband was responsible for supporting the family, so a woman did not need to work and at most it would be an exception; a woman should not have to work side by side with men and, indeed, co-education was also not possible and the separation between men and women in social life in general had an enormous influence on the whole life of the individual) and to the financial status of the family. So the number of employed women in the traditional, rich Gulf States was very low although work was available and foreign workers were even being brought into the country. Other factors were labour migration (which resulted in households where women were the head and in a kind of feminisation of the family) and the rate of male unemployment which, because of the traditional conception that men were the bread winners, quickly led to men being given preference without taking into account that there were also women who needed to support their families. Even when women went to work, it would often be observed that they were overqualified for the work they were doing. Here, their more limited mobility naturally played a part, indeed, in many societies women had no independent mobility at all. And that in turn affected class mobility; ultimately, social advancement for women was not possible by education or their own achievements but only through marriage. Moreover, the improvement of the social and economic status of women in certain urban families only resulted in women being shut in as status symbols, similar to the way in other societies in which men saw the non-employment of their wives as a demonstration of their own status. These were all examples of the widespread practice of using women as symbols of the status of a family, often with negative consequences for their participation in social and economic life. But even generally women were at a disadvantage for payment, promotion and further education in both the state and the private sector.

Legislation in this connection existed only on paper; the prevailing social system set the tone in which the leadership role in the power structures belonged to men and in which this kind of discrimination against women was current practice. Women who were employed in the informal sector in households and agriculture were usually not taken into account at all. With regard to salaries, again there was the idea that women did not need to earn money so urgently; for them it was merely extra income added to the income of their husbands so they could also work for less money, apart from the fact that they could not work regularly because of bringing up the children. Therefore the salary scale was generally lower in areas in which only women were likely to work. All in all, society but also the mass media, films, literature and school books reinforced the traditional image of the wife and mother who did not go out to work, even when it was emphasised on the Islamic side that women and men had the same rights and duties and therefore there was nothing against a woman exercising a profession. However, quite a number of Arab countries have only very partially ratified the international conventions concerning women's work because they partly contradict Islamic law – a very interesting point with which the speaker regrettably did not deal in greater detail, not even when she demanded in conclusion that these conventions should be ratified in so far as they did not contradict Islamic law. Politically, women were under-represented in practice. In a number of countries, the laws did not allow them to vote or be elected, sometimes their husbands prevented them from doing so and the prevailing value system in general treated them more as subjects than as citizens. The address closed with a series of recommendations on what should be done, even including the right of women to vote, but also aids which could help to combine professional and family life, namely all-day care for children and children's day centres as well as comprehensive encouragement of part-time work. It was a long list which was no surprise when one looked as the preceding description of the current situation.[105]

It is interesting that this address from the Muslim side saw the traditional distribution of roles also with regard to work in a much more critical way. It was less a matter of contributing supposedly typical feminine values to society and of strengthening the role of families again, and more a question of practically strengthening the role of women in a male society. The reason for this may be that it was not looking at problems in a worldwide setting, as the Christian address had done, but only at a very small and traditional section of the Arab world. This was a clear example of something that had already been indicated in the first Muslim address, namely that the theoretical Islamic equality of rights between men and women was eroded in practice by the distribution of roles according to nature and by the separation of the sexes, and turned into the opposite in a way that was unique in the world, and that there was still a long way to go before that could be overcome. It would be interesting to know where the borders really lie that are drawn not by social tradition but by Islamic law, but these were unfortunately only mentioned in general and no details were given at any point.

7.3.4 How could things continue?
7.3.4.1 What would a Muslim model for the future look like?
The next Muslim speaker, Dr. Samira Khawaldeh, began her view of the future with a description of present conditions in the Islamic world, in a sense, with a summary of the issues dealt with in the preceding address, and this summary is as brief as it is shattering. "The woman is still in the stage of awakening. And she is still a branch on the tree of man deprived of personal independence which

could enable her to make her own decisions and to interact in society on her own. Many phenomena in our society are deceptive. They indicate that woman is financially and personally independent, that she is beginning to play a growing political role as a member of government, of the parliament and of some political parties. They also indicate that girls surpass boys in number at the university level. But, as it has been mentioned, these phenomena will be found deceptive and contradictory after scrutiny."[106] The women's movement in the Muslim world had assumed only a superficial form as rebellion against Islamic dress and replacing it with Western dress. Otherwise, the interests of women were still focussed on such superficial things as cookery recipes, flat decorations and the latest fashion, at least if one were to believe the women's magazines. Intellectual independence had hardly been achieved yet; it was very rare to find a woman who held an opinion different from that of her husband or father, and that was surely less a consequence of intellectual harmony in a successful marriage that simply of dependency. The economic independence which seemed apparent in the increasing professional activity of women was in reality only an extension of the serving function of women in the home. They now had double duties in the home and at work whereas the men were still only working for an income. Moreover, the money which women earned did not belong to them. Sometimes the father made an agreement with the future husband that they would divide the income between them and the woman then only retained some pocket money. And in politics there was also no really active role of women. In Jordan, for example, not a single woman had been elected to parliament and the ministries which were entrusted to women (social development and information) could only be considered peripheral. As far as the relatively high number of women university graduates was concerned, that could also be attributed to the fact that women simply did not have much choice and therefore made studying their hobby, whereas young men, who were also allowed to go out, frequently preferred a sweet life to the university. All of this could only be described as unfair on women. In her opinion, it was wrong that the women's movement had so much emphasised the particularities of women and thus overlooked their essential nature as thinking beings endowed with reason. The women should have become involved alongside the men in dealing with national problems instead of fighting over questions of dress. She also apparently considered the struggle for the absolute equality of the sexes to be similarly senseless, but without saying how exactly she understood the relationship between the sexes and the nature of women and what the consequences of that should be in her view. "There is also the radical demand for absolute equality between the / p. 99 sexes, and opening up every opportunity for woman regardless of her psychological and physiological differences. Some even claim that those differences are no more than the outcome of ages of sex-discrimination which ultimately drove her to the back rows when it comes to serious business and responsibility."[107] In her opinion, the struggle against polygamy, which she mentioned in the same breath as the struggle against the right to divorce, was superfluous, a waste of effort, for the simple reason that it was against Islamic law and hence against religion. However, on the other side, the cause given for the regrettable situation of women was also the discrepancy between Islam and Muslims, especially the Muslim reformers who rejected any participation of women in public life as a forbidden mixing of the sexes and hence un-Islamic, but that did not reflect the real meaning of the prohibition.

This analysis therefore shows somewhat more clearly the points which had already come to light in the two preceding addresses. It was a question of precisely what Islamic law determined and what it did not, because that alone decided the exact position and the precise rights of women and not their wishes, however understandable (e.g. the abolition of polygamy), and at this point it was also not necessarily a matter of further developments but simply of who had the right or real interpretation. And in practice, at least, as became clear time and again, women have worse chances there.

That was also one aspect of the challenges which women faced in the present and future, and the speaker also included modernity among them although that was also a challenge for the men, indeed, for human beings and humanity in general. But the challenges also included the question of the true nature and identity of women. What made them different or special, and were these characteristics part of their essence or rather coincidental and a result of their environment? There were plenty of definitions and views of women and femininity. A final challenge was modern science and the possibilities which it offered and which, if one thought of the new technologies for conception, for example, completely revised the traditional understanding of motherhood, just as social changes had already done for the educational role of mothers. So this was a wide and not always promising area. As a possibility offered by Islam in view of these challenges, the speaker began by outlining a realm of unchangeable values where faith came first. Separation from and sin against the Creator were the real causes of the modern plight of humankind. What people today lacked was not material wellbeing but rather good intentions and peace with God, with themselves, with other people and with the universe. The Muslims had used the same experimental scientific methods as today and produced a marvellous civilisation without disfiguring the earth or polluting the environment. With its eternal values and unambiguous moral standards, Islam offered a better possibility for taking the right decisions. If she believed in the values of Islam, a woman could do herself, future generations and humanity the greatest service. As a person before God, a woman in Islam was absolutely equal to a man. In Islam she also had all the civil rights and could naturally commission someone to exercise her rights and occupy any post in the government apart from that of head of state, but recently there had even been discussion about that. However, the way for women into politics had been blocked for centuries by representatives of religion and the Islamists still restricted women to a very limited realm today, even though the first martyr in Islam was a woman, and this was possible only because she had not limited herself to the context of her home. Islam recognised the woman's independence and free will and made no distinction between man and woman in relation to financial transactions. But, on the other hand, it was a fundamental principle of Islam that men and women had different natures. That was a heavenly law. The woman would always be the one who bore children and gave birth to them and who also cared for them. Islam gave her the possibility to do so by providing for the care of or regular payments for each newborn child. At this point, the speaker stated concretely that many people were unable to reconcile the equality of man and woman with the different nature of the woman. She saw motherhood as the role and profession intended for the woman by her femininity, as the special responsibility which should have priority for her. In addition, women were physically weaker and their psychology was somewhat different from that of men; for example, they tended to be more forgetful than men. For both points she referred to passages of the Quran and Sunna, for forgetfulness naturally the one which gives the reason why two women are needed as witnesses compared with one man.

In addition to these unchangeable principles, however, there were others that related to particular times and were subject to interpretation. Here, it was especially important to avoid extremism when it was a matter of adopting precautionary measures, concretely not pushing women into oblivion away from the men and depriving them of any possibility of social activity in order to prevent sin and social problems. This was also the reason why many people made the veil a requirement and prevented Muslim girls from attending mixed universities. In part, this was naturally a reaction against the Western conception of completely free association between the sexes, and against the stereotype of the Western woman which the media had propagated. But the great confidence which God had placed in his servants with his mandate was much reduced by those who set themselves up as the protectors of religion – an impression which outsiders had also already gained. A further example which the speaker gave was that divorce was made unnecessarily difficult for women on the grounds that they had too little comprehension and religion and might demand a divorce for trivial reasons. Her argument was that Islam was divine and therefore not in danger, whereas Muslims were, namely in danger of becoming the curators of a historical museum. Overall, Islam had given women a central position in society which they could also regain in the future. Specifically, this meant being equal to men as human beings, having full rights as wives and mothers, being politically active, economically independent, leading their own lives and taking their own decisions, participating effectively in public life, being immune to contemporary corruption and libertinism and, above all, having mental stability and spiritual peace.[108]

What is anthropologically most interesting about this address and conception is certainly that here, for the first time, a concrete model was presented for mediating between the fundamental equality of man and woman and the simultaneous difference of their two natures by declaring motherhood the profession which nature had bestowed on women. Of course, at the same time, this raises questions as least for a Christian Western listener. Is motherhood then a kind of compulsion for a meaningful life for women (as had been mentioned as an idea in the Christian keynote address but as an example of the negative distortion of a concept) and what should then be said of women who perhaps against their will never become mothers? Do the protective provisions of Islamic law for a mother with children really exclude alternative ways for women to live as being unfeminine from the very beginning? Particularly in a dialogue with Catholic partners, this is a question of very great importance.

7.3.4.2 On the Christian side: where the divisions lie is clearer than the extent of the common views

The concluding address with a view to the future on the Christian side was given by Maurice Borrmans and followed the line of the concluding message of the Second Vatican Council addressed to women which stated, among other things, that the church was proud of having honoured and liberated women and of having underlined their fundamental equality with men. The time would come when this would be fully realised and women would enjoy influence in the world and a power that they had never had before. Perhaps under the influence of this approach and formulation, he looked at the situation of women in the world of today, including the Arab countries, in a quite different way from the previous speaker. School education was available today for all girls and all professions, not only housework and agriculture, were open to women; they had as much access to the press, radio, cinema and television as men. Women were perceived quite differently nowadays and had a quite different relation to political responsibility. Now they had a voice (and here he quoted an essay by Catherine Leroudier) which was also heard outside the home. Women had begun to demand the same rights as men in the political, social and civil spheres and this had revealed how male dominated these structures of society were and to what extent society had been shaped by men. In this way, the struggle had become a struggle of women against male domination. This had spread into other fields, for example to housework for which women no longer wanted to be responsible alone, and to religion where women dominated sociologically but men exercised power hierarchically. By demanding recognition for themselves and their abilities, women had raised the question of their own identity and of how they differed from men. In what followed, where it was a matter precisely of the question whether femininity existed or whether it had been created by men, it was striking that Borrmans really only used quotations – quoting women and their various views on the question, perhaps in order consciously to remain in the background as a man himself. The first whose words he quoted was Simone de Beauvoir who had thrown down the gauntlet with her statement that you were not born a woman but you were made into a woman – justifying this by saying that the place of the woman had always been determined by the fact that she lived in a male world. What appeared to be typically feminine was not the result of natural factors such as hormones and still less of the supposedly mysterious nature of women. That myth only helped to reinforce the existing situation. Men had always endeavoured to restrict women to a special feminine world, to the home, family and marriage. So the aim was to be truly human transcending any distinction between male and female. A woman could only really become free if she were able to keep her distance from the factors which tried to enclose her in her femininity and prevented her from being a human being, namely really free. Theories of this kind had also been supported by the research of such anthropologists as Margaret Mead.

But the author was in fact more inclined to other attitudes which, as he put it, appeared wiser and more suitable, namely that there really was an essential difference between men and women and an unchangeable female nature. Women were just as efficient as men, as he quoted from Suzanne Villeneuve, but they had their own way of living and carrying out activities. Their great difficulties in the contemporary world resulted less from their inability to occupy particular posts than from the compulsion to do so according to a male model. He then quoted Edith Stein in great detail who, after long meditation on the Genesis text, had come to the conclusion that one should not see marriage and motherhood as the only possible calling for women (in this, a fundamental line became visible which ran opposite to the basic line evident on the Islamic side). According to Edith Stein's conviction, the human species was developing in a dual way made up of man and woman. The character of the human being, which had to have some kind of essential feature, developed in both but along two different paths, dually, as it were, just as the whole human being was dual in structure. In addition, Edith Stein was concerned about the vocation and profession of women and the relevant ethics. A woman was attracted by everything that was alive and personal and she had a tendency to perceive this globally, as a concrete whole. Her natural need was authentically motherly. Abstraction was alien to her by nature. However, in addition to the profession of mother as such, Edith Stein also considered other professions which required motherly qualities even including the natural sciences. Indeed, every professional field could benefit from women provided they were allowed to follow their feminine professional ethic, although the latter was not defined more precisely, at least in the passages quoted. And although the difference between men and women had played a part right from the creation story, both, according to Edith Stein's emphasis, had been given the same threefold duty of being God's image, procreating and dominating the earth, but the woman had a feminine way of carrying out this duty. Edith Stein saw the influence of the woman as companion to the man mainly as helpful and harmonious, and as keeping the man from an activity that could completely consume him. The precise passage referred to from Vatican II was somewhat more cautious in its wording and less fundamentally anthropological; it only spoke in a rather flowery way about women always having had the lot of protecting the home, loving the beginnings and understanding cradles. But then the document went on to demand that women reconcile men with life and keep watch on the future of the human race. It also emphasised that not even families could live without the help of those who had no family, and naturally stressed the importance of the purity, selflessness and spirituality of nuns in a world in which egoism and the lust for profit seemed to be becoming a law. In statements on the dignity of women, Pope John Paul II had also emphasised that today scientific and technological progress were threatening to reinforce human pride and its sense of self-satisfaction to the point of self-destruction in one way or another, and that it was the dignity of the woman to help the man to become humble and wise in the use of his intelligence. Overall, Borrmans found a whole series of values which had already been mentioned by the Council and were already, and could become still more, the basis for dialogue and cooperation between Christian and Muslim women. "Transmit life, defend it in all its aspects; respect the continuity of tradition while enriching it through creativity; promote service to others as a response to the value of self-giving; fulfil religious consecration in order to honour God, the 'first served'; show maternal or sisterly compassion to those who are suffering or dying – these are the essential values which the Council, in its message, deems to belong particularly to /p.114 woman's vocation. In fact it is in these domains that there are many excellent examples of collaboration between Christian and Muslim women."[109] It is striking that, in the section dealing with the conference discussion, clear distinctions were also made concerning areas in which no cooperation was possible such as polygamy, dismissal and divorce, and it was emphasised that it was very difficult to distinguish between nature and culture, and that the faithful often attributed things to God's law and to nature as God had created it, whether behaviour or forms of discrimination, whereas in reality they were only consequences of living in different societies.

This is also where the weak point of the whole issue lay, if one considers the contributions in total and the little that can be deduced about the conference discussion. Are there natural or God-given differences between men and women and, if so, what exactly are they and what consequences are connected with them? There was a clear tendency on both the Christian and the Muslim side to say that differences did exist and to relate them to potential motherhood and a motherliness fundamentally linked with that and hence also to link them with what was natural to women or reflected their proximity to nature. But, even on the question of what else existed (potential threats to and weaknesses of women) or what social and religious consequences that had or should have, there was disagreement not only between Christians and Muslims but also diachronically within the religions themselves. Particularly in Islam, this was evidently a disputed point that was very explosive and to which the modern, Western image of women had added considerable fuel.

7.3.5 An elegantly worded joint declaration

There were, according to the final report, three working groups (as the addresses also made clear) in which agreement was reached at the end on the following points. Both religions agreed that God had created human beings and given them a special dignity in which men and women shared equally. The differences between men and women were also seen by both as given by God. Then both religions emphasised the status of the family as the nucleus of human society and appreciated the contribution which parents, and mothers in particular, made for the next generation. In addition, the latter also had an influence on other activities in the educational, social and economic realms but it was often not sufficiently appreciated or supported. This led to the interesting statement, "While participants emphasize the essential role of woman in the family and appreciate the woman's work at home, they also recognise her full right to engage in any other activities in keeping with her capabilities and circumstances, within the principles and guidelines given by each religious tradition."[110] This of course leaves much scope for interpretation and it was also observed that a lot of wrong practice existed which undermined the dignity of women and limited their possibilities. That was the final result of cumulative social traditions and needed to be corrected by a right and non-oppressive understanding of the scriptures – as had also become clear in the addresses. The only thing that was mentioned as negative was the exploitation of women which violated their dignity, especially in the entertainment industry and in advertising, something which naturally hardly concerned the Islamic realm that was the main focus of this question. In addition, there were legal gaps on certain issues such as employment, payment, education and maternity leave. Religion was seen as generally of great importance for forming a balanced personality with the highest values, and hence the promotion of religious values for everyone would also protect the dignity of women and help them to fulfil their role. It was not yet possible to agree on a theme for the next conference but it was to take place in January 1994 in Amman.

7.4 Amman 1994: nationalism

From *18–20 January 1994 the fourth dialogue conference* took place in Amman on *"Nationalism Today: Problems and Challenges"*. On the Catholic side there was an effort, on the one hand, to ensure a certain continuity among the participants and, on the other, also always to involve experts on the subject under discussion. The latter, in particular, was expected to be a double advantage: for the Muslims, the contributions by leading Catholic experts and, for these experts, their sometimes first contact with Muslims.

7.4.1 The welcoming addresses
7.4.1.1 A subject of special interest for the Hashemites
In the welcoming address by Prof. Nassir El-Din El-Assad it already became clear that the theme for 1994 was not just hanging in the air but related to concrete, political problems of the present, especially to the prominent position which the Hashemite royal family had in the Islamic world and with regard to the holy sites in Jerusalem. The two together apparently provided the motivation for a conference on precisely this subject, combined with the recognition that there was in fact a certain potential for conflict between religion and nationalism and that one could not simply say that the relation between an individual and his/her people and the society in which he/she lived was just natural and human and

somehow related to an instinct for survival and procreation. It was true that the religions had not simply abolished nationalities but they had purified them (which could almost have been a Catholic formulation), e.g. with regard to human values, and they had kept them away from conflict, discord and fanaticism.[111]

The way was therefore to move from contemporary problems to more fundamental issues in order then, it was hoped, to find better solutions for the contemporary problems.

7.4.1.2 Catholic view: nationalism between individual identity and general human solidarity

The second welcoming address by Cardinal Francis Arinze was certainly more substantial; he immediately gave a comprehensive but brief presentation of the Catholic point of view. It was a question of the relation between religious identification and national movements; that was his definition of the problem. The climax of his statements was, ""[T]he Christian view is to confirm a healthy patriotism, which serves to build and strengthen the personal values of self-identity and self-worth with /p. 14 the social goods of solidarity, self-sacrifice, and discipline for the good of the whole, while distinguishing it from a destructive nationalism or chauvinism, which seeks to oppose, dominate, and destroy others and to make one's nation absolute."[112] This introduced into the discussion at the same time the two key anthropological points which had to be dealt with adequately, namely one's own identity and sense of personal value, on the one hand, and the identity of others and their sense of personal value, on the other, or, to put it differently, the unity and solidarity of the whole of humankind. Both were justified and meaningful in their own context and had their roots in the Bible and tradition. In this way, Arinze introduced a distinction, albeit *expressis verbis* only fairly near the end, which John Paul II had made in a message to religious leaders in Sarajevo, namely the distinction between patriotism (positive) and nationalism (negative). His interpretation of patriotism here was very generous and even included support for one's own national team in international competitions, and he tried at the same time to find traces of this attitude of unabashed pride in the Bible. Patriotism became negative and ungodly if this attitude led to a conceptual or practical devaluing of others. It was relatively simple to prove this on the basis of the New Testament because there religious belonging clearly came before all other kinds of human belonging, just as God could always claim more obedience than a person or a human entity. To claim the same for the whole of the Hebrew Bible was more difficult, as he at least hinted, before then concentrating on the passages in the prophetic tradition which were intended to remind Israel that God was the God of all people and all nations belonged to him. In conclusion, Arinze voiced his hope that the conference would succeed in identifying the elements in each case which led to division, disorder and destruction – which was perhaps a simplified view of a complex problem where the limits are not always so clear.

7.4.2 The historical perspective(s)
7.4.2.1 Islam as the fulfilment and abolition of nationalism?

That immediately became clear in the first address which attempted to give a survey of the history of the relation between nationalism and religion from a Muslim perspective. It began with the fact that it was not at all easy, or was indeed almost impossible, to give a generally valid definition of nationalism which could apply equally to very different states at quite different times. But a majority view was that nationalism was a social concept referring to the bond that kept the members of a nation together, and that nationalism was a creative, stimulating force without which a nation would not come into existence at all. That was very abstract but history showed that the bonds which resulted in a nation and kept it together did not necessarily have to be of the same origin. Nationalism acquired its extreme significance and importance relatively late, in the nineteenth century in Europe and America and in the twentieth century in Asia and Africa, while its philosophical roots were to be found in the renaissance or, more precisely, in Macchiavelli's famous main work. It was interesting that the passages quoted by the speaker placed loyalty to one's own country and its freedom above all moral considerations, which, considering the reflections previously presented by Cardinal Arinze, should in fact really rule out nationalism as a concept to be tolerated on the Christian side. Understandably, Al-Sartawi did not deal with this idea but only spelled out how the calculation that nationalism led to peace because nations with self-determination did not go to war did not hold good, as was well known, and that in fact many individuals had been sacrificed to nationalist ideas. Following an excursus on the Marxist criticism of nationalism as serving the particular class in power, he analysed the nationalist movements in Asia and Africa in greater detail; they also affected many Muslim states which had come into being through them (Libya, Egypt, Pakistan, Saudi Arabia, Jordan. Syria, Iraq, Indonesia). Here too, he again offered various kinds of explanations but concentrated on the role of Islam in this context to which he attributed a key position in mobilising nationalist sentiments, as could be observed in individual figures, but it was more important that the Muslims had not accommodated themselves to colonial lords but rather preserved an independent character which was also very uniform, probably as a result of the influential force of the Quran. But finally he supported the definition of Muhammed Baqir Al-Sadr according to which nationalism was merely a purely historical or linguistic bond and not a philosophy or belief of its own; it had a neutral world view and needed to acquire a philosophy in order to shape a civilisation, social order or whatever. The Islamic message and, in line with that, the Islamic civilisation were worldwide, not nationalistic; they related to and liberated all people so that they then only served God. All other previous civilisations were described as civilisations of slavery whereas Islam offered brotherliness, justice, compassion and tolerance, a free, brotherly, friendly life, at least as long as it followed its own rules. In this respect, the so-called Arab (instead of Islamic) nationalism had been an error which had only led to fragmentation and to people again falling victim to imperialism and Zionism.[113]

At least it was clear where the political relevance and sensitivity of this presentation lay. The question also arises whether nationalism is in fact so formal and neutral in its world view as Al-Sadr and, following on from him, Al-Sartawi claimed and can therefore be used in the service of a good cause, namely Islam, or whether nationalism, in the sense of the Macchiavelli quotation, does not also at the same time use Islam to serve its territorial interests. Moreover, Al-Sartawi was very critical of Western nationalism which for him often meant almost the same as imperialism, but there was no basic criticism, such as that expressed by Cardinal Arinze, of their own sense of superiority. The world can be cured by and become free and brotherly with an Islam which imbues all the nationalisms of this world and subjects them to its organisational system; that would, as it were, be the ideal world for every person, although this is naturally a very Islamic view of the situation and of people.

7.4.2.2 Arab Christians: fighters for secularism and nationalism

It is very interesting to compare this with the Christian address, the theme of which was quite concretely the role of Christian Arabs in the formation of early Arab nationalism (nineteenth century); this naturally led to a quite different type of presentation with a completely different approach which nevertheless offered some interesting interconnections. To begin with, Adrian Musallam underlined that he did not only understand the term "Christian" in a religious sense but also socially and psychologically because an individual always lived and acted only in a human setting. His further remarks then made clear that he was referring here to the specific situation of Christians (and other minorities) in the Ottoman Empire with its (Islamic) millet system which was also one of the main reasons why Christians had played such an important part in the development of Arab nationalism. More precisely, it had developed in the area of Greater Syria as a plea for secularism because there were so many religious sects and ethnic groups there that it was the only thing which could unite them all. In the Muslim Ottoman Empire, on the other hand, belonging to a religion was the decisive factor. As long as the Christians paid their tribute, did not threaten the state by any kind of coalitions or violate the rules imposed upon them by Islam, the state refrained from intervening in their internal affairs and these were ordered by their own leader who had to collect the tribute and was otherwise the authority on practically all questions whether these related to worship, education or welfare or to finances, administration and legal affairs. According to Musallam (and thus practically in direct contradiction of the preceding speaker), the Arab Christians had brought about the liberation from this closed system and thus a cultural renewal of the whole nation and contact with a larger world which was already a world of nations, although they did this naturally primarily in their own interest. Firstly, (from the sixteenth century onwards), the church contacts with Rome were helpful in this (and led among other things to the establishment of schools from which many reformers then came) and then also the commercial contacts which were gradually taken over by the indigenous Christians (who thus obtained direct knowledge of European life), even including such details as that the first Arabic works were printed in Syrian characters until there was also an Arabic printing press in the Near East – important for spreading the new thinking, and naturally both on the initiative of Christian clergy. Later, Protestant missions came in as well which led among other things to the founding of the American University of Beirut. What they all had in common was that they took the Arabic language and culture very seriously and, at the same time, opened up new horizons with foreign languages such as English and French, promoted scientific thinking and offered critical curricula – for Christian Arabs. These academic endeavours separated gradually, however, from this foreign and also one-sidedly confessional background and for the first time, at least since the Ottoman rule over Syria, brought the different competing faiths together in an active partnership for a common aim. An additional external point was that, in the Ottoman Empire itself, there had also been changes in the direction of modern Western systems: protection of life, of honour, of citizens' property, equality of citizens irrespective of their faith. And within the millet system the citizens also gained in importance.

But the decisive stimulus towards a secular national system came from the massacre of 1860 which had almost eradicated the whole Christian community and of which the Turkish authorities did not gain control; indeed, on the contrary, Turkish soldiers and even the governor of Damascus participated in it. This lesson was also learnt by intellectuals like Butros al-Bustani who from then on insisted that everyone must recognise that all religions were finally equal. All people had a human nature, were descended from their ancestors and (only in third place) worshiped the same God. Laws had to be just, the same, adapted to the times – and they had to be based on a separation of the two realms, the religious and the worldly. Love of one's homeland was an act of faith and had to take precedence over all other religious ties. Overall, this is the clearest formulation of the Christian Arab conception of secularism and a fundamental battle cry against the social system which was and is traditionally described as Islamic. Taken out of context, the strongest statements made it sound as if Christians would get into difficulties because of this, cf. the statement by Cardinal Arinze on the issue of final loyalties. But in fact it was the particular circumstances, as the speaker made clear, which determined the form and then also the understanding of the protest. In general, Christian Arab intellectuals had played a major part until the beginning of the twentieth century, indeed up to the First World War, even though the real leadership passed gradually to the Muslim Arabs who, however, at least before a Christian audience, adopted the nationalist in contrast to the religious line of argument. The majority of Muslim Arabs, on the other hand, and that was also something one had repeatedly to recall, were not aware that it was a question of national unity. Especially the idea of secularism, the cornerstone of the whole affair for the Christian Arabs, and thus the separation of religion from life, which Islam normally rejected vehemently, did not enter their consciousness. They had also not experienced that following a religion could be a source of danger and that it should therefore not be the basis for citizenship or for equality concerning rights and duties.

These two addresses are a good example of an encounter not only between two different religious approaches but still more between a predominantly ideal approach of one religious type and a very real historical approach of the other. The results can be said to be typical. The one ends up with Islam as the protection and liberation for all people and the other with secularism as the only protection and only liberation and the way to equal rights for religious minorities. But both, it should be noted, were speaking about the same historical situation, although it should be added (information provided by the Christian contribution) that the first to point to the danger of Jewish nationalism were the Christian Arabs – but without any religious connotation and only based on the fact that two nationalisms were claiming the same territory and this laid the ground for a permanent conflict.[114] *Hardly anything could better demonstrate how markedly different concepts and experiences can determine the perception of one and the same situation, although one might get the impression that the Muslim perception was completely impermeable to even massive experiences of the opposite. It is extremely regrettable that the conference documents do not indicate how the discussion following the two addresses went.*

7.4.3 Nationalism: contemporary problems and challenges
7.4.3.1 Nationalism – a fundamental clarification of the term from the Christian side

The second round of addresses dealt with the present-day problems and challenges of nationalism although references back to history and general clarification of terms were unavoidable so that the link with the preceding addresses was more or less automatic and there was merely a shift of emphasis. The address from the Christian side was given by Dr. Jørgen Nielsen from the Selly Oak Colleges in Birmingham. He began with the major changes which had taken place in the evaluation of nationalism in the past three decades or so as a result of the economic and political situation, including the fact that, following the collapse of communism and thus of a whole world order, ethnic, nationalist and religiously exclusive movements worldwide were appealing to the lowest instincts of humankind – and with regrettable success, just when nationalism and the old concept of nations had been declared superseded or appeared inadequate. Nielsen also added a clarification of the term showing that what is called nationalism today came into existence in Europe from the seventeenth century onwards when old identities and authorities such as the church, feudal rule or tradesmen's guilds lost their formal power and were driven out mainly as a consequence of technological progress. Before and beyond this historical context, simple comparisons were not possible because there nationalism was at the most a European export. The whole presentation repeatedly returned to this when it attempted to identify generally valid trends and particular European conditions. Apart from this historical limitation, Nielsen also tried to limit the term in linguistic history, especially in view of the fact that, although there was reference to nationalism throughout Europe, on the one hand, in practice one spoke about this or that people. This could be traced back to two different ways of defining how an individual belonged to a community with common social and political interests: as a blood relationship or by common residence or belonging to a common class in the broadest sense (both were found already in ancient Rome which had then supplied the linguistic basis, at least in the Romance languages). At the level of a people, common descent was important although as a rule this was mythological in character – the myths of romanticism had had an especially strong influence in forming nations in Central Europe in the 19th century, but similar founding myths were also found in other European nations and they still provided identification even in the present.

Special features of European nationalism
It was especially European that nationality and citizenship practically always coincided, something that was and is not the case on other continents or at other times. The next criterion (and this had also already been mentioned in an earlier address) was language. In Northern and Western Europe it was not usually known for human communities to be normally bi- or multi-lingual rather than monolingual. This was connected with the fact that romanticism and linguistics, as they developed in the nineteenth century, were so much drawn into the process of forming nations (here again especially in Germany) that language and blood became almost identical in the public awareness. Even today one spoke of Germanic, Semite, Arian and Arab peoples, thus defining tribes linguistically. But what had developed in Europe did not correspond to the major part of human reality but rather obscured it and made people unable to perceive it, although the Anglo-Saxon tradition, whether in Great Britain or the USA, was fundamentally more open, and that in turn also led necessarily to other definitions of nation and citizenship. For the Arabic speaking realm, Nielsen stated that the corresponding vocabulary had first needed to evolve because the potentially suitable terms were either markedly religious *(umma, milla)* or insignificant, although in the latter case distinctions could again be observed along the lines of descent or territory. It was also interesting that, as a result of the political constellations of the period, it was precisely the German variant which had the decisive influence on the nationalisms that were developing at the beginning of the twentieth century in the Islamic realm from the Arab world via Mustafa Kemal Atatürk's Turkey to Muhammad Iqbal and Pakistan. Other more open definitions of nation, such as that of Grundtvig in Denmark, that anyone who felt they belonged to a particular people was thus part of that people (more or less also the definition of the Arab League for Arabs), had had next to no effect on political practice.

Concrete and potential effects of nationalism
It can be observed worldwide, however, that economic and technological modernisation have led to the wider spread of literacy and education and to working outside of traditional structures, and thus to the formation of a new identity, namely that of the nation. Education in particular had a key function in overcoming regional, minority or immigrant mentalities but could also become tendentious and even transmit specific hostile images. In addition, the formation of nations had almost never come about without violence towards the inside and/

or the outside. In a modern state, the bond which linked the individual with the state was citizenship. Normally, that included the right to residence and access to the various benefits which the state provided such as education, health and social care and the expectation to be able to find work and keep the family together. Irrespective of specific conditions, persons or a community without the full status of citizens were at a disadvantage. As far as the European Union was concerned, the main criticism of it in connection with human rights was the chaos relating to citizenship and the criteria for it, an inheritance of the history and tradition of the various member states, not to say their founding myths. These founding myths (worldwide) could be fundamentally different interpretations of one and the same historical event and this made such myths which provided identity so explosive and raised the question how they could be surmounted and new myths created that would promote unity. Similar items could be derived from international private law which was known both in Western societies and in the Islamic realm but was based on different criteria, or the underlying concept of the family which was instinctively always at the centre of a collective identity, irrespective of whether a culture defined itself otherwise as religious, secular or based on descent. If the Muslim minorities in Europe were increasingly to make more demands concerning family law, which was certainly conceivable, that could easily result in a mobilisation of some kind (national, ethnic or religious). All of this made clear that something which at first sight seemed so simple as belonging to a nation could rapidly, on closer examination, turn into collective pathology. That could also possibly lead to self-destruction as civil wars had shown, and experience had demonstrated that these were more emotional and hence more brutal than conflicts between foreign or different states. And, in the collective memory at least, the role of Christianity and Islam here was negative. The long series of disputes was generally known and could quickly be mobilised politically, whereas the long periods of peaceful coexistence and interchange were not remembered and could not offer a counterweight. On the contrary, conflict even seemed to be exported to areas which were not familiar with this kind of Christian-Muslim dispute. A contributing factor in part was that it was easier to obtain sympathy and support from outside if one could sell as religious an ethnic conflict which was otherwise of no interest. In this connection, the frequent appeal to the Christian character of Europe was interesting, although the almost more interesting question was which elements of its Christian character were actually cited. Spiritual leaders were also born into these collective pathologies in which very different strains were inseparably intertwined and they always remained part of them. In principle, it was neither Christian nor Muslim to support one's own brothers in the faith against every foreigner, especially not when it was a matter of exploitation or material or physical injustice. The major challenge which both religions faced in the contemporary situation could only be met jointly; that was the only way to liberate religion from the prison of such circumstances and then reintroduce the generally accepted values (in which Nielsen believed).[115]

This address showed more clearly that all those that preceded it how complicated the circumstances were, how very many factors were involved and how quickly an individual person or group of people, or even a religion, could fall captive to circumstances and constructs from which it was then difficult to escape. Because one was always also a part of a system which could develop very destructive forces (and had frequently done so), common efforts were the only means for finding a new path that both sides could follow.

7.4.3.2 The address by the minister of justice – a classical Islamic approach

The address from the other side, given by the Jordanian minister of justice, dealt with the situation in the Near East and, when one compares it with the two preceding addresses, tended more to simplification, not to say glossing over matters, and this could naturally have related to the person giving the address in perhaps three respects.

His perception as a Muslim was clearly marked by the Islamic ideal of society and the state; deviations from this were perceived less strongly and judged quite differently from a Christian who was at home in the same society; moreover, he did not come from university research and he occupied a very high office in society – that could also have contributed to his seeing the situation in somewhat less detail and calling for a more positive view.

But this made the anthropological statements clearer and simpler and also much easier to identify that in a context where belonging to a nation was seen as a complicated and in part almost imaginary network in which a wide variety of factors influenced the identity of a person, and it could be difficult to define all these factors and their share, let alone find a simple solution when conflicts arose. That was all much simpler as presented by Dr. Taher Hikmat. Islam was a universal or international, human religion so every person can be a Muslim. Naturally, the Arabs had a special role and standing within Islam because God had chosen an apostle from among them and had then chosen them, and because the union of this nation was a religious union or, more precisely, a religio-political one, intended to give it security, support, victory and a basis for living. Again he emphasised how important the community was over and above any little distinguishing criteria; this was typically Islamic, and significantly the markedly Islamic term *"umma"* was used for nation, initially in the form of the Medina agreement which, in addition to Muslims, also included the non-Muslims who had joined them. This nation was based on the principles of equality and cooperation. Later definitions looked quite different; membership of the *"umma"* required the recognition of certain things such as the creation of the world, the oneness of its creator, his eternity, justice and wisdom and that he lacked anthropomorphisms, Muhammad's prophetic role for all people and his message as right and something one had to stand up for, the Quran as the source of Islamic law and the Kaaba as the direction for prayer. That was no small number of statements of faith on which belonging to the nation depended but anthropologically the same can also be expressed very positively. Muslim citizenship was very easy to acquire; one did not need any special permission or minimum stay in a country but only the faith of the heart and the confession of the mouth. Strange and stranger were not something natural and unchangeably inborn in Islam like colour, race or language, but merely a personal matter which is subject to the will of the individual, and it was repeatedly emphasised strongly that, for Islam, race was not a criterion for distinguishing or discriminating. Thus it was clear and was also stated that the Muslim state was not a nation state in the Western or modern sense, and that the shifts mentioned naturally always had to be taken into account. The Near East had become a pan-Islamic state in which religion had replaced all tribal and other bonds and at the beginning this had been unproblematic. It was only with the rise of Muslim sects and with the domination of political disunion over the essential, doctrinal unity of the faith which was religious, political and unified, that the problems had also arisen. The problems of those who did not fit into the unity of faith were significantly enough not mentioned.

Islam and nationalism – a Muslim view
Nationalism in the modern sense only appeared in the Arab world in the twentieth century apart from a few literary references in the nineteenth century (this was confirmed in a striking way by the mainly historical address by Musallam, since Arab nationalism had been a Christian concern in the nineteenth century that found literary expression but was otherwise not recognised by the Muslim public which seemed to be all that Hikmat had in mind here). Turkish nationalism could be traced back to European influences (on this Nielsen was more precise) and Arab nationalism to oppression by imperialist capitalism (although the European colonial powers and the Ottoman Empire in its late phase were practically lumped together here) and this seemed to justify the latter (for Islam). It was also emphasised that Arab nationalism had never been anti-religious; on

the contrary, it was always very much connected with Islam because Islam was the greatest that the Arabs had even produced (and, *vice versa*, Islam was always connected with the Arabs because the Quran was in the Arabic language and the Arabs were the first Muslims). Hence Islam played a major role in the struggle against recent forms of hegemony and this completed the circle. Hikmat identified two safety valves in the current system of the Near East. The one was Islam which defused national tensions (e.g. between Arabs and Berbers or Arabs and Kurds) and the other was nationalism which linked Christians and Muslims (although he underlined that the Lebanese civil war was finally a political and not a religious conflict). Both could help to keep people together, but he also dealt in great detail and almost aggressively with the idea that an opposition between Christians and Muslims had been constructed but did not really correspond to the truth and rather served immoral, economic interests. Christians and Muslims had to commit themselves to these values and they could do so."[W]e, both Muslims and Christians, have a moral legacy and an ethical religious heritage sufficient to ward off the most flagrant violations of the freedom of man who is honored by all religions."[116]

So right at the end one can observe clear parallels to Nielsen's address but the tone has shifted. The shift in perception towards conflict rather than peaceful coexistence was the fault of the West alone and cooperation over common values was seen as quite simple and positive (as if, for example, a lack of religious liberty in Islam was not a flagrant violation of a human right to freedom, at least according to a non-Islamic view of humankind). One could call it naivety or the consequence of a policy in which there was no room for shades of meaning and uncertainties or even for self-criticism.

7.4.4 The role of the faithful
7.4.4.1 Humankind as a nation in the Muslim view
The last two addresses dealt with the role of the faithful in relation to questions of nationalism and attempted, as it were, to look into the future but nevertheless always on the basis of the past. In the case of the Muslim speaker, it was striking from the very beginning that, although he spoke about the need for a Christian-Muslim dialogue precisely on these questions in order to come to a common understanding, and said that the question of secularism in the past and the future was the decisive issue between Islamic and nationalist thinking (as had also become clear in Christian Arab thinking), he also quite naturally assumed an Islamic Arab identity, and equally naturally presupposed that Islamic Arab unity was the goal of both Islamic and Christian thinking; he could really

have been sure that Christian Arabs would contradict him on both points. But, like the Muslim speaker before him, he saw a great harmony between nationalist thinking, Islamic thinking, inter-religious dialogue and Arab unity as a factor which – according to Muslim logic – repeatedly came to light. His conceptual and anthropological starting point for this was Sura 2:213 according to which humankind was created as a single nation and the differences came afterwards, and therefore God sent prophets with a book so that it could be the judge in all things where they differed, and the people of the book were only different from one another because of their recalcitrance. His interpretation of this was that the differences between the individual religions could be traced back to their links with particular nations, although Islam was very tolerant towards the other monotheist religions in which he saw a likeness, not only in their belief in God but also in eschatology and in some sense in ethics, and sometimes more tolerant of them than towards certain of its own schools of law or even sects. With regard to the relation between society and religion in general and Islam in particular, the speaker then spelled out the idea that the human individual as such and in association with others faced two major challenges: either to survive or to be exterminated by civilisation. Here, frequently psychological factors predominated and caused a consolidation of the feeling of human brotherliness. That should indeed be the case and control individual and social thinking and action because otherwise there could be neither peace nor justice. The monotheist religions had rooted this principle of human brotherhood firmly in their creeds and had thus contributed directly to the great intellectual and behavioural achievements of human civilisation and to social relationships. Certain thinkers even believed that it was religion which led people to reflect and thus stimulated them to develop wisdom and visions for the future. One could summarise the essence of all the monotheist religions as human brotherliness resulting from the fact that all people were created from the same life by the same creator, with altruism as the starting point for social reforms and finally the general reform of one's self in order to be able to bring about a reform worth mentioning. Two of these three points are decidedly anthropological. Of course, the calls for social justice and human brotherhood and equality had always had an individual human tinge, but self-sanctification, the invitation to charitable and just action as well as the achievement of social justice, security and stability belonged to the fundamental aims of all religions. They all put the highest ideals before the human conscience and urgently called on people to maintain the bonds of brotherliness, love and peace. His statements in this connection were very far reaching. "[A] true believer is the nearest to perfection regardless of the faith he believes in. If people turn to religions as a means of rapprochement and amity through their sublime vocation, love and peace in society will be attained. The most outstanding characteristic of the message of religions is perhaps the universal love based on the general moral code of humanity that aims at openness to world cultures and thoughts and the common elevated notions of humanity and human progress."[117]

It is very interesting to see how far he went at this point with equality on an anthropological basis, namely as far as a moral code common to all people. That this is not only finally a more or less theoretical entity that is logical and can be explained by creation, but is also a present reality in the world of all religions and not merely in the monotheist ones as stated above, is something which indeed cannot just be taken for granted even in the Islamic understanding. Despite the conception of innate Hanif-hood, he set great store by the necessity of the revelation of God's will.

It was only in the following sub-section that Jarradat concretely took up the question of nationalism or, as he put it, the problem of contemporary Arab societies, namely that a confusion had arisen about the concept of Islam and the concept of nation. He understood the latter as the concept derived from Rousseau of a social contract between citizens that led finally to an ethnological nationalism or chauvinism and that had conflicted from the very beginning with the spirit of religion in general and of Islam in particular. For him, Islam was the core of the Arab civilisation. It had idealistic and humane aims and was in every respect universal, especially in its approach to life. And its language and *Sitz im Leben* were originally Arabic whereas, according to these criteria, Christianity was not European in the same way. So, as a faith and a culture, it had become the message of the Arab people. Jarradat's remarks came to a climax in a sentence which even linguistically was not quite correct. "So it is religious doctrine which forms the national ethos while the array consisting of (religion, language, culture and geographic and economic unity) may form the basic constituents of the national concept which does not conflict with the Islamic concept."[118]

It would appear that a very open concept was being used here in the practical Arab context to present the religion of Islam with its collection of beliefs and especially of regulations as the only natural basis for society and the nation. What is even praised as non-contradiction when seen from another, namely Christian Arab point of view, is more the taking over of the Arab nation by Islam, without paying attention to the fact that there is a more or less large Arab minority which neither views nor wishes to have Islam as the solution to the problem and which has evaluated its universality and humanity in a quite different way. So there are indications here that the initially so open formulation of an Islamic anthropological basis for common ethics for humanity can also become a dangerous trap from the point of view of non-Muslims, namely if it is simply equated with the factual Islamic system and then declared the aim of all Arab people (if not of all people in general). Anyone who dares to see things differently is in danger of being declared a person who does not live up to the highest values of Arab culture or even of human existence in general, and that in a society in which religion (and hence also a religious judgement) is the determining factor.

Values and goals for humankind
But Jarradat started from the harmony between the Islamic and national concepts which Islam had created by bringing the Arabs onto the international stage. This connection would also be positively decisive for achieving unity, freedom, social justice and democracy in the future. The civilising Islamic heritage belonged to both Muslims and Christians and both had had common roles in building it up and still did. In an Arab society which had yet to be established, Jarradat saw national Arab and civilising Islamic forces at work; Islam was needed because of its values such as social justice, social cohesion and unity, and nationalism because the social and emotional ties of the individual with his/her nation had to be strengthened by a national feeling which benefited both spiritual and material development in life. This last point was an idea which had already come up in a similar way right at the beginning in Cardinal Arinze's remarks. The line of argument became anthropologically more fundamental when dealing with the role of inter-religious dialogue in these contexts. "This makes interreligious dialogue a necessity dictated by a civilizational challenge for contributing to the one human civilization. Such contribution comes from the points of similarity and agreement among mankind in search of a better future for humanity. Through this dialogue monotheistic religions will contribute to the enhancement of good, the values of faith and the principles of brotherhood amid the humanity which Almighty God has created and through the natural disposition he has deposited in man leading him towards belief in the Creator." What was especially important to Islam in practice here was justice, social cohesion, consultation (shura) or democracy, as well as moderation or the middle way. Justice was necessary to protect the individual and society from injustice and aggression. Social cohesion also belonged to the prerequisites for a human life in dignity and here the connection between the spiritual, intellectual and material dimensions of life was also included. Consultation/democracy was the possibility of obtaining different views and opinions on an issue and achieving solidarity and freedom in society so that everyone could express their opinion irrespective of the religion they followed. Finally, moderation concerned the whole of life whether it were a matter of the common good or differences of opinion, tolerance or even worship life. These in turn were all necessary prerequisites

for human development, social progress and the promotion of moral values. The monotheist religions agreed on all of these points and especially Islam and Christianity.

Tensions and problems – only a concern for non-Muslims
Critical points in the dialogue were found, *inter alia*, in historical studies, especially when it was a matter of links with and the support of missionary activities by imperialist hegemony. But even in these sensitive areas positive or more objective trends could be identified. The question of secularism was mentioned as in special need of clarification through inter-religious dialogue, but this could probably only mean explanation for non-Muslims because Jarradat made his own point of view clear with unusual decisiveness and directness: Secularism was only something for societies in which religion was merely part of their social structure and essential basic building blocks, and where secularism was the prerequisite for unity, progress, freedom and liberation in society. In the Arab Islamic setting (the two are basically equated here as was already shown above), however, religion was the basis of human social life, nationalism and also of science and scientific thought. Anything else would be extremist in an Arab context. The reasons for the genesis of secular thinking in Arab society were of no interest to him; they had run their course and secularism in the Arab society of today was thus also reactionary in his view. Where, in reverse, secularism had been forced upon Muslims, they had rejected it and were unable to assimilate it. The absolutely logical conclusion for him, although it could only be a slap in the face for every Christian Arab and especially for the person who had spoken at the conference on the role of Christian Arabs in the formation of Islamic nationalism, was: "If secularism does not suit Muslim peoples in general, it does not suit the Arab nation in particular; because religion, and particularly Islam, is the basic constituent of this nation and the message it carries to humanity."[119] But the speaker himself had no feeling for this; he saw no tension at all between regional, national (or pan-Arab), civilisational, Islamic and global forms of belonging which related to the whole of humankind. He also saw no tension with Christianity on this point (nor with real Christians) and also not with democracy, quite the contrary. And right at the end, at least, he again emphasised where this attitude came from. Islam saw itself as the final valid divine message for humankind – and this message was in harmony with the natural human disposition and also showed the other monotheist religions and their messages where they belonged. That was precisely what he had demonstrated in his address.

7.4.4.2 From the Christian side: nationalism as a concept of "imagined communities"

The corresponding Christian address by Ian Linden did not start, however, from a specific Christian concept but from the concept of "imagined communities" in Benedict Anderson's analysis of nationalism which saw the rise of nationalism as a development in the human consciousness, namely that a new "imagined community" came into being in contrast to the religious community and based on other conceptions of space and time from the latter. This development was linked, among other things, with the formation of a national written language and it enjoyed a sweeping victory, so that after not quite 200 years it was seen worldwide as natural and essentially human to have a nationality. He also distinguished between popular nationalism based on Herder's sense of belonging, which he defined as the cry of the oppressed and as hope for the reestablishment of well-being in particular cultural forms, and official nationalism which manipulated the symbols and images of the former. As a rule, this nationalism received much more attention whereas the other, the genuinely anthropological aspect of the phenomenon which was connected with various forms of human solidarity, was really hardly discussed; and, with regard to nationalism, the political power of the phenomenon seemed in any case to be in inverse proportion to its intellectual, philosophical inadequacy. Even when Linden assumed (and he was very correct about indicating in each case what assumptions he was making) that the possibility of cultural self-determination was the prerequisite for the development of a truly human Christian spirituality, he still specified clear conditions for the involvement of Christians in nationalist movements (option for the poor). In this context, it was explicitly possible to discuss conflicts between the demands of religious and national communities and to criticise manipulations, transcending of limits and misuses of nationalism and religion, which he also did.

For the previous speaker, in contrast, (and for other Muslim speakers in a similar way) it was more a matter of putting into practice an ideal conception of nationalism which already existed in principle, namely the Islamic one (or perhaps still or even the Arab Islamic one). This considerably limited the scope for criticism, especially for basic criticism of the system, because one was firmly bound to one single system.

What is "imagined" in national entities?
For Linden, on the other hand, there was plenty of room for basic examinations of all the phenomena in the realm of nationalism beginning with how an "imagined community" should really be understood as far as "imagined" was con-

cerned. It was the exact opposite of private and passive, nor was it somehow a pure invention where one could imagine an inventor manipulating unsuspecting subjects into ethnic groups. It was much more a matter of widespread, collective fictions which were intensely discussed and repeatedly re-invented. In the mind of every representative of a nation there was a conception of commonality with other representatives of that nation although, in practice, he/she would never have the opportunity of entering into a personal relationship with them. In this sense, anything which went beyond a village community could only be an "imagined community". Here, a second definition came into play, which was also derived from Anderson, namely the nation as a political community that was perceived simultaneously as sovereign and limited. This conception came about when, under the pressure of the French revolution and the enlightenment, old systems such as dynasties or religious conceptions of hierarchy, space and time broke down or at least faded. That was precisely what produced a sharp contrast between nation and religious community. The latter was geared to community with the holy and was therefore best described as vertical and tending towards the centre, whereas the former was more horizontal and directed to borders. Especially in the Christian-Muslim context, the claim to religious leadership had been expressed mainly by control over the interpretation of holy texts in a particular, especially "true" language. Space was experienced and conceivable in pilgrimages, whether of the faithful or of the religious leader (in the Christian context, this definition naturally applied especially to the Catholic Church) and there was a certain timelessness between the people of today and a divine event long ago. In this respect, the development of printed everyday language, for example, (or the question of the legitimacy of translations in the religious realm,) was a serious challenge for the religious communities. And often these linguistic communities finally developed into nations, or the limits of a nation were determined by geographical or technical, administrative factors. Finally, nationalism was also a challenge to religious identity because it aroused strong feelings related to belonging. People loved the community and even died for it, and because this national identity was presented ("imagined") as natural, as something in the blood, people also had the feeling that they had no choice about the sacrifices that might be demanded of them. Even within a religious community, this could lead to different evaluations of the same situation depending on whether the national or the international level was affected. Such differences had arisen, e.g., between an English cardinal and the Pope with regard to silent approval versus public denunciation of the Gulf War.

Herder – a philosophical foundation for nationalism?
As far as a philosophical foundation for nationalism was concerned, the most plausible source was Johann Gottfried Herder who considered a sense of belonging fundamental to a human conception of wellbeing. He believed in a people's or a national spirit as a form of cultural self-determination and was an opponent of the universalism of the enlightenment, but he also strongly rejected any form of domination of one group over another. On the contrary, the negative aspect of the fact that people lived in unique cultures was that they separated themselves from those who robbed, excluded or belittled them. That was, in a way, the anthropological description of nationalism, while the historical evidence itself made clear that it was obviously not a natural category of the human after all, since the coexistence of human beings had been determined by quite other factors for most of their history and, particularly when one considered the Muslim realm, these quite obviously still determined it today or at least were supposed to do so. Perhaps that was also the reason for the difficulties about a philosophical foundation for and description of the phenomenon, since Johann Gottfried Herder was not really a philosopher but a poet and therefore fitted quite well into the context of "imagined communities".

Nationalism in modern times – a Christian approach
But it was also important to remember, as Linden emphasised time and again, that this popular nationalism preceded the official nationalism to justify a nation state and its rulers (sometimes referring to pre-national dynasties) and had to be distinguished from it, even though the latter used symbols from the former – never exactly the same but always similar, also in the Near East. Nationalism was, in a sense, an artistic product of human society. In contrast, a religious community existed in quite different dimensions and, when it was a world religion, it always had a worldwide dimension. Linden considered a national, religious identification no longer conceivable or even dangerous in modern times. Moreover, there were too many competing world religions and also millions more people who preferred a secular state of whatever form. He considered a moral, spiritual criticism of the state community by the religious community, of whatever kind, to be legitimate but not a fundamentally theocratic one. This naturally also involved the fact that Islamic provisions were considerably more concrete than Catholic ones and hence the scope for human Christian spirituality was broader. He then discussed the problem that cultural self-determination, which he considered as much part of human wellbeing as material possessions, was almost automatically understood today as the right to one's own state, which often created political problems and also starting points for manipulation. On the question of a genuinely Christian concept in this context, he ap-

pealed over against official nationalism for a radical openness to strangers and for radical justice, because for him the roots of the cry for cultural identity lay in oppression, fear and poverty. But it was always necessary to be self-critical about one's own structures and not only about state structures. He certainly recognised that, on the Islamic side, Islam was put forward as the solution to the problems and injustices mentioned. In contrast to Islam, Christianity did not simply offer a solution that could be applied like a blueprint. He himself saw his proposal as only one possible Christian approach and he was aware how strange this basic approach must seem to Muslims. But his concern was that societies should preserve human dignity, solidarity, diversity and culture in the future as well, and he believed that was possible by virtue of the love of God which embraced us all.[120]

Overall, this address can be seen as a key contribution, the only one which really attempted to clarify and explain the phenomenon of nationalism on a fundamentally anthropological level, whereas the other analyses remained more on the surface. It also became clear here why Christians and Muslims seemed not to be communicating on this question throughout the conference – they were arguing, so to speak, on different levels of an anthropological development and the Christians were historically more advanced and, above all, a stage further on in accepting this development because their community model was also much less rigidly established. It is not too clear whether this basic problem can be resolved; that can only be hoped in the interest of worldwide coexistence as Linden did.

7.4.5 The concluding comments – different in every respect
At the end of the meeting comments were made by both sides but of very different sorts. The one from the Christian standpoint written by Lambert Ejifor from Nigeria was as extensive as the largest earlier contributions and hence by nature almost a new contribution, because he made absolutely no direct reference to the previous speakers and introduced a wealth of completely new ideas which – being compiled from a different national point of view – defined the Christian approach completely differently and, at the same time, since he did not really want to accept a national perspective, claimed an absoluteness in exactly the same way as the Islamic side had done earlier. In a sense, he presented a new paradox and it is certainly noteworthy to see with which arguments and how Ejifor developed them.

7.4.5.1 A Christian voice from Nigeria: nationalism

After a very short analysis – if one can call it that – indicating some situations, he came to the conclusion that nationalism was a phenomenon and factor deeply rooted in human motivation and of unparalleled duration. He saw nationalism as strongly ideological, namely as a comprehensive complex of values which made total demands on the life and actions of human beings just like religion. Religion and nationalism also had many other things in common. Both were simultaneously abstract and concrete, mysterious and practical, had enormous energy, were immeasurable and sometimes incomprehensible. They both relied on mysticism and the power of feelings. Although here he saw religion more directly linked with human beings, because atheism and agnosticism were demonstrably difficult to maintain precisely in times of crisis. This and the innumerable ways of worshipping God all around the world were a sign, for him, of humankind's profound consciousness of God and of the need for religion as the soul of life and of a meaningful human existence. In contrast, nationalism was more artificial. Ejifor also made a distinction – perhaps to some extent in line with Linden or his authorities although he did not say so explicitly – between a genetic or anthropological and a political significance of nationalism. On the first level, it was a feeling or an emotional surrender to one's own origin, reinforced by an unshakable identification with everything derived from that origin.

Nations, wars and treaties

He defined "nation" as the commonality of shared origins and hence nationalism also as enthusiasm for, devotion to or cultivation of one's own nation or of the principle of nationality in general. The genetic aspect, which was to be taken very literally, clearly predominated in his view; a common genetic origin, when one was aware of it, inevitably led to an alliance with one another and against others, where the identification somehow became a psychological reflex for the integration of the human being and the perfection of the ego and for everything that guaranteed a person's self-preservation. In addition, there were non-human factors such as land, myths, values, common plans and goals for defence against hostile attacks, disorganisation and disintegration. The individual or the group defended this togetherness of relatives (although, linguistically at least, account was taken only of the male line) as if a larger self was being attacked. This original nationalism, which however normally wanted to possess its own state (at least according to a quotation he used which was broader than his own definition; another quotation even stated that the common features did not need to be real at all and could be a construct, literally "imaged", which naturally clearly recalled Linden's "imagined communities"), could also be derived from a common language, a common religion, the tradition of a common origin or the belief in it. According to his view and statements, this original nationalism would stand no dissolution from inside or outside. It was spontaneous and jealously possessive and could also evolve from patriotism to chauvinism. In any case, it was always fundamentally at the expense of its surroundings, and here he gave a large number of historical and contemporary examples from colonialism to present-day fragmentation, and he did not even hesitate to include the Jews and their Zionism after the fall of Jerusalem. In contrast to Linden, he certainly also found philosophical terms for this indivisible group spirit which was quite different from a pragmatic gathering of individuals. He saw Hegel's words about the soul of the race and Kant's pan-psychism as appropriate. Surprisingly enough, and this was apparently his own private hypothesis, he saw this original nationalism as the cause of innumerable wars – although these took place between secondary nation states. Ejifor's view of this was that the excessive hatred and exclusiveness, which fuelled (primary) nationalism, were ideal material for manipulation by propaganda, indoctrination and unobtrusive prejudices. Historically, he saw a kind of turning point in the Congress of Vienna where, for the first time, values and virtues were no longer a national monopoly and the civilised world came together to think about and feel for the well-being of humankind on a grand scale. From that time onwards there had been treaties, conventions and multilateral negotiations on an international level right through to the Charter of the United Nations, thanks to which there were now new fundamental freedoms and rights of individuals and moral persons, the observance of which could also be effectively supervised, although he at least admitted that a law which could not be broken was not made for human beings. But irrespective of whether he perhaps judged the role and authority of the international organisations too positively before he officially came to the problems at all, it certainly became clear that he already viewed nationalism in its original, anthropological form as natural but not ethically neutral and, on the contrary, considered it unavoidably negative in two respects. This was – in some sense surprisingly – the most negative evaluation presented thus far.

The present-day problems caused by nationalism
Irrespective of what effects it was having in practice, whether in large imperialist aggregates or in atomisation, frequently known politically as balkanisation, for him the key problems of nationalism were sovereignty and territoriality and those were massive problems. In all, he produced a list of almost 20 problems from his point of view, some of which belonged to the primary, anthropological level and had already been touched on indirectly. Nationalism by nature prescribed self-interest and the discrimination of others and emphasised the differences between people, thereby fragmenting the world, and the present problem was

precisely linking up to form a system of sovereign states, that is, the possibility of determining or destroying values by means of legislation. Nationalism was not responsible to anyone outside, also not to a universal or divine or natural law but, on the contrary, was ideological, hence to a large extent a system of beliefs which served the national élites in each case. Nationalism wanted to expand both on the primary and on the secondary level, naturally at the expense of others (which inevitably led to conflicts and even war) and it condemned any form of intervention that was seen as a challenge to its own power. But, since nationalism today had recognised that religion was indestructible, it had adopted an instrumental, pragmatic attitude to it, indeed, it was inclined to choose the religion which best corresponded to its aims. But nationalism contained elements of religious discrimination and persecution. Because it was aware of the principle of universality and of crimes against humanity, it avoided abolishing them by bloodshed in favour of a bloodless death. In addition, nationalism had been intellectualised and refined to such an extent that it at least partly took account of the needs of people with a religious consciousness and thus made positive religion unnecessary or at least partly replaced it. What Ejifor constructed here on the basis of a so-celled primary, anthropological nationalism was a massive front against religion. This was really much more deeply and fundamentally rooted in people but, according to their worldwide value system, it opposed nationalism and for this reason there was bound to be a battle between the two which, in the meantime, was no longer being conducted openly on the side of nationalism but through internal erosion and manipulation. A kind of nationalist substitute satisfaction was being offered for people's religious needs. Naturally, these were serious accusations, especially because they were not directed against excesses of the system, as might be the case with others, but against purported or real inner necessities. He saw the problems and challenges for religion mainly in three points: a power struggle between two hierarchies (his formulation is striking that religious ones could be formed either by ordination or by election) and hence automatic tensions when the religious and the secular came up against one another, at least with regard to the solutions which still seemed possible today. He again underlined the insight that nationalism had proved to be one of the strongest factors in modern psychology, and went on to regret that, whereas the secular world was looking at the major questions of humankind, the religious world – and here he must have meant particularly the Catholic and/or the Western Christian world – was becoming increasingly divided by its efforts at acculturation. Here, and again this is reminiscent of Linden's remarks, he spoke about different concepts of values but first and foremost about the languages of the people, the prevalence of which also in the religious realm was tantamount to the Fall, also in the Christian-Muslim dialogue. His arguments and choice of words on this point could hardly be more strident. "As people splinter away in their vernacular they tend to add more doctrines by sheer dint of rolling home. This fission phenomenon has become institutionalised in the Western and Christian world, where state Constitutions delight in 'secular state' axioms and Christian religion is basking in fissiparous acculturations and adaptations. It is now an anachronism to advise peoples to contemplate and worship through a shared linguistic medium. Eccentrics are practically inaugurating a tower of babel in Christianity. With religion sundered from the secular state and immersed in a liturgy of incommunicable tongues, most of the Western and Christian world cannot be disposed to dialogue with fused societies. The indivisibility and totality of the man in nationalism rooted and cultured in positive religion is unthinkable aberration from enlightenment, development and scientific rationality. This is the major stumbling block to sincere dialogue with the Muslim world which still strives to maintain unity of belief with unity of religious worship. Idiomatic communication has consequently suffered a fatal blow."[121]

Reintegration of nationalism into religion?
Against this background, it was surprising that he saw any chances at all for reconciliation between the two. He found them, to some extent following Linden (but again without stating it explicitly), in his view of nationalism today as a protest against inequality and injustice. If only everyone followed the international declaration on human rights (and some other things along those lines), as he repeatedly emphasised, then the problem would be solved; here, he quoted Cardinal Arinze who had said that there could not be different classes of citizenship – but without saying that this statement could also be explosive for religious systems, not least for the Islamic one. For Ejifor like for Linden, the consequence for religion was an option for the poor to obtain distributive justice and counteract political and economic factors of which the main factor was secularism again. In particular, religion should exercise its prophetic qualities when it was a matter of denouncing racial discrimination and discrimination against minorities which was a-religious or the greatest of all crimes. But he did not completely discourage cooperation if it served his aim of reconciling religion and nationalism in the sense that the integration of nationalism was re-established which, for him, constituted the life of human beings in its totality. Here, he again spoke very clearly and even poetically.

"A nation ruled by godly dictators is better than one ruled by ungodly democrats."[122]

One can imagine that this commentary caused much more confusion than clarity, especially among the Christians themselves. Unfortunately, the conference report contains no article on how the discussion went although that could certainly have been one of the most exciting and enlightening parts of the whole volume.

7.4.5.2 The Muslim commentary – a clear question about the future

In comparison, the Muslim commentary provided by Ms. Hanan Ibrahim from the University of Amman was very short. Among all the contributions, she particularly saw that of Ian Linden as positive because of its analytical depth and comprehensive approach, and she attempted to apply it linguistically and in content to the Arab realm although she did not quite succeed, and this led in turn to her main point of criticism against that approach. It was too one-sidedly Western, i.e. nationalism was seen too narrowly as only relating to states. She saw Arab Islamic nationalism more on the primary level in Linden's concept, namely that of the sense of belonging to a particular culture as an anthropological necessity, but she emphasised that there had always been room in this framework for other ethnic groups and religions and their cultural contributions. However, she admitted that the situation in the West had been different, possibly because it was composed of Christian and Hellenistic elements which were based on the separation between religion and state, and hence religion had not been in a position to shape national and political entities from the very beginning or as a whole; that she considered neglectful. She also saw that, in addition to Latin, there had always also been mother tongues which were able to develop and on the basis of which various cultures could evolve – not to mention the fact that the churches themselves had contributed directly to divisions. For her, the real questions – and here she was clearer and more emphatic than all the others – lay in the future. What would become of nationalism in the future in view of the economic pressure to which all were equally exposed, in view of growing religious movements which wanted to put religious unity in the place of national and regional unities (an amazing comment from this direction!) and in view of the extremely rapid progress of communication media which meant people were being overwhelmed by the values of the West? She saw better chances of renewal if nationalism was rooted in the primary level, and more opportunities of influence for religion if the latter was linked with nationalism as in the Arab realm. All the contradictions between national and religious or between national and regional belonging (although it remained unclear what exactly she meant by the latter) did not exist; there were only different historical models of equilibrium between Islamic and local demands between which people could somehow choose.[123]

Even in this commentary, one can still sense that, despite all the listening to one another, there was no real communication, at least between Western Christianity and Islam, because they were unable to agree on a common role which religion should play for people in the social and political realm. They were far removed from a common view of the situation and still more from a common solution.

7.4.6 The concluding remarks – surprisingly conciliatory

The meeting ended with a reception given by the then Crown Prince of Jordan at which Cardinal Arinze again expressed his thanks for the meeting and appreciated it as a step towards the right solutions. That could, however, have been a specifically Christian view of things which simply generously included Islam at this point, although Islam perhaps saw itself as being more directly at work and not only via the detour of influencing human hearts, as Arinze put it. "The religions do not pretend to have all the answers to world problems in matters political, economic, social, cultural and otherwise. But the religions can motivate, convert, heal, conscientize, reconcile, suggest, and dispose human hearts towards the correct solutions. Our colloquium which is about to conclude its fourth session is a step in that direction."[124]

Then the Crown Prince of Jordan, Hassan, went into greater detail and was more concrete, apart from the fact that he expressed his hope that the time would come when a centre for inter-religious studies could be established in the region. He also asked Cardinal Arinze to convey thanks to Pope John Paul II for his continuous condemnation of acts of violence, especially against Muslims, in the former Yugoslavia. The Pope had done so even more clearly than the representatives of the Muslim world themselves. Crown Prince Hassan also repeatedly emphasised the importance of young people (invited to this conference as well) if what had been started was to continue; in this respect, his thinking was similar to what John Paul II had made so clear when speaking to Muslim youth in Casablanca. He also set great store by human sensitivity and common respect for values, precisely when it was a matter of fundamental decisions about other people or about nature, and he went so far as to coin a new term "anthropolitics, politics for people"[125], for him the key term for the twenty-first century.

The final communiqué does mention that there had been lively participation in the discussion but only lists the points on which agreement could be reached which, considering the material in the various contributions, seem rather banal. A distinction had to be made between a natural healthy love of one's country and a destructive, chauvinistic nationalism that attempted to exclude, humiliate and subjugate others. There was no contradiction between religion and nationalism. According to both Islamic and Christian doctrine, before God, no race or people was to be considered of greater value than any other – before God, people were judged only by their faith and obedience. Followers of both religions recognised that the positive values of nationalism could contribute to forming and strengthening identity and to the sense of one's own value. Therefore religions should support these values and, in reverse, condemn and work to prevent national impulses becoming means for dominating or destroying others.

7.5 Rome 1996 – using the resources of the earth rightly

Together with these results it was announced that the next joint colloquium was to take place in Rome during the second half of the year 1995. The subject was to be religion and the environment with such sub-headings as religious bases for dealing with environmental concerns, the just distribution of the gifts of creation and the role of the faithful in influencing decisions related to the environment and the sharing of resources. In the end, this meeting took place from *18–20 April 1996* with 30 participants from both sides and the official title *Religion and the Use of the Earth's Resources*. As usual, there were opening addresses by the two presidents in which, in contrast to that by Prof. Nasir El-Din El-Assad, the one by Cardinal Francis Arinze provided a lot of content and, above all, anthropological depth; it started immediately by saying it was our duty as human beings to recognise God as the creator of all that existed, the visible and the invisible. Everything apart from God needed God in order to exist. It was God's will that the resources of the earth were entrusted to human beings so that they could use them, care for them and, especially, praise God through this administrative service and also share them with their fellow human beings, since the nature of human beings, according to God's will, was social and the benefits of creation were intended for everyone. The reason for their predominant position (described symbolically in the Bible by humans receiving from God the task of giving the animals names) was particularly the intelligence of human beings, their rational structure, the fact that they had an intellect and will and consisted of body and soul, matter and spirit. This background alone was enough to make clear that, in their practical action, some individuals and countries did not accept that all human beings were brothers and sisters in the common pilgrimage of life, and that they had probably not grasped that, one day, they would have to give account before God's judgement throne for their administrative service with the gifts created by God.[126]

This, in essence but naturally not in all the details, already said what had to be said about the position of human beings in relation to the resources of the earth from the Christian point of view in the wide sweep from creation to divine judgement.

7.5.1 Statements on the use of resources
7.5.1.1 The use of resources – a major theme for Islam
The first round consisted of addresses and responses on what the two religions said about using the resources of the earth. It was striking that the volume of the Muslim statements on this amounted to three or four times the number of pages for the Christian contributions. The Muslim response, for example, was almost as long as the Christian address itself. That showed that this was obviously a subject that was of special importance for Islam and Muslims and on which they had a lot to say. Another reason for this was that Islam, much more than Christianity, did not just lay down bases and basic lines but rather regulated the aspects of everyday life in great detail so that a speaker on an issue of this nature could refer to a broadly diverse and detailed legal tradition. The Muslim speaker, Abdul-Salam Al-Abbadi, divided up his address accordingly: the general Islamic attitude to the universe, life and humanity was only the starting point; his second point discussed the fact that the question of resources was especially important to Islam which endeavoured to regulate the relation between humans and resources in every detail; then the third point dealt with the most important, specific rules for the relation between humans and their natural environment, whether for the individual, a community or the whole of humankind; and the fourth, concluding point gave examples of the historical implementation of these rules. So it was hardly surprising that this produced an address of nearly 40 pages which was a prime example of how one can govern the world and human behaviour very thoroughly – or, on the other hand, leave many things open in this respect. The participants were also aware of this.

The vicarious role of human beings – the comprehensive approach of Islam
But the Muslim speaker was aware of this with special pride since he wrote at the beginning of his remarks, "In this context, it has to be noted that the most significant characteristic of this religion is that it has a comprehensive outlook towards existence, life and man, which had the greatest effect on all Islam's moral orientations and legal principles. It is a perfect outlook that explains all that is related to human life, clarifies the numerous facts of existence and answers all the questions that may occur to man's mind about his existence and the dimensions of such existence. This outlook is different from all the views and envision-

ments known to humanity about this existence in the past as well as in the present."[127] Particularly with regard to the theme of this dialogue conference, the basic point was naturally that God had created all living beings (from water) and made human beings his representatives, equipping them with everything they needed to fulfil their duties, namely developing the earth and establishing the complete lordship of God on earth. The whole earth was to serve God in every way. For this reason, human vicarious authority was also not absolute but subject to principles and rules laid down in Islamic law. Its only aim was the right service of God, namely persisting in all of God's requirements. If people deviated from them, they did not fulfil the conditions for being representatives and thus disqualified themselves. So people should act rightly and justly, exactly as God had prescribed, so that their possessions were not taken away from them again and entrusted to others – because they were not the owners; God alone who had created everything was the owner. People were also responsible to God and he would punish them for wrong actions. This vicarious responsibility is also not permanent but for a specific period of time determined by God, just as God had also determined the lifetime of people on earth and thus made this life a transition period, a test phase for the other life after death. God had honoured human beings by creating them in the best way and given them qualitatively more than the major part of creation so that humans were also capable of their tasks. This included understanding and reason and the possibility of acquiring quite new knowledge in this way. No other system took so comprehensive an account of the true nature of human beings as Islam. "[I]t is clear that man, according to Islamic outlook, consists of body, mind and spirit, each of which has its requirements and needs that have to be satisfied and met to enable man to perform the duties linked with his vice-regency on this earth. Hence various Islamic institutions have organized and regulated human life in a way that gives due observance to this human many-sided and harmonious innate character (*fitrah*), under which man has led a happy, serene and secure life. For the institutions that survive in the regulation of human reality and achieve happiness, rest and equanimity for man are institutions that pay due attention to man's innate nature. Such an outlook is one of the many characteristics that distinguish Islamic from other ancient and modern institutions."[128]

Matter and spirit – no contradiction in Islam
According to the Muslim understanding, it was also the Quran itself (Suras 10:101 and 7:185) which had led to the development of the experimental method and of scientific research and discoveries in general. For Islam, that was a form of worship for which humans received a great reward. It was not appropriate to maintain that Islam was against the natural sciences; on the contrary, in this respect

the Europeans had taken over everything from the Arabs; Islam was only concerned about the moral guidance of the natural sciences away from anything that could disturb the order and harmony which God had given to the whole cosmos. Human beings also had the right to use everything that God had created and to enjoy it, but naturally they also had to thank God for it. This Islamic approach was also directed against any claims that matter and spirit could not be combined, and was therefore vehemently against Western materialism which was a spiritual impoverishment and a crisis of values and morality. Wise people in the West were truly yearning for Islam, at least indirectly. "Wise people living under the umbrella of this civilization complain loudly because of individual as well as social deviation, chaos and problems as a result of the absence of spiritual and moral values in politics, economics, and mutual relations of people whether as individuals or as states. They yearn for an order that combines spiritualism with materialism and coordinates between them in a manner that ensures human elevation in material as well as in spiritual spheres. They want an order that maintains the advancement achieved so for *(sic!)* by humans while it insures human progress in the world of values and spirit. Islam, in fact, combines materialism with spiritualism through a special outlook that makes a person who performs all the requirements of his life on this earth in terms of benefitting and enjoyment, in accordance with what God /p. 24 has prescribed, a true worshipper of Almighty God for which he deserves the most generous reward. This will also provide him with the means that make his life on this earth a path of man's welfare and happiness in this life and in the hereafter."[129] Up to this point we are dealing with the realm of generalities and basic principles.

Resources – some concrete rules from Islamic law
Then he entered the realm of more precise regulations which could naturally not all be mentioned because practically everything relating to possessions had been regulated. But it was clear that there was a duty to develop the land, namely to make it suitable for fields, fruit plantations and buildings. But naturally the regulations of special interest were those which prevented one generation from intervening in the rights of another. In Islam, environmental protection was, in a sense, a matter of divine law; an offence against the environment was simultaneously an offence against God's commandments and laws. Indeed, the Quran prohibited excesses in the use of possessions, namely extravagance, luxury and unnecessary waste, and demanded moderation. It could be stated generally that the exploitation of the environment had to be dependent on security aspects, especially that no harm was done to others; the golden rule of Islamic law was considered to be the saying of Muhammad that Islam did not allow

harm or hindrance and therefore the whole of Islamic law aimed at bringing about good deeds and preventing harm and, in comparison, the latter was even the more important. In Islam the totality, the community, was very important, as had already been emphasised several times, and therefore public interests took precedence over private interests; so private harm was allowable in order to prevent public harm. For example, the Islamic tradition prohibited the use of bodies of water, roads and shady places to relieve oneself. On the contrary, roads were to be cleaned. And e.g. residential and industrial areas were to be separate; in practice, special places were even assigned to the individual trades in the markets. For trades such as butcher, water seller, baker or chemist there were special, precise hygienic regulations. In general, Islam considered attractive surroundings very important (even specifying that people should wear decent clothes and clean shoes) because that comforted the human soul and in this way motivated people to put their hearts into their work and work harder. Regulations applied even in the case of war: nobody should be (deliberately) maimed, children, women and old people should not be killed, palm trees and fruit trees should not be felled, wells not destroyed and no kind of burnt earth policy should be followed. So, also in the view of outsiders, the Arabs were the fairest conquerors in history, just as Islam was concerned to have a just society in general. "In reality, Islam has been careful to set up its society on bases of mutual dependence, cooperation, love, benefaction, mercy, justice and combatting injustice and all forms of aggression and harm against others."[130]

7.5.1.2 The Christian response: similar but less detailed

The Christian response by Jamal Khader strongly emphasised the common approach to creation and its protection which also made cooperation for the well-being of humankind possible. The biblical account of creation included that God had a plan which human beings should consequently also respect. Humans were to have dominion and cultivate the garden, naturally within the framework of observing divine law in justice and holiness and with respect for their having been created in God's image. In that way God was to be exalted throughout the earth (as in Ps.8:7+10). In this, human beings were God's representatives, not his substitutes, as Islam also claimed. But, as far as the relation between material and spiritual reality was concerned, the church saw it differently; it believed in the autonomy of earthly affairs. As a result of creation itself, the material world had a stability, truth and excellence of its own; it had its own orders and laws which humans had to respect and also follow. But these were not defined further and it was left open what was human autonomy and what was God's law. The problem today was that people wanted more to possess and enjoy that to be and grow, and that they therefore exploited the resources of the earth and also their own lives excessively and in a disorderly way. Instead of carrying out their task as God's fellow workers in creation, human beings set themselves up in God's place. For the sake of justice, however, the rich nations should share with the other members of the human race and all cooperate to protect nature and preserve it for future generations as God willed and according to his laws. Technical progress as such was seen as less valuable than progress towards greater justice, brotherhood and a more humane social environment. There were also remarks on work and social responsibility for property and its relation to the common good but these were not very clear either.[131]

Despite all the emphasis on human beings as creatures and the very similar tasks and responsibilities entrusted to them in Islam and Christianity, it again became clear that there were major conceptual differences regarding practical implementation in the sense that Christianity offered a lot of freedom but this left a number of practicalities unclear, whereas Islam claimed to offer a balanced, ingenious overall conception of divine origin and could provide a considerable collection of regulations.

7.5.1.3 The bases of a Christian ethics of creation

The official, fundamental Christian contribution by Msgr. René Coste also started with the theology of creation, moved on from there to human beings and the resources of the earth and only then provided a few points of reference within the framework of a theological ethics of creation. Of course, he also emphasised that the Abrahamic religions had the belief in creation in common and that one could go on from there to the question of how this creation should be treated ecologically in its dogmatic, ethical and spiritual dimensions and generally in many different forms of cooperation. God the Creator was not seen as overwhelming but rather as attractive. He was the fulfilment of humankind and the expression of its freedom. Christians were moved by grateful admiration and adoration for him because they were convinced that everything came from him, themselves and all that they possessed. When thanking, one had to recognise oneself as a creature and the earth and its resources as a gift from God. God had created out of love and very well, so a degradation of nature was also an offense against the creator. In addition, the outstanding dignity of humans, of each human being, even the most despised, was fundamental biblical teaching and anthropology, as was the equal dignity of man and woman because both were created in the image of God. Human dignity was participation in the dignity of God, the

Creator. "The affirmation of the fundamental dignity of a person, inherent in each human being as God's image, is one of the essential elements of the biblical conception of man. Its consequences are of capital importance for the ethical principles regarding the use of the earth's resources: not only no human being must be excluded, but each must have the same fundamental right to participate as well."[132]

The human being's standing in creation
The Bible also emphasised the unity of humankind in God; human beings were never seen as isolated individuals but as members of a large family who should relate to one another like brothers and sisters. They were called to live together in peace and solidarity, beginning by refraining from violence and crimes against one another. Humankind should be our horizon and we should see every person as a member of humankind. As far as resources were concerned, ethics demanded solidarity in time and space, especially in relation to the poor and all who lacked the essentials for living. At this point, the speaker also referred to the accusation that a so-called biblical and Christian anthropocentrism had made people not only into privileged beings but also absolute lords of the earth and the centre of the universe. Western Christianity, in particular, had had to face the accusation that it had taught that it was God's will for human beings to exploit nature for their own interests. This view had then decisively influenced the aggressive development of science and technology in relation to nature and was therefore responsible for the present degradation of nature. That attitude was most emphatically rejected. Certainly, human beings were at the top of the hierarchy of created beings, but in the sense that they had the greatest responsibility and naturally remained God's creatures and should act as God's servants. "We should not speak here of anthropocentrism but rather of theocentrism with regard to the whole creation and therefore for man himself in its midst."[133] However, he did not deny that many Christians had contributed to the current enslavement of nature and used the Bible to justify themselves. But the choice of language in Genesis on this point was no basis for tyrannical rule or exploitative misuse. This was where the love of the creator for his creation and human beings being created in God's image came in because it implied that human beings should also act lovingly in relation to creation. Yes to personal human initiative but at the same time also a subordinate and necessarily obedient role in relation to the creator who wished solidarity and peace for humans and also expected the same from them. The true role of human beings was to serve in relation to God and to creation – even though they had often not carried out this role. Moreover, all people had the basic right to benefit equally from the resources of the earth.

This was why the Bible constantly criticised injustice towards the poor because that deprived them of this right.

Key elements of a Christian theology of creation
A theology of creation required an ethics of creation but this had to be accessible for everyone, including non-believers, because ecological and environmental problems affected the whole creation. This type of ethics was a speciality of Catholic social doctrine. Reason demanded a doctrine of morality and a monotheist faith could only reinforce such teaching. An ethics of creation had to define the responsibility of human beings in a specific context and spell it out concretely. The speaker attempted to set up some pointers in that direction. In the light of belief in God, the Creator, no property right was absolute because God alone was the true owner and we humans were always responsible to him and to our fellow brothers and sisters. A practical definition of the concept of property always had to include an ecological approach and the universal purpose of possessions. Any use that we made of the resources of the earth had to presuppose that we would not ignore the rights and needs of other people in the process. This also led to the principle of the right of future generations in an ethical and juridical sense because, from a theological and philosophical point of view, the earth belonged as much to them as to us by right of birth. The ecological approach was also in the interest of the economy itself which had to be an economy for the whole person and for all people. The so-called "option for the poor" had been born within the Catholic Church but in the meantime become an ethical demand of the Christian churches in general. It was clear evidence of the universality of the church and its mission. The values which had to be fundamentally respected were respect for life and especially for human dignity, justice and solidarity on the worldwide level, as well as love for one's neighbour as a specifically Christian as well as basic and global value which also gave rise to a radical demand for justice. In conclusion, Msgr. Coste made a few more remarks on the primacy of ethics in the field of science and technology, mainly quoting documents and papal speeches, and emphasised that people's happiness and health depended less on material possessions and more on the gifts of nature and on other creatures, on human relationships and the relationship with God. He concluded by repeating a papal appeal for a different style of life, for greater simplicity, modesty, discipline and a sacrificial spirit, so that modern society would be able to solve the ecological problem.[134]

Overall, it is striking that, parallel to the main Muslim emphasis, here, too, much value is placed on understanding creation as brotherliness in space and time and also on the community, the system as a whole and primary care and attention for it. Msgr. Coste also made clear that it had not always been possible to take that for granted but that it corresponded to the true sense of the biblical message and to God's will for humankind. Here one can speak about great or growing proximity between the Christian and Muslim standpoints although the stage of their expression is naturally different.

7.5.1.4 The Muslim reaction – far-reaching agreement

The Muslim reaction to that was generally affirmative before going into a wealth of Islamic details. "If I am required to comment on this presentation from an Islamic standpoint, I will readily say that I have found that these foundations and teachings are essentially almost the same in terms of their ideological reference and jurisprudential teachings starting from God's aim of creation, through man's relation with God and the universe, and ending up with the creation system which necessitates that man should look at life with respect, dignity and integrity and within the framework and the responsibility that man assumes in existence."[135] But the speaker naturally did not miss the opportunity to spell that out again in a detail which, as already stated, was almost as long as the address itself. The amazing thing was the extent to which, when dealing with the very first Islamic point of departure, human dignity, he came close to the Christian formulations and even to speaking about being in God's image. "Thus in addition to the honor bestowed by God on man in creating the latter in His own Image and the blessings and bounties He conferred on him, He gave him / p. 60 preference through making him learned and civilized, and apt to acquire more knowledge and achieve development and progress in all aspects of his life, which is lacking in animals or other creatures."[136] And, in addition, humans were God's representatives and Islam focussed especially on the community and appealed for coexistence and cooperation. Two quite new aspects were that Islam saw society realistically and perceived that there simply were rich and poor people because, according to Sura 16:71, God was more generous in his gifts to some than to others. Nor did Islam preach the abolition of poverty or wealth but attempted to reduce the gap between the two and to do away with humiliation, exploitation, violence, tyranny, anger and their grounds. Muslims were a community of the centre, totally balanced, which relied on work as the basis for affluence and on nobody being deprived of their rights and rewards, especially not orphans and daily wage earners. Extremes were prohibited such as extravagance and miserliness, expenditure on worthless trivialities, and this was also taken to mean any illegal profit such as interest, games of chance, bribery and deception as well as monopolies and hoarding possessions. The reason for this was that Islam tried to reduce the harsh effects of capital. This was done, for example, by the alms tax and various expiatory payments, but Islam demanded generally that pious foundations be established and money be spent for religious purposes. This money was seen as a kind of loan to God which God would pay back several times over. Muhammad had also laid down that the essentials for living (without which the service of God would also be unthinkable), such as water, grazing land and fire or energy, should be common property on which no price was to be set. Such provisions could also prevent a class struggle because, if a society was endangered when people did not carry out their duties, it was also in danger if they did not get their rights (food, security, justice, wellbeing, happiness, dignity, etc.). That, too, caused social instability. The contrast between oppressors and oppressed, hungry or overfed or secure and anxious should not exist in a society in which social peace and faith reigned. Muhammad had already said that a person was not a true believer if he spent the night with a full stomach while his neighbour went hungry and he was aware of his situation. Gratitude to God really included all that automatically, but humans had a tendency to ingratitude and generally to being negative and destructive and to departing from moral values in their thinking and action.[137]

Again, the fundamental points of agreement became very clear but also that Islam is much more concrete and intervenes more directly to regulate but in the process (inevitably) compromises a bit on the ideals, something which Muslims praised as the so-called intermediate attitude.

7.5.2 Protection of resources
7.5.2.1 For Catholics: an ethical question
Then the addresses turned to the more practical question of what the two religions said about protecting the resources of the earth, although the Catholic speaker, Kevin Irwin, underlined that he could only speak for the Catholic part of Christendom and that for them it was more important to see what was not or only cryptically said by the magisterium. The statements on this question always adopted an ethical or anthropological approach (human responsibility, being in God's image, the dignity of human beings and of human work); they saw human beings at the centre of creation and responsible for respecting the autonomy of creation and for the just, brotherly distribution of its benefits. The human control of creation had to be conducted so that an economic order came into existence which respected the dignity of all people, namely that the earth and its resources were preserved because they were also important for the life of future generations. Human beings had to keep watch over creation.

Their incomparable dignity and their special gifts corresponded to their great responsibility. As farmers, for example, but also generally through their work, they were fellow workers with the creator as they carried out their tasks. But with their free will they could also misuse their gifts, not carry out their tasks, destroy, spread poverty and fear and finally harm themselves. John Paul II had stated explicitly that environmental protection was an ethical question and that the people of today had to make provision for those who would come after them. He, too, had emphasised the task – albeit limited – of humankind (from the very beginning there had also been divine prohibitions) and the importance of a just distribution of the goods God had given, especially between industrialised and developing countries, and had called for a different style of life also because human beings had a spiritual nature and a transcendent purpose. And finally he himself had declared responsibility for creation also to be a concern for ecumenical and interreligious cooperation. He had also made the link with human rights because development was necessary in order for human rights to be implemented. On the other hand, it was an anthropological error if people forgot that the resources of the earth were a gift from God and believed that they could use them just as they liked. Regrettably, people had more desire to own things than to associate them with the truth (this connection had already been mentioned several times at central points in the Islamic addresses but with the aim of demonstrating how well disposed Islam was to science; on the Catholic side the speaker described it as a sacramental view of reality). Again it was the anthropological axis, as the speaker called it, of the outstanding position and responsibility of humans around which everything else rotated, combined with a special emphasis on global and also interreligious responsibility. To summarise in the words of the speaker, "John Paul II has been very concerned to support the intrinsic worth of the human person and concomitantly to articulate responses human beings owe to God their creator and to fellow human beings. It is in this connection that contemporary papal teaching stresses responsibility for creation. This teaching can be characterized as overwhelmingly theocentric and anthropological. Creation is God's good work; humans must respect it and preserve it for future generations. (...) This anthropocentric vision grounds a contemporary rereading of the scriptural theme of stewardship. Here 'dominion' means responsibility for creation. (...) [T]he problem with destruction of nature is not that nature itself is injured but that it is a symptom of our lack of obedience to God, of sin, of the perversion of what *our* human nature ought to be – rational, sensible, humble. Given the fact that creation itself has traditionally had a privileged place in Catholic theology and belief and that in Islam due acknowledgement is paid to God the Creator, it would be helpful if this appropriate anthropological foundation could be broadened to include respect for the world's resources precisely because they are resources from the world as created and blessed by God as 'good'."[138]

Additional Christian lines of reasoning
The speaker also indicated other possible lines, such as the covenant with Noah or, as something specifically Catholic, the sacramental view of the world, namely that human beings experience God in and through the world in which they live, and that God can be and really is experienced there, perhaps all the more if the environment can be liberated again in part from its degradation. The parallel to Islam already mentioned above was specified here quite explicitly, although the term "sacramentality" was naturally extremely un-Islamic. He also took up the question of creation and redemption and the fact that their precise relationship still needed to be clarified theologically in reference to a theology of creation and ecology. God the creator redeemed and God's redemption extended to the whole creation. The image of pilgrimage was also very appropriate for reminding us that we should preserve the resources of the earth for future generations because we were all here only for a limited time. Finally he also referred to the Benedictine tradition of monasticism with its emphasis on human work as well, especially in connection with making land productive, and to Francis of Assisi as an example of amazement in his approach to and sacramental view of creation which rightly made him the Catholic patron of ecology.[139]

As a whole, this was more a collection of elements which repeatedly revolved by inner necessity around specific points than a fully developed theology on this question, but to that extent it was of special significance within the framework of this investigation because it was therefore still very anthropocentric and anthropological in its approach.

7.5.2.2 Islam – very well suited for cooperation on matters of nature and ecology

The Muslim response by Mai Muthaffar also began by strongly emphasising the extremely important task of human beings (which everybody else had rejected) and their equally outstanding characteristics and that the earth had been created for humankind. She saw common features in Islam and Christianity at this point and also the possibility of acting together precisely in the present situation in which human beings, to whom the earth had been entrusted, were ruining it by destruction and pollution and the violation of human values, because they had abandoned all morality when they turned away from God. Human beings were no longer able to stand up to this destruction alone; international cooperation was necessary. Islam and Islamic law were again seen very positively in this connection; among other things it had been Islam which had made the nomads sedentary. Indeed, faith brought with it the possibility of shaping external things because this life itself demanded sacrifice and self-denial. In brief, Islam was very well suited for this cooperation in which it was a matter of developing ethics of nature and ecology and of cooperating for more justice and a fairer distribution of the possibilities for living; that was seen as one of the really big problems.[140] All in all, this was a very positive response with hope for the future.

7.5.2.3 The basic Islamic question: who is a true believer?

The Muslim address on this issue, which was given by Sufian El-Tall, began by trying to clarify fundamentally what a believer was and how the earth was understood by Islam, in order then to be able to relate the two. Inevitably, many aspects which had also played a central part earlier were repeated and thus underlined. To begin with, the criterion for believing was very demanding. Faith had not only to be so deeply rooted in people's psyche that it governed the will and feelings of the persons concerned and their perception of the world, life and humankind; it had also, according to Sura 49:15, to reflect such a degree of decisiveness and certainty that doubt was no longer possible. Faith also led to action because believers could not bear any discrepancy between the image of faith in themselves and the real image round about them without always being shocked and hurt. Faith determined the role of human beings in the world, and here Islam had always emphasised people's freedom and their responsibility for themselves and their actions. Believers loved and respected nature because God had created it for them. But God had also made nature subject to them for them to fulfil their task; they were entitled to use it, enjoy it, etc., provided they also protected it which was their responsibility. Thus, special emphasis was placed on a rational treatment of the environment because the right to use it belonged just as much to the coming generations. Believers loved all creatures and all people irrespective of their skin colour and race. They considered them brothers who were created together with them from one person. "His faith does not permit him to get involved in ethnic, racial or gender conflicts, or to envy him or fanatically side against him; because he believes / p. 99 that all men descend from Adam and that Adam is made from the soil of this earth."[141] In addition, believers believed in work, loved it and carried it out without compulsion from above because faith was inconceivable without work. This work was connected with right deeds which were often mentioned by the Quran. By way of a short summary, one could say, "[A] believer according to Islam believes in God, His Books and Apostles, and makes no distinction between one and another. He is unfamiliar with fanaticism. He loves his fellow-men and respects their freedom and beliefs. He likes and seeks righteous work. He loves nature and its creatures and is contented with the universe and life."[142]

The Islamic view of the world – from water to environmental pollution
All of this was said in an almost apodictic way which became even more marked when it came to the Islamic view of the world. By means of combining Quranic texts and statements from scientific sources, special emphasis was placed on the equilibrium of the world as a habitat and on the fact that everyone had a share in it, including future generations. Each generation had only been lent the world; it belonged to God. In this connection, the saying of the prophet was quoted again that water, grazing land and fire (in the sense of energy and energy sources, cf. above) belonged to everyone, and also his saying that uncultivated land belonged to the person who made it cultivable. This was seen as a stimulus to make all land cultivable but without going into the question whether that might not in fact be harmful from a modern ecological point of view. The fact that the worldwide efforts for environmental protection had so far not led to an improvement was attributed to (personal) interests being stronger than a faith which emphasised that one also had to take account of the interests of others. He then dealt with specific areas of resources beginning with human beings themselves. Their role as God's representatives meant, above all, that they had to obey God and protect the earth; development only came after that. And in this they should show love and not envy to their fellow humans even though God in his omniscience had distributed his gifts in different ways (Sura 17:30 and 59:9). But there was some tension between that and the vehemence with which he had previously attacked poverty and called for justice (not charity) in view of the fact that many nations consumed considerably more resources and caused greater environmental pollution than others. As far as water was concerned, he emphasised that the Quran saw it as the origin of life from which God had created everything, that it was also of religious importance to Islam (ritual cleansing) and that, as already stressed, Muhammad had particularly protected its purity. Later lawyers had developed that further even to the point of prohibiting weapons of mass destruction. Animals, like people, were God's creatures, witnesses to his creative power and in part there to serve human beings, at least in the case of farm animals. On the other hand, people also had the duty to protect animals – again even quoting Muhammad's saying that one should leave brooding birds in peace. As far as the plant world was concerned, the cultivating and protective function of human beings was also emphasised which applied even in wartime, as had already been mentioned. With regard to the air and air pollution, the speaker applied the well-known Is-

lamic legal provision that the prevention of harm took precedence over the promotion of interests or, in plain language, no industrialisation if that entailed air pollution. With regard to food, there was naturally discussion of the foodstuffs Islam prohibited or considered harmful, of the provisions which required that part of the harvest be given to the poor and needy, and of the Islamic prohibition of wasting food which is still very relevant. With their weakness and lack of faith, human beings had begun to harm the world even to the point of endangering the human race itself, and had forgotten that God had given them the earth in good condition and commanded them not to spoil it. This was the point at which the speaker's own religious understanding played a major part. "Modern science and technology have been unable to stop corruption or to protect the earth's resources from wastage and depletion, while we can see from what has been said above that belief according to the Islamic faith is sufficient to tackle the problem and stop the onslaught of corruption in this earth and protect its resources from mishandling and deterioration. There are Islamic legislative principles that can be taken as starting points for the protection of the earth's resources. These principles and points of departure will be based on the enhancement of Islamic religious awareness, emphasis on the creation of faith within people's souls and hence commitment to Islamic concepts in dealing with nature, environment and the earth's resources. These concepts are based on avoidance of extravagance, wastage, ruining, depletion and pollution of resources. Public welfare should, of course, be given priority to individual interests. Private or individual harm can be tolerated in order to avoid public harm or damage. Harm cannot be removed through causing a similar harm, and the State is entitled not to license the projects which do more harm than good (...). It is our duty, therefore, to give the believers their chance to tackle the problem from the Islamic perspective."[143]

In the face of this massive self-confidence, it would have been desirable to see at least a brief critical comment on the extent to which these provisions had really been or are implemented, or whether many believers did not necessarily act in that way and were therefore not as faithful as they should be in an ideal world. In addition, there is a (frequent) danger that one compares one's own marvellous theory with the bad practice – in this case on the secular side – which is not really fair.

7.5.2.4 Christian comment: ethical agreements and theological differences

The official Christian response dealt quite generally and positively with the fact that great value had been attached to religion (in this case Islam, of course) as a psychologically deeply rooted and living, inspiring force for commitment in the area of environmental protection, and that the same was true for the Catholic side where religion led people to feel responsible before God for the environment and its protection, the evidence of which could simply be repeated. But then Bernard Przewozny made a well-informed and interesting comparison of the details of the two systems and their approaches to the issue. Whereas on the Islamic side it was precisely the absolute unity of God which led to God's being the only one who can regulate human relations with creation, in Christianity it was the conception of the loving, inner-trinitarian relationship that became visible outwardly in creation and was the key to a loving relationship with the latter, visible in humans being created in the image of God. Here, the Islamic conception was that of a watchman or representative which was frequently connected with the rational nature of human beings and with the human duty to preserve God's work (his interpretation of *fitra*). Here, he demonstrated very well that the Christian and Muslim conceptions were not just very close to one another as far as their practical consequences went, but that, if one then really investigated their theological grounding, they were worlds apart, although this was not so evident on the ethical level. Therefore Przewozny also offered a quite critical analysis of the industrial revolution (which was conceptually based on the enlightenment) that had (had) a tendency to abandon values and ethical norms and instead to concentrate on material things, and this had led to an increasing secularisation of life. It was necessary to build up a new relation between religion and modern science, technology and economics. As expressions of human creativity, they should not be seen as negative as such, but also as means and not ends in themselves; hence they should be used so that, among other things, an overall view of human dignity was preserved and also the possibilities of future generations were not lost from sight. Development should be defined in more spiritual and intellectual terms because, as John Paul II had also said earlier to the United Nations, material things alone could not satisfy people.[144] It was only surprising that at this point he did not simply refer to Jesus' saying that man does not live by bread alone.

7.5.3 The just distribution of resources
7.5.3.1 The Islamic role of human beings: the importance of work
A last round dealt concretely with the role of believers in the just distribution of the earth's resources which was clearly demanded by both Christianity and Islam on the basis of God's having given the earth to all people as his creatures and as brothers and sisters to one another. It was the Islamic address given by Abdul-Aziz Khayyat which began by taking up in great detail the marvel of this gift of God to human beings as his representatives, starting with the day for work

and the night for people to rest. He strongly emphasised the importance and blessing of work for human beings, illustrated by quoting a tradition about Muhammad according to which seven deeds of a person were sufficient for them to receive the reward after death: teaching profitable knowledge, constructing a waterway or enabling a river to flow, digging a well, planting plants and cereals, building a mosque, leaving a copy of the Quran to someone and leaving descendants who could pray to God for their forgiveness after they died.

The Islamic logic of sharing
Before then going into detail about distribution, he underlined two things which were important to Islam: that the earth's resources and the environment remain clean and suitable for life and that resources be used sparingly so that they were not exhausted; in other words, a rationalisation of consumption (which again related to the rational qualities of human beings that had already been mentioned several times in this connection by the Muslim side). The worst form of wastage, according to Muslim tradition, was unreasonable, unjustified spending. Distribution itself took place on three levels – between states, within states and between individuals. It was first assumed that the resources of the earth were sufficient for its inhabitants irrespective of how much they multiplied, because the resources were also renewable and would increase. The task of the believers was to ensure, by means of the necessary management and appropriate distribution, that there was sufficient food for all living beings. Here again there was reference to the regulation about certain things being common property where all mineral resources were assigned to an area of common property called fire. For this reason, it was legitimate for such resources to be state property and they could be used by the nation states. A clear line was drawn here e.g. against any form of colonisation or post-colonial exploitation of the raw materials of weaker countries by stronger ones. In other areas relating to nation state sovereignty over specific things, there was general agreement with international law and the corresponding treaties were recognised. The profit from the exploitation of national raw materials should go into the state treasury and be spent on public interests and the welfare of the state in accordance with a principle of Islamic law dealing with public welfare. Parts of public property could, however, also be given to private individuals for their use. And land could be distributed to poor inhabitants, or the state could assign unused public land to those who cultivated it – in line with the traditional saying of Muhammad often mentioned before.

Dealing with the state treasury and alms tax
Money from the state treasury or the alms tax could also be given directly to needy people or lent free of interest in order to promote development and coun-

teract poverty and hunger. Here there were five categories. The first to receive support were the poor who could not work, such as the handicapped, widows or orphans, for example. They should not only receive a minimum for survival, namely food and drink, clothing and housing, but also for a certain standard of living depending on the conditions determined by time and place, according to the legal principle which stipulated that necessities should be calculated in relation to the situation. In this connection, Muslim lawyers had mentioned suitable accommodation but also, e.g., enjoyment as an important aspect of human needs. Secondly, those whose expenditure exceeded their income were to receive what they absolutely needed from the alms tax fund and the necessary extras, namely marriage, promotion, horses to ride and suitable furnishings. Thirdly, those who had lost their fortune because of a catastrophe or bankruptcy should receive from the alms tax enough money to be able to get their business or farm running again. Fourthly, inheritance law was mentioned which prohibited the appointment of only one heir because the inheritance had to be distributed to the family. Fifthly, the state had to intervene in order to re-establish a balance if one class dominated another. The example given was a tradition stemming from Muhammad that someone who owned land must either work it or hand it over to his brother. The aim of all of this was that everyone got what they needed even if this meant that then everybody only had the bare essentials.

The practical bases of the Islamic distribution procedure
Distribution between the believers took place on a voluntary basis through the donations which Islam required, especially for the poor and needy, relatives, orphans, travellers and all who were in difficulty. Then there was also the possibility of disposing freely of up to one third of an inheritance. Finally, the exchanging of gifts was also counted as part of this process. In addition, there were due payments. These included payments for sustenance within the family, payments for offenses against divine law (here there is a whole series of very precise regulations), financial substitute payments, for example, if someone was unable to observe the fast, and finally foundations for charitable purposes such as the needy, orphans, mosques, schools or hospitals. Here again a tradition from Muhammad was quoted that, with the death of a person, everything came to an end that he/she had achieved in their life with three exceptions: continuing charity, useful learning and a pious child who prays for the deceased person.[145] This again showed the characteristics of Islamic law that it really does not leave any realm of human life without usually very detailed regulations.

7.5.3.2 The Christian commentary: far-reaching agreement?

The Christian commentator, Ignazio Musu, particularly noted the points of agreement which indicated the common roots of Christianity and Islam, especially God's love for all people in the past, present and future (although Muslims would probably never use the term "love", which Musu uses here, for the relation between God and man). He emphasised again that, according to the Christian tradition, God gave the earth to all humankind without excluding or favouring anyone, but this did not reflect the real state of things, and therefore the struggle against poverty and the promotion of social justice were very necessary. Social responsibility for property was also mentioned again in this connection. Otherwise, he followed the line of argument in the address and only observed that more emphasis should have been given to the effects which transcended the generations and made environmental protection so necessary. At certain points he perhaps saw the Muslim speaker as his ally too quickly and easily. "[A] very important point made by Prof. Khayyat is that God has created the world having in mind, so to speak, the needs of nourishment and adequate living for every human being. Hence it is man's responsibility to preserve the balance between what is available and what has to be used, through an accurate management to make resources sufficient for all living beings. The conclusion is immediate: we should not reject new born human beings for fear of poverty."[146]

This is certainly an example of how one mainly hears what is specially important to oneself in what another person says, whether or not it was the central point for the other person.

7.5.3.3 A description of the problem from a Christian point of view

The official Christian address on the subject was given by Anis Muasher who naturally also started from the statement that God had placed human beings above the other creatures with their wisdom and various abilities to use nature, and had entrusted nature to them, but precisely not to one person or one generation but to all people and all generations. It was an obvious injustice that a few persons had amassed an abundance of possessions and wasted the available resources in the process while many people were hardly able to survive. The threat of ecological collapse was a lesson on how much individual and collective greed and self-interest ran counter to the order of creation which relied on mutual dependency including all people and also future generations. "Man is not permitted to exploit his private capability which the Creator had granted to him, to aggress nature and other creatures and to cause an imbalance or to appropriate and monopolize them."[147] But in fact the way people had treated nature had become worse and worse from generation to generation, and at the expense of future generations in proportion to the extent to which they had developed their ability to bring about fundamental changes on the planet. As far as the question of equality was concerned, all people were God's children and hence brothers and sisters. In the Lord's Prayer, Christians were called upon to be satisfied with their daily requirements (give us this day our daily bread), not to be greedy or wasteful, not to amass things, not to consume excessively and thus make exaggerated demands on natural resources. The threat to the environment that had been described was the direct consequence of human society's deviating from this Christian way to such an extent that excess production had become a feature of Western society, in particular, as the speaker put it. Economic prosperity had become the measure replacing genuine need and just distribution, and this had upset the whole balance especially for coming generations.

A diversity of church statements on just distribution
Here there was again reference to John Paul II's call for greater modesty and simplicity, also in the interest of world peace, but many biblical references were also quoted which underlined that all people are brothers and sisters, that love knows no limits, that justice also requires that the poor be given a share of the earth's resources and that it is a serious violation of the rights of the poor not to help them, as well as being a religious and social crime, right back to the behaviour of the early church. Statements by the church fathers also went along these lines from Basil via Ambrose and John Chrysostom to Augustine. But modern industrial societies had cast all of that aside at the expense of poor people and underdeveloped countries. This structural problem of poverty, as John Paul II had also said, had to be tackled if one wanted to solve the ecological problem and also the problem of peace and war – indeed the three largest problems of humankind at the present time. But John Paul II was not the first to have taken up this issue repeatedly; John XXIII had called for the responsibility of the rich nations towards the poor countries whose inhabitants were victims of poverty and had no human rights, and had given the reason that wealth was not based on a surplus of possessions but on their really being distributed to individuals. When dealing with questions of how wealth, education, science and technology could be distributed more justly, one should also not forget spiritual values, because often the increase in scientific and technological progress and in material wealth was accompanied by an increase in excesses. The real danger for society was the tyranny of material work and the desire for profit over the soul and values. In view of the major social inequalities, Vatican II had also underlined the fundamental equality of human beings even though their physical, mental and ethical abilities were certainly not equal. An international economic

system needed to be created which could put an end to exploitation, the yearning for profit and the desire for political domination and replace these with friendship and brotherliness. Every person should be able to live in dignity with food, clothes and healthy accommodation and with possibilities of choice for their own life, establishing a family, to work and to practise their religion in peace. As had already become clear, John Paul II had spoken often on the question and also emphasised human dignity and brotherly unity which, as conceptual goals for humankind, went much farther that the multiplication of material production. Earthly goods existed for the whole human family and could not be the privilege of an individual while others did not even have a hope of the basic essentials for a life in dignity. Therefore it was necessary to enlighten people about their limitations and the possible consequences of industry. On certain fundamental issues, the rich industrialised countries contradicted the teaching of the church and the law of love and justice which the church had stood for from the beginning. Human beings had to respect the commandments of the creator; otherwise the yearning for profit and power so as to impose one's will on others would grow automatically. But conversion was possible at any time. Solidarity was seen as a Christian virtue which voluntarily brought about justice in the light of faith, forgiveness and reconciliation. However, economic regulations which encouraged greed, the amassing of money and possessions and also wastage needed to be changed.

A special aspect of the problem: expenditure on armaments
The address devoted a separate section to expenditure on armaments which in a sense caused double destruction; firstly, they cost money which could be better spent otherwise and they also destroyed existences directly. In summary, "If man adhered to the teachings of the Creator and behaved according to His will and partook the morsel of life and the surplus of his need with others, if the North cooperated with the South, if humans cooperated with one another and spread science and culture especially among women, if they were content with their need and avoided the surplus which has become a burden on them and if they were committed to recycle and use materials several times instead of transforming them to garbage and neglecting them and making of their disposal a burden on people, if they stopped manufacturing the instruments of war, destruction and fighting, if they partook the yield and production of the earth, gift of God to all mankind, human society would alienate the specter of destruction from this planet, and the human society would become as God wanted it, a peaceful welfare society."[148]

7.5.3.4 The Muslim commentary: more different agreements

Again, as so often on this issue, the author of the response saw major, fundamental agreements, in this case starting from how God had created humankind. "In many instances Mr. Anis Mouashir starts from entrenched human principles, uncontested by none of the human race, regardless of their positions of their religious beliefs. From the totality of these principles comes the love of peace, the love of others, justice in the distribution of wealth (of which the resources of the earth are part), the love of helping those in need and the poor, and the maintenance and conservation of the comprehensive environmental balance. Such principles represent the innate disposition of mankind, the moral constitution that Allah (God) created man upon."[149] There was emphasis on the freedom of choice between good and evil, gratitude and unfaithfulness, which God had given to humanity. Those were the characteristics which distinguished human beings from God's other creatures. The fundamental principle concerning the just distribution of goods and the relationship with the poor was also the same, although the statements on poor and rich in Islam did not seem to be quite so clear and unequivocal. On the one hand, it was emphasised that, at the beginning, there had been neither poor nor rich but, on the other, it was stated that God had created human beings with different abilities and different wealth as a sign of earthly pluralism but he had commanded justice as one of the tests for humankind. There was a certain tension here which could probably not be resolved. As far as the attitude to the environment was concerned, he again saw fundamental agreements with one big difference. In the Christian address, it was never completely clear which mechanisms were to be used for the necessary redistribution. In Islam that was resolved quite clearly and simply by the alms tax, charity and the use of the resources of the earth and, in addition, the state had the possibility of guaranteeing the implementation of the good principles. Here again there was a difference between the two religions concerning the degree to which human life was to be regulated or not (although the response also underlined that Muslims naturally did everything out of conviction and not by compulsion). The response ended with a final, somewhat abrupt agreement which the speaker believed could be observed. "This excessive waste in production and consumption of resources and products is pushing our natural resources to their limits. This flow / typhoon cannot be stopped except when mankind gives up its Darwinism, and undoubtedly this is where Islam and Christianity meet again."[150] *It could be that this reflects the same thing as was seen in reverse in the last Christian response, namely that the wish dominated the thought or the agreement observed.*

7.5.4 Crown Prince Hassan of Jordan: a very balanced address

A special highlight was the public lecture on the subject of Islam and the environment given by HRH the Jordanian Crown Prince in the Gregorian University. He, too, emphasised that Islam had a lot to say about the right relationship between humans and the environment and mentioned in this connection the principle of vicarious representation and the related responsibility of human beings to reject greed, wastage and pollution. He, too, viewed the role of the Enlightenment and the environmental damage connected with it critically but not in so absolute a way as the speaker just mentioned. Nor did he see the role of Islam in preserving nature in so glorious a light as many of the other Muslim presentations but recognised that there had also been quite other factors involved. "The little share that Muslims have had so far in spoiling the earth has been due to the prevalent underdevelopment of the Muslim countries rather than the contrary. The fact that much of the Muslim world remains preindustrial in character, (sic!) has helped preserve much of the environment, but for how long? As a Muslim in a position of responsibility, I fear the consequences of development as much as I am aware of its urgent need."[151] Rapid developments in this field had led to strain on the environment and human alienation, as the past had shown. A change of approach in the industrialised countries was required as well as a transfer of environmental technology. Probably the energy question would be decisive. He referred with great appreciation to the address of Pope John Paul II to the United Nations where the Pope had underlined that each human culture strove to answer the fundamental questions of existence and that deserved recognition and respect. Crown Prince Hassan also emphasised the many principles and values which Christians and Muslims had in common and that these human values were the key to the environmental problem. This can certainly be seen as a good concluding statement for the whole meeting.

7.5.5 At the end: not much theology but many practical recommendations

In the official recommendations, it was agreed that God had created the earth for the well-being of everyone and that it still had sufficient resources to feed everyone, even though in some countries less was being produced in order to keep prices stable, whereas the inhabitants of other countries did not have enough to eat. This led to a joint appeal for all types of resources, both renewable and non-renewable, to be used reasonably and responsibly and a corresponding style of life to be encouraged. Christians and Muslims alike needed to commit themselves to this individually and as communities; food producers and especially farmers should be encouraged to produce food to meet the needs of all people in a way compatible with the environment and that was ethically correct.

Land that was not in use should be made usable with resources from religious foundations, and politically all nations should be given access to the natural resources needed for survival in order to guarantee a minimum standard of living; anything else was immoral. As far as environmental protection in general was concerned, it was seen as a matter of cooperation right up to the international level; in particular, governments had to keep a check on industry and on the correct disposal of dangerous waste. Immense natural and especially human resources were destroyed by wars and violence so these were strongly opposed. The production and sale of armaments should be limited in favour of development and the emphasis put on dialogue to achieve just and peaceful solution of conflicts. Mutual dependency in the world was seen as positive; it reflected the unity of the human race as the creator had intended it. However, the developed countries should not tie their financial support for developing countries to the introduction of types of family which were not in line with God's will for humanity or to the adoption of forms of government alien to their tradition. The countries themselves should be able to determine the prices of raw materials and the use of the income from them. Reference was also made to the question of colonialism, stating that at the end of the century a complete or considerable cancellation of debts should take place, taking account of the idea that, as a counterpart, land should be designated a nature reserve. Finally, and this related more directly to persons, it was pointed out that the political leaders of developing countries had the duty to set the right priorities, to use the available means and resources wisely, to combat corruption at all levels and themselves to set an example of uprightness and efficiency combined with a genuine love of their fatherland. Reference was again made to the importance of education which, in line with the wording of UNESCO, was based on respect for nature, human dignity and the future, and on the will to create a quality of life open for all to participate in it. This began with the parents, especially with mothers, who were often sensitive to such questions and could therefore more readily influence their children accordingly, and ran through all the stages of formal education, but also included private associations, religious institutions, the media and the governments. The latter, in particular, had important tasks of awareness building or of making funds available to this end.[152]

The use of resources is generally a subject where human beings as such are central because of their responsibility, their involvement in the totality of the human family and in creation as a whole, but where they are of less importance at the same time. This has to be recognised finally and, above all, put into practice and this is also an important aspect for both a Christian and an Islamic anthropology at the beginning of the twenty-first century.

The next conference, it was decided, was to be held at the end of 1997 in Amman and to deal with the issue of human dignity as such.

7.6 The last dialogue conference for the time being – Amman 1997

The *sixth* and, at least for the time being, last *joint colloquium* took place from *3–4 December 1997* in Amman on the subject *Human dignity in Christianity and Islam*. It dealt with three aspects of this broad issue: the concept of human dignity, a historical review of Christian and Muslim attitudes to human dignity and, finally, future prospects.[153] (A generally positive thing about these conferences, as already mentioned, is the fact that they were published in both English and Arabic and are thus available to a wider public and can have further effects.) However, the conference on culture and religion in society, which had been planned for 2–4 July 2001 in Rome, was cancelled by the Muslim side on the grounds that there had been changes in the leadership and policy of the foundation.[154] This is regrettable in any case because the joint conferences with the Al Albait Foundation are the only ones which have been documented in six conference volumes and thus make it possible to follow the course of the conversations from 1989 to 1997, namely over nearly ten years, relatively comprehensively and easily at the same time.

7.6.1 The words of greeting
The change had in fact begun to be apparent at this last conference where the three speeches of greetings at the beginning (by the directors of the two institutions and by Crown Prince Hassan of Jordan himself) dealt less directly with the issue of human dignity and more with the question what the real purpose of these dialogue conferences was and to what extent it could be achieved.

7.6.1.1 Crown Prince Hassan of Jordan: what does not yet exist
Thus, Crown Prince Hassan underlined that they were not there to promote an international ethos (which could perhaps endanger the identity of the individual) but rather together to draw up a common code for human behaviour. Or more concretely: "We assemble in the spirit of convenience to speak to each other to identify what aspects of tradition are relevant to each other while retaining our particular identity. In the context of the rules of conduct discussed in the interfaith meeting, in one of the working groups' proposals the following attracted my attention: firstly, the importance of emphasising the association between the-

ology and practicality – to begin with commonality; to take into account the Enlightenment tradition; to embrace the principle of 'No Coercion'; to uphold the right to proclaim one's own religion; to reconsider the contempt of education. (...) May I entreat you all to be courageous in looking afresh at first our own, and secondly each other's texts, heritage and history – to develop a **framework for disagreement.** If we claim to be living, at the end of this millenium and the beginning of the new century, according to civilised norms – let us develop a respectful framework for disagreement. To accept responsibility for words and for actions at all levels, and (...) to recognise the political and economic dimensions of interfaith dialogue. (...) / p. 8 (...) I think it is crucially important not to allow our value system to be trammled, tainted and misused by vested interests, whether those vested interests are concerned with the selling and buying of weapons or concerned with the selling and buying of people's human dignity."[155]

As far as the question of the right to human dignity was concerned concretely, irrespective of a person's social, economic or political standing, he was convinced that Islam and Christianity completely agreed on this and the only differences were how this fact was expressed theologically, although the basis was the same, namely God's interest in human beings which demanded our loving interest in our neighbours and in the whole of humankind. In Christianity, that could be summarised in the formula of the commandment of love, whereas Islamic theology emphasised the creation of humans as free beings able to take independent decisions as the prerequisite for their task as God's representatives. Differences of rank between people were, so to speak, a divine test for people. Being God's vicarious representatives required that people be treated justly and equally and that the weak should not be oppressed. The cause of the weak and oppressed was equated with God's cause, both in the Bible and the Quran. In this connection, Crown Prince Hassan pointed out that all the laws relating to refugees were connected with martial law and that no law on peace existed, and therefore there was really no code of conduct for after a conflict; there were only human rights but (as yet) nowhere a code of human duties and of how one should establish peace concretely in a particular situation; in a sense this was a real but very sad irony and a stimulus from his side for reflection on the practical future of human beings and human dignity, as he set store by and emphatically underlined concrete and rapid implementation irrespective of all the theological reasoning. The president of the Jordanian foundation supplemented these remarks on human dignity with the thought that God had forbidden the killing of a person and had equated it with killing a whole people, but otherwise he dealt more with questions of dialogue as such.

7.6.1.2 Cardinal Arinze: God's rights do not contradict human rights

Cardinal Arinze tried to find related points in earlier dialogues, for example, the one on the question of the rights of children, but otherwise mainly gave a brief summary of what could be said on the question of human rights from the Christian side. Perhaps these differences in emphasis already indicated that the Jordanian side, following a series of dialogues also with other Christian denominations, was now looking for a somewhat different way of working. But it is at least worth mentioning the brief and telling expressions used by Arinze.

> "Human dignity is therefore an honour which comes from the fact of being man. It is God's gift. It is not a concession coming from other human beings, or associations or the State. It is part and parcel of being human, whether that human being be learned or unlettered, rich or poor, healthy or sick, young or old. This dignity is the foundation of the right of every human being to life, to freedom of religion, to work, to possession of property, to self expression and to the founding of a family. (...) Human dignity carries with it corresponding responsibilities. The human being has the duty to know God and to worship him. Every human being should recognize and respect the dignity of other human beings and their rights. This recognition begins in the family and extends to the whole of one's country and indeed to the world. (...) When human dignity is properly recognized and respected, God is honoured. The rights of man are in no way opposed to the rights of God. It is /p. 17 God who gave man all the rights that he or she has. When we respect our fellow human beings, we are giving glory to God who created them."[156]

It is evident that this was not simply a Christian basis for and formulation of human dignity but rather one in contrast to Islam, especially at the point where it clearly related human and divine rights positively to one another whereas, when it was a matter of the UN human rights convention and its scope and validity, these were frequently played off against one another on the Islamic side by indicating that they were not identical and so naturally God's rights took priority.

7.6.2 The basic features of human dignity
7.6.2.1 Three vehement delimitations on the Muslim side

It was only after this that the main addresses were given which further expanded and substantiated the basic lines sketched out before. On the Islamic side (Ammar El-Talebi), the initial starting point was that human beings only existed at all because God had breathed his breath into them and the human soul which resulted from this was considered self-evident. But then, in the course of a very detailed linguistic examination of the use of dignity and related terms in the Quran (root *karam* which really has to do with noble descent and can also be used e.g. for horses, camels and trees), it became evident that this soul was very closely connected with the knowledge which enabled human beings to be

God's representatives, but also with correct, pious behaviour which finally pointed to Islamic law. And this led in turn (and was now far from new) to long explanations about there being no separation in Islam and certainly no conflict between humans and nature and that, on the contrary, there was a profound link between faith and reason and the recognition of God was innate or created in human beings. Science, understood rightly, belonged to the nature – *fitra* – of humans and was even the secret of their dignity and superiority over the other creatures. Almost more interesting that these predominantly well known, positive statements were the negative delimitations related to them. The first was directed against the theory of evolution, and that in the name of all monotheist religions which had surely already said everything about the origin, nature, rise, place, purpose and destiny of humanity. They were all agreed that human beings were different from and superior to the other living beings. Above all, this concerned their reasonable souls and their intelligence, but also further physical elements such as language, work and walking upright and finally the true knowledge which God had taught them and only them (Sura 2:31). Free will also came in here without which there could be no responsibility although at the same time the Quran very much emphasised the servant status of human beings. The life and possessions of people were sacrosanct (cf. above). The second clear delimitation was directed against humans being created in God's image. There was a tradition of Muhammad's that God had created Adam in his image but this was merely a metaphorical expression for God's having breathed his spirit into humans, thus giving them intellect and access to knowledge (cf. above) and still other characteristics, especially strength of will, which belonged absolutely to God and only relatively to human beings and not at all to other living beings. Moreover, the "being" in the image could be applied most meaningfully to Adam himself in the sense that all people were like him. And Genesis 1:27 had been understood metaphorically by Christian and Jewish philosophers, although the speaker provided no concrete evidence of this. The last very clear delimitation was from the West which saw humans and nature quite differently and put itself in the place of God. Since in Islam all areas of knowledge were totally linked with the principle of monotheism, Muslim countries had rejected the purely materialist science of the West. His review of European philosophical history led to the conclusion that really nothing and nobody was so balanced as Islam and almost all the Muslim scholars. According to Western science, human beings were in a hopelessly bad situation. So, in order to regain their humanity and dignity, human beings should first rescue the world and science from their break with God and it was high time to do so.

7.6.2.2 Response from the Christian side

The Christian response (Roman Stäger) agreed that, indeed, many people in the Western world had denied the dependency of the creatures on the creator, but he raised the critical question to what extent the leading philosophers mentioned (including Nietzsche and Freud) could be considered representatives of Christianity. He also saw a rather narrow approach to the dimension of knowledge as the basis for human dignity which ignored other things such as human relationships, compassion, etc. He could only agree with the importance of the prohibition of killing but also asked critically about the practical consequences (which were lacking). "This seems to be a very modern statement not only in relation with preservation of nature, but more important with events in Algeria, Egypt, Somalia, Burundi, Bosnia, etc., as Christians as well as Muslims are concerned. If thus violence /p. 38 kills human dignity in how far has any conquest of power in the name of a religion, whatever its name is, the right to exist? And in how far has modern science the right to experimentation, in modern technology, in biotechnics?"[157] This raised important questions which were far from easy to answer.

7.6.2.3 Human dignity and human rights – not dependent on Christianity

The Christian fundamental address on human dignity, given by Ignazio Sanna, professor at the Lateran University, was striking – also in the context of the previous Christian statements – because of its decidedly Christological approach which also interpreted the statements on creation theology in a strictly Christological way, even subordinating them to Christology. But firstly he undertook the task, which the Muslim side considered very commendable, of distinguishing between the various sources of the concept of human dignity and human rights and of explaining them in their historical inter-relation or contradiction. In addition to Christian ethics, ideas from classical, namely Greek, philosophy and European humanism had played a part. The contribution from the Christian message consisted of two inter-related ideas: the natural equality of all people and the idea that, for God, each individual incorporated the dignity of the whole of humanity. However, a kind of secularisation of the religious understanding of human dignity had taken place. But here there was substantial agreement on important anthropological principles between the contributions mentioned from different backgrounds. However, there had always been different concepts of human rights so that, even when the Universal Declaration of Human Rights was accepted in 1948, someone said that they agreed but on condition that nobody asked why (even then it was obviously considered "rather Western and generally middle class"[158]). However, Sanna considered this refusal to participate in a common justification of human rights and human dignity to be a dangerous shortcoming with regard to its real validity. Its concrete realisation presupposed a common belief in its universal, absolute validity and thus the reasons for it became an absolute prerequisite for its realisation. In some way he saw a natural justification for human rights. ""[T]he concept of human dignity does not depend in a particular or indeed exclusive manner on any one historical cultural movement which may have left its mark on modern culture. Instead, like the modern ethos of human rights in general it is still compatible with all the philosophico-religious bases which have in some way contributed to its formation. The personal dignity of man and his specific natural vitality can be recognised without any need to resort to God or to have recourse to the answer given to this question by a historical revealed religion like Christianity. This dignity is already founded on the original parity of all members of a community based on the rule of law, and founded as well on man's experience of himself as a moral being, even before being rooted (...) in a specifically theological basis. If one makes a distinction between the historical genesis and the normative /p.42 validity of a particular reality, and one applies this distinction to the relation between Christianity and the modern ethos of human rights, one can affirm that the idea of human dignity, which is the basis of our democratic culture, has definitively been decisively influenced by Christianity, but this historical influence does not establish a permanent systematic dependence, in the sense that accepting this idea should somehow bind one to the intellectual assumptions of the Christian faith. From the point of view of a pluralistic society, this means that the idea of human dignity, even though its roots are embedded in a religious vision of life, is part of the undeniable core of a rational natural ethos that lays claim to universal recognition. On the other hand, the idea of man as 'corresponding to God', a modern translation of the biblical and patristic theme of the 'image of God', is, in the final analysis, present in the heritage of religion generally and can be accepted also within purely rational and philosophical vision of man, because the starting point of the experience of the sacred and of a consequent anthropological concept based on it is common to all people from all civilizations and cultures, whereas the conceptualization of this experience can develop in diverse ways and reach different destinations, according to whether the conceptualization takes place within an immanentist or a transcendent vision of reality. The West has contributed to the formation of this natural ethos, for example, by promoting the understanding that the human body belongs to the individual and is not to be manipulated by anyone, neither by the State, nor by the Emperor nor by any charismatic leader."[159] So the doctrine of humans being in the image of God was seen explicitly as inclusive; it did not exclude other religious or cultural justifications of the fundamental dignity of human beings.

Christian grounds for human rights – the long beginnings
On the other hand, the theological grounds for human rights were also not simply a superstructure forced upon modern human rights. This was where the theological and, above all, the Christological line of argument began. The final justification for human dignity was that, according to Eph. 2:18 and 2 Peter 1:4, they had a share in the nature of God; in other words, it was based on human beings being in the image of God which was also found in the New Testament. In philosophical terms, it was based on the fact that human beings developed and found their realisation naturally by believing in someone. Neither fate not the forces of evolution but the Holy Spirit made them into living beings and for their self-awareness they needed not the cosmos but God himself. "Man is not a little cosmos, a microcosmos, but a little God."[160] The idea here was particularly the intelligence of human beings, their conscience and their freedom. It was the special (indirect) anthropological conviction of John Paul II "that the Christian is the successfully realized person". Theology alone led to a correct anthropology which understood humans in both their dimensions and not only in their biological one. For this reason it was also meaningless to ask when a person began to be or stopped being a person. The beginning and the end vanished in God's eternal will and were beyond historical definition by human science. The real origin of the concept of the person was thus a theological one; it linked the person's being a person with God's being a person (here he quoted e. g. Albertus Magnus). It was only afterwards that the concept of person was gradually secularised and taken over by philosophy and law. Since God had created humans as ends and not as means, they had absolute value in themselves which God had given them and which depended on nothing else. Humans were unique and unrepeatable. Christ had confirmed this by dying for each one. This value was also made visible by the liturgy linkung baptism with unction, a ritual which had previously been reserved for kings. Here he also quoted the saying from the Talmud that the killing or rescuing of a single person was as valuable as the killing or rescuing of a whole people (a statement also known to and emphasised by the Muslim tradition, cf. above). Then Sanna's arguments turned to the Christological core following the statement of Tertullian that God had thought of Christ as a model when he created human beings. "[I]n the actual order of the history of salvation, man was constituted a person from the very beginning because he was created in Christ, the perfect image of God. Hence he is not simply a being that is open to the transcendence of the mystery of God, but is instead, first of all, a Christocentric being because he is historically defined as

person on account of his relationship to Christ." So the real dignity of human beings was to be persons in Christ. It was their calling and task to become persons in Christ. Sanna then discussed humans being created as man and woman in order to point particularly to humans not only as individuals but also as social beings. This was related to the inner-trinitarian relationships and that in turn indicated that social relationships which were built on possessions, power or privileges did not correspond to them but only such that were based on mutuality and inner meaning. Then the address took a further step to the concrete contributions of the church to the defence of human rights. This was demonstrated particularly by the fact that humans were never allowed to have control of human life, also not in conflict situations, because they were always supported by God's immeasurable power and because they were responsible before God; indeed, without God they could not really know who they were. Then Sanna briefly mentioned that the Catholic Church under Pius VI, Gregory XVI and Pius IX had been very critical about human rights, especially the freedom of religion and civil liberties, a fact which was attributed to the anti-ecclesiastical aspects of the Enlightenment which had taken up the cause of human rights and to the individualistic context (he did not mention that here there were convergences with the Islamic criticism of human rights that was quite typical nowadays).

The Catholic Church and human rights today
Pius XII was the first to provide a list of human rights in connection with human dignity – significantly enough, in his Christmas radio broadcast in 1942 – and he became one of the most influential figures in the new era which led up to the adoption of the Universal Declaration of Human Rights by the United Nations. The Council and popes who came after him followed this line further, especially John Paul II. Sanna saw the work of the church as a sentinel of humanity which offered the world an anthropology of the person that respected human values and was open to transcendence. The latter was emphasised particularly in view of the developments of modern times with their technological rationality which only saw the physical reality and had thus made humans firstly into creators and then into their own objects. One example of this was genetic engineering and also the fact that, although there was no higher life expectancy and lower infant mortality, at the same time the uniqueness of the unborn child and of old people in the face of death had been deleted from people's memory. In line with John Paul II, rational ethics and Christian revelation were necessary. At the end, Sanna again gave a summary of his thoughts on the subject. "Human dignity, in itself, is not an absolute, since it is not connected to any quality or value that man possesses in himself, like his soul or his intelligence or his virtues, but derives from the fact that man is always considered on the basis of

his relationship with God, a relationship that can either be positive or negative. Man is great not so much as God's creature, but rather as His personal interlocutor, in freedom and love. His dignity does not consist exclusively in being the image of God, but in being His child, who shares, starting down here and continuing forever, in divine life. Man in God's image is a manifestation and revelation of God in a role which surpasses all visible creation. Whereas other created things exist only as a result of a relationship of causal dependency on God and therefore manifest His power to the utmost, yet not His nature, man is a reflection of the mystery of God Himself, and in the profoundest essence of his spirit he renders visible who and how God is, i.e. pure person in perfect love and freedom is the only place in the visible world in which God is recognizable as personal spirit, because he indicates God not only in his existing, but also in his being such. The relative autonomy of the human spirit in the likeness of God is the highest form, although indeed still obscure, concealed and incomplete, of the natural revelation of God."

It is easy to see that this went far beyond anything that had been said on the Christian side in the dialogues thus far where, in practice, they had restricted themselves to interpreting the texts of Genesis and where, in fact, being in the image of God had been brought relatively close to the Islamic concept of being vicarious representatives. Here the theology which had been avoided so resolutely and successfully up to this point came in in its highest expression on the condition already mentioned: without theology no anthropology or, more precisely, without Christology no anthropology.

7.6.2.4 The Muslim response: in practice we are better

It is interesting and a good example to read the Muslim response. It ended by stating that the speaker did not agree with the previous speaker on a number of points but he did not spell this out further. The theological criticism that could have been expected was thus not stated and the criticism remained on the practical level. "However, I wish that the writer fully explained to us the meaning of dignity and how the image of God was in man particularly with the regard to the relative and moral outlook. Mgr Sanna did not, either, show clearly how this dignity can be safeguarded within the international community while he did not condemn the violations perpetrated by strong powers against weak peoples. It would certainly have been better had he stated that Almighty God has called for liberation of all peoples without any discrimination. To mention Israel more than once is not without significance (...)."[161] One gets the impression that here two people had failed to communicate, and especially that

Prof. Khayyat then took advantage of the opportunity to give a second address which was almost of the same length, pointing out that all those things had been stated precisely and dealt with in the Quran and by Islam, from the creation of human beings in the best form via their being equipped, above all, with understanding, their exalted position above creation, their vicarious responsibility and finally the legal regulations, naturally demonstrated not by theological considerations but by direct quotations from the Quran, all of which showed and proved that God had honoured human beings without making any distinction of race, skin colour or gender. Then he went on to spell out the regulations of Islamic law over many pages with special emphasis on the humane Islamic martial law which, it would appear, did not ignore any normative precedent. To quote the original, "In order to safeguard human dignity, Islamic *Shari'a* (Canon Law) called for respecting man's rights, and securing his liberties and banning any encroachment on these liberties at any time, unless such a person commits a punishable crime, in which case he will receive due punishment that deters him as well as others without any molestation done to his dignity, person and rights that are not related to the crime he has committed. Islam has guaranteed for man his dignity, liberty and belief."[162] This can be described as an elegant euphemism for punishments which directly affect the physical inviolability and even the life of human beings. However, "Many are the rights secured and guaranteed for man by Islam at all times and places and under all circumstances."[163] And as far as the oppressed were concerned he even came to the conclusion, "In this outlook, Islam is a long distance ahead of the present human laws and conventions in defining humanity while they are, as a whole, in conformity with the Islamic humanitarian outlook."[164] This constituted a very clear example of orthodoxy contrasted with orthopractice, probably reinforced with a good dose of Near East conflict, so that no real communication took place between them.

7.6.3 Human rights concretely in history
7.6.3.1 Human rights implemented concretely in Christianity
The next round was devoted to the practical, historical expressions of these concepts of human rights and was naturally even more extensive. For this reason, the Christian address (given, incidentally, by a layman, an engineer and journalist) was only reproduced in a shortened version. Its starting point was again a double one: on the one hand, the accessibility of human dignity for everyone including unbelievers and, on the other, the specifically Christian approach. Perhaps the former is also the more interesting because it had not been dealt with in such detail before and is probably the more decisive in a global context.

George Ajaiby formulated his conviction right at the beginning of his remarks in the following way. "It is possible for every human being, be he a believer or an unbeliever, who is genuinely open to reality and good through the light of reason and internal action of bliss, to discover in the natural law inscribed in hearts the sacred value embedded in human life from its beginning and until it come to an end. He can also emphasize the sublime human dignity enjoyed by each and every human being. Undoubtedly respect of human life and dignity constitutes two mainstays that are indispensable for human coexistence and social stability of all communities."[165] For a believer, however, all people were children of God without distinction; God created them in his image and loved them more than anything else. So they were all brothers and sisters, a point which Christianity especially emphasised but also shared with many other religions. The belief that God was the source of humanity was the greatest possible confirmation of human dignity, and precisely that was the biblical view and grounds for the inviolable human rights of all people irrespective of any other difference there might be between them. Christ had confirmed this again and even set human beings above religious systems (Sabbath). Neither Christ nor the apostles had made this into a system with precise laws for a particular form of society. But every community which wanted to last organised itself gradually in harmony with its message for the world. The message and mandate he formulated for the church was to bring about within itself the unity of humankind transcending ethnic, cultural, national, social or other barriers (another way of formulating the view of the church as the sacrament of unity as in Vatican II), and the aim was to achieve the perfection of the individual and of humankind as a whole. By doing this, the church promoted the fraternal coexistence of all peoples and it had also evolved a special form and a special content of the promotion of human dignity which was most obvious in the saints and martyrs and other upright people.

Implementation of human rights in the course of church history
The address then proceeded by epochs, starting with the first three centuries of Christianity which were naturally not so easy to identify. It seemed clear that Christians had resisted particular forms of behaviour which ran counter to the Gospel such as contempt for life, fornication, love of luxury and wealth. There was emphasis on the giving of alms, especially for widows and orphans, and on visiting and caring for brothers and sisters in the faith who were in prison. Slavery was not rejected but Christian slaves were seen as brothers and sisters and maltreatment of slaves was rejected and even punished within the church. Liberated slaves were able to become priests and even bishops. With the Constantinian era, overall conditions changed completely, but what remained and

even spread to society as a whole was charity which became social work for everyone. In this, the monasteries were especially prominent. The church had an influence on family legislation. From that time onwards, sexual relations with slave women were prohibited as was murdering children. Divorce was made more difficult but not abolished completely. The prison system was also made more humane. Prison supervisors were no longer allowed to let prisoners starve, and prisoners also had to be brought into the daylight once a day. Visits of clergy to prisons were officially allowed. But a prohibition of gladiator fights did not achieve much and slaves continued to be considered indispensable for the economy even though the church continued to work for an improvement of their living conditions. Many of the church fathers also took a stand on such matters. Their social teaching could more or less be summarised as follows. Social and economic relationships had to be subject to the criteria of justice and the good. Public interest had to take precedence over private interest (which could also be found as a principle of Islamic law). All people were equal irrespective of their social standing. God wanted the inequality, which resulted from natural differences and human freedom, to be balanced by the development and growth of social life. Every effort should be put into serving others. With regard to the Middle Ages, it was first emphasised that the church had really been the only instrument which could and did commit itself to human welfare, first and foremost the monasteries and monks as the embodiment of the Christian ideal. But there had also been a great gap between the Gospel ideal and the practical reality experienced by the broad mass of the faithful. From a modern point of view, it was difficult to explain how a church which valued the gospel highly was able at the same time to burn alive those who rejected its teaching. That was a consequence of the spirit of the time for which torture and the death penalty were normal, but also of the close interrelationship between religion and society. Religious oppression, the author judged in conclusion, was a sign of civilisatory backwardness resulting from a misunderstanding of religion and it was justified by the need to protect the community from heresies. This phenomenon had in no way been limited to Catholicism or Christianity. He then discussed the age of discoveries and the new upsurge of slavery connected with it. At least on the church side there had been a few shining examples of resistance to slavery such as the Dominican, Bartolomeo de las Casas, and others. The last historical stage was the French revolution which accused the church sweepingly of intolerance and of supporting all forms of tyranny. Initially, the church had tried to defeind itself in the traditional way, e.g. by censorship, intervention or written defences, but finally it had come out of this trial completely changed. Most of the church property had been secularised, religious liberty became part of the law and only the pope retained secular power as well. An efficient clergy came into existence and bishops were not also kept busy by civil affairs. Theology was renewed as well and there was a more tolerant and reconciliatory approach to the Protestants. Finally, Ajaiby dealt with the modern social teaching of the church which had started with *Rerum Novarum* of Leo XIII (1891) who had spoken out for the rights of the workers in the industrial society. He listed the documents of the magisterium which had dealt with these issues since then and had spoken in favour of the implementation of social justice and the defence of the dignity, rights and freedoms of human beings.

7.6.3.2 The Islamic judgement: the double standards of Christians

The Muslim response, presented by Ms. Mai Muthaffar, naturally also dealt with the central role of the Quran when Muslims were faced with so difficult and (in her view) multi-faceted an issue as human dignity, because it was the Quranic statements which determined the concrete definition of human dignity for Muslims, the most exalted model of which was the Prophet Muhammad himself. In Islam, humans were honoured by God as long as they maintained the values of their religion and believed in God's prescriptions. Faith was the criterion for ethics and human dignity was a moral concept that led to rights and duties which had in turn to become social conventions and norms, for example, mutual solidarity irrespective of faith, race or religion. All the divine scriptures agreed about the value of human beings and that the latter should abide by God's prescriptions but, in all monotheist religions, reality had often looked quite different in practice. Despite the differences in the practical interpretation, she finally saw many common features as far as the consequences in the civilisatory and humanitarian realm were concerned. With regard to the West, however, she observed critically that, in connection with human rights, it had produced a double standard through materialism and, as a consequence, through competition and exploitation. "[S]o we have a society that deserves honor and elevation while another society is subjugated to starvation and humiliation all at one and the same age and in accordance with one and the same rationale!!"[166]

This is indeed a question and an accusation which is not only raised by Islam but also by the countries of the "Third World" in general, and it shows that, if one tries to transfer a concept from one system to another, it not only has to be examined for the global acceptance of its validity but also for its global implementation by those who stand for it – also when this costs them something. That that could be and had been the case had just been underlined by the Christian historical presentation and was also repeatedly stated by the Muslim side.

7.6.3.3 A large amount of disagreement over a Muslim address – typical?

The Muslim counterpart address and the discussion of it evident in the reactions and responses (the Muslim speaker, a Lebanese Shiite, Saud El-Mawla, was unable to attend the meeting himself because of a car accident beforehand and could therefore not take account of the questions on the original document so he had written a separate response) are a prime example of how profound the differences are despite all the efforts at dialogue, and how quickly what one says fails to communicate and the one party does not understand the basic concepts of the other. This began with the Muslim speaker not understanding at all what was intended by the title "Attitude of Muslims towards Human Dignity, A Historical Review" and, instead of giving the address intended on the historical development and application of the concept of human dignity in Islam, deliberately presented another basic address which certainly contained some new, interesting and important considerations but not those that should have been discussed. The (basically self-critical) question about the historical implementation of theory in practice was not asked; instead, the ideal Islamic conception was presented again, based mainly on oft repeated quotations from the Quran and on their Shi'ite interpretation – which was formally new but not really in content. This bloc was contrasted with the Christian and Western understanding where the latter was much more clearly in evidence than the former, and then an Arab (both Muslim and Christian) understanding was invoked. The speaker considered this a constructive approach to Christian-Muslim dialogue on the question of human dignity and human rights and was puzzled when it appeared to the officer for Islam of the Pontifical Council for Interreligious Dialogue, Msgr. Khaled Akasheh, to be a confused, supercilious, un-dialogical and frequently ill-informed failure to address the subject. But precisely this criticism of the criticism really demonstrated the degree of misunderstanding because it confirmed more or less everything that had been said in the address and adopted the position that the critic should have known that he was really a self-critical person and convinced about Christian-Muslim dialogue but that that was not the issue here. This raises the question whether this was really only an unfortunate misunderstanding or perhaps a proof that Islam is immune to a large extent to criticism in general and especially from outside because it already has the only right and perfectly developed solution for all possible questions. If others have anything positive to contribute, that is really already present in Islam as well and the numerous shortcomings must be attributed to others, especially to the atheist and materialist West and even more to Israel.

From a decidedly Muslim point of view: human dignity in Islam and in the West
El-Mawla began by emphasising that dignity – in contrast to preference – in the Quran related only to characteristics which were given to people, on their own and not only to a certain extent. In human beings, these characteristics were reason and hence the ability to learn, as well as the will and freedom of decision. Humans were, on the one hand, hermaphrodites between dust and clay and, on the other, divine spirits; but this made them God's representatives on earth before whom even the highest beings, the angels, had to bow down. In practice, human beings were created for work and the acquisition of knowledge and finally for the adoration of God and thus for the world to come as the real goal of creation. What raised people so high above all other creatures was that they surrendered only to God, hence absolute monotheism. Human beings were religious by nature; this natural religion which was in fact identical to Islam was called *fitrah*, derived from the Arabic word for creation *fatr*. The message of all the prophets including Muhammad appealed to this natural religion. (On the other hand, the word *din* for religion referred only to Islam and was therefore not found in the plural in the Quran.) In his comparison, he came to the following conclusion:

> "In Islam, human dignity, like all moral values (truth, learning, knowledge, beauty, good, justice, virtue, creativity, originality, love and worship), are innate and, consequently, constant, absolute and common to all people and at all times in terms of their origins. Their ramifications, on the other hand, are transformed into what we call 'rights'. If we divest these values from constancy and absoluteness, they will be part of good manners and become a code of relative rules like laws. Constant and absolute moral values precede laws and Arabic language, therefore, distinguishes justice from law. This is unlike other languages; because societies not based on Islam regard justice to be law and whatever law may be, it is justice; whereas in Islam, the basic concept and ruler is justice and not equality or rights. Human beings, for example, are not equal but brethren, because equality is a legal term while brotherhood is more profound and comprehensive, because it emphasizes the same nature and substance. Women and men in Islam are not equal but made of the same substance because 'rib' in Hebrew as well as in Arabic means the same nature and the same mix. Woman was created from man's rib i.e. from his nature and mix.
>
> What the West views as a right (like the right of life and the right of learning) which can be relinquished by its possessor, is viewed by Islam as a /p.91 divine precept and legal obligation which should not be compromised even by its own holder even if he had to struggle and die for it. Man's dignity is embodied in his inborn search for knowledge; in his commitment to high morals, virtue or good; and in his propensity towards absolute physical as well as moral beauty, ending with love and adoration.
> (...)
> In the Western philosophical perspective, on the other hand, man is a given being, existent in this way, in this world without roots, without source, without ties linking him to the universe, nature and the course taken by life therein, and in human kind i.e. a vain creature. Man's perfection here is realized through the acquisition of control over nature, environment and his fellowmen. The most perfect being is, thus, the strongest being who

is capable of maintaining his existence (according to Darwin, Nietzsche, Freud and Marx). The relation of this human being to others is one of eviction, expulsion, usurpation and predominance; for it is the fittest, i.e. the most powerful, who survives. The theory of material and physical indulgence and enjoyment in this life is the dominating one according to this concept. Even learning or science is good only because it is a means to attain power which, in turn, is the source of enjoyment."[167]

Two completely different anthropologies: Islam and the West
So there were two different viewpoints, indeed, two different people, the religious one and the secular atheist. Which should one begin with in order to investigate the question of human dignity? For the one, everything – politics, society, the economy – was imbued with faith. The other believed nothing or believed that faith really had and should have nothing to do with all that, or was not at all interested in these relationships because all that counted were the economic, social, political and material advantages. This was followed by a detailed comparison of the two persons, some important extracts of which will be reproduced here. "A person without faith either because he does not care for that or because he absolutely does not believe in that, can mould his life in accordance with various visions. But the person who starts his life from a philosophical and spiritual point of departure usually takes a homogeneous course in his life. The former human being may be just and unjust within the same setting. He can be moral and at the same time a savage beast. The latter human being, on the other hand, cannot but be the same under all circumstances. (...) The former is torn among divided loyalties, various wishes and different norms; while the latter follows a clear and sound direction. The former is confused, perplexed and a victim of whims and caprice while the latter is rightly guided. We believe that man within the cultural perspective cannot be the human who is an outcome of social reflections according to some psychological schools of thought; nor can he be the human who is the product of chemical and physiological reactions that build up his psychological, mental and emotional systems as behaviourists think; nor the man that has resulted from the Oedipus complex as the psychoanalysis school holds, nor the person who is equivalent to his socio-economic output according to materialist Marxism. This is not real man. He is a distorted one. The cultural perspective of such man cannot be but of the same material of which this person has been made, and of the same womb from which he has been born. Under such conditions, we find no suitable cultural environment for such a man except alienation from / p.93 the world, from himself and from others in the fullest sense of the term 'alienation' and all the spiritual and psychological torment, shallowness and aridity it entails."[168]

Islamic and Jewish-Christian worldview – a sharp contrast
With that, he came to his antithesis to human dignity, namely alienation associated with the lack of a balanced combination of self-knowledge, knowledge of the world and knowledge of God which – as had frequently been underlined – was so characteristic of Islam. "Islam is a conception of the universe and life, supported by logic and science and at the same time characterized by faith and a target. Human dignity lies in freedom and knowledge and in justice and piety, all of which are embodied in religion i.e. inner nature; i.e. human dignity lies in the complementarity and sublimation of religion's line between what is permissible and what is taboo between this world and the next and between science and faith."[169] In contrast, in the Jewish-Christian tradition the basis for the contradiction between religion and science had already been laid fundamentally in original sin because the forbidden tree had been the tree of knowledge. Thus people had been faced with the choice between religion and knowledge which El-Mawla equated with the dispute between Zeus and Prometheus. So the devil had been the voice of reason and knowledge. Knowledge was seen as holy and divine and ignorance as human. In Islam, the forbidden tree, on the other hand, was the symbol of greed, that is the symbol of animality and not of the humanity of human beings. Knowledge was precisely what God had given to humans when he taught them the names of all things and with which he had marked them out in particular. "Hence religion and science or learning in Islam are synonymous while atheism is rejection and refusal to comply with learning and knowledge. In the Judeo-Christian Western civilization separation or conflict between science and religion has led to a division of history into the age of religion and faith and the age of learning or science. The former is the age of darkness and ignorance and the latter is the age of enlightenment and knowledge. In Islam, meanwhile, they did not separate: they flourished together and declined together."[170]

The Western concept of human rights: accused of hypocrisy
The latter point and its possible causes were however not discussed at all; instead, he drew very far-reaching conclusions from his own interpretation of the fall. "The concept of man and human dignity in the contemporary Western cultural perspective carries the meaning of deification; violence, sensual pleasures, visual sensory culture, and direct interests. It is this type of man who produced the colonial age, the Marxist devil and the vanity and meaninglessness in existentialism. This man is a distorted image of human beings. It is indeed paradoxical to talk about dignity and human rights in accordance with this perspective of man. What is being said about dignity and human rights is an outcome of the religious and spiritual dimension of man. It is the product of a moral world

and of a philosophy which sees in the creation of the universe, life, the world, nature and man, reason and the aim. (...) /p.96 (...) Therefore what may explain the failure at a global international or national level, results from the assumption that the systems charged with the implementation of these rights are distorted ones which are not born from the same womb that gave birth to these rights, but from another womb. From here we see that human dignity and rights are being violated by the institutions and states which advocate these rights and punish other nations and states violating them. We notice practical examples in the international attitude towards Zionism and Israel on the one side and in Iraq, Libya, Iran and Sudan on the opposite side. Human rights with regard to these powers or to national, regional or international institutions, are being exploited as a political weapon used or overlooked according to needs and circumstances."[171] The surest and simplest for human dignity and human rights would therefore be to return to religion, meaning to Islam; then all the problems would be solved. "Dealing with others from a Muslim perspective may be exercised on the basis of justice, beneficence, righteousness and balance and as has been commonly held by many researchers, on the basis of tolerance. For tolerance is something supplementary which is annexed to another thing which is primary, and it means that the other is to be under the tolerant party's mercy, and that the Muslim should have done him a favor. No!! Islam treats others just as they deserve and not on sufferance, or as a favor or condescension or charity. We all treat one another on the basis of human dignity."[172] That he had just denied this to the second, the other, the distortion of human beings, apparently did not strike the author, and one can only speculate about the treatment that these alienated rather than worthy persons would deserve. For him, it was sufficient to purify the Arab way of thinking and living from everything un-Islamic in order to have the ideal solution to all present and future problems from environmental protection via genetic engineering to nuclear technology and biochemical weapons. Arab Christians were generously included in this solution. "Speaking about human dignity in Islam and Christianity would lead us to asking questions about the cultural change required of the Arab-Muslim world in order that it may get a good-conduct certificate in the spheres of human rights, democracy and other components of genuine human dignity. We may ask here: Is the Arab world required to change its cultural constants and the character of its culture that emanates from its human belief, be it Christian or Muslim, in favor of a primitive conception of man which confines complete humanity to particular people while depriving all other humans of it pursuant to the mythical doctrine of the 'chosen people of God' and in accordance with the fabulous precept which claims that the deity has selected and given preference to one particular people to the exclusion of all other peoples and ethnicities? Is the Arab world and the

Arab individual to cast aside their cultural mainstays in order to provide an opportunity for coexistence with another human of this type who is not content with mere coexistence but wants to impose his hegemony over the region, its destiny, inhabitants and their fate?"[173]

7.6.3.4 Heated discussions and profound misunderstandings

That was very clear and it was equally clear that the reaction of Msgr. Akasheh, an Arab himself, was extremely negative. Among other things, he called for a revision of history teaching so that other religions and civilisations would also be able to recognise themselves in it. Obviously, he could not find himself in the generalisations of the speaker and especially not in his definition of original sin. The Jewish tradition would certainly also not have recognised itself in his definition of election and its relationship with human dignity but was unable to defend itself due to its absence. The speaker felt that the recommendation to check in advance with the persons concerned (in this case the reference was to the Christians in connection with original sin) was generally offensive and that he had been deprived of his right to the free expression of interpretations in the dialogue. He emphasised his critical view of Islamic conquests and of an aggressive and not purely defensive Holy War, but critical historical science and the difficulties of the *sine ira et studio* appeared to be as foreign to the speaker as *summum ius, summa iniuria* as an established element of the Western tradition even before the Arab Islamic world was able to formulate anything of the sort. The infinite self-confidence of his arguments, which led Msgr. Akasheh to quote Luke 6:42, made the sweeping nature of his statements including obvious gaps in his knowledge all the more obvious. "Finally I do not agree with Rev. Khaled's call 'to remodel history teaching so that it may be acceptable to other religions and cultures'... and I cannot understand how he wants to twist history in such a manner as to make it acceptable to the West and to Israel, for example. History is history and we can neither change nor modify it."[174]

Anyone who makes things so easy for themselves need not be surprised if finally arguments which have some justification are also not heard. However one may look at the affair, it is evidence of a profound misunderstanding between two ways of approaching an important question and unfortunately shows characteristics which are more typical than one would wish.

7.6.4 Challenges for the future
7.6.4.1 Once again: human beings as seen by Islam
The last round of discussion dealt with the challenges for the future which existed for both religions in the field of human dignity. Here again, at least for the Muslim address by Muhammad El-Sammak, the presentation of basic issues occupied just as much space as the main issue for discussion. Human beings were described, in line with Quranic statements, as people who had been faced with temptation, committed a sin and had then shown penitence. Their present existence after these events could be described as follows. Human beings were not considered guilty for something they had not committed, i.e. they were not born guilty but had an innate character which predetermined them to seek God and to believe in him. Human beings were much respected as God's representatives on earth; hence everything on earth was subordinate to them for them to develop but not to destroy. God had given people dominion over nations and nature, namely the environment. For this reason, during their creation people had been endowed by God with the necessary understanding and hence knowledge. Human beings were, in a sense, the supervisors, controllers and watchmen over themselves, the judges or arbiters of their own deeds and intentions, while here, according to a tradition of Muhammad, the latter were more significant. So human beings were more important than nature, i.e. they were more important and greater than the sun, moon, fire, wind or similar passing phenomena. They were directly connected with God and obedient to him alone, not blindly but with reason, learning and reflection. According to Sura 2:256, faith was seen as something which could not result from compulsion but was given by God. This again pointed to the great honour of human freedom which had been bestowed on human beings. For this reason, in Islam there was no mediator between God and human beings, so absolute monotheism also revealed the high standing of human beings. On questions of personal faith, there was only the authority of human beings in this world and God's authority in the world to come. So apostasy as such was not punished in Islam. What was punished was abandoning the Muslims and going over to join their enemies, because this was inner-worldly treachery for which inner-worldly punishments were provided by law and were imposed by the competent juridical authorities. Islamic legislation, for its part, was based – as was well known – on the Quran and Sunna and on rational considerations in harmony with the tradition according to specific criteria. This Islamic legislation also laid weight on the protection of human dignity, especially the equality of people and the protection of human life. The very possibility of making an independent legal judgement (the technical term used was *ijtihad*) was an outstanding sign of the respect for human dignity in the concrete exercise of human freedom of thought, opinion and research. This freedom

went so far as to recognise different opinions about the understanding of the Quran itself. Hence the various legal schools in Islam – within a certain academic and religious framework, variations could even be considered healthy. Islam had not imposed itself by force and also not forced its way into hearts and heads by miracles, because Muhammad had done nothing supernatural, namely not healed any sick people, raised the dead or turned a staff into a snake. His only miracle was the Quran from which infinite numbers of meanings could be derived in order to do justice to human development at all times and in all places. It was also important that human dignity was absolute; it did not depend on belonging to a particular group, race, skin colour, language or religion; it did not even depend on whether a person believed in God or not.

Islamic reservations about the Human Rights Declaration of 1948
Against this background, the speaker took up the two questions which constituted the greatest challenges in his view: human rights, on the one hand, and pluralism and the rights of ethnic and religious minorities, on the other. Islam was accused both of neglecting human rights and of not accepting pluralism. Within Islam, this was often seen as a deliberate campagne against Islam as the new enemy, and the reaction and defence of Islam was correspondingly emotional and unthinking. The speaker approached the subject in a very practical way with the question which Islamic country had not consented, or only consented with reservations, to which declaration and on what grounds. Among the member states in 1948, the year the UN Human Rights Declaration was adopted, only Egypt had expressed reservations, namely in relation to Articles 16 and 18 which dealt with the free choice of one's marriage partner and the freedom to change one's religion. Both conflicted with Islamic law which did not allow a Muslim woman (in contrast to a Muslim man) to marry a Jew or a Christian because those religions did not recognise Islam, and the reverse was also true. In addition, Islamic law considered every Muslim who changed their religion to be an apostate and therefore freely changing one's religion was not permitted.

Reservations about the rights of children and a personal conception of the matter
As far as the convention of 1989 on the rights of children was concerned, a number of Islamic states had clearly expressed reservations on certain points, for example on Article 14 on children's freedom of thought, conscience and religion, Article 16 which prohibited disturbing the private sphere of children and private correspondence, Article 17 according to which the child should have access to information of national and international origin with special linguistic consideration for minorities, Article 20 on adoption, Article 29 on respect for other cultures e.g. of their country of origin, or Article 30 which included granting the

children of religious minorities the right to public confession of their faith and the practice of its rituals. Again, it was conflicts with Islamic law which had led to these reservations and could lead to the claim that Islamic law despised the rights of children or was not interested in them at all. For this reason, the Islamic countries had drawn up their own statement on children's rights which emphasised, in the first place, that social values and principles were based on divine revelation and that the fundamental solution was for individuals, societies and governments to refer to these heavenly values and not to others that were forced upon them from outside. In practice, the rights of children in Islam began (long) before birth. "They begin with restricting the sexual relation between man and woman to legal marriage (which non-Muslim communities lack). Islam also forbids adultery, sexual perversity and concubinage. Islam, furthermore, urges that the spouse be chosen from those who have good character and a sound faith and calls for freedom from hereditary diseases to safeguard the child even before birth."[175] Reference was also made to the prohibition of abortion, to the right of the unborn child to possessions and inheritance and to the fact that an expectant mother is relieved of many religious duties and should therefore also be relieved of certain civil and practical duties. The newborn had as a divine gift an absolute right to life and to descent from a particular father, hence adoption was prohibited but the care of orphans was spelled out precisely. The child had the right to be breast fed by its own mother and to be brought up in its own family, to receive care and protection and even to play. In return, the child owed its parents obedience, in particular. The document also contained special regulations for physically and mentally handicapped children, for refugee children, and for illegitimate, homeless, begging, working and stateless children. The Muslim states were to incorporate these regulations in their legislation.

A special political difficulty: rights of minorities
With regard to the declaration of 1993 on the rights of minorities, on the one hand, Muslim minorities benefitted from it but, on the other, non-Muslim minorities in Muslim and non-Arab minorities in Arab countries should also benefit from it, because one could not demand something for one side and refuse the same thing for the other. As far as the rights of the Peoples of the Book to religious liberty in a Muslim society were concerned, they had always been legitimate. In the case of the Arab countries, the speaker saw the question of minority rights as a pretext for intervention in their internal affairs which Europe had practised since the middle of the sixteenth century when it won the Christians in the Arab world for its cause in order, so to speak, to get a foot in the door. The defence of minority rights was mixed up with colonial and imperial politics

and it still was today when America and Europe defended the rights of Muslim non-Arab or Arab non-Muslim minorities in the new world order. For this reason, and not because Islamic law was against minority rights, a large number of Arab and Muslim countries had serious reservations about the declaration on the rights of minorities. The human rights made by human beings distinguished between political and civil rights, on the one hand, and social, cultural and economic rights, on the other. Islam, for its part, divided human rights into three categories: the rights of God (related to religious observance), human rights and a combination of the two. "Islamic Shari'a reflects the high value accorded to man basically with regard to both liberties and rights, at individual and societal levels, and in delicate balance between religious, moral and interactive behavorial controls of these liberties and rights. Such controls include avoidance of excess in exercising rights, and avoidance of encroachment on the rights and liberties of others."[176] All of this was adopted unanimously by all Muslim countries in the Cairo Declaration on Human Rights in Islam (4 August 1990). In conclusion, the speaker referred positively to the UN Human Rights Declaration and the UN declaration on minority rights as the basis for a new, peaceful world order, but warned about double standards in their implementation, especially in the Middle East, which was again using human rights as a political means or even an instrument of political oppression and subjugation. Human rights and human dignity were a task for all societies, not only for the Muslim and Arab ones.

7.6.4.2 From the Christian side: mainly contradiction

The Christian commentary was limited to the second part of the address and agreed that the secular view of human beings had demonstrated its limits, particularly in its inability to protect the unborn, the old and the weak. From a Christian and particularly a Catholic point of view, Sandra Keating naturally also spoke about the quesion of divorce and monogamy. In the Christian view, it was not a loss of status to be an adopted child, because every child remained a child of God and all parents were only guardians of the gift which God had given. There was sharp criticism of the prohibition against a Muslim woman marrying a non-Muslim on the grounds of the non-recognition of her religion. It was not correct to say that Islam recognised Christianity when it did not recognise two of its fundamental dogmas, namely the Trinity and the incarnation of Christ. Therefore the whole argument on this question was flawed and the prohibition simply violated the right of a person to choose a partner and found a family with him. The speaker also objected to the insinuation that minority rights were merely a means for intervening in Arabian government affairs. In addition, she doubted whether mutual respect between two religions was possible if only one of these religions was allowed to proclaim its message openly because, although Muslims were allowed almost everywhere in the world to spread the teachings of the Quran without limitations, Muslims in Muslim countries were not permitted to take an interest in Christianity.[177]

It could also be added here that there was no reference at all to other religions apart from Judaism and Christianity, and the awkward question of what religious liberty followers of polytheist religions enjoyed in Muslim countries was not mentioned with even one syllable, let alone answered.

7.6.4.3 The dignity of human beings – a Christian concept

The first striking thing in the Christian address given by Andrea Pacini was that the fundamental considerations were presented much more briefly and he concentrated mainly on two quite different questions for the future, namely on bioethics and the model of society in view of increasing globalisation. As the basis for human dignity in Christianity, he naturally stated that human beings were created in God's image which was visible concretely in human intelligence, spirituality, will, conscience and freedom. It also meant that the purpose of human beings was to be God's partners. Humans were the only beings that had a free and intelligent subjectivity and were therefore called to enter into dialogue with God and into a relationship of obedience to him and his will. Jesus Christ was the model for humans as God had intended them, that is, in agreement with God's original will and, still more, Christ made it possible for humans to enter again into a real relationship with God. All of this was summarised in a philosophical concept, the concept of person, which comprehended the whole dignity of human beings. The concept of person had two dimensions: firstly, that of individual subjectivity (*individium subsistens*) which could not be reduced to anything else nor be instrumentalised because it had freedom and conscience. "Person" was the expression for the transcendental dimension of humans in contrast to all other creatures. Hence, humans had always to be treated as an end and never as a means, as Kant put it in secular terms. But the category of person did not only comprise this free, conscious subjectivity but also the rationality of human beings. Precisely because they had freedom, conscience and intelligence, humans were also open for others, for God, for humans, for creation as a whole. There were not isolated monads but rational, responsible beings. They had a subjective and a social, relational dimension and precisely that made them so unique according to God's will. Their dignity which was based on free reason placed great responsibility on them, namely the responsibility repeatedly to create conditions which made it possible for everyone to live in dignity. Human beings also had duties towards God, themselves and others and towards creation. But that human beings were persons meant also that they were bearers of fundamental rights which were not imparted to them by society but belonged to them as objective rights based on their being persons. The human rights formulated in international documents were not specifically Christian in origin but expressed the dignity of human beings in a legally binding way for the public sphere. In the Christian view, these rights were always linked with duties because the rights of another person implied duties for myself.

Reproductive biology – on the way to instrumentalising human beings?
Against this background, two problem areas were then discussed, firstly that of bioethics which the speaker saw as divided into the questions of reproductive biology and of genetic engineering. Here, the basic question was always whether a concrete or often only potential practice was in harmony with human dignity. As far as reproductive biology was concerned, the Catholic Church, as had been seen previously in the dialogue on the rights of children, agreed unreservedly with all purely curative measures but equally firmly rejected any separation of the act of procreation from marital intercourse, and particularly the practice of fertilising more egg cells than necessary and then destroying them irrespective of their dignity as embryonal human life. The general idea behind this was also that precisely the random nature of natural conception, apart from the aspect of the mutual, free love of the parents, further reinforced the character of person and gift of the developing person and thus counteracted any instrumentalisation of persons which would make them interchangeable products. Otherwise there was a danger that a society would no longer be in a position to grasp the dignity of human beings as a whole or to respect it at later stages. Moreover, for Christians children were a gift from God and not the right of human beings, and they were also more than the fruit of the will of their parents. Here there was a danger that the image of the family would be changed quite existentially and here, in the realm of bioethics, human beings had the task of shaping society in such a way that human dignity would also be safeguarded in the future.

Genetic engineering – a common ethical challenge
In the field of genetic engineering, therapeutic measures were also explicitly accepted but everything was rejected that reduced people to means and was not compatible with their dignity. This required special ethical training for the people in this field but there was also positive mention of the establishment of national ethics committees. And it was not surprising that this question had be-

come a popular subject for interreligious and intercultural dialogue in order together to define what was harmful for human dignity.

Human dignity in the light of globalisation – an additional challenge
The second challenge was concerned less with the physical and more with the social life of human beings. Every transition created ethical and social problems and hence so did increasing globalisation which was accompanied by growing urbanisation and new forms of work. In traditional societies, human dignity had been safeguarded by traditional forms of solidarity, above all by the extended family and the local community. The processes of change which even extended beyond the competence of national governments also needed to be controlled so that human dignity was not only preserved but even strengthened by a sort of ethical management. This related, on the one hand, to the more cultural and spiritual level of training the conscience but also, on the other, to the level of structural action, namely the level of laws and politics. In connection with what had been said before, special value was attached to the interplay between rights and duties. In recent decades, duties had been somewhat neglected in the West and the societies in the East had denounced this rightly in part, although there was certainly also a secular ethic of duty for which Kant was an example. On the other hand, there were also good historical reasons for this emphasis concerning the protection of the human person in society and the state. In the current transition to modernity, there was a danger that human dignity might become secondary, e.g. compared with economic development. In the Christian view, there was no alternative to human rights for the protection of human dignity in modern societies, precisely to defend individuals and their freedom on the intellectual level (freedom of conscience, right to freedom of expression, political rights) and on the economic and social level (workers' rights, social rights) within the broader framework of an ethic of duties. Rights and duties were complementary, expressed in Christianity by an ethic of rights and an ethic of responsibility. And again it was a matter both of the conscience of the individual as the main moral actor and of the creation of models of society which reflected human dignity.

7.6.4.4 The Islamic commentary: ethical agreements and many areas of work
The Islamic commentary noted far-reaching agreement particularly with the development of Islamic law, referred, in view of the challenges from genetic engineering for example, to a change in the innate nature of humans (*fitrah*), and again emphasised the elements which Islam saw as safeguarding human dignity, namely the status of God's representatives, submission to the will of God, the exalted position of humans above the rest of creation, the subjugation or instru-

mentalisation of the other creatures and responsibility before God, before oneself, before one's fellow human beings and before all other creatures. It also named other subjects which could be dealt with in greater depth in this area, such as the question of the relation between predestination and free will (according to Muhammad, one should believe in predestination with qualified freedom of choice, without getting too much involved in this issue) or the question of the necessary divine support which Pacini had only mentioned generally and in which Balarabi included the sending of prophets with one and the same message (which, he believed, could and should be discussed in greater detail), the subordination of the universe to human beings (cf. above on questions of the environment) and also ethics and duties. The exclusion of revelation as a source of knowledge and science in the West had led, in his opinion, to naturalist methods and a concentration on the rights of people while simultaneously neglecting duties and ethics. This was contrasted with the Islamisation of knowledge with Islamic science as its outrider, which saw revelation as a source of knowledge and science and thus arrived at a complex system of rights and duties in society and for the individual. Other questions which he considered still open and to be discussed were the relation between the divine and the human will in view of humans being God's representatives and image, the relation between reason and revelation in view of human responsibility, the relation between ethics and science and the broad realm of economic ethics.[178]

Perhaps the presentations as a whole were really somewhat too smooth or too positive and did not take account of the fact that certain things were also seen differently theologically and ethically even within one's own religion, quite apart from secular viewpoints. In particular, the realm of the assumed and real practical freedom of human beings between reason and revelation is an exciting question as other discussions have already demonstrated.

7.6.5 A positive summary of all dialogues with the Al Albait Foundation
The joint report reflected the structure of the conference in some detail whereas it contained rather little on the content. The dialogue was considered successful and the subject of human dignity one which had already been fundamental in the earlier dialogues (which was indeed true and has inevitably led to a number of repetitions in this presentation as well which can however be seen as reinforcements). Christians and Muslims shared the view, above all, that human beings were created by God and God had given them their dignity. This dignity was absolute and independent of race, skin colour, language or conviction. There was an urgent appeal not to confuse values and ideals with the reality experienced

by persons and societies which often failed to live up to the ideals. Special attention needed to be paid to those who – as individuals or as a society – had to suffer injustice and the deprivation of these divine human rights, namely the poor, the oppressed and, in particular, abandoned children, persecuted minorities and refugees. But reference was also made to the challenge of scientific progress and the hope was expressed for a new ethical world order which respected human dignity everywhere.[179]

Looking back, this conference was a collection of many anthropologial aspects which had been discussed in earlier dialogues so it formed a logical conclusion to some extent although this had not initially been planned and is also most regrettable. No other dialogue has been so well documented as this one or can therefore be so well followed with the insights and benefits it offers.

8 Contacts with the Secretariat for Inter-religious Dialogue, Tehran

8.1 The beginnings still under the Shah

The first contacts with Iran, i.e. with Shi'ite Islam, date back to the time of the Shah. From 1–7 June 1976, at the invitation of the deputy prime minister, Cardinal Pignedoli, the secretary, Rossano, and Father Abou Mokh were in Tehran for a comprehensive programme of visits with numerous encounters. The meetings with the leading figures of Shi'ite Islam, in which the Secretariat was particularly interested with a view to dialogues (in Rome or in Iran), took various courses. The leader of the Shi'ites worldwide at that time, Ayatollah Chari'at Madari, was very hesitant, whereas Ayatollah Taqui-Al-Coumi, the general secretary of the house for the rapprochement between the various Muslim sects with its headquarters in Cairo, was very open to the idea so that it was already possible to discuss potential themes. The concerns of the Secretariat were above all youth, social justice and the religious phenomenon. On the state side, the interest in dialogue was also seen positively and it was emphasised that in Iran every religion enjoyed complete religious liberty provided it did not interfer in state concerns. On the Secretariat's side, the concern was not to have several different partners in a dialogue but rather a counterpart which agreed on what had been understood and accepted as the main concern.[180] On 3–4 December 1977 there was then a meeting at the Vatican between the Secretariat and an Iranian delegation during which, despite its brevity, an interesting exchange on anthropological questions took place. It was emphasised that faith in God led Christians and Muslims to a similar view of various aspects of human beings, their duties and the ideal for which they should strive. However, the Christians saw a certain tension here between religion as a preserving element and religion as a critical, prophetic element which aimed at strengthening the rights of human beings. It was asked whether that was also found in Islam and the answer was fairly clearly negative. On the contrary, the theory of human rights could prove to be a trap. Since it was spread by Western movements, it was a theory which had evolved outside of Islam. Islam did not recognise such an ideology. In Islam, humans had no innate rights; on the contrary, they had a radical duty to God and everything else was just a consequence of this relationship. The Catholic participants saw a greater ontological density, as they called it, in the Bible and in Christianity than in the Quran. But they were still able to agree on stating that the basis of human rights lay in human beings' relation to God (quite the opposite of the Marxist and communist theories which were clearly a major motivation on the Iranian side for seeking a dialogue with the Catholic Church which was certainly well received). It was asked whether they could define a common ideal of human beings, and the initial practical suggestion was the question of dialogue or, more generally, of the relation of humans to God, and the question of fundamental human rights as understood in the Christian or Muslim revelation. The aim was to have a conference organised jointly, also with youth participants and possibly ecumenical and even Jewish representatives. Naturally, the results were to be adequately published. On the spot in Iran they wanted particularly to deal jointly with such questions as mixed marriages. What was immediately agreed, however, was a more symbolic exchange of students with the Gregorian University.[181]

8.2 A dialogue on the theological evaluation of modernity

Later on (after extensive preparations also with the local church and changes in both the date and the place for the start), there were various dialogues between the Pontifical Council for Interreligious Dialogue and the Secretariat for Inter-Religious Dialogue, Organisation for Islamic Culture and Communications, Tehran, Islamic Republic of Iran, the first from 30 October – 2 November 1994 in Tehran on the theological evaluation of modernity. The participants were a Vatican delegation, Muslim scholars from Tehran and Ghom and about 60 other participants in addition. The approaches to the theme varied considerably. Many addresses on the Muslim side concentrated on the question of modernity and a religious society or were of a very basic theological nature, whereas the Christian addresses included sociological, historical, cultural and spiritual perspectives as well and also focussed specifically on the situation in Asia and Africa. The documents were supposed to be published in Persian and English by the Center for International Cultural Studies in Tehran and so only one of the addresses is accessible, that by Felix Machado, one of the addresses from a Christian point of view. In Rome, there was general satisfaction that there had been a dialogue with Tehran at all but not with the way it went: too large an audience and too little time for discussion. (But it was hoped that these problems could be solved for future dialogues.)[182]

8.2.1 The development of modernity in the West

In his address, Felix Machado linked the concept of modernity with the concept of humanity. The age which had begun with the Enlightenment in the eighteenth century had intended to safeguard a peaceful social existence for all people and, in addition, the greatest dignity of each individual. At that time, human beings with their dignity and rights were the central focus; the human being as a whole was to be liberated. But the Catholic Church reacted negatively to the Enlightenment and also to the humanism which it produced, although it should naturally also be recognised that reason was absolutised and applied to the political and religious realm as well, and the period was far from harmonious or peaceful but rather marked by terrible conflicts and atheism. In addition, Machado mentioned the dispute of 1524–25 between Luther and Erasmus over human freedom as the beginning of modernity. Modernity as a whole was a Eurocentric phenomenon with Christian roots, set in motion by a sense of necessity to rework the language of the Christian tradition and of scholastic thinking. But this had also led to a sharp contrast between faith and reason. By chance, the beginning of European modernity had coincided with the beginning of the colonial period so that Asia had also been affected by Western modernity. On the other hand, Asia had always been marked by a diversity of cultures and religions, a phenomenon also reflected in modernity. But the religious harmony of Asia could not simply be equated with Western secularisation, a phenomenon of modernity. It was the ideal of the Asian ethos and could better be compared with what is known today in the West as religious pluralism. However, this ethnic, cultural, religious and linguistic diversity of Asia had always been felt to be a unity, indeed honoured as an authentic witness to the ultimate mystery of life. In contrast, secularism was often equated with materialism. For modern people who had emancipated themselves, God seemed no longer relevant. Religion was no more than opium for the people and therefore also alienated from the social, political and cultural life of society, the moral and ethical behaviour of which was then dependent only on its own authority. Modernity did indeed celebrate the death of God but human beings had not become so divine that they could take God's place.

8.2.2 Asia and modernity

With its colonial and post-colonial structures, this system had also influenced Asia and the primacy of reason and the central role of the individual and his/her rights as a free subject had fundamentally changed religious practice in Asia. The concept of democracy and also the separation of religion from social, political and cultural life had already affected Asia and this had to be taken into account when discussing so complex a phenomenon as modernity. But this did not mean that today's Asia was less religious; on the contrary, the ties with one's own religion had become stronger in each case as well as the criticism of the West. People in Asia continued to believe that religion was an answer to their practical needs and gave them the necessary motivation and ultimate meaning in life. So it could be said that Asia had avoided the conflict between holy and secular, faith and reason, state and religion. On the other hand, the influence of modernity was not only negative; it had also encouraged devotion to this world, humanism and social awareness. The priority of the human person was not only recognised but highly valued. Despite all the traditions which oppressed people (modernity often saw these people as prisoners of a holy universe), human rights were becoming important and popular movements were winning support. The exploited groups had now become political forces themselves. Machado also noted challenges for the future. Now that it was possible to harvest the fruits of modernity (scientific and technological, political and cultural), the question of justice was especially important. Now human beings were no longer subjected to nature and tradition; reason and freedom had become the norms. It was also clear in the meantime that modernity was not unproblematic, that science and technology could not simply be adopted as neutral but had effects on society, religion and culture. Modernity presupposed demythologisation and rationalisation whereas in the past there had been a mythical understanding of a human-God-nature reality, an anthropotheocosmos. Both the fundamentalist and revolutionary counter-currents refused to acknowledge that Asia had changed under the influence of modernity and saw modernity only as destructive of tradition, as a violent event for humankind and as an enemy of the Asians. On the contrary, modernity was welcomed without limits and uncritically by the Asians themselves. The consequence was complete confusion, the collapse of the family and society, indifference to religious traditions, loss of cultural identity, dangerous ecological problems, unemployment, the concentration of power and wealth, and violence in the political realm. But it was important to recognise that a reshaping of the mentality had already taken place and that modernity was a global and not a purely Western phenomenon. The signs of modernity included disorderly urbanisation, the new economic system and the changes brought about by technology. "Modernity is here to stay in Asia."[183] That had to be recognised and, on that basis, the future had to be shaped in the light of the positive influence of modernity. A good example of the combination of modernity with Asia was Japan. Machado particularly emphasised that science and technology could not be separated from their underlying philosophy and ideology. But the characteristics of modernity were, "primacy and importance accorded to the individual and his rights; separation between the public and private spheres of society, object and subject, and priority given to subjectivity; pluralism and ideology; linear conception of history; relation between science and technology and priority given to its own rationality; political reorganization of society"[184]. Machado certainly saw problems in that the emancipated Asian was more at home in Europe or America than it his/her homeland and that the political, economic or even religious institutions from the West were very fragile in Asia, but he emphasised nevertheless that everyone had to accept these challenges and work together for a common human future. It is a great pity that the plan to publish all the documentation was not implemented and therefore this address, which concentrated mainly on Asia, cannot be supplemented by other contributions more directly related to Christian-Muslim dialogue.

8.3 A dialogue on religious plurality – continuation needed

Then, from 7–23 June 1999, the secretary of the Pontifical Council for Interreligious Dialogue and the officer for Islam paid a visit to Iran in order to re-establish contact with the persons responsible with a view to preparing a new dialogue. This second dialogue subsequently took place from 7–9 February 2000 in Rome on the theme: Islam and Christianity Confronted with Religious Plurality. A total of five addresses were given by each side on fundamental teachings but also on more recent reflections on religious pluralism and historical experiences of it, on the rights of religious minorities in the Islamic Republic of Iran and, in line with the most recent teaching of the Holy See, on the present situation and the future prospects for coexistence and cooperation. It was agreed that nowadays religious pluralism attracted much attention because of its new dimensions which were related in turn to greater human mobility and progress in the field of communication media. Apart from that, differences of content were naturally noted in relation to the Islamic Republic of Iran compared with countries with a Catholic tradition, a number of subjects still needed closer study in further meetings,[185] and generally an exchange of professors and students from Catholic and Islamic universities could contribute a lot to better and deeper mutual understanding. That was a path which the Pontifical Council for Interreligious Dialogue had already followed in diverse ways through the partnership with the University of Ankara and the *Nostra Aetate* foundation. Finally there was also a wish to publish the results of the dialogue jointly in Italian and Persian which would certainly have made a real evaluation from an anthropological point of view possible.

8.4 A dialogue on youth and many old questions

From 11–12 May 2000 there was then another preparatory meeting in Rome to determine the programme and practical details for the next dialogue. This third meeting took place from 18–20 September 2001 in Tehran. It dealt with youth, their identity and religious education. Both sides presented addresses on the sub-themes, namely the identity crisis of youth, their momentary hopes, religion and identity, and finally youth and religious education. Worthy of mention (and in some sense typical) was that the Christian addresses had been prepared in writing and were thus available whereas the Muslims used notes which were then not made available to others. Each address was followed by lively discussion during which the Christian side appealed for the so-called religious minorities in Iran to be able to publish their holy books and to be allowed to introduce their religion to others as was the case in Lebanon and in most of the other countries (although the choice of Lebanon, which was multireligious and very liberal for the Near East, and the collective mention of other countries did not seem very convincing even for an outsider). It was also generally significant that a Christian-Muslim dialogue, which was really being conducted on a quite different subject, again ended up with the question of the rights of religious minorities and hence with the different definitions of religious liberty on the two sides. In addition, however, the matter of broader university cooperation was taken further with an offer from the Pontifical Council for Interreligious Dialogue to establish contacts for students of religion at Qom with one or more Catholic universities. In general, there was agreement on the importance of religious education in schools in order to promote religious and ethical values among young people and to awaken an authentic sense for religion. In order to help young people in their search for God, spirituality was very important. They were also able to agree on the need to promote people's rights to their personal faith in a spirit of respect and dialogue (whatever that might or might not mean in practice) and that they wanted to continue on the path of dialogue in order to encourage understanding and more cooperation between Christians and Muslims[186]. It would certainly be extremely interesting to be able to follow the details of the course of the dialogue and analyse them but this is unfortunately not possible for the reasons mentioned.

9 Islamic-Catholic Liaison Committee

9.1 The creation of the Islamic-Catholic Liaison Committee

After the informal Liaison Committee, which, in addition to the PCID, had also comprised the WCC, the World Muslim League, the World Muslim Congress and the World Islamic Call Society, had more or less petered out, in February 1993 the PCID made another attempt in Jeddah and met the leaders of the World Muslim League and also of the organisation of the Islamic Conference. This resulted in a small conference of the parties involved in June 1994 to discuss a draft for the Conference on Population and Development in Cairo. This in turn led to a wish for further contacts and on 22 June 1995, on the occasion of the dedication of the mosque in Rome, an agreement to appoint a so-called Islamic-Catholic Liaison Committee was signed; it formed a link between the Pontifical Council for Interreligious Dialogue and four important international Islamic organisations: the World Muslim League, the World Muslim Congress, the Conseil Islamique Mondial pour la Da'wa et le Secours Humanitaire and the I.S.E.S.C.O which is the Islamic branch of UNESCO. It was established that this committee was to meet at least once a year, alternately at the invitation of the one or the other side, to discuss a subject which was important for Christians and Muslims and to examine the state of relationships between Christians and Muslims worldwide. At the end in each case there was to be a joint press release in English and Arabic. Theoretically six persons from each side were to participate with a tendency to invite more Muslims. From the Catholic side, four representatives of the Curia were to participate, in practice from the Pontifical Council for Interreligious Dialogue and the second section of the Secretariat of State, as well as one expert and a representative of the local church in question. It can therefore justifiably be said that the undertaking was a high-ranking and representative affair on both sides. The Vatican side explicitly called on the members of the PCID at its next assembly to state the things about which they wished to inform the Muslims so that they could then be passed on to the newly established Liaison Committee. The very next day there was a meeting with the World Muslim League, World Muslim Congress, International Islamic Council for Da'wa and representatives of the Al-Azhar on the issue of the position of women in society in preparation for the UN conference on women in September 1995 in Beijing. There was both agreement and disagreement on the opinions voiced. In general, there was affirmation of the necessity to recognise the dignity, role and rights of women in society and to adopt appropriate measures and it was regretted that the initial draft for Beijing did not take account of the positive role of religions in this connection.

9.2 The first meeting in Cairo – issues can only be touched on

The first meeting of the Liaison Committee took place on 30 May 1996 in Cairo and was organised on the Muslim side by the International Islamic Council for Da'wa and Humanitarian Relief. It dealt with three subjects all of which at least touch on anthropological questions: the relation between justice and human dignity (and here it is striking that justice was mentioned first before human dignity, apparently a result of the influence of Muslim priorities), the environment and human security and finally poverty and humanitarian aid. It is easy to understand that these issues could not be dealt with exhaustively in one day and led only to general resolutions. It was agreed that everything belonged to the competence of both state and religious institutions and that joint studies and cooperation were needed in order to deal effectively with these problems, e. g. to provide the poor with basic necessities such as housing, health and food. Experts were to prepare specialised meetings related to these themes. On this occasion, all the participants also expressed their concern at the murder of the seven monks in Algeria and vigorously condemned all forms of violence in the name of religion.

9.3 The second meeting in Rabat: many delicate points

The next high-ranking meeting from 18–20 June 1997 in Rabat was organised by the Catholic side deliberately in an Islamic country and was deliberately two days longer albeit for only two issues: on how Muslims and Christians speak about one another and on minority rights (which was to be dealt with both conceptually and practically). The discussion covered a large number of delicate points such as freedom of conscience, mutuality and mixed marriages, although the concluding press release avoided all these questions and was therefore described as "very anodyne"[187].

9.4 The third meeting: very anthropological issues

The third meeting of the committee (the fourth if one counts the signing in Rome) took place from 17–18 July 1998 in Cairo and was again organised by the Muslims. It dealt with clearly anthropological questions, again in a very typical form and sequence: 1. Duties of man and woman in family and society, 2. Human rights and duties, 3. The rights of the child in family and society. That in this field the question of duties came first (and not e.g. that of rights) clearly demonstrates the Islamic influence, further reinforced by specifying man and woman and also family and society, which already suggests that the duties of men and women will be different in both areas, and this is indeed a basic Islamic premise. One could say that only when this question has been clarified can one ask or be allowed to ask about human rights and those too are supplemented by people's duties. In the case of children it was only a matter of rights – probably because here both religions do not or do not yet demand very much and, above all, because on this point the danger in modern society seen as a kind of common counterpart was that children would be neglected either in families or in society as a whole. But this can only be deduced from the subjects for the papers submitted and not proved because the press release speaks only very generally and more or less in passing about divergences and then points emphatically to common values, three of which were particularly underlined: human dignity which came from the Almighty God and was the source of human rights and duties. This statement was immediately followed by the sentence, "These are correlated and complementary for the realization of the Divine will."[188] In addition both saw the family founded on marriage as the basic unit of society and for both the child had a right to life, to a good family setting, to education and to religious instruction. The press release closes with an appeal to the media to respect and promote religious and moral values and with the more or less usual declaration that the work would continue, in this case even including the aim of concrete cooperation at local or international level in order to put these principles into practice as well. In contrast to the official declaration, the view of the officer for Islam in the Pontifical Council for Interreligious Dialogue is significant. "There was not much exchange on these subjects, and even less on questions of common interest, as had been foreseen. The atmosphere was however relaxed, and there was a growth of mutual understanding, respect and friendship. This is perhaps the real, if humble, contribution of these meetings."[189] That could be said to be progress on the human level between the highest representatives on both sides, even though progress on the anthropological issues did not seem possible.

9.5 The fourth meeting: the culture of dialogue and religious values

The fourth (or according to the other reckoning fifth) meeting was held in Paris from 1–3 July 1999 on the questions of how one could develop a culture of dialogue in the present generation and what could be done jointly to support lasting religious values in a changing world order; in other words, it was a question of "the third way". As far as the first issue was concerned, it was important initially to be clear about the nature of dialogue itself and to recognise that it comprised all forms of encounter which promoted mutual understanding and mutual respect. A culture of dialogue had to be built on increasing mutual trust in order to be able to bring about justice, promote peace, safeguard respect for human dignity and develop the secure coexistence of human society as a whole. To this end, a culture of dialogue had to be based, among other things, on the implementation of (divine) ethical principles. In the case of religious values, they should guide political leaders, in particular, in their endeavour to create a world order which safeguarded the best for all people. Once again, both sides appealed at the end to the media in reference to promoting religious values and a culture of dialogue.[190] This approach to the theme certainly also comprised anthropological elements because it was a question of dialogue between people and of building up a culture of coexistence in which certain values were considered to be of lasting importance.

Apparently, both the promotion of dialogue as such and the promotion of ethical values were less disputed, because they belonged to the practical level, than the question exactly which these values were for both sides and where there was not only overlap but also divergence, particularly in connection with the ideal image of human beings and their behaviour.

9.6 The further development: less issues, more discussion

From 4–5 July 2000 in Cairo the issue was the rights and duties of citizens according to Christianity and Islam. From 21–22 February 2001 an extraordinary meeting in Rome became necessary due, on the one hand, to the situation in the Holy Land and, on the other, to the situation of Muslims in Europe. At the next regular meeting from 3–4 July 2001 in Rome, the subject was the role of religion in the dialogue of cultures in an era of globalisation. The addresses were given by Fr. Joseph Ellul o.p. and Dr. Hamid Al-Rifaie, the president of the International Islamic Forum for Dialogue. At the end of the discussion it was agreed that civilisation was a common heritage of humankind. Its positive

elements had to be preserved and its advantages made available to everyone. They should also be developed and promoted in the interest of security and the wellbeing of human society as a whole. In addition, both parties affirmed that religious values should be the starting point for leading humankind to maintain human dignity and promote peaceful coexistence between the nations and the protection of the environment. There was also emphasis on the importance of dialogue between civilisations for better mutual understanding and for common activities in harmony and peace in order to safeguard human societies from catastrophes, poverty, ignorance, moral decline, the disintegration of families, wars and the effects of weapons of mass destruction. They rejected the hypothesis of an unavoidable conflict between civilisations and of social conflict. Globalisation and its advantages were to be recognised as important, but it was necessary also to direct attention to their dangers in the realisation of a just world order, indeed to the acceptance of just criteria for achieving the well-being of all and the respect of religious and cultural values in human societies. They thus intended to work together to spread a culture of dialogue more widely and to encourage a spirit of responsibility for society, in order to resist the consumer society, to preserve human dignity and human rights, to prevent aggression, oppression and injustice, to safeguard the rights of refugees to return to their countries of origin and to counteract any form of discrimination against human beings. All of these principles needed to be communicated by the mass media, by educational and cultural institutions and through other available channels. At the invitation of the Muslims, a meeting was planned for 12–13 July 2002 in London or Stockholm. In addition, on the initiative of Msgr. Michael Fitzgerald following the attacks of September 11th, there was a joint statement drawn up by the two secretaries condemning these acts of violence. In general in the course of the years, as was also evident here, it had been resolved to devote more time to discussion and therefore to limit the meetings to one subject with one address each from the Christian and Muslim sides. As could be observed after individual meetings, the discussion had been open and friendly as a rule but not without its difficulties. They were united when it was a matter of cooperating to defend the major values of humankind but tensions arose when dealing with specific situations in countries where Christians were persecuted. There were also difficulties when they tried to work in greater detail on questions of principle such as creating a culture of dialogue, or on fundamental issues like the freedom to change one's religion. Despite these difficulties, there was a growing awareness of dialogue and also of the effort it required, and at the same time the exchange between the members had become freer and deeper.[191]
It would appear that a lot of patience and humility are needed simultaneously in order to deal with the difficult situation in which the large and probably very pro-

found differences in the conceptions of human beings and their freedoms can really not be denied.

9.7 Surprising unanimity on the Holy Land: "two state" solution

But there were also situations in which there was obvious unanimity as the joint statement of 21 April 2002 on the situation in the Holy Land showed. Everyone could not but be interested in bringing about just and lasting peace. Concretely, it was a question of an immediate ceasefire in order to save human lives, especially innocent children, women and elderly people. There was also a joint appeal that people's basic living conditions should no longer be destroyed, for example by cutting down trees in plantations. Even during an armed conflict people should have access to water, food, medical care and other essentials for living. Basically, it was a plea for two independent states accompanied by the reasoning that violence would only produce violence and dialogue was the only way out of the present situation on the way to freedom, security and peace for all in the framework of a "two state" solution. As often in other places, the practical action for peace was seen in parallel to prayer for peace which was, after all, primarily a gift from God.[192] *So, in a sense, the theological component was superimposed on the anthropological one but without eliminating it.*

9.8 Human dignity and human rights in armed conflicts

The ninth meeting with the four Islamic organisations, World Muslim League, World Muslim Congress, International Islamic Council for Da'wa and Relief and Islamic Educational Scientific and Cultural Organisation, took place from 19–20 January 2004 in the Vatican. The leader of the Muslim delegation was Hamid bin Ahmad Al-Rifaie, the president of the International Islamic Forum for Dialogue in Jiddah. The subject of the meeting was human dignity and human rights in armed conflicts. There it was agreed that human dignity was a gift from God. Therefore both sides appealed for persistent prayer for peace and affirmed that justice and peace were the basis for relationships and any form of action between people. They also appealed jointly for an immediate end to all conflicts and aggression against the security and stability of nations. Both affirmed the right of nations to self-determination so that human life would be spared, especially that of innocent people, children, women, elderly and the handicapped. Both called for full respect for humanitarian law and the rights of the civilian population and of prisoners in armed conflicts, and also demanded that nobody should be prevented from having access to water, food, medication and medical care. They also appealed for the preservation of the infrastructure, namely possessions, houses, trees, animals and the essentials of life. The basis for this appeal was religious values and the need to observe international conventions. Places of worship were also to be respected and there should be a right to worship in peace and in wartime. On the basis of the conviction that violence would only produce violence, the spiral of violence had to be broken by a culture of dialogue. That was the only way to make justice and peace between people and societies a reality, which was why the Islamic-Catholic Liaison Committee encouraged it.[193] Here again, there are points of agreement with other dialogue bodies of a similar type.

10 Joint Dialogue Committee with the Al-Azhar University

10.1 The first contacts with Al-Azhar

The first contacts with Cairo were made very early on, namely in December 1970, when the Supreme Council for Islamic Affairs in Cairo came to Rome on an official visit which was returned by Cardinal Pignedoli and two of his colleagues from 9–16 September 1974. The visit dealt among other things with peace, cooperation and the new climate of dialogue, but explicitly not with theological questions and also not with politics, the Palestine question in particular, but it was certainly concerned that faith did not just consist of words but also of actions. In fact, there was a lot of discussion on young people's relation to religion (it was still the period of the youth rebellion), and how these problems and critical questions could be handled positively so that young people did not succumb to atheism. It became evident that the situation was considerably less critical on the Islamic side (faith in God was too deeply rooted in every Muslim for a "God is dead" theology to be conceivable) and Islamic spiritual leaders were dealing considerably less with such current questions but were all the more worried. The Catholic partners in the dialogue were at least partially self-critical in recognising that young people were not rebelling against faith in God itself but against many of the things which human beings had wrongly connected with it, and that it was important to find a form which spoke to them and their spiritual needs. Here the Muslims saw true faith in the true God at risk much more quickly and directly and the cause particularly in families – without an intact family life, young people could not have an intact faith. But the conversations themselves took place generally in an atmosphere of great openness and great respect, indeed even cordiality, friendship and brotherliness, as it was later described.

Right at the first conversation in Rome, there was a clear intention to cultivate regular contacts with one another from then onwards on all questions concerning the relations between Christians and Muslims and to appoint a representative for this from each side. So it is no surprise that from 11–14 April 1978 Cardinal Pignedoli and a Vatican delegation paid another visit to Cairo. There was an encounter with the religious and academic dignitaries of the Al-Azhar University and for the first time a joint statement on the total of three sessions. Once again, the search for a cure for materialism and atheism was the prime focus, but cooperation against racial discrimination and generally against all crimes against human beings as such was also considered necessary. Particularly in the struggle against atheism and for peace there was emphasis on common religious values, especially justice, love, respect for the right to life and human rights. Only the way of obedience to God could help to solve the social, economic and political problems. In the Islamic view, God had created human beings expressly so that they could help one another in all good things and in justice in order to bring about peace. Fanatical tendencies as a threat to peace were not recognised at all at that time; their full attention was directed to the danger of atheism, and alienation from religious values was seen as the reason for the typical problems of modern societies such as hectic living, disorder and a lack of orientation. Both Christians and Muslims with their specific religious values could work well together and contribute to fulfilling the wishes of humankind for wellbeing, peace and happiness. To this end, closer contacts and academic cooperation were envisaged. One of the first official acts of Cardinal Pignedoli's successor, Msgr. Jean Jadot from Belgium, was a visit in Cairo to renew these contacts with the Al-Azhar University. And his successor, Cardinal Arinze, also paid a visit at the beginning of February 1989 and in March 1995 to Sheikh Al-Azhar.[194] After that, there were contacts between the Pontifical Council for Interreligious Dialogue and the renowned Islamic Al-Azhar University, or more precisely its Permanent Committee for Dialogue with the Monotheist Religions, with a view to establishing a mixed committee for dialogue. Representatives of Al-Azhar were also present at the meeting in June 1996 in Rome. During the meeting of the Catholic-Islamic Liaison Committee in Cairo in 1996, the newly appointed Sheikh Al-Azhar, Dr. Mohammed Sayed Tantawi, came in person to the office of the papal nuncio to meet Cardinal Arinze, the president of the Pontifical Council for Interreligious Dialogue, and inform him of his consent to the establishment of such a mixed committee which had already been discussed under his predecessor.

10.2 The founding agreement

The agreement on the establishment of this committee was finally signed by both sides on 28 May 1998 in Rome. The agreement comprises a good two pages and is subdivided into eleven short paragraphs. The point of departure of the document and of the whole process was the necessity for each to have precise knowledge and a right understanding of the content of the faith and the practice of the other religion. It was also important, and came second on the list, for the religions to be able to play their part in human societies by promoting brotherliness, solidarity, cooperation, justice and peace, and by solving problems related to the welfare of the whole of humankind and the common fight against religious fanaticism. The document also goes into great detail about its own prehistory from the visit of a delegation from the (then) Secretariat for Non-Christians to the Al-Azhar via the exchange of correspondence, some of which is described in detail, to the dialogues and colloquia jointly attended. Now these relations were to be further strengthened and the results produced thus far supplemented. The areas of work for the mixed committee were research on common values and the promotion of justice, peace and respect for religions. It was to promote interchange on issues of common interest such as the defence of human dignity and human rights and the encouragement of mutual knowledge and respect for one another – goals also of a clearly anthropological nature. For all of this, the committee was to develop its own method of cooperation; all that was specified for the time being was its maximum size, composition and its moderation, together with the related necessary information process in each case, and at least a yearly meeting for which the travel and accommodation costs would be borne by the delegations themselves. But additional meetings, for example in order to establish the agenda for the yearly meetings, were possible. Binding rules on publicity were also laid down from the start. At the end of each meeting there would be a press release. Without the consent of both sides, no information would be made public on the papers presented to the committee.[195] This naturally leads one to conclude that information in general might be rather sparse and a number of interesting and controversial issues might remain confidential.

10.3 The development of the cooperation: better and better despite difficulties

The first committee had prominent members; from the side of the Pontifical Council for Interreligious Dialogue its president and vice-president, and from the side of the Al-Azhar the retired assistant to the Great Imam and the vice-president of the Permanent Committee for Dialogue with the Monotheist Religions. From 13–14 April 1999, the yearly meeting took place in Cairo and the next on 28 March 2000 in Rome, with an extraordinary meeting on 6 July 2000 in Cairo. After the historic visit of Pope John Paul II to the Al-Azhar, it was decided that the date of that visit, February 24, would be declared the "day of dialogue" and of the annual meetings. So, starting in 2001, the annual meetings were held on 24 February In 2001 in Cairo a sort of review was conducted of what had been achieved positively so far (especially a joint declaration on the Balkans and particularly on Kosovo), and at the end a joint declaration was produced on the situation in the Near East. The mixed committee took stock of the humanitarian situation which had resulted from the conflict in the Holy Land and expressed its horror at the increasingly numerous sacrifices of human lives, the wounded, the material damage, the destruction of institutions and the essentials of life and other forms of suffering for the population (especially, it was emphasised, in the illegally occupied areas); it voiced its sympathy and solidarity with all the victims and their families and expressed hope for peace founded on justice and international law. In this connection, the committee also appealed to the religious leaders, stating that the true basis for peace was justice and mutual respect. Any attempt to exploit the situation (especially in relation to the Holy Sites in Nazareth, meaning the mosque, on which there had already been an exchange of correspondence judged internally to be extremely positive) in order to sow discord and division between Christians and Muslims was condemned, as was any attack on religious liberty, and here again it was especially a matter of Jerusalem and its Holy Sites and of access to them[196] – all in all a traditionally difficult subject on which this joint declaration follows lines which are also familiar from papal statements, particularly including the concern about human suffering. *The special features of this statement with Muslims are the explicit effort to ensure that division or disunity between Christians and Muslims is not encouraged and the explicit mention that the occupied territories were illegally occupied.* Further declarations on the terrorist attacks of September 11th followed later. Regret for the victims was expressed and condolences for the families bereaved. That was not the way to peace; one declaration spoke of inhuman acts of terrorism. The next official meeting in Rome on the culture of dialogue in education was surprisingly fixed already for 23 February 2002 (and then also took place on exactly that date). After a few meetings, a general evaluation showed their great similarity to the meetings of the Catholic-Islamic Liaison Committee. The atmosphere at the meetings was relaxed, cordial, full of trust and respect; there was listening on both sides and this improved from one meeting to the next despite certain difficulties. And, on each occasion, they had been warmly received and encouraged by the Great Imam.[197] The difficulties are not spelled out further but left to the conclusions of the reader.

10.4 The witness of the Al-Azhar at the prayers for peace in Assisi in 2002

In between, there was the Day of Prayer for Peace at Assisi on 24 January 2002 to which the Pope had invited representatives of all religions but especially Muslims in view of the situation. This meant of course that the Muslim witness for peace carried a special weight at this event. It had been written by Sheikh Al-Azhar Mohammed Tantawi and was read by his deputy, Dr. Ali Elsamman. It dealt with the (five) stimuli which, according to Tantawi, the Muslim faith provided for creating a better world. Firstly, it was pointed out again that God had created all people and from only one father and only one mother. Moreover, all monotheist religions could agree on respect for values and this was to the benefit of humankind. The religions proclaimed ethical values such as honesty, justice, peace and prosperity and also the exchange of good deeds, cooperation between the nations for well-being and spirituality (and not for attacks and aggression). Then God had created people as tribes and peoples so that they could get to know one another, and according to God's criterion the most honourable was the most pious. In addition, all monotheist religions called upon people to honour justice and the law by assisting legal owners to obtain their rights. (In order to avoid any doubt about what that meant, Tantawi then thanked the Vatican State for its noble support for the Palestinian people.) And his last point was that there was no compulsion in religion and that Christians and Muslims had lived peacefully together in Egypt since the fourteenth century with equal rights and duties. Because of its direct and inseparable link with justice, according to the closing sentence, the Ulemas of the Al-Azhar agreed with conviction with the appeal for peace.[198] *Those were definitely strong words although it would have been good to hear the Coptic brothers on the same rights, but this was naturally not possible because this was not intended to be a dialogue.*

10.5 Meeting in 2002 – a start with activities against religious extremism

The two dialogue committees then met again, as already mentioned, on 23 February 2002, the day preceding the anniversary of the now historic visit of Pope John Paul II to the Al Azhar in 2000. The meeting took place in the offices of the Pontifical Council for Interreligious Dialogue and dealt with the subject of religious extremism and its effects on humankind. Each side gave an address on this. The hypothesis presented by the Christian side was that fanaticism was derived from an attitude of "divisiveness"[199] explained as meaning that anyone who was not explicitly in favour was considered an opponent. It was necessary for the sake of peace, cooperation and harmony between religions, for Christians and Muslims to take up the challenge of religious plurality. The speaker from the Muslim side, Sheikh Zafzaf, emphasised the generosity and moderate approach of Islam. The discussion which followed dealt with the following points: extremism had to be condemned because it contradicted the teachings of both religions. Extremists, especially religious extremists, could certainly sometimes have serious intentions, but they had the tendency to consider themselves the only ones who were right and to be intolerant towards others who did not agree with them, namely not accepting them with their differences and also violating their rights. Dialogue could be helpful as a counterweight to extremism but with the reservation that the conditions had to be such that a positive result could be guaranteed. (It would be good to know how the positive outcome of a dialogue could be guaranteed but unfortunately the sources say nothing on this; nor do they reveal exactly what was proposed – if anything.) But dialogue alone was not enough anyway; the fundamental aspects of society had to be included, namely family life, education, social development, the mass media, and the promotion of justice and solidarity in the different countries and at international level. Independently of one another, both bodies resolved to continue to follow the path of dialogue and to influence public opinion so that it would reject extremism. *That was certainly all well and good but, in view of so explosive a subject, it was also very vague and far from practical.*

The joint declaration of the meeting from 24–25 February 2004 in the Vatican was somewhat more precise in relation to the difficulties. It was a question of rejecting generalisations and of the importance of self-criticism on which both sides gave addresses. The address from the Muslim side was given by Sheikh Fawzi al-Zafzaf himself, the president of the Permanent Committee of Al-Azhar for Dialogue with Monotheistic Religions. He stated that both religions rejected generalisations when judging people. If an individual or a community committed a sin, they alone were responsible. Both religions also promulgated self-criticism both at the level of the individual and of communities, as well as examination of conscience and asking for forgiveness which could serve as an example for others. In this sense, the Joint Committee appealed for the avoidance of all generalisations and only calling those to account who had really done wrong so as not to punish innocent people for the misdeeds of others. It also appealed to everyone to examine their consciences and, wherever possible, to admit their guilt and then to resume the right behaviour. In that way, the Committee hoped to spread justice, peace and love to everyone.[200]

11 Regional dialogues

11.1 The regional dialogues: an activity of the whole Church from the very beginning

There were also a number of dialogues which were not connected with a fixed counterpart but which involved a whole region in each case. This type of dialogue was very important but was rarely conducted directly from Rome which often only provided the impetus. That applied especially to the dialogue of life; the first phase of a Christian-Muslim encounter had already taken place in their respective home countries so it was not necessary to start by discussing problems with one another to build up trust; on the contrary, common problems could be tackled directly. According to the person responsible, Cardinal Arinze, this role of local churches in promoting interreligious dialogue had existed particularly since 1980. For this, the decisive role belonged to the diocesan bishop because it was in his town or country that there were to be fruitful relationships between the faithful. Therefore the Pontifical Council preferred not to enter into contact with the religious leaders before they had held dialogues with the local churches. And invitations were also normally only accepted if the local church was involved or at least approved. The motivation behind this was relatively clear. Dialogue was not a matter of purely diplomatic activities of the Holy See but of activities of the church whether local or universal. As such, dialogues started where the people lived. The Pontifical Council, as an office of the Holy Father, the worldwide shepherd, encouraged, made proposals, cooperated, reminded and sometimes urged but it did not replace the diocese, the congregation, the monastery or the bishops' conference. Where Christians were in the minority, the visits or other activities of the Pontifical Council always tried to respect the role of the local church and to strengthen it in the eyes of the non-Christians.[201] The special project of local Christian-Muslim dialogues in fact provided the first stimulus for dialogue work in a number of countries.

11.2 North Africa in 1988 – also possible stimuli for others

The first such conference, which was prepared and continued regionally but conducted with the help of what had become the Pontifical Council for Interreligious Dialogue, took place from *24 – 28 October 1988* in Assisi and was simultaneously the event to mark the second anniversary of the Prayer for Peace. The working title was "*Co-existence in the midst of Differences*"; the 25 participants came from North Africa (Mauritania, Morocco, Algeria, Tunisia, Libya and Egypt – unfortunately Sudan was not possible) where the bishops' conference had also suggested the meeting and decided on it with the relevant bodies in the Secretariat, although with a view to other parts of the world, e. g. Europe, where the Christian majority certainly needed more mutual knowledge of, respect for and cooperation with Muslims. The overall evaluation on the Catholic side was that interreligious dialogue in general was making only slow progress and required a lot of patience and generosity. In the Christian-Muslim dialogue, some time since the Second Vatican Council with its new approach, new considerations of the differences and hence some degree of coolness and even disillusionment had come about, but these were now being replaced by new energy based on a realistic, balanced appreciation. The range of participants at the dialogue in Assisi was very broad and had been selected only on the basis of their honest interest in Christian-Muslim dialogue and not with regard to their official functions; indeed, the official religious institutions were reticent anyway about the dialogue or even openly opposed it. What united the participants in practice was the spiritual path they had followed – from whatever kind of negative view of the others to genuine knowledge. Faith in the one Creator led to greater openness (and was also strengthened) to the point where it was said to be a temptation to see oneself as the only possessor of truth and goodness whereas really all people were merely simple witnesses to and fellow workers with God. From an anthropological point of view, it should also be mentioned that there was a lot of emphasis on the need for a respectful attitude, whether in the religious education of the family or in school books and teacher training (and also beyond their official training, giving attention, above all, to the correlation between family influence and formal education), on social communication in general – and here great value was attached to the possibilities of encounters between young people and still more to an appropriate presentation in the media, and finally on the aim of promoting justice and peace among all people. The starting point for these reflections was the reasons for the right to differ in the Bible and the Quran. What that really meant was the right to religious differences and hence to freedom of religion and conscience. This also primarily implied respect and love. In addition, it was interesting that Christians and Muslims from the different countries had prepared this conference and, on their return, became core groups which continued to cooperate. The conference itself was strongly marked by prayer; there were no addresses but only joint reflection on the scriptures and opportunities for practical interchange.[202]

11.3 West Africa in 1991 – the decisive point: democracy

The second meeting from *4–8 August 1991* brought together two dozen Christians and Muslims from Ghana, Gambia, Nigeria and Sierra Leone, that is from English speaking West Africa, at Ibadan in Nigeria. The meeting began with an exchange on the situation in the different countries and then divided up into workshops on the issues of development, democracy and relations with the government, mixed marriages and the influence of the media on mutual relationships. The greatest amount of agreement, particularly on the anthropological level, was found in connection with development. It became evident that Christians and Muslims had a very similar conception of human development, namely the development of the whole human being in all of his/her aspects – economic, socio-cultural, political, emotional, moral and spiritual – and all this in harmony with the aim of the creation of humankind. Human beings were created and raised above the other creatures in order to worship God and glorify the name of their Creator, to live in the world according to his requirements of love, prayer, good neighbourliness, peace and compassion and thereafter to be with him. Development in all its facets had to be for everyone, both poor and rich, and should also respect nature. However, it was also noted that the motivation for such involvement could vary according to the religion and, above all, the conceptions of political development differed because Christians and Muslims had different views about the relation between religion and the state. In the field of development there was already a lot of cooperation even including mutual assistance in the construction of mosques and churches. But it would be possible and desirable to extend this cooperation e.g. in the area of work with women, in promoting life and health, in the fight against drugs and against excessive expenditure on funerals. In contrast, the approach to the question of mixed marriages was more pragmatic. There was a lack of mutual knowledge, especially about the religions' understanding of marriage. In general, religious leaders should encourage people to marry within their own religion. But the choice of one's partner for life was a basic human right and therefore mixed couples should be supported, especially with regard to education, financial support and the custody of children. Marriage preparation courses were essential particularly for mixed marriages. Finally, with regard to democracy, that was seen as the decisive point connected with understanding and cooperation between Christians and Muslims. But it was precisely here that there was the greatest difference between the church, which did not identify with any political system on principle, and Islam, which considered religion and politics inseparable because the Quran was also a law book which regulated the life of all Muslims in all the realms of human existence and human activity, also in the realm of government. In practice, in Ghana, Gambia and Sierra Leone, state and religion were separate with the consequence that the religious leaders did not interfere in state affairs. In Nigeria, the situation was different and for this reason many people on both sides saw politics mainly as a means of obtaining power for their own religious group.

11.3.1 A whole series of recommendations
The conference also adopted recommendations; firstly, a general one to work for the overcoming of prejudices, to avoid describing the followers of the other religion as unbelievers and rather to treat them with respect. On the basis of common attitudes of faith about the meaning of human existence, Christians and Muslims were encouraged to cooperate in the field of development for the whole population, but especially for the poor in such areas as education, health, social welfare and particularly on problems like drugs, defence of the values of the family, and the fight against unnecessary and excessive expenditure (cf. above). As far as the sensitive area of democracy was concerned, it was emphasised that democracy comprised respect for the rights of every person irrespective of their origin, religious conviction or gender. It was particularly emphasised that every citizen had the right to religious liberty and that religious communities had the right to freedom of action. The particularly sensitive areas that gave rise to conflict and were mentioned as needing especially careful treatment were the appointment and promotion of civil servants, the allocation of land for religious purposes and the access to and use of the media. In connection with mixed marriages, it was also underlined that the religious liberty of both partners should be respected. The media, including the secular media as far as possible with their wider influence, should be used to promote understanding and harmony. Here, too, it was resolved to hold similar dialogues in the future in the various countries themselves.[203]

11.4 South East Asia in 1994

11.4.1 A very positive tribute to Cardinal Arinze
The third of these dialogues took place between Christians and Muslims in South East Asia, namely from *1–5 August 1994* at Pattaya, Thailand, on the very Asian or even Far Eastern subject *"Harmony among believers of different faiths"*. The meeting opened with an address by Cardinal Francis Arinze. The cardinal again emphasised that only the arrangements for the meeting were the responsibility of the Pontifical Council while the meeting itself was to be shaped by the

participants themselves. The meeting brought together Christians and Muslims from the Philippines, Indonesia, Malaysia and Thailand but unfortunately not from Singapore or Brunei. The cardinal described religiosity as a characteristic of Asian people. Asia had produced many different religions and cultures but also harmony and tolerance between them. In that sense, it could be described as a pioneer of inter-religious relationships from which the whole of humankind benefited. That should be further encouraged, particularly in view of the growth of materialism, exploitation, the decline of values, injustice, poverty and egoism among individuals and groups. For dialogue, mutual respect and, to back that up, especially respect for religious liberty were required. Dialogue then grew through cooperation and Asia with its many inter-religious groups at the local level was exemplary in that and therefore there was a wish to learn from Asia. Harmony had an effect also on cooperation for the wellbeing of people, the family, society and creation as a whole. The grounds given for this, in analogy with the Catholic tradition, were the unity of creation according to God's plan. God was the creator, redeemer and goal of every person, and every person had something of this image of God within them and was created for the same purpose of seeing God in heaven as he was. The differences between Christians and Muslims faded in view of these fundamental and decisive common features, an expression which was again reminiscent of John Paul II. Arinze even went so far as to say that not only did both Christians and Muslims believe it was necessary to submit the details of their lives to God but that both considered prayer, fasting and alms giving necessary. After listing the various types of dialogue and their possibilities in each case, he again emphasised the importance of religious liberty, irrespective of whether it related to the majority, a minority or an individual person. The fruits of an interreligious dialogue required patience and here the behaviour of Christian-Muslim groups during the Gulf War in 1991 was mentioned as a positive example. On the other hand, increasing materialism and consumerism were again mentioned as challenges, as well as the egoism of the rich who saw the poor increasingly as disturbing intruders – living in religious indifference and as if God did not exist. His definition of dialogue was interesting. "Dialogue is an invitation to believers of different religions to research together the wholeness of Truth. It is gathering together as a family to listen to the voice of God speaking through every believing person and through every religious tradition."[204]

11.4.2 A Christian plea for Asian values
Unfortunately, only one of the addresses given was published in Pro Dialogo, the one by Archbishop Fernando Capalla from Davao, Philippines. It is striking that

Christians and Muslims were not really mentioned in it directly and it dealt more with the role of the other religions in the region and what could be learnt from them for creating harmony or, more precisely, for a process of modernisation that was both humane and Asian. He emphasised that the unity of Asia was attributed particularly to its diversity of religions but they had one thing in common: that religion determined the social goals pursued by economics, politics and society. Because the culture of South East Asia comprised religious values, it played a humanising role. In addition to Hindu and Chinese sources for this kind of inclusion of diversity, mention was also made of the practice of the Muslim ruler Akbar (1542–1605). Modernity had come with Western expansion and brought with it industrialisation, urbanisation and commercialisation and, as a consequence, secularisation and individualism. This caused serious tensions and disturbances which the governments tried to counteract by creating an artificial, enforced unity, often on the religious level, which somehow stabilised relative values again by compulsion. Dialogue was seen as the way in which religious groups could help societies in South East Asia to rediscover their identity in this complex, pluralist situation. This meant a dialogue which critically examined the processes mentioned, but also a dialogue which looked for moral and spiritual resources in the religious and cultural heritage. He mentioned specifically the Asian understanding of secular and rational which differed from the European understanding. According to the Western understanding, secular meant the autonomy of temporal things which could develop according to their internal logic independently of traditional, religious claims in this realm. But, according to the Asian or Eastern understanding, secular meant that everyone, including all religious groups, had their place in society and none dominated the others. There were similar differences in the area of the rational. According to the modern, Western understanding it meant control, theory and projects of a logical, mathematical type, whereas Asia distinguished between rational and reasonable (here again the example is Confucianism and not Islam). Reasonable had also always to be rational but rational had also to be reasonable in order not to become inhuman. Everything in this contribution pointed to a specific, South East Asian spirituality which could be expressed in Christianity, in Islam, but also in Hinduism and Buddhism. Fundamentally, this old attitude was seen as a hope for the growth of human values and a basis for Christian-Muslim dialogue. And the Philippines, the mainly Christian context where the author worked, needed to link up with this again for the sake of human survival and well-being, both economic and political.[205]

11.4.3 The final report: a typical appeal for solidarity

The final report particularly emphasised that the result should also be a dialogue of action. Modernisation was seen as a special challenge. Technological progress and economic growth brought about rapid change both outwardly in society and inwardly in the way in which people saw and understood their mutual relationships and the purpose of their lives in general. The approach to the problem of modernisation had to be creative and not entirely negative. However, what was seen very negatively was the process of secularisation. It tended among other things to destroy humanness. A very pronounced individualism was considered new, namely the rights and fulfilment of the individual at the expense of the family and society as a whole. In contrast, in the past solidarity and harmony (which was the theme) were traditional values of culture and models for common life; now there was more polarisation, and among religions as well. This development was also viewed more or less clearly as negative. The quite practical problems were that less people could enjoy modern wealth whereas many were exploited or marginalised instead. This resulted among other things in a wave of emigration which caused the break-up of many families and exposed the emigrants to new exploitation. One particularly repulsive and despicable form was the exploitation of women and children for labour or sex. Corruption and bribery were an additional problem and there was also an ecological crisis in the region. The mass media were seen as an additional problem because they had further encouraged the developments described above which led to a breakdown of solidarity. The final report spoke out clearly in favour of the family over against an exaggerated individualism, and against all attempts to place homosexual relationships on the same level as heterosexual ones. The responsibility and rights of parents concerning the sexual education of their children in accordance with their religious convictions were explicitly emphasised. The participation of women in dialogues should be encouraged (unfortunately the document does not indicate how high the proportion of women was at this conference). The phenomenon of fear and its varied causes (historical and sociological during the colonial period, the minority phenomenon) were also discussed. The participants were concerned about exclusive and extremist religious trends. They wanted mutual respect, not mistrust.[206] That seems to have succeeded because at the general assembly of the Pontifical Council for Interreligious Dialogue it was stated explicitly that, following the conference, the contacts between Christians and Muslims at local level had been renewed, especially in Malaysia. So plans were made immediately for a meeting of East African countries at the beginning of 1997.

It became clear here how highly Christians and Muslims in general, or at least in this region, value mutual respect and solidarity and that both, despite all the expressions of openness for modern developments, support a more traditional image of the family and society and in some way form a joint opposition to many developments of modernity and secularisation – at least as long as both are negatively affected in the same way by the latter and are able to conjure up a contrasting image of a harmonious society not only of people but also of religions from their regional past. This demonstrates most clearly, in my view, the regional character of this dialogue which, under other conditions, could well have produced quite different results.

12 Future prospects: difficult but still hopeful

A good 40 years after Nostra Aetate and after a large number of dialogues it was time for the Pontifical Council for Interreligious Dialogue to look back as well. Khaled Akasheh undertook this officially with regard to the Christian-Muslim dialogue and asked what had changed positively or negatively in the meantime and what might be the prospects for the future. Akasheh observed an increasingly critical attitude of the Muslims over these 40 years, and especially after September 11th 2001, in relation to their heritage and their religious traditions. Religious extremism and violence in the name of religion had become issues and were hence less taboo than in the past. Human dignity and human rights, including the right to religious liberty, had also become issues that were discussed in the intellectual circles of Muslim majority societies. One could also observe generally that Muslims were increasingly taking the initiative for dialogues, although naturally often to correct the image given by a particular press or by religious minorities that used violence in the name of religion. On the other hand, the anti-terror initiative of the West with its anti-Muslim aspects had produced a feeling of humiliation among the Muslims that was hard to overcome. All in all, religion and politics remained a difficult subject. A degree of healthy secularism seemed hard to achieve in countries with a Muslim majority. There was a real danger of the politicisation of religion and thus of mixing religion and politics. Islamic political parties were working to establish Islamic states and to introduce *Sharia*, although Islam as *the* solution was more of a slogan than a political project. Countries which had adopted Islamic law had not thus found the solution to their problems that had been announced before. Moreover, this kind of politics did not encourage genuine cultural and religious plurality; nor did it result in respectful relationships with those who were really "other". Here Akasheh mentioned as a particularly delicate issue school textbooks which dealt with religious and historical questions. They still raised questions related to their role in the spread of Islamic extremism and even violence. It was not yet possible to discuss such sensitive historical matters as the Islamic conquests, the Islamisation of the countries conquered and the crusades. For the future, in view of the risk of a clash of cultures, Akasheh saw the task of the Christian-Muslim dialogue particularly as safeguarding humankind from anything that could lead to new world wars. In addition, means had to be sought jointly for an equal distribution of resources, because countries with a Muslim majority were often at the top of the so-called "Second World" in the southern hemisphere. But one should also not lose sight of the fact that both Christians and Muslims worshipped the one God. As far as the range of dialogue participants was concerned, Akasheh was convinced that everyone would join in who did not support terrorism or engage in it themselves. In a world that was becoming constantly more global, every religion would have to rely on others at some point. Moreover, Christian-Muslim dialogue was also an expression of brotherly love.[207] *So in general the propects were positive except for the people who caused the biggest headache for society because they wanted to destroy it for the sake of a religious ideal image which then did not keep its promise in practice.*

However, the final word should be a personal experience from the scholarship programme of the *Nostra Aetate* foundation – a young Muslim from Tunisia who, following his studies at the Ezzitouna, also studied Christianity and Judaism with a scholarship at the Gregorian Pontifical University and was able during that time to live in the main monastery of the White Fathers with the same rights and duties. He described the discussions with and also among the White Fathers in the following way. "I once discussed Qur'anic and Biblical anthropolgy with an elderly priest. I spoke of man as God's vicar in the Qur'an while he spoke of man as God's likeness in the Bible, and at a certain point we discovered that we were / p. 305 expressing the same concept of humankind and the same human condition both in the Qur'an and in the Bible."[208]

Notes

1 As a rule, reference is simply made to this change; in retrospect, it appears more or less to have been the compulsory next step (cf. Note 2). But here one should certainly not underestimate the role of Cardinal Pignedoli and his personality. Already when he came into office he made clear, despite all appreciation for the previous work, that there had to be and would be a further development and that this would follow the line of more real encounter. He was seeking support for this, cf. Pignedoli, Sergio, Lettre du nouveau Président, BSNC 22 (1973), p. 12–16 or idem, Perspectives du Secrétariat, BSNC 23/24 (1973), p.88. A further impression of this is given by Shorter, Aylward, Cardinal Pignedoli: A Brief Memoir, Pro Dialogo 72 (1989), p. 377–

378. Arnulf Camps, Two Eminent Secretaries: Pierre Humbertclaude and Pietro Rossano (1964–1982), Pro Dialogo 72 (1989), p.379 makes the point clearly in retrospect: "Cardinal Pignedoli knew how to make friends and his thoughts were always a little in advance of many of us as he considered friendship to be a better instrument of communication than sharp theological positions." It is also very well put in his letter for the tenth anniversary of the Secretariat (in which, incidentally, he also discussed its unfortunate negative name), in Pignedoli, Sergio, Pentecostal Letter of the President of the Secretariat, BSNC 26 (1974), p.93: "There is another thing about which we are really convinced. It is that the religious man who accepts the Other in his life is, among all types of men, the very one who has within him, who is most capable of building up lasting relationships with others. Every religion, in fact, opens the heart of men to their brothers for the very reason that it brings man to God. It is not possible to ignore this interrelation of values." Pietro Rossano also writes in retrospect that he was not particularly interested in theological questions but had suffered when doubts were expressed about salvation being for non-Christians as well and when other religions were shown so little appreciation (10th Anniversary of the Death of Cardinal Sergio Pignedoli, Pro Dialogo 74 (1990), p.174, cf.also p.172f). The sermon on the occasion of the mass for the tenth anniversary of his death is also very moving: Shirieda, John, Don Sergio, Witness to God's Goodness, Pro Dialogo 74 (1990), p. 175–177. In this connection, the review by Cardinal Francis A. Arinze is very interesting: Prospects of Evangelization, with Reference to the Areas of the non- Christian Religions, Twenty Years after Vatican II, BSNC 59 (1985). It makes clear that the Pignedoli era was the only one to date during which the Secretariat actively approached non-Christians and organised dialogues instead of, as was later the case, only being indrectly involved or accepting invitations either to conferences in a broader framework or to an interchange with specific partners (p. 118–124). From a purely chronological point of view, the list here of the activities in 1974–1978 (p.121) is of interest; unfortunately not all of those relating to Islam are reflected to the desirable degree in the accessible publications of the Secretariat. The most balanced is probably still the evaluation of Pignedoli in Fitzgerald, M[ichael] L[ouis], The Secretariat for Non-Christians Is Ten Years Old, Islamochristiana 1 (1975), p.91, as he also writes in Fitzgerald, Michael L[ouis] / Borelli, John (eds.), Interfaith Dialogue, A Catholic View, London/Maryknoll 2006, p.91: "To my mind the experience of dialogue, meeting people whose sincerity and goodness cannot be denied, is the best way of broadening one's horizons and of coming to recognize God's action in people." An appreciation of his successor, Archbishop Jadot, can also not avoid appreciating Pignedoli; on the contrary, it contains this evaluation of the outstretched arm of the Holy Father, as in Michel, Thomas, Growing towards "dialogue in community", a Reflection on Archbishop Jadot's Tenure at the Secretariat for non- Christians, Pro Dialogo 103 (2000), p.68. However, it is clear that we cannot deal with every mutual visit nor with every local or regional dialogue and also not with all the individual meetings with members and advisers – that would go far beyond the scope of this study. But all of these encounters and activities have been documented and commented on with great care and are available to those interested in the house's own bulletin. A general review of the first 25 years is also provided by Fitzgerald, Michael L[ouis], Twenty-five Years of Dialogue, The Pontifical Council for Inter-Religious Dialogue, Islamochristiana 15 (1989), p. 109–120, naturally not forgetting Number 74 (1990) of the Secretariat's bulletin which deals with the silver jubilee and all the events, appreciations and reviews. Number 78 (1991) of the Secretariat's bulletin is dedicated, following his early death, to the equally eminent figure and secretariy already mentioned, Pietro Rossano, who later became a bishop. Another review can be found in Akasheh, Khaled, Considerations on Forty Years of Religious Dialogue with Muslims (A Report), Pro Dialogo 116/

117 /2004), p. 195–204, which provides an equally brief review of the events with a wealth of material. He is the only person who includes a discussion (p.196) of Islamic-Christian meetings "at Church level", as he calls it, even before the Second Vatican Council, namely in 1964 in Egypt, in Lebanon and once also in the Philippines, all countries in which Christians and Muslims were living together and, in certain cases, practically ever since Islam came into being. Akasheh observes that these countries were also preferred for dialogue meetings later on. He then honours the preparatory role of Pope Paul VI who had received important Muslim personalities after the Council. Value must certainly also be attributed to the cooperation in the university setting in the form partly of colloquia and also in an exchange of professors; but this field can also only be mentioned in passing here and not analysed further because it would go far beyond our scope. In general, however, a tendency to institutionalisation can be observed which was also consciously perceived as such. "Dialogue will remain precarious where commitment to it does not take an organized form." (according to Fitzgerald, Michael L[ouis] / Borelli, John (eds.), Interfaith Dialogue, A Catholic View, London/Maryknoll 2006, p.92). On the conference in Luxembourg mentioned separately cf. Canivez, Jean, **Luxembourg:** Les Non-Chrétiens en Europe – Brève relation sur la rencontre au Luxembourg (13–14 Mars 1974), BSNC 26 (1974), p. 145–147. In a quite different way in relation to Europe, there is an interesting article by Gjergij, L., La situation religieuse des Albanais en Yougoslavie, Bulletin Pontificium Consilium pro Dialogo inter Religiones 75 (1990), p. 286–289, which deals with the Christian-Muslim dialogue between the Albanians in Kosovo. The published documentation on Bamako/Mali 18–20 June 1974 is considerably more detailed, beginning with the list of participants: Participants à la réunion de Bamako, BSNC 28/29 (1975), p.3. However, it is difficult to get a real impression of that meeting because the two published reports, if one can call them that, could hardly be more different. The one by Cardinal Pignedoli, as one would expect, was really enthusiastic, cf. Pignedoli, Sergio, Relation de S.E. le card. Sergio Pignedoli, BSNC 28/29 (1975), p. 12–15, whereas that by the specialist, Josef Stamer, is rather devastating, cf. Stamer, Josef, Compte Rendu: L'Eglise d'Afrique Occidentale en dialogue avec les musulmans?, BSNC 28/29 (1975), p. 4–11, starting with the disastrous participation of precisely those persons who should have been the most interested and ending with the consistently defensive attitude of the Christians who did not recognise the need for a dialogue with Muslims or took this to absurd conclusions. It is amazing that Cardinal Pignedoli saw the dialogue with the Muslims locally as having almost achieved its aim while Father Stamer had the impression that it was necessary here for the insights of Vatican II about the religious value of Muslims first to be received. With only these two reports, it will certainly hardly be possible to clarify who was more or less right, but it is an example of how much evaluations can differ from one another and how cautiously reports should be treated.

2 According to Rossano, P[ietro], Bilan d'une rencontre Domus Mariae – 1972, BSNC 21 (1972), p. 9–15. The later report by Akasheh mentions (p.199) that there were also multi-lateral dialogue conferences from 1971 onwards, and he situates the high points of the dialogue during the presidency of Cardinal Arinze because it was then that the dialogue was institutionalised. "During the second part of the pontificate of Pope John Paul II and under the presidency of Cardinal Arinze (1984–2002), this dialogue experienced unprecedented success; it broadened and simultaneously became institutionalised. These were times during which talks and meetings multiplied. The institution of permanent liaison committees between the great Islamic organisations and this Pontifical Council (with four large international organisations and with al-Azhar), by holding regular meetings, allowed bonds to be strengthened, ideas to be exchanged and the signing of common declarations for national or international events or situations."

3 Mokh, Francois Abou, Le dialogue dans le monde arabe, BSNC 30 (1975), p.224, cf.also p.223 + 225f, and the passage is of special interest (the Lebanese civil war had only just begun) where (p.226) he quotes a Lebanese Muslim leader. "Nous avons accepté à contre coeur le contrat national, au moment où nous étions pauvres et divisés. Aujourd'hui, la situation est changée, et il nous est impossible de vivre dans les mêmes conditions. D'ailleurs, les musulmans, selon leurs traditions, ne peuvent être gouvernés par des chefs non musulmans". The other basic contributions are: Rossano, Pietro, Introduction, BSNC 30 (1975), p. 211–213 and Camps, Arnulf, Le dialogue interreligieux et la situation concrète de l'humanité, BSNC 30 (1975), p. 315–318, whose judgement of the work of the World Council of Churches is as precise as it is theologically damning, especially with reference to the Protestants who were no longer able to see the usefulness of religious dialogue. "Même le département pour le dialogue du COE à Genève n'a pas réussi à trouver une base ouverte et chrétienne pour le dialogue. On a l'impression qu'on évite le vrai problème en commençant depuis bien des années des dialogues concrets en différentes parties du monde (Le Liban, Sri Lanka, etc.) sans toucher le problème théorique et religieux. Le sujet des dialogues était toujours l'unité de l'humanité, la paix, le développement etc. Je sais bien qu'il y a des exceptions parmi nos frères protestants, mais ils sont peu nombreux à en être et ils n'ont pas la force de se faire entendre." (p.315). More attention is paid to the situation of the Muslims in sometimes very limited contexts by de Souza, Achilles, Christian-Muslim Dialogue with Reference to Pakistan, BSNC 30 (1975), p. 218–222, Anawati, Georges C., Le dialogue islamo-chrétien en Égypte aujourd'hui, BSNC 30 (1975), p.227f, Cuoq, Joseph M., Notations sur la situation du dialogue au Maghreb, BSNC 30 (1975), p. 234–236, Lelong, Michel, La communauté chrétienne et les musulmans en Tunisie, BSNC 30 (1975), p. 237–240, idem, Le secrétariat de l'église de France pour les relations avec l'islam, BSNC 30 (1975), p. 249–252, Borrmans, Maurice, Quelques informations sur la Libye et l'église, BSNC 30 (1975), p.241, Roest Crollius, Ary A., Four Notes on Dialogue (More Especially Regarding Islam), BSNC 30 (1975), p.246f, Fitzgerald, Michael L[ouis], Christian-Muslim Dialögue in Indonesia and the Philippines, BSNC 30 (1975), p. 214–217 (which as always constitute a special case), cf. also later Michel, Th[omas], Religion in Indonesia today, BSNC 44 (1980), p. 213–220. In this connection mention must also be made of the report published considerably later by Henri Teissier, Dialogue Islamo-Chretien *(sic!)* au Maghreb et en Algerie *(sic!)*, BSNC 39 (1978), p. 198–204, the basic tenor of which is certainly positive despite all the difficulties that existed there. He saw the main problem for dialogue as being that the Muslim society in North Africa was on a different psychological level compared with Christian-Western society. For this reason, a conversation on dogmas did not yet make sense; it was necessary firstly to be clear about the frames of reference on each side and have a discussion on the conceptions of revelation, inspiration, the Book, religious language and the like. This, in my opinion, expresses an insight which is valid also beyond North Africa and should still be taken into account today. A report by Sister Odile de La Fortelle on the dialogue situation in West Africa must also be mentioned here: **Ghana:** Christian-Muslim Dialogue, BSNC 39 (1978), p. 248–251. She came to the conclusion, precisely in view of the difficult minority situation of the Christians there, "I realized still more how much, when addressing the Christians to open them to the real values of another faith, I cannot but strengthen them in their own faith" (p.251). Dalmais, P., L'Église et l'Islam au Tchad, BSNC 44 (1080), p. 221–236, also belongs to this context. As demonstrated by Secretariatus pro non Christianis, Questionnaire, Dialogue: Situation and Problems of 24 April 1979 from the archives of the WCC, DFI Box 37, there were repeated questionnaires of this kind about the progress of inter-religious dialogue. And, for other reasons as well, the Bulletin repeatedly examined individual geographical areas and asked how things stood, for example, for Europe in BSNC 60 (1985). Especially interesting here is Sabanegh, E[douard] S[ami] Martin, Les directives de la Hiérarchie catholique de l'Europe de l'Ouest concernant les Musulmans Étrangers, BSNC 60 (1985), p. 291–302, which gives a review of how Muslim migrants in Western Europe were treated by the Catholics, although it is also stated that, after 30 years and with a second generation, one could no longer speak about immigrants. The same volume contains an obituary for him as the Officer for Islam at the Secretariat: Basterrechea, José Pablo, Avant la messe de funérailles du Frère Martin Sabanegh (3 juillet 1985), BSNC 60 (1985), p. 337–338.

4 In this connection, the report given a little later by Michael Fitzgerald on the occasion of the visit of an Indonesian delegation to the Vatican is interesting, cf. Fitzgerald, Michael [Louis], The Secretariat for Non-Christians and Muslim-Christian Dialogue, BSNC 37 (1978), p. 9–14. There the local dialogue is mentioned first and the other dialogue activities which assist it (p.9); then the emphasis is on basic openness even at the expense of being misunderstood, which refers relatively clearly to Tripoli in particular (p.9f, cf. the chapter below). Then he goes into greater detail about the dialogues of the Secretariat as an organ of the Vatican always being top-level affairs and therefore particularly needing to be supplemented and continued at local level (p.10). It is also interesting that the need for small, continuous groups is mentioned. "The continuous research could be expected / p.11 to produce a far from negligible growth in openness and understanding." (p. 10–11). It is also stated very clearly that dealing with areas of human experience rather than dogmatics and theology is a better service to understanding and avoiding difficulties, even though this does not exclude all structural problems, e.g. the question how exactly a social order should be based on faith (p.11). He identifies a wide field for possible cooperation ranging from a new economic order, via promoting peace, human rights and the struggle against racial discrimination to religious liberty, and the issue of prejudices is also discussed in relative detail. "More accurate knowledge will show up the weakness of current clichés concerning Christians and Muslims." (p.13) It can be said that much of what is mentioned here certainly does not only apply to the dialogues of the Secretariat in question but is important in addition for Christian-Muslim dialogue in general. On the other hand, the corresponding view from the Indonesian side, Inter-Religious Dialogues and the Development of Religious Harmony in Indonesia, BSNC 37 (1978), p. 14–24, is mainly interesting as evidence of the familiar special Indonesian situation and the conception of harmony which is also considered very important in other places in South East Asia. It is certainly interesting that the first potentially disturbing factor mentioned here is the missionary dimension of religions, even before ignorance, fear, suspicion or lack of communication at leadership level (p.20). At least it is possible to point out that considerable success has already been achieved by dialogues although, or precisely because, they had not relied principally on representatives of the various local communities or of the state. : "At the outset, participants of the dialogues came mostly from the universities in the hope that more objectivity could be expected." (p.22) Even though it was neither a matter of a Christian-Muslim dialogue nor of a conversation on the present state of Christian-Muslim dialogue, the interested reader is referred to BSNC 48 (1981) which constitutes a sort of volume of issues concerning Islam, and to BSNC 49/50 (1982), a volume of issues concerning Africa which also deals with African Islam in great detail. BSNC 51 (1982), on the other hand, is a volume of issues concerning Asia but it naturally also contains some interesting articles from Asia on Islamic questions. That the Catholic conception of anthropology, which is discussed here in contrast to the Islamic conception, can be said to vary depending on the point of view from which it is considered, is very well demonstrated by Fitzgerald, Michael L[ouis], Mission and Dialogue: Reflections in the Light of Assisi 1986, BSNC 68 (1988), p. 113–120.

5 cf. Rossano, Pietro, The Secretariat for Non-Christian Religions from the Beginnings to the Present Day: History, Ideas, Problems, BSNC 41/42 (1979), p. 95–100+105, cf. idem, Le cheminement du dialogue interreligieux de "Nostra Aetate" à nos jours, Pro Dialogo 74 (1990), p.135 and (for the further differences in the Muslim reaction to the new conception of dialogue) p.139, and p.140f concerning the perhaps unspectacular but nonetheless perceptible progress which was made in the next few years precisely in the realm of practical cooperation; Arinze, Francis, Reflections on the Silver Jubilee of the Pontifical Council for Inter-religious Dialogue, Pro Dialogo 72 (1989), p.314f (here it is stated in retrospect that the inter-religious encounters helped the Christians to clarify the understanding of inter-religious dialogue and were in some way a test of the principles which had been established in the guidelines for dialogue – as was very well demonstrated in the preceding chapter specifically for Christian-Muslim dialogue and its guidelines; a reference must also be allowed here to Number 77 (1991) of the Secretariat's own bulletin which is devoted to the document "Dialogue and Proclamation"), Masson, Joseph, Souvenirs et perspectives, Pro Dialogo 72 (1989), p. 334–342 (which gives background information based on his experience about why real dialogues were not possible earlier, what could be achieved – and was also achieved – and why, for example, the change in name was very important to the partners in the dialogue; on this last point cf. also Arinze, Francis, Editorial, BSNC 69 (1988), p.185, for whom it was also a final recognition) and Rossano, Pietro, **Rome:** Restitution of the Visit of the WCC Commission for Dialogue to the Secretariat for Non-Christians (7 March 1974), BSNC 26 (1974), p.140: "The contents and the spiritual values of man expressed in various religions should be the fundamental bases from which dialogue operates." This, moreover, was an observation made expressly in agreement with the corresponding body of the WCC, cf. later Taylor, John B., **Geneva – Switzerland:** Third Working Meeting between the Secretariat for non-Christians and the World Council of Churches' Sub-unit on Dialogue with People of Living Faiths and Ideologies – Ecumenical Institute Bossey (March 13th-14th 1983), BSNC 52 (1983), p.85: "In any future planning for actual dialogues between Christians and people of other faiths it was recognized that it could be very useful to centre the agenda round common human concerns such as threat of nuclear war, the crisis of the familiy life or the violations caused by racism." In any case, the interchange and cooperation between these two institutions can almost be described as an established fact in the background which required less mention and documentation the longer it continued because it was such a normal thing for both sides. The best document on this relationship, its extent and significance is found in Ucko, Hans, Pontifical Council for Interreligious Dialogue 40 Years Testimony, Current Dialogue 45 (2005), p. 13–15. At the same time, however, this document also shows that there had been very few really concrete projects on which there had been cooperation and that these were never related exclusively to Christian-Muslim dialogue but always included at least one more if not several other religions. The joint Christian theological conference on the significance of Islam, to which Taylor's report refers as being envisaged for 1985 or 1986 (Taylor, p.85: "might promote"), has so far never come about, which indicates how difficult such a fundamental theological question is, perhaps the most difficult in the whole of inter-religious dialogue. With regard to the continuing problems in Christian-Muslim dialogue mentioned, cf. in greater detail and in a more balanced form Cordeiro, Joseph / Teissier, Henri, Rapport du Groupe "Islam", BSNC 41/42 (1979), p.186: "Le Secrétariat ne saurait oublier, enfin, que le difficile dialogue avec les Musulmans aura toujours de plus ou moins graves implications politiques, se heurtera sans cesse au problème de la 'représentativité', discutable mais réelle, des instances officielles islamiques et devra accepter que beaucoup de musulmans y poursuivent le dessein à eux déjà assigné par le Coran, à savoir conduire les Non Musulmans à l'Islam par les 'voies les meilleures du dialogue' ".20 years later in Report

on the Activities of the PCID: Relations with Muslims, Pro Dialogo 101 (1999), p.219, Khaled Akasheh wrote: "Christian-Muslim dialogue is without doubt the most difficult of dialogues, but at the same time the most necessary and perhaps the most promising. It constitutes a sign of hope for the future, but also a challenge. The PCID, strengthened by divine assistance, by the encouragement of the Holy Father and by the collaboration of Members and Consultors, hopes to succeed in this mission to bring about a future of peace and collaboration not only between Christians and Muslims, but also between all believers and men and women of good will." It is also interesting that the advisers of the Secretariat mentioned here were very much at the centre during the initial phase up to 1971 in research, publications and planning, but then receded into the background and thus became real "advisers", according to Zago, Marcello, The Life and Activity of the Pontifical Council for Inter-religious Dialogue, Pro Dialogo 74 (1990), p.151. On p.154, Zago also gives a very detailed, clear explanation of why the change of name to Pontifical Council for Inter-religious Dialogue was long overdue. For him, dialogue stands for all positive, constructive inter-religious relations with individuals or communities that are aimed at mutual understanding and mutual enrichment. To describe these expressly as inter-religious recognised "the other person's identity, the value of his experience and of his religious traditions." So here the name is not indifferent but rather *nomen est omen*; in this connection cf. also auch Shirieda, John, Editorial, Bulletin Pontificium Consilium pro Dialogo inter Religiones 70 (1989), p.1. The first issue of the Secretariat's bulletin with the new name gives him the opportunity to emphasise that this new name itself is a renewed confirmation of the mandate from the silver jubilee. There is a similarly positive reaction to the appointment of the secretary of that time, Michael Louis Fitzgerald, as the titular bishop of Nepte; cf. Shirieda, John, Editorial, Bulletin Pontificium Consilium pro Dialogo inter Religiones 79 (1992), p.1. A little later, the bulletin was renamed "Pro Dialogo" which again conveyed something more positive about the content than the previous name, according to Fitzgerald, Michael L[ouis], Editorial, Pro Dialogo 85/86 (1994), p.1. This double issue is also interesting for anyone who wishes to have information about new reflections in the field of the theology of religions although that is expressly not our subject here. One of the most recent external deveopments (since May 1998) is the Groupe Interdicastériel pour l'Islam (GII), which has nothing directly to do with either dialogue or research but is intended to follow the contacts at the level of the Roman Curia with the various Islamic institutions and to create a uniform voice even with identical wording for the curia, nunciatures and bishops' conferences. The two-monthly meetings started in December 1998 and included the second section of the State Secretariat, the Congregation for the Oriental Churches, the Congregation for the Evangelisation of the Peoples, the Pontifical Council for Migrants and Refugees, the Pontifical Council Justice and Peace, the Pontifical Institute for Arab and Islamic Studies (PISAI) and naturally the Pontifical Council for Inter-religious Dialogue. A document on terminology, which is intended to avoid both offensive vocabulary and vocabulary overloaded with Islamic ideology, has already been produced, as well as an Arabic glossary on Islam with an Italian index which might become the subject for a larger study. The last subject which was tackled during the initial phase was the implementation of Islamic law and its implications for Muslims and, above all, for non-Muslims. It is possible that there will be an exchange on this issue between the curia, the nunciatures and the bishops' conferences. The idea has also been mooted of an introduction to Islam for which it would be possible in part to have recourse to "Cheminer ensemble" of the Pontifical Council for Inter-religious Dialogue, and of a group that would work on the level of the religious orders and communities where these are affected by Islam. The GII, according to Akasheh, is a support for the Pontifical Council for Inter-religious Dialogue in dealing with the challenge which Islam constitutes for the Church as a whole, although it could certainly be further expanded; cf. Akasheh, Khaled, Bureau pour l'Islam Rapport d'activités: novembre 1998-octobre 2001, Pro Dialogo 109 (2002), p.99f. Less structured and less regularly, i.e. with only two or three meetings a year, there had already been meetings earlier on with the intention of keeping members of other departments of the curia informed about the various current aspects of Christian-Muslim relations; cf. Fitzgerald, Twenty-five Years of Dialogue, p.113. As far as general developments and problems are concerned, reference can also be made to Machado, Felix A[nthony], A Summary of Reports by the Members: Interreligious Dialogue Promoted by the Church, Pro Dialogo 109 (2002), p. 128–130. It is significant that Georges C. Anawati had already written in an essay in 1978: "Mir will scheinen, daß auf katholischer Seite das Anliegen des islamisch-christlichen Dialoges zum Durchbruch gekommen ist, zumindest der Intention nach" (It seems to me that, on the Catholic side, the issue of Christian-Muslim dialogue has made a breakthrough, at least in its intention. Zur Geschichte der Begegnung von Christentum und Islam, in: Bsteh, Andreas (ed.), Der Gott des Christentums und des Islams, Beiträge zur Religionstheologie 2, Mödling 1978, p.33, whereas on the Muslim side he mainly mentions individuals (p.33f, the footnotes in particular provide interesting information about persons who do indeed repeatedly appear in prominent roles in the discussion). It is also very significant that Robert Caspar, in his essay on Islam and Secularisation, BSNC 15 (1970), p.156, speaks about secularisation as the fulcrum for Christian-Muslim dialogue. There were often invitations to establish a kind of common front of religions against atheism – something which again became evident e.g. in the dialogue of Tripoli. Caspar also sees clear tendencies in the political realm (p.155): "I only note the almost constant fact that many political leaders before independence resolutely opted for a secular State, but once in power, not only did they yield to public pressure and admit Islam as the State religion, but themselves hardly seem to want now to come back on this point, conscious as they are of maintaining religious tendencies within the framework and the options of their internal and international politics."

6 According to Report of Mgr. Rossano's Journey to Iraq, Pakistan, Bangladesh, Northern India, BSNC 34/35 (1977), p. 62–65; the details of the report on the journey can be found on p. 52–62. On the very positive consequences of the stay in Bangladesh, cf. also Orlando, G., **Bangladesh:** Reports on the Activity of the "Brotherhood Commission" for Muslim and Christian Encounter, BSNC 36 (1977), p. 195–197. A list of all visits and of the conferences organised by the Secretariat is provided by Cardinal Pignedoli in Pignedoli, Sergio, Letter of Cardinal President of the Secretariat to the Bishops, BSNC 36 (1977), p. 90, where he also emphasises that dialogue is particularly the concern of the local churches and that the Secretariat can only provide assistance. The concrete dialogue depended on the circumstances and the counterparts, and (p.91) general reference was made to common goals on the human, spiritual and moral levels. Difficulties are mentioned, indeed relatively clearly, and top of the list is political intervention, especially when one is dealing with religions which do not make a strict distinction between worldly and spiritual spheres; this certainly is a reference to Islam. The person responsible for Islam sounds much more enthusiastic in a report on a conference in which he participated when he states that, after twelve years of effort, some rapprochement had come about; cf. Mokh, François Abou, **Vienne:** Rencontre des chrétiens et des musulmans à Saint-Gabriel (Mödling – Vienne) 31 mai – 5 juin 1977, BSNC 36 (1977), p. 192. However, he also repeats that the Secretariat is aiming at meetings which go beyond politics. An excellent example of the Muslim visit to the Secretariat and of the non-intervention of the Secretariat in the affairs of the local bishops is found in Shirieda, John, Report of the Conversation between the Indonesian Delegation and that of the Secretariat for Non-Christians, BSNC 37 (1978), p. 5–8. The main issue in the conversation with the Indonesian minister, Mukti Ali, was mission (or what the Indonesian Muslims considered to be mission, even going as far as the psychologically disturbing presence of Christian buildings such as churches, schools and hospitals) or of proselytism, although every form of church aid suffers from this suspicion unless it is distributed by the state. There were further visits of this kind later on as well, also to strengthen people on difficult ground, e. g. **Alger – Algérie:** Visite du Cardinal F. Arinze, Président du Secrétariat pour les non-Chrétiens (28 nov. – 5 déc. 1987), BSNC 67 (1988), p. 87–88. With regard to the regional dialogues of the Secretariat and their effects cf. Machado, Felix, Interreligious Dialogue in the Various Regions of the World, p. 268, where Machado deals with Catholic dialogues in general and not only those organised centrally. Generally, anti-ecumenical and fundamentalist tendencies prove to be obstacles to dialogue, irrespective of whoever expresses them. It is obvious that Christian-Muslim dialogue is especially frustrating in countries with a Muslim majority (p. 269). Fitzgerald also mentions on the same occasion (assembly of the Pontifical Council for Inter-religious Dialogue 20–24 November 1995) that many people initially associated Islam with fundamentalism and extremism and questioned whether dialogue was possible with Muslims at all. His answer was the counter-question about the alternative – perhaps the option of continuing the crusades? (Fitzgerald, Michael L[ouis], Report on the Activities of the PCID: November 1992 – November 1995), Pro Dialogo 92 (1996), p. 169).

7 cf. Cordeiro, Joseph / Teissier, Henri, Rapport du Group "Islam", BSNC 41/42 (1979), p. 183–186, the early version of which with insignificant variations can be found as the "Compte rendu du groupe Islam" in the archives of the World Council of Churches, DFI Box 37. There was agreement that, on the Christian side, more knowledge and preparation were needed, but Francis Arinze, at that time still archbishop in Nigeria, doubted that a Muslim lecturer would be the most suitable for this, because the latter would be anything but objective and naturally could also not say what was positive or negative about Islam from a Christian point of view; cf. Arinze, F[rancis], Means Necessary for Developing Dialogue with Non-Christians, BSNC 41/42 (1979), p. 169. Cardinal Pignedoli, on the other hand, referred generally for dialogue to something that Louis Gardet had stated in an article in Islamochristiana 3 (1977) about the three dangers of Christian-Muslim dialogue: "a combative apologetic, an apologetic of insufficiency, and a practical syncretism"; cf. Pignedoli, Sergio, Introduction of the Cardinal President to the Plenary Meeting of the Secretariat for Non-Christian Religions, 1979, BSNC 41/42 (1979), p. 86. A little later, on 19 September 1981, the Islamic Council of Europe published a new Islamic declaration on human rights which was discussed critically in the Bulletin by the Asia officer of the PCID at that time, Thomas Michel, particularly because of its limitations (it is more of an Islamic than a truly universal declaration of human rights as far as its background and its commitment to Islamic law are concerned, a secular legal system is not envisaged, with regard to its validity it cannot speak for all Muslims and nor does it claim that its content is being implemented in all Muslim countries). Nevertheless, Michel considers it important enough to be made the content of studies and discussions among the Muslims themselves and also in their dialogue with non-Muslims; cf. Michel, Thomas, New Muslim Declaration on Human Rights, BSNC 48 (1981), p. 248f. Reference must also be made to an article or lecture by the same author on God's Covenant with Mankind according to the Qur'an, BSNC 52 (1983), p. 31–43, even though this was not given at an event of the PCID. The same can be said of the lecture originally given in Italian by Thomas Michel on Sin, Forgiveness and Reconciliation in Islam, BSNC 53 (1983), p. 115–134.

8 A good example of this is Borelli, John, Recent Muslim-Catholic Dialogue in the USA, in: Fitzgerald, Michael L[ouis] / Borelli, John (eds.), Interfaith Dialogue, A Catholic View, London/Maryknoll 2006, who on p. 105f lists ten points which he has learnt from his experience of dialogue with Muslims. These can and must not all be quoted here but some of them are so fundamental and important that they cannot be left out.

"1. For too many centuries, from the beginning of the encounters between Arab Muslims and the Christians outside of the Arabian Peninsula, Christians have made outlandish statements about Islam, Muhammad and Muslims in general. Muslims feel compelled to lecture Christians about the basics of Islam in order to correct our mistaken views. (...) / p. 106 (...)

"6. Muslims are particularly eager to tell Christians about their respect for Jesus, and cannot understand why Christians can be negative or distrustful towards them.

"7. The word 'mission' functions in the same way among Muslims as the word 'jihad' does among Christians. Both words have beautiful meanings but they carry connotations of violence, intolerance and disrespect. The problem for Christian-Muslim conversation is that in a thorough discussion these words are difficult to avoid using. (...)

"10. Christians and Muslims often judge one another by their extremists. This can happen between any two groups, but because of the particular history they have had and the way strife has been promoted as a way of dealing with one another, they each make the mistake of judging the other's worst by their own best. They often let the extremists do the talking and thus capture public attention."

In addition, this volume provides other interesting and important information on Christian-Muslim dialogue to which the whole of the second part of the book is devoted. Theological dialogue is considered to be difficult and the formal dialogue between representatives of the two religions will most probably be concerned frequently with social questions. The difficulties are stated openly and clearly, namely that there are different understandings of the relationship between human beings and God and also a different understanding of religious liberty and (at least as Christians see it, this in turn is related to the fact that things which are different may or must also be treated differently) women are discriminated against in Islam. In addition to this, it should not be forgotten that there were radical elements in Muslim societies under which the Muslims themselves also suffered. Following 11th September 2001, the Muslims themselves had felt the need to make clear that Islam and terrorism were not simply interchangeable (cf. p.131, 138, 140 f, 147). Inter-religious dialogue as a whole probably always moves between the extremes of naivity and distrust, both of which must be avoided: "ingenuousness which accepts everything without questioning and on the other hand a hypercritical attitude which leads to suspicion" (p. 168). In this connection, the chapter on modern religious fundamentalism is also well worth reading (p. 170–179). Concerning Jadot cf. Fitzgerald, Twenty-five Years of Dialogue, p. 114 and especially p. 115 on his remarks to the Roman bishops' conference in October 1983. "He explained that dialogue is pre-eminently the task of the local Churches. The role of the Secretariat is to conscientize, encourage and collaborate, by being an agent of communion and information. He also stated that Islam merits special attention because it is a monotheistic religion, because it is so widely spread and is witnessing a re-awakening, and also because of its socio-political implications." (Michael Fitzgerald became the secretary of the Secretariat for Non-Christians in 1987; cf. his "opening speech": Greetings from the New Secretary, BSNC 64 (1987), p. 5–7).

Another very special area, which has become interesting because of the major problems that have arisen in the meantime, is the question of Christian-Muslim dialogue among the working class in France; cf. the very noteworthy article by Serain, Michel, Chrétiens et Musulmans en classe Ouvrière, BSNC 63 (1986), p. 251–261, who comes to the conclusion (p. 261): "Bien que les chrétiens et musulmans ne partagent pas nécessairement le même sens de l'homme et n'aient pas la même vision d'une société idéale, ils sont confrontés ensemble aux mêmes problèmes. Les réponses communes dépendront de la qualité du dialogue entre partenaires."

For Jadot's evaluation cf. Jadot, J[ean], The Growth in Roman Catholic Commitment to Interreligious Dialogue since Vatican II, BSNC 54 (1983), p. 205–220, an essay which is also very interesting because it paints an unusually clear picture of the historical developments before the Council and because it highly appreciates the ecumenical dimension. In this connection, mention can be made of an article which appeared in the bulletin of the Secretariat and which in the briefest form reflects precisely the anthropological approach of the Eastern Church: Yannoulatos, Anastasios, Relations between Man and Nature in the World Religions, BSNC 47 (1981), p. 139–141. For the later deveopments in the Secretariat, reference can be made to P.C.I.D. Dialogue with Muslims since the last Plenary, Bulletin Pontificium Consilium pro Dialogo inter Religiones 82 (1993), p. 34–45. It points out, e.g., (p.43) that the Organisation of the Islamic Conference (OIC) was interested in common projects. Although this organisation was really responsible for inter-state relations, whereas the Secretariat was responsible for relationships with Muslims as individual believers, an attempt would be made to find a common way. (At that time, a meeting was envisaged for the beginning of 1993.) This was the first assembly after a long period which actually took the time to enter in greater detail into questions of Islam (and other religions), according to Shirieda, John, Editorial, Bulletin Pontificium Consilium pro Dialogo inter Religiones 82 (1993), p. 1–2. Good insight into the general work of the Pontifical Council and its involvement in Vatican and other bodies is provided by Arinze, Francis, Meeting Other Believers, Bulletin Pontificium Consilium pro Dialogo inter Religiones 82 (1993), p. 17–22.

9 cf. Fitzgerald, Michael L[ouis], Report on the Activities of the PCID: November 1992 – November 1995, Pro Dialogo 9 (1996), p. 170.

10 According to Fitzgerald, The Secretariat for Non-Christians Is Ten Years Old, p. 90 f, who also does not consider all visits worthy of specific mention, but this one (in addition to Cairo) was. A more precise report will be found in Visite des Uléma Saoudiens au Secrétariat pour les Non-Chrétiens, BSNC 28/29 (1975), p. 181–185. This report contains an infinite number of details but unfortunately not what was said on human rights from a Christian or Muslim point of view. But this is such a sensitive issue precisely in connection with Islam in general and Saudi Arabia in particular.

11 Fitzgerald, Michael L[ouis], Tolerance – Respect – Trust, Forum Mission 6 (2010), p. 33.

12 This whole section is based on Report on the "Seminar on Islamic-Christian Dialogue" Held in Tripoli (1st – 5th February 1976), BSNC 31 (1976), p. 5–7+10; for the exact programme cf. p. 7–10 which sometimes give small details that are not found in Borrmans; the internal guideline for the whole conference on the Catholic side was (according to p.7): "[T]he uniquely religious competence of the Delegation is confirmed, certain recommendations of the Cardinal Secretary of State are notified, it is decided that we will listen a lot and give collective witness of unity, charity, prayer, welcome and spiritual brotherhood." As far as the composition of the delegation is concerned, p. 6 states: "[W]e must note immediately that the Copt Patriarchate of Alexandria 'cannot' be present, and the WCC of Geneva 'does not consider it necessary' to delegate anyone." In addition, reference can be made to an American report: Islamic Christian Dialogue, Tripoli, Libyan Arab Republic, The Link 9,1 (1976), p. 1–8.

13 'Utba, Abd ar-Rahmân, quoted from Borrmans, Maurice, Le séminaire du dialogue islamo-chrétien de Tripoli (Libye) (1–6 février 1976), Islamochristiana 2 (1976), p. 142, cf.also p. 135–141.

14 Chullikal, Anthony, quoted from ibid., p. 143.

15 According to ibid., p. 143–154; special emphasis must be given to p.154, note 31, where there is also mention of a contribution from the former president of the Republic of Venezuela, Cordeira, who gave a very detailed report on how much the church's social doctrine had helped him in his office.

16 ibid., p. 155, cf.also p. 156, where he speaks of a "climat indéfinissable". The Report on the "Seminar on Islamic-Christian Dialogue" Held in Tripoli (1st-5th February 1976), p. 10/11 summarises it as follows: "The *Moslem Delegation* carried out its 'dialogue' by systematically attacking the Catholic Delegation: our faults, according to them, are: Conscious falsification of the Holy Scriptures (particularly the Gospels), a monotheism which is not authentic, continual faults in history (from the Crusades to colonialism, up to the 'exculpation' of the Jews), / p. 11 scorn and misunderstanding of Islam both on the level of faith and on the part of the orientalists, proselytism and conversion hunting, theological and ethical inferiority, lack of doctrine and social ideology, separation of life from the norms of religion etc. Islam appeared constantly sure of itself, a perfect religion which has nothing to learn, aware that today it is in possession of the prestige weapon of raw materials, avenger of rights too long ignored, proud of its religious purity. But with reference to this, it seems only right to note that the religious profession of the participants seemed to be rather a juridical fact than a true inner dimension." To what extent the latter remark is perhaps a direct consequence of the highly praised structure of Islam (at least to a certain degree) would also be an interesting anthropological question although difficult to answer. Very much later it was interpreted as a lack of tolerance and respect; cf. Fitzgerald, Tolerance – Respect – Trust, p. 34: "The Muslim speaker, however, adopted a different method. Taking the encouragement to frankness to its extreme, he pronounced a diatribe against Christianity for its misdeeds against Islam, particularly in connection with colonialism and neo-colonialism. In the ensuing exchange (...) a Catholic bishop from Africa appealed for an end to simplistic slogans and demanded *respect* for minorities everywhere."

17 Some decisive quotations from the beginning of this declaration, quoted from Borrmans, p. 159, note 42 (p. 159/160):

"I – Nous avons pris conscience que notre même foi en Dieu créateur et juge des hommes et de l'histoire a été et reste une révolution dans le devenir humain.

C'est cette foi qui a fait surgir l'homme dans sa dignité de 'vice-régent' de Dieu sur terre; elle implique ainsi et exige le respect de la dignité et de la liberté de l'homme.

Cette même foi, tout en demeurant au-dessus des idéologies, en constitue la source vivifiante, qui donne à l'homme les motivations et les énergies qui lui permettent de se situer parfaitement vis-à-vis de Dieu, de ses frères et du cosmos.

Sans cette même foi en Dieu, il est impossible d'assurer à l'homme toute sa liberté et toute sa dignité.

"II – Il est du devoir de tous les Fils d'Abraham de faire en sorte (...) que leur vie et leur travail les animent à collaborer comme frères au service de la famille humaine.

"III – La foi dans le Dieu unique nous engage à reconnaître tous les hommes comme nos frères, à faire notre autocritique devant toutes les formes d'injustice qui existent dans nos sociétés respectives et à nous convertir personnellement aux exigences de la justice, à l'appel des pauvres et des opprimés.

(...)

Quel que soit le niveau auquel nous plaçons l'intervention de nos textes sacrés – c'est là où nous avons des conceptions différentes – nous sommes pratiquement d'accord sur toutes les exigences concrètes de la justice sociale."

The second part of this declaration dealt with the obstacles to dialogue and how they could be ovecome. Again, a few extracts which are of anthropological relevance (p. 160, note 42):

"3) être assez loyal de part et d'autre pour garantir à toutes les minorités religieuses tous les droits et tous les devoirs reconnus à la majorité;

"4) reconnaître à chacune des deux religions le 'devoir de l'apostolat' dans la pureté du témoignage qu'il faut porter, et dans le respect des libertés humaines, ce qui implique la condamnation de tout prosélytisme."

One can imagine that such such suggestions would have come up against difficulties on the Muslim side if it had been possible to make them at all which was unfortunately not the case.

18 ibid., p. 161, note 43 (the note which attempted to give an approximate description of what had taken place).

19 ibid., p. 160, note 43; the final events are best described in Report on the "Seminar on Islamic-Christian Dialogue" Held in Tripoli (1st-5th February 1976), p. 12. (p. 11 listed the participants from the Catholic side: Abou Mokh, Roest Crollius, Borrmans and Lanfry. Msgr. Rossano had taken his leave relatively quickly when it became clear that the final document was to be negotiated between two Arabic speaking commissions). "At 4 p.m. the Congressmen begin to fill the hall; at 5 p.m. Fr. Abou Mokh arrives with the first half of the copy of the Arab text. When questioned by Mgr. Rossano and the Cardinal he replied that the work had gone ahead in total harmony and he and Fr. Borrmans wrote a personal note to the Cardinal, at the opening of the sitting in which he was invited to congratulate the two Commissions publicly for the total agreement in the draft of the final communiqué. At 5.15 p.m. begins the reading of the long Arab text, with simultaneous but defective translation into French. In the assembly there is applause, various members of the Catholic Delegation applaud, then the meeting breaks up hurriedly. We must admit that the greater part of the assembly was not interested in the points which were called into question and which provided an immediate bomb for the journalists. But when the meeting broke up the Cardinal did not hide his disapproval of what he had heard, and Mgr. Rossano, surrounded by journalists, expressed his dismay at what had happened".

20 Here some of the sections that are important from an anthropological point of view will be quoted as they appear in the Bulletin in English translation to make it easier for the reader: Text of the Final Declaration of the Tripoli Seminar on the Christian-Muslim Dialogue, BSNC 31 (1976), p. 14–21. Right away in the introduction fraternity and the concerns of humankind are mentioned and naturally the importance of religion to this end: "*Under the sign: (...) 'Let us seek, therefore, what strengthens peace and brotherhood'*. In an atmosphere of hope and mutual trust, with a common awareness of responsibility for the future of man (...)" (p. 14). The overall aim is (p. 15) "to create an atmosphere that will assist understanding of the material and moral crises that modern man is enduring, in order to find practical solutions for them. (...) Man finds himself today, it is true, in a state of restlessness, anxiety, spiritual exile, estrangement from all inner peace and happiness. He struggles in a hell of misfortunes, caused by the oppression of materialism, which draws the world away from the source of good, justice and piety; sources of which religion is the real and authentic origin. The commitment for the liberation of man from all forms of ignorance, injustices, tyranny and exploitation, finds its foundation in Religion. (...) There are priorities that no 'Heavenly Religion' can sacrifice, or neglect to affirm, including: the dignity of man, his right to life, freedom, justice and equality." Then there is a brief review of the results on the sub-themes of the conference and the individual points follow under the heading (p.16): "The two parties agreed to begin a new page, based on respect, cooperation and joint action for the good of Humanity." § 4 (p.17) is very clearly influenced by Islam (Religion as true guidance): "The organization of life cannot be accomplished outside Religion, which constantly guides Humanity along the right way and the straight path; consequently, the two parties affirm that Religion is the basis of just legislation, and that all legislation established by man is unable to reach perfection". This is more clear than the following paragraph which does speak about human dignity, well-being and freedom, but neither gives reasons for not a more precise definition of these. There is more substance in the brief § 6 (p.17): "In respect of man's dignity, the two parties reject and denounce racial discrimination in all its forms, because it is a degradation of man, who is honoured by God". The following two paragraphs deal, on the one hand, with hunger and the underpriviledges of many which are described as a "disgrace to Humanity" (p. 17, § 7), and, on the other, with religious liberty with a clear tendency against religious persecution, presumably meaning communist doctrine and regimes. § 9 (p.18) may prove to be ambiguous on closer examination (especially after the simultaneous, practical Libyan interpretation: "The two parties aspire to the realization of Peace on the basis of Justice and respect for Rights. They call upon the countries that possess destructive arms to stop manufacturing them". Explosive material can be contained in sentences such as these (p.19, § 16), especially when the reproaches of the conference are still ringing in your ears: "The patrimony of civilization and culture belongs to the whole of Mankind. It is the right of Mankind to receive this patrimony in a correct, just way." Even apart from the two disputed paragraphs, it is clear that the Muslim side was more successful in getting things included which were important to it and that these had considereble consequences, even though Christians are not used to this as a rule because they do not practise and are not familiar with this simple, direct application of religion to every kind of activity.

21 Report on the "Seminar on Islamic-Christian Dialogue" Held in Tripoli (1st-5th February 1976), p. 13; cf.also Borrmans, p. 157–164; the Arabic wording of the declaration then follows on p. 164–170 with the two non-accepted paragraphs correctly only on p.169 in note 1. In his article, Borrmans also refers to various press reports on the conference. He kindly made part of this file (some of which has been translated into French from other languages) available to me as a photocopy so that it is possible for me to provide a small selection of considerations on certain issues. First of all, the official version of why the press conference, which could have immediately clarified the disagreements concerning the final declaration, never took place (Utba, Abd al-Rahman, in: Holtz, Bruno, Une voie tracée pour la compréhension, La Liberté (18 February 1976), p. 9): "La conférence de presse, d'ailleurs, n'a pas été reportée parce qu'il y avait des difficultés, comme vous le croyez probablement, mais parce que M. Shahati, vice-président de la délégation islamique, désigné pour diriger la conférence de presse a été retenu à la présidence. Je peux même vous en révéler les raisons: le colonel Mu'ammar Kadhafi a présenté au cardinal Pignedoli la demande d'établir des relations diplomatiques entre le Saint-Siège et la République arabe libyenne. En plus, il a annoncé son accord pour l'ouverture d'une église catholique à Benghazi, deuxième ville du pays." The question arises whether one could not simply have postponed the press conference or whether some other person could have been asked to preside over it if that was really the only reason. As an example of the Jewish reactions, we can mention from this collection the (originally English) telegram from the general secretary of the International Council of Christians and Jews, Rev. P.W.W. Simpson, to the Vatican State Secretariat and to Father Pierre de Contenson, the secretary of the Vatican Commission for religious relationships with the Jews, published in The Tablet (14 February 1976), p. 170. Le Monde (13 February 1976), p. 3 (signed S.J.), speaks of an affair "qui détruit le mythe d'une 'machine' vaticane parfaitement huilée." Another commentary in the same issue of Le Monde and even on the same page, headed 'Paradoxes à Tripoli' and written by André Mandouze, a professor at the Sorbonne and one of the observers, refers to the "complexité religieuse d'une intrigue sans commune mesure". His very sharp commentary deserves to be quoted, at least in part. "À vrai dire, il faut reconnaître d'emblée que seuls ont été pris au dépourvu – aussi bien parmi les spectateurs que parmi les acteurs eux-mêmes – ceux qui n'ont pas encore admis l'inévitable dimension politique de tous les événements religieux de quelque ampleur. Le problème n'est de savoir si le Vatican a de moins bons diplomates que naguère ou si le dynamisme du jeune Kadhafi l'a emporté. Dans cette affaire, la faute des chrétiens est originelle, s'ils n'ont pas accepté d'avance que discuter de religion avec un partenaire, pour lequel la séparation du spirituel et du temporel est proprement incompréhensible, comportait inévitablement non point un simple risque mais une chance certaine de 'faire de la politique'. Si les chrétiens sont des enfants de choeur, qu'allaient-ils faire dans cette galère voguant en pleine tempête israélo-palestinienne sur des eaux forcément mêlées?" He then goes on to speak about the paradoxes of the dialogue of Tripoli: "Le premier paradoxe concerne donc les chrétiens et leur impréparation politique, ici comme ailleurs, aux tâches évangeliques." More lay people would have been helpful here, in his opinion, and in addition: "De plus, trop occidentale et nullement oecuménique, cette délégation trop 'romaine' ne pouvait que donner abusivement aux musulmans l'impression qu'elle détenait directement du pape le pouvoir de décider en toute matière." But he goes still further: "Mais c'est alors qu'apparaît le second paradoxe, lequel concerne cette fois-ci l'orientation dominante des musulmans présents à ce colloque. Loin de moi l'idée de porter le moindre jugement dépréciatif sur la qualité spirituelle ou sur la valeur intellectuelle de tel ou tel. D'une part, cependant, il n'a échappé à personne que certains grands centres de la pensée islamique n'étaient point représentés." So the Muslims had theoretically enjoyed great freedom but had not used it in practice. As for fundamentalism, according to the author, the two partners were equal despite their different behaviour and their different opinions on political issues. This could easlity have led to a joint crusade against the modern world of unbelievers – but had not done so, and that, for Mandouze, is the third paradox. Since he rejects the explanation of economic dependence on the Soviet Union, he tries to find a religious solution. "Je préfèrerais, quant à moi, résolument, un autre genre d'explication, même si, encore une fois, cette explication repose sur un nouveau paradoxe, le quatrième et dernier, mais le plus important et que je résume ainsi: même s'il y a eu beaucoup de facteurs contradictoires à l'origine de ce 'Séminaire du dialogue islamo-chrétien', celui-ci a été sauvé par la dynamique interne de la foi dans un même Dieu." But here he does not succumb to the illusion that this common root would be enough for everything without further efforts such as had indeed been required; on the contrary, there too his view is very sharp and very critical, this time more in relation to the Muslims. "En dépit des droits que donne aux musulmans le fait d'avoir souffert de la part des chrétiens, il eût sans doute été plus constructif de la part de la délégation musulmane de ne pas céder constamment à la facilité de critiquer l'autre." His (borrowed) final thought: "Pour reprendre un mot d'un des délégués musulmans, les deux parties savent maintenant que les roses de la compréhension réciproque sont protégées par beaucoup d'épines." The article 'Relations christianisme-islam: Un travail scientifique s'impose' by Mohammed Arkoun, a Muslim professor at the University of Paris, published in Le Figaro (27 February 1976), partly follows the same line and partly another (page numbers are lacking in the collection). He calls for a sociological approach to religious life and a demythologising view of religious matters. That would automatically solve many problems. "Il n'y aurait plus place pour une question aussi dérisoire et anachronistique que celle qui consiste à demander aux chrétiens, comme on l'a fait à Tripoli, de reconnaître Muhammad comme prophète. Il n'y aurait plus lieu non plus pour le chrétien de consentir à une auto-accusation publique pour toutes les blessures et toutes les incompréhensions dont le monde chrétien s'est effectivement rendu coupable à l'égard de l'Islam: l'histoire nous enseigne en effet que celui-ci n'a pas toujours été particulièrement compréhensif à l'égard du christianisme." Borrmans also expresses himself on this question in an interview with Michel Bavarel published in La Liberté (18 February 1976), p.9, under the title 'Muhammad: un prophète?' But what is more interesting is how he sees the possibility of future communication between Christians and Muslims: "Je pense qu'il faut être patient. Malheureusement, j'ai peur que l'homme moderne, parce que les moyens de communication sont rapides, s'imagine que 13 siècles, je ne dirai pas d'incompéhension mais d'ignorance réciproque, peuvent être effacés en quelques années et par quelques coups de baguette magique. Quand on voit déjà combien difficile est le rapprochement entre les Églises chrétiennes, comment pouvons-nous croire que demain ou après-demain tout sera clair, tout sera simple et tout sera résolu entre chrétiens et musulmans? (...) Il arrive de leur côté ce qui arrive de notre côté : (*sic!*) les hommes de dialogue sont une infime minorité. Ils ont conscience, les uns et les autres, d'être et de demeurer demain et après-demain encore incompris de beaucoup. Parce que chez eux, comme chez nous, ils entendront souvent leurs coréligionnaires leur dire: à quoi cela sert-il ? (*sic!*)" Peter Hebblethwaite writes something similar in his (originally English) article 'Un dialogue islamo-chrétien' in The Tablet (14 February 1976), p. 174 f: "Malgré cela, les chrétiens sont considérés comme les plus proches des 'croyants', et il vaut la peine d'insister sur cette vérité qui n'a pas précisément été à l'avant-plan de la conscience occidentale durant les 14 derniers siècles. Il y a un long héritage de suspicion et d'agressivité qui ne peut être dépassé d'un seul coup. L'apprentissage du dialogue est long, et il y aura des retours en arrière." Significantly enough, he also observes that certain Arab Christians join in energetically when it is a question of running down Israel. He contrasts this in a second (naturally also originally English) article 'Enfin les croisades sont terminées ! (*sic!*) in The Times (14 February 1976) – no page numbers given – with the extraordinarily balanced attitude of Gadaffi: "Il fit allusion, avec une discrétion étonnante, au problème palestinien. La guerre contre Israël ne peut être un 'jihad', une guerre sainte, parce que les juifs, eux aussi, croient en un seul Dieu, et que la guerre sainte ne peut se faire que contre les incroyants." But communication remained difficult and pitfalls were waiting everywhere: "Mais en fait, en bien des points du monde chrétiens et musulmans ont appris, comme le dit le Cardinal Pignedoli 'que, pour les relations entre chrétiens et musulmans, le mot clé est l'amitié'. Malheureusement c'était une citation d'un livre sur Charles de Foucauld que les Arabes regardent comme un espion français. Ce sont là des pièges d'un dialogue sur de nouvelles frontières. Mais, au moins, les Croisades sont finalment terminées." It is put very well at the beginning and end of the article 'La voie difficile' by Michel Lelong, Le Figaro (11 February 1976), also without page numbers. "Faudrait-il conclure, après la confusion des dernières heures, que le dialogue islamo-chrétien est décidément impossible, ou, du moins, si difficile qu'il vaut mieux y renoncer? Nombreux sont ceux, dans l'une et l'autre communauté, qui se refusent à le croire. (...) Ce qui unit le plus profondément les chrétiens et les musulmans – et ce qui les unit aux croyants du judaïsme – c'est la certitude que l'homme trouve son véritable bonheur, sa véritable liberté dans la lumière de Dieu. C'est aussi une commune espérance qui donne à l'existence humaine et à l'histoire leur signification. C'est en se situant dans une telle perspective que peut et doit se poursuivre le dialogue islamo-chrétien." In contrast, the article by the Muslim judge, Tayssir Kana'an, entitled 'Non ... non ... non ... o sainteté le pape!' speaks volumes along the opposite line; it appeared in Arabic in the paper 'Al Quds (21 February 1976), page numbers missing: "Et que mon frère arabe chrétien me permette d'assurer, en son nom, que les liens qui l'unissent à moi sont mille fois plus forts et plus fermes que ceux qui l'unissent au chrétien belge, kényan ou italien. Notre nationalisme est un, notre langue est une. Nos traditions sont identiques et notre destin est le même. Malgré les différences de rites, les chrétiens endurent avec nous les mêmes souffrances et partagent avec nous les prisons et les camps de réfugiés. (...) Sainteté, Si vous considérez que le rejet, par les arabes chrétiens, des idées et des positions du Saint Siège au sujet de la souveraineté arabe sur Jérusalem constitue un 'péché', alors ils sont tous 'pecheurs' et viendront au Saint Siège pour se confesser. En même temps, ils attendent et espèrent que vous bénirez leur péché. Sainteté, Le prélat de l'Église, Hilarion Capucci, arabe chrétien : (sic!) porteur d'un passeport du Vatican, vous envoie son salut de sa prison." Where the Arab Christians had stood from the very beginning in this unfortunate disagreement at the end of the conference in Tripoli and for which reasons there can be very little doubt after reading this. Report on Mgr. Rossano's Journey in South-East Asia, BSNC 31 (1976), p. 49–61, also shows very well how strong and in part united was the resonance with which the dialogue and declaration of Tripoli were received in the non-Arab Islamic world as well, e.g. p.50: "The Tripoli meeting, whose final session had already been translated into Malaysian by Mr. Kamaruddin, was spoken of with satisfaction; the Great Mufti made vows for the restitution of the Cordova mosque, and for Jerusalem, 'the third holy city of Islam', asking me whether Christians were able to go there and what was the Vatican position regarding it: I replied simply that all Christians wished for respect and freedom of access to the holy places of all the religions."

22 Akasheh, Considerations on Forty Years of Religious Dialogue with Muslims (A Report), p. 196; his superior put it in the following way (Fitzgerald, Tolerance – Respect – Trust, p. 35): "The seminar had achieved the opposite of what it intended, for it had contributed to creating a climate of suspicion towards dialogue." In the next sentence he describes it as "[t]his unfortunate event".

23 According to Balić, Smail, Worüber können wir sprechen? Dialog der Religionen 1 (1991), p. 72. The report by Heinz Gstrein, Die Lage der christlichen Kirchen im arabischen Maghreb, Wort und Wahrheit 27 (1972), p. 463–469, forms an interesting background; it is concerned mainly with the situation in Libya and does not paint a flattering picture of the country in the past or the present and also not in religion or politics – and, indeed, even at that time, no flattering picture of the behaviour of the Catholic Church in these entanglements. Concerning the fact that at least the opening of a church was achieved, cf. Fitzgerald, Michael L[ouis], Twenty-five Years of Dialogue, p. 111. It is also clear that the Catholic Church desired further meetings and thus getting to know one another better but in a religious framework far removed from politics, cf. Mokh, François Abou, **Vienne:** Rencontre des chrétiens et des musulmans à Saint-Gabriel (Mödling – Vienne) 31 mai – 5 juin 1977, BSNC 36 (1977), p. 192.

24 cf. Rossano, P[ietro], Convergences Spirituelles et Humanistes entre l'Islam et le Christianisme, BSNC 40 (1979), p. 15–19.

25 cf. Fitzgerald, Twenty-five Years of Dialogue, p. 119, Visit of Cardinal Arinze to Libya (16–20 March 1989), Islamochristiana 15 (1989), p. 220, Rome: **Meeting between the World Islamic Call Society and the Pontifical** (sic!) **Council for Interreligious Dialogue** (14–15 February), Pro Dialogo 74 (1990), p. 178–180, Muslim-Christian Meeting at the Offices of the Pontifical Council for Interreligious Dialogue with Representatives of the World Islamic Call Society (Tripoli, Libya) (14–15 February 1990), p. 294–295.

26 cf. Arinze, Francis, In Respect and Openness, in: Pontifical Council for Interreligious Dialogue / World Islamic Call Society (eds.), Co-existence between Religions, Reality and Horizons, Vatican City, cf.also, p. 7 f. Sharif, Muhammad Ahmad, Understanding Tolerance, in: Pontifical Council for Interreligious Dialogue / World Islamic Call Society (eds.), Co-existence between Religions, Reality and Horizons, Vatican City, cf.also, p. 10 f. This contribution is general and in various ways rather exaggerated in that it maintains the theory that Christians were seen as Muslims because they believed in the same God, and he also writes (p.10): "We have contributed to human understanding by making tolerance a cornerstone in human civilization that emanated from both Christianity and Islam. This was a great development in the progress of mankind."

There is also an interesting remark in the welcome by the local Maltese representative of the World Islamic Call Society, Mukhtar Aziz (Words of Welcome, in: Pontifical Council for Interreligious Dialogue / World Islamic Call Society (eds.), Co-existence between Religions, Reality and Horizons, Vatican City, cf.also, p.5), where he praises God as the one who sent prophets in order to direct people away from worshipping one another. That is certainly an interesting and important aspect of the Islamic view of human beings and of the history of humanity. A short report on the meeting, the two opening addresses and the final declaration can also be found in **Paola – Malta:** *Christian-Muslim Meeting*, Bulletin Pontificium Consilium pro Dialogo inter Religiones 76 (1991), p. 101–108.

27 Ghwel, Ibrahim, Islamic Perception of Christianity, in: Pontifical Council for Interreligious Dialogue / World Islamic Call Society (eds.), Co-existence between Religions, Reality and Horizons, Vatican City, cf. p. 20, cf. also p. 16–19; for the Christian view mentioned cf. Ellul, Joseph, A Christian Understanding of Islam, in: Pontifical Council for Interreligious Dialogue / World Islamic Call Society (eds.), Co-existence between Religions, Reality and Horizons, Vatican City, p. 12–15.

28 Ghwel, p.21.

29 El-Saadi, Muhammad, Religious Tolerance in Islam, in: Pontifical Council for Interreligious Dialogue / World Islamic Call Society (eds.), Co-existence between Religions, Reality and Horizons, Vatican City, p. 25, cf. p. 22–24.

30 cf. Michel, Thomas, Tolerance: A Christian Perspective, in: Pontifical Council for Interreligious Dialogue / World Islamic Call Society (eds.), Co-existence between Religions, Reality and Horizons, Vatican City, p. 27–31

31 Borrmans, Maurice, Christian-Muslim Cooperation: Building a Modern Pluralist Society, in: Pontifical Council for Interreligious Dialogue / World Islamic Call Society (eds.), Co-existence between Religions, Reality and Horizons, Vatican City, p. 34, cf.also p. 32f.

32 ibid. p.38, cf. p. 35–37.

33 Kawtharani, Wajih, Practical Application of the Spirit of Tolerance: an Islamic Perspective, in: Pontifical Council for Interreligious Dialogue / World Islamic Call Society (eds.), Co-existence between Religions, Reality and Horizons, Vatican City, p. 42, cf.also p. 39f, and Symposium on "Co-existence between Religions: Reality and Horizons" (22–23 November 1990), Islamochristiana 17 (1991), p. 234–236. Here are a few more quotations from Kawtharani's address which make his approach still clearer: "In the North they are called upon to take stands on the state of increasing impoverishment in the South, on the state of forced usurpation of the centre of world decision-making, on the colossal network of economic and military hegemony over the sources of energy and raw materials, on the sav-/ p. 41 age aggression against Palestine, on racial discrimination in South Africa, and on religious and national discrimination against the Muslims in Greece, Bulgaria and the republics of the Soviet Union. (…) The Islamic awakening – even if sometimes wrought with tension – is an expression of a sharp world wide division between North and South. And if Christian awakening in Eastern Europe is an expression of the state of rejection of the one-party system and the one ruling ideology, the Islamic awakening is also a state of rejection of injustice and hegemony whether it came from the outside or from within. In both cases, this is manifested in the bad distribution of wealth and bad utilization of it, and in the unequal sharing of international decision-making and the arbitrariness of the stands taken from the crises of the Islamic world and its countries. (…) And at a time when we are discerning a notable growth in Europe of racism, which expresses itself in some European Countries (*sic!*) by varied forms of infringement upon the individual liberties of the Muslim immigrants, we do not notice any offical (*sic!*) Christian condemnation of this wave of racism. On the contrary, the Islamic world expressed amazement at the stand taken by the representative of the Vatican in the United Nations when that representative raised questions about Islamic *sha-ri'a* in the context of racism in a meeting of the Committee of social affairs. (...) / p. 42 (...) [W]e maintain that, from an Islamic viewpoint of the concept of tolerance, the weak and helpless, whether Christian or Muslim, cannot separate his tolerance from his just and human causes. Is it reasonable to demand from the Muslim or Christian Palestinian to be tolerant while he is being annihilated every day and expelled every day only because he is demanding his right as a human being, his right to a culture and his right to a state? (...) Are the Arabs and Muslims destined to pay for the misdeeds of racist Europe past and present? In the past when Europe repented for its antisemitism by the Balfour Declaration and subsequent Western policies, and at present when it attempts to compensate for its oppression of freedoms in the Soviet Union and Eastern Europe by unleashing Jewish migrations to Palestine and other Arab Lands (*sic!*) (South Lebanon and the Golan Heights)? Must the price for the freedom of the European peoples which we support be the colonization of Muslims and their subordination to Zionist hegemony?" It is evident that here two different agendas have collided.

34 cf. Magro, Sylvester, Religious Tolerance in Libya, in: Pontifical Council for Interreligious Dialogue / World Islamic Call Society (eds.), Co-existence between Religions, Reality and Horizons, Vatican City, p. 47–50; the Final Statement can be found in: Pontifical Council for Interreligious Dialogue / World Islamic Call Society (eds.), Co-existence between Religions, Reality and Horizons, Vatican City, p. 43–46.

35 cf. LIBYA *Symposium on Religion and Mass Media (Tripoli, 3–6 October 1993)*, Islamochristiana 20 (1994), p. 242 and Michel, Thomas, **Tripoli – Libya:** Seminar on the Media and Presentation of Religion (3–6 October 1993), Bulletin Pontificium Consilium pro Dialogo inter Religiones 84 (1993), p. 308–310. The addresses of special anthropological interest from the Christian side were given by Josè Martinez and Hans-Peter Röhlin. The Muslim participants did not only come from Libya but also from Morocco, Lebanon, Syria and the USA. The Palestinian imam from the Islamic Centre of Malta was also present. However, what clearly did not happen although it had been announced, was the publication by the PCID of the revised addresses. The meeting had been planned in October 1992 by representatives of the W.I.C.S. together with Fr. Michel, according to Michel, Thomas, P.C.I.D. Dialogue with Muslims since the Last Plenary, p. 40, and the original title was to be: "Media and the Presentation of Religion."

36 cf. Press communiqué of the Workshop on "The Media and Religion", Jointly Organized by the World Islamic Call Society (Libya) and the Pontifical Council for Interreligious Dialogue (Vatican) (Vienna, 7–8 October 1994), Islamochristiana 21 (1995), p. 142–143, also Press Communiqué, Pro Dialogo 87 (1994), p. 272f; on the event in general cf. Fitzgerald Michael L[ouis], **Vienna – Austria:** Workshop on "Religion and the Media" (7–8 October 1994), Pro Dialogo 87 (1994), p. 271f., on both events cf. Fitzgerald, Michael L[ouis], Report on the Activities of the PCID: November 1992 – November 1995, Pro Dialogo 92 (1996), p. 170

37 cf. Akasheh, Report on the Activities of the PCID: Relations with Muslims, p. 216 and Colloquium on "Islamic Da'wah and Christian Mission in the Next Century", Jointly Organized by the World Islamic Call Society (Libya) and the Pontifical Council for Interreligious Dialogue (27–30 April 1997), Islamochristiana 23 (1997), p. 201–202. The most exact description of the event is found in: Troll, Christian W., **Vatican City:** Christian Mission and Islamic Da'wa, 27–30 April 1997, Pro Dialogo 96 (1997), p. 394f. The subject also comes up elsewhere, e.g. at a meeting of the members and advisers from the Near East and North African region (30 November -2 December 1997 in Amman) where a lack of mutuality was noted especially with regard to freedom of conscience and religion. Here it is stated, certainly in reference to the minority of Arab Chris-

tians in the area who are also threatened by emigration and self-marginalisation, that neither mission nor *Da'wah* should become proselytism that takes advantage of a situation of poverty, weakness and the ignorance of people, cf. Akasheh, Khaled, **Amman – Jordanie:** Réunion des Membres et des Consulteurs du Moyen Orient et du Nord de l'Afrique, 30 novembre-2 décembre 1997, Pro Dialogo 99 (1998), p. 337. Indeed, it was only on 10 March 1997 that the Holy See and Libya had established full diplomatic relations, although on that occasion it was naturally emphasised how old the relationships really were, maintained especially by the Franciscans, cf. Diplomatic Relations Are Established between the Holy See and the Socialist People's Libyan Arab Jamahiriya (10 March 1997), Islamochristiana 23 (1997), p. 221.

38 cf. Akasheh, Bureau pour l'Islam Rapport d'activités: novembre 1998-octobre 2001, p. 107, **Tripoli – Libya:** "A Culture of Dialogue in an Era of Globalization", 16–18 March 2002, Pro Dialogo 110 (2002), p. 229–231.

39 cf. Les musulmans en Europe, BSNC 33 (1976), p. 337–340, cf. also from the French side Dufaux, Louis, Le rôle de l'islam dans les problèmes de l'immigration, Bulletin Pontificium Consilium pro Dialogo inter Religiones 76 (1991), p. 36–38.

40 Borrmans, Maurice, Welcome to the Participants, Islamochristiana 11 (1985), p. 4; for further information cf. Colloquium on Holiness in Islam and Christianity, Rome, 6–7 May 1985, Islamochristiana 11 (1985), p. 1–3, and Michel, Thomas, **Rome – Italy:** Colloqium on "Holiness in Islam and Christianity", BSNC 60 (1985), p. 320–322; cf. also for the content Michel, Thomas, The Idea of Holiness in Islam, Pro Dialogo 92 (1996), p. 220–240.

41 Faruqi, Ziaul Hasan, The Concept of Holiness in Islam, Islamochristiana 11 (1985), p. 10, cf. p. 7–9 and his reference to Shah Waliullah p. 12.

42 ibid., p.15, cf. p. 10–14.

43 ibid., p.18, cf. p. 16–17.

44 ibid., p.24. For the rest of the presentation cf. p. 19–23 + 25 f. The response by Paul Jackson does not have a title of its own but is subsumed under Faruqi's contribution and can be found on p. 27 following the French summary of that contribution.

45 Nizami, K[haliq] A[hmad], Models of Holiness for Muslims, Islamochristiana 11 (1985), p. 53, cf. p. 51f+ 54–57 on the significance of Muhammad, with its climax in a hadith (p.57) "that ten virtues should be cultivated for moral and spiritual health: (1) truthfulness, (2) curtailment of worldly involvements, (3) sharing food with the hungry neighbour, (4) fulfilling the need of the beggar, (5) reciprocating kindness by kindness, (6) honesty in trust, (7) kindness towards kith and kin, (8) kindness towards friends, (9) kindness towards guests, (10) the root of all these virtues being modesty and penitence (*haya*)." There then follows on p. 57–66 the description of the lives of individual saints. The information about Sheikh Fariduddin (1175–1265) is interesting, e.g. p. 61: "His private life was a perfect mirror of his public life and he never said or did different things in public and in private. There was a complete harmony between his thought, words and actions." At the end (p.67) there is a response by the Franciscan sister L. Vagogne from Pakistan who asks, on the one hand, about living examples and spiritual leaders (where the latter were distinguished less by their knowledge than by their nearness to God) and also about the problem that there are absolutely no official criteria for canonisations in Islam and so they are something like plebicites. This entailed the problem that, despite all dogma, performing miracles counted more than setting an example. Finally, the question also arose as to whether holiness could somehow be inherited and whether then the ordinary person born outside of this stream of holiness was not at a disadvantage.

46 Futrell, John Carroll, The Concept of Christian Holiness, Islamochristiana 11 (1985), p. 31, cf. p. 29 f and for what follows p. 32–34. The speaker initially focussed very strongly on the kingdom of God, cf. p. 31 where the quotation continues: "The Kingdom will come when all human persons have formed an interpersonal community of love with one another in communion with Father and Son and Holy Spirit, so that the reign of God is universally real and visible, in all peoples' social and political and economic organization of human society and in their use of all material things, finally transforming the universe in beauty (Genesis 1) rather than polluting it." The response by Riffat Hassan can be found, again without a separate heading and also only in an indirect summary, according to the French summary on p. 35f, although p.36 deals more generally with the difficult points of the 'trialogue' between Jews, Christians and Muslims.

47 cf. Le Gal, Suzanne, Models of Holiness for Christians, Islamochristiana 11 (1985), p. 37–48, the summary of the response by Riffat Hassan, who clearly saw herself in this connection as a feminist theologian, is to be found on p.49f. It is interesting that Suzanne Le Gal, in contrast to the previous Christian speaker, particularly emphasised that the kingdom of God was a gift which could also not be acquired by good deeds (p.38).

48 Gaudeul, Jean-Marie, A Christian Critique of Islamic Holiness, Islamochristiana 11 (1985), p. 80, cf. p. 69–79.

49 ibid., p.82, cf. p.80f.

50 ibid., p.83.

51 ibid., p.88, cf. p. 84–87.

52 Ayoub, Mahmud [M.], A Muslim Appreciation of Christian Holiness, Islamochristiana 11 (1985), p. 96/97. His Islamic approach then also becomes very clear (he was born in Ayn Qana in Lebanon): "It is my conviction that Christianity began to lose its power of sanctification as it lost its Eastern home and character. It was in this Eastern piety and spiritual dynamism of the holy desert fathers that Islam was born and nourished. It was not dogma but holiness, victory against the demonic spirits of uncleanness which spoke to the needs of men and women. We dismiss as a bit of Eastern superstition that Jesus cast out unclean spirits. Yet it was this piety of healing and sanctification, whose ultimate source Jesus was, which played an important role in the life of the society of the ancient Near East, and which can once again rejuvenate the materialistic society of our world today." (p. 96) On the subject of Protestantism he also made some interesting remarks (p. 95): "It is fortuitous that the Reformation has not produced a true mystical tradition with its particular spiritual discipline and philosophical outlook. Holiness is a Divine gift of grace mediated by the people of God. The lack of a mystical tradition in Protestantism may be due to the fact that there are no saints to mediate this gift of sanctifying grace to the people. It is for this reason that so many young Christians who are earnestly seeking have been obliged to look outside their churches for spiritual guidance." There is also a short Christian commentary (p.98) by Suzanne Le Gal, who was mentioned before, referrig to the two sacraments, baptism and the eucharist, and to the communion of saints being based on and brought about by them.

53 cf. Roest Crollius, Ary, Rome 10–13 mai 1989, **"Communiquer les valeurs religieuses à la jeunesse aujourd'hui"**, Pro Dialogo 73 (1990), p. 79–81, Séminaire académique de théologiens musulmans et chrétiens à l'université Pontificale Grégorienne: "Communiquer les valeurs religieuses à la jeunesse aujourd'hui", Islamochristiana 15 (1989), p. 242–243. For the bases of this cooperation cf. Signature d'un accord de coopération académique entre l'université d'Ankara et l'université grégorienne de Rome, Islamochristiana 12 (1986), p. 222–223. In the context of this cooperation, Fr. Thomas Michel SJ gave lectures on Christian subjects at the universities of Ankara (1987), Izmir (1988) and Konya (1989) while Prof. Yurdaydin, on the other hand, was teaching in 1987–88 at the Gregoriana and the PISAI, cf. At the Ankara University: Christian Lessons to Muslims, Islamochristiana 13 (1987), p. 217, und Fitzgerald, Twenty-five Years of Dialogue, p. 118. Another form of academic communication which will certainly have a broad effect, although it cannot be taken into account by a study like this, is the foundation *Nostra Aetate* established on 5 April 1993 which e.g. over the period from 1991 to 1999 awarded scholarships to 17 Muslim students for Christian studies in Rome, combined with accommodation in a Christian house and with a monthly meeting where these students met Christians and Muslims from the city who were active in the field of dialogue, not to mention the contacts with the Pontifical Council for Interreligious Dialogue itself where the person responsible for Christian-Muslim dialogue also held monthly meetings, according to Fitzgerald, Michael L[ouis], Report on the Activities of the PCID: November 1998-October 2001, Pro Dialogo 109 (2002), p. 80f, and also Akasheh, Considerations on Forty Years of Religious Dialogue with Muslims (A Report), p. 198. This is a means of also promoting a practical understanding of Christianity which could hardly be achieved in another way; cf. Akasheh, Report on the Activities of the PCID: Relations with Muslims, p. 219 and another practical example Tabbara, Nayla, A Roman Experience, Pro Dialogo 108 (2001), p. 393–396 and Demiri, Lejla, My Experience of Dialogue, Pro Dialogo 114 (2003), p. 375–378, both of whom especially praise the combination of theory and practice which gave them insights which they would not have gained from the one or the other alone. The official report by Lejla Demiri also comes to this conclusion but is not so personal or impressive: **Rome, Italy:** 15 March 2003 A Report for "Nostra Aetate" Foundation, Pontifical Council for Interreligious Dialogue, by a Muslim Student who Had Received a Study Grant, Pro Dialogo 114 (2003), p. 409–420. It is interesting that a study report presented by a man to the general assembly of the Pontifical Council from 14–19 May 2001 mentions the value of common life as greater than purely academic studies as well but also discusses the image of Islam in the West and in Christianity which consists mainly of violence, fanaticism and intolerance. From the viewpoint of his home country, Tunisia, he sees the roots of violence, terrorism and conflict in poverty, injustice and alienation. He himself, on the basis of his personal experience of dialogue, saw the conflict of cultures as a war of ignorance. A final interesting reference in this report relates to the famous home university of the student, the Ezzitouna, and its agreements with the Gregorian Pontifical University, the Pontifical Institute for Arabistics and Islamic Science and the theological faculty of Sicily, according to Merchergui, Ahmed, Testimony (Muslim, Tunisia), Pro Dialogo 116/117 (2004), p. 304–306. On 15 June 2000, the president of the department for religious affairs in Turkey offered the Pontifical Council for Interreligious Dialogue an exchange of scholarships between Catholic and Muslim students. As Akasheh, Khaled, Bureau pour l'Islam Rapport d'activités: novembre 1998-octobre 2001, Pro Dialogo 109 (2002), p. 112 (cf. p.111) wrote: "Notre Conseil a exprimé sa satisfaction pour cette initiative de réciprocité et sa disponibilité pour l'étude d'un protocole d'accord, mais a insisté sur la nécessité du dialogue au niveau local." In this context mention should also be made of the John XXIII Ecumenical Centre in Ankara that was to encourage understanding between Christians and Muslims with a library and dialogue activities and was inaugurated in February 1990 with distinguished participants, cf. Inauguration of the John XXIII Ecumenical Centre (22 February 1990), Islamochristiana 16 (1990), p. 280. On John XXIII and the whole range of dialogue activities in Turkey, cf. Akasheh, Khaled, Some Reflections on Islam and Christian-Muslim Relations in Certain Countries, Pro Dialogo 116/117 (2004), p. 254f.

54 cf. Yilmaz, Mehmet Nuri / Arinze, Francis, A STATEMENT OF INTENT between the Pontifical Council for Interreligious Dialogue (Vatican City) and the Presidency of Religious Affairs, Prime Minister's Office, Republic of Turkey, Pro Dialogo 110 (2002), p. 193; for an evaluation see Akasheh, Considerations on Forty Years of Religious Dialogue with Muslims (A Report), p. 198.

55 cf. Arinze, Francis, Greetings to the Participants, in: Religious Education and Modern Society, Acts of a Muslim-Christian Colloquium Organized Jointly by the Pontifical Council for Interreligious Dialogue (Vatican City) and the Royal Academy for Islamic Civilization Research Al Albait Foundation (Amman), Vatican, p. 5–7, El-Assad, Nassir El-Din, Inaugural Speech, in: Religious Education and Modern Society, Acts of a Muslim-Christian Colloquium Organized Jointly by the Pontifical Council for Interreligious Dialogue (Vatican City) and the Royal Academy for Islamic Civilization Research Al Albait Foundation (Amman), Vatican City, p. 8–11. For previous developments see Sabanegh, E[douard] S[ami] Martin, L'Église catholique au sein du monde arabo-musulman, BSNC 57 (1984), p. 301; the article is also interesting for its analysis of the situation in various Arab states. One special emphasis in the cooperation with the foundation is questions related to the family. In addition, in this context, reference can be made to a publication which does not relate directly to Christian-Muslim relations and is therefore not a subject for this study but could nevertheless be of interest for the reader: Pontifical Council for Interreligious Dialogue / Pontifical Council for the Family (eds.), Marriage and the Family in Today's World, Interreligious Colloquium Rome, 21–25 September 1994, Vatican City 1995. On the specific question of Christlian-Muslim marriages, reference can be made e.g. to Salama, Andraos, Les mariages entre chrétiens et musulmans approches pastorales, Bulletin Pontificium Consilium pro Dialogo inter Religiones 70 (1989), p. 59–76.

56 cf. Kamel, Abdul Aziz, Identity and Openness: Religious Education in a Pluralistic World, in: Religious Education and Modern Society, Acts of a Muslim-Christian Colloquium Organized Jointly by the Pontifical Council for Interreligious Dialogue (Vatican City) and the Royal Academy for Islamic Civilization Research Al Albait Foundation (Amman), Vatican City, p. 15–36

57 Galenzoga, Rodulfo M., Response to the Paper of Dr. Abdul Aziz Kamel on the Theme: Identity and Openness: Religious Education in a Pluralistic World, in: Religious Education and Modern Society, Acts of a Muslim-Christian Colloquium Organized Jointly by the Pontifical Council for Interreligious Dialogue (Vatican City) and the Royal Academy for Islamic Civilization Research Al Albait Foundation (Amman), Vatican City, p. 110, cf. p. 109 + 111.

58 Sabbah, Michel, Identity and Openness: Religious Education in a Pluralistic World, in: Religious Education and Modern Society, Acts of a Muslim-Christian Colloquium Organized Jointly by the Pontifical Council for Interreligious Dialogue (Vatican City) and the Royal Academy for Islamic Civilization Research Al Albait Foundation (Amman), Vatican City, p. 39, also p. 37f.

59 ibid. p.40.

60 ibid.

61 ibid., p. 42, cf. p. 41; Al-Abbadi, Abdul Salam, Response to the Paper of S.B. Mons. Michel Sabbah on the Theme: Identity and Openness: Religious Education in a Pluralistic World, in: Religious Education and Modern Society, Acts of a Muslim-Christian Colloquium Organized Jointly by the Pontifical Council for Interreligious Dialogue (Vatican City) and the Royal Academy for Islamic Civilization Research Al Albait Foundation (Amman), Vatican City. p. 112–113 give a summary of the bases of Christian education as he had heard and understood them in the address, and he regretted that the author had not indicated any original sources because a number of relationships had remained unclear – which demonstrates very well that, for a listener of the same religion, many things can be omitted or put much more briefly and still appear comprehensible, whereas this does not apply to a listener or reader from a different religious background. The following is his brief Muslim summary especially of the anthropology of Christian education in response to the address discussed.

"A. Education aims at the bettering of man.

B. Sound human education is what enables men to relate to God and accept the Creator's design.

C. Man's dignity stems from God Himself.
D. Man is made of matter and soul; balance between his spiritual and material potentialities is of paramount importance. The soul is at the basis of any identity, and it is through the soul that man relates to God (...) / p. 113
F. Love is at the foundation of any attempt to direct man's behaviour. While worthy of trust, man remains prone to error and inclined to evil, at least as much as he is capable of goodness. His essence was flawed at the outset by Original Sin, but, through the Redemption, he has regained his ability to do good."

62 Sabbah, p.45, cf. p.43f.
63 Al-Abbadi, p.113, cf. p. 112–114.
64 ibid., p.116, cf. p.115.
65 ibid., for what follows cf. p.117f. Al-Abbadi also quotes an interesting hadith on p.118: "'All creatures are children of God, and God loves most those of them who are most useful to His children' (Related by Abu Ya'ala, Al-Bazzar, and Tabaran.)", but also Sura 8,58 which deals with the cancellation of alliances when one fears treachery, and it does not encourage much confidence when it is added that this has influenced the behaviour of Muslims in times of war and peace.
66 Sheridan, Rosaleen, Science, Faith and the Concerns of Youth, in: Religious Education and Modern Society, Acts of a Muslim-Christian Colloquium Organized Jointly by the Pontifical Council for Interreligious Dialogue (Vatican City) and the Royal Academy for Islamic Civilization Research Al Albait Foundation (Amman), Vatican City cf. p.50, cf.also p.49.
67 ibid., p.51/52, cf. a sentence like p.51: "Self transcendence and religious openness characteristic of each man and woman find a special intensity in Jesus of Nazareth" or the summary of Teilhard's four statements of faith about God and the universe, including the third: "I believe that in man spirit is fully realised in person."
68 ibid., p.56, cf. p. 53–55+57–59, including p.58: "In conclusion, it must be stated that both faith and science search for truth which in the former is arrived at through love and in the latter is arrived at through analysis and enquiry. Youth seeks truth which is revealed gradually over time through faith, science and personal experience" and as the really last sentence p.59: "The human mind is propelled to grow through energy generated by tension created by extraverted activity directed toward the external world and by introverted activity directed towards the internal world of values, beliefs and thoughts."
69 Abedin, Zainul, Response to the Paper of Sr. Rosaleen Sheridan on the Theme: Faith and the Concerns of Youth, in: Religious Education and Modern Society, Acts of a Muslim-Christian Colloquium Organized Jointly by the Pontifical Council for Interreligious Dialogue (Vatican City) and the Royal Academy for Islamic Civilization Research Al Albait Foundation (Amman), Vatican City, cf. p.120.
70 Innocent, The Misery of Human Condition, quoted from ibid., p.121. Unfortunately neither the name nor the reference are given more precisely (unlike the quotations from Augustine) which is naturally a serious academic omission, however admirable it may be that a Muslim is so familiar with the Christian tradition. For the further remarks cf. p.122.
71 Abedin, p.122.
72 El-Zuhaili, Wahbah, Science, Faith and the Concerns of Youth, in: Religious Education and Modern Society, Acts of a Muslim-Christian Colloquium Organized Jointly by the Pontifical Council for Interreligious Dialogue (Vatican City) and the Royal Academy for Islamic Civilization Research Al Albait Foundation (Amman), Vatican City, cf. p 64, also p. 60–63.
73 ibid., p. 67–68, cf. p.65f.

74 According to Aquilina, Oscar, Response to the Paper of Dr. Wahbah El-Zuhaili on the Theme: Science, Faith and the Concerns of Youth, in: Religious Education and Modern Society, Acts of a Muslim-Christian Colloquium Organized Jointly by the Pontifical Council for Interreligious Dialogue (Vatican City) and the Royal Academy for Islamic Civilization Research Al Albait Foundation (Amman), Vatican City, cf. p. 123 – 127.

75 El-Khayyat, Abdul Aziz, Religious Education in the Modern Society: Islamic Education, in: Religious Education and Modern Society, Acts of a Muslim-Christian Colloquium Organized Jointly by the Pontifical Council for Interreligious Dialogue (Vatican City) and the Royal Academy for Islamic Civilization Research Al Albait Foundation (Amman), Vatican City, cf. p.77, cf. also p. 71–73+83. On the question of Muslim technical sciences, there are differing legal opinions at present ranging from complete rejection to complete agreement but the tendency is towards acceptance provided that no Islamic rules are violated, cf. p. 79 – 81.

76 ibid., p.87, cf. p. 74 – 76+85 – 86.

77 According to Root, Howard, Response to the paper of Dr. Abdul Aziz El-Khayyat on the Theme: "Religious Education in Colleges and Universities: Present Situation and Aspirations for the Future", in: Religious Education and Modern Society, Acts of a Muslim-Christian Colloquium Organized Jointly by the Pontifical Council for Interreligious Dialogue (Vatican City) and the Royal Academy for Islamic Civilization Research Al Albait Foundation (Amman), Vatican City, cf. p. 128 – 129.

78 According to Duminuco, Vincent J., Religious Education in Colleges and Universities: Present Situation and Aspirations for the Future, in: Religious Education and Modern Society, Acts of a Muslim-Christian Colloquium Organized Jointly by the Pontifical Council for Interreligious Dialogue (Vatican City) and the Royal Academy for Islamic Civilization Research Al Albait Foundation (Amman), Vatican City, cf. p. 89 – 105 (in contrast, cf. Sabanegh, [Edouard Sami] M[artin], Pour une meilleure compréhension de la législation matrimoniale en Islam, BSNC 44 (1980), who stated right at the beginning (p.157) that that which no influence of modern thinking had been able to change so far in Islamic countries was family law in the broadest sense).

79 According to Aydin, Mehmet, Response to the Paper of Rev. Vincent Duminuco, S.J. on the Theme: "Religious Education in Catholic Higher Education: Present Situation and Aspirations for the Future", in: Religious Education and Modern Society, Acts of a Muslim-Christian Colloquium Organized Jointly by the Pontifical Council for Interreligious Dialogue (Vatican City) and the Royal Academy for Islamic Civilization Research Al Albait Foundation (Amman), Vatican City, cf. p. 130 – 133.

80 cf. Fitzgerald, Twenty-five Years of Dialogue, p.114, Jarrar, F[aruk A.] / Fitzgerald, M[ichael] L[ouis], Islamo-Christian Colloquium on: "Religious Education in Modern Society" (With Particular Emphasis on Colleges and Universities), Pro Dialogo 73 (1990), p. 82– 83, Jarrar, Faruk A., The Royal Academy of Jordan for Islamic Civilization Research: A Continuing Dialogue, Islamochristiana 16 (1990), p.151, Christian-Muslim Consultation on "Religious Education in Modern Society" (6 – 8 December 1989), Islamochristiana 16 (1990), p. 290 – 292, Colloquium: "Religious Education in Modern Society", in: Religious Education and Modern Society, Acts of a Muslim-Christian Colloquium Organized Jointly by the Pontifical Council for Interreligious Dialogue (Vatican City) and the Royal Academy for Islamic Civilization Research Al Albait Foundation (Amman), Vatican, cf. p. 134 –135. The papal address can be found in John Paul II, Address to Partecipants *(sic!)*, in: Religious Education and Modern Society, Acts of a Muslim-Christian Colloquium Organized Jointly by the Pontifical Council for Interreligious Dialogue (Vatican City) and the Royal Academy for Islamic Civilization Research Al Albait Foundation (Amman), Vatican City, cf. p. 136 – 137. It is interesting that diplomatic relations between the Holy See and Jordan

were only established later, see: The First Ambassador to the Holy See Is Received by the Pope (19 November 1994), Islamochristiana 21 (1995), p. 188–190.

81 cf. H.R.H. Crown Prince Hassan, Inaugural Address, in: Pontifical Council for Interreligious Dialogue / Royal Academy for Islamic Civilization Research Al Albait Foundation (eds.), The Rights and Education of Children in Islam and Christianity, Acts of a Muslim-Christian Colloquium, Vatican City, cf. p.3+5–6, Arinze, Francis, The Child Is Precious, in: Pontifical Council for Interreligious Dialogue / Royal Academy for Islamic Civilization Research Al Albait Foundation(eds.), The Rights and Education of Children in Islam and Christianity, Acts of a Muslim-Christian Colloquium, Vatican City, cf. p.7. The last word of greeting (El-Assad, Nassir El-Din, Inaugural Address, in: Pontifical Council for Interreligious Dialogue / Royal Academy for Islamic Civilization Research Al Albait Foundation (eds.), The Rights and Education of Children in Islam and Christianity, Acts of a Muslim-Christian Colloquium, Vatican City, cf. p. 9–12), is less interesting in content but does provide a review of what had already been done on this issue on the Muslim side, and also states fundamentally that at some point an evaluation should be conducted of what had come out of the dialogue conferences thus far (many people were in fact asking this question) and this basis should serve for further planning. A short summary of the meeting and its results can be found in **Amman – Jordan:** *Christian-Muslim Colloquium* **"The Rights and Education of Children in Islam and Christianity"**, Bulletin Pontificium Consilium pro Dialogo inter Religiones 76 (1991), p. 109–111.

82 According to Arinze, Francis, The Rights of the Unborn Child, in: Pontifical Council for Interreligious Dialogue / Royal Academy for Islamic Civilization Research Al Albait Foundation (eds.), The Rights and Education of Children in Islam and Christianity, Acts of a Muslim-Christian Colloquium, Vatican City, p. 15–22, El-Abbadi, Abdul Salam, Muslim Response, in: Pontifical Council for Interreligious Dialogue / Royal Academy for Islamic Civilization Research Al Albait Foundation (eds.), The Rights and Education of Children in Islam and Christianity, Acts of a Muslim-Christian Colloquium, Vatican City, p. 24–32.

83 El-Samerra'i, Faruk, The Rights of the Embryo, in: Pontifical Council for Interreligious Dialogue / Royal Academy for Islamic Civilization Research Al Albait Foundation (eds.), The Rights and Education of Children in Islam and Christianity, Acts of a Muslim-Christian Colloquium, Vatican City, p.34/35, cf. p.33f.

84 ibid., p.40. cf. p. 41–57.

85 cf. Tannous, Nemra, Christian Response, in: Pontifical Council for Interreligious Dialogue / Royal Academy for Islamic Civilization Research Al Albait Foundation (eds.), The Rights and Education of Children in Islam and Christianity, Acts of a Muslim-Christian Colloquium, Vatican City, p. 60–64.

86 According to Khayyat, Abdul Aziz, The Pre-school Child: Rights and Education, in: Pontifical Council for Interreligious Dialogue / Royal Academy for Islamic Civilization Research Al Albait Foundation (eds.), The Rights and Education of Children in Islam and Christianity, Acts of a Muslim-Christian Colloquium, Vatican City, p. 67–93, and Rasmussen, Nete, Christian Response, in: Pontifical Council for Interreligious Dialogue / Royal Academy for Islamic Civilization Research Al Albait Foundation (eds.), The Rights and Education of Children in Islam and Christianity, Acts of a Muslim-Christian Colloquium, Vatican City, p. 94–97.

87 'Ayesh, Husni, Muslim Response, in: Pontifical Council for Interreligious Dialogue / Royal Academy for Islamic Civilization Research Al Albait Foundation (eds.), The Rights and Education of Children in Islam and Christianity, Acts of a Muslim-Christian Colloquium, Vatican City, p.113, cf. p. 110–112+114. He referred to the previously summarised address by Eisele, Marie-Paule, Rights and Education of the Pre-school Child, in: Pontifical Council for Interreligious Dialogue

/ Royal Academy for Islamic Civilization Research Al Albait Foudation (eds.), The Rights and Education of Children in Islam and Christianity, Acts of a Muslim-Christian Colloquium, Vatican City, p. 98–109.
88 cf. Fahim, Amin, The Rights and Education of Children in Schools, in: Pontifical Council for Interreligious Dialogue / Royal Academy for Islamic Civilization Research Al Albait Foundation (eds.), The Rights and Education of Children in Islam and Christianity, Acts of a Muslim-Christian Colloquium, Vatican City, p. 117–126, El-Sheikh, Omar, Muslim Response, in: Pontifical Council for Interreligious Dialogue / Royal Academy for Islamic Civilization Research Al Albait Foundation (eds.), The Rights and Education of Children in Islam and Christianity, Acts of a Muslim-Christian Colloquium, Vatican City, p. 127–129.
89 Rahman, Hani Abdul, Rights and Education of Children in Schools, as Related to Government and Society, in: Pontifical Council for Interreligious Dialogue / Royal Academy for Islamic Civilization Research Al Albait Foundation (eds.), The Rights and Education of Children in Islam and Christianity, Acts of a Muslim-Christian Colloquium, Vatican City, p.137, cf. p. 130–136.
90 ibid. p.143, cf. p. 138–142 and, for what follows, p. 138–150.
91 According to Naber, Antun, Christian Response, in: Pontifical Council for Interreligious Dialogue / Royal Academy for Islamic Civilization Research Al Albait Foundation (eds.), The Rights and Education of Children in Islam and Christianity, Acts of a Muslim-Christian Colloquium, Vatican City, p. 152–157.
92 According to The General Report of the Muslim-Christian Colloquium on the Theme: "The Rights and Education of Children in Islam and Christianity" (Amman, Jordan, 13–15 December 1990), Islamochristiana 17 (1991), p. 220–222, also as General Report, in: Pontifical Council for Interreligious Dialogue / Royal Academy for Islamic Civilization Research Al Albait Foundation (eds.), The Rights and Education of Children in Islam and Christianity, Acts of a Muslim-Christian Colloquium, Vatican City, p. 158–160.
93 Arinze, Francis, The Importance of Women's Role in Society, in: Women in Society According to Islam and Christianity, Acts of a Muslim-Christian Colloquium Organized Jointly by the Pontifical Council for Interreligious Dialogue (Vatican City) and the Royal Academy for Islamic Civilization Research Al Albait Foundation (Amman), Vatican City, p.9, cf. p. 5–8.
94 El-Assad, Nassir El-Din, Woman the Mainstay of Human Life, in: Women in Society According to Islam and Christianity, Acts of a Muslim-Christian Colloquium Organized Jointly by the Pontifical Council for Interreligious Dialogue (Vatican City) and the Royal Academy for Islamic Civilization Research Al Albait Foundation (Amman), Vatican City, p.13, cf. p.14.
95 Khayyat, Abdul Aziz, Women's Status in Islam, in: Women in Society According to Islam and Christianity, Acts of a Muslim-Christian Colloquium Organized Jointly by the Pontifical Council for Interreligious Dialogue (Vatican City) and the Royal Academy for Islamic Civilization Research Al Albait Foundation (Amman), Vatican City, p.17.
96 ibid.
97 ibid., p. 24–25, cf. p. 18–23.
98 ibid., p. 27–28, cf. p.26.
99 ibid., p. 28–29, cf. also p.32.
100 Cantore, Stefania, Woman in Christianity, in: Women in Society According to Islam and Christianity, Acts of a Muslim-Christian Colloquium Organized Jointly by the Pontifical Council for Interreligious Dialogue (Vatican City) and the Royal Academy for Islamic Civilization Research Al Albait Foundation (Amman), Vatican City, p.36, cf. p,33–35.
101 ibid., p.43, cf. p.46: "Mary is the original beauty who reappears in history." For the other comments see p. 37–42..

102 ibid., p.45.
103 ibid., p. 46–47.
104 According to Barbieri Masini, Eleonora, Contemporary Problems and Challenges that Women Have to Face Specifically as Christian Women, in: Women in Society According to Islam and Christianity, Acts of a Muslim-Christian Colloquium Organized Jointly by the Pontifical Council for Interreligious Dialogue (Vatican City) and the Royal Academy for Islamic Civilization Research Al Albait Foundation (Amman), Vatican City, p. 52f., 55, 58f. + 62–68.
105 cf. Dahhan, Omaymah, Muslim Women in the Arab World, in: Women in Society According to Islam and Christianity, Acts of a Muslim-Christian Colloquium Organized Jointly by the Pontifical Council for Interreligious Dialogue (Vatican City) and the Royal Academy for Islamic Civilization Research Al Albait Foundation (Amman), Vatican City, p. 71–93.
106 Khawaldeh, Samira F., Future Opportunities and Prospects for Women in Islam, in: Women in Society According to Islam and Christianity, Acts of a Muslim-Christian Colloquium Organized Jointly by the Pontifical Council for Interreligious Dialogue (Vatican City) and the Royal Academy for Islamic Civilization Research Al Albait Foundation (Amman), Vatican City, p.97.
107 ibid., p. 98–99.
108 According to ibid., p. 99–106.
109 Borrmans, Maurice, Christians and Muslims Working Together for the Improvement of Woman's Condition: Possibilities and Prospects, in: Women in Society According to Islam and Christianity, Acts of a Muslim-Christian Colloquium Organized Jointly by the Pontifical Council for Interreligious Dialogue (Vatican City) and the Royal Academy for Islamic Civilization Research Al Albait Foundation (Amman), Vatican City, p. 113/114, cf. p. 107–112+116. The comments on the discussion are found on p.117f, including on p.118 an unambiguous statement about the share of Mary in redemption which is significantly described as Christian and not merely as Catholic. "For us Christians, if Eve took the initiative of sinning while the man Adam was in immediate agreement (thus, equality in sin), Mary participated with Jesus her son in the work of universal redemption (equality in the redemption) and became the first human model of holiness for all."
110 General Report of the Muslim-Christian Colloquium of 24–26 June 1992 on the Theme "Women in Society according to Islam and Christianity", Islamochristiana 18 (1992), p.319, cf. p.318+320; the report can also be found under General Report, in: Women in Society According to Islam and Christianity, Acts of a Muslim-Christian Colloquium Organized Jointly by the Pontifical Council for Interreligious Dialogue (Vatican City) and the Royal Academy for Islamic Civilization Research Al Albait Foundation (Amman), Vatican City, p. 119–121, and under Rome – Italy: **Women in Society** (24–26 June 1992), Bulletin Pontificium Consilium pro Dialogo inter Religiones 81 (1992), p. 378–380. The address by the Pope at the reception for the conference participants was also included in the volume on the conference: H.H. Pope John Paul II, Address of H.H. Pope John Paul II to Participants in the Colloquium "Women in Society According to Islam and Christianity", in: Women in Society According to Islam and Christianity, Acts of a Muslim-Christian Colloquium Organized Jointly by the Pontifical Council for Interreligious Dialogue (Vatican City) and the Royal Academy for Islamic Civilization Research Al Albait Foundation (Amman), Vatican City, p. 123–124.
111 cf. Fitzgerald, Report on the Activities of the PCID: November 1992 – November 1995, p.170 and El-Assad, Nassir El-Din, Inaugural Address, in: Nationalism Today: Problems and Challenges, Acts of a Muslim-Christian Colloquium Organized Jointly by the Pontifical Council for Interreligious Dialogue (Vatican City) and the Royal Academy for Islamic Civilization Research Al Albait Foundation (Amman), Rome 1994, p.6f.

112 Arinze, Francis, Inaugural Address, in: Nationalism Today: Problems and Challenges, Acts of a Muslim-Christian Colloquium Organized Jointly by the Pontifical Council for Interreligious Dialogue (Vatican City) and the Royal Academy for Islamic Civilization Research Al Albait Foundation (Amman), Rome 1994, p. 13–14, cf. p. 9–12.

113 cf. Al-Sartawi, Mahmoud, Nationalism and Religion: An Historical Overview, A Muslim Perspective, in: Nationalism Today: Problems and Challenges, Acts of a Muslim-Christian Colloquium Organized Jointly by the Pontifical Council for Interreligious Dialogue (Vatican City) and the Royal Academy for Islamic Civilization Research Al Albait Foundation (Amman), Rome 1994, p. 17–26.

114 cf. Musallam, Adnan, Nationalism and Religion: An Historical Overview, A Christian Perspective, in: Nationalism Today: Problems and Challenges, Acts of a Muslim-Christian Colloquium Organized Jointly by the Pontifical Council for Interreligious Dialogue (Vatican City) and the Royal Academy for Islamic Civilization Research Al Albait Foundation (Amman), Rome 1994, p. 27–44.

115 cf. Nielsen, J[ørgen] S., National Ties Today: Problems and Challenges, A Christian Viewpoint, in: Nationalism Today: Problems and Challenges, Acts of a Muslim-Christian Colloquium Organized Jointly by the Pontifical Council for Interreligious Dialogue (Vatican City) and the Royal Academy for Islamic Civilization Research Al Albait Foundation (Amman), Rome 1994, p. 47–61.

116 Hikmat, Taher, National Ties Today: Problems and Challenges, A Muslim Viewpoint, in: Nationalism Today: Problems and Challenges, Acts of a Muslim-Christian Colloquium Organized Jointly by the Pontifical Council for Interreligious Dialogue (Vatican City) and the Royal Academy for Islamic Civilization Research Al Albait Foundation (Amman), Rome 1994, p.68, cf. p. 63–67. One really has the feeling of listening to a politician entreating the masses when one reads the concluding sentences that followed: "The Middle East, and particularly our country, is the birthplace and radiation focus of monotheistic religions. It is the beacon of tolerance and love and will never be a new Yugoslavia, Bosnia-Herzegovina or Serbia. / p. 69 Let our slogan be: 'inter-communion of civilizations not conflict between civilizations'."

117 Jarradat, Izat, The Role of Believers in Dealing with Questions of Nationalism, A Muslim Viewpoint, in: Nationalism Today: Problems and Challenges, Acts of a Muslim-Christian Colloquium Organized Jointly by the Pontifical Council for Interreligious Dialogue (Vatican City) and the Royal Academy for Islamic Civilization Research Al Albait Foundation (Amman), Rome 1994, p.77, cf. p. 73–75, and for the specifically Islamic view on religion and society p.76: "For from an Islamic angle, the call of religions for belief represents the invitation of man to maintain his human dignity and feeling equal to others in terms of human and value distinctive qualities and his deep conception of the possibility of intellectual, social and cultural understanding and agreement at both individual and collective levels." Although the formulation is less elegant, it follows the same lines as the quotation in the main text.

118 ibid., p.78.

119 ibid., p.83, cf. p. 78–82+84–86. In his summary on p.85 there are such sentences as: "(a) Islamic Arab civilization is the message of the Arab nation. (b) The civilizational content of Arab nationalism is Islam and not secularism. (c) The Arab national feeling, far from being an extremist, is a humane one. (...) (f) Fighting the danger of atheism is a common target for Islamic as well as for national thought."

120 cf. Linden, Ian, The Role of Believers in Dealing with Questions of Nationalism, A Christian Viewpoint, in: Nationalism Today: Problems and Challenges, Acts of a Muslim-Christian Colloquium Organized Jointly by the Pontifical Council for Interreligious Dialogue (Vatican City) and

the Royal Academy for Islamic Civilization Research Al Albait Foundation (Amman), Rome 1994, p. 87–99.
121 Ejifor, Lambert, A Christian Comment, in: Nationalism Today: Problems and Challenges, Acts of a Muslim-Christian Colloquium Organized Jointly by the Pontifical Council for Interreligious Dialogue (Vatican City) and the Royal Academy for Islamic Civilization Research Al Albait Foundation (Amman), Rome 1994, p.114, cf. p. 103–113. The essay by Charles Amjad-Ali, Towards a New Theology of Dialogue, Al-Mushir 33 (1991), p. 57–69, follows a similar line, again from a Christian but this time not Western but Pakistani point of view and dialogue experience. He too appeals for the prophetic and political dimension of religion precisely in order to safeguard the incarnation, and vehemently attacks the philosophical, theological insights since the Enlightenment as too narrow and as unsuitable in another context, meaning the dialogue with Muslims; see p.64: "So dialogue is not just a concern for setting relationships in motion with people of other faiths, but much more significantly it is a challenge to the epistemological foundations upon which we have operated at a very fundamental level."
122 ibid., p.118, cf. p. 114–117.
123 cf. Ibrahim, Hanan, A Muslim Comment, in: Nationalism Today: Problems and Challenges, Acts of a Muslim-Christian Colloquium Organized Jointly by the Pontifical Council for Interreligious Dialogue (Vatican City) and the Royal Academy for Islamic Civilization Research Al Albait Foundation (Amman), Rome 1994, p. 119–121.
124 Arinze, Francis, Concluding Discourse Together for Better Times, in: Nationalism Today: Problems and Challenges, Acts of a Muslim-Christian Colloquium Organized Jointly by the Pontifical Council for Interreligious Dialogue (Vatican City) and the Royal Academy for Islamic Civilization Research Al Albait Foundation (Amman), Rome 1994, p.123.
125 H.R.H. Crown Prince Hassan, Concluding Discourse, in: Nationalism Today: Problems and Challenges, Acts of a Muslim-Christian Colloquium Organized Jointly by the Pontifical Council for Interreligious Dialogue (Vatican City) and the Royal Academy for Islamic Civilization Research Al Albait Foundation (Amman), Rome 1994, p.125, cf. p.126f; for the following cf. Nationalism Today: Problems and Challenges Final Communiqué, 20 January 1994, in: Nationalism today: Problems and Challenges, Acts of a Muslim-Christian Colloquium Organized Jointly by the Pontifical Council for Interreligious Dialogue (Vatican City) and the Royal Academy for Islamic Civilization Research Al Albait Foundation (Amman), Rome 1994, p.129f; for a short summary of the whole meeting cf. also JORDAN *Final Communiqué of the Muslim-Christian Colloquium of 18–20 January 1994 on the Theme "Nationalism Today: Problems and Challenges"*, Islamochristiana 20 (1994), p.239, or Nationalism Today: Problems and Challenges, Final Communiqué, Pro Dialogo 87 (1994), p.266f; for a very short summary cf. **Amman – Jordan:** Christian-Muslim Colloquium on "Religion and Nationalism" (18–20 January 1994), Pro Dialogo 87 (1994), p.266.
126 cf. Arinze, Francis, For a Responsible Use of the Earth's Resources, in: Religion and the Use of the Earth's Resources, Acts of a Muslim-Christian Colloquium Organized Jointly by the Pontifical Council for Interreligious Dialogue (Vatican City) and the Royal Academy for Islamic Civilization Research Al Albait Foundation (Amman), Rome 1996, p 5–6; the other address can be found under El-Assad, Nasir *(sic!)* El-Din, Inaugural Address, in: Religion and the Use of the Earth's Resources, Acts of a Muslim-Christian Colloquium Organized Jointly by the Pontifical Council for Interreligious Dialogue (Vatican City) and the Royal Academy for Islamic Civilization Research Al Albait Foundation (Amman), Rome 1996, p. 7–9, but its statements more generally reflect creation theology and are of less specific interest for anthropology.

127 Al-Abbadi, Abdul-Salam, Religious Teachings on the Use of the Earth's Resources, A Muslim Perspective, in: Religion and the Use of the Earth's Resources, Acts of a Muslim-Christian Colloquium Organized Jointly by the Pontifical Council for Interreligious Dialogue (Vatican City) and the Royal Academy for Islamic Civilization Research Al Albait Foundation (Amman), Rome 1996, p.15, cf. p.13 f.
128 ibid., p.19, cf. p. 16–18+26.
129 ibid., p. 23–24, cf. p. 20–22.
130 ibid., p.34, cf. p. 28–33, 35 + 37–41.
131 cf. Khader, Jamal, A Christian Commentary on Dr. El-Abbadi's Paper, in: Religion and the Use of the Earth's Resources, Acts of a Muslim-Christian Colloquium Organized Jointly by the Pontifical Council for Interreligious Dialogue (Vatican City) and the Royal Academy for Islamic Civilization Research Al Albait Foundation (Amman), Rome 1996, p. 43–45.
132 Coste, René, Religious Teachings on the Use of the Earth's Resources, A Christian Perspective, in: Religion and the Use of the Earth's Resources, Acts of a Muslim-Christian Colloquium Organized Jointly by the Pontifical Council for Interreligious Dialogue (Vatican City) and the Royal Academy for Islamic Civilization Research Al Albait Foundation (Amman), Rome 1996, p.50, cf. p.48 f.
133 ibid., p.51.
134 cf. ibid., p. 52–57.
135 El-Jerrari, Abbas, A Muslim Commentary on Mgr Coste's Paper, in: Religion and the Use of the Earth's Resources, Acts of a Muslim-Christian Colloquium Organized Jointly by the Pontifical Council for Interreligious Dialogue (Vatican City) and the Royal Academy for Islamic Civilization Research Al Albait Foundation (Amman), Rome 1996, p.59.
136 ibid., p. 59–60.
137 cf. ibid., p. 61–67.
138 Irwin, Kevin W., Religious Teachings on the Protection of the Earth's Resources, A Christian Point of View, in: Religion and the Use of the Earth's Resources, Acts of a Muslim-Christian Colloquium Organized Jointly by the Pontifical Council for Interreligious Dialogue (Vatican City) and the Royal Academy for Islamic Civilization Research Al Albait Foundation (Amman), Rome 1996, p.83, cf. p. 71–82.
139 cf. ibid., p. 85–90.
140 cf. Muthaffar, Mai, A Muslim Commentary on Fr. Irwin's Paper, in: Religion and the Use of the Earth's Resources, Acts of a Muslim-Christian Colloquium Organized Jointly by the Pontifical Council for Interreligious Dialogue (Vatican City) and the Royal Academy for Islamic Civilization Research Al Albait Foundation (Amman), Rome 1996, p. 91–94.
141 El-Tall, Sufian, Religious Teachings on the Protection of the Earth's Resources, A Muslim Point of View, in: Religion and the Use of the Earth's Resources, Acts of a Muslim-Christian Colloquium Organized Jointly by the Pontifical Council for Interreligious Dialogue (Vatican City) and the Royal Academy for Islamic Civilization Research Al Albait Foundation (Amman), Rome 1996, p.98/99, cf. p.96 f.
142 ibid., p.99.
143 ibid., p.116, cf. p. 99–115.
144 cf. Przewozny, Bernard, A Christian Commentary on Dr. El Tall's Paper, in: Religion and the Use of the Earth's Resources, Acts of a Muslim-Christian Colloquium Organized Jointly by the Pontifical Council for Interreligious Dialogue (Vatican City) and the Royal Academy for Islamic Civilization Research Al Albait Foundation (Amman), Rome 1996, p. 119–122.

145 cf. Khayyat, Abdul-Aziz, The Role of Believers in Ensuring an Equitable Distribution of Earth's Resources, A Muslim Point of View, in: Religion and the Use of the Earth's Resources, Acts of a Muslim-Christian Colloquium Organized Jointly by the Pontifical Council for Interreligious Dialogue (Vatican City) and the Royal Academy for Islamic Civilization Research Al Albait Foundation (Amman), Rome 1994, p. 125–141.
146 Musu, Ignazio, A Christian Commentary on Prof. Khayyat's Paper, in: Religion and the Use of the Earth's Resources, Acts of a Muslim-Christian Colloquium Organized Jointly by the Pontifical Council for Interreligious Dialogue (Vatican City) and the Royal Academy for Islamic Civilization Research Al Albait Foundation (Amman), Rome 1994, p.146, cf. p.145+147f.
147 Muasher, Anis, The Role of Believers in Ensuring an Equitable Distribution of Earth's Resources, A Christian Point of View, in: Religion and the Use of the Earth's Resources, Acts of a Muslim-Christian Colloquium Organized Jointly by the Pontifical Council for Interreligious Dialogue (Vatican City) and the Royal Academy for Islamic Civilization Research Al Albait Foundation (Amman), Rome 1996, p.150, cf. p.149.
148 ibid., p.165, cf. p. 151–164.
149 Hani, Abdul Razak Bani, A Muslim Commentary on Dr. Muasher's Paper, in: Religion and the Use of the Earth's Resources, Acts of a Muslim-Christian Colloquium Organized Jointly by the Pontifical Council for Interreligious Dialogue (Vatican City) and the Royal Academy for Islamic Civilization Research Al Albait Foundation (Amman), Rome 1996, p.167.
150 ibid., p.171, cf. p. 168–170.
151 His Royal Highness Crown Prince El Hassan Bin Talal of the Hashemite Kingdom of Jordan, Islam and Environment, in: Religion and the Use of the Earth's Resources, Acts of a Muslim-Christian Colloquium Organized Jointly by the Pontifical Council for Interreligious Dialogue (Vatican City) and the Royal Academy for Islamic Civilization Research Al Albait Foundation (Amman), Rome 1996, p.174, cf. p. 175–178.
152 cf. Christian-Muslim Colloquium on Religion and the Use of the Earth's Resources (Rome, 18–20 April 1996), Islamochristiana 22 (1996), p. 231–233, or Recommendations, in: Religion and the Use of the Earth's Resources, Acts of a Muslim-Christian Colloquium Organized Jointly by the Pontifical Council for Interreligious Dialogue (Vatican City) and the Royal Academy for Islamic Civilization Research Al Albait Foundation (Amman), Rome 1996, p. 181–183.
153 cf. Akasheh, Report on the Activities of the PCID: Relations with Muslims, p.215. It is perhaps also interesting that this dialogue conference was also used in order just afterwards to hold a meeting of the regional staff and advisers, according to Akasheh, Khaled, **Amman-Jordanie:** Réunion des Membres et des Consulteurs du Moyen Orient et du Nord de l'Afrique, 30 novembre-2 décembre 1997, p.337f. In the discussion naturally the well-known subject of the lack of reciprocity came up again, especially with regard to freedom of conscience, religion and witness. But in fact the main addresses dealt with spirituality in the dialogue and a number of interesting ideas were voiced. cf. Couvreur, Gilles, Spiritualité du Dialogue dans un Contexte Pluraliste, Pro Dialogo 99 (1998), p.324, against the background of the French experience, the confirmation that there was a link between faith in God and encountering the poor. Commitment to justice was not simply optional for someone who believed in God. The consequences could be very concrete. "Les enjeux sont importants pour l'avenir: les jeunes musulmans des banlieues seront-ils les seuls à dénoncer l'exclusion dont ils sont victimes? Ne risquent-ils pas, alors, d'être conduits à nommer 'Société occidentale' la source de leurs maux et de rêver à une contre-société qui serait un état islamique?" Sabbah, Michel, Spiritualité dans un milieu arabo-musulman, Pro Dialogo 99 (1998), p. 312–316, on the other hand, was more steeped in an almost Islamic Christian spirituality, although he also raised the question of justice and injustice, adding wisely that their

definition was related to the major interests of the nations and their rulers and to different worldviews. Another striking statement was (p.314): "Accepter l'autre dans sa différence, c'est reconnaître la dignité de tout être humain et sa vocation à la vérité de Dieu. Dans le Christ, il y a la plénitude de la divinité. Tout être humain est en voie vers cette plénitude, mais selon la grâce que Dieu lui donne." It is noteworthy that, on the Catholic side, the common conception of human dignity had been mentioned very early on (1976) as a special reason for the great importance of Christian-Muslim dialogues. cf. Report on Mgr. Rossano's Journey in South-East Asia, p.56: "To which I answered that I believe that the dialogue between Islam and Christianity is the urgent task of the years ahead because their prophetic and historical monotheism offers the most solid spiritual guarantee for the dignity of Man, his inviolability, sacral character, equality and liberty against the impending menace of materialism, of totalitarianism and of technocracy."

154 cf. Akasheh, Bureau pour l'Islam Rapport d'activités: novembre 1998-octobre 2001, p.107.

155 Prince el Hassan bin Talal, Inaugural Address, in: Human Dignity, Acts of a Muslim-Christian Colloquium Organized Jointly by the Pontifical Council for Interreligious Dialogue (Vatican City) and the Royal Academy for Islamic Civilization Research Al Albait Foundation (Amman), Rome 1999, p.7/8, cf. p.5f+9f.

156 Arinze, Francis, Inaugural Address, in: Human Dignity, Acts of a Muslim-Christian Colloquium Organized Jointly by the Pontifical Council for Interreligious Dialogue (Vatican City) and the Royal Academy for Islamic Civilization Research Al Albait Foundation (Amman), Rome 1999, p.16/17, cf. p.15; cf. also El-Assad, Nassir El-Din, Inaugural Address, in: Human Dignity, Acts of a Muslim-Christian Colloquium Organized Jointly by the Pontifical Council for Interreligious Dialogue (Vatican City) and the Royal Academy for Islamic Civilization Research Al Albait Foundation (Amman), Rome 1999, p.13; but the fundamental reflections on the dialogues are almost more interesting, p.11/12: "The aim was quite obvious to us from the start, namely, to organize symposiums for a cultural civilizational interlocution that embraces the value system and social issues as seen by the faithful followers of both religions in order to make known such values and issues and propagate knowledge through the attitude of both faiths towards them. Other aims include discernment of contemporary challenges that face the followers of these two religions in those topics and an attempt to have a future envisionment of how to deal with them, and to coordinate and cooperate to make this envisionment materialize as much as possible. We all know that these aims involve correction of the image of each party in the eyes of the other. /p.12 (...) The principle of matchability and equivalence, it must be said, between the two parties has been achieved without requiring direct verbal statements. It has been achieved in practice by the very holding and frequency of these meetings. Had it not been for that, meetings would not have been held in such frequency while ideas, information and experience would not have been so actively exchanged. Reasons why these two points have been so clearly and emphatically underlined include avoidance of discussing issues of doctrine and worship, keeping away from missionary and preaching practices and restriction of dialogue to social values and matters that concern both Muslim and Christian sides".

157 Stäger, Roman, The Concept of Human Dignity According to Islam, A Christian Comment, in: Human Dignity, Acts of a Muslim-Christian Colloquium Organized Jointly by the Pontifical Council for Interreligious Dialogue (Vatican City) and the Royal Academy for Islamic Civilization Research Al Albait Foundation (Amman), Rome 1999, p.37/38, also El-Talebi, Ammar, The Concept of Human Dignity According to Islam, in: Human Dignity, Acts of a Muslim-Christian Colloquium Organized Jointly by the Pontifical Council for Interreligious Dialogue (Vatican City)

and the Royal Academy for Islamic Civilization Research Al Albait Foundation (Amman), Rome 1999, p. 21–36.
158 Sanna, Ignazio, The Dignity of Man, A Christian Perspective, in: Human Dignity, Acts of a Muslim-Christian Colloquium Organized Jointly by the Pontifical Council for Interreligious Dialogue (Vatican City) and the Royal Academy for Islamic Civilization Research Al Albait Foundation (Amman), Rome 1999, p.37/38, also El-Talebi, Ammar, The Concept of Human Dignity According to Islam, in: Human Dignity, Acts of a Muslim-Christian Colloquium Organized Jointly by the Pontifical Council for Interreligious Dialogue (Vatican City) and the Royal Academy for Islamic Civilization Research Al Albait Foundation (Amman), Rome 1999, p.40, cf. also p.39.
159 ibid., p. 41–42.
160 ibid., p.43.
161 Khayyat, Abdul Aziz, A Muslim Comment, in: Human Dignity, Acts of a Muslim-Christian Colloquium Organized Jointly by the Pontifical Council for Interreligious Dialogue (Vatican City) and the Royal Academy for Islamic Civilization Research Al Albait Foundation (Amman), Rome 1999, p.53.
162 ibid., p.57, cf. also p. 54–56+58–63.
163 ibid., p.58.
164 ibid., p.57.
165 Ajaiby, George, Attitude of Christians towards Human Dignity, A Historical Review, in: Human Dignity, Acts of a Muslim-Christian Colloquium Organized Jointly by the Pontifical Council for Interreligious Dialogue (Vatican City) and the Royal Academy for Islamic Civilization Research Al Albait Foundation (Amman), Rome 1999, p.67, cf. p. 68–79.
166 Muthaffar, Mai, Comment, in: Human Dignity, Acts of a Muslim-Christian Colloquium Organized Jointly by the Pontifical Council for Interreligious Dialogue (Vatican City) and the Royal Academy for Islamic Civilization Research Al Albait Foundation (Amman), Rome 1999, p.82, cf. p.81+83.
167 El-Mawla, Saud, Attitude of Muslims towards Human Dignity, A Historical Review, in: Human Dignity, Acts of a Muslim-Christian Colloquium Organized Jointly by the Pontifical Council for Interreligious Dialogue (Vatican City) and the Royal Academy for Islamic Civilization Research Al Albait Foundation (Amman), Rome 1999, p.90/91, cf. p. 85–89; for the criticism see Akasheh, Khaled, Comment, in: Human Dignity, Acts of a Muslim-Christian Colloquium Organized Jointly by the Pontifical Council for Interreligious Dialogue (Vatican City) and the Royal Academy for Islamic Civilization Research Al Albait Foundation (Amman), Rome 1999, p. 101–102.
168 El-Mawla, p. 92–93.
169 ibid., p.94.
170 ibid., p.95.
171 ibid., p. 95–96.
172 ibid., p.97.
173 ibid.
174 El-Mawla, Saud, Explanatory Remarks, in: Human Dignity, Acts of a Muslim-Christian Colloquium Organized Jointly by the Pontifical Council for Interreligious Dialogue (Vatican City) and the Royal Academy for Islamic Civilization Research Al Albait Foundation (Amman), Rome 1999, p.105.
175 El-Sammak, Muhammad, Muslims Faced by the Challenges of Human Dignity, in: The Pontifical Council for Interreligious Dialogue / The Royal Academy for Islamic Civilization Research

Al Albait Foundation (eds.), Human Dignity, Acts of a Christian-Muslim Colloquium, Vatican City 1999, p.118.
176 ibid., p.120.
177 cf. Keating, Sandra, Comment, in: The Pontifical Council for Interreligious Dialogue / The Royal Academy for Islamic Civilization Research Al Albait Foundation (eds.), Human Dignity, Acts of a Christian-Muslim Colloquium, Vatican City 1999, p. 123–125.
178 cf. Pacini, Andrea, Christians Facing the Challenge of Human Dignity: Prospects for the Future, The Pontifical Council for Interreligious Dialogue / The Royal Academy for Islamic Civilization Research Al Albait Foundation (eds.), Human Dignity, Acts of a Christian-Muslim Colloquium, Vatican City 1999, p. 127–139 (it is perhaps also significant that here the active form was chosen whereas the Muslim contribution used the passive form, possibly as a symbolic expression of the challenges being seen more as imposed than as something which one could approach openly and with some degree of confidence) and Balarabi, Abdul-Hafeez, Comment, in: The Pontifical Council for Interreligious Dialogue / The Royal Academy for Islamic Civilization Research Al Albait Foundation (eds.), Human Dignity, Acts of a Christian-Muslim Colloquium, Vatican City 1999, p. 141–146.
179 General Report, The Pontifical Council for Interreligious Dialogue / The Royal Academy for Islamic Civilization Research Al Albait Foundation (eds.), Human Dignity, Acts of a Christian-Muslim Colloquium, Vatican City 1999, p. 153–155; for the criticism of the methods cf. El-Assad, Nassir El-Din, The Closing Address, The Pontifical Council for Interreligious Dialogue / The Royal Academy for Islamic Civilization Research Al Albait Foundation (eds.), Human Dignity, Acts of a Christian-Muslim Colloquium, Vatican City 1999, p.150 f, who was particularly critical when the authority over interpretation of the religion in question, namely of the partner in dialogue, was violated, but also positively underlined the importance e. g. of the dialogue on the rights of children which had filled in a genuine gap of knowledge on the Muslim side.
180 Mokh, François Abou, Voyage d'une délégation du Saint-Siège en Iran, BSNC 33 (1976), p. 319–326. During this journey, decisive contacts were also established for a visit from 15–17 September 1976 to another predominantly Shi'ite country, the Yemen. The reception there was extremely friendly; the government even made a military aeroplane available, but for the time being the concern about a dialogue was not taken up, according to Mokh, François Abou, Visite de Son Éminence le Cardinal Pignedoli au Yémen, BSNC 33 (1976), p. 327–330. And when one looks through the list of non-Christian visitors during the Holy Year, it can be noted that Muslims were weakly represented by very select persons (Turkey and Indonesia) and that they only approached the Secretariat and did not seek a papal audience; for example, cf. List of the Non-Christian Visitors During the Holy Year, BSNC 31 (1976), p. 78–79.
181 cf. Rossano, Pietro, Report of the Conversation between the Iranian Delegation and that of the Secretariat for Non-Christians, BSNC 37 (1978), p. 25–28.
182 cf. Michel, Thomas, IRAN *In Tehran, the Muslim-Christian Colloquium on "A Theological Evaluation of Modernity" (30 October – 2 November)*, Islamochristiana 21 (1995), p.172, idem., **Teheran – Iran:** Christian-Muslim Colloquium on "A Theological Evaluation of Modernity" (30 October – 2 November 1994), Pro Dialogo 87 (1994), p.275 f (the first part of these reports is identical while the report in Pro Dialogo adds a section "Meetings and visits", p.275 f), Fitzgerald, Report of the Activities of the PCID: November 1992 – November 1995, p.170; on the preparation, idem., P.C.I.D. Dialogue with Muslims since the Last Plenary, p.43. At that time a general preparatory meeting had been planned for November 1992. There seemed to be a general interest among the Shi'ites to establish contact with the PCID, especially on questions of the Near East. When Cardinal Arinze had visited Syria, the spiritual head of the Lebanese Hizbollah had travelled per-

sonally to Damascus to meet the cardinal and underline this. And that confirms the first hypothesis which Michel put forward at the end of his report (p.44). "In its relations with Muslims, the PCID works on the principle that dialogue should also be carried out within situations of ambiguity and even conflict. Dialogue cannot wait until good relations with Muslims are established throughout the world." Although a little later he reflected openly on what the issue really was about Islamic fundamentalism: was it exaggerated by the media, was it a passing phenomenon in Muslim history or an expression of "true" Islam? But contacts with Iran, in particular, had already existed considerably earlier according to Cuoq, J[oseph M.], Iran: (sic!), BSNC 14 (1970), p.103 f, who reported on his travels and on contacts with religious and university authorities interested. What he described as decisive was a friendly and positive attitude to Islam and the Muslims which would create an atmosphere of brotherliness and cooperation. In addition, cf. Gardet, L[ouis], A Few Notes on My Journey, BSNC 15 (1970), p. 175–178, who also gives an impressive description of the Christian ecumenical dimension of his encounters.

183 Machado, Felix, A Theological Evaluation of Modernity: An Asian Point of View, Pro Dialogo 89 (1995), p.181, cf. p. 176–180.

184 ibid., p.182, cf. p.183.

185 cf. Akasheh, Bureau pour l'Islam Rapport d'activités: novembre 1998-octobre 2001, p.107 f, and idem., **Rome:** Islam and Christianity Confronted with Religious Plurality, 7–9 February 2000, Pro Dialogo 104/105 (2000), p. 228–229.

186 cf. Akasheh, Bureau pour l'Islam Rapport d'activités: novembre 1998-octobre 2001, p.108, and idem.**, Tehran, Iran:** The Third Colloquium, 18–20 September 2001, Pro Dialogo 108 (2001), p. 406–407.

187 Akasheh, Report on the Activities of the PCID: Relations with Muslims, p. 214, cf. p.213, and idem., Bureau pour l'Islam Rapport d'activités: novembre 1998-octobre 2001, p.105; also Press Statement of the 1st Meeting (22 June 1995), Islamochristiana 21 (1995), p.178; on the pre-history, Fitzgerald, Report of the Activities of the PCID: November 1992 – November 1995, p.171, where at another point (p.180) he described the effects of interreligious relations on ecumenical relations as follows: "It can surely be said that relations with people of other religions is a spur to ecumenism, both in so far as it brings out the urgent need for overcoming the divisions among Christians, and also as it leads us to see in a clearer light the really essential elements of our faith." On the other meetings, Press Statement of the 2nd Meeting (28 (sic!) June 1995), Islamochristiana 21 (1995), p.178, EGYPT *The 2nd Meeting of the Islamo-Catholic Liaison Committee (30 May 1996)*, Islamochristiana 22 (1996), p.209, MOROCCO *The Third Meeting of the Islamo-Catholic Liaison Committee (Rabat, 18–19 June 1997)*, Islamochristiana 23 (1997), p. 223–224.

188 **Cairo-Egypt:** Annual Meeting of the Islamic-Catholic Liaison Committee, 17–18 July 1998, Pro Dialogo 99 (1998), p.342.

189 Akasheh, Report on the Activities of the PCID: Relations with Muslims, p.214.

190 cf. **Paris, France:** Islamic-Catholic Liaison Committee, 1–3 July 1999, Pro Dialogo 102 (1999), p.345 f, and Akasheh, Bureau pour l'Islam Rapport d'activités: novembre 1998-octobre 2001, p.106.

191 cf. Akasheh, Bureau pour l'Islam Rapport d'activités: novembre 1998-octobre 2001, p.105 f; Michael Fitzgerald personally spoke more openly at a meeting called in reaction to September 11th by the International Islamic Forum for Dialogue together with the World Muslim Congress: "To condemn and to show sympathy is however not enough. There is a need to analyse the causes and the consequences of the events of 11 September. If this can be done together, the better it will be. Hence the importance of our meeting these days." It was a question, as the agenda showed, of the need to protect human life, the definition of terrorism compared with legitimate

self-defence, the necessity of justice as a foundation for peace, the role of the media in influencing society, and of the possibilities for cooperation in the humanitarian sphere which was obviously very important to him. Here he was naturally concerned to protect religion from manipulation, but also about education for mutual respect and, above all, joint service to humankind: conflict prevention, joint action to limit the purchase of weapons and the development and use of chemical and biological weapons, and possibly in programmes of aid for refugees, especially in reconstruction in war zones and even more in strenghtening civil society, because the rebuilding of trust in one another was the most important prerequisite for everything else (Fitzgerald, Michael L[ouis], **Cairo-Egypt:** Cairo Encounter, 28–29 October 2001, Pro Dialogo 109 (2002), p.156, cf. p.157). For the regular meeting in July, see Al-Sharif, Kamel / Arinze, Francis, Religion and the Dialogue of Civilizations in an Era of Globalization (Islamic-Catholic Liaison Committee), Pro Dialogo 108 (2001), p.325 f.

192 cf. Fitzgerald, Michael L[ouis] / Al-Rifaie, Hamid A., Islamic-Catholic Liaison Committee: Joint Declaration on the Present Situation in the Holy Land (21st April 2002), Pro Dialogo 110 (2002), p.192.

193 cf. Fitzgerald, Michael L[ouis] / Al-Rifaie, Hamid bin Ahmad, Vatican City: 20 January 2004 **Joint Declaration of the Islamic-Catholic Liaison Committee,** Pro Dialogo 115 (2004), p. 37–38.

194 cf. Fitzgerald, The Secretariat for Non-Christians Is Ten Years Old, p.90 and, in even greater detail, idem., Twenty-five Years of Dialogue, p.111+113+119; also Visit of Cardinal Arinze to Egypt (3–10 February 1989), Islamochristiana 15 (1989), p.190. A contemporary report of that first meeting is Lanfry, J., **Rome-Vatican:** Muslim Delegation at the Secretariat, BSNC 16 (1971), p. 41–44, with the Christian reflection on themselves (p.44): "when we do act, let us remain men of faith. We must not be believers only when we pray, and mere technicians and administrators when it comes to doing something." The return visit has been described in detail in Visite du Secrétariat pour les Non-Chrétiens au Conseil Suprême pour les Affaires Islamiques, BSNC 28/29 (1975), p. 174–180. The joint statement can be found under Rencontre islamo-chrétienne du Caire: 12–13 avril 1978, BSNC 38 (1978), p. 157–160.

195 cf. Akasheh, Report on the Activities of the PCID: Relations with Muslims, p.214 f, and **Cité du Vatican:** Accord entre le Conseil Pontifical pour le Dialogue Interreligieux et le Comité Permanent d'Al-Azhar al-Sharif pour le Dialogue avec les Religions Monothéistes, 28 mai 1998, Pro Dialogo 99 (1998), p. 338–340.

196 cf. Déclaration commune d'Al-Azhar et du Conseil pontifical pour le Dialogue interreligieux sur la situation au Moyen-Orient, Pro Dialogo 107 (2001), p.193, and Akasheh, Bureau pour l'Islam Rapport d'activités: novembre 1998-octobre 2001, p.106 f; cf. also Arinze, Francis, **Visit to Jerusalem:** 26 Feb.–3 March 2001, Pro Dialogo 107 (2001), p. 264–266, which, like the dialogue conferences in Jerusalem (cf. below), demonstrates the extensive commitment of the Pontifical Council for Interreligious Dialogue to the Holy Land, but also shows how necessary and difficult this is when, for example, it is stated openly here that the proportion of Christians in the population has sunk to a mere 1.5 per cent; this should make it clear to everyone that the number of Christians in the country is no longer in any proportion to the significance of the Christian sites in the Holy Land and hence to the "Christian feeling" of the country. That this commitment was not without problems, and that precisely the dialogue with the Muslims was in danger from the very beginning of being exploited/used by the other side for political ends, is demonstrated by the commentary in the Saudi newspaper *Oukaz* of 23 April 1974 on the occasion of the visit of Cardinal Sergio Pignedoli to that kingdom: "Il n'y a pas de doute que l'établissement d'un dia- /p.204 logue entre les musulmans et les chrétiens renforcera les requêtes islamiques dans les rencontres internationales et aboutira à une prise de position commune dans les discussions sur les lieux saints de Jérusalem" (quoted from Rencontres avec les non-chrétiens, Proche-Orient Chrétien 2 (1974), p.203/204).

197 cf. Akasheh, Bureau pour l'Islam Rapport d'activités: novembre 1998-octobre 2001, p.106 f; Declaration of the Joint Committee of the Permanent Committee of Al-Azhar for Dialogue with the Monotheistic Religions and the Pontifical Council for Interreligious Dialogue, Pro Dialogo 108 (2001), p.332, Al-Rifaie, Hamid Ahmad / Fitzgerald, Michael L[ouis], **Declaration of the Joint Islamic-Catholic Liaison Committee** (12 settembre 2001), Pro Dialogo 108 (2001), p.333; Fitzgerald/Borelli, p.93: "It is therefore a young body, one which has still to find its way."

198 cf. Sheikh Al-Azhar Mohammed Tantawi (Islam), Pro Dialogo 109 (2002), p.148 f; cf. also DAY OF PRAYER FOR PEACE IN ASSISI (24th January 2002), Pro Dialogo 109 (2002), p.135.

199 **Rome – Italy:** Annual Report of the Committee for Dialogue of the Pontifical Council for Interreligious Dialogue and the Permanent Committee of al-Azhar for Dialogue with Monotheistic Religions, Pro Dialogo 110 (2002), p.228, cf. also p.227; as far as the date is concerned which had been specified to the exact day, cf. Akasheh, Considerations on Forty Years of Religious Dialogue with Muslims (A Report), p.198.

200 Al-Zafzaf, Fawzi / Fitzgerald, Michael L[ouis], Vatican City, 24–25 February 2004 **Joint Committee of the Permanent Committee of al-Azhar for Dialogue with Monotheistic Religions and the Pontifical Council for Interreligious Dialogue,** Pro Dialogo 118 (2005), p.52 f.

201 cf. Machado, Felix, Harmony among Believers of the Living Faiths: Christians and Muslims in Southeast Asia, Pro Dialogo 87 (1994), p.216; on regional dialogues in general, cf. Arinze, Francis, Forty Years of Developing Theological and Dialogical Engagement, Pro Dialogo 116/117 (2004), p.285 f.

202 cf. Jean Paul II reçoit les participants de la rencontre d'Assise, Islamochristiana 15 (1989), p.245, and Arinze, Greetings to the Participants, p.6; also especially Arinze, Francis, Believers Walking and Working Together, Bulletin Pontificium Consilium pro Dialogo inter Religiones 70 (1989), p. 43–45 (Opening and Welcome Address of 25 October 1988). The most comprehensive report is that of Peteiro, Antonio, Chrétiens et Musulmans constructeurs de la Paix, Pontificium Consilium pro Dialogo inter Religiones 70 (1989), p. 37–42 – in fact really for the Spanish bishops' conference. In the official report Rapport de Synthèse, Bulletin Pontificium Consilium pro Dialogo inter Religiones 70 (1989), p.42, the decisive passage reads as follows: "Les Écritures Saintes insistent sur l'importance du respect de l'Autre (...). Nous devons donc approfondir l'esprit des Livres Sacrés et reconnaître le droit de l'Autre à la différence religieuse, au nom de la liberté des consciences, afin de vivre en paix et en bonne entente." Individual Muslims went considerably further in their statements, e. g. Rahal, M. Redouane, Réponse d'un participant musulman, Bulletin Pontificium Consilium pro Dialogo inter Religiones 70 (1989), p.46: "Notre engagement découle de la foi en l'homme. (...) Nous croyons au même Dieu qui a fait l'Homme à son image c'est-à-dire plein de mansuétude et de concorde. (...) Ces Écritures en tant que parole divine ne visent-elles pas la promotion de l'homme qui est notre préoccupation majeure? (...) Que Dieu nous inspire pour travailler dans la sérénité et aboutir ainsi à des conclusions qui seront des jalons dans l'édification d'une communauté humaine tolérante où les valeurs morales, Essence de nos Écritures, auront la première place." In addition, note should be taken of the article which accompanies these reports in the Bulletin, Michel, Thomas, Christian and Muslim Minorities: Possibilities for Dialogue, Bulletin Pontificium Consilium pro Dialogo inter Religiones 70 (1989), p. 47–58. The final evaluations (p. 56–58) indicate a good and up to date analysis and lead to this conclusion (p. 58): "The alternatives to dialogue – political pressures, power strategies, armed struggle – all simply defer the problem to a later age. It is only when Christians

and Muslims agree to live together in a relationship of mutual trust and respect for the persons and rights of the other that the situation of minorities can be hoped to improve. This slow and painful process is the path to which the Holy See, the World Council of Churches, and many Muslim individuals and organizations are committed." As far as prayer was concerned, some sentences can be quoted here from another context (Shirieda, J[ohn], Editorial, Bulletin Pontificium Consilium pro Dialogo inter Religiones 80 (1992), p.129): "On the occasion of the recent hospitalization of Pope John Paul II to undergo serious surgery we received many messages, letters and telegrams, from religious leaders round the world. They expressed their concern and assured the Holy Father of their prayers for his prompt and complete recovery. (...) It is a great comfort and joy to know that those who believe in God, the Giver of life and health, are united in prayer in times of suffering and difficulty. Our brothers and sisters of other religions can be sure of the Holy Father's prayers for them." Another good example is mentioned in **Vatican City:** Visit to the Vatican of Sh. Ahmed bin Mohamed Zabara, Grand Mufti of the Yemen (28 April – 5 May 1993), Bulletin Pontificium Consilium pro Dialogo inter Religiones 83 (1993), p.199: Father Thomas Michel, the Secretariat's officer for Islam, accompanied the guest to the Friday prayers in the mosque in Rome.

203 cf. Report of the Christian-Muslim Dialogue Meeting in Ibadan (4–8 August 1991), Islamochristiana 17 (1991), p. 249–252.

204 Arinze, Francis, In Openness and Collaboration: Opening Address to Christian-Muslim Seminar from ASEAN Countries Gathered at Pattaya, Thailand, 2 August 1994, Pro Dialogo 87 (1994), p.228, cf. p. 224–227+229 f.

205 cf. Capalla, Fernando R., The Role of Religious Groups in Modern Pluralistic Societies – Ideals and Realities; Traditions of Harmony in Southeast Asia as a Basis for Christian-Muslim Dialogue, Pro Dialogo 87 (1994), p. 232–239; reference can also be made to Machado, Felix A., A Summary of Reports by the Members: Interreligious Dialogue Promoted by the Church, Pro Dialogo 109 (2002), p.126, who points to the forum between bishops and ulama in Mindanao which had worked so convincingly, effectively and successfully that it was now officially related to the Inter-Religious Federation for World Peace.

206 cf. Christen und Muslime in Südasien, Schlussbericht eines regionalen Seminars, Weltkirche 8 (1994), p. 242–244; in English, Christian-Muslim Seminar on "Harmony among Believers of Living Faiths: Christians and Muslims in Southeast Asia" (1–5 August 1994), Islamochristiana 20 (1994), p. 271–273, also as Final Report, Pro Dialogo 87 (1994), p. 219–223. On the dialogue situation in Asia, Number 71 of the Pontifical Council's Bulletin is interesting and also contains some reports on Christian-Muslim dialogues, for example in Pakistan, Bangladesh, India, Malaysia, Indonesia. and the Philippines.

207 cf. Akasheh, Considerations on Forty Years of Religious Dialogue with Muslims (A Report), p. 202–204; as far as the estimate of terrorism is concerned, in the background there is probably the report by the Latin Patriarch of Jerusalem, Michel Sabbah, which he had given to the plenary of the Pontifical Council from 14.–19 May 2004 and which very strongly underlined the connection with oppression and poverty which could turn a religion into aggressive violence and war, something which governments should also pragmatically take into account; see Akasheh, Some Reflections on Islam and Christian-Muslim Relations in Certain Countries, p.252f. Another report which he includes here (p.256f) could also have played a part, namely the report of Archbishop Jan Pawel Lenga from Kazakhstan, which began by describing how their common situation of persecution under communism had resulted in a kind of tacit solidarity. And the visit of Pope John Paul II from 22.–25 September 2001 had been received as a witness of genuine spirituality and holiness which overcame religious barriers. But the new situation of freedom also had its dark sides. The view could be found especially among the religious leaders that belonging to an ethnic group should also determine the religion to which one belonged, and this implied that anyone who considered changing was denied freedom of conscience and religion. The description in the same report of the situation in Algeria (p.253f) was problematic in a similar but different way where Islam had become an important political factor in the new state through the war of independence. Officially, the constitution recognised the right to freedom of conscience but in practice this was restricted to freedom of worship for foreigners. So the dialogue was also mainly conducted by foreigners, especially by Christian students from Africa south of the Sahara who, with their professors, had opened up the way for dialogue for the local priests and bishops. But, in addition to social and contemporary questions, a spiritual dialogue was especially important and had gained a sad reputation in 1996 with the murder of the monks of Tibhérine. It becomes clearer here than elsewhere that the Christian church felt itself swept along by Islam from the disadvantages of Islamic family law in mixed marriages to the 19 deaths in their ranks. Although the situation appeared safe at the moment, it was viewed very critically among Algerians who were definitely critical of Islam and a new Islamisation evident, for example, in the veiling of girls. The situation in Bosnia-Herzegovina was described as especially problematic (p. 257–259). Following a history of occupation and Islamisation from 1463–1878, after the collapse of the Yugoslav totalitarian regime a violent conflict took place from October 1991 to December 1995 which destroyed Christian-Muslim relationships. Ethnic demarcation lines (cf. Kazakhstan above) became religious ones; and Muslim guerilla fighters came from outside, some of whom remained through marriage, and a small number still formed a terrorist element in the society today. Financial assistance came from the oil states, especially Saudi Arabia, for the organsation of the Muslim community. But this Islamic reawakening with foreign aid had negative effects on Christian-Muslim relations in the form of anti-Catholic books and statements by fanatical groups. The World Conference of Religions for Peace had contributed to a law on religious liberty and the legal status of religious communities that was adopted in 2003. Pope John Paul II had deliberately visited Sarajevo from 12.–13 April 1997, but the general judgement of the top religious leaders locally, who had good relations in spite of everything, was that dialogue did exist despite the challenges and because of the past and the present.

208 Merchergui, p.304/305.

IV Evaluation

1 The statements of the magisterium

There are an infinite number of statements concerning human beings in everything related in the broadest sense to the Christian-Muslim dialogue of the Vatican from its beginnings at the Second Vatican Council up to the death of Pope John Paul II. Here, I only wish to recapitulate the most important lines. Naturally, it is clear, both in time and in content, that a hierarchy exists, that developments were set in motion by the Second Vatican Council and were continued especially by Paul VI and John Paul II at the highest level, but also at the next level down in a more executive form by the Secretariat for Non-Christians or, as it was later called, the Pontifical Council for Interreligious Dialogue. Only when the preparatory theological work had progressed to a certain stage did dialogues actually take place, and these dialogues endeavoured, for their part, to stick to the lines laid down up to that moment. But it is a peculiarity of dialogue that two partners come to a dialogue with different interests and no party can simply determine everything in advance. Therefore the biggest changes in the original agenda came about with the move from the preparations to the actual dialogues.

1.1 The Second Vatican Council and its time

To start with the basic intentions, as far as the Second Vatican Council was concerned, the main statements on followers of other religions are undoubtedly found in *Nostra Aetate* and they present a clear theological anthropology. All nations and hence all people have God as their common origin and common goal and therefore form one single community of which God himself is the original source. According to statements in *Ad Gentes* and *Gaudium et Spes*, Christ is the model for a renewed humanity and thus the Church is the connecting link between different human communities and nations, and cooperation between all people is an explicit imperative. It is also worth mentioning the official recognition of religious liberty in *Dignitatis Humanae*. The emphasis there is on people's conscience and their duty to seek the truth, as well as on people's social nature. A refusal of religious liberty constituted a violation of the human person and of the divine order for human beings, and religious liberty had to relate explicitly to the practice of religion in community. At that particular time, these were very new and clear statements however much they may have become generally accepted in the meanwhile. The period of the Second Vatican Council was also the time when the Secretariat for Non-Christians was established by Pope Paul VI with the double task of studies to make progress on the issue, and also of maintaining real contacts with non-Christians, while special value was attached to the latter. During the Council, the encyclical *Ecclesiam Suam* was also issued which attempted to provide a theological foundation for the new term "dialogue". Through Jesus Christ in the Holy Spirit, God offered the Church a dialogical relationship, indeed, the whole of God's revelation could be understood as dialogue. Therefore the Church should also establish such a dialogical relationship with humanity. In more anthropological terms, dialogue was the best form of relationship with people in the modern world.

1.2 The official statements by the magisterium

The new catechism appeared already during the pontificate of John Paul II. It also emphasised that all people had the same origin and the same goal which was why they had a yearning for God in their hearts that was expressed in prayer but also in sacrifice, worship and meditation. So it was quite right to describe humans as religious beings. Here, the Church explicitly defended the ability of human reason to recognise God, and saw that also as a basis for dialogue with followers of other religions. A possibility for the salvation of non-Christians, if it was not their own fault that they did not know Christ and the Church, was seen in their obeying their consciences nevertheless. Freedom of conscience and religion were also mentioned as an aspect of human dignity. These statements were completely in accordance with earlier encyclicals of John Paul II. Apart from that, as his encyclicals show, his understanding of dialogue was certainly broad. Interreligious dialogue had always to be concerned with moral values, social justice, peace and freedom, as well as the protection of life and the family. The mutuality or problem of religious liberty in the Christian-Muslim context was also mentioned frequently. But, for John Paul II, interreligious dialogue was not merely a strategy for the peaceful coexistence of the nations. He saw the Church, in a much more profound sense, as the sacrament of unity for the whole of humankind, and prayer and contemplation in interreligious dialogue were especially important to him – hence, also, the interreligious encounter for prayer in Assisi as early as 1986. Both aspects are also clearly evident in the encyclical *Vita Consecrata* in which he particularly commended interreligious dialogue to the religious orders across the whole span from spirituality to a common commitment to justice, peace and the integrity of creation.

1.3 The magisterium of Paul VI

Naturally, there is a large number of statements of lesser importance from the magisterium. In the addresses of Paul VI for the anniversaries of the founding of the Secretariat for Non-Christians, for example, there is the statement that every person participates in the unfathomable mystery of God because, as God's creatures, they are images of God and represent the humanity of Christ. The meeting point with Christ's humanity was precisely the search for being God's image which had left traces in all people and also provided the urge for the unity of all. After the Council, Paul VI even spoke of the religious genius of humankind. In addition, there was naturally a lot of praise for the work done thus far as a visible and institutional sign of the dialogue with non-Christians, recognisable in the new climate between the Catholic Church and other religions, and the appreciation and interest at the highest possible level provided encouragement. On the occasion of the transfer of the Institute of the African Missionaries for Arab Studies to Rome in the form of the Pontificio Istituto di Studi Arabi e d'Islamistica, Paul VI described cultural differences as divine providence which should serve to supplement and enrich. The more such addresses were directed to broader Church circles, the more appealing but sometimes also almost apologetic was the tone adopted by Paul VI. Other people also had moral and religious values and it was important to awaken this sediment of goodness. But this should definitely not lead one to any false conclusions. The unity of humankind corresponded to the unity of the truth, and an objective religion corresponded to people's subjective religiosity. In the presence of or when speaking to Muslims, however, such statements were made in a weaker form, if at all. Many of the declarations by Paul VI were adapted to the specific, practical situation, and there was really only one permanent theme: freedom of religion or, more precisely, freedom of proclamation for minorities, which was usually recognised in theory but not implemented in practice. The pontificate of John Paul I was simply too short to merit a mention here.

1.4 The magisterium of John Paul II

Among the many speeches of John Paul II, special emphasis must be given to his address to Muslim youth in Casablanca on 19 August 1985 which has acquired a leading role in theological history and ranks within the Catholic Church *de facto* on a level with *Nostra Aetate*. Right from the start, John Paul II made clear that he wanted to witness to his faith in God and to the human values which originated with God and were important especially for young people facing decisions about life. His conceptual approach and his wording attempted to communicate with young people and with Muslims. He stated that faith witnessed through its respect for other religious convictions because every person expected to be respected for what they were and what they believed according to their conscience. Then he tried to link that to the Christian understanding of religious liberty. It was a question of voluntary tribute to reason and the heart which constituted the dignity of human beings and which respected God and humans simultaneously. Nor could we call upon God if we did not behave in a fraternal way to people who were all in God's image. The reasons for respect, love and help for every person were based on their being God's creatures and, in a certain sense, God's images and representatives. This was a juxtaposition of the Christian and the Muslim conceptions of human beings which were considered from the start to be open to interpretation which made this statement less pointed. The dignity and freedom of human beings were present throughout the address and young people were seen as allies in the struggle for peace and a more just world. There was reference to the mutual recognition of religious values but also to it being possible to accept theological differences with humility and respect in mutual tolerance. He, John Paul II, believed that God was inviting us to change our old habits, to respect one another and to stimulate one another to do good works along God's path, and he was also convinced that, along this path, a world could come into being in which men and women with a living faith would try to build up a human community corresponding to God's will. Here, it is striking how central elements of Catholic doctrine were linked with Muslim theological thinking, and how they were not just doctrine but personal conviction and witness to faith. What he said was also meant and being lived out at that moment, and, behind it, was the highest representative of the Catholic Church who, at the same time, was an ordinary believer. One can sense this even when reading his words many years later and it constitutes the strength of this address which gave it a value far beyond its official standing.

John Paul II gave a similar but much less famous address in 1990 at Bamako in front of 1,500 Christian and Muslim young people. Whether people were described as God's images or representatives, they were in any case a sign that pointed to God. Human rights were also an expression of God's will. But humans were also marked by dependency on God and this gave meaning to their lives. Christians and Muslims had different motives and means for putting this ideal into practice. In any case, it contrasted with ideologies and ephemeral happiness. This was again a fundamentally anthropological speech and it was again addressed to young people, probably deliberately because young people, in particular, needed guidance, and because one had to start with them if one wanted to create a new world.

Speaking to the participants at a Christian-Muslim meeting on religious education in modern society, John Paul II then also made clear that this was one of the issues he wanted discussed in the dialogue. Young people had to rediscover the social nature of human life or, more precisely, the inalienable rights and duties of human beings. He was concerned about a better transmission of religious values, especially respect for others and openness towards all God's children. He was also critical of modern affluence which could result in a refusal to recognise God as creator and law giver. But he did not totally reject progress in view of considerations related to the theology of creation (God's mandate for humankind).

John Paul II's remarks about the Secretariat for Non-Christians followed on from those of Paul VI. The Secretariat was an expression of the will of the Church to establish contact with the many who were looking to non-Christian religions for something that would give their lives meaning and direction. The aim stated for dialogue, as well as for all believers and all religions, was for every person to be able to make their authentic faith a reality and to reach their transcendent goal. When the representatives of the World Council of Churches were also present, he emphasised that the vast number of ordinary believers needed to understand followers of other faiths and accept them as brothers and sisters. He also emphasised prayer as the best means for uniting humankind. In general, this pope set great store by spiritual interchange in order to give dialogue quality and depth and to guard it against the danger of mere activism. A genuine spirituality of dialogue also had to expect misunderstandings and prejudices and that an outstretched hand could be turned down. The encyclical *Pastor Bonus* changed the name of the Secretariat to Pontifical Council for Interreligious Dialogue but did not really change its task. It can also be noted that the priorities directly set by the Pope, namely studies and encounter, were followed and the subjects he wanted really did come up as issues in the dialogues. The problem of peace was a constant concern. The fundamental basis of dialogue was mutual respect which would lead to religious liberty. Speaking to a specialised audience in the broadest sense and in view of current cases, the Pope also discussed problems such as Muslim fundamentalism. In order to resolve that problem, it was necessary also to tackle its root causes because, in many cases, religion was exploited to detract from social or economic difficulties. In addition, the Pope distinguished between the phenomenon of fundamentalism, its effects and the persons who perpetrated it. The attempt to force upon other people what we considered to be the truth was not only an offence against human dignity but ultimately an offence against God himself whose image was in these people. Therefore fundamentalism was basically an attitude diametrically opposed to belief in God. So terrorism did not only exploit people but, in fact, God as well, because terrorism made God into an idol which could be misused to satisfy one's own needs.

As far as the Near East was concerned, naturally the peace issue combined with the question of human rights and especially of religious liberty were the main focus. It becomes clear that the crisis situations were known to John Paul II in detail, and also that they were often presented in a simplified or biased way, whereas the Pope endeavoured to have a balanced approach. At these points, too, it is clear that John Paul II was less concerned about elaborate theological reasoning, although it was present in the background, and more about bringing a few fundamental things to bear in a rather tangled political situation with all his institutional and personal weight.

Religious liberty generally in states with a Muslim majority remained a major issue in the Pope's addresses and certainly a feature of his argumentation. Human rights were important in order to build up just, stable societies which could meet the people's wishes for dignity and freedom. Speaking to the Catholic congregation in Ankara, the Pope emphasised in the traditional way the Church's appreciation of the religious values of Muslims. A new age had begun for Christians and Muslims in the sense that the spiritual bonds which linked the two had to be developed for the sake of moral values, social justice, peace and freedom. The belief in creation constituted a firm foundation for mutual comprehension and peaceful dialogue, for dignity, brotherliness and people's freedom because, as God's creatures, human beings had inviolable rights, but they were also bound by the law of good and evil which had its origin in the divine order. These universal, unchangeable norms, he added in another speech, were written on the human heart and thus formed the real meeting point between persons of different religious traditions – a more practical expression of being in God image as the meeting point for all people.

In Europe, on the other hand, John Paul II encouraged people to dialogue with Muslims but also called for human rights as protective rights for foreigners. In his view, the attitude of Muslims to God expressed the existential state of every person before God. He even spoke of a joint pilgrimage to eternity, visible in prayer, fasting and charity, as well as common efforts for peace and justice, for human progress and environmental protection. However, he also spoke of refraining from any kind of violence.

An address, which was an equally good example and which pursued many of the aspects already mentioned, was his speech in 1995 on the occasion of the 50th anniversary of the United Nations. Fundamental similarities existed between the nations because cultures were really only different ways of expressing the transcendent dimension of human life. For John Paul II, this implied respect for cultures and thus logically also respect for religious liberty on which, in turn, all human rights and thus every really free human society rested. To abolish the differences would rob human life of the depth of its mystery. Nevertheless, he still maintained that there was an unchangeable truth about human beings, but this truth was so complex that each culture could teach something about one dimension or another of this truth by means of respectful dialogue. Unfortunately, he did not enter into practical details. The address of 1999 for the 50th anniversary of the recognition of human rights is also of similar interest. There, human dignity was presented as a transcendent value which related back to creation. Persons' rights were not just granted to them by the Declaration of Human Rights; it only proclaimed them (and that is exactly how it understands itself). According to John Paul II, all human rights were endangered if one of them was violated. He recognised a special threat to the rights to life and to religious liberty. Where it was a question of threats to religious liberty, he was convinced that faith and violence had nothing to do with one another. Hints of divergences from Islam were very evident here when John Paul II demanded the right to change one's religion, to the public exercise of religion and to the abolition of discrimination against religious minorities.

These points were not always developed so extensively; often there was simply reference to the religious or spiritual dimension of human beings or also, on that basis, to the unimpeachable dignity and hence inalienable rights of people irrespective of practical benefit, and to human solidarity transcending all religious borders. Time and again, the predominantly materialist approach of modern people was cited as a common negative image. But it is interesting that the anthropology of a relatively early speech (1981), which otherwise referred explicitly to Paul VI, started not from creation, in the usual way, but from Christology. Every individual and every nation had become children of God through the cross and resurrection of Christ and thus shared in the divine nature and were heirs to eternal life. In Christ, every person received the full measure of their dignity and their final purpose. The need for prayer was mentioned as particularly linking the religions because human spirituality was directed to an absolute – an idea which recurred frequently.

In general, it can be said that Muslims occupied a special place in the papal addresses compared with Hindus and Buddhists, but that did not make any major difference on the anthropological level. However, the difference became all the more evident when it was a question of religious liberty. Then his argument repeatedly and emphatically used the image of a reasonable, free and hence responsible person, and stated that God could surely not love any religious observance which was imposed from outside. On the occasion of the renewal of a cooperation agreement between the University of Ankara and the Gregorian University, John Paul II referred to freedom of conscience and religion as the fundamental condition for this cooperation. With regard to the equality of men and women, however, he did not speak so clearly when addressing Muslims. The differences between men and women were not to be used to oppress the ones, nor to declare the superiority of the others. But, also on the sensitive issue of mission, respect for the inalienable dignity and freedom of human beings created and loved by God was important. He referred repeatedly to respect, knowledge and acceptance but also to insight, dialogue and cooperation. In view of people's indifference to tragedies, such as the many years of civil war in Lebanon, there was also a very strong emphasis on God as the judge of humankind. Time and again, it was a matter of peace, and not merely of outward peace but of people's inner peace which was connected with knowing where one came from, why one was on earth and to whom one would one day return.

During the encounters with Muslims on his pastoral journeys, human dignity and human rights were a main issue for John Paul II from the very beginning. Children were to be introduced to this by religious education, and dignity could certainly also be formulated in Muslim terms as belonging to the servants of God, because that was the real meaning of calling people brothers and sisters in the faith. God had set every person on earth so that they could make a pilgrimage of peace, each in his/her own situation and culture. Here, dialogue was seen as a possibility for making modern society more human, more just and more respectful. Interestingly enough, in this context prayer was also mentioned several times as helping to prepare people to meet their neighbours, and for relationships of respect, understanding, appreciation and love free of all discrimination. The dignity of human beings was open for transcendence. In practice, that could lead to greater discipline and uprightness in private and public life and to abolishing every form of discrimination. Respect and brotherliness should even be a motivation for abolishing poverty. In addition, for John Paul II, human dignity was the criterion for deciding which elements of contemporary technology and science should be integrated into society. But, on the other hand, John Paul II was also able (when addressing the president of the Sudan) to talk about humans being rational beings who were capable of solving all conflicts when there was a will to do so.

To conclude the discussion of the addresses by John Paul II, something must now be added about his statements to diplomats accredited with the Holy See. There again, we find the same theological sequence spelled out repeatedly: faith in God the Creator – the dignity of human beings as God's creatures – human rights, especially passive and active religious liberty – the effects on the level of coexistence, and common action such as justice, solidarity and peace. Religious liberty strengthened the moral integrity of a nation. The importance of natural law was also underlined. It had inspired the law of the nations and was still able to bring the unity of the human race to light. In this context, he was naturally able to deal with individual cases in a very variable way and definitely to argue very skilfully and diplomatically. It is striking that, in his very last addresses, the importance and benefit for religious liberty of the separation between the state and religion, but also for religious and state organisations and for society in general, was very strongly emphasised. That shows, in conclusion, that John Paul II made a decisive contribution at many points to furthering theological anthropology in dialogue precisely with Muslims, and was always able skilfully to adjust his statements on these issues to his specific audience and thus also to different political situations. His theological and practical commitment to interreligious dialogue in general, and especially between Christians and Muslims, was profoundly spiritual, but this did not prevent him, at the same time, as had been the case to a lesser extent with his pre-predecessor Paul VI, from being a good ambassador for human rights, particularly for religious liberty, or from being active to this end in the political realm. In addition, he was concerned to win all people and especially young people to a spiritual and hence profoundly human life transcending all religious boundaries.

1.5 Documents from the Curia

In conclusion, some ideas on this subject must be mentioned which come from three documents of the Curia. The document on the *Attitude of the Church Towards Followers of Other Religions* was published in 1988 by the Secretariat for Non-Christians. It states that the Christian doctrine on man was spelled out in reference to the biblical revelation but also in permanent confrontation with the hopes and experiences of the nations. In their origin and future, all people were equal but there were also differences, even in their moral abilities. However, the similarities were seen as so great that one could draw up common declarations, conventions and international agreements, for example on the defence of human rights. In this, the aim was to reach an agreement on legislation and the moral law by democratic means, and here it was imperative that the same law applied to everyone in a state. Respect for human beings was seen explicitly as respect for their basic rights, dignity and fundamental equality. Local churches should stand up for the rights of minorities.

Dominus Iesus explicitly underlined that parity between the partners in a dialogue, which was the prerequisite for dialogue, only applied to their equal personal dignity, and the document was also critical of how interreligious dialogue had so often given preference to the subject of creation. The overall impression of this document is not that of a clear, independent viewpoint on the basis of which one could approach others in one's own right and seek commonalities, as had been the case with the new movement since Vatican II. Now, the same arguments appeared to be disputed by the development they had undergone. The vanguard positions had become defensive bastions.

The last document worthy of mention is *Erga Migrantes Caritas Christi* which has a separate chapter on Muslim migrants and was published after decades of intensive dialogue activity. It mentions that, for both Christians and Muslims, God plays the decisive role in the life of humankind for all the reasons of creation theology already discussed. But there is also a lot of detail about the differences over the question of what human dignity and rights look like precisely and what is expected of them. Against the background of bitter experiences as well, the hope could only be expressed that, in connection with human rights, Muslims would increasingly understand that it is impossible to give up fundamental rights to freedom, inviolable personal rights, the equal dignity of men and women, democratic principles of government, a healthy laïcité of the state and an appropriate autonomy of creation. This document expresses the real disagreement between Christians and Muslims more clearly than all the other statements of the magisterium. It becomes clear here just how much was involved in practice behind the pronounced papal commitment to human rights.

2 The Secretariat's own publications

2.1 The basic starting point

The Secretariat for Non-Christians understood its establishment by Pope Paul VI as the outward sign of the interest of the Church in followers of other religious traditions. Part of its task was to awaken this interest within the Church and to pass on knowledge. In this situation, many things were still unclear despite the Council's new anthropology. It appeared that universal human brotherliness could provide a basis for dialogue, combined with the importance of natural law, the traces of a primitive revelation and the grace of God which God did not deny to people of good will. The intention was to expand this ideological basis further, building on the bible, the church fathers, the Council and naturally the papal instructions. In fact, the initial phase of the Secretariat was marked almost exclusively by preparing publications of which the more internal ones are naturally of particular interest.

2.2 Publications of the Commission for Religious Relations with Muslims

The publications of this commission, which was set up in 1974, served to work with a group of international specialists on Islam on particular issues, initially of a very disparate nature. The ways of publishing them varied but their character is more internal, in the sense of information for decision makers.

2.2.1 Harmony and conflict
2.2.1.1 Truth and violence in Islam
This study dealt in depth with a subject which has lost nothing of its relevance. The decisive issue was the spiritual inheritance which influenced the mentality of the Muslim community. As on many other issues, with regard to violence and tolerance the Quran was fairly balanced, almost neutral. But the revealed text provided a wealth of material which gave Muslims a feeling of superiority. The further one moved away from the original documents, the clearer the tendency to violence became. The intellectual tradition had then deprived Islam of its original innocence and begun to interpret it in a decidedly one-sided way. There were passages of the Quran which justified arbitrary violence against unbelievers and the majority claimed their unrestricted validity. A fundamental idea behind this was that only a powerful God could convincingly be called God. That strength then also had an effect on the Muslim community. It was the best community because it prescribed what was good and prohibited the bad, and did both in a reliable way, namely by battle. The uniqueness of God left no room for anyone else and hence not for human rights either.

In conclusion, an attempt was made to describe some examples of ways frequently chosen for dealing with violence, three of which will be mentioned here. One way was to close one's eyes to one's own violence. As a first step, an attempt was made to show that *jihad* had always been only defensive. In the next step, what should be and what is were equated. Since the Quran forbade any compulsion in faith, it simply could not exist. As a last step, one's own violence was overcome for ever by pointing to the much greater violence of the Peoples of the Book, or by questioning or simply ignoring all Muslim authors who had argued for the legitimacy of an aggressive *jihad*, although there were more than a few of them and they were not insignificant. But there was also a conscious approval of violence to bring about a new divine world order, represented, for example, by the leading thinker of the Muslim brotherhood. For him, *jihad* was total revolution against every kind of absolute sovereignty of human beings, whether governments or laws. This approach was also found in many school books. According to another approach, recourse to armed violence was an arbitrary and illegitimate war (*harb*) because it lacked divine justification. Islam alone could practice *jihad* and had even turned it into the main expression of submission to God. It was simply the implementation of the divine right to legitimate correction.

2.2.1.2 The attitude of the Franciscans
The Franciscans derived from the second Vatican Council the duty of a "godly" behaviour for Christians and the conviction that followers of other religions could give Christians a ray of truth. But, since the Council, it had been hard for this idea to find acceptance in the Church or in Islam. At the same time, it was linked with Francis himself who had visited the Sultan of Egypt in the midst of the crusades. This tradition was officially revived to mark the 800th birthday of Francis of Assisi in 1982 with the Letter from Assisi.

2.2.1.3 Change of perspective
The last contribution also drew on experiences from East Asia and put forward the hypothesis that the concern of dialogue was to understand similarities in differences, and that a genuine dialogue needed both harmony and conflict. Depending on the mentality, the emphasis was more on the infinite nature of truth or on the right grasp of the truth. So some people were more interested in an inclusive solution while others thought the truth had to be confessed clearly and one had to have the courage to defend it. It was in this context that one should also understand the development of something like a Holy War, and that applied to all three monotheist religions. Among the three major similarities, and hence good subjects for dialogue, the author listed the question of self-denial, the experience of modernism and the modern technologies which resulted in mental processes of assimilation.

2.2.2 Reference work on Islam – special case of Algeria
The reference work on Islam (1992/93) was intended as information for a group of decision makers. It was compiled by specialists who were very well informed. From the point of view of anthropology, the single volume on Algeria is interesting because it also discusses the image of human beings behind the conflicts and bloodshed there, for example that of the great Algerian philosopher, Malek Bennabi, and his efforts, in response to the social trauma of the colonial period, to create a complete human being who comprised an Islamic tradition that was true for him together with the autonomy of reason. Another new variety was rep-

resented by the modernists or reformists who appealed for a return to seeing humans as reasonable beings and were therefore particularly committed to human rights and the equal status of women. One could speak of a search for an enlightened Islam which adapted its spiritual heritage to the new historical conditions from which political and personal rights to freedom have evolved. At the heart of the debate on modernity there is the conception of the rights of the individual. The core of this conception, namely various political rights and democracy, was seen as a universal achievement. A public debate on religious liberty had existed only for a relatively short time. Here it became evident how much was involved, how human beings were understood on the Muslim side, and that there was no longer one single view; on the contrary, the different views conflicted and this sometimes led to bloodshed. The importance and difficulties of the question had already become clear when the matter of human rights was repeatedly raised as the prime issue by the popes when addressing Muslims.

2.2.3 Common prayer

The question of common prayer was naturally also of practical importance. The first Muslims apparently had no hesitation about performing their ritual prayers also in Christian places of worship (despite icons and statues) and, *vice versa*, the Christians also prayed in the mosque of the Prophet at Medina. But, nowadays, Muslims had rarely taken the initiative for common prayer. Common liturgical prayer was not considered possible because it would violate the Christian identity. But common prayer could also give dialogue initiatives a special dimension because it could focus the dialogue on God and on the search for his will. Therefore more spirituality was demanded on both sides.

2.2.4 Religion and politics

The collective volume on religion and politics, a work with very condensed information compiled over the period 1995–1999, deals with the question of authority as far as anthropology is concerned. What authority belongs to God and what authority to human beings as individuals or as a group with regard to legislation and the shaping of society and of their own lives? Do human beings have some leeway for their judgement, in which areas and under which conditions? The Muslim answers were rarely formulated in a directly anthropological way; they were political or, more precisely, juridical. But the way in which humans were understood in relation to divine authority determined what they should, or were allowed to, do in practice. What if Muslims had to adapt to the values of others in the realm of public institutions, whether these be academic, cultural or social? How did one deal with these competing claims to authority? On the theoretical level, this raised the question to whom a Muslim was responsible and whether different authorities fitted into a Muslim framework at all. Indeed, according to the Muslim understanding, authority was based on sovereignty which belonged to God alone, and on God alone being the law giver. If humans were sovereign in the form of the people, this would be diametrically opposed to that view. Surprisingly often, in this connection, the name of the Hanbalite, Ibn Taymiyyah, was mentioned who claimed that Islam was inherently theocratic. But overall there were very many different Islamic concepts of authority and so it was impossible to find any uniformity.

2.2.4.1 Internal disagreement

A contribution which was rather different from the norm dealt with the dispute between the Maulânâ Sayyid Abû'l A'la Mawdûdî, who provided the framework for the modern discussion of terms such as Islamic revolution, Islamic state and Islamic theology, and his former follower Maulânâ Wahîduddîn Khan. In this case, it was a matter of the key to understanding religion. For Mawdûdî, the key was found in the order of life. Whereas, as Wahîduddîn understood it, that implied taking a consequence for the whole. Religion was the relation between God and his servant and that was more than obedience. It was the full expression of human nature. Mawdûdî, in contrast, represented more ideology than spiritual depth. The outcome was an extremist attitude with an unwillingness to contribute actively to peaceful and just coexistence in pluralist and secular societies. But the author of the contribution also pointed out three major weaknesses in Wahîduddîn's reasoning. Muhammad, as the example for Muslims, had himself understood Surah 9:32f as a call for a war against polytheism and for the subjection of the religions of the Book. Wahîduddîn had also claimed that one had to stand up for religion, but it remained unclear whether secularism was to be understood as part of the confrontation with non-religion. The question of human rights, and of equal rights in a situation with a Muslim majority and non-Muslim minorities, was equally unclear. Violence to obtain the outward victory of good and to prevent evil seemed then to be allowed. So it was clear also in this internal dispute that Islam had a relatively concrete conception of what it meant specifically for society that God was the absolute Lord and humans were subordinate to God. Even for a Muslim minority, it is hard to spiritualise this conception, and all the more so for a Muslim majority.

2.2.4.2 Journal of Muslim Minorities' Affairs

The Journal of (the Institute of) Muslim Minorities' Affairs formed the basis for analyses of *Europe* and West Africa from an internal Muslim viewpoint. The former editor, Syed Zayn al-Abidin, was of the opinion that Muslims could live comfortably as a minority provided that certain minimal requirements were met. Therefore the journal supported a European Islam. When comparing the European countries, Germany was even mentioned as best for the survival of the Muslim diaspora. The German state was secular but not secularist, and possibly Muslims had more individual and political rights there than in Muslim countries. However, Muslims generally had the impression of never quite being welcome in Europe. Eastern Europe continued to be a special situation. Only in Poland had the situation been really good for Muslims. Otherwise, there had sometimes been massive anti-Muslim propaganda including the prohibition of circumcision. The still more exceptional case was that of the Croatian Muslims because they had adopted a number of Christian customs and practices, so they drank alcohol, ate pork and allowed their women also to marry non-Muslim men.

As far as *West Africa* was concerned, it was pointed out, firstly, that Islamisation on a large scale had only taken place during the colonial period, and this did not fit into the basic Islamic scheme according to which religious liberty was directly opposed to Islamic order. The articles examined offer interesting insights into the religious characteristics of the Muslims and their consequent reactions. A number of questions, such as the dispute over the introduction of Islamic law, really affected all the countries of the world today, because now there were Muslim minorities practically everywhere. With the modern state, the colonial rulers had introduced a model of society into the region which was totally alien to Islam. Here, there were problems of a fundamental nature so now a clear conception was needed of what Muslim life should look like today, but the sole common feature was often only an extreme aversion to everything which came from the West. If one wished to describe the Muslim minorities in West Africa in general, that was best done by saying they were minorities between emigration and struggle. So the really important question was whether pragmatism could be more than just a temporary solution. There was a typical to-ing and fro-ing between the desire for a pure, authentic, Muslim state and constant compromises with a much deeper African reality. Was it possible at all to cooperate with political institutions whose laws set themselves above Islamic law? It seemed intolerable if Muslim life were to be ruled by an administration described as Christian. This fundamental rejection of plurality also formed the background to the reasons for most of the Muslims' political demands in West Africa, often presented in the name of religious liberty and democracy. In all, they were markedly more extensive than in Europe or North America, even where there was partial overlap, and they boiled down to a society marked by Islam in the quite traditional style. This ideal image was deeply rooted and could not be reconciled with any other model. One had at least to expect a debate on these questions and that was probably also the reason why a collective volume on this subject was considered necessary.

2.2.4.3 Regional studies

The last section of the volume on politics is devoted to regional studies, this time from the viewpoint of a Christian observer, and again beginning with *Europe*. The legal situation in Europe was as varied as the legal situation of the churches in Europe. But it was evident that, with the switch from a provisional to a lasting presence, it had become important for the Muslims to ensure that they could live out their religion completely. This raised two decisive questions. Was Europe able to accept on its territory the existence of communities of a religion which refused to be private and made demands not only in the cultic but also in the social realm? And could the Muslim communities, for their part, adapt to the minority situation? The Christians required the Muslims in this situation to interpret their tradition anew in the light of the demands of modern life and with an open heart and mind.

The situation in *Turkey* was also analysed in reference to Islam and politics and two phases came to light. Initially, Islam had been relegated to private life because all sovereignty rested with the people. Things began to change in 1945 when the multi-party system was introduced in order to meet the requirements of the United Nations. The elections demonstrated that the reforms had not filtered down to the level of the ordinary population and had led to an erosion of laïcité. In its foreign policy, Turkey had officially been an Islamic state since May 1976 anyway. Now, there were also official voices in favour of Islamic law; indeed, God's laws were more important than human laws. With a Muslim population of 99.5 per cent, the Islamists were the strongest party at 21 per cent. So the adaptation of Islam to the twenty-first century remained an urgent and difficult problem in Turkey as well. The decisive point at issue was again whether people would now adapt to God's provisions, or whether they would adapt their faith and religion to their lives; in other words, how and in which Islam the realm of God and that of human beings would be delineated separately and what freedom of decision then remained for the humans.

The central questions for Islam of political power and social organisation had also been answered traditionally in the past by *Indian Muslims* as an either/or matter. Under British colonial rule, the question had arisen whether Muslims could be loyal subjects of the British crown. The answer in the majority of cases was that cooperation was possible provided that Islamic values were not violated and the Muslims' religious and cultural character was preserved. But there were also four more recent approaches: pragmatic, modernist, fundamentalist-missionary and legally constructive. Pragmatic meant being primarily a Muslim only in matters of the faith. The modernist view relied on re-interpretation where the basic problem was seen to be that Islamic law combined law and religion within itself, where the law had to adapt whereas religion was unchangeable at heart. So the question was how a valid form of consensus could be reached. The fundamentalist-missionary approach saw theocracy or theo-nomocracy as an essential prerequisite for a full Islamic life and as contradicting national sovereignty, democracy and secularism. It was a matter of a kind of theo-democracy. Muslims were only to use the different system and bring about a peaceful change of the system. Legally constructive meant in practice that traditional Islamic law would be changed significantly. But Muslims had been set apart by God himself by special right guidance and had to remain so with their own social and cultural system and Muslim personal law. To make concessions in this field would be the equivalent of rejecting Islam. With all these restrictions, the question certainly arose whether there could then still be a unity of the country in economics, education and culture. But, legally, no republican country with a non-Muslim majority would today be considered a "house of war". Secularism had to be accepted pragmatically in order to organise an extremely pluralist state, and possibly to achieve the full goals and rights of the Muslims. The results of this investigation made clear that major progress is still needed in Islamic legal and theological thinking in order for Muslims really to participate in the life of the democratic and secular Republic of India. The same applies generally to the whole relationship of Islam to democracy and secularism.

The concluding part of the volume focuses on *Africa*, beginning with the position of the Catholic Church in *Kenya* as a representative of the whole of *East Africa*. Politics had to move within the framework of morality and be directed to the wellbeing of people in general. Both were lacking in the region. The key issues of the moment were identified as human rights, justice and peace, as well as abuse of power and corruption. It is really a spectacular view – in line with the Second Vatican Council – of what Christianity and Islam could do for humankind on the basis of their common anthropological characteristics. Unfortunately, the position of the Muslims was less clear. The development in *West Africa* showed certain parallels with Turkey. There, too, there had been a secular period during which the growing political freedom (multi-party system) had led to a gradual undermining of secularism in favour of traditional Islam. With the help of the Arab Muslim countries and the international Islamic organisations, Islam had been able to make extensive use of the newly established political leeway to present itself as the solution to the crisis. The new freedom of speech allowed certain Muslim currents to demand the Islamisation of the life and structures of countries in the name of the democratic majority principle. There was a real danger of a radical questioning of laïcité. The latter was always presented as a negative value imported from the West and synonymous with atheism. But, in the preceding centuries, African Muslims had developed a great ability for living under non-Muslim governments without major problems – and their religious scholars and religious leaders had shown a corresponding ability to justify this situation. During the colonial period, Islam had in fact grown. The situation today in West Africa was more or less that African laïcité granted religious communities an important place in society. Islam and Arab culture had become factors which strengthened individual personalities, on the one hand, and, on the other, expressed opposition to foreigners and their culture.

Overall, it is clear that the situation of Islam and politics is complicated and diverse. But, at the same time, there is a clear tendency for Islam to see the overall political conditions not as completely indifferent and interchangeable but rather as connected with human subordination to God. It is hard to judge how strong this tendency is in detail, but the ideal of a separate Muslim model, expressed concretely or symbolised by Islamic law, is very influential and repeatedly in tension with democracy and secularism. It is also clearly present in Muslim minority situations. That is certainly a consequence of its being profoundly rooted in Islam's theological conception of human beings. Political and legal leeway depend on the fundamental leeway which humans have in relation to God, and there again there is a tendency to strengthen God's position over against humans and not the other way round.

2.2.5 Experiences of dialogue on the eve of the third millennium

On the eve of the third millennium, the commission raised the question how the Church had lived out the Christian-Muslim dialogue since the Second Vatican Council and what dialogue Christians and Muslims would wish to have in the third millennium. A short evaluation of this inquiry has been published. On the Christian side, there had been a wide range of different experiences. One discovery was how many human values Christians shared with Muslims. But, in the West, there was still ignorance and a superiority complex, whereas in the East there was compulsory coexistence even when no relationship was desired. The evaluation recognised a gradual progression from relationships in life to theological reflection and to questions concerning the issue of evangelisation. Young, educated Christians were the most ready to recognise the necessity of Christian-Muslim relations. Young Muslims were also considered more open. There was dialogue at all levels. Dialogue at the personal level was easy. More frequently it was an exchange on social and human issues but sometimes there was also theological dialogue. In general, dialogue efforts could also be frustrating, at least for the Christian side.

The parallels on the Muslim side are striking. Christians, too, were simply human beings. The Muslims associated theological dialogue with Europe and daily common life with Asia. The mass of Muslims was sceptical. Interreligious solidarity for a just world was seen as the motivation for dialogue. Dialogue was also a good weapon against violence and hatred. To this end, one needed to start with the education of children and to make the results of the dialogue thus far widely known. However, the results might also be compromised by an enthusiasm for conversion. There was a call for an explicit declaration by the Church that it had no hidden motives in dialogue, and for clarification of the terms "dialogue" and "proclamation".

So, on both sides, the human level was considered very important. This confirmed the direction taken by the Council in placing the emphasis on common humanity and common values. Reference had also been made to the question of a Church dialogue initiative without hidden intentions because of the suspicion that this would not be believed. The repercussions in the realm of evangelisation had not been expected. In any case, the importance of Christian-Muslim dialogue for a common peaceful future was affirmed despite all the difficulties.

2.2.6 Religious liberty
2.2.6.1 Legal bases
The next and, in this case, final publication of the Commission was *Religious Liberty: A Theme for Christian-Muslim Dialogue*. So it dealt with a subject that had always been very strongly emphasised on the papal side in relation to Muslims and did so in a comprehensive and profound way. The subject is of special interest because Christians and Muslims understand freedom differently and therefore the practical implementation differs both at the level of the individual and of society.

The first contribution to the publication discussed the development of religious liberty as the human right which had been recognised the longest (Edict of Nantes). One article of the Universal Declaration of Human Rights concerned the right to religious liberty, namely the right to change one's religion and to practise one's religion alone or in community. This had made religious liberty an international customary law for all states. It was a fundamental personal and hence indestructible and inviolable right. This meant religious liberty was an absolute, holy right and every person a *sanctum sanctorum*. Religious liberty was based on the innate dignity of human beings. Human dignity resulted from persons' reason and free will which led to personal responsibility. Religious liberty had to be enshrined in the constitution. Every nation had the duty to recognise the diversity of views about the last things. Because of people's personal responsibility, psychological freedom was also necessary. What had, in fact, given rise to disagreements since 1948 was the right to change one's religion, but the essence of freedom was precisely the possibility of change.

The overall evaluation of this and further documents emphasised that it had never been intended to produce a totally uniform practice. In 1999, the Arab human rights movement had rejected every attempt to exploit civilisational or religious peculiarities in order to dispute the universality of human rights. Since John Paul II had seen the Church as having an apostolate of human rights, the Catholic Church was committed to a binding human rights convention with supervisory mechanisms. Since this convention was still lacking, there was also still no international organisation to supervise religious liberty although that was urgently needed.

2.2.6.2 Catholic grounds for human rights

The second contribution dealt with how the official Catholic view of religious liberty had evolved historically. The concept had always been present but in changing ideas, culture and legal sources. Religious liberty did not mean that human beings could be their own highest authority or that all religions were of equal value. From the Catholic point of view, each person had the right to build up his/her own relationship with God according to his/her conscience, and to be protected from any kind of outward compulsion in the process. So, in the Catholic understanding, people were not free from the commitments which follow from religion; on the contrary, human freedom consisted of obeying one's conscience in religious matters. Freedom of religion was freedom of conscience. The conscience had absolute priority. Even if it were in error, a person had to obey it. Human dignity and freedom were respected only if there was totally free consent. Religious liberty was part of the act of faith and, for this reason, the missionary command was one of the strongest motivations for the Church to stand up for freedom of religion. But this only became official under Pius XI with his encyclical on *The Church and the German Reich* and the statement that any laws which suppressed or hindered the confession and practice of faith contravened natural law. So it was a matter of the existence of an objective moral order. The Church based religious liberty on rational principles of the personal dignity of individuals. Religious liberty was presented as an unbroken doctrine of the Church. The Gospel truth had been at work like yeast in society as far as freedom was concerned. Thus, theological anthropology was again clearly in evidence as the foundation for religious liberty, although it must also be stated that this line of argument was only spelled out in the official papal declarations in the course of the twentieth century. As the available documentation shows, that was also how it found its way into significant documents for the Christian-Muslim dialogue.

2.2.6.3 Religious liberty in modern Islamic thought

The next article was devoted to looking in the opposite direction, to a Catholic view of religious liberty in modern Islamic thought, and it began with the long and stony path in the Catholic Church towards the recognition of human rights and religious liberty, because many of the questions which the Catholic Church came up against along this path were obviously very similar to those which Islam was encountering today.

For the majority view in traditional Islamic teaching on human rights, two points were absolutely binding. It was the fundamental duty of human beings to accept Islam. And Muslims had to remain true to this faith and law at all costs because they had committed themselves for ever and also to its community, the *umma*. The latter's unity and wellbeing were closely connected with the faith and faithfulness of the Muslims. So the reason for depriving apostates of their rights was their battle against the believers, e.g. in the form of temptations which diverted them from their faith. The Islamic Charter for Germany had, however, recognised the right to change one's religion or to have no religion, because the Quran prohibited any compulsion in religious matters. Anthropological considerations had gained new significance because, as a result of Western influence, people were at the centre, something which was alien to pre-modern Islamic culture. The new state of Islamic thought was that the concepts of human dignity and of individual freedoms had greater influence. Freedom, with its concrete political and legal implications, had become increasingly attractive for Muslims.

But, up to the present, that had not succeeded in changing either the attitude or doctrine of any of the great, traditional institutions of Muslim learning, nor of any noteworthy Muslim movement. Freedom as understood by modern Europeans was a term unknown to Muslims up to 50 years ago. For them, freedom was merely the opposite of slavery. The modern Arabic expression for human dignity related to the Quran's statement that God had honoured the children of Adam. The statement that God had appointed human beings as vicarious representatives was similar to the Jewish and Christian conceptions of humans as God's image. The most important and most quoted passage of the Quran, when it was a question of a justification for human dignity, was Surah 33:72 according to which God had entrusted heaven and earth to humankind because all others had refused. The Quran's proofs all claim that human dignity was created by God himself and was not granted to humans on the basis of religion, so it can also not be lost.

But there were obstacles: the questions of slavery, of women and of corporal punishment. The Quran presupposed slavery but nowhere did it state that it was essential. However, this very distinction between social prerequisites and lastingly valid norms was not accepted by the more conservative theologians as far as the traditional discrimination of women was concerned, and it could not be applied at all to the corporal punishments laid down in the Quran. As a rule, no connection was made between the arguments for human rights and human freedom. At best, freedom was derived secondarily from the duty of human beings to make use of the reason God had given them. But there were also three different ways of linking human dignity with freedom which were connected with the role of humans as divine representatives. The first saw God's freedom as the decisive point in this representation, the second creativity and the third autonomous action. In addition, it was mentioned that human freedom was the prerequisite for witness. Enforced obedience would not only be morally worthless and unworthy of human beings but even less worthy of God.

So then how could human freedom be related to the absolute duty of human beings to obey the will of God? Was there a limit to freedom marked out by religious law, as in classical, traditional Islamic thinking? What was the relationship between rights to freedom and religious duties? Among conservative and fundamentalist theologians one can observe a priority of duties over rights. Only those who have fulfilled their duties could insist on their rights. So no freedom existed which was as fundamental as the duty to carry out God's will. For conservative Muslims, this consent to religious law had to be given at least by those who were already Muslims. There could be no rights to freedom that counteracted religious duties or acting according to the sharia. Rights to freedom were a consequence only of following the sharia and the revealed principles of social ethics. Freedom belonged to the desirable consequences of faithfulness to one's God-given duties, but was precisely not the way willed by God for carrying out religious duties. The fundamentalist position was in line with the modern ethos of freedom to the extent that the revealed norms could only be followed by means of personal, free devotion, but, in practice, it still ended up in the attitude described above. However, in the meantime, there were also quite different approaches which did not consider the confession of Islam to be automatically identical with committing oneself to the shari'a. This meant that religious law belonged to the realm of secular control of the body and not to the realm of the spiritual control of souls.

That then raised the following questions. What is the relation between the duties of the individual and those of the state? What is the purpose of the legal order and of the state at all? The ideal Islamic state was organised on the basis of an identity claimed between the people of the state and the community of faith. According to Islamist thinking, the responsibility for fulfilling God's will lay mainly with the state which had to lead individual Muslims back to the path of the true faith, if necessary by force. Since 1970, there had repeatedly been discussions about what should be created first, an Islamic society or an Islamic state. Even those in favour of the latter solution had come to a clearer recognition, in the meanwhile, of the value of freedom as a basis for religious commitment and for democratic ways to promote Islamic values. Nowadays, the conception of human beings as responsible representatives of God was mentioned in this connection. The question needed to be asked how God really wanted our social conditions to be. The historicity of human beings was a reality even among Islamic thinkers. The most important Islamic bulwark for the defence of religious liberty was seen in the Quran's statement that judgement belongs to God alone.

The final opinion on this important point was that an Islam existed which can be reconciled with human rights including religious liberty, but at present it was represented by a small, modern minority and definitely not in the countries with a Muslim majority. That, and the Islam there, would change considerably if religious liberty were really to prevail there. A realistic estimate of the possibilities was a matter of personal opinion. It was only clear that the struggle concerned a conception of theological anthropology of enormous consequence.

2.2.6.4 Religious liberty in practice

The other contributions are somewhat less fundamental and more of a practical evaluation, beginning with *Pakistan* and its founder, Muhammad Ali Jinnah. He had been an enthusiastic supporter of religious liberty and wanted everyone to be citizens with equal constitutional rights and equal responsibility. Jinnah's dream was of a democratic, secular Pakistan. He died in the autumn of 1948 and, in the spring of 1949, Islam was declared the basis of the new state. According to the constitution of 1973, Islam is the state religion, and religious liberty is subordinate to law, order and morality which means Islamic law. Under Zia ul Haq, Pakistan became a rigid, Islamic state in which non-Muslims classically became second-class citizens. Religion dominates the national assembly. Extremist religious and political parties have a strong influence and, in certain provinces, have enforced the introduction of strict Islamic laws such as the blasphemy law which has been abused to take private revenge on Christians. The overall judgement was that a weak state seeks refuge in religion.

Pakistan's education system is governed by Islamic ideology; every subject is taught in the light of Islam. Religious tolerance or concord are not found in the curriculum. The wording refers literally to an education like among the Taliban in Afghanistan. There is no education for religious liberty. This can be clearly demonstrated statistically by the mosque schools to which the masses from the lower classes flock in large numbers. The situation was worst just after September 11th. The ordinary people sympathised with the Muslims in Afghanistan against their own government. But perhaps that was also a turning point even if there is still a long way to go. The regulations in force and also the constitution are to be adapted to the UN human rights' charter and Pakistan is to sign the relevant human rights' conventions. The education system is to be revised and a place given to interreligious dialogue. But, above all, the discriminatory laws are to be revised.

The second example was the largest Christian-Muslim nation in the world, *Nigeria*, although it had simply changed its observer status in the Organisation of the Islamic Conference into membership. In Nigeria, British colonial policy had made religion into an instrument of conflict. Later, military rule made the ethnic and religious tensions still more acute. In the Muslim north, religion was the only reason for fear among the non-Muslims, but the mutual mistrust between religious representatives in general was so profound that they were unable to work together. And the whole discussion about shari'a in Nigeria was more a statement about the failure of the state and about the corruption which had brought it to its knees. It was important that non-Muslims emphasised that they did not oppose sharia because they were in favour of the weaknesses which the supporters of sharia wanted to combat. The role and place of religion in a democracy and of religious liberty in a plural state remained open questions. In Nigeria, the present situation was not very positive either, but hope for the future was seen in democracy and dialogue.

The last example was *France*. With two million French citizens, Islam was the second largest religion in France. Only between five and ten percent of the Muslims in France were practising Muslims. The strengths and weaknesses of adherence were very similar when comparing the Christians and Muslims in France, which led one to conclude that religious liberty causes a softening of religious attitudes. Education was organised in such a way that there was no kind of religious instruction in public schools. Dalil Boubakeur, the rector of the largest mosque in Paris, considered, in line with Surah 4:62, that it was necessary to obey those who were the rulers in a particular place. By means of political negotiations, the Muslims had succeeded in getting a direct mention of the right to change one's religion deleted from the declaration in which they expressed their agreement with the French constitution. The document deals in relative detail with the question of the veil. The veil was an expression of a social system in which men and women did not have the same social status. A commission had investigated whether there were more such cases in which secular principles and the equal rights of women were being violated in suburbs, schools and hospitals. It revealed an undermining of the secular nature of these institutions and especially of the status of women. As a result, a new law was passed prohibiting all visible religious symbols in primary and secondary schools and especially the veil. Only a small minority demanded the veil for girls in school. Feminist Muslim associations even wanted a still stricter approach by the school authorities and demonstrated for this on the streets. But there were also hostages taken and death threats. The author of the article believed that it was basically a matter of demonstrating difference at some point while being able to agree on everything else.

Islam and Christianity were increasingly equal in France and were reflecting together on the role and visibility of religion in the pluralistic society of tomorrow; laïcité needed to take on a more neutral form. Public opinion and the media no longer listened to individuals; it was necessary to make joint statements. Islam was a religion of French citizens or of foreigners in the integration process. In practice, there were repeatedly moments of tension and conflict between public order in France and ways of life which came from other countries. But finally it was less a matter of laïcité as an ideological conflict and more of laïcité as a possibility for a positive interchange between all involved.

2.2.6.5 Supervision of religious liberty

The third and last major section dealt with supervising the practice of religious liberty. Who does that and how? Non-governmental organisations had recently discovered their special interest in the complex issue of religious liberty. How much freedom of thought and action was a particular tradition prepared to grant within the framework of how it understood itself? Was someone with a different religious approach restricted or even aggressively attacked? The old but permanently complicated key question here was who speaks for a religion? All of these reflections and considerations showed that religious liberty and its supervision were not as simple as might appear at first sight.

The efforts for religious liberty thus far had been made by religious organisations and their human rights' activists who had noted religious persecution world-wide. What was the most effective way of counteracting it? In this connection, it was decided to link up with the standards which the individual nations themselves had recognised. This involved the American Congress, a specially appointed, small, two-party commission, a newly established office for international religious freedom under the American Department of State and a special post of ambassador who was to present an annual report on religious liberty in 194 countries. The ambassador for religious liberty was one of the main advisors to the President and the Secretary of State. The combination of a moral initiative with foreign policy naturally presented difficulties and led to the accusation that a western understanding of religion lay behind the definition of religious liberty. In fact, international norms and treaties were necessary because the problem of religious liberty was of global urgency. The most comprehensive possibilities for supervision were still those of the USA and the United Nations which generally agreed on who had violated religious liberty most seriously. And there was not a single supervisory body for issues of religious liberty in which the USA had no say.

The reports of the UN and the USA varied to some degree as far as the situation of Christians in *Iraq* was concerned where the USA saw attempts to undermine the identity of Christians. That naturally encouraged the suspicion that the content of the reports was connected with American politics. But one of the major problems was that there was no precise definition of such terms as terrorism, extremism in the name of religion, or radical or militant religious-extremist groups. For the USA and its Department of State, a policy in favour of religious liberty was a means in the fight against terrorism. That was considered one of the most effective and lasting counter-measures against a potential clash of cultures. To deprive someone of religious liberty meant to deny that they were human. At the heart of freedom was the right to raise the very fundamental questions about the origin, nature, value and significance of human life, and then to live accordingly. The Americans emphasised individual human rights in contrast to cultural norms. The UN report noted a tension between the fight against terrorism and democratic processes. It provided a definition of extremism and demanded that it be combated according to a minimum of common rules of behaviour.

The overall conclusions from all these reflections were that state organisations were perhaps not adequate for dealing with interreligious problems, while non-governmental organisations lacked the power to get their proposals implemented. The author saw supervision on a religious basis as a corrective to political supervision (alien to many countries), although it was positive that documents had been produced and improvements were being urged.

2.2.6.6 Conscience and religious liberty

The final contribution raised the question of what was going on in the conscience of a believer who demanded freedom of choice. In both Christianity and Islam, there were conservatives who denounced human rights as anti-religious. But there was equally a new awareness that saw religious liberty as a means for serving God better. Even the most credulous were no longer able to believe a message without consciously choosing it. Thus an inner desire for religious liberty had developed and this had been recognised by Vatican II. The desire in Europe and North America to be liberated simultaneously from absolute monarchies and from abusive religious institutions had previously paralysed reflection. Was it possible to embrace a freedom which rejected submission to God? Only after the church had again become the object of persecution, like in its early days, had it rediscovered that one can also demand religious liberty in order to be more faithful to God.

Without harm and tragedies, as the Muslims themselves said, there could be no transition to a tolerant, pluralist society. The article provided examples of the democratisation of religious institutions in both religions. The lay people had, in a sense, grown up and religion existed for people, not *vice versa*. That could be an opportunity for Islam and result in honouring God in human beings by instilling new life into old values such as respect and tolerance. Muslims themselves felt that the ethical vision of the Quran had been lost. Many of the laws and institutions of mediaeval Islam had been revived but, in the process, ethically there had been only a gradual decline. In many areas, it was no longer possible to build one's life on the rules which had been handed down by a religious tradition from the distant past. Some degree of convergence about ethical values was emerging in all religions. Truth, in the opinion of the author, was automatically winning intelligence to be its follower. For centuries, people had belonged to a religion just as you belong to a family or a nation. There was political pressure and deviating personal decisions were prevented. Changing one's religion became treason. The religious authorities of all religions had urgently to face the fact that they were losing a large number of followers with each generation who were then only Catholics or Muslims in name. The author prophesied that, in the future, the only people who would become or remain believers were those whose education had prepared them to be non-conformists. Only with sufficient inner freedom would it be possible to withstand the indoctrination techniques which were becoming increasingly subtle. In Islam, that attitude was most readily found among the Sufis according to whom a person born a Moslem, when he or she became an adult, should also recite the creed of their own free will as if he/she were reciting it for the first time.

The desire to belong to a community of faith certainly existed, but unity could no longer be enforced simply in an authoritarian way or by social pressure. That gave rise to the temptation to cement a group together by hatred of another group. On the other hand, people felt they had been left alone in face of major existential decisions and less and less equipped to deal with the vicissitudes of life. Answers to birth and death, sickness and suffering could no longer be found merely in social conformity. Today, people were sceptical, on the one hand, and, on the other, yearned for everything reflected in the existential decisions of more or less famous persons. Doctrine had to become witness or it would have no effect.

The new type of religious person was more spiritual than ritualistic, more mystical than dogmatic and more uncertain. This was a general phenomenon although at different speeds. The religious specialists on both sides were servants who were sometimes approached for their advice or knowledge, but whose intervention was felt increasingly to be indiscretion. Serving people did not mean enslaving them. Their yearning for religious freedom should not be understood merely as a refusal to believe in God. God himself had taken upon himself the risk of addressing each individual so that they could respond to him individually. It was a matter of the extraordinary freedom of God who guided each person as he willed (a veiled quotation from the Quran). It was a question of a new anthropology; awareness and freedom were constitutive elements of human nature. God did not manipulate us. The Most Exalted One, the Almighty, took people's freedom seriously. Human beings were also respected in their freedom to say "No". So how could we attempt to enslave those whom God himself wished to address as partners? It was quite inconceivable that we should prevent persons from giving their own, free answers. God himself caused the tremendous yearning for freedom to well up in human hearts, and education and preaching should not combat the freedom which existed. These considerations had gained new force among thousands of believers. It seemed quite logical and simple (although often very difficult in practice) that God himself had given people freedom so that they could seek God of their own free will. And this demonstrated very clearly that, after a large number of historical, juridical and practical remarks, the issue of religious liberty leads one back to a decidedly theological anthropology which was already basically present in the Second Vatican Council and could also serve as a basis for further dialogue with Muslims.

2.3 Additional publications of the Pontifical Council for Interreligious Dialogue

The publications of the Pontifical Council for Interreligious Dialogue relating to Islam covered a much wider area than can be discussed in this brief evaluation. With this breadth, they varied necessarily and positively, whether one considers basic approaches or individual problems, strictly theological anthropology or the endeavour to grasp the needs of the other party and take them seriously. The real effort to do so can be recognised clearly in each case, but the Council's own approach was equally clear and naturally followed a line influenced by the popes, especially by John Paul II. It is almost more than obvious how much emphasis was placed on the common elements of the understanding of human beings (God as creator and goal of humankind) when Muslims were addressed directly, particularly in contrast to a purely materialist view of the world. That was naturally right and also justified in view of practical cooperation for the wellbeing of real people which was indeed the stated aim. But it could also easily ignore the fact that there were differences on this human, anthropological level as well which were certainly recognised on other occasions. The more the documents were designed to serve as internal information, the more clearly the differences were discussed. Even though both religions saw human beings as God's creatures, it still made a major difference whether humans were in God's image or God's vicarious representatives. The freedom of human beings and the inherent dignity which they had, independently of their correct behaviour in relation to their creator, were greater, on closer examination, as God's image than as God's representatives. Although neither should act independently of God and his instructions for humans and the world, the creative freedom which the biblical, Christian tradition attributed to the image of God was indeed great, whereas for God's representatives, as viewed by the Islamic tradition, freedom was geared first and foremost to carrying out the very concrete, both comprehensive and detailed, will of God. And precisely at this point, where it was less a matter of anthropology as such than of its (necessary) effects on society and the state, time and again the major difficulties arose to which there seemed to be no obvious solution. Of course, one should not over-emphasise the difficulties when one wished to achieve something important together, but it would also be misleading and, in a sense, dangerous to brush the difficulties completely under the carpet and, at the same time, repeatedly to emphasise the common front against materialism and secularism. Precisely the attitude to the secular world reveals differences today between Christianity and Islam, so a common front really cannot be built up as simply as it might seem when Muslims are being addressed. Naturally, it is difficult to find exactly the right formulations and the right tone, but good cooperation can only be developed on a very solid basis, and therefore it is a real pity that the book which had been planned, on the view of human beings as seen in the other religions, never materialised. It would have been ideal in relation to the subject discussed here.

3 The Christian-Muslim dialogues

3.1 The beginnings

The first attempt to prepare for Christian-Muslim dialogues was to invite two non-Christians to the second assembly of the consultants in Rome in 1972, including Prof. Muhammad Talbi from Tunisia. A confrontation took place over the question of the relation between the two religions and the modern world. The final conclusion was that contemporary culture led the representatives of religions to take a look at anthropology. Religions were devoting more and more of their resources to human beings and their social advancement. Among its short-term aims, the Secretariat was to deal with subjects which were simultaneously religious and of human interest – such as human rights, for example. This mention shows, on the one hand, that the popes with their emphasis on this issue shared in a consensus that transcended religions and, on the other, that this double task was being taken seriously and tackled in detail.

The major change in the Secretariat came about in 1973. Publications notably took second place and, instead, there were many more direct contacts of very different kinds, including fully fledged dialogue conferences. There was reference to the hand of the Holy Father being extended to followers of other faiths. This was accompanied by a more precise description of dialogue. It was the outcome of an anthropology for which human beings as God's image and the object of God's love were the central focus. In dialogue (cf. above), the discussion did not only deal with religious matters but also with things that interested people on a religious level.

A more decisive role than the conference in Rome in 1972 was that of the conference of consultants in 1975 at Grottaferrata. By means of a questionnaire, the current situation of dialogue in various areas had been identified in order to decide how to proceed. The majority of the reports gave an up-to-date picture of the dialogue situation relating to the overarching question of what caused mistrust or disquiet among the Muslims. The Christian-Muslim dialogue was still at an initial stage, or was seen as a matter for specialists and not for congregations and churches. The report which was most critical of Muslims was given by the Secretariat's officer for Islam, Abou Mokh, who came from a Muslim context.

He described the situation as schizophrenic. Arab Muslims, in particular, expected nothing of Christians or of dialogue. Islam alone could save the world and bring it peace and justice. One could ask oneself whether, under such circumstances, there could ever be a dialogue on an equal footing, indeed, any kind of dialogue worthy of the name, and what it would look like. For the Catholic side, dialogue meant the acceptance of others and their rights but, on comparison, that was clearly not the case for the Muslim side. The programmatic limitation to *homo religiosus* and the reticence in relation to practical implementation were considered problematic for Islam because there the link between religion and the worldly order was considerably more direct. Religion and practice were so closely interwoven that it would make no sense to separate them. Religion or one's world view was known to be the key to one's behaviour. One example of this in Islam was popular fatalism. If someone did not believe that human life could be improved before people died, it seemed rather pointless to attempt to improve peoples' earthly life, unless this was combined with a religious dialogue about the relation between God and humankind.

When the dialogues really started, they opened up a new and wide field, particularly from an anthropological point of view. Justice, peace and human development were very high on the priority list of the partners to the dialogues. However, the differences between the individual Muslim organisations and countries were quite considerable and, for this reason, the dialogues and their results varied a lot. (In the course of the years, for example, it was seen as more and more important to root out religious fundamentalism and terrorism.) Attention had always to be paid to ensuring that the dialogues were not exploited by the other party and that one did not compromise one's own cause. But the main task of the Secretariat was to stimulate and support dialogues at regional and local level, without being their main organiser. In addition, there were visits by representatives of the Secretariat who were seen somehow as emissaries of the Pope. These visits could have a catalytic effect because the Christian churches were weak and not strongly rooted in their Muslim contexts, and that made dialogue very difficult. Freedom to confess the faith was also very important as far as social expressions of belief were concerned. Materialism and secularism were seen as a danger by the Muslims as well; they would be very happy to cooperate with Christians on this front. And there was concern about the young people. The proposals were all anthropological issues against a theological background: juridical, ethical and philosophical cooperation on a theistic basis.

At the first assembly of the Secretariat in 1979, Christian-Muslim dialogue was assessed as difficult, starting with the question of representation. In addition, one had to accept that many Muslims followed the scheme laid down in the Quran of using the best dialogues to lead non-Muslims to Islam. For Muslims, dialogue also always had a more or less strong political slant; one has only to remember the many attempts to re-establish Islamic law. One would wish that Islam had a less juridical-political approach and could see itself more as a way to God. So far, that approach was mainly found among Muslim minorities in non-Islamic countries.

Msgr. Jean Jadot, as Pro-President of the Secretariat, gave it the encouragement to participate in dialogues organised by others instead of conducting its own. This was a line which was followed continuously from then onwards, dividing the work into three parts, namely studies of Islam (Commission for Religious Relations with Muslims), the role of animator for dialogues at local level and, finally, the Secretariat's own involvement in dialogues which was far less pronounced than the other two and was restricted, above all, to the international level. But there was a whole series of dialogues with specific partners, the most important of which will be mentioned again here.

3.2 Dialogues with Libya

3.2.1 The spectacular beginning in 1976

The first major dialogue conference took place from 1–6 February 1976 in Libya at the invitation of the head of state, General Muammar al-Gaddafi. It turned out not to be the calm dialogue between experts expected by the Vatican, but more of a show on a large stage. Cardinal Sergio Pignedoli, as President of the Secretariat, had expressed readiness for this dialogue provided that it dealt with a religious and not a political subject. But the practical preparation was in the hands of the secretary of the Commission for Islam, Fr. Abou Mokh, and a representative of the Libyan embassy in Rome. The preparatory commission promised the Vatican that political questions would not be discussed. However, before the meeting started, it was not even known who the Muslim speakers would be. This gave rise to the suspicion that it would not become an academic conference. At the meeting itself, the Muslim participants and observers were considerably more representative than those on the Christian side. But there was no one representing the famous Islamic universities, especially the Al-Azhar, nor Saudi Arabia. Among the Christian observers, there were many representatives of the Orthodox churches whose reaction was strongly anti-Jewish. The hosts determined the line for the meeting to follow: to honour Islam and to work for the liberation of humankind from all forms of injustice and slavery.

The first real issue (requested by the Muslims) was religion and theology. The approach to this question was immediately linked to anthropology on the Muslim side. The question whether ideologies were capable of responding to all of people's needs was presented as the decisive criterion. Ideologies never had a comprehensive view to offer of people, society and the universe. Religion alone was the source of order for life and society, and naturally Islam in particular. What this implied was Islamic law which also governed interpersonal relationships strictly and flexibly at the same time. The Christian side, too, recognised the inability of ideologies to respond to all the problems of people as persons. Religion was the only protection for the inalienable freedom of human beings. In all, the main point of the discussion was that religion gave Christians guidance but did not provide them with ready made solutions once and for all. The Muslims, for their part, saw a dangerous autonomy of reason in a temporal order for earthly polity and were very critical of laïcité. They asked typically whether Christians were not too often torn between the spiritual demands of their religion and the confusion which they felt in relation to the political options of a social order which did not follow religious lines.

Next, there was discussion of a subject requested by the Christian side, namely about what Christianity and Islam had in common. For Christians, this meant the Creator who loves human beings and their fulfilment, who is merciful and shows mercy, and who raises people up from the dead. Commitment to human beings had theological roots. Faith in the living God led Christians to respect life; faith in a just God led to combating every form of discrimination, and faith in a free God led to defending rights to freedom. Because God was peace, Christians had to strive for an international, fraternal community. The Muslims were invited to a dialogue about values. On the Muslim side, there was reference to a natural religion which was marked by a dichotomy between Creator and creature, by the communication of the Creator with the creatures through revelation, by the gift to the creatures of the ability which enabled them to carry out the plan of the Creator in the world, by implementing moral values and thus, finally, by responsibility and sanctions. The second part of the contribution, however, consisted of polemics.

The next issue suggested by the Libyans was faith and justice, the occasion for an excellent presentation on Islamic jurisprudence in the economic field. It was possible to construct a whole social system on the principles of Islam. In this context, human beings were God's slaves and this liberated them from any other, alienating slavery. Fraternity between all people and their benefitting together from the whole of creation were derived from this. According to the Quran, non-Islamic cultures were unable to protect people. But Islamic culture entrusted human beings with the task of thanking God by their work and never being unjust. Human beings were God's representatives on earth. God had made everything subject to them. There was an abundance of resources, enough for all people to live human lives. To this end, it was sufficient to obey Islamic law. The corresponding Christian address tried to demonstrate what special link with justice the monotheist religions had in common. Social justice had something to do with faith in the one God. Faith in a personal God, a positive evaluation of creation, a linear conception of history which was moving towards a kingdom of peace and justice according to God's plan, and an awareness of the social nature of human beings – all of this promoted the development of justice. But, for the Muslims, it was a deficit that the Gospel only made proposals and did not contain any precise divine law. It was a shortcoming of Christianity that it rejected theocracy. Here, too, it was again clear that, for Islam and the Muslims with their all-embracing basic conception, is was much more difficult to exclude any realm at all as not being religious, not to mention politics. The comparison between the issues suggested by the Christians and the Muslims is very telling in itself. The subjects requested on the Christian side were more on a theological level; on the Muslim side, they all also had a practical, legal, social, political emphasis. In the eyes of its followers, that is precisely the specificity of Islam and its superiority. To exclude anything would be like an amputation. So the difficulties were to be expected from the start.

The last major issue, again suggested by the Christians, concerned how it was possible to counteract divisive prejudices and misunderstandings. The discussion contained few directly anthropological statements. The Muslim response consisted of classical accusations against Christians and of political demands that Christians should reject an oppressive and unjust international order. What this meant was, firstly, to condemn Zionism and abandon the papal demand for a special status for Jerusalem and, only secondly, a just distribution of the material and technological wealth of the West.

Against this background of obviously fundamental differences and conflicts of interest, and the obsession of the Muslims with political problems, a joint final statement then had to be drawn up. In the end, it contained paragraphs on human dignity and mobilising it to combat every form of injustice, on the rejection of racial discrimination, on religious liberty, but also on the distinction between Judaism and Zionism (describing Zionism as an aggressive, racist movement without reference to Palestine), and on the national rights of the Palestinian people, such as the right of the Palestinians to return and Jerusalem as an Arab city. Late in the night, it was agreed that the last two paragraphs should be considered a Muslim statement which would be submitted to the Vatican State Secretariat responsible for political matters. The latter naturally promptly rejected it. But feelings had already risen very high, especially on the Jewish side. This all-but-catastrophe led the Catholics and also some Muslims to reflect on the difficulties of the dialogue and the requirements for its preparation. The résumé on the Catholic side was that dialogue with Muslims was difficult. The reason was the discrepancies in history, culture, social development and religious mentality which had only now been fully recognised. The mental attitude of dialogue needed first to be born in the Muslims. The dialogue which had taken place was considered a political misuse of dialogue. Moreover, dialogues which brought together hosts and invited guests had no chance of success. But the Christians also needed to develop in order to be able to think their way into Islam. The terms used most frequently, such as dialogue, prayer, monotheism, revelation or the will of God, had different meanings for Christians and Muslims. For the time being, the best method seemed to be a welcome: a friendly community in which each sought only the best for the other.

It was only after a break of more than a decade that regular contact was again established with mutual visits and encounters. Then, too, anthropological elements played a part; for example, there was emphatic interest in promoting human development and liberation. Respect for human dignity and freedom was necessary. The points considered problematic, on the other hand, were compulsion, seduction, manipulation and mockery, as well as materialism as a problem of modern society and migration. The question of the mutuality of religious liberty was also raised.

3.2.2 Coexistence and tolerance

Then, in 1990, there was a symposium in Malta on the coexistence of religions or, in other words, on the question of tolerance. In the rounds of addresses, it was firstly a matter of how each viewed the other, and only afterwards of specific conceptions of tolerance.

The Muslim side emphasised how Islam viewed humankind in general and hence also people's religions or rather religion. Islam was peaceful submission to laws which resulted in calm and security. The rest of creation obeyed these laws in any case; only human beings had the possibility of disobeying. Departure from these laws of nature really constituted the moment when human existence was born. It was expected that human beings would voluntarily live in harmony with the whole universe. They needed (true) guidance in order to become part of the harmonious universe. Humankind was seen as a family, the diversity of which was an enrichment, a sign of God's hallowed, creative power. Every tribe and every nation had its own land and received its own appropriate message. In Islam, religious tolerance was firmly anchored in the duties of a Muslim. God had created human beings free and given them the freedom to choose their faith. That was precisely the reason why there were different attitudes of faith. People would only be called to account for this at the Last Judgement. The so-called "holy war" had been prescribed to protect religious liberty. Hence, according to Muslim conviction, tolerance and coexistence had sufficient safeguards. The small problems, such as dealing with polytheists, the *dhimma* status, the factual legal and practical treatment of church construction, and the parallel between a lack of religious liberty for Muslims and the prohibition of mission, received no mention whatsoever.

The Catholic view of Islam was a summary of statements from the Council and from *Ecclesiam Suam*, supplemented by the Casablanca speech of John Paul II. The aim of the Catholic contribution was to present how the Catholic Church under John Paul II had been working for a peaceful society. In the process, it was certainly recognised that, in practice, monotheist belief was not always a secure foundation for the dignity, fraternity and freedom of individuals. That was traced back to a narrow conception of religion. In contrast, otherness had to be recognised with appreciation, and more tolerance and trust had to be brought into religious culture and thus also into culture in general. That could only be built up slowly.

The striking thing at this point is the imbalance of the arguments. The Quran and Sunna both start from situations in which a certain tolerance was possible and prescribed among Muslims. But persecuted, biblical Christianity was far removed from a situation in which it would have made sense for it to call for tolerance towards people of a different faith. The problem lies less in the many great exceptions to the principle of tolerance, as in Islam, and more in the way people understand their own tradition and handle it as soon as they become masters of the situation. The Catholic contribution also presented a deliberately positive view, but at least included the recognition that there were problems as well. That was probably easier that admitting that the problems were of a structural nature and not simply false developments, even if they were very long lasting.

However, the question of practical implementation was also discussed directly. In the modern world, people of very different religious backgrounds lived side by side everywhere and they had to be treated equally. The old type of society, which attributed a clearly limited standpoint to followers of certain other religions, was no longer conceivable in the modern world, according to the Christian conviction, because it did not respect the rights of all. For Christians, the question was whether they really felt at ease and could develop themselves in a country which tried to implement Islamic ideals and Islamic law. For Muslims, the question was the reverse – whether they could accept to live out their Islam as a personal religion in a society in which the state left the responsibility for religious practice and its organisation to its citizens. Could the Human Rights Declarations serve as a common charter for such new, pluralist states? The Casablanca speech of John Paul II was quoted again. The speaker identified some points of agreement, as far as equality and justice for all were concerned, with the Malikitic principle of the five essential interests which had to be safeguarded for every person: religion, physical integrity, family descent, material possessions and intellectual abilities. The basic prerequisite was that a society organised pluralism on all levels and provided the possibility for deciding according to one's conscience. The practical aspects were repeatedly brought back to the one fundamental point that separation between religion and state was absolutely necessary. Combined with human rights, that did not lead to a secularist but to a secular state which respected and valued religion.

The Muslim participants, on the other hand, objected that, as long as Muslims were treated unjustly, no one could demand any tolerance from them. The speaker did not take up the point about the separation of religion and politics. Religion was the sanctuary of human beings when they were tortured. Religious liberty within the Muslim community was obviously not a problem as long as it coincided with the traditional views of Islamic law. Again it was evident that the differences between the approaches were so great that it was difficult to mediate between them, so that the same subject became two quite different subjects, giving the impression that people were speaking past rather than to one another. It can certainly be almost equally difficult to define concretely, and then also to agree, what (divine) justice is in a situation, as to come to an agreement on a definition of monotheism. This conference did not produce any joint statements.

3.2.3 Religion and mass media

In 1993, a conference took place on religion and the mass media, including the positive role of the media in forming real people. The media were able to instil tolerance and respect for people irrespective of their religion or race. It was important to promote common human values such as uprightness. It was agreed to set up a joint Christian-Muslim committee which was to observe closely how religion was presented in the media, either with appreciation or to take action against distorted presentations. In addition, educational institutions were important to counteract prejudices and extremism.

The issue of religion and the media was dealt with further in another conference which analysed the situation and how it could be influenced positively. On that occasion, it was agreed to make suitable material available to news agencies and to act jointly to deal with cases of misunderstanding, distortion and prejudice.

3.2.4 The problematic case of mission

The question of mission was taken up again at a later conference but all that was achieved was to clarify the positions on each side and to identify agreements and differences when comparing the fundamental conceptions. It was common to both sides that they addressed their message to all nations and all people. A clear contrast could certainly be observed between missionary and colonial attitudes and strategies on the Christian side. With a view to the future, human dignity had to be respected and this included the others' religion when speaking or writing about it, as well as freedom of conscience as part of religious liberty. There should be no compulsion about religion, according to the Muslims. Economic emergencies should not be exploited but injustice and exploitation should be combated jointly. The key questions which arose included the conception and practice of dialogue and mission, *diakonia* (social service) and mission and the possible misuse of *diakonia*, as well as the spread of secular ethics. And there was a discussion on religious liberty including the free choice of and changing one's religion, but also on whether it was a possible aim of *da'wah* to reinforce the system of sharia. All in all, the fronts had hardened somewhat.

3.2.5 The prospects for this dialogue

Nevertheless, there were further dialogues, for example on the culture of dialogue in globalisation. There was a need to evaluate the Christian-Muslim dialogue process thus far and to outline an approach for the future. The dialogue itself was seen as especially urgent in view of the hypothesis of a clash between cultures. There was a call for precise knowledge of the standpoint of the others, for honesty and a spiritual form of approach, for mutual acceptance precisely in otherness, as well as freedom of belief and religious practice. Dialogue was not merely intellectual discussion; it included positive relationships in everyday life and the will to work together in the service of humankind. There was a new necessity to cooperate to help the poor, the weak and the needy, to promote development and strive for justice and peace. Programmes of education for dialogue were required. Greater continuity was desired for the dialogue meetings thus far and a still greater exchange of information in advance. The two organisations could work together to reduce tensions between Christians and Muslims. To this end, a Coordinating Committee was established that was to meet regularly once a year.

3.3 The European situation in the dialogue

Another section of the dialogue was directed to Muslims in Europe, initially in 1976 in Vienna. The Secretariat for Non-Christians invited the bishops' conferences of Western Europe, the World Council of Churches and a number of leading Muslims resident in Europe to a total of three study and reflection days. They began by taking stock of the number and origin of the approximately nine million Muslims in Europe, of whom almost two thirds were immigrants and a good third lived in what was Yugoslavia at that time. The immigrant workers in Europe had come into a secular society which offered them no guarantee for a secure human and religious life. Their human dignity and their cultural and religious identity were considered to be constantly under threat. The inalienable rights of the immigrants were work, security, a just wage and culture. Christians needed to learn more about Islam (cf. Vatican II). The Muslims expected a lot of the Church. The opening up of the Catholic Church towards Islam since the Second Vatican Council was expressly acknowledged, and this led to special trust in relation to the Catholic Church in a situation in which their brothers in the faith were considered second class citizens in Europe, if not even as new slaves. They wished to be allowed to maintain their faith and practise it freely, to be recognised as a community under public law and to have Islamic instruction for Muslim children.

3.4 Dialogue on holiness

A special dialogue was devoted to the subject of holiness in Christianity and Islam. Indeed, several individual Muslims had also repeatedly shown interest in the question of spirituality. However, on the Muslim side, with one exception, the participants in this dialogue all came from the Indian sub-continent. The idea of the conference had also come up in India at a seminar on Islam for the bishops there, connected with visits to Islamic institutions, buildings and pilgrimage sites. Now the intention was to offer Muslims a similar Christian experience in the footsteps of St. Francis and the Benedictine order. All the practical organisation was done by the Secretariat for Non Christians. The focus was on conceptions and examples of holiness and each was even able to criticise the holiness of the other side. The subject of this dialogue was particularly productive as far as anthropology was concerned. It showed a specially deep interest in who human beings are, or should be and become, in their relation to God and to other beings. It was a matter of the created thirst for God. One good example was the contribution by a specialist on the history of Sufism in India. It dealt with the non-rational level which can certainly be awakened in people but not taught. The beginning and end of things remained hidden from intellectual capabilities; human beings depended instead on God, the Creator, and on intuitive experience. In the visible world, there were enough divine signs to enable people to realise that there must be a Creator behind them. But there were always also people who closed their minds to this insight and persisted in unbelief. Intuition came from a deeper, non-rational level of human nature which reflected an inner urge for contact with the heavenly world. The angels were the link between God and human beings. They influenced individual people and humankind as a whole in more or less the same way that the spirit influenced the body. Humans themselves had immense rational and spiritual capacities. God had formed people from earth and breathed his breath into them which enabled them to hear, see, feel and understand. God taught his representatives on earth the names of everything, their inner nature and their characteristics and feelings. That was the higher aspect of human beings. There had been a bond between God and humans when humans had another form of existence, a kind of divine existence. Everything else – knowledge, feelings, will power, understanding and the power to differentiate – were added to this and the total constituted the singularity of human beings. Being human was a reflection of God's characteristics. (Ghazali even said God created Adam in his image.) So the perfect human being was the one whose life was more and more an expression of the divine characteristics. The human task was to purify their whole being and surrender to God in order to be able to enter into contact with the invisible. Humans were created in the best possible form, they had a proximity to the divine soul and their free will was limited. But they were also impatient, arrogant, ungrateful and easily inclined to all evil. Humans were in a constant tension between the mutually exclusive elements of their nature. So their worldly existence was a state of alienation from their real selves. On the material level, life was merely transitory, absurd and full of worries. Therefore the soul was eternally yearning restlessly. There were three stages of the soul: on the lowest level it was inclined to evil and destruction. Then the lower level had to be brought under control by taking refuge in God as the protector of the faithful. Finally, people heard the divine call and set aside all earthly lusts. For the Sufis, the love of God in the double sense was the unifying bond between humans and the divine.

3.5 Dialogues with the Royal Academy for Islamic Civilization Research, Jordan

3.5.1 Religious education

The best documented dialogues took place with the Royal Academy for Islamic Civilization Research, part of the Al Albait Foundation. Here too, at the first dialogue conference in 1989, there was reference back to the Council's declaration, *Nostra Aetate*, which had suggested to Christians and Muslims that they should overcome the conflicts of the past and find a way to cooperate. Especially for young people, as was emphasised at the beginning, it was important to know how they could lead a life in harmony with the will of God. Therefore, initially, there was a discussion of religious education in relation to identity and openness. The aim in mind was peaceful coexistence in a context of tolerance and mutual respect. In this connection, Islamic identity was described as two circles around the centre of monotheism, the prophetic circle with Muhammad and the widest possible circle of *fitrah*, generally described as a natural knowledge of God but here as intuition. Creation and its unchangeable divine order were, in a sense, God's open book. The classical *dhimma* status was interpreted to mean that the only friends Muslims were not allowed to have were those who warred against them for religious reasons or had at least helped to expel them. Otherwise, freedom of worship and protection were appropriate because all people were part of the large human family which God had created and endowed with special dignity. There was also an ethical law common to all religions. Followers of different religions could cooperate in many areas, especially in connection with that ethical law. In summary, Islam could thus unite mentalities within a flexible framework of identity and pluralism.

The Christian response by the Latin patriarch of Jerusalem contained more stringent anthropological reasoning that the opening address but offered no practical conclusions. Pluralism had always been a reality because it was a feature of the human race. Every believer had to learn to deal with these influences without aggression and also without spiritual alienation. The question how religious education dealt with this led, firstly, to the question of how identity and openness were understood. Identity was defined, first and foremost, by the freedom, self determination and independence which God had given to each person and which every other individual and every social organisation had to respect. The primary identity of human beings was their immediate relationship with God. Human identity was defined by their superiority to the material world. It was the basis for self-determination, freedom and dignity. In addition, each person's identity resulted from their links with a particular society. Here, nationality and religion were fundamental. Mentality, behaviour, customs and traditions depended on them. Each person had the right and the duty to preserve their identity with these two dimensions. One of the tasks of education was to respect and promote this identity. Openness was, primarily, openness to the way in which God viewed his creation. All of God's creatures were encountered in God, their Creator. Consequently, any closing off of oneself from others ran counter to the authentic definition of identity. Openness through interaction with all that was good in human beings (one can hear echoes of the formulations of Vatican II) was one thing; separation and scepticism were something different. Openness did not mean breaking off the relationships which rooted one in a particular milieu (family, nationality, religion) which was seen as a divine gift. Education had to counteract a break by developing a personality which was aware of its own identity, had values and could distinguish between right and wrong.

Education consisted of helping people to be human. Similarly to the Muslim view, humans/humaneness and God were seen as being ultimately in harmony. In this sense, education equipped a person to enter into a relationship with God. Education had to awaken awareness for the God-given dignity of persons and to preserve and develop it. The educators themselves had to be aware of this dignity and to be respectful and conscientious in their dealings. Education had to maintain the balance between mind and matter. Body, intellect and soul had to be able to grow. Relationships with God and with other people had to grow and be interlinked. One should love one's neighbours as God loved them, even if they perhaps had a different understanding of truth. So education was a matter of coming to share the view which God had of one's neighbours and of oneself. In short, if God placed so much trust in people, the educators should do so also. Education was based on freedom and self-determination. Without these, there was no education at all. Even for the sake of truth, no one had the right to question the dignity of another person or to violate it. God did not compel people to accept the truth or to behave rightly. The faithful had to be liberated from aversion to followers of other faiths – another reference to Vatican II. This, indeed, suggested an anthropological foundation for education in a generally and religiously pluralist world. One can note a strong Catholic influence (as one would expect) on the interpretation of the doctrine of being in God's image, as well as the influence of the Second Vatican Council. The emphasis on the milieu (nationality, etc.) was clearly shared by the Muslims, and indeed the involvement of human beings in the whole of creation and society was stressed. The strongest link between the two sides was that human beings were seen as brothers because of creation. Particularly in direct comparison with the Muslim document, it is striking how the idea of creation, of the fraternity of human beings and of the world as a whole as mirroring God, can certainly be described as common. On the Muslim side, naturally, the idea of humans being in God's image was absent.

The Christian address on faith and science drew heavily on Teilhard de Chardin. Human beings should hear and feel the love of God in creation and voluntarily accept or reject this offer. Humans had to be people who desired this mutual love. The readiness to allow oneself to be changed by love was the core aspect of human existence. Christ as the pinnacle of creation was always present in every person from the start. God's message in Christ explained to all people what happened in their lives and what their task was. Other religions fitted into this well with their striving for perfection, truth and beauty and their concern for basic human problems and experiences. The common point with other religions was the grateful relation with the Creator.

This address was the object of strong Muslim criticism voicing one of the most typical reproaches of Islam towards Christianity: that it had too negative an image of humanity and the world (original sin) and was too much concerned about the world to come, that it did not keep the right balance between this and the next world and between the various elements in the nature of human beings, whereas Islam covered all of that.

For the Muslim speaker on the topic, faith and science were the two essential truths which led to God, the Creator. Faith was an innate component of human nature and disclosed to people that they were weak and incapable and in need of God, the Creator. Scientific knowledge recognised its own limitations in respect of the mystery of life and proved that God was the Creator. It gave one peace and certainty to leave the things to God which were beyond human capabilities. For young people, in particular, faith had only positive consequences. As soon as a young person, especially, believed in the oneness of God, he/she refused to be dominated by other people and only feared the almighty God. As a result, young people had positive characteristics and were in a position to achieve aims. The dignity which was acquired by religious belonging, and not by striving for riches, was especially important. It could be very helpful to assist young Muslims in getting rid of the Western intellectual invasion. So here identity was emphasised far more than openness. In the real situation of young people, perhaps not every reality leads to Islam quite so automatically.

At the end, there was also a discussion of religious education in colleges and universities. In this connection, it was emphasised on the Muslim side that lifelong learning was a duty, but that it was sufficient if certain persons performed this duty vicariously for everyone. Religious education took place in any case because human beings were God's representatives on earth. Humans were honoured with reason, intellectual abilities and an inclination to learn. Everything existed to serve human beings; in return, God demanded that humans fill the earth with love, faith and good works. The prerequisites were a firm will, the possibility to choose, the knowledge of good and evil, God's right guidance and teaching through the prophets and thus, naturally, a religious education. This corrected people's behaviour and made them good members of society. However, the question of the extent to which Islamic education really implemented this very broad claim was never even raised.

On the Christian side, it was stated that education did not automatically result in more humaneness and Christian spirit. Against this background, the speaker spelled out a vision of the future for Catholic universities based on a theological anthropology. The so-called "value free" education had proved to be a myth. Values gave meaning to life. They gave a person an identity, a character. Important human problems and concerns had to be investigated at Catholic universities and any partial or distorted view of humans had to be rejected. Catholic education had always been concerned about the holistic intellectual development of each individual. The leaders of tomorrow had to be confronted with the questions of the consequences of progress and the moral implications of supposedly financial decisions. A Catholic university without this comprehensive conception of human beings would not even be conceivable. In this connection, a Catholic university would attach special importance to the realm of the contemplative capacity right at the centre of human existence. On the Muslim side, this was understood as an objection to secularism and they explicitly emphasised the myth of value-free education. That again confirmed where the emphases lay on each side and how that could influence what one hears.

In the end, all the participants at the dialogue conference agreed that God had created human beings with dignity, freedom, rights and duties. Religious education should not only emphasise the development of a religious identity but also contribute to a better understanding of common values and to mutual respect.

3.5.2 Education and children's rights
The next dialogue conference did not only deal with education but also with the rights of children under the headings of unborn children, pre-school children and school children. One of the three opening addresses was given by the (then) Crown Prince Hassan of Jordan who considered that the conference was concerned with the duties adults had in regard to children, also when there were problems. For his part, Cardinal Arinze emphasised during the opening session that the attitude to children born and unborn said a lot about a society; in that way, a society signed its own judgement.

Arinze himself also gave the Christian keynote address on the rights of the unborn child. From the very beginning, a human was a person, namely a unity of body and soul, and had to be treated as such. Human life was a gift from God. It began with the union between the ovum and the sperm cell and God alone had the right to determine its beginning and end. But human beings had the right to life and dignity from conception to their natural death. The Catholic Church had always been against abortion and now saw its view affirmed by modern biology. Although the state authorities in many countries were legalising abortion, humans did not have the power to make laws against God, the Creator. Even if the law declared an embryo a non-person, in the eyes of God it remained a human being with the rights of every human person. *In vitro* fertilisation could not be reconciled with human dignity, in the Catholic view. Since human life was the life of a person, it could not be a technological product.

The criticism from the Muslim side was that the Catholic presentation had only mentioned the right to life, dignity and health whereas, in Islam, the unborn child had far more rights and could, for example, inherit or receive endowments.

According to the Islamic tradition, the first right of the foetus was to determine the duration of the pregnancy. (This is derived from Surah 13:8 which states that pregnancy and breast feeding last thirty months and the minimum duration stipulated for pregnancy is six months. No maximum duration is given.) In Islam, the rights of the embryo started with the choice of the mother. This included marriage prohibitions and the right to dissolve a marriage if the mother were sick and, in particular, mentally handicapped. Optimum opportunities for development were to be given to the child from the very beginning. To this end, prayers should be offered for the child and the parents should behave better to set the child an example. The Quran should be read even to the unborn child in order to confirm the innate nature of the human being. A woman who was pregnant or breast feeding did not have to fast, nor to bother about her financial support, because work was always connected with annoyance which could be harmful for the child and even cause a miscarriage. Islam also prevented modern harm to unborn children from alcohol abuse and AIDS because it prohibited extra-marital sex and alcohol consumption. Only at the very end did the address also deal with protecting the life of the unborn child. The Muslim conceptual approach was much more juridical and the protection of unborn life less absolute than on the Catholic side. Under specific conditions, abortion was possible and, even where it was prohibited, the punishment for it was much less severe. Unborn life also had its price (blood money, compensation). That made it valuable but, at the same time, to some extent, a commercial object. Despite all the dignity given by God to humans, the price could definitely vary depending on gender, age and status. This again made clear that there were fundamental differences between the Christian and the Muslim sides in the nature of their argumentation – more philosophical-theological or more juridical-theological.

On the rights of the pre-school child, the Islamic contribution again went into infinite detail and was, in practice, a brief compendium of Islamic law on the subject. The rights of children were the duties of society. In progressive states, these rights were normally well respected, but worst in Oriental, African and Muslim countries. The question of the reason for this extreme discrepancy was not even raised. On the contrary, it was regretted that many people had failed to recognise the positive influence of the Quran, Sunna and the Islamic legal tradition on this issue. The child had the right to be brought up and educated but an infant had far more rights: firstly, to hear God's word. As a boy, he had the right to circumcision. The child should receive the best and the care of its own mother was the best. Every Islamic education had six basic principles. The first was education in the faith; according to Muhammad, children should be led to three things: love of their prophet, love of his household and the recital of the Quran. The second was moral education. That was followed by physical education, mental education, gradual social education and, right at the end, practical education. Children also had nine concrete rights: the right to maintenance by their parents, the right to equal treatment with their siblings, the right to sympathy and genuine love, the right to playing and fun (children should learn to swim, shoot and ride), the right to care, the right to paternal guardianship, where appropriate the right to guardianship by other persons, the right to legitimate descent by birth within marriage. Only the ninth and last right was the right to life, meaning the child must be protected from hunger, aggression and other forms of death and must also not be sold into slavery or bondage. Education revolved around the core of actionable rights. Here, four points were important to the speaker: the good example of the parents, good company (well brought up children of the same age from good families), and friendliness to children (violence destroyed what we today would call social competences). The last point was that children should be taught the characteristics which Islam demanded, also including manliness (namely, not becoming accustomed to comfort and luxury, not having every wish fulfilled). So one could say that Islamic education wants to create a good person first and then a good citizen.

The Christian side was also concerned about practice. What was not merely the ideal situation like, with which Muslims usually answered every question? What were the real conceptions and how were they being implemented? Perhaps this is a typically Western Christian approach, but it was an important point reflected in many addresses and many dialogues. There is often the danger of comparing a Christian actual situation with an Islamic ideal situation, and this results in an imbalance of argument and sometimes also of feelings because, despite formally dealing with the same subject, in fact it is a comparison of apples with oranges.

The Christian speaker, for example, approached the same issue on the basis of her own practice as a mother and a Christian. That a child had rights meant that the child was a person, an actor. This was obvious to people who believed in God as Creator. Since the human being was both a material and a spiritual being, it was also important to develop this second aspect. Practical tests had demonstrated that children needed individual attention and regularity as far as time, place, persons and education were concerned. That was most readily found within the family. Education should produce free, responsible adults capable of taking their own decisions. The child must be respected and would then gradually learn to adapt to other persons and circumstances. The child's body must also be respected. But that implied setting limits, teaching him/her responsibility and respect for others and also giving him/her an awareness of his/her spiritual side, his/her relationship with God. Authority was necessary. "Must" can become "should" and, finally, the aim of education was autonomy. Today, in particular, quiet and inwardness were necessary as a help for finding one's own purpose and ultimately God. The child should also be assisted to overcome the roots of fanaticism, violence and aggression. The child should marvel at things and learn idealism and selflessness. Having no other aim than profit deprived people of any nobleness and led to frustration. The child should learn to forgive. He/she should learn from living examples; that would also enable adults to correct themselves. Among the rare common points in this case, both main addresses mentioned that parents should or could improve themselves because they automatically functioned as examples.

The final round of discussion dealt with the rights and education of school children. The Christian speaker emphasised that religions, including Islam, had promoted the UN Convention on the Rights of the Child. A theoretical discussion on human rights was not yet possible with primary school children but their rights could be practised in school, e. g. by refraining from punishment by beating, out of respect for the child. A concern of the speaker's was how uniformity could be achieved on this issue between government, family, school and places of worship, as well as the question how children could be educated to peace, love of neighbour and the practice of human rights, particularly in areas where violence and religious extremism were dominant. The speaker came from Egypt and described the path as stony if one had the courage to tackle the discrepancy between the ideal and the real situation. Children did have the right to learn and to a suitable school education, but the school education of Egyptian children did not correspond to what the UN Convention understood by an education suited to the personality, freedom, abilities and future tasks of children – and this situation was found throughout the Arab world. It is striking that both Christian contributions raised the same (self-)critical questions.

The Muslim address, in contrast, offered reflections on the theoretical substructure. To be good was an inherent human characteristic. What mattered was to produce good human citizens. To this end, a selective process for dealing with values and trends was required. That was no problem in a society with a clear identity and stability but could become difficult and even dangerous in societies subject to rapid social change, which had no clear cultural vision and in which individuals had plural loyalties. The main concern was harmony and equilibrium between all components. But, in practice, in the Arab world the exact opposite was the case. Of special importance for an education system was the philosophy behind it, its consistent implementation and its agreement with the specifics of the society concerned, because usually values were invisible, meaning unconscious, and so they were maintained all the more rigidly. It was important to bring reason to bear on situations which had been ruled thus far by impulse and emotion. Nor was it possible to combine eclectically elements of very different origin which had contradictory philosophies. Naturally, the speaker relied strongly on the long neglected Islamic charter which was to bring about the practical implementation of values such as responsibility, self-giving, uprightness, trust, steadfastness and productivity. Islam emphasised the protection of motherhood and childhood in a more profound way than modern society because its ethical norms were connected with rewards in this world and the next. In Islam, learning, in particular, was a duty of both men and women and its non-observance was punishable. The Islamic state was the first to provide free education for all, as was still evidenced by the religious schools where students received financial assistance for their livelihood. Strictness and beating were certainly included but playing was also prescribed – after all, Islam was a realistic and therefore perfect religion. What was important was a good example and everything that promoted a child's courage and freedom and avoided crises of identity among children.

The final report mentioned pre-natal care, for example, because life was a gift from God. Genetics and pre-natal technology needed a moral framework based on divine provisions and ethical principles. Mothers should have the time and means to take care of the development of their children in their first few years. The sanctity of marriage was strongly emphasised because both parents were important. Disadvantaged and handicapped children had to receive special assistance. The parents had the primary right to determine the education of their children. All children had a right to religious education, as well as a right to an education which respected their particular faith and values. But children should also learn to understand the faith of others. No education was free of values. Peace, justice, development and health care were the best way to safeguard the rights and education of children.

3.5.3 Women in society
The third dialogue conference in Rome discussed women in society, their fundamental status, contemporary challenges and problems, together with future prospects and possibilities, and on this occasion the majority of speakers were also women. What would have happened if the women had been left to discuss the subject on their own? Cardinal Arinze described women as the givers and sustainers of life. But a more far reaching, spiritual motherhood was also required. Both kinds of mothers taught children to respect older people and people of their own age, to serve those in society who were weaker and less fortunate, as well as moderation, patience, generosity and loyalty. God had also used the image of the mother to describe God's faithful, passionate love for humankind – and the Arabic word for "merciful" was derived from the Arabic term for a mother's womb. So a mother's love and protective care conveyed the very first impression of what God was like.

The Muslim opening address, however, saw the meeting mainly as a possibility for correcting false impressions of the wretched position of women in Islam, significantly enough under the title *Distortion of the Image of Woman's Status in Islam*. Women were seen as human beings just as much as men without any distinction apart from what nature determined. Women were prepared to bring up children and to care for them. These were duties which followed from their physical nature in line with creation and with reward in the world to come. Anthropologically, all the details related to the one factor that, according to Islam, men and women were equal with regard to origin and purpose, but otherwise the gender difference, namely the woman as mother, predominated. In this role, women were highly respected and should be well protected, but in such a way that this protection tended to undermine their rights to participate in the life of society because that could not be expected of their feminine sensitivity.

The corresponding Christian address was given by a woman and adopted a historical approach beginning with humankind being created as man and woman. Beyond and above all their differences, they were marked by their resemblance to God. Men and women were relational beings created by God with the same dignity in perfection and harmony. But this was not reflected by the present situation. Humans had rejected their status as creatures and were now no longer in harmony with themselves. Moreover, women were frequently subordinated to men. The problem was that men and women were apparently unable to return to their original mutual behaviour, even though they always yearned for it with a kind of nostalgia. Jesus had restored this dignity of human beings that had been overshadowed by sins and injustices and, contrary to the traditions of his time, had done so equally for men and women. Mary was the archetype of women. Women were no longer subordinate but free to serve. Despite all the differences, an amazing agreement can be observed here. The serving role was attributed mainly to women as their natural role corresponding to their nature.

The addresses which followed dealt with the challenges and problems for women. Women had generally adopted the role of supporting the weaker members of society, had committed themselves successfully to environmental concerns and in a variety of self-help groups, both in industrialised and developing countries. Women were the ones who could bring about or destroy tolerant communities as far as education for mutual respect was concerned. Their values were predominantly those of a pre-industrial society less interested in accumulating material possessions. This is definitely an anthropological description of women but it makes little reference to religion. However, an empirical analysis seems to point to promoting the life of the next generation and to a generally serving function.

The Muslim address described the situation of Muslim women in the Arab world because there, despite economic and political differences, the historical, cultural, social and religious conditions were similar. There were five problem areas, beginning with the large number of female illiterates. The employment rate for Muslim women was the lowest in the world. And, according to tradition, a woman did not have to work, or only in exceptional circumstances. A woman should not have to work alongside men. Ultimately, social advancement was possible for women not by means of education or their own achievements but only by marriage. In the predominant social system, the leading role belonged to men. The media and school textbooks also reinforced the traditional image of women as non-employed housewives and mothers. Politically, women were practically not represented at all. The Islamic theory about the equal rights of men and women was, in practice, eroded and turned into the opposite in a way unique in the world by the distribution of roles according to nature and the separation of the sexes, and this was viewed very critically.

The Muslim view of the future assumed that the woman would continue to be a branch of the man's tree, unable to take any independent decisions or to act autonomously in society. However, the female speaker did not say what the consequences should be in her opinion. She considered the struggle against polygamy to be superfluous for the simple reason that it was against Islamic law and hence against religion. So it was again a question of what exactly was laid down by Islamic law and what not, because that alone determined the precise standing and the exact rights of women and not their perhaps understandable wishes. The challenges also included the question of women's true nature and identity. What made women different and special – and were these characteristics part of their being or merely conditioning? As a human being before God, in Islam a woman was absolutely equal to a man. Women also enjoyed all civil rights. But, on the other hand, it was a fundamental principle of Islam that men and women had different natures. Women gave birth to children and also cared for them. The speaker saw motherhood as the role and profession assigned to women by their femininity. By declaring motherhood to be the natural profession for women, a practical model was provided to mediate between the fundamental equality of men and women and the simultaneous difference between the nature of the two.

Maurice Borrmans linked his address on future prospects to the final message from the Second Vatican Council. There it had been stated that the Church was proud to have honoured and liberated women and to have identified their fundamental equality with men. According to Suzanne Villeneuve, the major difficulties for women in today's world resulted less from their inability to occupy certain positions than from the pressure to do so according to a male model. Women were particularly attracted to whatever was living and personal, and they tended to perceive such things as a concrete whole. Their natural desire was authentically motherly. By nature, abstraction was alien to them. Men and women had been given the same threefold duty to be the image of God, to procreate and to dominate the earth, but women had a feminine way of carrying out this duty – helpfully and harmoniously. Women should reconcile men with life and keep watch over the future of the human race. The message of the Council as a whole was to pass life on and defend it in all its aspects. No cooperation was possible in the case of polygamy, repudiation or divorce.

But the decisive question of the whole conference was: are there natural, meaning God-given, differences between men and women and, if so, what exactly do they comprise and what effects are related to them? The clear tendency was to say "Yes" and to derive differences from motherliness (and a close relationship to nature) because of potential motherhood.

At the end, agreement was reached on the following points, among others: that God had created human beings and given them a special dignity in which men and women shared equally. The differences between the two were God-given. The status of the family was emphasised and the achievements honoured which parents, and especially mothers, contributed to future generations. Women had the right to activities that corresponded to their abilities and circumstances within the guidelines laid down by their particular religious tradition.

Religion was considered generally to be of the greatest importance for a personality of the highest value, which meant that promoting religious values among all people would also safeguard the dignity of women and help them. Everything else, namely the possible weaknesses of women and their social consequences, was not only a matter of dispute between Christians and Muslims but also within the religions themselves and, especially for Islam, an extremely explosive point of disagreement, particularly when facing the modern, western image of women.

3.5.4 Nationalism

The next conference in Amman dealt with a completely different subject, namely the problems and challenges of nationalism today. This certainly had to do with the position of the Hashemite royal family. At the opening, the hosts emphasised that, although religions had not simply abolished nationalities, they had purified them – which could almost be a Catholic formulation. So the approach should lead from contemporary problems to the more fundamental ones in order then, it was hoped, to find better solutions to the contemporary problems. The second welcoming address by Cardinal Arinze provided a brief summary of the Catholic point of view. It was a question of the relation between religious identification and national movements. Christians were concerned about healthy patriotism, in contrast to destructive nationalism or a chauvinism which attempted to dominate and destroy others and set up its own nation as absolute. This introduced some anthropological considerations: one's own identity and one's sense of one's own value, as well as the identity of others and their sense of their own value, in fact the unity and solidarity of humankind as a whole. John Paul II had made a distinction between (positive) patriotism and (negative) nationalism. Whatever led to a mental or practical devaluation of others was ungodly. God was the God of all people and all nations belonged to him.

The first main address attempted to give a general review of the history of the relationship between nationalism and religion from a Muslim point of view. The speaker analysed the nationalist movements in Asia and Africa which had also led to the formation of many Muslim states. Islam played a key part in mobilising nationalist sentiments. It was important that the Muslims had not accommodated themselves to the colonial rulers but rather maintained an independent character which was very uniform, probably because of the strong influence of the Quran. The speaker agreed with a definition according to which nationalism was a purely historical or linguistic bond with a neutral worldview and that it needed a philosophy before it could form a social order. The message of Islam liberated everyone so that they were able to serve God alone. All other civilisations were described as civilisations of slavery. But an

Islam, which filled all the nationalisms of this world and subjected them to its system of order, could heal the world and make it free and fraternal. That in a sense would be the ideal world for every person.

The Christian address, for its part, discussed the role of the Christian Arabs in the development of early Arab nationalism in the nineteenth century. The Christians had had a dominant role in the development of Arab nationalism. It developed in the region of Greater Syria as a plea for secularism, because there were so many religious sects and ethnic groups there that this was the only thing which could unite them all – in contrast to religious affiliation which was the decisive factor in the Ottoman Empire. Almost directly contradicting the view of the previous speaker, Christians had brought about liberation from this closed system, and thus cultural renewal and links with the wider world. The decisive impetus in this direction had been the massacre of 1860 which had almost wiped out the whole Christian community and which the Turkish authorities were unable to control. On the contrary, Turkish soldiers participated in it, even including the governor of Damascus. So what was important was for everyone to recognise that ultimately all religions were equal. All people were human by nature, originated from primordial parents and (only as the third point) worshipped the same God. There had to be a division between the two realms, religious and secular. Love of one's homeland had to supersede all religious ties. This was a major challenge to the traditional Islamic social system. But the conception of secularism, and hence of the separation between religion and life, did not normally enter the consciousness of Muslim Arabs. And that was also connected with the experience of whether religion could be a source of danger or not.

The second round of addresses took up contemporary problems and challenges of nationalism. The speaker for the Christian side was a Dane who was lecturing in England. He said that nationalism had developed in Europe in the seventeenth century. On a world scale, one could observe how economic and technological modernisation had led to the formation of a new identity, namely that of the nation. Education was a key factor here for overcoming regional, minority and immigrant mentalities. In Arabic speaking areas, the necessary vocabulary had first needed to be found because the potentially suitable terms had strong religious overtones (*umma, milla*). The criteria for citizenship were a heritage of the founding myths. On the world level, these could be fundamentally different interpretations of one and the same historical event; this constituted the brisance of such myths as the basis for identity and raised the question how they could be overcome and replaced by new myths which could bring about unity. Something similar could be said of private law and the concept of the family behind it which is instinctively always at the heart of a collective identity. If the Muslim minorities in Europe were to make an increasing number of demands related to family law, which is certainly conceivable, this could easily result in some kind of (national, ethnic, religious) mobilisation. Something like national belonging, which appeared so simple at first sight, could rapidly become a collective pathology when examined more closely. In Christianity and Islam, the collective memory was negative. Disagreements could quickly be mobilised for political ends, while the long periods of peaceful coexistence and interchange were not remembered and could not offer a counterweight. On the contrary, it was easier to win sympathy if an ethnic conflict was presented as religious.

The address from the other side was given by the Jordanian minister of justice on the situation in the Middle East. An anthropological statement is easier than when national belonging is seen as a complicated and, to some extent in fact, imaginary tangle. For the minister of justice, Islam was a universal, human religion. Every person can be a Muslim. In a typically Islamic way, he again emphasised how extensively this community (*umma!*) transcended all minor distinguishing criteria. Muslim citizenship was very easy to acquire; one only needed the faith of the heart and the confession of the mouth. Foreign and foreigner were not unchangeable innate factors but merely subject to the will of the individual.

The last two addresses dealt with the role of the faithful and the prospects for the future. For the Muslim speaker, the question of secularism in the past and the future was the decisive issue between Islamic and nationalist thinking. He quite naturally assumed an Islamic Arab identity. Like the previous Muslim speaker, he saw a major harmony between nationalist thinking, Islamic thinking, interreligious dialogue and Arab unity as a given fact which – according to Muslim logic – repeatedly came to the fore. The anthropological starting point was that humankind was created as a single nation and the differences only came afterwards. The differences between the individual religions could be traced back to their links with particular nations. A possible summary of the essence of all monotheistic religions could be human fraternity as a consequence of the fact that all people were created by the same Creator from the same clay, altruism as the starting point for social reforms and, finally, self-reformation in general. Two of these points were thus clearly anthropological. All religions urgently called upon people to maintain the bonds of brotherliness, love and peace. So equality on an anthropological basis also included a moral code common to all people. The speaker assumed a harmony between Islamic and national conceptions. That was positively decisive in order to bring about unity, freedom, social justice and democracy in the future. Islam was needed for its values such as social justice, social cohesion and unity, and so was nationalism because the social and emotional ties between individuals and their nation needed to be strengthened by national feelings. The following were important to Islam: justice, social cohesion, consultation (*shura*) or democracy, as well as moderation or the middle way. Social cohesion was also one of the preconditions for a human life in dignity, together with the connection between the spiritual, intellectual and moral dimensions of life. On all these points, the monotheist religions agreed, especially Islam and Christianity. Secularism was mentioned as needing special clarification by means of dialogue. The speaker considered that secularism was good only for societies in which religion was just one part of their social structure and essential foundation stones. In the Arab Islamic region (the equation again), religion was the basis of human life in society and of nationalism. Anything else would be extremist in the Arab context. The reasons for the development of secular thinking also in Arab society were no longer relevant. Secularism did not suit Muslim nations. Right at the end, the speaker again underlined where this approach came from. Islam understood itself as the final divine message for humankind – and of course this message was in line with people's natural predisposition. The common ethic so often put forward as the Islamic anthropological basis could therefore become a disastrous trap if it were identified with the real, comprehensive Islamic system, so one could easily miss the highest value of human existence if one looked elsewhere.

The Muslim speakers were concerned to see an ideal nationalism which already existed, namely the Islamic one, put into practice. The leeway for criticism, especially basic criticism of the system, was limited. The corresponding Christian address started by discussing *imagined communities* (Benedict Anderson). The rise of nationalism was the development of a new *imagined community* in contrast to the religious community and based on different conceptions of space and time compared with the latter. The idea of the nation as a political community evolved when, under pressure from the French revolution and the enlightenment, old systems such as dynasties, and also religious conceptions such as hierarchy, space and time, were collapsing or at least declining. That was precisely what produced the sharp contrast between nation and religious community. For this reason also, nationalism was a challenge to religious identity because it aroused strong feelings of belonging and even demanded sacrifices. Within a religious community as well, it could lead to varying evaluations of the same situation. The speaker named clear conditions for the involvement of Christians in nationalist movements (option for the poor). In this connection, one could talk about conflicts between the demands of the religious and the national community and denounce manipulation, over-stepping of limits and misuses of nationalism and religion. National-religious identification was considered by the speaker to be no longer conceivable in modern times, indeed dangerous. There were too many competing world religions and millions of people who preferred a secular state. He considered it legitimate for a religious community, of whatever kind, to engage in moral-spiritual criticism of the state community, but not in fundamentally theocratic criticism. Nevertheless, the Christian leeway was greater compared with the considerably more specific Islamic provisions. On the question of a genuinely Christian conception of nationalism, the speaker appealed for radical openness to what was alien and for radical justice, because he believed the roots of the cry for cultural identity resided in oppression, fear and poverty. On the Islamic side, Islam was increasingly cited as the solution to the problems and injustices mentioned. Christianity did not simply offer a solution which could be applied as a blueprint. The proposal was only one possible Christian proposal and necessarily alien to Muslims because of its basic approach. This was the only contribution which attempted to clarify the phenomenon of nationalism on a fundamentally anthropological level. Throughout the conference, Christians and Muslims seemed to speak past one another because, in a sense, they were arguing on different levels of the anthropological development just described.

Official agreement was possible only on banal points. A distinction had to be made between what was healthy or destructive but there was no contradiction between religion and nationalism. Neither Islam nor Christianity should consider one nation higher than another. The positive values of nationalism could contribute to forming and strengthening identity and the sense of one's own worth. For this reason, both sides had to support these values and work to prevent nationalist thrusts becoming means of domination.

3.5.5 Environment

The next dialogue conference dealt with the environment. Naturally, it discussed the religious basis, just participation and influencing decisions. The opening address by Arinze had an anthropological depth as such. It was a human duty to recognise God as the Creator of all that existed. It was God's will that the earth's resources had been entrusted to human beings so that they could care for them, use them and share them with their fellow beings because, according to God's will, human nature was social. This status of human beings was manifested in intellect, will, soul and spirit. But individual persons and countries did not realise that they had to give account of their responsibility. That was, essentially, all that needed to be said from a Christian point of view about the situation of human beings in relation to the earth's resources.

3 The Christian-Muslim dialogues — 585

To begin with, there were addresses (and responses) on what the two religions say about the exploitation of the earth's resources. The Muslim presentations were three to four times longer than the Christian ones. This was obviously a subject very dear to Muslim hearts but, again, it also reflected the detailed Islamic legal tradition governing everyday life. The general Islamic approach to the universe, life and humanity was the starting point; the second issue was the question of the resources; the third point dealt with the most important, actual rules and the fourth and last point gave examples of how they had been implemented in the course of history. Islam's comprehensive view of existence, life and humankind was considered its most important characteristic. This approach explained everything related to human life and answered all the questions that could arise. God had made human beings his representatives and equipped them to this end with all that they needed to carry out their duties. These were to develop the earth and to establish God's complete lordship. So this vicarious responsibility was subject to principles and rules which were laid down in Islamic law. The aim was to persevere in all that God had prescribed. Anyone who deviated from this disqualified themselves. God alone was the owner, human beings only proprietors. God determined the duration of their vicarious responsibility. God had honoured human beings by creating them in the best way so that they were also capable of their tasks. That included understanding and reason and the possibility of acquiring new knowledge by those means. No other system took such careful account of humans' true nature as did Islam. Islamic institutions had organised everything to take adequate account of the innate human character (*fitrah*). But, naturally, the provisions of special interest were those which prevented one generation from intervening in the rights of another. For Islam, environmental protection was a matter of divine law. The Quran forbade excesses in the use of resources. The exploitation of the environment was dependent on security aspects. The golden rule of Islamic law was considered to be Muhammad's saying that Islam did not permit harm or hindrance. Islamic law as a whole was geared to producing good deeds and preventing harm.

According to the Christian view, on the other hand, as a result of creation the material world itself had its own stability, truth and excellence. It had its own orders and laws. What was human autonomy and what was divine law remained undecided. The common points were that human beings were created and the very similar task and responsibility given to them by both Islam and Christianity; the great differences related to the practical implementation. Christianity provided great freedom and some lack of clarity on the practical side, whereas Islam, with its claim to an overall concept of divine origin, had a vast compendium of rules.

From the Christian point of view, the fundamental dignity of human beings as God's image had very important consequences for the ethical principles concerning the use of the earth's resources. With regard to these resources, solidarity was required in time and space, especially in relation to all those who lacked the basic necessities of life. The real role of humans was to serve. A theology of creation had also to be accessible to non-believers. This kind of ethics was a speciality of Catholic social doctrine: reason required moral teaching and a monotheist faith could only reinforce this teaching. The ethics of creation had to spell out what human responsibility implied. Some of the main elements were: no property right was absolute because God alone was the true owner. A definition of the concept of ownership should always also include an ecological approach to and the universal purpose of resources. The rights and needs of other people should not be ignored. This led to the principle of the rights of future generations. In the meantime, the "option for the poor" had become an ethical requirement in the Christian churches in general. The values which had to be observed at all costs were respect for life and especially for human dignity, justice and solidarity on the world scale, as well as love of one's neighbour which resulted in the radical demand for justice. Similarly to the Muslims, creation was understood as brotherliness in space and time. That was generally affirmed by the Muslim side, even including the mention of being God's image.

The meeting then turned to the more practical question of the protection of resources. John Paul II had emphasised the importance of a just distribution of the resources given by God, especially between industrialised and developing countries, and called for a different style of life because human beings had a transcendent destiny. He had also related this to human rights because development had made the exercise of human rights possible. Everything revolved around the (literally) anthropological axis, around the outstanding position and responsibility of human beings. Contemporary papal teaching underlined responsibility for creation. This teaching could be described as overwhelmingly theocentric and anthropological, according to the speaker. This anthropological perception was the basis for the present new interpretation of the biblical theme of stewardship. In this case, ruling meant responsibility. It would be helpful if this appropriate anthropological basis could be expanded to include respect for the world's resources. The parallel with Islam was mentioned explicitly although the terms naturally differed. It was a matter of creation and redemption and the theological link with ecology which still needed clarification. We were all here only for a limited period, so the expression "pilgrimage" was fitting in order to remind us to safeguard the resources of the earth for future generations. All in all, this contribution had a strong anthropological approach.

The Muslim paper on the subject repeated many points. Because people's own interests were stronger than their faith, the world-wide efforts for environmental protection had not led to any improvement. Modern science and technology had been unable to safeguard the earth's resources whereas the Islamic faith could solve this problem. The Islamic conceptions of how to handle nature, the environment and the earth's resources were based on avoiding extravagance, wastage, ravaging, destruction, exploitation and the pollution of resources. It was therefore a duty to give the believers the chance to approach these matters from an Islamic perspective.

On the Catholic side, it was also true that people felt responsible before God for the environment. A well-informed and interesting detailed comparison of the two systems and their approaches to this question showed that, on the Islamic side, the absolute oneness of God led directly to God's being the only one who could govern human relationships with creation, while, in Christianity, the loving relationship within the Trinity was the real key to a loving relationship to creation. The two conceptions were very similar as far as their practical consequences were concerned, but worlds apart in the case of their theological bases, although this was hardly noticeable on the ethical level.

The last round dealt with the role of believers in the distribution of resources. On the Muslim side, it was assumed that there were sufficient resources on the earth for all its inhabitants. It was the task of the faithful to ensure, by means of proper management and suitable distribution, that there was enough food for all living beings. All mineral resources were subsumed under the general resource of fire and were therefore legitimately the property of the nation state. The profit from national raw materials had to be spent in the public interest and on state welfare. In addition, the state had to intervene to re-establish the economic equilibrium if one class dominated another. Everyone should receive what they needed even if that meant that everyone had only the basic essentials. Distribution between the faithful took place on a voluntary basis, for example through donations.

The Christian address again started by stating that God had exalted human beings above the other creatures with wisdom and various abilities for using nature, and had entrusted nature to them, but specifically to all people and all generations. In the Lord's Prayer, we were called upon to be satisfied with our daily needs. The threats described to the environment were the direct consequence of human society's departing from this Christian path. The speaker repeated the call of John Paul II for greater moderation and simplicity, and also quoted numerous biblical passages showing that the poor should be given their share of earthly goods. The Church Fathers had followed this line as well. Modern industrialised societies had completely upset all this at the expense of poor people and underdeveloped countries. The structural problem of poverty, as John Paul II had said previously, had to be tackled if one wanted to solve the ecological problem and also the problem of war and peace, namely the biggest problems of humankind today. But spiritual values should also not be forgotten, because excess often increased with scientific and technological progress and material wealth. The real danger for society was the tyranny of material work and of the desire for profit over the soul and values. If human beings were to behave according to the will of the Creator, then society would become as God wished – a peaceful welfare society.

In the Islamic view, there were major, fundamental agreements but one great difference. The Christian address had never made really clear what the mechanisms were for the necessary distribution. In Islam, this was regulated by the alms tax, beneficence and the use of the earth's resources, and the state had the possibility to guarantee the implementation of these good principles. Here again, the difference lay in the basic conception of the degree to which human living should be regulated.

The climax was the public address at the Gregorian University by the Jordanian Crown Prince on the subject of Islam and the environment. Islam had a lot to say on this. He mentioned the vicarious role and responsibility of human beings. He also recognised the role of the enlightenment and the scientific stimulus to which it had given rise, the progress it had brought about and the subsequent environmental damage, and he was critical about this but not absolutely. Nor did he consider the role of Islam in protecting nature to be quite so glorious. That had had more to do with under-development in the Muslim countries. As a Muslim in a responsible position, he was just as fearful of the consequences of development as he was aware of its urgent necessity. Past history had demonstrated that rapid development had resulted in harm to the environment and in human alienation. The Crown Prince emphasised the many principles and values which Christians and Muslims had in common and which were the key to environmental problems.

The official recommendations agreed, among other things, that God had created the earth for the well-being of all and it still had sufficient resources to feed everyone. All kinds of resources, both renewable and non-renewable, had to be used reasonably and responsibly. A corresponding way of living had to be encouraged. Food should be produced without damaging the environment and in an ethically correct way. Cooperation was necessary for environmental protection. Immense amounts of natural and, above all, human resources were destroyed by war and violence, so they spoke out strongly against that. It was necessary to rely on dialogue to find just and peaceful solutions. Mutual dependency in the world reflected the unity of the human race as willed by the Creator. Reference was made to the importance of education based on respect of nature, of the dignity of human beings and of the future, and to the will to create a quality of life which was open so that all could participate in it. Religious institutions, the media and governments had the task of promoting awareness of and/or providing the means for this.

3.5.6 Human dignity

The last of these dialogue conferences dealt with human dignity from three angles: conception, historical review and prospects. But the opening greetings raised the question what the purpose of these dialogue conferences was and to what extent it could be achieved. Crown Prince Hassan emphasised that there was no intention of promoting an international ethos but rather of building up a common code of human behaviour. The participants were speaking with one another in order to identify aspects of their traditions which were relevant for one another while maintaining their own identity. He was convinced that Islam and Christianity could agree completely on the question of the right to human dignity and would only differ in the way in which this fact was to be expressed theologically, and that the basis for this, God's concern for human beings, was the same. In Christianity, it was expressed in the formula of the command to love, while, for Islamic theology, it was the creation of human beings as free beings, able to take autonomous decisions. The cause of the weak and the oppressed was equated with God's cause in the Bible and the Quran. Despite all his theological reasoning, he particularly set store by and emphasised practical and rapid implementation.

The first main address from the Muslim side took as its starting point that humans only existed at all because God had breathed his breath into them. But the negative distinctions were almost more interesting. The first, voiced on behalf of all monotheist religions, was directed against the theory of evolution, because these religions had already said everything possible about the origin, nature, progress, place, purpose and destiny of human beings. They all agreed that humans were different from other beings and were superior to them because of their rational soul, intelligence, physical elements such as speech, work and upright gait, and the true knowledge which God had taught them alone. Free will, without which there could be no responsibility, was also part of this, although the Quran at the same time strongly emphasised people's serving status. The second clear distinction was humans being in the image of God. It was most meaningful to put "his" before image in the case of Adam himself. The last, very marked distinction was from the West which understood humankind and nature in a very different way. In order to regain humanness and dignity for human beings, science had to be saved from its breach with God.

The basic Christian paper adopted a strong Christological approach which also included a Christological interpretation of statements on the theology of creation. The contribution of the Christian message to the conception of human dignity and human rights combined two inter-related ideas: the natural equality of all people and the conception that, for God, every individual incorporated the dignity of humankind as a whole. However, a kind of secularisation of the religious understanding of human dignity had taken place. There was substantial agreement between these contributions from different backgrounds on some important anthropological principles. The speaker considered the failure of the Universal Declaration of Human Rights of 1948 to attempt a common justification of human rights and human dignity to be a dangerous shortcoming affecting its real validity. Its practical implementation presupposed a common belief in its universal, absolute validity which made such a justification an absolute condition for its implementation. He saw a natural reason for human rights. The conception of human rights, like the modern ethos of human rights, was generally compatible with all the basic philosophical-religious reasons that had made any sort of contribution to it. The individual dignity of persons could be recognised without reference to God. Dignity was based on persons' experiencing themselves as moral beings. The idea of human dignity had been decisively influenced by Christianity but this historical influence was not a basis for dependence, in the sense that the acceptance of this idea somehow bound one to the Christian faith. As viewed by a pluralist society, this meant that the idea of human dignity was part of a rational, natural ethos which claimed worldwide recognition.

The theological basis for human rights was also not simply a superstructure which had been imposed on modern human rights. The final argument in favour of human dignity was that humans shared in God's nature (being in the image of God). In order to understand themselves, human beings needed God himself. It was particularly the anthropological conviction of John Paul II that Christians were persons who had been successfully realised. Only theology could lead to a correct anthropology which perceived human beings in both of their dimensions and not merely biologically. For this reason, it was also meaningless to ask when a human being started or stopped being a person. That was hidden in the eternal will of God. So the real origin of the concept of person was theological and linked humans being persons to God's being a person (Albertus Magnus). Since God had created human beings as goals and not as means, humans had absolute value as such. Christ had confirmed that by dying for every person. Tertullian had claimed that God had thought of Christ as the model when he created human beings. Humans were created in Christ, the perfect image of God. Humans were, first and foremost, Christocentric beings because they were defined historically by their relationship to Christ as a person. Humans were created as social beings and this reflected the relationships within the Trinity. The only thing which could correspond to this was social relationships based on mutuality and inner meaning, but not those which were built on possessions, power and privileges, The real contribution of the Church to the defence of human rights could be seen, *inter alia*, in the view that humans must never have control over human life. The very critical attitude of the Catholic Church to human rights was traced back to the anti-ecclesiastical aspects of the enlightenment which had written human rights large on its banner, and also to the individualistic context. The speaker did not mention that this showed convergences with the quite typical, contemporary, Islamic criticism of human rights. Pius XII, in 1942, was the first to become one of the most influential figures of the new era, together with John Paul II, in particular. The work of the Church was to stand guard over humaneness which offered the world an anthropology of the person that respected human values and was open to transcendence. Human dignity did not only consist of being in the image of God but also of being God's children who shared in divine life. Whereas other created beings only manifested the power of God, the human being was a reflection of the divine mystery itself and made visible who and how God is, namely a pure person in perfect love. The relative autonomy of the human spirit, in analogy to God, was the highest form of natural knowledge of God. This went far beyond what had been said previously by the Christians in the dialogues, where they had restricted themselves to the texts from Genesis and come relatively close to the Islamic concept of vicarious responsibility. The theology, which had thus far been avoided so resolutely and successfully, was in full swing here: without theology, without Christology, no anthropology.

Next, there was a discussion of the practical, historical expressions of these conceptions of human dignity. The starting point for the Christian address was twofold: human dignity as accessible for every person and the specifically Christian approach. With the light of reason and inner enlightenment, it was possible to discover, in the natural law which was written on people's hearts, the holy value which was encapsulated in human life from start to finish. Undoubtedly, respect for human life and respect for human dignity were two important pillars that were essential for humans to live together and for the social stability of all communities. All human beings were brothers and sisters, as Christianity emphasised, but it shared this insight with many other religions as well. Neither Christ nor the apostles had turned this into a system or drawn up precise laws for a specific form of society. But every community, which wished to endure, had gradually organised itself in line with their message. The message and mandate of the Church was to bring about the unity of humankind within itself, irrespective of ethnic, cultural, national, social or other boundaries (the Church as sacrament of unity). All people were equal independently of their social status. God wished the inequality, which resulted from natural differences and human freedom, to be levelled out by the development and growth of life in society. Religious oppression was a sign of backward civilisation derived from a misunderstanding of religion. According to Muslim conviction, all monotheist religions agreed on the value of human beings but the practical reality had often appeared quite different. So a conception of human dignity had not only to be measured by the global acceptability of its basis but also by its global implementation by those who supported it.

The corresponding Muslim address, response and answer to this were prime examples of how profound the differences were in and despite all the efforts of dialogue, and how quickly one can fail to communicate and one person does not understand the basic conceptions of the other. The address given was another basic presentation of the ideal conception of Islam but the question of the historical implementation of theory in practice was not asked. The speaker considered this a constructive approach for a Christian-Muslim dialogue on human dignity and human rights. But Akasheh from the Pontifical Council for Interreligious Dialogue saw it as a confused, supercilious, non-dialogical and frequently ill-informed misunderstanding of the subject. And precisely the criticism of this criticism clearly revealed the true degree of misunderstanding, because it affirmed the address and adopted the view that the critics should have known that the speaker was really a self-critical person and convinced about Christian-Muslim dialogue, but that that was not the point here. Perhaps this was indeed evidence of the fact that Islam is, to a high degree, immune to criticism in general and especially from outside, because it has the only right solution already perfectly formed for dealing with all questions. If anyone else had anything positive to contribute, that was really already present in Islam, and the numerous shortcomings were all the fault of others, especially of the atheist, materialist West and still more of Israel.

Dignity was related to characteristics which were given to people individually. In the case of human beings, this meant reason, the ability to learn, the will and freedom to decide and, above all, that they subjected themselves to God alone. Human beings were religious by nature, *fitrah*, derived from the Arabic word for creation, *fatr*, which was, in fact, identical with Islam. For Islam, human dignity, like all moral values, was innate and hence unchangeable, absolute, and common to all people at all times. In Islam, the fundamental, determining concept was justice and not equality or rights. What the West considered a right (the right to life, the right to learning) was seen by Islam as a divine precept. From a Western, philosophical perspective, human perfection came about by humans gaining control over nature, the environment and their fellow human beings. Consequently, the most perfect being was the strongest being, as argued by Darwin, Nietzsche, Freud and Marx. The relationship between this human being and others was a relationship of more or less legal expulsion, of illegal appropriation and of domination because the fittest survived. Even learning and science were only good as means to power which, in turn, was the source of pleasure. Such a life was dominated by material and physical licentiousness. So there were two different ways of seeing things, indeed two different people, one religious and the other secular and atheist. Which should be taken as the starting point for the question of human dignity?

This was followed by a detailed comparison of the two persons. A person without faith could base their life on various approaches in contrast to a person with a spiritual starting point. The former was the victim of whims and caprices whereas the latter was rightly guided. Muslims believed that a person who was the product of social reflexes could not be human. Nor could persons be human who were the product of chemical and physiological reactions which determined their psychological, mental and emotional systems, as the behaviourists believed, nor persons who were equated with their socio-economic output as in Marxist materialism. Those were not really human beings. That was alienation, the antithesis to human dignity, defined by the lack of a balanced combination of knowledge of self, of the world and of God, such as was exemplified by Islam. Human dignity was found in the complementarity between this world and the next, between science and faith. In contrast, in the Jewish-Christian tradition the basis for the contradiction between religion and science was rooted in the conception of original sin. At the tree of knowledge, humankind had been confronted with the choice between religion and knowledge. Knowledge was seen as divine and ignorance as human. In Islam, on the contrary, the forbidden tree was the symbol of lust, namely the symbol of animality and not of the humanness of humans. God had given humans knowledge by teaching them the names of everything. Therefore, in Islam, religion and science or learning were synonyms. In the Jewish-Christian, Western civilisation, the separation and conflict between science and religion resulted in dividing up history into the age of religion and faith and the age of learning and science. In Islam, on the other hand, they flourished together and also declined together, although this was not further discussed. The lessons which the speaker drew from his interpretation of the fall were very broad. That was the type of person who had produced the colonial age, the devil of Marxism, and existentialism with its lack of meaning and significance. Indeed, it was paradoxical to speak of dignity and human rights with this conception of human beings. Those systems were faulty and did not spring from the same womb as human rights.

Looked at in this way, according to the speaker, one could see that human rights were being used as a political weapon. The most reliable and simple thing for human dignity and human rights would be to return to religion, meaning Islam, and then all the problems would be solved. Islam treated others as they deserved. One had to ask what must be changed in the Arab Muslim world so that it could be given a certificate of good behaviour for human rights. Did it have to switch to a primitive conception of human beings which limited being fully human to particular people, while all other people, nations and ethnic groups could not be God's chosen people?

In this connection, Akasheh called for a revision of history teaching so that other religions and civilisations could recognise themselves in it (e.g. original sin). The speaker felt that the recommendation firstly to inquire of those affected (the Christians) deprived him of his right to the free expression of his interpretations in the dialogue.

The last round of the discussion examined the challenges for the future. On the Muslim side, two questions were identified as major challenges. Islam was accused both of neglecting human rights and of failing to accept pluralism. The free choice of one's marriage partner and freedom to change one's religion contradicted Islamic law. As far as the rights of minorities were concerned, religious liberty had always been legitimate for the peoples of the Book. Minority rights were an excuse for intervening in internal affairs, as Europe had been doing since the middle of the 16th century when it won over the Christians in the Arab world to its side. That was mixed up with colonial and imperialist policy even down to the present when America and Europe defended the rights of Muslim non-Arab or Arab non-Muslim minorities. Excess in the exercise of rights should be avoided. In the Christian view, the Islamic prohibitions of marriage simply conflicted with people's basic right to choose their partner and start a family with him/her. There was also an objection to the insinuation that minority rights were merely a means for intervening in Arab governmental affairs. In addition, it was doubted that mutual respect between two religions was possible if only one of these religions was allowed to proclaim its belief openly.

The Christian address by Andrea Pacini focussed on two quite different questions for the future, bioethics and the form of society in the context of increasing globalisation. The basis for human dignity was their being in God's image. Humans were the only beings who had free, intelligent subjectivity and were called to enter into dialogue with God and into a relationship of obedience to God and his will. The concept of "person" comprised the whole dignity of human beings as such: individual subjectivity which could not be instrumentalised because it had freedom and conscience and the relational nature of human beings. Human beings had the responsibility repeatedly to create the conditions in which it was possible for every person to live in dignity. Being a person also meant being the bearer of fundamental rights which were not bestowed on one by society but belonged to the person as objective rights based on their being a person. According to the Christian understanding, these rights were always related to duties because the rights of another person implied duties on my side.

Bioethics consisted of both reproductive biology and genetic engineering. The coincidental nature of natural conception itself reinforced the personal character and the gift character of a developing human being and counteracted their instrumentalisation. Otherwise, there was a danger that society would no longer be in a position to grasp the dignity of humans as a whole and thus also to respect it at later stages. The second challenge had less to do with the physical than the social life of human beings. In traditional societies, human dignity had been safeguarded by traditional forms of solidarity (extended family and local community). Processes of change needed to be steered in such a way that human dignity was not only preserved but even further strengthened. It was a matter of developing awareness but also of structural action in legislation and politics to produce forms of society which reflected human dignity. Special significance was attached to the interplay between rights and duties. In the West, duties had been somewhat neglected. But this emphasis had good historical reasons, namely the protection of the human person. At the moment, there was a danger that human dignity might become secondary compared with economic development, for example. As Christians saw it, to protect human dignity in modern societies there was no alternative to human rights.

From the Muslim point of view, the exclusion of revelation as a source of knowledge and science had led in the West to a concentration on people's rights while simultaneously neglecting duties and ethics.

According to the joint report, Christians and Muslims particularly shared the view that human beings were created by God and had received their dignity from God. This dignity was absolute and independent of race, skin colour, language or convictions. There was a strong call not to confuse values and ideals with the reality of human life which often did not live up to them. Special attention had to be paid to those who had to suffer injustice and the deprivation of these God-given human rights: the poor, the oppressed, abandoned children, persecuted minorities and refugees. Hope was expressed for a new, ethical world order which would respect human dignity everywhere. The conference resembled a collection of many anthropological aspects which had been discussed in the previous dialogues and thus formed a logical conclusion to some extent.

3.6 Dialogue with Iran

From 1976 onwards, there had been contacts with Iran on the question of future dialogues. The concerns of the Secretariat were youth, social justice and the phenomenon of religion. In 1977, there was an interesting exchange on anthropological questions. It was emphasised that faith in God led Christians and Muslims to have a similar view of different aspects of human beings, their duties and the aims they should achieve. In tension with this, Christians saw religion as critical and prophetic with the aim of strengthening people's rights. The question was whether this also applied to Islam. The answer was a fairly firm denial. On the contrary, the theory of human rights could prove to be a trap because it had been drawn up outside of Islam. Islam saw it as an ideology which it could not accept. In Islam, human beings had no innate rights; on the contrary, they had a radical duty to God and everything else was simply a consequence of that relationship.

The first conference with the Iranian Secretariat for Interreligious Dialogue took place in 1994 in Tehran on the theological evaluation of modernity. The address by Felix Machado is the only one available. He linked the concept of modernity with the concept of humanity. The enlightenment had wanted to guarantee a peaceful social existence for all people. At the present time, human beings with their dignity and rights were absolutely central. The Catholic Church saw this as negative. After all, reason was also applied to the religious realm. A sharp contrast had developed between faith and reason. Modernity was a Eurocentric system with Christian roots. The primacy of reason and the central position of the individual and his/her rights had fundamentally changed the practice of religion in Asia. But people in Asia were still of the opinion that religion was an answer to their practical needs and gave them the necessary motivation and ultimate meaning for life. The conflict between sacred and secular, faith and reason, state and religion had been avoided. But the influence of modernity had also encouraged devotion to this world, humanism and social awareness. The main characteristics of modernity were the importance and priority given to individuals and their rights.

In 2000, the second dialogue dealt with Islam and Christianity in relation to religious plurality. Religious pluralism was connected with greater mobility and progress in the realm of the communication media. In 2001, the third dialogue dealt with youth and their identity and religious education or, more precisely, the identity crisis of young people and their momentary hopes. It was significant that a Christian-Muslim dialogue, which was actually conducted on a quite different subject, still ended up dealing with the issue of the rights of religious minorities and the different definitions of religious liberty on each side.

3.7 Islamic-Catholic Liaison Committee

On 22 June 1995, on the occasion of the dedication of the mosque in Rome, the agreement to establish a so-called Islamic-Catholic Liaison Committee was signed. It was to meet at least once a year to discuss a subject that was important to both Christians and Muslims, and to take stock of the status of the relationships between Christians and Muslims worldwide. For both parties, it was a high-ranking, representative affair. Right at the beginning, it raised the question of the standing of women in society. The need to recognise the dignity, role and rights of women in society was strongly affirmed. At the first meeting, the Islamic Council for Da'wah and Humanitarian Relief was also represented. The meeting dealt with three issues, all of which have at least something to do with anthropological questions: justice and human dignity, the environment and human safety, as well as poverty and humanitarian aid. The poor should be provided jointly with the basic necessities such as housing, health care and food. In addition, all violence in the name of religion was very strongly condemned. At the next meeting, it was mainly a question of minority rights and of the way in which Muslims and Christians talked about one another. The third meeting concentrated on decidedly anthropological issues: initially, the duties of men and women in the family and society, then human rights and duties and, finally, the rights of the child in the family and society. Obviously, a danger was recognised here that children would be neglected in modern society. Divergences were mentioned before coming back emphatically to underline common values: human dignity which came from the almighty God and was the source of human rights and duties, the family based on marriage, the child's right to life, to a good family environment, to education and religious teaching. The atmosphere of the meeting was relaxed. It was noted that mutual understanding, respect and friendship had grown. Perhaps the most important contribution of these meetings was the progress on the human level between the highest representatives of both sides. The next meeting dealt with a culture of dialogue and with lasting religious values, the so-called "third way". It stated that dialogue embraced all forms of encounter which promoted mutual understanding and respect. The basis should be increasing mutual trust in order to be able to bring about justice, promote peace, to safeguard respect for human dignity and to make secure coexistence in human society a reality. The culture of dialogue should be based on the fulfilment of ethical principles. The furthering of ethical values was less disputed than the question what exactly those values might be for both sides, and where there were not overlaps but differences as far as the ideal image of human beings and their behaviour was concerned. The next meeting took up the question of the rights and duties of citizens For both sides, it was important that religious values and a spirit of responsibility were the starting point, and these had to be encouraged in order to safeguard human dignity and promote coexistence between the nations. The idea of a clash of civilisations was rejected. After September 11th, a joint declaration was produced. There was always agreement when it was a matter of cooperating to defend the high values of humankind, of tensions in the situation of countries in which Christians were persecuted, or of fundamental issues of a culture of dialogue or questions of principle such as changing one's religion. The differences in the view of human beings and their freedoms could not be denied. The ninth meeting discussed human dignity and human rights in armed conflicts. In armed conflicts, humanitarian law and the rights of the civilian population and of prisoners had to be fully respected. In addition, no one should be refused access to water, food, medicine or medical care. The infrastructure, namely everything necessary for living, had to be preserved. This appeal was based on religious values and international conventions. Places of worship should also be respected and there should be a right to conduct worship in peace and war.

3.8 Dialogues with the Al-Azhar University

The first contacts with the Al-Azhar University concerned peace, cooperation and the new climate of dialogue, and also faith which consisted not only of words but also of deeds. The relation of young people to faith was important. This issue was less critical on the Islamic side – faith in God was too deeply rooted in every Muslim for a "God is dead" theology to be conceivable. In the Catholic view, young people were not rebelling against faith in God itself but against many things that people had made of it. It was important to find a form that spoke to their spiritual needs. On this point, the Muslims saw true faith in the true God threatened much more quickly and directly. As soon as family life was no longer intact, young people were unable to have an intact faith. There was an immediate intention to maintain regular contact with one another from then onwards on all questions affecting the relations between Christians and Muslims and to appoint a representative from each side to this end. During a further visit of a Vatican delegation in Cairo, a joint statement was drawn up with the religious and academic dignitaries of the Al-Azhar University. It dealt with the quest for a cure for materialism and atheism and with cooperation against racial discrimination and generally against all crimes against humanity. In this connection, it emphasised common religious values: justice, love, respect for the right to life and human rights. Only the way of obedience to God could help to solve the social, economic and political problems. In the Islamic view, God had explicitly created human beings so that they could help one another especially in the area of justice in order thus to achieve peace. Alienation from religious values was generally the reason for the typical problems of modern societies: hectic, disorder and a lack of orientation. Both Christians and Muslims could work well together by virtue of their own religious values and contribute to the fulfilment of humankind's desire for well-being, peace and happiness.

Thereafter, there were contacts with a view to setting up a joint committee for the dialogue. The agreement to establish this committee was adopted on 28 May 1998 in Rome. The starting point of the whole process was the need for precise knowledge and a right understanding of the contents of the faith and of the practice of one another's religions. The second point was that religions should be able to play their part in human society as far as promoting brotherliness, solidarity, cooperation, justice and peace was concerned, and also for the well-being of humankind as a whole and in the common struggle against fanaticism. The results achieved thus far were to be augmented. Work was to be done in the areas of research on common values and the furthering of justice, peace and respect for religions. Subjects of common interest were to be dealt with such as the defence of human dignity and human rights and encouraging respect for one another – aims that were clearly anthropological. There was to be a meeting at least once a year and each meeting should conclude with a press communiqué. No official information would be released by the committee about the papers submitted without the consent of both sides; this leads one to conclude that a number of interesting and controversial points might be kept under lock and key. After the visit of Pope John Paul II to the Al-Azhar, the date of his visit, February 24, was declared the "day of dialogue" and the date for future meetings. In total, there was a joint declaration on the Balkans, one on the situation in the Near East and, later, one on September 11th. Any attempt to exploit a particular situation and any attack on religious liberty was rejected. This joint declaration also followed the lines were which were otherwise familiar from papal statements.

After a number of meetings, the general evaluation was very similar to that of the meetings of the Islamic-Catholic Liaison Committee: the atmosphere was relaxed, full of trust and respect, and the participants were increasingly listening to one another. For 24 January 2002, the Pope had invited people to Assisi for a day of prayer for peace in view of the situation of Muslims, in particular. On this occasion, special importance had been attached to the Muslim witness to peace. God had created all people from one father and one mother. All monotheist religions could agree on respect for such values as uprightness, justice, peace and welfare, an exchange of good deeds, cooperation between nations for well-being and on spirituality. God had created people as tribes and nations and here, according to God's standard, the most pious was the best. All the monotheist religions called upon people to esteem justice and law by helping legal owners to regain their rights – the practical meaning of this was support for the Palestinian people. There was no compulsion in religion. In Egypt, Christians and Muslims had lived peacefully together for fourteen centuries with equal rights and duties. On the basis of the reference to justice, the Al-Azhar agreed with the appeal for peace.

At the next regular meeting of the committee, the subject was religious extremism and its effects on humankind. The Christians explained fanaticism as considering every person to be an enemy who was not explicitly for something. It was necessary for both Christians and Muslims to take up the challenge of religious plurality. The Muslim participants underlined Islam's moderate approach. In the discussion, extremism was condemned because it was not in harmony with the teachings of either religion. Extremists might have serious intentions but had the tendency to believe that they alone were right and to be intolerant of others, not accepting them with their differences and also violating their rights. Dialogue could be helpful as a counterweight to extremism. Society had to be included as well, namely family life, education, social development, mass media, promotion of justice and solidarity. The meeting in 2004 dealt with rejecting generalisations and the importance of self-criticism. Both religions rejected generalisations when judging people. The responsibility rested solely with the individual or community concerned. Only those who had really done something wrong could be held to account. Innocent people should not be pilloried for the misdeeds of others. Both religions were also in favour of self-criticism, examination of conscience and asking for forgiveness. Everyone should engage in the examination of conscience, wherever possible admit their guilt and thus return to the right behaviour. In this way, it was hoped to spread justice, peace and love.

3.9 Regional dialogues

The project of regional dialogues was very important. But they were rarely conducted directly from Rome. Above all, they were concerned with the dialogue of life. The first phase of encounter had already taken place. It was no longer necessary to built up trust and so common problems could be tackled. The diocesan bishop had the decisive role. The motives behind this approach were clear. What mattered in these situations was not purely diplomatic activity of the Holy See but rather activities of the Church. The Pontifical Council for Interreligious Dialogue also exercised pressure sometimes but could not take the place of other levels or institutions. Where Christians were in the minority, the visits or other activities of the Pontifical Council always tried to respect the role of the local church and to strengthen it in the eyes of the non-Christians. Moreover, in certain countries, the project of regional dialogues provided the first impetus for dialogue at all.

3.9.1 North Africa

The bishops' conference for North Africa had proposed a dialogue meeting and it was agreed together with the competent bodies of the Pontifical Council, although with a focus on Europe where the Christian majority needed more mutual knowledge of, respect for and cooperation with Muslims. The overall Catholic view was that interreligious dialogue generally progressed only slowly and required a lot of patience and generosity. After a certain time lapse since the Second Vatican Council, new reflection on the differences between Christians and Muslims had come about, and also some measure of cooling off and disillusionment, and this now needed to be replaced by a new impetus based on a realistic, balanced appreciation. The official religious institutions were holding back in regard to dialogue or even rejected it outright. The spiritual path, which the participants in dialogue had followed, united them, moving from some kind of negative view of one another to genuine knowledge. It was even described as a temptation to see oneself as the sole owner of the truth and the good. Everyone was simply a witness to and a fellow worker of God. In an anthropological sense, the necessity of a respectful attitude was strongly emphasised, from religious education in the family to text books and teacher training, and special attention had to be paid to reconciling family influences and official education. The starting point was the reasons found in the Bible and the Quran for the right to religious difference, and hence to freedom of religion and conscience which was what was really meant by "respect and love". After returning home from the conference, the participants formed core groups which continued the cooperation.

3.9.2 West Africa

The second conference of this kind brought together participants from English-speaking West Africa in 1991. It comprised workshops on development, democracy and relationships with government, mixed marriages and the influence of the media. The greatest agreement was reached in the realm of development. Both Christians and Muslims had a very similar conception of human development as the development of the whole human being in all his/her aspects: economic, socio-cultural, political, emotional, moral and spiritual. Human beings had been created and raised above all other creatures in order to worship God and glorify the name of their Creator, to live according to his precepts of love, prayer, good neighbourliness, peace and mercy in the world and to be with him thereafter. Development should benefit all, both poor and rich, and should respect nature. However, the motivation for this varied according to one's religion, because Christians and Muslims had different conceptions of the relation between religion and the state. There was already much cooperation in development but it would be desirable to have more. On mixed marriages, the views were more pragmatic. There was a lack of knowledge, especially in connection with the understanding of marriage in the two religions. Religious leaders should encourage people to marry within the limits of their own religion. But the choice of one's marriage partner was a basic human right. Democracy was seen both fundamentally and practically as the decisive point as far as understanding and cooperation between Christians and Muslims were concerned. Here, there was a major difference between the Church which, as a matter of principle, did not identify itself with any political system, and Islam which considered religion and politics inseparable because the Quran was, after all, also a law book which regulated the life of all Muslims, also in the realm of government. The conference recommended overcoming prejudices, neither side describing the other party as unbelievers, and especially working together for the poor in the struggle against exaggerated expenditure. Democracy included respect for the rights of every person irrespective of origin, religious affiliation or gender. It was particularly emphasised that every person had the right to religious liberty, and religious communities had the right to act freely. The following areas were mentioned as especially sensitive and riddled with conflict: the nomination and promotion of state officials, the designation of land for worship purposes, and the access to and use of the media. Here, too, the participants stated their intention to organise similar dialogues themselves in their various countries.

3.9.3 South East Asia

In 1994, there was a regional dialogue in South East Asia (excluding Singapore and Brunei). Cardinal Arinze described religiosity as a characteristic of Asian people. Asia had produced many different religions and cultures but also harmony and tolerance between them. This should be further strengthened. For dialogue, mutual respect and especially respect for religious liberty were required. This could lead to cooperation for which Asia was again an excellent example. God was the Creator, Saviour and goal. Human beings were God's image and created with the aim of seeing God in heaven. In this light, the differences between Christians and Muslims paled – a statement which also referred back to John

Paul II. Christians and Muslims wanted to subject the details of their lives to God. Both considered prayer, fasting and alms giving to be necessary. Patience was needed for interreligious dialogue to bear fruit. As a positive example, Arinze mentioned the behaviour of Christian and Muslim groups during the Gulf War in 1991. The challenges were the growth of materialism and consumerism, the egoism of the rich who saw the poor increasingly as disturbing intruders, and, in general, living as if God did not exist. In dialogue, it was important to seek the fullness of truth together. Of the addresses given, unfortunately only one was published in which Christians and Muslims were practically not mentioned directly. The final report emphasised the dialogue of action. Technological progress and economic growth brought about rapid change also in the way in which people saw their relations with one another and the purpose of their lives in general. The process of secularisation was viewed very negatively. It tended to eradicate humaneness. A highly developed individualism – the rights and self-fulfilment of the individual at the expense of the family and society – was considered new. In the past, in contrast, solidarity and harmony had been the traditional cultural values and the model for community life. Now there was more polarisation, also between religions. Few people were able to enjoy modern affluence and, instead, many were exploited and marginalised. The exploitation of women and children by work or for sex was particularly repulsive. The mass media had further encouraged the breakdown of solidarity. There was clear preference for the family rather than an exaggerated individualism, and objection to all attempts to set homosexual relationships on the same level as heterosexual ones. There was also explicit emphasis on the responsibility and rights of parents in relation to the sexual education of children in accordance with their religious convictions. There was concern about extremist religious trends. What was needed was mutual consideration, not mistrust.

3.9.4 East Africa

During the planning for a regional dialogue in East Africa at the beginning of 1997, it became clear how much value was attached to mutual respect and solidarity among Christians and Muslims in general, or at least in this region, and that, despite all the declarations of openness toward modern developments, both stood for a more traditional image of family and society and found themselves, as it were, united in opposition to many developments of modernity and secularisation.

3.10 After 40 years of dialogue and September 11th

The official review by the Pontifical Council for Interreligious Dialogue asked what had changed for the better or the worse in 40 years, and noted a more critical attitude of the Muslims to their inheritance and religious tradition following September 11th 2001. Extremism and violence in the name of religion had become issues and were thus less taboo that in the past. Human dignity and human rights, including the right to religious liberty, had also become subjects that were being discussed in intellectual circles of societies with a Muslim majority. Muslims were increasingly taking the initiative for dialogue in order to correct their image. The anti-terror initiative in the West with its anti-Muslim aspects had given rise among Muslims to a feeling of humiliation which was hard to overcome. Religion and politics remained a difficult subject. A healthy degree of secularism seemed difficult to achieve in countries with a Muslim majority. There was a real danger of politicising religion and thus confusing religion and politics. Islamic political parties were working to create Islamic states and to introduce Islamic law. A particularly sensitive issue was school books which dealt with religious and historical questions and their role in spreading Islamic extremism and even violence. It was not yet possible to discuss such sensitive historical issues as the Islamic conquest, the Islamisation of the countries conquered or the crusades. For the future, in view of the risk of a clash of cultures, Akasheh identified the task of protecting humankind from anything which could lead to new world wars. In his opinion, everyone would join this cause if they did not support or practise terrorism.

4 Final remarks

The wealth of detail, especially in the dialogues related to anthropology, was and is very great. That was less a result of the anthropology of the Roman Catholic Church, which is based strictly on creation theology with few exceptions, as can be seen in the remarks on the statements of the magisterium, and more because anthropology has to be seen in a much broader way in the Islamic context. With regard to the theological basis for anthropology, Christians and Muslims are not so far removed from one another. Both maintain that God created human beings, gave them their tasks and will require an account from them. The differences lie in the application. From the Christian point of view, the image of God allows more leeway in determining the practical accomplishment of these God-given tasks. Whereas, for God's vicarious representatives in Islam, Islamic law specifies this accomplishment fairly precisely and in considerable detail, something which was recognised very early on as a difficulty for the dialogue. So the real freedoms are very different and considerably greater in the Christian context. But, from the point of view of the Muslim participants in the dialogues, this Christian freedom does not appear desirable but rather frightening. The provisions of Islamic law are felt more to be the strength of their own religion which has solved all the problems in advance that way. This approach had a strong effect on the very choice of subjects for the dialogues; the Catholic side would have liked them to be more theological. The consequence was that Christian practice was often compared with an Islamic ideal and this naturally produced an imbalance in the dialogue as a whole. For this reason, it can be said that the dialogues clearly identified the common features already present in the anthropology of the two religions but did not bring about any convergence concerning the content of the differences and probably could not do so. However, there was some human convergence as a result of the actions of Paul VI and John Paul II thanks to the diverse activities of what is now the Pontifical Council for Interreligious Dialogue, in particular through the dialogues. It was certainly a stony path at times, but knowledge, understanding and trust had grown enormously, nevertheless. That alone should make it worthwhile to continue along this path.

Bibliography

Abedin, Zaineh, "Response to the Paper of Sr. Rosaleen Sheridan on the Theme: Science, Faith and the Concerns of Youth", in: *Religious Education and Modern Society, Acts of a Muslim-Christian Colloquium Organized Jointly by the Pontifical Council for Interreligious Dialogue (Vatican City) and the Royal Academy for Islamic Civilization Research Al Albeit Foundation (Amman)*, s.l.s.a., p. 120–122.

Abou Mokh, François, "'Le dialogue dans le monde arabe", BSNC 30 (1975), p. 223–226.

Abou Mokh, François, "**Vienne:** Rencontre des chrétiens et des musulmans à Saint-Gabriel (Mödling – Vienne) 31 mai – 5 juin 1977", BSNC 36 (1977), p. 191–193.

Abou Mokh, François, "Visite de Son Éminence le Cardinal Pignedoli au Yémen", BSNC 33 (1976), p. 327–330.

Abou Mokh, François, "Voyage d'une délégation du Saint-Siège en Iran", BSNC 33 (1976), p. 319–326.

"Abuja – Nigeria: 22 March 1998, **Meeting with Muslim Leaders at the Apostolic Nunciature on the Occasion of the Pope's Second Pastoral Visit to Nigeria to Beatify Fr Cyprian Michael Iwene Tansi**", BPCDIR 99 (1998), p. 276–278.

"Abuja – Nigeria: 23 March 1998, **Discourse of the Pope at Meeting with the Episcopal Conference of Nigeria in the Apostolic Nunciature** (Excerpt)", BPCDIR 99 (1998), p. 278.

Abu-Rabi, Ibrahim, "John Paul II and Islam", in: Sherwin, Byron L. / Kasimow, Harold (edd.), *John Paul II and Interreligious Dialogue*, Maryknoll 1999, p. 185–204.

"Accra. Greeting to Leaders of the Muslim Community in Ghana", in: Segretariato per i non cristiani (ed.), *Chiesa e islam*, s.l. 1981, p. 19–20.

Ajaiby, "George, Attitude of Christians towards Human Dignity, A Historical Review", in: *Human Dignity, Acts of a Muslim-Christian Colloquium Organized Jointly by the Pontifical Council for Interreligious Dialogue (Vatican City) and the Royal Academy for Islamic Civilization Research Al Albeit Foundation (Amman – Jordan)*, Rome 1999, p. 67–79.

Akasheh, Khaled, "**Amman – Jordanie:** Réunion des Membres et des Consulteurs du Moyen Orient et du Nord de l'Afrique, 30 novembre – 2 décembre 1997", BPCDIR 99 (1998), p. 337–338.

Akasheh, Khaled, "Bureau pour l'Islam Rapport d'activités: novembre 1998 – octobre 2001", Pro Dialogo 109 (2002), p. 98–112.

Akasheh, Khaled, "Comment", in: *Human Dignity, Acts of a Muslim-Christian Colloquium Organized Jointly by the Pontifical Council for Interreligious Dialogue (Vatican City) and the Royal Academy for Islamic Civilization Research Al Albeit Foundation (Amman – Jordan)*, Rome 1999, p. 101–102.

Akasheh, Khaled, "Considerations on Forty Years of Religious Dialogue with Muslims (A Report)", BPCDIR 116–117 (2004), p. 195–204.

Akasheh, Khaled, "Report on the Activities of the PCID: Relations with Muslims", BPCDIR 101 (1999), p. 213–219.

Akasheh, Khaled, "**Rome:** Islam and Christianity Confronted with Religious Plurality, 7–9 February 2000", BPCDIR 104/105 (2000), p. 228–229.

Akasheh, Khaled, "Some Reflections on Islam and Christian-Muslim Relations in Certain Countries", BPCDIR 116/117 (2004), p. 252–259.

Akasheh, Khaled, "**Tehran, Iran:** The Third Colloquium, 18–20 September 2001", BPCDIR 108 (2001), p. 406–407.

Akasheh, Khaled, *"Une espérance nouvelle pour le Liban* Lecture de l'exhortation apostolique post-synodale à la lumière des relations islamo-chrétiennes", dans: Isizoh, Chidi Denis (ed.), *Milestones in Interreligious Dialogue, A Reading of Selected Catholic Church Documents on Relations with People of Other Religions, Essays in Honour of Francis Cardinal Arinze*, Rome/Lagos 2002, p. 246–254.

"A Letter Sent by the Holy Father to Archbishop Vinko Puljić of Sarajewo "Change Violence into Acceptance" (2 February 1993)", Islamochristiana 19 (1993), p. 212–213.

"A Rome (Vatican), le Pape reçoit les Evêques du Tchad en visite "ad limina" (27 juin 1994)", p. 269–271.

Al-Abbadi, Abdul-Salam, "Religious on the Use of the Earth's Resources, A Muslim Perspective", in: *Religion and the Use of the Earth's Resources, Acts of a Christian-Muslim Colloquium Organized Jointly by the Pontifical Council for Interreligious Dialogue (Vatican City) and the Royal Academy for Islamic Civilization Research Al Albait Foundation (Amman – Jordan)*, Rome 1996, p. 13–41.

Al-Abbadi, Abdul Salam, "Response to the Paper of S.B. Mons. Michel Sabbah on the Theme: Identity and Openness: Religious Education in a Pluralistic World", in: *Religious Education and Modern Society, Acts of a Muslim-Christian Colloquium Organized Jointly by the Pontifical Council for Interreligious Dialogue (Vatican City) and the Royal Academy for Islamic Civilization Research Al Albait Foundation (Amman)*, s.l.s.a., p. 112–119.

Al-Rifaie, Hamid Ahmad / Fitzgerald, Michael L[ouis], **"Declaration of the Joint Islamic-Catholic Liasion Committee** (12 settembre 2001)", BPCDIR 108 (2001), p. 333.

Al-Sartawi, Mahmoud, "Nationalism and Religion: An Historical Overview, A Muslim Perspective", in: *Nationalism Today: Problems and Challenges, Acts of a Muslim-Christian Colloquium, Organized Jointly by the Pontifical Council for Interreligious Dialogue (Vatican City) and the Royal Academy for Islamic Civilization Research Al Albait Foundation (Amman)*, Rome 1994, p. 17–26.

Al-Zafazaf, Fawzi / Fitzgerald Michael L[ouis], "Vatican City, 24–25 February 2004 **Joint Committee of the Permament Committee of al-Azhar for Dialogue with Monotheistic Religions and the Pontifical Council for Interreligious Dialogue**", BPCDIR 118 (2005), p. 52–53.

"**Alger – Algérie:** Visite du Cardinal F. Arinze, Président du Secrétariat pour les non-Chrétiens (28 nov. – 5 déc. 1987)", BSNC 67 (1988), p. 87–88.

"Allocution de S.M. le Roi Hasan II à l'occasion de la rencontre du Pape avec la jeunesse marocaine", Islamochristiana 11 (1985).

Amjad-Ali, Charles, "Towards a New Theology of Dialogue", Al-Mushir 33 (1991), p. 57–69.

"**Amman – Jordan:** Christian-Muslim Colloquium on "Religion and Nationalism" (18–20 January 1994)", BPCDIR 87 (1994), p. 266.

"**Amman – Jordan:** *Christian-Muslim Colloquium* **"The Rights and Education of Children in Islam and Christianity"**",BPCDIR 76 (1991), p. 109–111.

Anawati, Georges C., "Le dialogue islamo-chrétien en Égypte aujourd'hui", BSNC 30 (1975), p. 227–232.

Anawati, Georges C., "Zur Geschichte der Begegnung von Christentum und Islam", in: Bsteh, Andreas (Hrsg.), *Der Gott des Christentums und des Islams*, Beiträge zur Religionstheologie 2, Mödling 1978, p. 11–35.

Aquilina, Oscar, "Response to the Paper of Dr. Wahbah El-Zuhaili on the Theme: Science, Faith and the Concerns of Youth", in: *Religious Education and Modern Society, Acts of a*

Muslim-Christian Colloquium Organized Jointly by the Pontifical Council for Interreligious Dialogue (Vatican City) and the Royal Academy for Islamic Civilization Research Al Albait Foundation (Amman), s.l.s.a., p. 123–127.

Arinze, Francis [A.], "1984", p. 36–37,
- "1985", p. 38–39,
- "1986", p. 40–41,
- "1987", p. 42–43,
- "1988", p. 44–45,
- "1989", p. 46,
- "1990", p. 47–48,
- "1992", p. 52–53,
- "1993", p. 54–55,
- "1994", p. 56–57,
- "1995", p. 58–59,
- "1996", p. 60–61,
- "1997", p. 62–63,
- "1998", p. 64–65,
- "1999", p. 66–67,
- "2000", p. 68–69,
- "2000", p. 70–72,
- "2001", p. 73–75,

in: Pontifical Council for Interreligious Dialogue (ed.), *Meeting in Friendship, Messages to Muslims for the End of Ramadan (1967–2002)*, Vatican City 2003.

Arinze, Francis [A.], "Believers Walking and Working together", BPCDIR 70 (1989), p. 43–45.

Arinze, Francis [A.], "Concluding Discourse Together for Better Times", in: *Nationalism Today: Problems and Challenges, Acts of a Muslim-Christian Colloquium Organized Jointly by the Pontifical Council for Interreligious Dialogue (Vatican City) and the Royal Academy for Islamic Civilization Research Al Albait Foundation (Amman)*, Rome 1994, p. 123–124.

Arinze, Francis [A.], "Editorial", BSNC 69 (1988), p. 185–186.

Arinze, Francis [A.], "For a Responsible Use of the Earth's Resources", in: *Religion and the Use of the Earth's Resources, Acts of a Muslim-Christian Colloquium Organized Jointly by the Pontifical Council for Interreligious Dialogue (Vatican City) and the Royal Academy for Islamic Civilization Research Al Albait Foundation (Amman)*, Rome 1996, p. 5–6.

Arinze, Francis [A.], "Forty Years of Developing Theological and Dialogical Engagement", BPCDIR 116/117 (2004), p. 283–291.

Arinze, Francis [A.], "Greetings to the Participants", in: *Religious Education and Modern Society, Acts of a Muslim-Christian Colloquium Organized Jointly by the Pontifical Council for Interreligious Dialogue (Vatican City) and the Royal Academy for Islamic Civilization Research Al Albait Foundation (Amman)*, s.l. s.a., p. 5–7.

Arinze, Francis [A.], "In Faith and Gratitude", BPCDIR 74 (1990), p. 117–118.

Arinze, Francis [A.], "In Openness and Collaboration: Opening Address to Christian-Muslim Seminar from ASEAN Countries Gathered at Pattaya Thailand, 2 August 1994", BPCDIR 87 (1994), p. 224–231.

Arinze, Francis [A.], "In Promotion of Interreligious Dialogue in Africa", BPCDIR 87 (1994), p. 210–213.

Arinze, Francis [A.], "In Respect and Openness", in: Pontifical Council for Interreligious Dialogue / World Islamic Call Society (edd.), *Co-existence between Religions, Reality and Horizons*, Vatican City s.a., p. 7–9.

Arinze, Francis [A.], "Inaugural Address", in: *Human Dignity, Acts of a Muslim-Christian Colloquium Organized Jointly by the Pontifical Council for Interreligious Dialogue (Vatican City) and the Royal Academy for Islamic Civilization Research Al Albait Foundation (Amman)*, Rome 1999, p. 15–17.

Arinze, Francis [A.], "Inaugural Address", in: *Nationalism today: Problems and Challenges, Acts of a Muslim-Christian Colloquium Organized Jointly by the Pontifical Council for Interreligious Dialogue (Vatican City) and the Royal Academy for Islamic Civilization Research Al Albait Foundation (Amman)*, Rome 1994, p. 9–14.

Arinze, F[rancis A.], "Means Necessary for Developing Dialogue with Non-Christians", BSNC 41/42 (1979), p. 168–170.

Arinze, Francis [A.], "Meeting Other Believers", BPCDIR 82 (1993), p. 17–22.

Arinze, Francis [A.], "Preface to the English Translation", in: Pontifical Council for Interreligious Dialogue / Borrmans, Maurice, *Guidelines for Dialogue between Christians and Muslims*, Interreligious Documents 1, New York / Mahwali 1990, p. 5.

Arinze, Francis A., "Prospects of Evangelization, with Reference to the Areas of the non Christian Religions, Twenty Years after Vatican II", BSNC 59 (1985), p. 111–140.

Arinze, Francis [A.], "Reflections on the Silver Jubilee of the Pontifical Council for Inter-Religious Dialogue", BPCDIR 72 (1989), p. 313–318.

Arinze, Francis [A.], "The Child Is Prescious", in: Pontifical Council for Interreligious Dialogue / Royal Academy for Islamic Civilization Research Al Albait Foundation (edd.), *The Rights and Education of Children in Islam and Christianity, Acts of a Muslim-Christian Colloquium*, Vatican City s.a., p. 7–8.

Arinze, Francis [A.], "The Importance of Women's Role in Society", in: *Women in Society According to Islam and Christianity, Acts of a Muslim-Christian Colloquium Organized Jointly by the Pontifical Council for Interreligious Dialogue (Vatican City) and the Royal Academy for Islamic Civilization Research Al Albait Foundation (Amman)*, s.l.s.a., p. 5–9.

Arinze, Francis [A.], "The Rights of the Unborn Child", in: Pontifical Council for Interreligious Dialogue / Royal Academy for Islamic Civilization Research Al Albait Foundation (edd.), *The Rights and Education of Children in Islam and Christianity, Acts of a Muslim-Christian Colloquium*, Vatican City s.a., p. 15–22.

Arinze, Francis [A.], "The Way Ahead for Muslims and Christians", BPCDIR 91 (1996), p. 26–32.

Arinze, Francis, "**Visit to Jerusalem:** 26. Feb.–3 March 2001", BPCDIR 107 (2001), p. 264–266.

Arkoun, Mohammed, "Réflexions d'un musulman sur le "nouveau catéchisme"", Islamochristiana 19 (1993), p. 43–54.

Arnaldez, R., "Les valeurs de l'Islam", BSNC 9 (1972), p. 21–29.

"Assisi: 24th January 2002, **Address on the Occasion of the Day of Prayer of Peace**", BPCDIR 109 (2002), p. 45–48.

"Assisi: 24th January 2002, **Greetings on the Occasion of the Day of Prayer for Peace**", BPCDIR 109 (2002), p. 43–44.

"Astana International Airport, Kazakhstan: 22 September 2001, **Response to the Welcome of the President**", BPCDIR 108 (2001), p. 305–308.

"Astana, Kazakhstan: 23 September 2001, **Angelus Message**", BPCDIR 108 (2001), p. 311–312.
"Astana, Kazakhstan: 23 September 2001, **Homily at the Holy Mass**", BPCDIR 108 (2001), p. 308–311.
"Astana, Kazakhstan: 23 September 2001, **To the Ordinaries of Central Asia** (Extract)", BPCDIR 108 (2001), p. 312–313.
"Astana, Kazakhstan: 25 September 2001, **On the Occasion of the Parting Ceremony**", BPCDIR 108 (2001), p. 317–319.
"At the Ankara University: Christian Lessons to Muslim", Islamochristiana 13 (1987), p. 217.
Aydin, Mehmet, "Response to the Paper of Rev. Vincent Duminuco, S.J. on the Theme: "Religious Education in Catholic Higher Education: Present Situation and Aspirations for the Future"", in: *Religious Education and Modern Society, Acts of a Muslim-Christian Colloquium Organized Jointly by the Pontifical Council for Interreligious Dialogue (Vatican City) and the Royal Academy for Islamic Civilization Research Al Albait Foundation (Amman)*, s.l. s.a., p. 130.133.
'Ayesh, Husni, "Muslim Response", in: Pontifical Council for Interreligious Dialogue / Royal Academy for Islamic Civilization Research Al Albait Foundation (edd.), *The Rights and Education of Children in Islam and Christianity, Acts of a Muslim-Christian Colloquium*, Vatican City s.a., p. 110–114.
Ayoub, Mahmoud [M.], "Pope John Paul II on Islam", in: Sherwin, Byron L. / Kasimow, Harold (edd.), *John Paul II and Interreligious Dialogue*, Maryknoll 1999, p. 169–184.
Ayoub, Mahmud [M.], "A Muslim Appreciation of Christian Holiness", Islamochristiana 11 (1985), p. 91–98.
Aziz, Mukhtar, "Words of Welcome", in: Pontifical Council for Interreligious Dialogue / World Islamic Call Society (edd.), *Co-existence between Religions, Reality and Horizons*, Vatican City s.a., p. 5–6.
Balarabi, Abdul-Hafeez, "Comment", in: Pontifical Council for Interreligious Dialogue / Royal Academy for Islamic Civilization Research Al Albait Foundation (edd.), *Human Dignity, Acts of a Muslim-Christian Colloquium*, Vatican City 1999, p. 141–146.
Balic, Smail, "Worüber können wir sprechen?", Dialog der Religionen 1 (1991), S. 57–73.
Baltimore, Secrétariat pour les Relations avec l'Islam: Lettre n° 51 nov. 1995.
"Bangladesh, The New Ambassador to Holy See Received by John Paul II", Islamochristiana 12 (1986), p. 196
"BANGLADESH *The New Ambassador to the Holy See Is Received by the Pope (10 January 1992)*", Islamochristiana 18 (1992), p. 244–245.
Barbieri Masini, Eleonora, "Contemporary Problems and Challenges the Women Have to Face Specifically as Christian Women", in: *Women in Society According to Islam and Christianity, Acts of a Muslim-Christian Colloquium Organized Jointly by the Pontifical Council for Interreligious Dialogue (Vatican City) and the Royal Academy for Islamic Civilization Research Al Albait Foundation (Amman)*, Vatican City s.a., p. 51–69.
Basterrechea, José Pablo, "Avant la messe de funérailles du Frère Martin Sabanagheh (3 juillet 1985)", BSNC 60 (1985), p. 337–338.
Bayraktar, Mehmet," Le Pape Jean Paul II et la politique islamique de la papauté", Islamochristiana 15 (1986), p. 237–239.

Bellinger, Gerhard J., *Knaurs großer Religionsführer, 670 Religionen, Kirchen und Kulte, weltanschaulich-religiöse Bewegungen und Gesellschaften sowie religionsphilosophische Schulen*, Augsburg 1999.

"Bethlehem: 22 March 2000, **Homily** (Extract)", BPCDIR 104/105 (2000), p. 172.

"Bethlehem: 22 March 2000, **To the Chairman, Palestine Autonomous Territory, and Palestine Friends**", BPCDIR 104/105 (2000), p. 170–172.

Bianchi, U[go], "Le concept du bien et du mal dans les religions", dans: Secretariatus pro non Christianis (éd.), *Religions, Thèmes fondamentaux pour une connaissance dialogique*, Rome 1970, p. 407–438.

Blockhausen, Denise, "Les relations islamo-chrétiennes depuis Vatcian II et les perspectives de collaborations islamo-chrétienne *(sic!)* pour le troisième millénaire", BPCDIR 102 (1999), p. 332–340.

Blockhausen, Denise, "Pluralisme et laïcité", dans: Commission pour les rapports religieux (éd.), *Religion et politque: Un thème pour le dialogue islamo-chrétien*, Cité du Vatican 1999, p. 54–59.

Borelli, John, "Forty Years ago: Vatican II, Dialogue and Lay Leadership", in: Fitzgerald, Michael L[ouis] / id. (edd.), *Interfaith Dialogue, A Catholic View*, London/Maryknoll 2006, p. 13–24.

Borelli, John, "Recent Muslim-Christian Dialogue in the USA", in: Fitzgerald, Michael L[ouis] / id. (edd.), *Interfaith Dialogue, A Catholic View*, London/Maryknoll 2006, p. 97–108.

Borrmans, Maurice, "Christian-Muslim Cooperation: Building a Modern Pluralist Society", in: Pontifical Council for Interreligious Dialogue / World Islamic Call Society (edd.), *Coexistence between Religions, Reality and Horizons*, Vatican City s.a., p. 32–38.

Borrmans, Maurice, "Christians and Muslims Working Together for the Improvement of Woman's Condition: Possibilities and Prospects", in: *Women in Society According to Islam and Christianity, Acts of a Muslim-Christian Colloquium Organized Jointly by the Pontifical Council for Interreligious Dialogue (Vatican City) and the Royal Academy for Islamic Civilization Research Al Albait Foundation (Amman)*, Vatican City s.a., p. 107–118.

Borrmans, M[aurice], "Islam", in: Secretariatus pro non Christianis (ed.), *Religions in the World*, Città del Vaticano s.a., p. 85–118.

Borrmans, Maurice, "L'esprit de la déclaration *Nostra Aetate:* Paul VI et Jean Paul II en dialogue avec les Musulmans", dans: Isizoh, Chidi Denis (ed.), *Milestones in Interreligious Dialogue, A Reading fo Selected Catholic Church Documents on Relations with People of Other Religions, Essays in Honour of Francis Cardinal Arinze*, Rome/Lagos 2002, p. 47–66.

Borrmans, Maurice, "Le Pape Paul VI et les Musulmans", Islamochristiana 4 (1978), p. 1–10.

Borrmans, Maurice, "Le séminaire du dialogue islamo-chrétien de Tripoli (Libye) (1–6 février 1976)", Islamochristiana 2 (1976), p. 135–170.

Borrmans, Maurice, "Les discours de Jean-Paul II aux jeunes de Casablanca", Seminarium 26/1 (1986), p. 44–72.

Borrmans, Maurice, "Les évaluations en conflit autour de *Nostra Aetate*", Communio 25/5 (2000), p. 96–123.

Borrmans, Maurice, "Quelques informations sur la Libye et l'église", BSNC 30 (1975), p. 241.

Borrmans, Maurice, "Welcome to the Participants", Islamochristiana 11 (1985), p. 4–5.

Bouwen, Frans, "Colloquium on the Spiritual Significance of Jerusalem in Glion (2–6 May 1993)", Islamochristiana 19 (1993), p. 294–295.

Bühlmann, Walbert, *Alle haben denselben Gott, Begegnung mit Menschen und Religionen Asiens*, Frankfurt/Main 1978.

Buonomo, Vincenzo, "The Catholic Vision of Religious Liberty", in: Pontifical Council for Interreligious Dialogue / Commission for Religious Relations with Muslims (ed.), *Religious Liberty: A Theme for Christian-Muslim Dialogue*, Vatican City 2006, p. 33–54.

"Cairo, Egypt: 25 February 2000, **A la fin de la Sainte Messe**", BPCDIR 104/105 (2000), p. 166–167.

"**Cairo – Egypt:** Annual Meeting of the Islamic-Catholic Liaison Committee, 17–18 July 1998", BPCDIR 99 (1998), p. 342.

Camps, Arnulf, "Le dialogue interreligieux et la situation concrète de l'humanité", BSNC 30 (1975), p. 315–318.

Camps, Arnulf, "Two Eminent Secretaries: Pierre Humbertclaude and Pietro Rossano (1964–1982)", BPCDIR 72 (1989), p. 379–382.

Canivez, Jean, "**Luxembourg:** Les Non-Chrétiens en Europe – Brève relation sur la rencontre au Luxembourg (13–14 Mars 1974)", BSNC 26 (1974), p. 145–147.

Cantore, Stefania, "Woman in Christianity", in: *Women in Society According to Islam and Christianity, Acts of a Muslim-Christian Colloquium Organized Jointly by the Pontifical Council for Interreligious Dialogue (Vatican City) and the Royal Academy for Islamic Civilization Research Al Albait Foundation (Amman)*, Vatican City s.a., p. 33–47.

Capalla, Fernando R., "The Role of Religious Groups in Modern Pluralistic Societies – Ideals and Realities; Traditions of Harmony in Southeast Asia as a Basis for Christian-Muslim Dialogue", BPCDIR 87 (1994), p. 232–239.

"Cardinal Arinze Visits Pakistan, India, Malaysia, Lybia", BPCDIR 71 (1989), p. 285–289.

Caspar, Robert, "Islam and Secularization", BSNC 15 (1970), p. 147–157.

Caspar, Robert, "La recherche du salut dans l'islam", dans: Secretariatus pro non Christianis (éd.), *Religions, Thèmes fondamentaux pour une connaissance dialogique*, Rome 1970, p. 115–138.

Caspar, Robert, "La religion musulmane", dans: Henry, A[ntonin]-M[arie] (éd.), *Les relations de l'église avec les religions non chrétiennes*, Unam Sanctam 61, Paris 1966, p. 201–236.

Caspar, Robert, "Pour une théologie chrétienne du monde sécularisé: quelques perspectives récentes", dans: Groupe de recherches islamo-chrétien (éd.), *Pluralisme et laïcité, Chrétiens et musulmans proposent*, Paris 1996, p. 179–194.

"Castel Gandolfo: 13 September 2001, **To the New Ambassador of the United States of America to the Holy See** (Extract)", BPCDIR 108 (2001), p. 301–304.

"Castel Gandolfo: 30 août 2003, **Aux évêques de l'Eglise Catholique copte d'Egypte à l'occasion de leur visite "ad limina"** (Extrait)", BPCDIR 114 (2003), p. 343–344.

"Castel Gandolfo: 25 September 2003, **To the Bishops from the Philippines (Ecclesiastical Provinces of Cagayan de Oro, Cotabato, Daono, Lipa, Ozamis and Zamboanga) on the Occasion of their "ad limina" visit** (Extract)", BPCDIR 114 (2003), p. 347–348.

"Castel Gandolfo: 3 September 2004, To the Reverend Brother Cardinal WALTER KASPER, President of the Pontifical Council for Promoting Christian Unity", BPCDIR 118 (2005), p. 27–29.

"**Chad and Niger:** Study Sessions on Islam and Mission", BSNC 16 (1971), p. 49–50.

Channan, James, "**Pakistan – Karachi:** 9th General Assembly of the World Muslim Congress (30 March – 2 April 1988, Karachi)", BPCDIR 70 (1989), p. 120–122.

Channan, James, "State of Religious Freedom in Pakistan", in: Pontifical Council for Interreligious Dialogue / Commission for Religious Relations with Muslims (ed.), *Religious Liberty: A Theme for Christian-Muslim Dialogue*, Vatican City 2006, p. 83–103

"Christen und Muslime in Südostasien, Schlussbericht eines regionalen Seminars", Weltkirche 8 (1994), p. 242–244.

"Christian-Muslim Colloquium on Religion and the Use of the Earth's Resources (Rome, 18–20 April 1996)", Islamochristiana 22 (1996), p. 231–233.

"Christian-Muslim Consultation on "Religious Education in Modern Society" (6–8 December 1989)", Islamochristiana 16 (1990), p. 290–292.

"Christian-Muslim Seminar on "Harmony among Believers of Living Faiths: Christians and Muslims in Southeast Asia" (1–5 August 1994)", Islamochristiana 20 (1994), p. 271–273.

"Christian Prayer for Peace", BSNC 64 (1987), p. 123–124.

CIBEDO e.V. (Hrsg.), *Die offiziellen Dokumente der katholischen Kirche zum Dialog mit dem Islam*, Regensburg 2009.

"Cité du Vatican: 15 Janvier 1999, **Discours du Pape aux évêques de la Conférence épiscopale de la Bosnie-Herzégovine à l'occasion de leur visite "ad limina"** (extrait)", BPCDIR 102 (1999), p. 289–290.

"Cité du Vatican: 25 juillet 1999, **Télégramme du Pape à l'occasion du décès de Sa Majesté, Hassan II, Roi du Maroc**", BPCDIR 102 (1999), p. 312.

"Cité du Vatican: 28 août 1999, **Aux évêques du Côte d'Ivoire à l'occasion de leur visite "Ad limina"** (Extrait)", BPCDIR 103 (2000), p. 17–18.

"Cité du Vatican: 9 Septembre 1999, **Aux évêques du Tchad à l'occasion de leur visite "Ad limina"** (Extrait)", BPCDIR 103 (2000), p.

"Cité du Vatican: 27 Septembre 1999, **Aux évêques de la République Centrafricaine en visite "Ad limina"** (Extrait)", BPCDIR 103 (2000), p. 24.

"Cité du Vatican: 26 Octobre 2000, **Au nouvel Ambassadeur du Liban près le Saint Siege** (Extrait)", BPCDIR 106 (2001), p. 15–17.

"Cité du Vatican: 6 novembre 2000, **Message à Sa Béatitude Michel Sabbah**", BPCDIR 106 (2001), p. 18–19.

"Cité du Vatican: 13 janvier 2001, **Au Corps Diplomatique accrédité près le Saint Siège à l'occasion de l'échange des vœux du Nouvel An** (Extrait")", BPCDIR 106 (2001), p. 50–52.

"Cité du Vatican: 17 mars 2001, **Aux Evêques du Rite Latin de la Région Arabe (C.E.L.R.A.)** (Extraits), BPCDIR 107 (2001)", p. 162–164.

"Cité du Vatican: 16 mai 2001, **L'audience générale** (Extraits)", BPCDIR 107 (2001), p. 176–177.

"Cité du Vatican: 18 mai 2001, **Au nouvel Ambassadeur de Tunisie près le Saint-Siège** (Extrait)", BPCDIR 107 (2001), p. 187–189.

"Cité du Vatican: 18 mai 2001, **Aux nouveaux Ambassadeurs de Népal, Tunisie, Estonie, Zambie, Guinée, Sri Lanka, Mongolie, Afrique du Sud et Gambie** (Extrait)", BPCDIR 107 (2001), p. 189.

"Cité du Vatican: 12 novembre 2001, **Aux membres de l'"International Catholic Migration Commission" et de la Fondation "Migrantes"** (extrait)", BPCDIR 109 (2002), p. 15.

"Cité du Vatican: 6 december *(sic!)* 2001, **Aux nouveaux Ambassadeurs** (Extrait)", BPCDIR 109 (2002), p. 29.

"Cité du Vatican: 6 décembre 2001, **Au nouvel Ambassadeur du Mali près le Saint-Siège** (Extrait)", BPCDIR 109 (2002), p. 27–29.

"Cité du Vatican: 11 décembre 2001, **Aux évêques de l'Eglise chaldéenne (d'Iraq, l'Iran, du Liban, d'Egypt, du Sirie, de la Turquie e** *(sic!)* **des Etats Unis d'Amérique) en visite "ad Limina"**, extrait", BPCDIR 109 (2002), p. 30–31.

"Cité du Vatican: 10 janvier 2002, **Discours du Saint-Père aux Membres du Corps Diplomatique accrédité près le Saint-Siège** (Extrait)", BPCDIR 109 (2002), p. 39–41.

"Cité du Vatican: 3 Mai 2002, **Au nouvel Ambassadeur du Maroc près le Saint-Siège** (Extrait)", BPCDIR 110 (2002), p. 181–183.

"Cité du Vatican: 17 Mai 2002, **Au nouvel Ambassadeur du Soudan près le Saint-Siège** (Extrait)", BPCDIR 110 (2002), p. 184–186.

"Cité du Vatican: 13 janvier 2003, **Au Corps Diplomatique accrédité près le Saint-Siège à l'occasion de l'échange des Vœux du Nouvel An** (Extrait)", BPCDIR 112 (2003), p. 16–20.

"Cité du Vatican: 15 février 2003, **Aux évêques de la Guinée Conakry à l'occasion de leur visite "ad limina"** (Extrait)", BPCDIR 114 (2003), p. 309–310.

"Cité du Vatican: 22 février 2003, **Aux évêques de la Région du Nord de l'Afrique pendant leur visite "ad limina"** (Extrait)", BPCDIR 114 (2003), p. 313–316.

"Cité du Vatican: 24 Février 2002, **Au nouvel Ambassadeur de la Jordanie près le Saint-Siège** (Extrait)", BPCDIR 110 (2002), p. 183–184.

"Cité du Vatican: 24 Février 2002, **Aux Chefs d'Etat ou de Gouvernement**", BPCDIR 110 (2002), p. 171–172.

"Cité du Vatican: 17 Mai 2002, **Au nouvel Ambassadeur du Bélarus près le Saint-Siège** (Extrait)", BPCDIR 110 (2002), p. 188.

"Cité du Vatican: 15 may *(sic!)* 2003, **Au nouvel Ambassadeur de Syrie près le Saint-Siège** (Extrait)", BPCDIR 114 (2003), p. 329–330.

"Cité du Vatican: 17 juin 2003, **Aux évêques du Burkina Faso et du Niger à l'occasion de leur visite "ad limina"** (Extrait)", BPCDIR 114 (2003), p. 338.

"Cité du Vatican: 28 juin 2003, **Exhortation apostolique post-synodale, "Ecclesia in Europa"** (Extrait)", BPCDIR 118 (2005).

"Cité du Vatican: 10 octobre 2003, **Aux participants à l'Assemblée parlementaire de l'Organisation pour la Sécurité et la Coopération en Europe (O.S.C.E.)**", BPCDIR 115 (2004), p. 9–10.

"Cité du Vatican: 12 décembre 2003, **A l'Ambassadeur du Qatar près le Saint-Siège** (Extrait)", BPCDIR 115 (2004), p. 21–22.

"Cité du Vatican: 12 janvier 2004, **Aux membres du Corps diplomatique accrédités près le Saint-Siège, à l'occasion de l'échange des vœux du nouvel an**", BPCDIR 115 (2004), p. 26–31.

"Cité du Vatican: 21 février 2004, **Au nouvel Ambassadeur de Turquie près le Saint-Siége,** (Extrait)", BPCDIR 118 (2005), p. 10–12.

"Cité du Vatican: 27 mai 2004, **Au nouvel Ambassadeur du Mali près le Saint-Siége** (Extrait)", BPCDIR 118 (2005), p. 19.

"Cité du Vatican: 27 mai 2004, **Au nouvel Ambassadeur de Tunisie près le Saint-Siége** (Extrait)", BPCDIR 118 (2005), p. 15–17.

"Cité du Vatican: 1 août 2004, **À Sa Béatitude Emmanuel III Delly, Patriarche de Babylone des Chaldéens**", BPCDIR 118 (2005), p. 25.

"Cité du Vatican: 7 octobre 2004, **Aux membres de la Commission théologique pontificale internationale** (Extrait)", BPCDIR 118 (2005), p. 29–30.

"Cité du Vatican: 15 novembre 2004, **Au nouvel Ambassadeur de la République d'Irak près le Saint-Siége** (Extrait)", BPCDIR 118 (2005), p. 33–35.

"Cité du Vatican: 15 novembre 2004, **Aux chefs religieux d'Azerbaïdjan**", BPCDIR 118 (2005), p. 35–36.

"Cité du Vatican: 26 novembre 2004, **Au Président, S.E. Ali Abdullah Saleh, de la République du Yémen**", BPCDIR 118 (2005), p. 38–39.

"Cité du Vatican: 10 janvier 2005, **Au Corps diplomatique accrédité près le Saint-Siège** (Extrait)", BPCDIR 118 (2005), p. 42–48.

"**Cité du Vatican:** Accord entre le Conseil Pontifical pour le Dialogue Interreligieux et le Comité Permanent d'Al-Azhar al Sharif pour le Dialogue avec les Religions Monothéistes 28 mai 1998", BPCDIR 99 (1998), p. 338–340.

"Colloquium on Holiness in Islam and Christianity, Rome, 6–7 May 1985", Islamochristiana 11 (1985), p. 1–3.

"Colloquium on "Islamic Da'wah and Christian Mission in the Next Century", Jointly Organized by the World Islamic Call Society (Libya) and the Pontifical Council for Interreligious Dialogue (27–30 April 1997)", Islamochristiana 23 (1997), p. 201–202.

"Colloquium: "Religious Education in Modern Society"", in: *Religious Education and Modern Society, Acts of a Muslim-Christian Colloquium Organized Jointly by the Pontifical Council for Interreligious Dialogue (Vatican City) and the Royal Academy for Islamic Civilization Research Al Albait Foundation (Amman)*, s.l. s.a., p. 134–135.

Commission pour les rapports religieux avec les musulmans (éd.), *Religion et politique: Un thème pour le dialogue islamo-chrétien*, Cité du Vatican 1999.

"Concluding Ceremony", BPCDIR 109 (2002), p. 152–155.

Conseil Pontifical pour le Dialogue Interreligieux (éd.), *Le dialogue interreligieux dans l'enseignement officiel de l'église catholique (1963–1997), Documents rassemblés par Francesco Gioia*, s.l. 1998.

Pontifical Council for Interreligious Dialogue (ed.), *Interreligious Dialogue, The Official Teaching of the Catholic Chruch (1963–1995)*, Edited by Francesco Gioia, Boston 1997.

Pontificio Consiglio per il Dialogo Interreligioso (ed.), *Il dialogo interreligioso nel magistero pontificio (Documenti 1963–1993)*, Città del Vaticano 1994.

Cordeiro, Joseph / Teissier, Henri, "Rapport du Groupe "Islam"", BSNC 41/42 (1979), p. 183–186.

Coste, René, "Religious Teachings on the Use of the Earth's Resources, A Christian Perspective", in: *Religion and the Use of the Earth's Resources, Acts of a Christian-Muslim Colloquium Organized Jointly by the Pontifical Council for Interreligious Dialogue (Vatican City) and the Royal Academy for Islamic Civilization Research Al Albait Foundation (Amman)*, Rome 1996, p. 47–57.

Couvreur, Gilles, "Spiritualité du dialogue dans un contexte pluraliste", BPCDIR 99 (1998), p. 317–324.

Cuoq, J[oseph M.], "1967", p. 8–9,
    "1968", p. 10–11,

"1969", p. 12–13,
"1971", p. 14–15,
Cuoq, J[oseph] M., "1972", p. 16–17, in: Pontifical Council for Interreligious Dialogue (ed.), *Meeting in Friendship, Messages to Muslims For the End of Ramadan (1967–2002)*, Vatican City 2003.
Cuoq, J[oseph M.], "Iran:" *(sic!)*, BSNC 14 (1970), p. 103–104.
Cuoq, Joseph M., "Notations sur la situation du dialogue au Maghreb", BSNC 30 (1975), p. 233–236.
Dahhan, Omaymah, "Muslim Women in the Arab World: Contemporary Problems and Challenges", in: *Women in Society According to Islam and Christianity, Acts of a Muslim-Christian Colloquium Organized Jointly by the Pontifical Council for Interreligious Dialogue (Vatican City) and the Royal Academy for Islamic Civilization Research Al Albait Foundation (Amman)*, Vatican City s.a., p. 71–93.
Dalmais, P., "L'Église et l'Islam au Tchad", BSNC 44 (1980), p. 221–236.
"Damas Syrie: 6 mai 2001, **Homilie au Stade Abbassyine**", BPCDIR 107 (2001), p. 168–169.
"Damas Syrie: 6 mai 2001, **Rencontre avec les Patriarches et les évêques au Patriarchat grec-melkite**", BPCDIR 107 (2001), p. 169–170.
"Damascus, Syria: 5th May 2001, **Discourse upon Arrival at the International Airport** (Excerpt)", BPCDIR 107 (2001), p. 167–168.
"Damascus Syria: 6th May 2001, **Discourse during the Visit to the Umayyad Mosque**", BPCDIR 107 (2001), p. 170–172.
"Damascus Syria: 8th May 2001, **Discourse of Farewell at the International Airport**", BPCDIR 107 (2001), p. 175–176.
"DAY OF PRAYER FOR PEACE IN ASSISI (24th January 2002)", BPCDIR 109 (2002), p. 135–137.
"Déclaration commune d'Al-Azhar et du Conseil pontifical pour le dialogue interreligieux sur la situation au Moyen-Orient", BPCDIR 107 (2001), p. 193.
"Declaration of the Joint Committee of the Permanent Committee of Al-Azhar for Dialogue with the Monotheistic Religions and the Pontifical Council for Interreligious Dialogue", BPCDIR 108 (2001), p. 332.
Demeerseman, Gérard, "Echos de presse dans certains pays arabes et musulmans à propos de la visite du Pape", Islamochristiana 11 (1985), p. 229–230.
Demeerseman, Gérard, "Journée de prière pour la paix à Assise", Islamochristiana 13 (1987), p. 200–204.
Demiri, Lejla, "My Experience of Dialogue", BPCDIR 114 (2003), p. 375–378.
Demiri, Lejla, **Rome, Italy:** 15 March 2003 A Report for "Nostra Aetate" Foundation, Pontifical Council for Interreligious dialogue, by a Muslim Student who Had Received a Study Grant, BPCDIR 114 (2003), p. 409–410
"Diplomatic Relations Are Established between the Holy See and the Socialist People's Libyan Arab Jamahiriya (10 March 1997)", Islamochristiana 23 (1997), p. 221.
"Discourse to the Representatives of the Muslim Community (Davao Airport -Philippines)", in: Segretariato per i non cristiani (ed.), *Chiesa e islam*, s.l. 1981, p. 50–53.
Donahue, John, *Islam in the Middle East*, C.R.R.M. Reports Recent Trends and Movements in Islam 4, s.l. 1992
"**Dr. H. Munawir Sjadzali, Minister for Religious Affairs in the Government of Indonesia, Received the Pope in the Name of All Religious Believers in Indonesia with the Following Address**, 10 October 1989", BPCDIR 73 (1990), p. 9–11.

Dufaux, Louis, "Le rôle de l'islam dans les problèmes de l'immigration", BPCDIR 76 (1991), p. 36–38.
Duminuco, Vincent J., "Religious Education in Colleges and Universities: Present Situation and Aspirations for the Future", in: *Religious Education and Modern Society, Acts of a Muslim-Christian Colloquium Organized Jointly by the Pontifical Council for Interreligious Dialogue (Vatican City) and the Royal Academy for Islamic Civilization Research Al Albait Foundation (Amman)*, s.l. s.a., p. 89–105.
Duval, Léon-Etienne, "Brève réflexion sur le dialogue", BPCDIR 72 (1989), p. 319–322.
Ecclesia Catholica / Secretariatus pro non-Christianis (ed.), *Guidelines for a Dialogue between Muslims and Christians*, Roma [1969].
"EGYPT The 2nd Meeting of the Islamo-Catholic Liaison Committee (30 May 1996)", Islamochristiana 22 (1996), p. 209.
Eisele, Marie-Paule, "Rights and Education of the Pre-school Child", in: Pontifical Council for Interreligious Dialogue / Royal Academy for Islamic Civilization Reasearch Al Albait Foundation (edd.), *The Rights and Education of Children in Islam and Christianity, Acts of a Muslim-Christian Colloquium*, Vatican City s.a., p. 98–109.
Ejiofor, Lambert, "A Christian Comment", in: *Nationalism Today: Problems and Challenges, Acts of a Muslim-Christian Colloquium Organized Jointly by the Pontifical Council for Interreligious Dialogue (Vatican City) and the Royal Academy for Islamic Civilization Research Al Albait Foundation (Amman)*, Rome 1994, p. 103–118.
El-Abbadi, Abdul Salam, "Muslim Response", in: Pontifical Council for Interreligious Dialogue / Royal Academy of Islamic Civilization Research Al Albait Foundation (edd.), *The Rights and Education of Children in Islam and Christianity, Acts of a Muslim-Christian Colloquium*, Vatican City s.a., p. 23–32.
El-Assad, Nasir El-Din, "Inaugural Address", in: *Religion and the Use of the Earth's Resources, Acts of a Christian-Muslim Colloquium Organized Jointly by the Pontifical Council for Interreligious Dialogue (Vatican City) and the Royal Academy for Islamic Civilization Research Al Albait Foundation (Amman – Jordan)*, Rome 1996, p. 7–9.
El-Assad, Nassir El-Din, "Inaugural Address", in: *Human Dignity, Acts of a Christian-Muslim Colloquium Organized Jointly by the Pontifical Council for Interreligious Dialogue (Vatican City) and the Royal Academy for Islamic Civilization Research Al Albait Foundation (Amman – Jordan)*, Rome 1999, p. 11–13.
El-Assad, Nassir El-Din, "Inaugural Address", in: *Nationalism Today: Problems and Challenges, Acts of a Muslim-Christian Colloquium Organized Jointly by the Pontifical Council for Interreligious Dialogue (Vatican City) and the Royal Academy for Islamic Civilization Research Al Albait Foundation (Amman)*, Rome 1994, p. 5–7.
El-Assad, Nassir El-Din, "Inaugural Address", in: Pontifical Council for Interreligious Dialogue / Royal Academy for Islamic Civilization Research Al Albait Foundation (edd.), *The Rights and Education of Children in Islam and Christianity, Acts of a Muslim-Christian Colloquium*, Vatican City s.a., p. 9–12.
El-Assad, Nassir El-Din, "Inaugural Speech", in: *Religious Education and Modern Society, Acts of a Muslim-Christian Colloquium Organized Jointly by the Pontifical Council for Interreligious Dialogue (Vatican City) and the Royal Academy for Islamic Civilization Research Al Albait Foundation (Amman)*, s.l. s.a., p. 8–11.

El-Assad, Nassir El-Din, "The Closing Address", in: Pontifical Council for Interreligious Dialogue / Royal Academy for Islamic Civilization Research Al Albait Foundation (edd.), *Human Dignity, Acts of a Christian-Muslim Colloquium*, Vatican City 1999, p. 149–152.

El-Assad, Nassir El-Din, "Woman the Mainstay of Human Life", in: *Women in Society According to Islam and Christianity, Acts of a Muslim-Christian Colloquium Organized Jointly by the Pontifical Council for Interreligious Dialogue (Vatican City) and the Royal Academy for Islamic Civilization Research Al Albait Foundation (Amman)*, s.l. s.a., p. 11–14.

El-Hage, Youssef Kamal, "Religious Liberty: A Universal Human Right", in: Pontifical Council for Interreligious Dialogue / Commission for Religious Relations with Muslims (ed.), *Religious Liberty: A Theme for Christian-Muslim Dialogue*, Vatican City 2006, p. 11–32.

El-Jerrari, Abbas, "A Muslim Commentary on Mag Coste's Paper", in: *Religion and the Use of the Earth's Resources, Acts of a Christian-Muslim Colloquium Organized Jointly by the Pontifical Council for Interreligious Dialogue (Vatican City) and the Royal Academy for Islamic Civilization Research Al Albait Foundation (Amman – Jordan)*, Rome 1996, p. 59–67.

El-Khayyat, Abdul Aziz, "Religious Education in the Modern Society: Islamic Education", in: *Religious Education and Modern Society, Acts of a Muslim-Christian Colloquium Organized Jointly by the Pontifical Council for Interreligious Dialogue (Vatican City) and the Royal Academy for Islamic Civilization Research Al Albait Foundation (Amman)*, s.l. s.a., p. 71–88.

El-Mawla, Saud, "Attitude of Muslims towards Human Dignity, A Historical Review", in: *Human Dignity, Acts of a Christian-Muslim Colloquium Organized Jointly by the Pontifical Council for Interreligious Dialogue (Vatican City) and the Royal Academy for Islamic Civilization Research Al Albait Foundation (Amman – Jordan)*, Rome 1999, p. 85–100.

El-Mawla, Saud, "Explanatory Remarks", in: *Human Dignity, Acts of a Christian-Muslim Colloquium Organized Jointly by the Pontifical Council for Interreligious Dialogue (Vatican City) and the Royal Academy for Islamic Civilization Research Al Albait Foundation (Amman – Jordan)*, Rome 1999, p. 103–105.

El-Saadi, Muhammad, "Religious Tolerance in Islam", in: Pontifical Council for Interreligious Dialogue / World Islamic Call Society (edd.), *Co-existence between Religions, Reality and Horizons*, Vatican City s.a., p. 22–26.

El-Samerra'i, Faruk, "The Rights of the Embryo", in: Pontifical Council for Interreligious Dialogue / Royal Academy for Islamic Civilization Research Al Albait Foundation (edd.), *The Rights and Education of Children in Islam and Christianity, Acts of a Muslim-Christian Colloquium*, Vatican City s.a., p. 33–59.

El-Sammak, Muhammad, "Muslims Faced by the Challenges of Human Dignity", in: Pontifical Council for Interreligious Dialogue / Royal Academy for Islamic Civilization Research Al Albait Foundation (edd.), *Human Dignity, Acts of a Christian-Muslim Colloquium*, Vatican City 1999, p. 109–121.

El-Sheikh, Omar, "Muslim Response", in: Pontifical Council for Interreligious Dialogue / Royal Academy for Islamic Civilization Research Al Albait Foundation (edd.), *The Rights and Education of Children in Islam and Christianity, Acts of a Muslim-Christian Colloquium*, Vatican City s.a., p. 127–129.

El-Talebi, Ammar, "The Concept of Human Dignity According to Islam", in: *Human Dignity, Acts of a Muslim-Christian Colloquium Organized Jointly by the Pontifical Council for

Interreligious Dialogue (Vatican City) and the Royal Academy for Islamic Civilization Research Al Albait Foundation (Amman – Jordan), Rome 1999, p. 21–36.

El-Tall, Sufian, "Religious Teachings on the Protection of the Earth's Resources, A Muslim Point of View", in: *Religion and the Use of the Earth's Resources, Acts of a Christian-Muslim Colloquium Organized Jointly by the Pontifical Council for Interreligious Dialogue (Vatican City) and the Royal Academy for Islamic Civilization Research Al Albait Foundation (Amman – Jordan)*, Rome 1999, p. 95–118.

El-Zuhaili, Wahbah, "Science, Faith and the Concerns of Youth", in: *Religious Education and Modern Society, Acts of a Christian-Muslim Colloquium Organized Jointly by the Pontifical Council for Interreligious Dialogue (Vatican City) and the Royal Academy for Islamic Civilization Research Al Albait Foundation (Amman)*, s.l. s.a., p. 60–68.

Ellul, Joseph, "A Christian Understanding of Islam", in: Pontifical Council for Interreligious Dialogue / World Islamic Call Society (edd.), *Co-existence between Religions, Reality and Horizons*, Vatican City s.a., p. 12–15.

"*Extract from* **Adress to the President and the Civil and Religious Authorities Present at the Airport before Leaving Pakistan**", in: Segretariato per i non cristiani (ed.), *Chiesa e islam*, s.l. 1981, p. 44–45.

"*Extract from* **Discourse of Pope John Paul II to the Catholic Community in the Chapel of the Italian Embassy in Ankara**", in: Segretariato per i non cristiani (ed.), *Chiesa e islam*, s.l. 1981, p. 12–15.

"*Extract from* **Message to President of Pakistan**", in: Segretariato per i non cristiani (ed.), *Chiesa e islam*, s.l. 1981, p. 39–40.

"Extraits du discours du Saint-Père au Corps diplomatique (13 janvier 1990)", Islamochristiana 16 (1990), p. 292–294.

Fahim, Amin, "The Rights and Education of Children in Schools", in: Pontifical Council for Interreligious Dialogue / Royal Academy for Islamic Civilization Research Al Albait Foundation (edd.), *The Rights and Education of Children in Islam and Christianity, Acts of a Muslim-Christian Colloquium*, Vatican City s.a., p. 117–126.

Faruqi, Ziaul Hasan, "The Concept of Holiness in Islam", Islamochristiana 11 (1985), p. 7–27.

Festarozzi, F[ranco], "Le Dieu vivant dans la révélation chrétienne", dans: Secretariatus pro non Christianis (éd.), *Religions, Thèmes pur une connaissance dialogique*, Rome 1970, p. 365–403.

"Ferrari, Silvio, *Gerusalemme nei documenti pontifici*, a cura di Edmond Farhat, Città del Vaticano, Libreria Ed. Vaticana, 1987, 377 pp.", Islamochristiana 14 (1988), p. 362–364.

"Final Report", BPCDIR 87 (1994), p. 219–223.

"Final Statement", in: Pontifical Council for Interreligious Dialogue / World Islamic Call Society (edd.), *Co-existence between Religions, Reality and Horizons*, Vatican City s.a., p. 43–46.

Fitzgerald, Michael L[ouis], "2002", in: Pontifical Council for Interreligious Dialogue (ed.), *Meeting in Friendship, Messages to Muslims for the End of Ramadan (1967–2002)*, Vatican City 2003, p. 76–77.

Fitzgerald, Michael L[ouis], "**Cairo – Egypt:** Cairo Encounter 28–29 October 2001", BPCDIR 109 (2002), p. 156–157.

Fitzgerald, Michael L[ouis], "Chrétiens et musulmans en Europe", dans: Commission pour les rapports religieux avec les musulmans (éd.), *Religion et politique: Un thème pour le dialogue islamo-chrétien*, Cité du Vatican 1999, p. 103–118.

Fitzgerald, Michael L[ouis], "Christian-Muslim Dialogue in Indonesia and the Philippines", BSNC 30 (1975), p. 2143 – 217.
Fitzgerald, Michael L[ouis], "Editorial", BPCDIR 81 (1992), p. 265 – 266.
Fitzgerald, Michael L[ouis], "Editorial", BPCDIR 85/86 (1994). p. 1 – 2.
F[itzgerald], M[ichael] L[ouis], "Final Discussion of the Plenary Assembly", BPCDIR 82 (1993), p. 89 – 91.
Fitzgerald, Michael L[ouis], "Greetings from the New Secretary", BSNC 64 (1987), p. 5 – 7.
Fitzgerald, Michael L[ouis], "Joseph Cuoq (1917 – 1986)", BSNC 63 (1986), p. 339 – 341.
Fitzgerald, Michael L[ouis], "MAURICE BORRMANS, *Guidelines for Dialogue between Christians and Muslims*. Translated from the French by R. Marston Speight. (Interreligious Documents I) New York / Mahwah, N.J., Paulist Press, 1990, pp. vi-132.", BPCDIR 76 (1991), p. 149.
Fitzgerald, M[ichael] L[ouis], "MAURICE BORRMANS, *Orientamenti per un dialogo tra cristiani e musulmani*", BPCDIR 70 (1989), p. 127.
Fitzgerald, Michael L[ouis], "Mission and Dialogue: Reflections in the Light of Assisi 1986", BSNC 68 (1988), p. 113 – 120.
Fitzgerald, Michael L[ouis], "Muslims in Europe", in: Commission pour les rapports religieux avec les musulmans (éd.), *Religions et poliltique: Un thème pour le dialogue islamo-chrétien*, Cité du Vatican 1999, p. 63 – 78.
Fitzgerald, Michael L[ouis], "My Personal Journey in Mission", in: id. / Borelli, John (edd.), *Interfaith Dialogue, A Catholic View*, London/Maryknoll 2006, p. 1 – 12.
Fitzgerald, Michael L[ouis], "Other Religions in the Catechism of the Catholic Church", Islamochristiana 19 (1993), p. 29 – 41.
Fitzgerald, Michael L[ouis], "Report on the Activities of the PCID: November 1992 – November 1995", BPCDIR 92 (1996), p. 167 – 182.
Fitzgerald, Michael L[ouis], "Report on the Activities of the PCID: November 1998 – October 2001", BPCDIR 109 (2002), p. 73 – 80.
Fitzgerald, Michael L[ouis], "**Sarajewo:** A Christian-Muslim Colloquium, 13 November 1999", BPCDIR 103 (2000), p. 83 – 84.
Fitzgerald, M[ichael] L[ouis], "*Seminarium, Joannes Paulus II e Islamismus*, Januario – Martio 1986, Città del Vaticano, 240pp.", Islamochristiana 13 (1987), p. 275 – 277.
Fitzgerald, Michael L[ouis], "The Pontifical Gregorian University and Interreligious Dialogue", BPCDIR 107 (2001), p. 244 – 251.
Fitzgerald, Michael [Louis], "The Secretariat for non Christians and Muslim-Christian Dialogue", BSNC 37 (1978), p. 9 – 14.
Fitzgerald, M[ichael] L[ouis], "The Secretariat for Non-Christians Is Ten Years Old", Islamochristiana 1 (1975), p. 87 – 95.
Fitzgerald, Michael L[ouis], "Tolerance – Respect – Trust", Forum Mission 6 (2010), p. 14 – 49.
Fitzgerald, Michael L[ouis], "Toward a Christian Theology of Religious Pluralism by Fr Jacques Dupuis", BPCDIR 108 (2001), p. 334 – 341.
Fitzgerald, Michael L[ouis], "Twenty-five Years of Dialogue, The Pontifical Council for Inter-Religious Dialogue", Islamochristiana 15 (1989), p. 109 – 120.
Fitzgerald, Michael L[ouis], "**Vienna – Austria:** Workshop on "Religion and the Media" (7 – 8 October 1994)", BPCDIR 87 (1994), p. 271 – 272.

Fitzgerald, Michael L[ouis] / Al-Rifaie, Hamid [bin] A[hmad], "Islamic Catholic Liaison Committee: Joint Declaration on the Present Situation in the Holy Land (21st April 2002)", BPCDIR 110 (2002), p. 192.

Fitzgerald, Michael L[ouis] / Al-Rifaie, Hamid bin Ahmad, "Vatican City: 20 January 2004 **Joint Declaration of the Islamic-Catholic Liaison Committee**", BPCDIR 115 (2004), p. 37–38.

Fitzgerald, Michael L[ouis] / Borelli, John (edd.), *Interfaith Dialogue, A Catholic View*, London/Maryknoll 2006.

Fortelle, Odile de La, "**Ghana:** Christian-Muslim Dialogue", BSNC 39 (1978), p. 248–251.

"Forward", BSNC 18 (1971), p. 135.

Futrell, John Carroll, "The Concept of Christian Holiness", Islamochristiana 11 (1985), p. 29–34.

Galenzoga, Rodulfo M., "Response to the Paper of Dr. Abdul Aziz Kamel on the Theme: Identity and Openness: Religious Education and Modern Society", in: *Religious Education and Modern Society, Acts of a Christian-Muslim Colloquium Organized Jointly by the Pontifical Council for Interreligious Dialogue (Vatican City) and the Royal Academy for Islamic Civilization Research Al Albait Foundation (Amman)*, s.l. s.a., p. 109–111.

Gardet, L[ouis], "A Few Notes on My Journey", BSNC 15 (1970), p. 174–178.

Gardet, L[ouis], "En Islam: Dieu et le croyant en Dieu", dans: Secretariatus pro non Christianis (éd.), *Religions, Thèmes fondamentaux pour une connaissance dialogique*, Rome 1970, p. 343–364.

Gaudeul, Jean-Marie, "A Christian Critique of Islamic Holiness", Islamochristiana 11 (1985), p. 69–89.

Gaudeul, Jean-Marie, "Free to Serve God Better", in: Pontifical Council for Interreligious Dialogue / Commission for Religious Relations with Muslims (ed.), *Religious Liberty; A Theme for Christian-Muslim Dialogue*, Vatican City 2006, p. 187–212.

Gaudeul, Jean-Marie, "Islam in a Minority Situation: The Situation in France", in: Pontifical Council for Interreligious Dialogue / Commission for Religious Relations with Muslims (ed.), *Religious Liberty; A Theme for Christian-Muslim Dialogue*, Vatican City 2006, p. 125–147.

Gelot, J[oseph], "Education des chrétiens en vue du dialogue", BSNC 20 (1972), p. 63–68.

Gelot, J[oseph], "Le bien et le mal en Islam", dans: Secretariatus pro non Christianis (éd.), *Religions, Thèmes fondamentaux pour une connaissance dialogique*, Rome 1970, p. 523–553.

"General Report", in: A Historical Review, in: *Human Dignity, Acts of a Muslim-Christian Colloquium Organized Jointly by the Pontifical Council for Interreligious Dialogue (Vatican City) and the Royal Academy for Islamic Civilization Research Al Albeit Foundation (Amman – Jordan)*, Rome 1999, p. 153–155.

"General Report", in: Pontifical Council for Interreligious Dialogue / Royal Academy for Islamic Civilization Research Al Albait Foundation (edd.), *The Rights and Education of Children in Islam and Christianity, Acts of a Muslim-Christian Colloquium*, Vatican City s.a., p. 158–160.

"General Report", in: *Women in Society According to Islam and Christianity, Acts of a Muslim-Christian Colloquium Organized Jointly by the Pontifical Council for Interreligious Dialogue (Vatican City) and the Royal Academy for Islamic Civilization Research Al Albait Foundation (Amman)*, s.l.s.a., p. 119–121.

"General Report of the Muslim-Christian Colloquium of 24 – 26 June 1992 on the Theme "Women in Society According to Islam and Christianity"", Islamochristiana 18 (1992), p. 318 – 320.

"**Geneva:** Visit of the President and the Secretary of the Secretariat for Non-Christians to the Ecumenical Council of Churches (15 January 1974)", BSNC 26 (1974), p. 137 – 139.

Ghwel, Ibrahim, "Islamic Perception of Christianity", in: Pontifical Council for Interreligious Dialogue / World Islamic Call Society (edd.), *Co-existence between Religions, Reality and Horizons*, Vatican City s.a., p. 16 – 21.

Gioia, Francesco, "Presentation of the Book Il dialogo inter-religioso nel magistero pontificio: Documents 1963 – 1993", BPCDIR 87 (1994), p. 214 – 215.

Gjergij, L., "La situation religieuse des Albanais en Yougoslavie", BPCDIR 75 (1990), p. 286 – 289.

Gonçalves, Teresa, "Card. PAUL POUPARD, *Le Religioni nel Mondo*, I Edizione Italiana, Edizioni Piemme, 1990, pp. 122", BPCDIR 74 (1990), p. 205.

Gravrand, Charbel, "L'Afrique au Secrétariat", BPCDIR 72 (1989), p. 350 – 361.

Greco, J[oseph], "Le bien et le mal, Perspectives chrétiennes", dans: Secretariatus pro non Christianis (éd.), *Religions, Thèmes fondamentaux pour une connaissance dialogique*, Rome 1970, p. 555 – 596.

Greco, J[oseph], "The Activity of the Consultors", BSNC 12 (1969), p. 206.

Gstrein, Heinz, "Die Lage der christlichen Kirchen im arabischen Maghreb", Wort und Wahrheit 27 (1972), S. 463 – 469.

"Gruß des Papstes an die Muslime in Deutschland", in: Segretariato per i non cristiani (ed.), *Chiesa e islam*, s.l. 1981, p. 37.

Hani, Abdul Razak Bani, "A Muslim Commentary on Dr. Muasher's Paper, in: Religion and the Use of the Earth's Resources", *Acts of a Christian-Muslim Colloquium Organized Jointly by the Pontifical Council for Interreligious Dialogue (Vatican City) and the Royal Academy for Islamic Civilization Research Al Albait Foundation (Amman – Jordan)*, Rome 1996, p. 167 – 171.

Henkel, Willi, "Directions in Vatican Documents on Interreligious Dialogue", in: Daneel, Inus / Engen, Charles Van / Vroom, Hendrik (edd.), *Fullness of Life for All, Challenges for Mission in Early 21st Century*, Amsterdam / New York 2003, p. 235 – 242.

Hikmat, Taher, "National Ties Today: Problems and Challenges, A Muslim Viewpoint", in: *Nationalism Today, Problems and Challenges, Acts of a Muslim-Christian Colloquium Organized Jointly by the Pontifical Council for Interreligious Dialogue (Vatican City) and the Royal Academy for Islamic Civilization Research Al Albait Foundation (Amman – Jordan)*, Rome 1994, p. 63 – 69.

"Holy Father's New Appeal for Peace: "We Beg again, End this Terrible Slaughter!" (19 February 1994)", Islamochristiana 20 (1994), p. 216 – 217.

Honecker, Martin, "Religion und Politik: Zur Geltung der Menschenrechte in Christentum und Islam", in: Busse, Heribert / Honecker, Martin (Hrsgg.), *Gottes- und Weltverständnis in Islam und Christentum*, EZW-Texte 123, Stuttgart 1993, S. 15 – 27.

Humbertclaude, P[ierre], "Clarification of the Nature and Role of the Secretariat for non-Christians", BSNC 11 (1969), p. 76 – 97.

Ibrahim, Hanan, "A Muslim Comment", in: *Nationalism Today, Problems and Challenges, Acts of a Muslim-Christian Colloquium Organized Jointly by the Pontifical Council for*

Interreligious Dialogue (Vatican City) and the Royal Academy for Islamic Civilization Research Al Albait Foundation (Amman – Jordan), Rome 1994, p. 119–121.
"Inauguration of the John XXIII Ecumenical Centre (22 February 1990)", Islamochristiana 16 (1990), p. 280.
"INDONESIA The New Ambassador to the Holy See Is Received by the Pope (19 June 1995)", Islamochristiana 21 (1995), p. 169–171.
"International Airport, Amman, Jordan: 20th March 2000", BPCDIR 104–105 (2000), p. 168–169.
"Inter-Religious Dialogues and the Development of Religious Harmony in Indonesia", BSNC 37 (1978), p. 14–24.
"Introduction", in: Pontifical Council for Interreligious Dialogue / Commission for Religious Relations with Muslims (ed.), *Religious Liberty; A Theme for Christian-Muslim Dialogue*, Vatican City 2006, p. 7.
"Iraq, New Ambassador to the Holy See Presents Credentials", Islamochristiana 12 (1986), p. 208–209.
Irwin, Kevin W., "Religious Teachings on the Protection fo the Earth's Resources, A Christian Point of View", in: *Religion and the Use of the Earth's Resources, Acts of a Christian-Muslim Colloquium Organized Jointly by the Pontifical Council for Interreligious Dialogue (Vatican City) and the Royal Academy for Islamic Civilization Research Al Albait Foundation (Amman – Jordan)*, Rome 1996, p. 71–90.
Isizoh, Chidi Denis, **"Report of Meetings in Ghana:** 18–30 January 1998", BPCDIR 99 (1998), p. 343–345.
Isizoh, Chidi Denis, **"Yaoundé, Cameroon:** Workshop on the "Challenges of Interreligious Dialogue in Sub-Saharan Africa", 20–25 March 2001", BPCDIR 107 (2001), p. 253–255.
"Islamic/Christian Dialogue, Tripoli, Libyan Arab Republic", The Link 9,1 (1976), p. 1–8.
"ISRAEL, The New Ambassador to the Holy See Is Received by the Pope (10 April 1997)", Islamochristiana 23 (1997), p. 199–200.
Jacob, Xavier, "Islam et politique (Turquie)", dans: Commission pour les rapports religieux avec les musulmans (éd.), *Religion et politique: Un thème pour le dialogue islamo-chrétien*, Cité du Vatican 1999, p. 119–144.
Jacob, Xavier," L'islam turc et le dialogue", Islamochristiana 15 (1989), p. 237.
Jacob, Xavier, "Situation de l'islam en Turquie", dans: Commission pour les rapports religieux avec les musulmans (éd.), Religion et politique: Un thème pour le dialogue islamo-chrétien, Cité du Vatican 1999, p. 145–159.
Jadot, Jean, "1980", p. 32,
    "1981", p. 33,
    "1982", p. 34,
    "1983, p. 35,
    in: Pontifical Council for Interreligious Dialogue (ed.), *Meeting in Friendship, Messages to Muslims for the End of Ramadan (1967–2002)*, Vatican City 2003.
Jadot, Jean, "Preface to the 1981 Edition", in: Pontifical Council for Interreligious Dialogue / Borrmans, Maurice, *Guidelines for Dialogue between Christians and Muslims*, Interreligious Documents 1, New York / Mahwah 1990, p. 7.
Jadot, J[ean], "The Growth in Roman Catholic Commitment to Interreligious Dialogue since Vatican II", BSNC 54 (1983), p. 205–220.

Jadot, J[ean], "L'Église Catholique et le dialogue inter-religieux depuis Vatican II", BSNC 54 (1983), p. 98–114.

Jarradat, Izat, "The Role of Believers in Dealing with Questions of Nationalism, A Muslim Viewpoint", in: *Nationalism Today, Problems and Challenges, Acts of a Muslim-Christian Colloquium Organized Jointly by the Pontifical Council for Interreligious Dialogue (Vatican City) and the Royal Academy for Islamic Civilization Research Al Albait Foundation (Amman – Jordan)*, Rome 1994, p. 73–86.

Jarrar, Faruk A., "The Royal Academy of Jordan for Islamic Civilization Research: A Continuing Dialogue", Islamochristiana 16 (1990), p. 147–152.

Jarrar, F[aruk A.] / Fitzgerald M[ichael] L[ouis], "Islamo-Christian Colloquium on: "Religious Education in Modern Society" (with Particular Emphasis on Colleges and Universities)", BPCDIR 73 (1990), p. 82–83.

"Jean Paul II reçoit les participants de la rencontre d'Assise", Islamochristiana 15 (1989), p. 245–247.

"Jerusalem: 23 March 2000, **To the Christian, Jewish and Muslim Leaders**", BPCDIR 104/105 (2000), p. 175–177.

"Jerusalem: 25 March 2000, **To the Consul Generals of Jerusalem**", BPCDIR 104/105 (2000), p. 179.

"Jerusalem: 25 March 2000, **To the Participants and Representatives of the Christian Churches and Confessions Present in the Holy Land** (Extract)", BPCDIR 104/105 (2000), p. 179.

Jeusset, Gwenolé, "Conflits et harmonie", BPCDIR 81 (1992), p. 328–359.

Johannes Paul II, *Die Schwelle der Hoffnung überschreiten*, hrsg. v. Vittorio Messori, Hamburg 1994.

John Paul II, "Address to Partecipants" (sic!), in: *Religious Education and Modern Society, Acts of a Muslim-Christian Colloquium Organized Jointly by the Pontifical Council for Interreligious Dialogue (Vatican City) and the Royal Academy for Islamic Civilization Research Al Albait Foundation (Amman)*, s.l. s.a., p. 136–137.

John Paul II, "To Cardinal Francis Arinze President Pontifical Council for Interreligious Dialogue", in: *Religion and the Use of the Earth's Resources, Acts of a Christian-Muslim Colloquium Organized Jointly by the Pontifical Council of Interreligious Dialogue (Vatican City) and the Royal Academy for Islamic Civilization Research Al Albait Foundation (Amman – Jordan)*, Rome 1996, p. 179.

John Paul II, H.H. Pope, "Address of H.H. Pope John Paul II to Participants in the Colloquium "Women in Society According to Islam and Christianity"", in: *Women in Society According to Islam and Christianity, Acts of a Muslim-Christian Colloquium Organized Jointly by the Pontifical Council for Interreligious Dialogue (Vatican City) and the Royal Academy for Islamic Civilization Research Al Albait Foundation (Amman)*, Vatican City s.a., p. 123–124.

"JORDAN Final Communiqué of the Muslim-Christian Colloquium of 18–20 January 1994 on the Theme "Nationalism Today: Problems and Challenges"", Islamochristiana 20 (1994), p. 239.

"Nationalism Today: Problems and Challenges, Final Communiqué", BPCDIR 87 (1994), p. 266–267.

Jullundry, Rashid Ahmad, "Islam", BSNC 32 (1976), p. 186–208.

Kamel, Abdul Aziz, "Identity and Openness: Religious Education in a Pluralistic World", in: *Religious Education and Modern Society, Acts of a Muslim-Christian Colloquium Organized Jointly by the Pontifical Council for Interreligious Dialogue (Vatican City) and the Royal Academy for Islamic Civilization Research Al Albait Foundation (Amman)*, s.l. s.a., p. 15–36.

Kaulig, Ludger, *Ebenen des christlich-islamischen Dialogs, Beobachtungen und Analysen zu den Wegen einer Begegnung*, Christentum und Islam im Dialog / Christian-Muslim Relations 3 (zugleich kath. theol. Diss. Bochum), Münster 2004.

Kawtharani, Wajih, "Practical Application of the Spirit of Tolerance, An Islamic Perspective", in: Pontifical Council for Interreligious Dialogue / World Islamic Call Society (edd.), *Coexistence between Religions, Reality and Horizons*, Vatican City s.a., p. 39–42.

Kayitakigba, Médard, "Chrétiens et Musulmans aujourd'hui face à face", BSNC 62 (1985), p. 192–204.

Kayitakigba, Médard, "Chrétiens et musulmans en Afrique sub-Saharienne", BSNC (1984), p. 328–338.

Khader, Jamal, "A Christian Commentary on Dr. El-Abbadi's Paper", in: *Religion and the Use of the Earth's Resources, Acts of a Christian-Muslim Colloquium Organized Jointly by the Pontifical Council for Interreligious Dialogue (Vatican City) and the Royal Academy for Islamic Civilization Research Al Albait Foundation (Amman – Jordan)*, Rome 1996, p. 43–45.

Khawaldeh, Samira F., "Future Opportunities and Prospects for Women in Islam", in: *Women in Society According to Islam and Christianity, Acts of a Muslim-Christian Colloquium Organized Jointly by the Pontifical Council for Interreligious Dialogue (Vatican City) and the Royal Academy for Islamic Civilization Research Al Albait Foundation (Amman)*, Vatican City s.a., p. 97–106.

Khayyat, Abdul Aziz, "A Muslim Comment", in: *Human Dignity, Acts of a Christian-Muslim Colloquium Organized Jointly by the Pontifical Council for Interreligious Dialogue (Vatican City) and the Royal Academy for Islamic Civilization Research Al Albait Foundation (Amman – Jordan)*, Rome 1999, p. 53–63.

Khayyat, Abdul Aziz, "The Pre-school Child: Rights and Education", in: Pontifical Council for Interreligious Dialogue / Royal Academy for Islamic Civilization Research Al Albait Foundation (edd.), *The Rights and Education of Children in Islam and Christianity, Acts of a Muslim-Christian Colloquium*, Vatican City s.a., p. 67–93.

Khayyat, Abdul-Aziz, "The Role of Believers in Ensuring an Equitable Distribution of Earth's Resources, A Muslim Point of View", in: *Religion and the Use of the Earth's Resources, Acts of a Christian-Muslim Colloquium Organized Jointly by the Pontifical Council for Interreligious Dialogue (Vatican City) and the Royal Academy for Islamic Civilization Research Al Albait Foundation (Amman – Jordan)*, Rome 1996, p. 125–143.

Khayyat, Abdul Aziz, "Women's Status in Islam", in: *Women in Society According to Islam and Christianity, Acts of a Muslim-Christian Colloquium Organized Jointly by the Pontifical Council for Interreligious Dialogue (Vatican City) and the Royal Academy for Islamic Civilization Research Al Albait Foundation (Amman)*, Vatican City s.a., p. 17–32.

Keating, Sandra, "Comment", in: *Human Dignity, Acts of a Christian-Muslim Colloquium Organized Jointly by the Pontifical Council for Interreligious Dialogue (Vatican City) and the Royal Academy for Islamic Civilization Research Al Albait Foundation (Amman – Jordan)*, Rome 1999, p. 123–125.

"Kiev, Ukraine: 24 June 2001: **To the Illustrious Representatives of the All-Ukrainian Council of Churches and Religious Organizations** (Extract)", BPCDIR 108 (2001), p. 293–295.
Konrad-Adenauer-Stiftung e.V. (Hrsg.), *Vatikan und Ökumene, Die Debatte über die vatikanische Erklärung "Dominus Iesus"*, Sankt Augustin 2000.
Kreutz, Andrej, "The Vatican and the Palestinians: A Historical Overview", Islamochristiana 18 (1992), p. 109–125.
Kroeger, James H., "Milestones in Interreligious Dialogue", Studies in Interreligious Dialogue 7 (1997), p. 232–239.
Kukah, Matthew Hassan, "Religious Liberty in a Plural Society: The Nigerian Experience", in: Pontifical Council for Interreligious Dialogue / Commission for Religious Relations with Muslims (ed.), *Religious Liberty; A Theme for Christian-Muslim Dialogue*, Vatican City 2006, p. 105–124.
"La visite pastorale de Jean Paul II au Mali (28–29 janvier 1990)", Islamochristiana 16 (1990), p. 255–259.
Lahbabi, Mohamed Aziz, "Islam", BSNC 32 (1976), p. 155–173.
Lahbabi, Mohamed Aziz, "L'Islam et la paix", BSNC 32 (1976), p. 174–185.
Landousies, J., "Algérie: Rapport sur le dialogue islamo-chrétien", BPCDIR 73 (1990), p. 99–101.
Lanfry, Jacques, "A la mémoire du P. Joseph Cuoq, premier responsable pour l'Islam (1964–1974)", BPCDIR 72 (1989), p. 383–390.
Lanfry, J[acques], "**Rome – Vatican:** Muslim Delegation at the Secretariat", BSNC 16 (1971), p. 41–44.
"Le discours du Saint-Père aux membres du Corps Diplomatique accrédité près le Saint-Siège (11 janvier 1992)", Islamochristiana 18 (1992), p. 317–318.
Le Gal, Suzanne, "Models for Holiness for Christians", Islamochristiana 11 (1985), p. 37–48.
"Le nouvel ambassadeur de Tunisie près le Saint-Siège est reçu par le Pape", Islamochristiana 15 (1989), p. 236.
"Le nouvel ambassadeur de Turquie près le St-Siège est recu *(sic!)* par le Pape", Islamochristiana 14 (1988), p. 307–308.
Legarde, Michel, "Violence et verité", BPCDIR 81 (1992), p. 282–327.
Lelong, Michel, "La communauté chrétienne et les musulmans en Tunisie", BSNC 30 (1975), p. 237–240.
Lelong, Michel," Le secrétariat de l'église de France pour les relations avec l'islam", BSNC 30 (1975), p. 249–252.
"Les formes succesives de la déclaration", dans: Henry, A[ntonien]-M[arie] (éd.), *Les relations de l'église avec les religions non chrétiennes*, Unam Sanctam 61, Paris 1966, p. 287–305.
"Les musulmans en Europe", BSNC 33 (1976), p. 337–340.
"Les rencontres de Jean Paul II avec les divers groupes de catholique sénégalais", Islamochristiana 18 (1992), p. 289–291.
"*LIBYA Symposium on Religion and Mass Media (Tripoli, 3–6 October 1993")*, Islamochristiana 20 (1994), p. 242.
Linden, Ian," The Role of Believers in Dealing with Questions of Nationalism, A Christian Viewpoint", in: *Nationalism Today: Problems and Challenges, Acts of a Muslim-Christian Colloquium Organized Jointly by the Pontifical Council for Interreligious Dialogue*

(Vatican City) and the Royal Academy for Islamic Civilization Research al Albait Foundation (Amman), Rome 1994, p. 87–99.
"List of the Non-Christian Visitors During the Holy Year", BSNC 31 (1976), p. 78–79.
Machado, Felix A[nthony], "A Summary of Reports by the Members: Interreligious Dialogue Promoted by the Church", BPCDIR 109 (2002), p. 125–130.
Machado, Felix [Anthony], "A Theological Evalutation of Modernity: An Asian Point of View", BPCDIR 89 (1995), p. 176–183.
Machado, Felix [Anthony], "**Bangalore – India:** Report: Asian Consultors and/or Secretaries of the National Episcopal Dialogue Commissions from Asia, 5–8 July 1998"., BPCDIR 99 (1998), p. 345–348.
Machado, Felix [Anthony], "Harmony among Believers of the Living Faiths: Christians and Muslims in South East Asia", BPCDIR 87 (19994), p. 216–218.
Machado, Felix [Anthony], "Interreligious Dialogue in the Various Regions of the World", BPCDIR 92 (1996), p. 267–270.
Machado, Felix A[nthony], "Presentation", in: Pontifical Council for Interreligious Dialogue (ed.), *Meeting in Friendship, Messages to Muslims for the End of Ramadan (1967–2002)*, Vatican City 2003, p. 5–7.
Magro, Sylvester, "Religious Tolerance in Libya", in: Pontifical Council for Interreligious Dialogue / World Islamic Call Society (edd.), *Co-existence between Religions, Reality and Horizons*, Vatican City s.a., p. 47–50.
Marella, Paul, Présentation, dans: Secrétariat pour les non chrétiens (éd.), Orientations pour un dialogue entre chrétiens et musulmans, 2ième éd., Ancona 1969, p. 5–6
Marella, Paul, "Présentation", dans: Secretariatus pro non Christianis (éd.), Religions, *Thèmes fondamentaux pour une connaissance dialogique*, Rome 1970, p. 5–6.
"MAROC *Commémoration du Xème anniversaire de la visite au Maroc de Sa Sainteté le Pape Jean Paul II, effectué en août 1985 (15 octobre 1995)*", Islamochristiana 22 (1996), p. 243–251.
Masson, Joseph, "Souvenirs et perspectives", BPCDIR 72 (1989), p. 334–342.
Mbiti, John, "**Enugu, Nigeria:** The Contributions of Africa to the Religious Heritage of the World; 8–13 January 2001", BPCDIR 107 (2001), p. 255–260.
McAuliffe, Jane Dammen, "Monitoring for Religious Freedom: A New International Mandate", in: Pontifical Council for Interreligious Dialogue / Commission for Religious Relation with Muslims (ed.), *Religious Liberty: A Theme for Christian-Muslim Dialogue*, Vatican City 2006, p. 151–183.
McAuliffe, Jane Dammen, "Rendering Allegiance to the Word: *Qur'anic* Concepts and Contemporary North American Concerns", in: Commission pour les rapports religieux avec les musulmans (éd.), *Religion et politique: Un thème pour le dialogue islamo-chrétien*, Cité du Vatican 1999, p. 13–36.
McDonald, Kevin, in: Fitzgerald, Michael L[ouis] / Borelli, John (edd.), *Interfaith Dialogue, A Catholic View*, London/Maryknoll 2006, p. VII-IX.
Merchergui, Ahmed, "Testimony (Muslim, Tunisia)", BPCDIR 116/117 (2004), p. 304–306.
"**Message aux chrétiens du Maghreb de la Conférence des Evêques de la Region Nord de l'Afrique (C.E.R.N.A.),** Alger, le 8 juin 1990", BPCDIR 74 (1990), p. 155–163.
"Message de Jean Paul II aux Patriarches et aux Evêques sur le Moyen-Orient (4 mars 1991)", Islamochristiana 17 (1991), p. 287–289.

"Opening Discourse", in: *Rencontre de S.S. le Pape Jean-Paul II avec les patriarches des églises orientales et les évêques des pays impliqués dans la guerre du golfe, Meeting of H.H. Pope John Paul II with the Patriarchs of the Oriental Churches and Bishops of Countries Involved in the Gulf War, 4–6 March 1991*, s.l.s.a., p. 18–21.

"Message of His Holiness Pope John Paul II for the Celebration of the World Day of Peace (7 January 1999)", Islamochristiana 25 (1999), p. 240–242.

Michel, Thomas, "25 Years of Letters to Muslims for Id Al-Fitr", BPCDIR 84 (1993), p. 300–302.

Michel, Thomas, *1991 Estimated Muslim Population of Various Countries*, C.R.R.M. Reports Recent Trends and Movements in Islam 5, s.l. 1992.

Michel, Thomas, "Christian and Muslim Minorities: Possibilities for Dialogue", BPCDIR 70 (1989), p. 47–58.

Michel, Thomas, "Christlich-islamischer Dialog: Gedanken zu neueren Verlautbarungen der Kirche (Fortsetzung)", Christlich-islamische Begegnung – Dokumentationsleitstelle (CIBEDO) Dokumentation 2 (1988), S. 45–63.

Michel, Thomas, "God's Covenant with Mankind According to the Qur'an", BSNC 52 (1983), p. 31–43.

Michel, Thomas, "Growing towards "Dialogue in Community" a Reflection on Archbishop Jadot's Tenure at the Secretariat for non Christians", BPCDIR 103 (2000), p. 68–70.

Michel, Thomas, "IRAN, In Tehran, the Muslim-Christian Colloquium on "A Theological Evaluation of Modernity" (30 October – 2 November 1994)", Islamochristiana 21 (1995), p. 172.

Michel, Thomas, "Islamo-Christian Dialogue: Reflections on the Recent Teachings of the Church", BSNC 59 (1985), p. 172–193.

Michel, Thomas, "New Muslim Declaration on Human Rights", BSNC 48 (1981), p. 248–249.

Michel, Thomas, "P.C.I.D. Dialogue with Muslims since the Last Plenary", BPCDIR 82 1993), p. 34–45.

Michel, Thomas, "Pope John Paul II's Teaching about Islam in His Adresses to Muslims", BSNC 62 (1985), p. 182–191.

Michel, Th[omas], "Religion in Indonesia today", BSNC 44 (1980), p. 213–220.

Michel, Thomas, "**Rome – Italy:** Colloquium on "Holiness in Islam and Christianity"", BSNC 60 (1985), p. 320–322.

Michel, Thomas, Sin, "Forgiveness and Reconciliation in Islam", BSNC 53 (1983), p. 115–134.

Michel, Thomas, "**Teheran – Iran:** Christian-Muslim Colloquium on "A Theological Evaluation of Modernity" (30 October – 2 November)", BPCDIR 87 (1994), p. 275–276.

Michel, Thomas, "The Idea of Holiness in Islam", BPCDIR 92 (1996), p. 220–240.

Michel, Thomas, "Tolerance: A Christian Perspective", in: Pontifical Council for Interreligious Dialogue / World Islamic Call Society (edd.), *Co-existence between Religions, Reality and Horizons*, Vatican City s.a., p. 27–31.

Michel, Thomas, "**Tripoli – Libya:** Seminar on The Media and Presentation of Religion (3–6 October)", BPCDIR 84 (1993), p. 308–310.

Michel, Tom, "Islam in Asia Today", BSNC 57 (1984), p. 314–327.

"MOROCCO *The Third Meeting of the Islamo-Catholic Liaison Committee (Rabat, 18–19 June 1997)*", Islamochristiana 23 (1997), p. 223–224.

Muasher, Anis, "The Role of Believers in Ensuring an Equitable Distribution of the Earth's Resources, A Christian Point of View", in: *Religion and the Use of the Earth's Resources*,

Acts of a Christian-Muslim Colloquium Organized Jointly by the Pontifical Council for Interreligious Dialogue (Vatican City) and the Royal Academy for Islamic Civilization Research Al Albait Foundation (Amman – Jordan), Rome 1996, p. 149–165.

Musallam, Adnan, "Nationalism and Religion: An Historical Overview, A Christian Perspective", in: *Nationalism Today: Problems and Challenges, Acts of a Muslim-Christian Colloquium Organized Jointly by the Pontifical Council for Interreligious Dialogue (Vatican City) and the Royal Academy for Islamic Civilization Research Al Albait Foundation (Amman)*, Rome 1994, p. 27–44.

"Muslim-Christian Meeting at the Offices of the Pontifical Council for Interreligious Dialogue with Representatives of the World Islamic Call Society (Tripoli, Libya), (14–15 February 1990)", Islamochristiana 16 (1990), p. 294–295.

Musu, Ignazio, "A Christian Commentary on Prof. Khayyat's Paper", in: *Religion and the Use of the Earth's Resources, Acts of a Christian-Muslim Colloquium Organized Jointly by the Pontifical Council for Interreligious Dialogue (Vatican City) and the Royal Academy for Islamic Civilization Research Al Albait Foundation (Amman – Jordan)*, Rome 1996, p. 145–148.

Muthaffar, Mai, "A Muslim Commentary on Fr. Irwin's Paper", in: *Religion and the Use of the Earth's Resources, Acts of a Christian-Muslim Colloquium Organized Jointly by the Pontifical Council for Interreligious Dialogue (Vatican City) and the Royal Academy for Islamic Civilization Research Al Albait Foundation (Amman – Jordan)*, Rome 1996, p. 91–94.

Muthaffar, Mai, "Comment", in: *Human Dignity, Acts of a Christian-Muslim Colloquium Organized Jointly by the Pontifical Council for Interreligious Dialogue (Vatican City) and the Royal Academy for Islamic Civilization Research Al Albait Foundation (Amman – Jordan)*, Rome 1999, p. 81–83.

Naber, Anton, "Christian Response", in: Pontifical Council for Interreligious Dialogue / Royal Academy for Islamic Civilization Research Al Albait Foundation (edd.), *The Rights and Education of Children in Islam and Christianity, Acts of a Muslim-Christian Colloquium*, Vatican City s.a., p. 152–157.

"Nairobi: Greeting to Leaders of the Muslim Community in Kenya", in: Segretariato per i non cristiani (ed.), *Chiesa e islam*, s.l. 1981, p. 25–26.

"**Nairobi, Kenya:** Interreligious Dialogue Workshop Session, 24–31 July 1999", BPCDIR 102 (1999), p. 346–348.

"Nationalism Today: Problems and Challenges Final Communiqué, 20 January 1994", in: *Nationalism Today: Problems and Challenges, Acts of a Muslim-Christian Colloquium Organized Jointly by the Pontifical Council for Interreligious Dialogue (Vatican City) and the Royal Academy for Islamic Civilization Research Al Albait Foundation (Amman)*, Rome 1994, p. 129–130.

"New Delhi, India: 6 November 1999, **Apostolic Exhortation: Ecclesia in Asia** (Extract)", BPCDIR 103 (2000), p. 27–47.

Nielsen, Jorgen *(sic!)* S., "National Ties Today: Problems and Challenges, A Christian Viewpoint", in: *Nationalism Today: Problems and Challenges, Acts of a Muslim-Christian Colloquium Organized Jointly by the Pontifical Council for Interreligious Dialogue (Vatican City) and the Royal Academy for Islamic Civilization Research Al Albait Foundation (Amman)*, Rome 1994, p. 47–61.

Nizami, K[haliq] A[hmad], "Models of Holiness for Muslims", Islamochristiana 11 (1985), p. 51–66.

Nnyombi, Richard, "Christianity and Islam in the Political Life of East Africa: The Example of Kenya", in: Commission pour les rapports religieux avec les musulmans (éd.), *Religion et politique, Un thème pour le dialogue islamo-chrétien*, Cité du Vatican 1999, p. 178–196.

Onah, Godfrey I., "Reflections on *Dignitatis Humanae*, 2–4", in: Isizo, Chidi Denis (ed.), *Milestones in Interreligious Dialogue, A Reading of Selected Catholic Church Documents on Relations with People of Other Religions, Essays in Honour of Francis Cardinal Arinze*, Rome/Lagos 2002, p. 83–110.

Orlando, G., "**Bangladesh:** Reports on the Activity of the "Brotherhood Commission" for Muslim and Christian Encounter", BSNC 36 (1977), p. 195–199.

Pacini, Andrea, "Christians Facing the Challenge of Human Dignity: Prospects for the Future", in: Pontifical Council for Interreligious Dialogue / Royal Academy for Islamic Civilisation Research Al Albait Foundation (edd.), *Human Dignity, Acts of a Christian-Muslim Colloquium*, Vatican City 1999, p. 127–139.

"**Paola – Malta:** *Christian-Muslim Meeting*", BPCDIR 76 (1991), p. 101–108.

"**Paris, France:** Islamic-Catholic Liaison Committee, 1–3 July 1999", BPCDIR 102 (1999), p. 345–346.

"Paris: Message to Representatives of the Moslem Community in France", in: Segretariato per i non cristiani (ed.), *Chiesa e islam*, s.l. 1981, p. 33–34.

"Participants à la réunion de Bamako", BSNC 28/29 (1975), p. 3.

Pedretti, G., "À la mémoire du P. Pierre Humbertclaude S.M. (1899–1984)", BSNC 55 (1984), p. 114–115.

Peteiro, Antonio, "Chrétiens et Musulmans constructeurs de la paix", BPCDIR 70 (1989), p. 37–42.

Pignedoli, Sergio, "1973", p. 18,
  "1974", p. 19–20,
  "1975", p. 21–22,
  "1976", p. 23–25,
  "1977", p. 26–28,
  "1978", p. 29–30,
  "1979", p. 31,
  in: Pontifical Council for Interreligious Dialogue (ed.), *Meeting in Friendship, Messages to Muslims for the End of Ramadan (1967–2202)*, Vatican City 2003.

Pignedoli, Sergio, "Introduction of the Cardinal President to the Plenary Meeting of the Secretariat for Non-Christian Religions, 1979", BSNC 41/42 (1979), p. 85–87.

Pignedoli, Sergio, "Letter of Cardinal President of the Secretariate to the Bishops", BSNC 36 (1977), p. 89–92.

Pignedoli, Sergio, "Lettre du nouveau Président", BSNC 22 (1973), p. 12–16.

Pignedoli, Sergio, "Pentecostal Letter of the President of the Secretariat", BSNC 26 (1974), p. 91–94.

Pignedoli, Sergio, "Perspectives du Secrétariat", BSNC 23/24 (1973), p. 86–88.

Pignedoli, Sergio, "Preface", in: Secretariatus pro non Christianis (ed.), *Religions in the World*, Città del Vaticano s.a., p. 7–8.

Pignedoli, Sergio, "Relation de S. E. le card. Sergio Pignedoli", BSNC 28/29 (1975), p. 12–15.

Platti, Emilio, "Tendances dans l'islam d'aujourd'hui: Europe Occidentale", Dossiers de la C.R.R.M. Courants et mouvements dans l'islam contemporain 3, s.l. 1992.
Pontifical Council for Interreligious Dialogue (ed.), *Journeying together, The Catholic Church in Dialogue with the Religious Traditions of the World*, Città del Vaticano 1999.
Pontifical Council for Interreligious Dialogue (ed.), *Meeting in Friendship, Messages to Muslims for the End of Ramadan (1967–2002)*, Vatican City 2003.
Pontifical Council for Interreligious Dialogue (ed.), *Recognize the Spiritual Bonds which Unite Us, 16 Years of Christian-Muslim Dialogue*, Vatican City 1994.
Pontifical Council for Interreligious Dialogue / Borrmans, Maurice, *Guidelines for Dialogue between Christians and Muslims*, Interreligious Documents 1, New York / Mahwah 1990.
Pontifical Council for Interreligious Dialogue / Fitzgerald, H.E. Archbishop Michael L[ouis], *Constructing Peace Today, Message for the End of Ramadan 'Id al-Fitr 1424 A.H. / 2003 A.D.*, Vatican City s.a.
Ponticifal Council for Interreligoious Dialogue / Pontifical Council for the Family (edd.), *Marriage and the Family in Today's World, Interreligious Colloquium Rome, 21–25 September 1994*, Vatican City 1995.
Pontifical Council for the Pastoral Care of Migrants and Itinerant People, *Istruction Erga Migrantes Caritas Christi (The Love of Christ towards Migrants)*, Vatican City 2004.
"Pope John Paul's Address to the Delegation of the World Islamic League during the Audience of Thursday 28 January 1993", Islamochristiana 19 (1993), p. 316.
"Pope John Paul's Address to the Participants in this Colloquium during the Audience of Friday 26 June 1992:" *(sic!)*, Islamochristiana 18 (1992), p. 320–321.
"Pope's Address to the Diplomatic Corps Accredited to the Holy See", Islamochristiana 15 (1989), p. 251–252.
"Pope's Address to the Participants in the Plenary Assembly 1998", BPCDIR 101 (1999), p. 191–192.
"Preface", BSNC 32 (1976), p. 105–106.
"Press Communiqué", BPCDIR 87 (1994), p. 272–273.
"Press Communiqué of the Workshop on "The Media and Religion", Jointly Organized by the World Islamic Call Society (Libya) and the Pontifical Council for Interreligious Dialogue (Vatican) (Vienna, 7–8 October 1994)", Islamochristiana 21 (1995), p. 142–143.
"Press Statement of the 1st Meeting (22 June 1995)", Islamochristiana 21 (1995), p. 178.
"Press Statement of the 2nd Meeting (28 *(sic!)* June 1995)", Islamochristiana 21 (1995), p. 178.
Pruvost, Lucie, "From Tolerance to Spiritual Emulation", Islamochristiana 6 (1980), p. 1–9.
Pruvost, Lucie, "L'islam dans les cinq pays du Maghreb arabe", DOSSIERS de la C.R.R.M. Courants et mouvements dans l'islam contemporain 6, s.l. 1993.
Pruvost, Lucie, "Tendances et courants dans l'islam algérien", Dossiers de la C.R.R.M. Courants et mouvements dans l'islam contemporain 7, s.l. 1993.
Przewozny, Bernard, "A Christian Commentary on Dr. El-Tall's Paper", in: *Religion and the Use of the Earth's Resources, Acts of a Christian-Muslim Colloquium Organized Jointly by the Pontifical Council for Interreligious Dialogue (Vatican City) and the Royal Academy for Islamic Civilization Research Al Albait Foundation (Amman – Jordan)*, Rome 1996, p. 119–122.
"Qunaytra, Syria: 7th May 2001, **Prayer for Peace**", BPCDIR 107 (2001), p. 173–174.
Rahal, M. Redouane, "Réponse d'un participant musulman", BPCDIR 70 (1989), p. 45–46.

Rahman, Hani Abdul, "Rights and Education of Children in Schools, as Related to Government and Society", in: Pontifical Council for Interreligious Dialogue / Royal Academy for Islamic Civilization Research Al Albait Foundation (edd.), *The Rights and Education of Children in Islam and Christianity, Acts of a Muslim-Christian Colloquium*, Vatican City s.a., p. 130–151.

"Ramadan 1969", BSNC 13 (1970), p. 52–53.

"Rapport de Synthèse", BPCDIR 70 (1989), p. 42–43.

Rasmussen, Nete, "Christian Response", in: Pontifical Council for Interreligious Dialogue / Royal Academy for Islamic Civilization Research Al Albait Foundation (edd.), *The Rights and Education of Children in Islam and Christianity, Acts of a Muslim-Christian Colloquium*, Vatican City s.a., p. 94–97.

"Recommandations", in: *Religion and the Use of the Earth's Resources, Acts of a Christian-Muslim Colloquium Organized Jointly by the Pontifical Council for Interreligious Dialogue (Vatican City) and the Royal Academy for Islamic Civilization Research Al Albait Foundation (Amman – Jordan), Rome 1996*, p. 181–183.

"Recommandations of the Four Section *(sic!)* Addressed to the Secretariat", BSNC 18 (1971), p. 212–217.

"Relations avec les non-chrétiens", POC 2 (1974), p. 203–204.

"Religious Leaders Gather anew in Assisi (Italy) to Fast and Pray for Peace in Balkans and Europe (9–10 January 1993)", Islamochristiana 19 (1993), p. 250–252.

"**Rencontre** de S.S. le Pape Jean-Paul II avec les patriarches des églises orientales et les évêques des pays impliqués dans la guerre du golfe, **Meeting** of H.H. Pope John Paul II with the Patriarchs of the Oriental Churches and Bishops of Countries Involved in the Gulf War, 4–6 March 1991", s.l.s.a.

"Rencontre islamo-chrétienne du Caire, 12–13 avril 1978", BSNC 38 (1978), p. 157–160.

Renz, Andreas, *Der Mensch unter dem An-Spruch Gottes, Offenbarungsverständnis und Menschenbild des Islam im Urteil gegenwärtiger christlicher Theologie*, Christentum und Islam: Anthropologische Grundlagen und Entwicklungen 1, Würzburg 2002.

"Report of Mgr. Rossano's Journey to Iraq, Pakistan, Bangladesh, Northern India", BSNC 34/35 (1977), p. 52–56

"Report of the Christian-Muslim Dialogue Meeting in Ibadan (4–8 August 1991)", Islamochristiana 17 (1991), p. 249–252.

"Report on Mgr. Rossano's Journey in South-East Asia", BSNC 31 (1976), p. 49–61.

"Report on the "Seminar on Islamic-Christian Dialogue" Held in Tripoli (1st – 5th February 1976)", BSNC 31 (1976), p. 5–13.

Roest Crollius, Ary A., "Four Notes on Dialogue (More Especially Regarding Islam)", BSNC 30 (1975), p. 245–248.

Roest Crollius, Ary A., "Harmony and Conflict", BPCDIR 81 (1992), p. 360–377.

Roest Crollius, Ary [A.], "Rome 10–13 mai 1989, "**Communiquer les valeurs religieuses à la jeunesse aujourd'hui**"", BPCDIR 73 (1990), p. 79–81.

"**Rome – Italy:** Annual Report of the Committee for Dialogue of the Pontifical Council for Interreligious Dialogue and the Permanent Committee of al-Azhar for Dialogue with Monotheistic Religions", BPCDIR 110 (2002), p. 227–228.

"Rome – Italy: **Women in Society** (24–26 June 1992)", BPCDIR 81 (1992), p. 378–380.

"Rome: **Meeting between the World Islamic Call Society and the Pontificial** *(sic!)* **Council for Interreligious Dialogue** (14–15 February)", BPCDIR 74 (1990), p. 178–180.

Root, Howard, "Response to the Paper of Dr. Abdul Aziz Al-Khayyat on the Theme: "Religious Education in Colleges and Universities: Present Situation and Aspirations for the Future"", in: *Religious Education and Modern Society, Acts of a Muslim-Christian Colloquium Organized Jointly by the Pontifical Council for Interreligious Dialogue (Vatican City) and the Royal Academy for Islamic Civilization Research Al Albait Foundation (Amman)*, s.l. s.a., p. 128–129.

Rossano, Pietro, "10th Anniversary of the Death of Cardinal Sergio Pignedoli", BPCDIR 74 (1990), p. 172–174.

Rossano, P[ietro], "Bilan d'une rencontre Domus Mariae – 1972", BSNC 21 (1972), p. 9–15.

Rossano, P[ietro], "Contents of Dialogue with Non-Christians", BSNC 26 (1974), p. 143–144.

Rossano, P[ietro], "Convergences Spirituelles et Humanistes entre l'Islam et le Christianisme", BSNC 40 (1979), p. 15–19.

Rossano, P[ietro], "Exposition from the Catholic Point of View", BSNC 17 (1971), p. 103–108.

Rossano, Pietro, "Foreword", in: Pontifical Council for Interreligious Dialogue / Borrmans, Maurice, *Guidelines for Dialogue between Christians and Muslims*, Interreligious Documents 1, New York / Mahwah 1990, p. 1–3.

Rossano, Pie[t]ro, "Introduction", BSNC 30 (1975), p. 211–213.

Rossano, P[ietro], "Introduction", dans: Secretariatus pro non Christianis (éd.), *Religions, Thèmes fondamentaux pour une connaisance dialogique*, Rome 1970, p. 95–99.

Rossano, P[ietro], "L'homme et la religions", dans: Secretariatus pro non Christianis (éd.), *Religions, Thèmes fondamentaux pour une connaissance dialogique*, Rome 1970, p. 7–92.

Rossano, P[ietro], "Le bien et le mal dans les religions: vue rétroperspective", dans: Secretariatus pro non Christianis (éd.), *Religions, Thèmes fondamentaux pour une connaissance dialogique, Rome 1970*, p. 597–603.

Rossano, Pietro, "Le cheminement du dialogue interreligieux de "Nostra Aetate" à nos jours", BPCDIR 74 (1990), p. 130–142.

Rossano, P[ietro], "Le salut dans le christianisme", dans: Secretariatus pro non Christianis (éd.), *Religions, Thèmes fondamentaux pour une connaissance dialogique*, Rome 1970, p. 101–114.

Rossano, P[ietro], "Our Programme and Method", BSNC 26 (1974), p. 141–143.

Rossano, P[ietro], "Préliminaire", dans: Secretariatus pro non Christianis (éd.), *Religions, Thèmes fondamentaux pour une connaissance dialogique*, Rome 1970, p. 227–235.

Rossano, P[ietro], "Presentation of the Volume "Religions" Published by the Secretariat", BSNC 16 (1971), p. 36–40.

Rossano, Pietro, "Report of the Conversation between the Iranian Delegation and that of the Secretariat for Non-Christians", BSNC 37 (1978), p. 25–28.

Rossano, P[ietro], **"Rome:** Restitution of the Visit of the WCC Commission for Dialogue to the Secretariat for Non-Christians (7 March 1974)", BSNC 26 (1974), p. 139–143.

Rossano, Pietro, "The Secretariat for Non-Christian Religions from the Beginnings to the Present Day: History, Ideas, Problems", BSNC 41/42 (1979), p. 88–109.

Sabanagh, E[douard] S[ami] Martin, "L'Église catholique au sein du monde arabo-musulman", BSNC 57 (1984), p. 289–309.

Sabanagh, E[douard] S[ami] Martin, "Les Chrétiens d'Europe face aux travailleurs musulmans imigrés", BSNC 57 (1984), p. 310–313.

Sabanageh, E[douard] S[ami] Martin, "Les directives de la Hiérarchie catholique de l'Europe de l'Ouest concernant les Musulmans etrangers", BSNC 60 (1985), p. 291–302.

Sabanageh, E[douard Sami] M[artin], "Pour une compréhension de la législation matrimoniale en Islam", BSNC 44 (1980), p. 157–177.

Sabbah, Michel, "Identity and Openness: Religious Education in a Pluralistic World", in: *Religious Education and Modern Society, Acts of a Muslim-Christian Colloquium Organized Jointly by the Pontifical Council for Interreligious Dialogue (Vatican City) and the Royal Academy for Islamic Civilization Research Al Albait Foundation (Amman)*, s.l. s.a., 37–45.

Sabbah, Michel, "Spiritualité dans un milieu arabo-musulman", BPCDIR 99 (1998), p. 312–316.

Salama, Andraos, "Le Message du Secrétariat aux Musulmans pour la fin du Ramadan 1407/1987 et ses échos dans le monde", BSNC 66 (1987), p. 307–314.

Salama, Andraos, "Les mariages entre chrétiens et musulmans approches pastorales", BPCDIR 70 (1989), p. 59–76.

Sanna, Ignazio, "The Dignity of Man, A Christian Perspective", in: *Human Dignity, Acts of a Christian -Muslim Colloquium Organized Jointly by the Pontifical Council for Interreligious Dialogue (Vatican City) and the Royal Academy for Islamic Civilization Research Al Albait Foundation (Amman – Jordan), Rome 1999*, p. 39–52.

Schmidt, Stjepan, "Cardinal Bea and Interreligious Dialogue", BPCDIR 74 (1990), p. 143–149.

Secretariatus pro non christianis (ed.), *Le discours de S.S. Jean-Paul II lors de sa rencontre avec les jeunes musulmans à Casablanca (Maroc), le 19 août 1985, The Speech of the Holy Father John Paul II to Young Muslims during His Meeting with them at Casablanca (Morocco), August the 19th, 1985*, s.l.s.a.

Segretariato per i non cristiani (ed.), *Chiesa e islam*, s.l. 1981.

Segretariato per i non cristiani (ed.), *Religioni, Temi fondamentali per una conoscenza dialogica*, 4. ed., Fossano 1987.

"Séminaire académique de théologiens musulmans et chrétiens 'à l'Université Pontifical Grégorienne: "Communiquer les valeurs religieuses à la jeunesse aujourd'hui"", Islamochristiana 15 (1989), p. 242–243.

Serain, Michel, "Chrétiens et Musulmans en classe ouvrière", BSNC 63 (1986), p. 251–261.

Sharif, Muhammad Ahmad, "Understanding Tolerance", in: Pontifical Council for Interreligious Dialogue / World Islamic Call Society (edd.), *Co-existence between Religions, Reality and Horizons*, Vatican City s.a., p. 10–11.

Sheikh Al-Azhar Mohammed Tantawi (Islam), BPCDIR 109 (2002), p. 148–149

Sheridan, Rosaleen, "Science, Faith and the Concerns of Youth", in: *Relgious Education and Modern Society, Acts of a Muslim-Christian Colloquium Organized Jointly by the Pontifical Council for Interreligious Dialogue (Vatican City) and the Royal Academy for Islamic Civilization Research Al Albait Foundation (Amman)*, s.l. s.a., p. 49–59.

Sherwin, Byron L. / Kasimow, Harold (edd.), *John Paul II and Interreligious Dialogue*, Maryknoll 1999.

Shirieda, John, "Don Sergio, Witness to God's Goodness", BPCDIR 74 (1990), p. 175–177.

Shirieda, John," Editorial", BPCDIR 70 (1989), p. 1–2.

Shirieda, John, "Editorial", BPCDIR 79 (1992), p. 1.

Shirieda, J[ohn], "Editorial", BPCDIR 80 (1992), p. 129–130.

Shirieda, John, "Editorial", BPCDIR 82 (1993), p. 1–2.

Shirieda, John, "Report of the Conversation between the Indonesian Delegation and that of the Secretariate for Non-Christians", BSNC 37 (1978), p. 5–8.

Shorter, Aylward, "Cardinal Pignedoli: A Brief Memoir", BPCDIR 72 (1989), p. 377–378.

"Signature d'un accord de coopération académique entre l'université d'Ankara et l'université grégorienne de Rome", Islamochristiana 12 (1986), p. 222–223.

"Solemn Closing", in: Rencontre de S.S. le Pape Jean-Paul II avec les patriarches des églises orientales et les évêques des pays impliqués dans la guerre du golfe, Meeting of H.H. Pope John Paul II with the Patriarches of the Oriental Churches and Bishops of Countries Involved in the Gulf War, 4–6 March 1991, s.l.s.a., p. 22–25.

Souza, Achilles de, "Christian-Muslim Dialogue with Reference to Pakistan", BSNC 30 (1975), p. p. 218–222.

"St. Egidio Community, **Rome – Italy: The Visit of the Grand Mufti of Syria to Rome** (December 1985)", BSNC 62 (1986), p. 218–219.

Stäger, Roman, "The Concept of Human Dignity According to Islam, A Christian Comment", in: Human Dignity, Acts of a Christian-Muslim Colloquium Organized Jointly by the Pontifical Council for Interreligious Dialogue (Vatican City) and the Royal Academy for Islamic Civilization Research Al Albait Foundation (Amman – Jordan), Rome 1999, p. 37–38.

Stamer, Josef, "Compte Rendu: L'Eglise d'Afrique Occidentale en dialogue avec les musulmans?", BSNC 28/29 (1975), p. 4–11.

Stamer, Josef, "Les tendances actuelles de l'islam en Afrique de l'ouest", Dossiers de la C.R.R.M. Courants et mouvements dans l'islam contemporain, s.l. 1992.

Stamer, Josef, "Présentation des minorités islamiques et de leurs revendications politiques dans l'Afrique de l'Ouest à travers le "Journal of the Institute of Muslim Minorities' Affairs" (JIMMA)", dans: Commission pour les rapports religieux avec les musulmans (éd.), Religion et politique: Un thème pour le dialogue islamo-chrétien, Cité du Vatican 1999, p. 79–100.

Stamer, Joseph, "Prier avec les musulmans?", BPCDIR 96 (1997), p. 357–370.

Stamer, Josef, "Religion et politique en Afrique de l'ouest: Des voix musulmanes", dans: Commission pour les rapports religieux avec les musulmans (éd.), Religion et politique: Un thème pour le dialogue islamo-chrétien, Cité du Vatican 1999.

"Symposium on "Co-existence between Religions: Reality and Horizons" (22–23 November 1990)", Islamochristiana 17 (1991), p. 234–236.

Tabbara, Nayla, "A Roman Experience", BPCDIR 108 (2001), p. 393–396.

Tannous, Nemra, "Christian Response", in: Pontifical Council for Interreligious Dialogue / Royal Academy for Islamic Civilization Research Al Albait Foundation (edd.), The Rights and Education of Children in Islam and Christianity, Acts of a Muslim-Christian Colloquium, Vatican City s.a., p. 60–64.

"SYRIA The New Ambassador to the Holy See Received by the Pope", Islamochristiana 15 (1989), p. 234–235.

[Talal, El] Hassan [bin], H.R.H. Crown Prince, "Concluding Discourse", In: Nationalism Today: Problems and Challenges, Acts of a Muslim-Christian Colloquium Organized Jointly by the Pontifical Council for Interreligious Dialogue (Vatican City) and the Royal Academy for Islamic Civilization Research al Albait Foundation (Amman), Rome 1994, p. 125–127.

Talal, Prince El Hassan bin, "Inaugural Address", in: Human Dignity, Acts of a Christian-Muslim Colloquium Organized Jointly by the Pontifical Council for Interreligious

Dialogue and the Royal Academy for Islamic Civilization Research Al Albait Foundation (Amman – Jordan), Rome 1999, p. 5–10.

[Talal], H.R.H. Crown Prince [El] Hassan [bin], "Inaugural Address", in: Pontifical Council for Interreligious Dialogue / Royal Academy for Islamic Civilization Research Al Albait Foundation (edd.), *The Rights and Education of Children in Islam and Christianity, Acts of a Muslim-Christian Colloquium*, Vatican City s.a., p. 3–6.

Talal, His Royal Highness Crown Prince El Hassan bin, of the Hashemite Kingdom of Jordan, "Islam and Environment", in: *Religion and the Use of the Earth's Resources, Acts of a Christian-Muslim Colloquium Organized Jointly by the Pontifical Council for Interreligious Dialogue (Vatican City) and the Royal Academy for Islamic Civilization Research Al Albait Foundation (Amman – Jordan)*, Rome 1996, p. 173–178.

Talal, El Hassan bin, Sua Altezza Reale Principe di Giordania / Elkann, Alain, *Essere Musulmano*, Milano 2001.

Taylor, John B., "**Geneva – Switzerland:** Third Working Meeting between the Secretariat for non Christians and the World Council fo Churches' Sub-unit on Dialogue with People of Living Faiths and Ideologies – Ecumenical Institute Bossey (March 13th -14th, 1983)", BSNC 52 (1983), p. 84–85.

Teissier, Henri, "Dialogue Islamo-Chretien *(sic!)* au Maghreb et en Algerie *(sic!)*", BSNC 39 (1978), p. 198–204.

Teissier, Henri, "L'experience Missionaire de l'Eglise au Maghreb et la mission du Conseil pontifical pour le Dialogue inter-religieux", BPCDIR 72 (1989), p. 323–333.

"Text of the Final Declaration of the Tripoli Seminar on the Christian-Muslim Dialogue", BSNC 31 (1976), p. 14–21.

"The Final Communiqué", Islamochristiana 19 (1993), p. 295–296.

"The First Ambassador to the Holy See Is Received by the Pope (19 November 1994)", Islamochristiana 21 (1995), p. 188–190.

"The General Report of the Muslim-Christian Colloquium on the Theme: "The Rights and Education of Children in Islam and Christianity" (Amman, Jordan, 13–15 December 1990)", Islamochristiana 17 (1991), p. 220–221.

"The Holy Father Appeals for Middle-East Peace (16 June 1997)", Islamochristiana 23 (1997), p. 252–253.

"The Holy Father's Address to the Diplomatic Corps Accredited to the Holy See (12 January 1991)", Islamochristiana 17 (1991), p. 277–279.

"The Holy Father's Address to the President of the Sudan", Islamochristiana 19 (1993), p. 280–282.

"The Inauguration of the Mosque in Rome (21 June 1995)", Islamochristiana 21 (1995), p. 176–177.

"The New Ambassador of Pakistan Received by the Pope", Islamochristiana 11 (1985), p. 232–233.

"The Patriarchs of the Catholic Churches of the Middle East and the Presidents of the Bishops' Conferences of the Countries More Directly Involved in the Gulf War, Final Communiqué", in: *Rencontre de S.S. le Pape Jean-Paul II avec les patriarches des églises orientales et les évêques des pays impliqués das la guerre du golfe, Meeting of H.H. Pope John Paul II with the Patriarchs of the Oriental Churches and Bishops of Countries Involved in the Gulf War, 4–6 March 1991*, s.l.s.a., p. 26–28.

"The Pope Receives the Bishops of Pakistan on the Occasion of Their "ad limina" Visit (21 October 1994)", Islamochristiana 21 (1995), p. 196–198.

"The Responsibility of Jews, Christians and Muslims for Peace in Jerusalem, Thessaloniki (25–29 August 1996)", Islamochristiana 22 (1996), p. 224–225.

"The Speeches Pope John Paul Prepared for His Pastoral Visit to Sarajewo (8 September 1994)", Islamochristiana 21 (1995), p. 145–149.

"**Tripoli – Libya:** "A Culture of Dialogue in an Era of Globalization" 16–18 March 2002", BPCDIR 110 (2002), p. 229–231.

Troll, Christian W., "Divine Rule and its Establishment on Earth: A Contemporary South Asian Debate", in: Commission pour les rapports religieux avec les musulmans (éd.), *Religion et poltique: Un thème pour le dialogue islamo-chrétien*, Cité du Vatican 1999, p. 37–53

Troll, Christian W., "Islam and Pluralism in India", in: Commission pour les rapports religieux avec les musulmans (éd.), *Religion et poltique: Un thème pour le dialogue islamo-chrétien*, Cité du Vatican 1999, p. 160–177.

Troll, Christian [W.], *Islam in South Asia: Facts, Movements and Trends*, C.R.R.M. Reports Recent Trends and Moevements in Islam 2, s.l. 1992.

Troll, Christian W.," Religious Freedom in Modern Islamic Thought: A Catholic Perspective", in: Pontifical Council for Interreligious Dialogue / Commission for Religious Relations with Muslims (ed.), *Religious Liberty: A Theme for Christian-Muslim Dialogue*, Vatican City 2006, p. 55–80.

Troll, Christian W., "**Vatican City:** Christian Mission and Islamic Da'wa 27–30 April 1997", BPCDIR 96 (1997), p. 394–395.

"TUNISIE *Le nouvel ambassadeur près le Saint-Siège est reçu par le Pape (19 novembre 1994)*", Islamochristiana 21 (1995), p. 211–212.

Ucko, Hans, "Pontifical Council for Interreligious Dialogue 40 Years Testimony", Current Dialogue 45 (2005), p. 13–15.

"Une délégation parlementaire intercommunautaire du Liban en visite à Rome", Islamochristiana 11 (1985), p. 226–227.

"Vatican City: 29 May 1998: **Pope's Discourse to the Members of the Delegation from Al-Azhar.**", BPCDIR 99 (1998), p. 281.

"Vatican City: 1st January 1999, **Message of His Holiness Pope John Paul II for the Celebration of the World Day of Peace** (Excerpt)", BPCDIR 102 (1999), p. 281–282.

"Vatican City: 11th January 1999, **Holy Father's Address to the Diplomatic Corps:** *"The Time Has Come to Ensure that Everywhere in the World Effective Freedom of Religions is Guaranteed"*. (Excerpt)", BPCDIR 102 (1999), p. 283–285.

"Vatican City: 19th February 1999, **Greetings of the Holy Father to a Delegation from the "International Forum Bethlehem 2000"**", BPCDIR 102 (1999), p. 291.

"Vatican City: 5th May 1999, **General Audience: Religious Dialogue with Islam**", BPCDIR 102 (1999), p. 303–305.

"Vatican City: 3 September 1999, **To the Bishops of Zambia on the Occasion of their "Ad limina" Visit** (Extract)", BPCDIR 103 (2000), p. 19.

"Vatican City: 6 September 1999, **To the Bishops of Malawi on the Occasion of their "Ad limina" Visit** (Extract)", BPCDIR 103 (2000), p. 20.

"Vatican City: 16 December 1999, **To the New Ambassador of the Islamic Republic of Pakistan to the Holy See** (Extract)", BPCDIR 103 (2000), p. 57.

"Vatican City: 25th May 2000, **To the Ambassador of Kuwait to the Holy See**", BPCDIR 104/105 (2000), p. 182–183.

"Vatican City: 12 June 2000, **To the New Ambassador of the Republic of Indonesia**", BPCDIR 104/105 (2000), p. 190–192.

"Vatican City: 7 September 2000, **To the New Ambassador of the Arab Republic of Egypt to the Holy See** (Excerpt)", BPCDIR 106 (2000), p. 7–9.

"Vatican City: 14 December 2000, **To the New Ambassador of Eritrea to the Holy See** (Excerpt)", BPCDIR 106 (2001), p. 30–31.

"Vatican City: 6 January 2001, **Apostolic Letter, Novo Millennio Ineunte** (Excerpts)", BPCDIR 106 (2001), p. 43–48.

"Vatican City: 22 January 2001, **To the New Ambassador of the Islamic Republic of Iran to the Holy See** (Excerpt)", BPCDIR 106 (2001), p. 48–50.

"Vatican City: 19th February 2001, **To the Bishops of Turkey, Ad Limina Apostolorum** (Excerpts)", BPCDIR 107 (2001), p. 160–162.

"Vatican City: 28th April 2001, **To the New Ambassador of Iraq to the Holy See**", BPCDIR 107 (2001), p. 165–166.

"Vatican City: 15th May 2001, **To the Bishops of Bangladesh**", BPCDIR 107 (2001), p. 176.

"Vatican City: 18th May 2001, **To the New Ambassador of Gambia to the Holy See** (Excerpt)", BPCDIR 107 (2001), p. 184–185.

"Vatican City: 19th May 2001, **To the Bishops of Pakistan** (Excerpts)", BPCDIR 107 (2001), p. 189–191.

"Vatican City: 19th May 2001, **To the Participants in the General Chapter of the Society of the African Missions** (Excerpt)", BPCDIR 107 (2001), p. 191–192.

"Vatican City: 26 May 2001, **Letter to Bishop François Blondel of Viviers on the Occasion of the Centenary of the Ordination to the Priesthood of Charles de Foucauld in 1901** (Extract)", BPCDIR 108 (2001), p. 289–290.

"Vatican City: 16 June 2001, **To the Participants in the "Study and Reflection Days" on the Occasion of the Tenth Anniversary of the Death of H.E. Mgr. Piero Rossano**", BPCDIR 108 (2001), p. 291–293.

"Vatican City: 18 October 2001, **Message on the Occasion of the 88th World Day of Migrants and Refugees, 2002**", BPCDIR 108 (2001), p. 321–324.

"Vatican City: 9th November 2001, **To the Participants in the Plenary Assembly of the Pontifical Council for Interreligious Dialogue**", BPCDIR 109 (2002), p. 10–12.

"Vatican City: 6th December 2001, **To the New Ambassador of Bangladesh to the Holy See** (Extract)", BPCDIR 109 (2002), p. 20–22.

"Vatican City: 7th December 2001, **To the New Ambassador of Turkey to the Holy See** (Extract)", BPCDIR 109 (2002), p. 33–34.

"Vatican City: 13th December 2001, **To Authorities Concerned about Crisis in Holy Land** (Extract)", BPCDIR 109 (2002), p. 31–32.

"Vatican City: 1st January 2002, **Message for the Celebration of the World Day of Peace** (Extract)", BPCDIR 109 (2002), p. 37–39.

"Vatican City: 6th March 2002, **To the Most Reverend Pietro Sambi Apostolic Nuncio to Cyprus**", BPCDIR 110 (2002), p. 172–173.

"Vatican City: 11th April 2002, **To the New Ambassador of Yugoslavia to the Holy See** (Extract)", BPCDIR 110 (2002), p. 175–176.

"Vatican City: 20th April 2002, **To the Bishops of Nigeria on the Occasion of their Visit "ad Limina"** (Extract)", BPCDIR 110 (2002), p. 177–178.

"Vatican City: 15 February 2003, **To the Bishops of the Gambia, Liberia and Sierra Leone during their "ad limina" Visit** (Extract), BPCDIR 114(2003), p. 308–309.

"Vatican City: 15 May 2003, **To the New Ambassador of the Islamic Republic of Pakistan to the Holy See** (Extract)", BPCDIR 114 (2003), p. 332–333.

"Vatican City: 31 October 2003, **To the Ministers for the Interior of the European Union**", BPCDIR 115 (2004), p. 12–14.

"Vatican City: 10 November 2003, **To a Delegation of the Organisation for the Liberation of Palestine** (Extract)", BPCDIR 115 (2004), p. 14.

"Vatican City: 20 November 2003, **To the Participants in the V World Congress of the Pastoral Care of Migrants and Refugees** (Extract)", BPCDIR 115 (2004), p. 17–18.

"Vatican City: 10 January 2004, **To the Ambassador of the Republic of Indonesia to the Holy See** (Extract)", BPCDIR 115 (2004), p. 25–26.

"Vatican City: 17 January 2004, **On the Occasion of the Concert for Reconciliation**", BPCDIR 115 (2004), p. 31–32.

"Vatican City: 20 January 2004, **To the Participants in the Christian-Muslim Dialogue which Was Organized by the Islamic-Catholic Liaison Committee in Rome**", BPCDIR 115 (2004), p. 32.

"Vatican City: 2 April 2004, **To the New Ambassador of Lebanon** (Extract)", BPCDIR 118 (2005), p. 12–14.

"Vatican City: 27 May 2004, **To the New Ambassador of Nigeria to the Holy See** (Extract)", BPCDIR 118 (2005), p. 17–18.

"Vatican City: 27 May 2004, **To the New Ambassador of Sri Lanka to the Holy See** (Extract)", BPCDIR 118 (2005), p. 19–21.

"Vatican City: 27 May 2004, **To the New Ambassador of Yemen to the Holy See** (Extract)", BPCDIR 118 (2005), p. 22–23.

"Vatican City: 27 May 2004, **To the New Ambassadors (Suriname, Sri Lanka, Mali, Yemen, Zambia, Nigeria and Tunisia) to the Holy See** (Extract)", BPCDIR 118 (2005), p. 14–15.

"Vatican City: 1 July 2004, **Common Declaration Signed together with Ecumenical Patriarch Bartholomew I of Constantinople** (Extract)", BPCDIR 118 (2005), p. 24–25.

"Vatican City: 24 January 2005, **To the Bishops from Spain (1st Group) on the Occasion of their "Ad Limina" Visit (Extract)**", BPCDIR 118 (2005), p. 49.

"Vatican City: **Message of the Special Assembly for Asia of the Synod of Bishops 13 May 1998** (Excerpt)", BPCDIR 99 (1998), p. 287–288.

"**Vatican City:** Visit to Vatican of Sh. Ahmed bin Mohamed Zabara, Grand Mufti of the Yemen (28 April – 5 May 1993)", BPCDIR 83 (1993), p. 199.

"Visit of Cardinal Arinze to Egypt (3–10 February 1989)", Islamochristiana 15 (1989), p. 190.

"Visit of Cardinal Arinze to Lybia (16–20 March 1989)", Islamochristiana 15 (1989), p. 220.

"Visite des Uléma Saoudiens au Secrétariat pour les Non-Chrétiens", BSNC 28/29 (1975), p. 181–185.

"Visite du Secrétariat pour les Non-Chrétiens au Conseil Suprême pour les Affairs Islamiques", BSNC 28/29 (1975), p. 174–180.

"Wadi al-Kharror, Jordan: 21 March 2000, **To the People of Jordan**", BPCDIR 104/105 (2000), p. 169–170.

Yannoulatos, Anastasios, "Relations between Man and Nature in the World Religions", BSNC 47 (1981), p. 134–142.
Yilmaz, Mehmet Nuri / Arinze, Francis, **"A STATEMENT OF INTENT between the Pontifical Council for Interreligious Dialogue (Vatican City) and the Presidency of Religious Affairs, Prime Minister's Office, Republic of Turkey"**, BPCDIR 110 (2002), p. 193.
Zago, Marcello, "Le Magistère sur le dialogue islamo-chrétien", BSNC 62 (1986), p. 171–181.
Zago, Marcello, "Les documents du conseil Pontifical pour le dialogue interreligieux", BPCDIR 72 (1989), p. 362–376.
Zago, Marcello, "The Life and Activity of the Pontifical Council for Inter-religious Dialogue", BPCDIR 74 (1990), p. 150–154.

Index

atheism 35, 48, 139, 406, 452, 465, 475f., 496, 517, 546, 598

belief 34, 46, 48, 52f., 81f., 84, 107–109, 144f., 149, 152, 155f., 181f., 185, 194, 215, 220, 242, 244f., 249, 264, 295, 303, 318, 334, 337, 389, 398–400, 407–409, 417, 419, 422, 425f., 433, 441, 445f., 453, 512, 517, 533f., 559, 564, 566, 590, 594

catechism 14f., 64f., 254, 256, 530
charity 205, 217, 229, 237, 244, 425, 429, 433, 447, 453, 499, 534
children/youth 14, 23, 27, 33, 37, 43, 46, 48, 50, 57, 59, 69, 72, 86, 100, 105, 116, 123f., 126, 130, 132, 140f., 144f., 153, 161, 166, 172, 181, 188, 194, 196, 210, 232, 237, 239, 241, 246, 280, 300, 304f., 328f., 331f., 334, 336f., 344, 346–348, 350–365, 367, 369, 373–378, 381f., 411, 416, 431, 435, 438, 446f., 456f., 460, 463f., 468, 471, 474f., 483, 487, 512–515, 523, 531, 533, 535f., 547, 549, 567, 572–577, 579, 591, 595–597, 603
Christ 9f., 14, 19, 29, 37, 43, 56, 64, 78f., 88, 91, 96, 111, 147f., 201, 220–222, 225f., 228, 230, 264, 266, 312f., 325, 329, 341, 373, 442f., 446, 458f., 521, 527, 529–531, 535, 571, 590f.
– christology 1, 43, 440, 444, 535, 591
Christians (regional groups / confessions) 2, 7, 9f., 14–16, 18f., 21, 26f., 29–31, 37–39, 43, 50–52, 54, 58f., 61–64, 67f., 70–72, 76, 78, 83–85, 89, 91, 93–98, 104f., 107, 109–111, 113, 116, 126, 130, 132, 140–142, 148, 159, 161, 169–171, 173–175, 181, 186, 188–196, 198–208, 211–214, 222, 228, 232, 234–236, 242, 250, 252–254, 256, 259, 262–265, 268, 270, 272, 274, 277, 280f., 283–285, 287–289, 291, 293– 301, 304–306, 313–315, 319, 321, 328, 331f., 342, 345, 356f., 365, 389–391, 397, 400–402, 405, 409, 417f., 431, 440, 446, 448, 453f., 457, 460, 464, 470, 473, 477, 481, 490–494, 496– 499, 502, 504f., 508f., 516, 522–525, 529–531, 533, 537f., 540f., 544, 546f., 552, 554, 558f., 561–564, 567, 580f., 583, 590f., 594–598, 600f.
community 8, 13, 24f., 31, 34, 36f., 58, 80, 86–89, 99, 101, 103–105, 108, 110, 112, 116, 119f., 123, 126, 130, 133, 135, 138, 141, 143, 150–152, 156f., 167, 183, 186f., 190f., 195f., 199f., 202–205, 210, 219, 227, 231, 239, 248f., 263, 267f., 282, 284, 286, 296–298, 300, 305, 311f., 316, 320, 329, 355f., 358, 373, 390, 392, 394f., 402–405, 413, 416, 419f., 441, 444, 446f., 461, 480, 490, 509, 528f., 532, 539, 547, 549, 551, 556, 561, 563, 565, 567, 581–584, 591, 595, 600, 603
conversion 10f., 17, 57, 94, 106, 130, 145, 148, 169, 180f., 186, 189, 213, 275, 298, 302, 312, 315, 366, 432, 500, 547
culture 9, 11f., 16–18, 20, 25, 33, 35–37, 40f., 46, 50, 52, 78, 87f., 96, 109, 115, 122, 125f., 130, 135, 138, 145f., 152f., 160, 165, 167–170, 176, 182, 188f., 193–196, 199, 208, 210, 217, 242, 255, 261, 273, 277, 285, 291, 296, 303–305, 321f., 326, 329, 335, 337, 339, 341, 361, 365, 385, 390, 394, 399f., 404f., 410, 432, 434, 436, 441, 452–454, 456, 464–466, 472f., 475, 478, 485–487, 489, 502, 507f., 510, 534–536, 545f., 548f., 554, 558, 561, 563f., 566f., 597f., 602, 604

dialogue 1–7, 10–21, 23f., 26–35, 37, 39–42, 44–51, 54–87, 89–91, 93–98, 107, 112f., 116f., 130, 138, 140–142, 146, 157, 160, 163, 167, 172, 185–189,

191–194, 197–202, 204, 208–212, 214 f., 220, 229, 231, 233, 235, 241 f., 250–256, 258–265, 267–280, 282, 284, 288, 290–292, 294 f., 297–299, 302–304, 306, 317–320, 322, 344, 355, 365, 382, 384, 386, 397 f., 400 f., 408 f., 414, 435–438, 444, 449, 454, 459–465, 467–469, 471–482, 484–508, 510–531, 533–538, 540 f., 546 f., 549, 552, 556–563, 566 f., 569, 572, 574, 577, 582–584, 588 f., 591 f., 594–605
– Christians and Muslims   1 f., 6 f., 13, 16, 25, 28, 31 f., 34, 36 f., 40, 42, 44–47, 51, 57, 68–70, 73, 80, 84, 89 f., 93, 116 f., 129 f., 137, 140, 142, 163, 172, 179, 185, 188 f., 192 f., 196 f., 200 f., 203, 207–213, 235–239, 241, 252 f., 255, 262 f., 283, 289, 292, 298, 302–304, 319, 366, 385, 397, 405, 434, 462, 464, 469, 476, 478–480, 482–489, 491, 493, 495, 498, 504, 509 f., 516, 526 f., 532, 534, 537 f., 546 f., 552, 563, 566, 569, 580, 584, 588, 595–604
– Christians and non-Christians   95
– interreligious (in general)   1–4, 6 f., 9 f., 12, 14 f., 17, 20, 23, 26–28, 30, 33, 39, 44, 46, 48–51, 55–62, 65 f., 68–73, 79, 82–84, 89–92, 98, 111, 178, 220, 235, 250 f., 253–256, 258–263, 268–270, 318 f., 400, 422, 449, 461, 464, 467–469, 471, 476 f., 479, 481 f., 485, 487 f., 494, 496 f., 499, 505–507, 510–527, 529 f., 533, 537, 547, 552, 554, 557, 582, 592, 596, 600 f., 603–605

enlightenment   169, 216, 333, 403 f., 409, 427, 434, 437, 443, 452, 465, 518, 583, 588, 591, 596

family   16, 18, 30, 33, 35–39, 43, 48, 50, 66, 77, 79 f., 82 f., 85, 88, 124, 129, 136, 144, 162, 164, 181–183, 185, 188, 193, 199 f., 209 f., 222, 229, 231 f., 236, 239–241, 243, 245, 285, 294, 297, 301, 312, 321, 323 f., 329, 331 f., 334, 336, 339, 344, 346, 348 f., 351–355, 358, 362 f., 369 f., 374 f., 377 f., 383–386, 394, 418, 429, 432, 435, 438, 447, 457 f., 460 f., 466, 471, 475, 480, 482, 484 f., 487 f., 511, 513, 528, 530, 555, 563, 565, 569 f., 575, 579–582, 594 f., 597 f., 600 f., 603

force   38, 72, 100, 107 f., 110, 120, 126, 138, 143, 146, 148, 156, 161, 180, 185, 200, 205, 230, 249, 258, 288, 308, 316, 321, 323, 388 f., 395, 400, 426, 442, 456, 466, 492, 533, 551 f., 556
– force and Christianity   110, 148, 185, 258, 426, 533
– force and Islam   38, 100, 107 f., 110, 120, 126, 138, 156, 161, 180, 185, 205, 249, 308, 400, 426, 456, 556

god
– adoration   417, 450
– covenant   14, 144 f., 202, 221, 224, 423, 497
– creator   7, 22, 28, 31, 35, 38 f., 41, 43, 45, 47, 51 f., 54, 56, 80, 83, 89, 95, 104, 137, 190, 193, 202, 212, 216, 223 f., 229–231, 238–241, 246 f., 250 f., 282, 284–286, 292 f., 298, 307, 315, 324, 329, 331, 333 f., 336, 345, 355, 362, 372, 380, 396, 398, 400, 412, 417–419, 422 f., 430, 432, 435, 440, 443, 482 f., 485, 511, 533, 536, 557, 561, 568, 570 f., 573, 575, 582, 584, 588 f., 602
– divine judgement   413
– divine law   8, 108, 217, 226 f., 287, 292, 315, 415 f., 429, 562, 585
– divine mercy   212, 284, 311, 561
– god's will   15, 25, 36, 39, 45, 83, 147, 155 f., 202, 227, 231, 240, 399, 412, 418, 420, 435, 459, 532, 550 f., 584
– reign   509
– revelation   10, 55, 64 f., 95 f., 109, 115, 157 f., 183, 185, 212, 222–224, 226, 246 f., 284, 291, 302, 322, 329 f., 337, 345, 399, 443 f., 457, 462, 464, 492, 530, 537 f., 561, 563, 595
Gospel   9, 11 f., 47, 60, 95–97, 118, 147, 150, 188, 207, 230, 276 f., 282, 287, 340, 364, 373, 446 f., 500, 549, 562

holy places   321, 505

Holy Spirit   10, 12f., 66, 228, 312, 316, 442, 509, 530
hostilities (Christians and Muslims)   7, 222
human being
- conscience   8, 11, 14, 17, 22, 24f., 28, 31, 33, 36, 41, 43–46, 48–50, 53f., 61, 79, 81f., 87f., 92, 137, 143, 146–149, 151, 156, 164, 179, 184, 193, 196, 201, 211, 213, 216, 222, 229, 258, 264, 266, 283, 292f., 298, 302, 318, 359, 398, 442, 456, 459, 461, 470, 480, 482, 500, 504, 507, 520, 526, 528–530, 532, 535, 548, 555, 565f., 595, 600f.
- destiny   55, 73, 84, 89, 212, 339, 439, 454, 586, 589
- dignity   8, 11f., 14f., 22, 24f., 27, 30–37, 39, 41, 43, 45–48, 50–57, 60–63, 77, 79–82, 84–86, 88, 143, 146f., 150f., 153–155, 157f., 176, 187, 189, 199–202, 209f., 237f., 240f., 244, 249, 251, 284, 290, 292–297, 302, 304f., 321–323, 325, 328, 334, 341, 343–347, 349, 366, 372f., 384–386, 400, 405, 417–422, 427, 432, 435–446, 448–456, 458–463, 465, 470–474, 477, 488, 501f., 512, 517, 521–523, 530, 532–538, 548–550, 557, 562–564, 566f., 569–573, 577–580, 583, 586, 589–599, 604
- free will   8, 68, 119, 143, 147, 151, 153, 155, 182, 185, 221, 381, 422, 439, 462, 548, 555f., 568, 589
- human rights, declaration of   17, 24, 27, 36–39, 41–43, 50–52, 54–57, 59, 65, 70, 77–81, 85–88, 90, 104, 106, 115, 120, 136f., 139, 143f., 146–153, 158–161, 168f., 171f., 174–179, 197–199, 201, 206, 208f., 211, 226, 240, 258, 260f., 266, 273, 278f., 297f., 305, 318, 321, 328, 334, 340f., 358f., 361, 376, 394, 409, 422, 431, 437f., 440 443, 445f., 448f., 452f., 456, 458, 460f., 463f., 466, 471, 473f., 476f., 488, 493, 497, 499, 532, 534–539, 541f., 545, 547–555, 558, 564f., 575, 586, 590–599, 604
- imago Dei   264

- nature   7–9, 13, 17, 20, 25, 31, 33, 37, 40, 43, 54, 58, 61f., 77–79, 81, 84, 86–88, 91, 96, 108, 120, 124f., 127, 138, 143, 147, 150f., 156, 162, 165, 176, 184, 189, 192, 198f., 202, 206, 208, 215, 219, 221, 224f., 227, 240, 242, 245, 249, 266f., 269, 273, 275, 279f., 286, 288, 290, 293f., 296, 298f., 304, 307–309, 313f., 317f., 324–329, 331, 333, 336f., 339, 344, 346, 348, 366–369, 372f., 378–385, 391, 405, 407, 411–414, 417–419, 422–427, 430, 434f., 439f., 442, 444, 450, 452–455, 460f., 464, 466, 472, 477, 483, 499, 529, 533, 535, 539f., 542f., 553f., 556, 562–564, 568, 571, 573f., 577–579, 581, 584f., 587–590, 592f., 595, 602
- ratio   77, 283, 509
- responsibility   8, 27, 69, 71, 107, 115, 143, 148, 151, 154–158, 167, 188, 190, 202, 204, 211, 227, 230, 232, 236, 240, 247–249, 260, 285, 297, 325, 329, 331, 334, 336, 338, 340f., 353f., 356, 361–364, 380f., 383, 414, 417–422, 424, 430f., 434f., 437, 439, 445, 459, 461f., 473, 484, 487, 501, 548, 551, 561, 564, 575f., 584–586, 588f., 591, 595, 598, 600, 603
humankind   9f., 13, 18, 21, 24, 27, 33f., 37, 40, 43, 55, 59, 79, 81, 97, 109f., 114f., 118, 135, 139, 146f., 152–154, 157, 180, 184, 187f., 190, 194f., 200, 203f., 206, 208f., 211f., 214f., 219–221, 223–227, 231f., 236f., 239, 241, 244, 250, 275, 282–284, 292–295, 302, 304, 310, 312, 322, 324, 326f., 336, 338, 346, 366, 380, 387, 392, 397f., 400f., 406–408, 413, 416–418, 420, 422–424, 430–433, 437, 446, 466, 472f., 476f., 479, 483, 485, 489, 501, 525, 530f., 533, 536, 538, 545, 550, 557, 559f., 563, 566, 568, 577, 580, 582f., 585, 588–590, 592f., 598–600, 604

ideology   89, 119f., 135, 159, 192, 204, 237, 274f., 282, 306, 317, 336, 464, 467, 495, 500, 506, 542, 552, 596

644 — Index

immigration 122, 129, 259, 508
international law 13, 428, 478
Islamic Law 34, 38, 104, 108f., 115, 118–120, 124–128, 130, 132–135, 157, 159, 162, 174, 178, 203, 232, 246, 256, 277, 282, 286f., 299f., 310f., 314, 327f., 335, 337, 346, 348–351, 362, 378, 380, 382, 396, 414–416, 424, 428f., 439, 445, 447, 456–458, 461, 488, 495, 497, 543–546, 551f., 560–562, 564f., 574, 578f., 585, 594, 604f.
Israel 64, 83, 97, 138, 167, 170, 190, 256, 286, 288, 388, 444, 449, 453f., 504, 592

Jerusalem 13, 15f., 20f., 31, 40, 63f., 69f., 74, 82f., 86, 277, 288, 290, 323, 386, 407, 478, 505, 525, 527, 562, 569
Jews/Judaism 3f., 15, 18, 20, 25, 28, 30, 46f., 57, 60, 65, 80, 82, 98, 103, 105, 107, 109f., 119, 149, 166f., 174, 191, 222, 226, 230, 250, 275, 284, 290, 300, 328, 407, 459, 489, 500, 502, 509, 562

liberalism 215

marriage 12, 14, 33, 82, 92, 122–124, 144, 166, 174, 208, 210, 245, 294, 339, 345–349, 353, 362–364, 368, 370, 374, 377, 379, 383, 429, 456f., 471, 483, 511, 528, 573f., 576, 578, 594, 597, 602
– mixed marriage 33, 57, 63, 91, 207f., 210, 464, 470, 483f., 528, 601f.
materialism 22, 34f., 238, 243, 251, 268, 277, 285, 288, 292, 301, 415, 448, 465, 476, 485, 501, 521, 557, 559, 563, 593, 598, 603
media 17, 27, 32f., 42, 91, 123, 128, 141, 160, 167, 175, 177, 182, 268, 300f., 304f., 318, 320, 323, 336, 339, 363f., 367, 375, 377, 381, 410, 435, 467, 471–473, 480, 482–484, 487, 507, 524f., 553, 565, 578, 589, 596, 600–603
migrants 56, 85, 91, 178, 209, 493, 495, 538
mission 9–11, 21, 40, 48, 54, 59f., 62, 73, 96f., 103f., 107, 111, 125, 129, 134–137, 207f., 222, 252f., 255, 257, 263, 285, 292, 295, 301f., 330, 390, 419, 493, 495–499, 507f., 536, 564, 566
modernity 113, 115, 124, 127, 133, 135, 153, 188, 199, 272, 297, 380, 461, 464–467, 486, 488, 523f., 541, 596, 603
monks 32, 250, 447, 470, 528
moral/ethics 7, 11f., 16, 18, 20f., 27, 34f., 39, 47, 53, 55, 59, 66, 69, 77f., 86, 100, 103, 108f., 118f., 122, 130, 134, 136f., 144, 146, 149f., 154–156, 163, 171, 181, 189–193, 195, 199, 205f., 213, 218, 226, 228, 230, 237, 240, 244–246, 248f., 263f., 285, 302, 309, 316, 322, 327f., 339–341, 344–346, 352, 354, 356, 364, 368, 380, 384, 388, 397–399, 401, 404, 407, 413, 415, 417–419, 421, 424, 433, 440f., 443f., 448, 450–452, 458, 460–462, 465, 471, 473, 483, 486, 496, 501, 508, 526, 530f., 534, 536f., 548, 550, 554, 561, 566, 572, 574, 576, 582–584, 586, 590, 592, 595, 601
– ecology 340f., 423f., 586
– justice 7, 13, 15–17, 19, 22, 27, 31, 34–36, 38f., 43f., 47f., 52, 59, 64f., 77, 79f., 82, 84, 88, 92, 95, 99, 109, 111, 119, 125, 136f., 157, 161–163, 181, 189, 192, 198, 201, 205f., 209f., 217, 221f., 228, 240f., 244, 248, 256, 263, 270, 274f., 279, 282, 284–287, 295, 297, 299f., 304, 312, 322, 328, 340, 343, 352, 364–366, 389, 395f., 398, 400, 405, 409, 416f., 419, 421, 424f., 430–433, 447f., 450, 452f., 456, 463, 466, 470, 472, 474–480, 482, 495, 500–502, 520, 525, 530, 534, 536, 545, 559, 561f., 565f., 577, 582–584, 586, 592, 596–600
– solidarity 22, 25, 30f., 37f., 47, 52, 77f., 86, 89, 95f., 113, 138, 141, 180, 193, 195, 205, 212, 236, 239, 286, 356, 387, 400, 402, 405, 418f., 432, 448, 461, 477f., 480, 487f., 527, 535f., 547, 580, 586, 595, 599f., 603

peace 7, 9, 11, 13, 15–17, 19–22, 25–28, 30–36, 38–42, 44–46, 48f., 51–53,

Index — 645

64, 67–69, 71, 77–81, 84–86, 88–91, 95f., 99–101, 103, 106f., 110, 116, 136f., 143, 149, 151, 160, 167, 187, 189, 191–193, 201, 205, 210f., 218, 220f., 230, 237–241, 243f., 249, 255f., 260f., 268–270, 273–275, 282, 284, 286, 293, 295, 297, 299f., 302, 304, 310, 336, 359, 362, 364–366, 380, 382, 388, 398, 418, 421, 425, 431–433, 437, 472–483, 493, 495, 501f., 512, 525, 527f., 530, 532–534, 536, 545, 559, 561f., 566, 571, 575, 577, 582, 588, 597–600, 602
pluralism   61, 121, 124, 162, 168, 180, 205, 257–261, 270f., 283, 288, 298, 303, 321, 323, 328, 433, 456, 465, 467, 565, 569, 594, 596
Pontifical Council for Interreligious Dialogue   261
pope   10, 12f., 15, 17, 19, 22–31, 33, 36, 38, 40, 45, 48f., 60–64, 66–70, 72–75, 78–84, 87–91, 93–95, 97f., 132, 140, 144, 146, 148f., 198, 235, 238f., 241, 250, 267–269, 276, 279, 296–298, 306, 317, 331, 384, 403, 411, 434, 443, 447, 478f., 491, 514, 516, 527–530, 533f., 538, 541, 557–559, 599
power   12, 49, 65, 105, 109f., 117, 120, 132f., 136–139, 153, 156, 161f., 167, 169f., 178f., 187, 203, 205–207, 227f., 236, 239, 241, 243, 247f., 259, 264, 275, 285, 294, 314, 325, 345, 366, 373, 375, 377, 382f., 388, 392, 396, 402, 406, 408, 425, 432, 440, 443f., 447, 451, 453, 466, 484, 496, 509, 526, 544f., 554, 563, 568, 573, 591, 593
prayer (interreligious)   12–14, 17, 24–28, 31, 33, 39f., 46f., 50f., 68–71, 79f., 89f., 100, 112, 116f., 119, 123, 174, 186, 194, 196, 199, 203, 212f., 218f., 234, 239, 255, 268f., 291, 309, 311f., 315, 334, 348, 351f., 363, 371, 396, 431, 474, 479, 481–483, 485, 499, 527, 530, 533–536, 541, 563, 573, 587, 599, 602f.

religion
– freedom of religion   8, 42, 81, 87, 93, 143–148, 150, 152, 157, 167, 169, 179, 244, 318, 438, 443, 482, 531, 548, 601
– religious criticism 215
– religious orders 306, 312, 495, 530, 567
– religious supremacy 222

secularism   15, 89, 119–121, 127, 131, 134–136, 138f., 161, 247, 251, 277, 320, 342, 389–391, 397, 401, 409, 465, 488, 517, 542, 545f., 557, 559, 572, 581–583, 604
sin   8, 12, 60, 100, 113, 204, 218, 220, 222, 226–229, 232, 240, 247, 264f., 267, 270, 285, 315–317, 325, 331f., 371, 373, 380f., 422, 452, 454f., 480, 497, 512, 516, 571, 578, 593f.
society   2, 12, 25, 27, 31, 34–36, 38f., 45f., 48, 50–53, 60, 65, 68, 73, 77, 80, 82, 86, 88, 101, 106, 109, 114, 117, 119, 121, 123, 125f., 128, 133, 136–139, 142, 149–152, 156, 162, 167, 175f., 179f., 189, 192f., 196, 203, 206–210, 214, 218, 229, 239, 241, 246, 249, 251, 255, 261f., 278f., 282f., 292, 296–298, 300, 303, 305, 313, 315f., 318–320, 323f., 326–328, 330, 332, 334, 338–340, 344–346, 350, 352, 354, 356, 359f., 363–367, 369–371, 373–375, 377–379, 382f., 385f., 395, 398–401, 404, 416, 419–421, 431f., 436, 441, 446–448, 451, 457, 459–467, 469–473, 480, 485–489, 492, 505–507, 509, 511–513, 515–517, 525, 528, 533, 535–537, 541–544, 546f., 549, 551, 553, 555, 557, 561, 563–565, 567, 569f., 572, 574, 576–578, 583, 587f., 590–592, 594f., 597, 599f., 603
spirituality   16, 26, 28, 113, 117, 119, 182, 184, 194f., 198, 211, 217, 266, 277, 306, 309f., 347, 354, 384, 402, 404, 459, 468, 479, 486, 520, 527, 530, 533, 535, 541, 567, 599
state   16, 30, 34, 38, 50, 53, 55, 57f., 61, 64, 66f., 69, 71, 73, 75, 77, 79f., 87–89, 100f., 103f., 106, 108, 113, 118–124, 126–128, 130–132, 134–136,

138 f., 143–147, 149–151, 153–159, 161–166, 170–178, 186, 189 f., 193, 197, 201, 205, 207, 213, 218, 222, 228, 230 f., 233, 244–247, 249, 251, 256 f., 261, 265 f., 268, 279 f., 283 f., 288, 291, 297 f., 308, 316, 323 f., 327, 339, 345, 350 f., 353, 358 f., 362–365, 374, 377, 381, 388, 390, 394–396, 404–410, 415, 426, 428–430, 433, 438, 441, 453, 456 f., 461, 463, 466, 469 f., 474, 479, 483 f., 488, 493, 495–497, 499, 501 f., 506 f., 511, 514, 528, 534, 537 f., 542–545, 547, 549–552, 554, 557, 560, 562, 564 f., 568, 573 f., 576, 580, 584, 587 f., 596, 602, 604

submission   23, 39, 50, 119, 147, 155, 179, 201–204, 213, 217, 227, 230, 239, 286, 293, 305, 309, 313 f., 328, 461, 540, 555, 563

theocracy   17, 109, 134, 160, 287, 545, 562

tolerance   25, 36 f., 44, 51, 80, 83 f., 99 f., 102 f., 105 f., 110, 138, 144 f., 159 f., 171 f., 180, 190, 260, 293–297, 299 f., 320, 328, 362, 364 f., 389, 400, 453, 485, 499 f., 505–507, 517, 532, 539, 552, 555, 563–565, 569, 602

truth   8, 11, 21, 31, 33, 41, 43, 46, 51, 81, 88, 95, 98–100, 110–113, 137, 141, 147 f., 150 f., 157, 168, 181, 184, 187, 200 f., 212, 221 f., 225, 241, 244, 247–249, 253, 257, 268, 271, 285, 304, 320, 322 f., 325 f., 328, 330 f., 333, 354, 372, 397, 416, 422, 450, 482, 485, 512, 529, 531, 533, 535, 539 f., 549, 555, 570 f., 585, 601, 603

woman, position of   57, 130, 208, 282, 298, 328, 348 f., 362, 365–374, 376–379, 381–385, 417, 443, 450, 456–458, 471, 573, 577–579, 611 f., 618

www.ingramcontent.com/pod-product-compliance
Lightning Source LLC
Chambersburg PA
CBHW031321230426
43670CB00006B/206